MODERN CLASSICS IN ANALYTICAL CHEMISTRY

VOLUME II

A collection of articles from
ANALYTICAL CHEMISTRY
selected for their value
as supplementary reading about
modern chemical instrumentation
and analytical methods

Compiled and edited by
Alvin L. Beilby
Pomona College
Claremont, Calif.

An ACS Reprint Collection

AMERICAN CHEMICAL SOCIETY

WASHINGTON, D. C. 1976

Library of Congress CIP Data

Modern classics in analytical chemistry—II

Includes bibliographic references and index.

1. Chemistry, Analytic—Addresses, essays, lectures.
I. Beilby, Alvin L., 1932–   II. Analytical Chemistry.
III. Chemical and Engineering news.

QD75.2.M63   543'.008   75-125864
ISBN 0-8412-0332-6   (d.2)   1-314

Copyright © 1976

American Chemical Society

All Rights Reserved. No part of this book may be reproduced or transmitted in any form or by any means—graphic, electronic, including photocopying, recording, taping, or information storage and retrieval systems—without written permission from the American Chemical Society.

PRINTED IN THE UNITED STATES OF AMERICA

CONTENTS

1 Introduction Alvin L. Beilby

2 Development of Analytical Chemistry As a Science
Izaak M. Kolthoff
45 (1), 24A (1973).

Spectroscopy

11 Carbon-13 Nuclear Magnetic Resonance Spectroscopy
George A. Gray
47 (6), 546A (1975).

23 Atomic Absorption Spectroscopy—Stagnant or Pregnant?
Alan Walsh
46 (8), 698A (1974).

30 Multielement Flame Spectroscopy Kenneth W. Busch
and George H. Morrison
45 (8), 712A (1973).

40 Atomic Fluorescence Spectrometry J. D. Winefordner
and R. C. Elser
43 (4), 24A (1971).

52 Fourier and Hadamard Transform Methods in Spectroscopy
Alan G. Marshall and Melvin B. Comisarow
47 (4), 491A (1975).

61 Atomic Spectrochemical Measurements with a Fourier
Transform Spectrometer Gary Horlick and W. K. Yuen
47 (8), 775A (1975).

67 X-Ray Energy Spectrometry David E. Porter and
Rolf Woldseth
45 (7), 605A (1973).

74 X-Ray Photoelectron Spectroscopy William E. Swartz, Jr.
45 (9), 788A (1973).

82 Tunable Lasers in Analytical Spectroscopy
James R. Allkins
47 (8), 752A (1975).

90 Surface and Thin Film Compositional Analysis: Description
and Comparison of Techniques Charles A. Evans, Jr.
47 (9), 818A (1975).

99 Surface and Thin Film Analysis Charles A. Evans, Jr.
47 (9), 855A (1975).

Electrochemistry

109 The Renaissance in Polarographic and Voltammetric
Analysis Jud B. Flato
44 (11), 75A (1972).

118 Anodic Stripping Voltammetry T. R. Copeland and
R. K. Skogerboe
46 (14), 1257A (1974).

Chromatography

125 Foundations of Modern Liquid Chromatography
L. S. Ettre and C. Horvath
47 (4), 422A (1975).

138 The Development of Chromatography Leslie S. Ettre
43 (14), 20A (1971).

144 Process Gas Chromatography R. Villalobos
47 (11), 983A (1975).

Automation and Instrumentation

154 GC/MS/Computers
Francis W. Karasek
44 (4), 32A (1972).

164 Computerized Signal Processing Gerald Dulaney
47 (1), 24A (1975).

171 Microprocessors. Part I: Bridging the Gap
Raymond E. Dessy, Peter Janse-Van Vuuren, and
Jonathan A. Titus
46 (11), 917A (1974).

179 Microprocessors. Part II: Applications Raymond E. Dessy,
Jonathan A. Titus, and Peter Janse-Van Vuuren
46 (12), 1055A (1974).

186 Automated Reaction-Rate Methods of Analysis
Howard V. Malmstadt, Emil A. Cordes, and
Collene J. Delaney
44 (12), 26A (1972).

196 Instruments for Rate Determinations
Howard V. Malmstadt, Collene J. Delaney, and
Emil A. Cordes
44 (12), 79A (1972).

Measurement Techniques

205 Applications of Chemiluminescent Reactions to the
Measurement of Air Pollutants
Robert K. Stevens and J. A. Hodgeson
45 (4), 443A (1973).

210 Chemiluminescence and Bioluminescence in Analytical
Chemistry W. Rudolf Seitz and Michael P. Neary
46 (2), 188A (1974).

219 Immobilized Enzymes: Analytical Applications
Howard H. Weetall
46 (7), 602A (1974).

227 Ultrapurity in Trace Analysis James W. Mitchell
45 (6), 492A (1973).

234 The Other Face of the Measurement Base
Richard W. Roberts
47 (7), 648A (1975).

241 Measurement Analysis by Pattern Recognition
Bruce R. Kowalski
47 (13), 1152A (1975).

248 Comparison of Analytical Techniques for Inorganic
Pollutants R. F. Coleman
46 (12), 989A (1974).

The Analytical Approach

254 Urgent Production Problem Solved by Unique Capabilities
of Analytical Chemists Claude A. Lucchesi
46 (4), 433A (1974).

257 Effects of Raw Material Change in Manufacturing Process Resolved   John Mitchell, Jr.
46 (9), 804A (1974).

259 Analytical Chemists Vital in Commercialization of New Food Packaging Material   V. F. Gaylor
46 (11), 897A (1974).

262 Multitechnique Approach Solves Construction Materials Failure Problems   William G. Hime
46 (14), 1230A (1974).

265 Analytical Chemists in the U.S. Customs Service   Daniel W. Vomhof
47 (2), 228A (1975).

268 Recognition and Solution of Production Problems in Vitamin Manufacture
E. MacMullan, R. Hagel, and R. Gomez
47 (4), 473A (1975).

270 Finding the Flavor Component: A Job for the Analytical Chemist   W. Noel Einolf
47 (6), 579A (1975).

273 The Forensic Chemist. An "Analytical Detective"
Larry S. Eichmeier and Michael E. Caplis
47 (9), 841A (1975).

276 Industrial Analytical Chemists and OSHA Regulations for Vinyl Chloride   S. P. Levine, K. G. Hebel, J. Bolton, Jr., and R. E. Kugel
47 (12), 1075A (1975).

Art Conservation

280 Art Conservation: Culture Under Analysis. Part I
Ben B. Johnson and Thomas Cairns
44 (1), 24A (1972).

290 Art Conservation: Culture Under Analysis. Part II
Ben B. Johnson and Thomas Cairns
44 (2), 30A (1972).

299 Index Listing of "Report for Analytical Chemists" Articles in ANALYTICAL CHEMISTRY, 1970-1975.

300 Index Listing of "Instrumentation" Articles in ANALYTICAL CHEMISTRY, 1970-1975.

303 Subject Index

INTRODUCTION

THE PURPOSE of this second volume of "Modern Classics in Analytical Chemistry" is the same as that of the first volume — to provide a selection of reprints of articles which can serve as supplementary reading material for analytical chemistry courses. In addition it is hoped that this volume will serve as a convenient source of articles for anyone wishing to keep abreast of recent advances in analytical chemistry.

The articles have been chosen exclusively from the 1970–1975 A-page articles in ANALYTICAL CHEMISTRY — "Report," "Instrumentation," and "The Analytical Approach." In order to keep the cost as low as possible the articles have been reprinted exactly as they originally appeared. No typographical errors have been corrected, and the authors' affiliations are the ones that they had when the articles were first published.

The selection of articles was made entirely by the editor but was reviewed by the editors of ANALYTICAL CHEMISTRY. There were two main criteria for the selection of articles from among the many that have appeared during the past six years in the A pages of ANALYTICAL CHEMISTRY. The first criterion was to include topics which are discussed in advanced analytical chemistry and instrumental analysis courses so that this reprint volume can serve as a convenient source of supplementary reading for existing textbooks. The second criterion was to have a broad coverage of the field of analytical chemistry. It should be noted that the authors come from a wide variety of institutions — academic institutions, governmental laboratories, instrument manufacturers, and industrial and research laboratories. The articles include historical discussions, theoretical presentations, applied discussions, and solutions of analytical problems. In the following paragraphs the articles are discussed in more detail.

What could be more fitting for an opening article for a volume titled, "Modern Classics in Analytical Chemistry" than an article on the "Development of Analytical Chemistry As a Science" by Izaak M. Kolthoff. As soon as the editor saw this article he knew that it should be the lead-off article for this volume.

The first main group of articles deal with spectroscopic methods of analysis. Of special interest in this group is one by Alan Walsh, the developer of the atomic absorption spectrometer, in which he reminisces on its development and looks at the future. Also included in this group are two articles which deal with surface and thin film analysis. Next come two articles on electroanalytical methods of analysis, followed by three articles on chromatography. Since the historical development of analytical methods is generally not included in textbooks, the editor felt fortunate that several articles of a historical nature could be included in this volume, such as two of the three articles on chromatography.

The article on "GC/MS/Computers" was chosen since it illustrated how a combination of several techniques can lead to results greater than the sum of results with each technique alone. This article is followed by three articles dealing with the general use of computers with analytical instrumentation. The next nine articles cover a variety of topics which the editor chose to increase the breadth of topics to this volume.

When the feature, "The Analytical Approach," first appeared in ANALYTICAL CHEMISTRY, the editor knew that the articles for this feature should be included in the second volume of "Modern Classics in Analytical Chemistry." Claude Lucchesi, in his introduction to the feature, stated that he hoped that the articles would be useful to teachers of analytical chemistry who might use the examples presented to show their students the scope and excitement involved in the service aspect of professional analytical practice. It is a pleasure to be able to include all nine of "The Analytical Approach" articles that appeared by the end of 1975 with the hope that the inclusion will help to fulfill Dr. Lucchesi's wish.

Many areas could have been chosen as examples of where analytical chemistry is applied to the solution of problems. Some of the areas have been covered in "The Analytical Approach" articles. To end this second volume of "Modern Classics in Analytical Chemistry," the editor felt that the most fitting area to choose would be one that deals with another type of "classics." Hence Volume 2 ends with the two articles on "Art Conservation: Culture Under Analysis."

In addition to the articles, index listings for all of the "Report" and "Instrumentation" articles for the period 1970–1975 have been added with the hope that the reader of this volume will peruse the listings and find other articles of interest.

The editor would like to thank the authors for their approval to include their articles in this volume and for their earlier efforts in writing the articles. The editor also gives his thanks to Josephine M. Petruzzi, Managing Editor of ANALYTICAL CHEMISTRY, for her support and to the staff of the American Chemical Society for their work in the preparation of this volume.

Alvin L. Beilby

Seaver Chemistry Laboratory
Pomona College
Claremont, California 91711

DEVELOPMENT OF ANALYTICAL

Analytical chemistry: science or art? Today analytical chemists, together with chemists in other disciplines, originate methods and techniques on a sound scientific basis

BEFORE REVIEWING the development of analytical chemistry as a scientific discipline in the first quarter of this century, it is of interest to briefly consider the slow growth of chemistry as a science. It was not until the 16th century that experimental chemistry was born. Although considered by many a charlatan, the physician Aureolus Philippus Theophrastus Bombastus Paracelsus (1) von Hohenheim (1493–1541) objected to the alchemical conception of elements and recognized the importance of chemistry for medicine: "Chemistry solves for us the secrets of therapy, physiology and pathology; without chemistry we are trudging in darkness." The importance of present clinical chemistry more than justifies Paracelsus' prophetic statement. Undoubtedly, also thanks to Paracelsus, up to the 19th century, many physicians and pharmacists contributed so much to chemistry. To mention only one (Flemish) practicing physician, J. B. van Helmont (2), born in 1577 in Brussels, is considered one of the early great chemists. Among his many contributions we may cite his analytically important discovery that silicic acid, when dissolved in a base, is recovered quantitatively upon addition of an acid. Van Helmont and other scientists, like Boyle (3) in the 17th century, described several color and precipitation reactions which found use in qualitative analysis. In that period use was also made of the blowpipe (a narrow tube with which air can be blown into a flame) which had been in use by goldsmiths since antiquity and found general application in the glass industry by the 17th century. In qualitative analysis it was used to investigate the color of melts, particularly that of borax for the detection of several metals and, to a lesser extent, the melt of sodium ammonium phosphate. The Swede Torbern Bergman (4) (1735–1784), generally recognized as the founder of qualitative and quantitative inorganic analysis, summarized in his book published in 1779 all applications of the blowpipe, a gadget still in use in the beginning of the present century. Bergmann was the first to present a system of qualitative analysis, introducing hydrogen sulfide for separations and describing a host of color and precipitation reactions. His system, with many additions and modifications, is still in use today, although thioacetamide has been gradually replacing the ill-smelling gas. Not too many years ago, all students in chemistry took a course in "Qual," using the modified Bergmann system and thus learning quite a bit of factual chemistry. Today, more and more colleges are teaching less and less classical qualitative and quantitative inorganic chemistry.

Gases: Phlogiston Theory

Chemistry in its embryonic state was mainly analytical in nature. Wrong theories, handicaps in the development of chemistry as a science, were often proposed on the basis of philosophy or metaphysics instead of experimental facts. As an extreme example, the phlogiston theory was intended to explain the chemistry involved in the combustion of gases. The Greeks considered that nature was composed of four elements, one being air. When water was heated and the "air" formed from it was mixed with ordinary air, that part of the air that once had been water could be changed back into water by cooling the mixture. It was not until 1620 that van Helmont introduced for this fraction the name of "vapor." For air he reserved the Paracelsian term "chaos," pronouncing and spelling it "gas," to my knowledge a word common to all Indo-European languages. Van Helmont found that some vapors differed not only from air but from each other. By burning charcoal he obtained "gas sylvestre" (from wood) which he noticed can put out a candle flame, and "gas pingue" (gas from fat, actually hydrocarbons) which can be burned as air cannot.

Actually, Robert Boyle (3) (1627–1691), who discovered the gas law which carries his name and who is also well known for constructing the hydrostatic balance, may be considered the father of quantitative gas analysis. Famous for his contributions to gas chemistry and analysis is Henry Cavendish (1731–1810) who a century later discovered, for example, that the "inflammable air" which develops upon dissolution of a metal in acid can combine at an electric spark with oxygen to form water. Daniel

CHEMISTRY AS A SCIENCE

IZAAK M. KOLTHOFF, Department of Chemistry, University of Minnesota, Minneapolis, Minn.

Rutherford (1749–1819), who lived in the same period, is credited with being the discoverer of nitrogen, called by him "phlogisticated air," which was given by A. L. Lavoisier (5) (1743–1794) its French name "azote" (meaning "no life" since animals died in it), a name still recognized in English in "hydrazoic" acid, "azides," "azo," and "hydrazo" compounds. Joseph Priestley (1733–1804), a Presbyterian minister who later in Philadelphia became a Unitarian minister and whose hobby was chemistry, is credited with being the discoverer of oxygen (called by him "dephlogisticated air") (6). This again was recognized by Lavoisier as an element and named by him "oxygen" (meaning "producer of sourness" because he thought oxygen was an essential constituent of acids). It is a cruel irony of fate that Lavoisier (no doubt the most outstanding chemist of the 18th century), who pioneered accurate analysis as a prerequisite for the development of chemistry, ended his life in 1794 during the French Revolution under the guillotine because of his association with nobility, and his friend Priestley was chased from his English laboratory to America because of his sympathy with the French Revolution. (Laboratory equipment used by Priestley in his experiments dealing with air is on display in the Smithsonian Museum of History and Technology.) It is almost incredible that in the period in which valuable discoveries in chemistry were made, a fantastic theory could be proposed and generally adopted which has greatly handicapped progress in theoretical chemistry. I am referring to the "phlogiston theory" proposed by the German chemist Georg Ernest Stahl (1660–1737) (7): "A substance that burns or a metal which rusts contains a substance called 'phlogiston' (from the Greek "set on fire") and when it burns or rusts it gives up phlogiston." It is almost incomprehensible that this theory remained accepted for some 70 years, even though several properties of gases were discovered which were hard to explain by the theory. For example, one experiment was carried out by Joseph Priestley who described in 1774 that mercury upon heating in air forms a brick-red "calx" (it "rusts"). When he heated this "calx" in a container by use of a lens that concentrated sunshine on it, the "calx" broke down and mercury deposited in the upper part of the container. When he put a smoldering splinter of wood into the glass container, it burst into a brillant flame. He explained his results on the basis of the phlogiston theory: "Mercury contains phlogiston, which upon heating is given off, leaving phlogiston-free calx (HgO) behind." When the phlogiston-free substance is heated, it gives off a gas which he called "dephlogisticated air" (oxygen).

All very confusing! Lavoisier in the same year (1774) dealt the death blow to the phlogiston theory. Priestley visited Lavoisier in 1774 and demonstrated to him the formation and decomposition of mercury(II)oxide. Lavoisier had already run several quantitative experiments and had noted a partial vacuum upon combustion of a metal in a closed vessel filled with air and had concluded that air is composed of a mixture of gases. He also had proved experimentally that when a metal "rusts" and "gives off its phlogiston," it gains in weight. No substance was known which had a negative weight, but several phlogistonists believed this rather than give up the theory. It is extremely difficult to understand that Bergman, who may be credited with being the first to enrich the literature with an elementary outline of systematic gravimetric inorganic analysis, remained a firm believer in the phlogiston theory. Based upon the results of his quantitative experiments, Lavoisier, who recognized that Rutherford's "phlogisticated air" was nitrogen and Priestley's "dephlogisticated air," oxygen, postulated his "anti-phlogiston theory" which stated that "the quantity of a substance is the same before and after all operations, only changes and transformations are going on." With Lavoisier it was the Russian Lomonosov (from whom Lomonosov University in Moscow, U.S.S.R., derives its name), who corrected the famous but erroneous experiment by Boyle that the weight of a closed container of lead and air increases upon oxidation of the metal by heating. The contributions by Lavoisier, known as the "Father of Modern Chemistry," and to a lesser extent by Lomonosov, were the basis of the development of stoichiometry to which Richter (1762–1807) contributed so much. The names of Berthollet (1748–1822), Proust (1755–1826), and Dalton (1766–1844) are mentioned in this connection. Based upon quantitative relationships found in stoichiometric studies, Dalton discovered his well-known law of multiple proportions. Only because of progress in analysis, which became more and more accurate as time went on, could rational theories and laws be discovered in the 19th century. This, in turn, led in the 20th century to a rational development of scientific analytical chemistry. Classical gravimetric and volumetric analysis became well established in the middle of the last century. Important contributions were made by Heinrich Rose (1795–1864) (think of the Rose crucible), J. J. Berzelius (1779–1848), and Carl Remigius Fresenius (1818–1897), the latter being the founder of the famous *Z. Anal. Chem.*, which made its first appearance in 1862 and is still being published.

Titration Methods

Volumetric analysis was crudely practiced in the second half of the 18th century. In 1757 Home described the titration of alkaline earth ions in water with potassium carbonate and of chloride with silver nitrate until the clear point, the latter method being greatly refined some 100 years later by Gay-Lussac. Lewis in 1767 weighed the container from which he added standard hydrochloric acid to a solution of potash and thus used the first (primitive) weight buret. He detected approach of the end point by cessation of effervescence and then titrated until blue litmus paper turned red. In 1786 Descroizilles was probably the first chemist to carry out a titration in a graduated cylinder in which the volume could be read with relatively fair accuracy. In a publication in 1795 in an obscure journal entitled *Journal des Arts et Manufactures*, he described his titration of hypochlorite with indigo solution, a method which was practiced for many years. In a later publication (8) he also described his titration method of various acids with alkali and vice versa, with syrup of violets as indicator. A detailed history of volumetric analysis was given by Szavadváry (9). Great improvement in the refinement of titrimetric apparatus was made by Gay-Lussac (10) who described various titration methods. Well known is his detection of the clear point (referred to above) in the titration of silver with chloride or vice versa. Volumetric analysis became a generally accepted and popular method of analysis for inorganic and organic compounds after Friedrich Mohr (11) published in 1855 his excellent book on titration methods. The end point in a titration was determined visually, usually with an indicator. Litmus was the only indicator in acid-base titrations until Luck recommended in 1877 phenolphthalein for the titration of weak acids with a strong base, and Lunge in the same year recommended methyl orange for the titration of weak bases with a strong acid.

Elementary Organic Analysis

For the sake of completeness, it should be added that quantitative elementary organic analysis was also developed during the 19th century. Combustion of an organic compound in oxygen was used in 1784 by Lavoisier, who weighed CO_2 and H_2O formed. His cumbersome apparatus was responsible for his quantitative results being considerably in error. Gay-Lussac and Berzelius enriched the literature with descriptions of combustion apparatus, but in 1837 the famous organic chemist Justus Liebig (12) made the most important and popular contribution. In 1831 Dumas (13) described his well-known method for the determination of nitrogen. Some 50 years later in 1883, Kjeldahl (14), chemistry director of the famous Carlsberg Laboratory in Copenhagen, described his popular method for nitrogen. Organic and inorganic quantitative analyses were practiced on a macro scale, but it was not until 1912, when Fritz Pregl published his method for the micro determination of carbon, hydrogen, and nitrogen, that quantitative micro and later submicro elementary analysis became generally applied. It is fair to mention here the name of the late Benedetti Pichler as one of the pioneers of ultramicroanalysis. This, in a nutshell, is the development of classical qualitative and quantitative analysis until the very end of the last century.

Analytical Chemistry: Science or Art

At that time analytical chemistry was a highly developed art but not a science. Analytical chemists were ignorant of the revolutionary development of physical chemistry in the last quarter of the 19th century, and analytical chemistry lost its historic luster because analytical chemists continued to develop their subject entirely empirically. Academically, analytical chemistry became discredited. This situation may explain why no chairs in analytical chemistry were established in European universities until some 25 years ago and why teaching of analytical chemistry was confined mainly to technical universities and remained a major part in the education of pharmacists. At the end of the last century, the famous German chemist Wilhelm Ostwald (1853–1932) made an effort to establish analytical chemistry as a scientific discipline. In the preface to his stimulating booklet (15), "Die wissenschaftlichen Grundlagen der analytischen Chemie," he made it clear that analytical chemistry was doomed to continue to occupy its subordinate position of a maidservant (somewhat comparable to our present "service laboratories" where analyses are carried out routinely) for the other branches of chemistry, unless the analytical chemists would discontinue teaching and practicing chemical analysis solely as an empirical technique and art which requires skill and precision, and start making use of experimental and theoretical developments of physical chemistry. If I could have been a reviewer of Ostwald's book which was published in 1894, I would have praised it greatly for what it contributed but would have criticized it for its many omissions. Briefly, in his booklet of modest proportions, Ostwald first discusses the importance of properties made use of in analytical chemistry and then separations in two chapters. For the separation of solids from solids, he makes use of the difference in density, the difference in rate of sedimentation having already been in use in ore analysis. Use of differences in magnetic and electrification properties, and especially of differences in solubility, is also mentioned. For the separation of liquids from solids, filtration is treated in some detail, including the theory of effectively washing a precipitate. The so-called "Ostwald ripening" for the increase of particle size to facilitate filtrations is fully explained. Apparently, Ostwald was not acquainted with the Gooch crucible, even though this had been described in 1878 (16). Distillation is discussed in some detail and, more briefly, the separation of gases from one another. The theory of extraction is clearly explained. The section dealing with "adsorption" is unsatisfactory, although Ostwald deserves credit for emphasizing the importance of adsorption when dealing with colloidal precipitates. The more modern methods of separation are not mentioned, although some had been made use of when dealing with specific materials. For example, use of paper chromatography as a separation technique is found in the literature of the dye industry of the 1800's. The technique of capillary analysis described by Goppelsroeder in several publications around 1870 is close to that of modern adsorption chromatography. Columnar chromatography was discovered in 1897 by the American J. T. Day, who applied it to fractionating petroleum by adsorption on columns of Fuller's earth. Ion-exchange chromatography has been utilized in soil studies since the middle of the last century and has become a most important separation technique after the discovery of exchange resins. Returning to Ostwald's monograph, analytically the most important part of his book deals with solution chemistry (the mass action law discovered by the Norwegians Guldberg and Waage in 1867), the theory of electrolytic dissociation by Arrhenius, and the combined application of the mass action law and Arrhenius' theory to the derivation of solubility products and an interpretation of the properties of acid-base indicators. Ostwald's little book is a classic, but it could have had a greater impact on analytical chemists if he had outlined the analytical importance of the work of others, particularly Gibbs, van't Hoff, and Arrhenius. Together with Ostwald, they may be called the founders of physical chemistry and, indirectly, of scientific analytical chemistry. Van't Hoff and Arrhenius were geniuses; Ostwald was a great scholar and prolific writer. His books and research have greatly contributed to making chemists acquainted with the new discoveries.

Jacobus Henricus van't Hoff was born in Rotterdam in 1852 and died in 1911. In 1874 at the age of 22, he published in Dutch an 11-page paper dealing with organic stereochemistry in which the theory of the asymmetric carbon atom was mentioned. (Independent of van't Hoff, the Frenchman LeBel published a similar theory in 1875.) A French translation of van't Hoff's paper was published in 1875. The chemical world did not pay any attention to his theory of the asymmetric carbon atom until a year after publication of the French version when the organic chemist Wislicenus (then a professor in Würzburg, Germany) asked permission to have a German translation published entitled, "Die Lagerung der Atome in Raume" (in English, "Stereochemistry"). It was violently attacked by the famous German organic chemist Kolbe who wrote, "A certain Dr. J. H. van't Hoff, teacher at the Veterinary School in Utrecht, has no appetite for exact chemical research. He found it easier to mount his Pegasus (apparently borrowed from the School) to announce to the world that he saw on his flight on the horse 'how the atoms in a molecule are arranged in space'."

Physical Chemistry Relationships

In spite of this scathing criticism the theory of van't Hoff-LeBel soon became universally accepted. To outline its analytical importance in connection with determinations by optical rotation would be superfluous. At the age of 32 in 1884, van't Hoff published his famous book, "Études de dynamique chimique," which led to an explosive development of physical chemistry. Quantitative relationships dealing with reaction kinetics in liquids and gases and the effect of temperature on reaction rates and upon chemical equilibria were developed. Soon after this publication, van't Hoff published in 1885 and 1886 in the *Archives Neerlandaises* his classical papers on the osmotic laws in dilute solutions, concluding that the gas laws of Boyle and Gay-Lussac also hold for dilute solutions. Also, van't Hoff was the first to derive thermodynamically the mass action law. Again the chemical world did not pay any attention to his fundamental classical studies until after Svante Arrhenius published in 1885 a review in the *Nordisk Revy* (Uppsala) which closes as follows: "It has been the intention of the reviewer to call attention to the vast perspectives which the work of the author has opened to future research." Solutions of strong electrolytes did not follow van't Hoff's osmotic laws. It is not surprising that the "abnormal" behavior could be explained by the theory of electrolytic dissociation of Arrhenius who traveled to Amsterdam where van't Hoff had become a professor. In 1884 at the young age of 25, Arrhenius originated in his doctor's thesis his well-known theory of electrolyte solutions under the title, "Recherches sur la conductibilité galvaniques des électrolytes." Like van't Hoff, he met with much resistance in having his revolutionary theory accepted. Some of his professors considered it too bizarre and speculative. How could sodium atoms (ions) be in solution when the metal reacted violently with water? As a result of this opposition, Arrhenius neither earned the doctor's degree cum laude nor was he awarded the "jus docendi."

It took quite a while before the theories of van't Hoff and Arrhenius became generally recognized, and great credit must go to Wilhelm Ostwald for having propagated the revolutionary developments through his research, writings, and personal contacts with van't Hoff and Arrhenius. He recognized the importance of Arrhenius' doctor's thesis. After this thesis was submitted, Ostwald traveled to Uppsala to discuss with Arrhenius his own work dealing with solutions of electrolytes and to offer him a lectureship in Riga where Ostwald was then a professor before moving to Leipzig. Arrhenius declined the flattering offer and stayed in Sweden. The rapid dissemination of the new ideas was also greatly promoted by the *Zeitschrift fur physikalische Chemie*, founded by Ostwald and van't Hoff, and by the publication of van't Hoff's lectures at the University of Berlin (*17*). In 1901 van't Hoff was the first recipient of a Nobel prize (in chemistry), followed by Arrhenius in 1903 and Ostwald in 1909. Both van't Hoff and Arrhenius enjoyed the lighter side of life, whereas Ostwald was more the typical dignified "Geheimrat." In my young days as a student in Utrecht, lots of stories were in circulation about van't Hoff, especially in the laboratory named after him. One night, van't Hoff (at that time a professor in Amsterdam) found his usual relaxation in a nightclub. When he arrived home in the early morning hours, his wife would not let him enter the house but sent him to a public bath, from where he went relaxed to the university to deliver his regular lecture at 9 a.m. At the end of the lecture he said, "Will the student who lent me 25 guilders last night in a nightclub come to my office so that I can refund him."

Ernst Cohen, who became professor and head of the van't Hoff laboratory in Utrecht and was the favorite student of van't Hoff, has written a 638-page biography of his teacher (*18*) in which van't Hoff's personal life and his scientific contributions are presented in great detail.

It is hard to imagine the development of chemistry (and analytical chemistry) without van't Hoff's monumental contributions which every student in chemistry is familiar with even today.

Gibbs' Contributions

Returning to the omissions in Ostwald's book, a few words may be said about the work of another pioneer in science, Josiah Willard Gibbs, who was not an analytical chemist—nor even a chemist for that matter—but whose discoveries are of great importance to analytical chemistry. In a classical paper entitled, "Thermodynamic Principles Determining Chemical Equilibria," and published in the rather obscure *Journal of the Connecticut Academy* (*19*), Gibbs introduced the principles of the phase rule as well as the concept of the chemical potential. Use of both is made in analytical separations, an analytical topic par excellence. For example, the phase rule states the conditions necessary to achieve a state of equilibrium, both for homogeneous and heterogeneous systems. These conditions are of special analytical interest in the preparation and preservation of pure crystal hydrates and other solvates, in the quantitative heat transformation of a compound into another (e.g., of calcium oxalate into carbonate), in phase solubility methods, in checking the purity of slightly soluble compounds (solubility independent of amount of solid), in zone melting purification, and in the interpretation of thermal and differential thermal analysis (including melting and freezing point curves and mixed melting points). Today, many of the more theoretically inclined analytical chemists make frequent use of the principle of the chemical potential, especially in formulating a quantitative relationship between equilibrium constants, including electrode potentials, and medium activity coefficients in different solvents. The importance of Gibbs' theories to *analytical* chemistry apparently did not occur to Ostwald, because no reference to Gibbs and his work is found in his book (*15*). This omission is not made in a modest book by Herz (*20*) in which a chapter (pp 83–90) is devoted to the phase rule with special emphasis on melting point diagrams and transformation temperature. The book is hardly ever referred to in the analytical literature. The probable reason is that it is more an introduction to physical chemistry, and hardly any analytical applications are discussed.

Optical Methods

It is also not clear why Ostwald (*15*) did not mention optical analysis, although the fundamentals of emission spectroscopy and spectrophotometry had been known and used for quantitative purposes for more than two centuries. Briefly, spectra were investigated by several workers in the 16th and 17th centuries; in this connection Newton's name may be mentioned particularly. The real development of emission spectroscopy started in the beginning of the last century; the names of Herschel, Wollaston, and Fraunhofer are known to most chemists and physicists. Emission spectroscopy became an established method of inorganic analysis after the classical publications of Kirchhof and Bunsen in 1860.

Robert Wilhelm Bunsen (1811–1899) was most versatile and started as an organic chemist but soon also carried out research in inorganic chemistry in connection with his classical studies entitled, "Studies in the Cacodyl Series," in the period between 1837 and 1842. In 1852 Bunsen accepted a call to Heidelberg to succeed Leopold Gmelin, who is the founder of "Gmelin's Handbuch der anorganischen Chemie," comprising now more than 100 volumes with new editions still being published continuously. From all over the world, students came to study inorganic and later analytical chemistry under Bunsen. Among his many interests Bunsen became involved in photochemical studies, especially of the hydrogen–chlorine reaction in sunlight. The Bunsen burner, still known and used by all chemists, had been constructed by Bunsen in 1855 and gave a "nonluminous" gas flame in which "flame reactions" were studied. In 1860 Bunsen in cooperation with Gustav Kirchhof published his first classical paper on emission spectroscopy entitled, "Chemical Analysis Through Observation of the Spectrum." The spectrum was produced with a spectroscope which consisted of a prism, a cigar box, and two ends of otherwise unusable old telescopes. The two authors predicted that spectral analysis would be the means for discovering new elements which either occurred in extremely small quantities or which analytical chemists were not able to distinguish from other elements with similar chemical properties. One year after publication of their paper, this prediction was fulfilled with the discovery of rubidium and cesium in spring water from Dürkheim.

The theory underlying spectrography and its further development became possible only in the late 19th century when the structure and properties of atoms and molecules became better understood, whereas emission spectrography became a routine method of analysis in this century, thanks to the revolutionary development of modern instrumentation.

The theoretical fundamentals of spectrophotometry were developed in the middle of the 18th century by Lambert, extended to solutions in 1852 by Beer, and are usually referred to as Lambert and Beer's law. As Szavadváry (9) states in his book: "We could equally well call these the Bougouer-Bernard laws, as these two Frenchmen established them independently of their German colleagues. Moreover, Bougouer (1698–1758) long preceded Lambert, while Bernard was only a few months behind Beer."

Colorimetry in a primitive form was practiced in the 18th century. However, apparent deviations from Beer's law, e.g., by formation of different complexes with different molar absorptivities as in the system CrO_4^{2-}, $HCrO_4^-$, $Cr_2O_7^{2-}$, or complexes of Cu^{2+} with amines, and also so-called dichroism in solutions (alkaline forms of some sulfonephthaleins), were not understood by analytical chemists toward the end of the last century.

Spectrophotometry was practiced by physical chemists at the end of the last and the beginning of this century. However, it became a popular technique and routine method of analysis only after simple and also self-recording spectrophotometers became available which allowed a rapid and accurate measurement of absorption in the visible light, now often referred to as colorimetry. These instruments made the classical Duboscq colorimeters obsolete.

Nephelometry in a primitive form was first used in the nineties by the well-known professor of physical chemistry at Harvard, T. W. Richards, in his atomic weight work based upon the determination of chloride as silver chloride. In 1915 Richards received the Nobel prize for his atomic weight work. Refined nephelometry (used for the determination of the concentration of suspensions) of particle size and molecular weight could be perfected only after the laws of light scattering were known. Lord Rayleigh in 1871 had discovered the laws of light scattering by molecules of a gas, but nephelometry could not become an exact technique until after Mie in 1908 developed a quantitative interpretation of the effect of particle size upon intensity and angular distribution of scattered light. Analytical chemists remained ignorant of these theoretical developments and practiced nephelometry in a purely empirical way; it was only during and after the Second World War, when so many of the modern techniques based upon radiation were discovered and/or developed, that nephelometry became a technique used for obtaining quantitative information on the number and size of particles in a suspension. This sketchy review of the historical development of optical methods of analysis serves to illustrate that this kind of analysis certainly deserved discussion in Ostwald's booklet. The tremendous analytical use of known optical techniques and the discovery of many new optical and radiation techniques date from the end of the Second World War.

Solution Analysis

We will now briefly discuss the development of solution analysis in the first quarter of this century. All of the fundamental principles have been contributed by physical and biochemists. By recognizing the importance of these fundamental studies for analytical chemistry, I have played a modest role in the establishment of analytical chemistry as a scientific discipline. It has been my good fortune to have been acquainted with scientific aspects of analytical chemistry by my teacher, Nicolaas Schoorl, who early in this century taught analytical chemistry at the University of Utrecht, Holland. Having received his education from the famous trio, van't Hoff, Bakhuis-Rozeboom (who introduced Gibbs' phase rule in Europe), and Lobry de Bruyn, professors in the University of Amsterdam, it is not surprising that Schoorl stressed the scientific aspects of analytical chemistry, which he taught both as an art and a science. When I took my first course in quantitative analysis in 1913, I had great difficulty in understanding the selection of the proper indicators in acid-base titrations. Several years before the classical paper by Sörensen (21) appeared, Schoorl in 1904 had published a series of papers in the Dutch Chemisch Weekblad in which he interpreted the main characteristics of acid-base indicators and explained the choice of indicator in the titration of weak acids or bases. These papers were not easy reading at that time, and in an effort to understand them, I found it necessary to make a thorough study of the literature on the subject. This study acquainted me with the work carried out by several *nonanalytical chemists* who may be considered the pioneers of scientific analytical chemistry. When talking about acid-base indicators, it is only natural to refer in the first place to the contributions by the Danish *physiological* chemist, S. P. L. Sörensen, who in 1901 succeeded Johann Kjeldahl as director of the Carlsberg Laboratory. In 1909 Sörensen (21) published in French in *Compt. Rend. du Lab. de Carlsberg* a classical paper in which he originated the notation of pH and in which he presented a detailed account of the colorimetric and potentiometric determination of pH and a set of buffer solutions, some of which are still in use. The limitations of both colorimetric and potentiometric methods with the hydrogen electrode were described in detail. For years to come, this paper and its German translation in 1910 in the *Biochem. Z.* (21) served as the most authoritative source of information on the determination of pH. It inspired me to explore the tremendous significance of pH in analytical chemistry, part of which study was presented in my first monograph (22) published in 1920. An English translation by Charles Rosenbloom of the third German edition was published in 1937 by Macmillan. It is only proper to mention here the name of another *physiological* chemist, an American, the late William Mansfield Clark, who in 1917 introduced the series of sulfonephthaleins which are still in daily use and a set of buffer solutions which are also still being used. In addition, he published the first series of oxidation reduction indicators, giving their reduction potentials and optical properties. The latest edition of Clark's book (23) is still a most authoritative source of information on pH, redox indicators, and equilibria. Sörensen visited me in Utrecht in 1917, and on a visit to this country in 1924, I spent a day with Wm. M. Clark. Both men impressed me deeply with their modesty. In addition to the monographs on the hydrogen ion concentration and on redox reactions by the German *biochemist* Leonor Michaelis (24), I like to single out particularly a monograph by the famous Danish *physical* chemist Niels J. Bjerrum (25) in 1914, entitled, "Die Theorie der alkalimetrischen und azidimetrischen Titrierungen," which has been instrumental in the scientific development of titrimetry. In this book Bjerrum calculated the shape of neutralization curves and titration errors in the visual end-point determinations. The book inspired me to start developing a theoretical interpretation of all methods of volumetric analysis which ultimately resulted in 1926 in the publication of a two-volume book (26), "I. Die Theorie der Massanalyse" and "II. Die Praxis der Massanalyse," translated into English by my friend the late N. H. Furman. Later, in cooperation with V. A. Stenger [and R. Belcher (Vol III)], a three-volume revision was published. Returning to the development of titrimetry after Bjerrum's publication, acid-base titrations with a few empirical exceptions were confined to aqueous systems. Empirically, the biochemist Foreman (27) in 1920 had developed a method for the titration with sodium hydroxide of amino acids in a medium high in alcohol or acetone. Only after Niels Bjerrum presented his Zwitterion (hybrid ion) theory of amino acids was the reaction on which the titration is based understood. The structure of an amino acid like glycine is $CH_2NH_3^+COO^-$, not CH_2NH_2COOH, and it is the NH_3^+ and not the COOH group which is being titrated with strong alkalis. For a quantitative interpretation of aqueous systems, the classical Arrhenius theory of acids and bases, on which Bjerrum's monograph is based and which was amended and refined by the Debye-Hückel theory in 1923, is still entirely adequate. However, for an interpretation of acid-base titrations in nonaqueous media, the Arrhenius theory is entirely inadequate. The Brønsted-Lowry theory (28), which was developed independent of each other by the two *physical* chemists in 1923, has had a tremendous impact upon the development of nonaqueous titrimetry.

Again, it was not an analytical, but a *physical-organic* chemist, James Bryant Conant (professor at Harvard, later president of Harvard, then Ambassador to West Germany, and finally an outstanding educator) who in 1927 made the first quantitative application of the Brønsted theory in his pioneer work with Norris Hall on basicity in glacial acetic acid (29). Development of nonaqueous acid-base titrimetry has been slow since, until some 15–20 years ago when there was an explosive development of the practice and theory of such titrations to which American analytical chemists particularly have made fundamental contributions.

A brief review of the history of acid-base titrations in aqueous and nonaqueous solvents would be incomplete without mentioning the G. N. Lewis theory of acids and bases, first proposed in 1923 by this famous American *physical* chemist (30). According to Lewis, a base is a donor of electrons and an acid an acceptor of electrons. Bases have at least one pair of unshared electrons which they can share with an acid with the formation of a coordination (neutralization) product. Thus, for example, boron trichloride, a Lewis acid, reacts in an inert solvent with dimethylaminoazobenzene with the formation of a red neutralization product, comparable to the addition of a proton to the indicator to form the red acid form. In a paper published in 1944 in the *J. Phys. Chem.* (31), I pointed out that we should distinguish between two classes of acids, Brønsted and Lewis acids. In a strict sense, Brønsted acids are the *neutralization* products of a base with a proton and not acids in the Lewis sense. Lewis was in complete agreement with this proposal, and the above notations are now generally in vogue. In many publications distinction is made between Lewis and Brønsted bases, a distinction which does not serve any purpose because every "Lewis" base is also a Brønsted base. The broad concept of acids by Lewis implies that all complexation titrations involve Lewis acid-base reactions, both in aqueous and nonaqueous media. Complexometric titrations, even in aqueous medium, were hardly used in the beginning of this century. Well known and in general use was the complexometric titration of cyanide with silver, based on the water solubility of $Ag(CN)_2^-$ and the insolubility of silver cyanide. The method is still referred to as the Liebig method, published by him in 1851 (*32*) and later perfected by Dénigès.

Complexes

At that time no advantage was taken by analytical chemists of the Werner theory of the structure of complexes. Alfred Werner (who lived from 1866 to 1919 and became a professor at the University of Zürich at the age of 27) made his monumental contribution in his "Habilitationsschrift" in 1891 at the young age of 26, when he developed the coordination theory and introduced the terminology "coordination number" and "coordination sphere." Not only is Werner rightfully considered the founder of scientific inorganic chemistry, but his theory has proved to be of great analytical importance. This was particularly recognized by Gerold Schwarzenbach, the successor of W. D. Treadwell at the Eidgenössische Technische Hochschule in Zürich. In the forties Schwarzenbach started his theoretical studies of the complex formation of aminopolycarboxylic acids with inorganic cations. Schwarzenbach's studies led him to the development of the titration of calcium and magnesium with EDTA and to the introduction of the first metallochromic indicators. Schwarzenbach is rightfully considered the originator and developer of modern complexometry (chelometry, chelatometry), the development of which is recent history.

I am omitting a review of the development of electroanalytical chemistry as this was given in a recent paper (*33*). I only mention that the famous German *physical* chemist Walther Nernst without any doubt should be called the father of modern electroanalytical chemistry. Not only did he in 1889 enrich the scientific world with the fundamental Nernst equation, but he also explored the analytical applications of this equation. Potentiometry, potentiometric titrimetry, and electrodeposition at controlled potential originated in the 1890's in his laboratory. In 1887 Nernst became an assistant to Wilhelm Ostwald in whose laboratory he originated the Nernst equation. This makes it almost inexplainable that Ostwald in the first and later editions of his monograph on the scientific fundamentals of analytical chemistry ignored Nernst's pioneer work in electroanalytical chemistry. Not even in his book, "Development of Electrochemistry" (*34*), published in 1916, in which Ostwald discusses in two pages the impact of electrochemistry upon the "rationalization of analytical chemistry," does he mention the fundamental studies by Nernst in the field of electroanalytical chemistry.

In conclusion of this sketchy review, it is fair to state that it was in the first quarter of this century that analytical chemists started to recognize the analytically important contributions made by physical and biochemists. This period can be characterized as the beginning of the scientific development of analytical solution chemistry. Since around 1930, analytical chemistry has not been taught solely as an art. The scientific fundamentals on which analytical methods were based gradually became incorporated in courses in qualitative and quantitative analysis. Textbooks published from the thirties to date present the theory on which procedures and techniques are based and no longer resemble the cookbook type of texts of an earlier period. Today, analytical chemists, together with chemists in the other disciplines, originate and develop new and perfect old methods and techniques on a sound scientific basis. As far as solution chemistry is concerned, modern analytical chemists now determine equilibrium constants of acid-base, oxidation-reduction, complexation and precipitation reactions, standard potentials, optical constants like molar absorptivities, and many other data, all of which belonged to the domain of physical chemistry in the first quarter of this century. Solution analysis is no longer confined to water as a solvent but has been greatly expanded to nonaqueous solvents, including fused salts. We analytical chemists are no longer the maidservants of other chemists, but together we contribute to further progress in the whole field of chemistry.

Literature Cited

(1) P. T. B. Paracelsus, "Bücher und Schriften," Basel, Switzerland, 1589.
(2) J. B. van Helmont, "Ortus medicinae...," Amsterdam, Holland, 1652; from J. R. Partington in "Great Chemists," Vol I, p 83, E. Farber, Ed., Interscience, New York, N.Y., 1961; Partington gives 170 references in van Helmont's biography.
(3) R. Boyle, "Philosophical Works," London, England, 1725.
(4) T. Bergman, "Opuscula physica et chemica," Uppsala-Stockholm, Sweden, 1779–88.
(5) A. L. Lavoisier, "Traité élémentaire de chimie," 1789; "Oeuvres," Paris, France, 1854 (from Ref. 9).
(6) J. Priestley, "Observations of Different Kinds of Air, 1772 (from Ref. 9).
(7) G. E. Stahl, "Specimen Beckerianum," Leipzig, Germany, 1703.
(8) F. A. H. Descroizilles, *J. des Arts et Manufact.*, **1**, 256 (1795); *Ann. Chim.*, **60**, 17 (1806).
(9) F. Szavadváry, "History of Analytical Chemistry," pp 197–284, Pergamon Press, London, England, 1966.
(10) J. L. Gay-Lussac, *Ann. Chim. Phys.* (a) **26**, 162 (1824); (b) **60**, 225 (1835).
(11) F. Mohr, "Lehrbuch der Chemisch-Analytischen Titriermethoden," Vieweg & Sohn, Braunschweig, Germany, 1855.
(12) J. Liebig, "Anleitung zur Analyse organischer Körper," Braunschweig, Germany, 1837.
(13) J. B. Dumas, *Ann. Chim. Phys.*, **47**, 198 (1831).
(14) J. G. Kjeldahl, *Z. Anal. Chem.*, **22**, 366 (1883).
(15) W. Ostwald, "Die wissenschaftlichen Grundlagen der analytischen Chemie," Verlag W. Engelmann, Leipzig, Germany, 1894.
(16) F. A. Gooch, *Proc. Amer. Acad.*, **13**, 342 (1878).
(17) J. H. van't Hoff, "Vorlesungen über theoretische und physikalische Chemie; Vol 1, Die chemische Dynamik; Vol 2, Die chemische Statik; Vol 3, Beziehungen zwischen Eigenschaften und Zusammensetzung," Friedrich Vieweg & Son, Braunschweig, Germany, 1898. In the first volume there is even a discussion (p 217) of the effect of solvent on reaction velocities. As an example, we quote that the rate constant of the reaction $(C_2H_5)_3N + IC_2H_5 \rightarrow (C_2H_5)NI$ at 100° is 0.00024 in hexane and 0.13 in benzylalcohol.
(18) E. Cohen, "Jacobus Henricus van't Hoff. Sein Leben und Wirken," Akadem. Verlagsges. m.b.H., Leipzig, Germany, 1912.
(19) W. Gibbs, *Trans. Conn. Acad.*, **3**, 108, 343 (1874–78).
(20) W. Herz, "Physikalische Chemie als Grundlage der analytischen Chemie," in series "Die Chemische Analyse," 114 pages, B. M. Margosches, Ed., Ferdinand Enke, Stuttgart, Germany, 1907.
(21) S. P. L. Sörensen, *Compt. Rend. du Lab. de Carlsberg*, **8**, 28 (1909); *Biochem. Z.*, **21**, 159 (1909).
(22) I. M. Kolthoff, "Der Gebrauch von Farbindikatoren," Julius Springer, Berlin, Germany, 1920; 4th German ed., 1932.
(23) W. M. Clark, "The Determination of Hydrogen Ions," 1920; "Oxidation Reduction Potentials of Organic Systems,"

Williams and Wilkins Co., Baltimore, Md., 1960.
(24) L. Michaelis, "Oxidations-Reduktions Potentiale," Part II, "Die Wasserstoffionenkonzentration," Julius Springer, Berlin, Germany, 1916.
(25) N. Bjerrum, "Die Theorie der alkalimetrischen und azidimetrischen Titrierungen," Sammlung Herz, Verlag Enke, Stuttgart, Germany, 1914.
(26) I. M. Kolthoff, "I. Die Theorie der Massanalyse," "II. Die Praxis der Massanalyse," F. Springer, Berlin, Germany, 1927.
(27) F. W. Foreman, *Biochem. J.*, **14**, 547 (1920).
(28) J. N. Brønsted, *Rec. Trav. Chim.*, **42**, 718 (1923); *Chem. Rev.*, **5**, 231 (1928).
(29) T. M. Lowry, *Chem. Ind. (London)*, **42**, 43 (1923); *Trans. Faraday Soc.*, **20**, 13 (1924).
(29) J. B. Conant and N. F. Hall, *J. Amer. Chem. Soc.*, **49**, 3047, 3062 (1927).
(30) G. N. Lewis, "Valence and the Structure of Atoms and Molecules," Chemical Catalog Co., New York, N.Y., 1923.
(31) I. M. Kolthoff, *J. Phys. Chem.*, **48**, 51 (1944).
(32) J. Liebig, *Lieb. Ann.*, **77**, 102 (1851).
(33) I. M. Kolthoff, *J. Electrochem. Soc.*, **118**, 5C (1971).
(34) W. Ostwald, "Entwicklung der Elektrochemie," p 176, Verlag J. A. Barth, Leipzig, Germany, 1910.

Based on talk given at the Fisher Award Anniversary Symposium in Boston, Mass., at the National ACS Meeting, April 9 to 14, 1972.

Izaak M. Kolthoff, *professor emeritus at the University of Minnesota, has been a leader in the field of electroanalytical chemistry and continues to be an active researcher. His current interests include studies of acid-base equilibria in nonaqueous solvents. He is the author or coauthor of nine books and more than 850 publications and is an editor of "Treatise on Analytical Chemistry." He is also a commission chairman for the International Union of Pure and Applied Chemistry. Professor Kolthoff has received many awards and honors, and in June 1972 a new chemistry building at the University of Minnesota was dedicated as Kolthoff Hall in his honor.*

CARBON 13

Nuclear Magnetic Resonance Spectroscopy

George A. Gray
Varian Instrument Division
25 Route 22
Springfield, NJ 07081

Since its discovery in the 1940's, nuclear magnetic resonance (NMR) spectroscopy has had a spectacular growth and acceptance in chemistry. Until recently, sensitivity considerations had limited its application to high abundance magnetically active nuclei such as ^1H, ^{19}F, and ^{31}P for which a conventional field or rf frequency sweep can elicit strong resonances, even on very small samples. Progress in instrumentation has centered on: increasing the applied magnetic field to increase spectral dispersion and simplify resonance patterns; improving the homogeneity of the applied magnetic field for greater resolution of spectral lines; stabilization of frequency field ratios for drift-free operation; sensitivity improvement; multiple-resonance capability (decoupling); and "other nuclei" observation. These developments in conventional swept (continuous wave or CW) NMR have resulted in the availability of low-cost routine proton NMR spectrometers and their consequent widespread use both in research and practical endeavors. The low sensitivity (Table I) and often small natural abundance of nuclei "other" than ^1H, ^{19}F, and ^{31}P have retarded their use until recently.

Given a choice, a chemist would certainly consider the NMR spectrum of carbon to be of fundamental and desirable importance, while recognizing that the practical development of the field has been guided primarily by what has been permissible technologically. And while it is now possible to obtain ^{13}C NMR spectra, the very factors that have retarded its development are actually now recognized to be advantageous. This seeming contradiction is easily explained. The low natural abundance of the stable magnetically active nuclide of carbon (^{13}C, 1.1% natural abundance, spin = $\frac{1}{2}$) makes the chances of two ^{13}C's being in one molecule very small, obviating any complications of *homo*nuclear spin-spin (J) coupling. (If desired, extended time-averaging or isotopic enrichment does make this observation possible.) Upon complete proton decoupling (see below), therefore, the ^{13}C NMR spectrum consists of single lines, one for each chemically shifted carbon. The routine application of this technique has depended on several technological advances such as broad-banded proton decoupling, time-averaging, and pulsed Fourier transform NMR.

Table I. NMR Nuclei

Isotope	Spin	Abundance, %	Rel sensitivity	Frequency (MHz) at 23 kG
^1H	1/2	99.98	1000	100
^2H	1	0.015	9.65	15.4
^{13}C	1/2	1.11	15.9	25.1
^{14}N	1	99.63	1.01	7.2
^{15}N	1/2	0.37	1.04	10.1
^{17}O	5/2	0.04	29.1	13.6
^{19}F	1/2	100	833	94.1
^{31}P	1/2	100	66.3	40.5

Proton Decoupling in ^{13}C NMR

Since the proton is a spin-½ nucleus, it can spin couple to ^{13}C and produce a "coupled" spectrum. For example, a methyl carbon would appear as a quartet with intensities of 1:3:3:1. If an intense ^{1}H rf decoupling field is used to irradiate the protons of the methyl group, collapse will occur, giving a narrow line centered at the chemical shift of the methyl carbon. In comparing the areas of the coupled and decoupled spectra, it is usually noticed that one obtains up to almost three times as much area as expected from simple multiplet collapse. This "extra" intensity is the nuclear Overhauser enhancement. It results in considerable sensitivity enhancement since a factor of three greater signal saves a factor of nine in time-averaging in attaining a desired signal-to-noise ratio.

The above decoupling experiment can be termed "coherent single-frequency" decoupling since only one type of proton in the proton spectrum was irradiated. For a complicated molecule this approach is impractical, and one resorts to application of broadbanded decoupling in which the coherent proton rf is modulated with "white" noise. If the frequency spread of this noise is greater than the range of proton frequencies and high enough power is employed, the various ^{13}C–^{1}H multiplets are collapsed, giving single lines for all carbons formerly coupled to protons.

Further advantage is realized since no detailed consideration is necessary for selection of the ^{1}H rf frequency.

Time-Averaging

Single-scan NMR spectra have provided and still provide the vast majority of proton NMR spectra. Weak nuclei such as ^{13}C and low-level proton spectra require time-averaging (multiscanning) for extraction of the signals from within the noise. In the CW mode this requires very long periods of time since each scan usually takes ~500 sec. This large expenditure of time has discouraged its routine use for signal-to-noise enhancement in the CW mode. One principal advantage of the pulsed Fourier transform method is that each scan requires only on the order of 1 sec. Hence, 100 scans can give up to a factor of 10 in signal-to-noise improvement and 10^4 scans up to a corresponding 100-fold improvement (the signal-to-noise ratio improves as the square root of the number of scans). The latter improvement alone can overcome the unfavorable natural abundance factor for ^{13}C.

Pulsed Fourier Transform NMR

In a nuclear spin system at equilibrium the steady state and impulse response form a Fourier transform (FT) pair. That is, the time response of the spins can be calculated from their frequency domain spectrum and vice versa. This is extremely useful for obtaining spectra quickly so that multiscan averaging can be performed. The basic FT experiment is illustrated in Figure 1. Rather than a continuous, weak (mW) rf field, we employ a powerful (kW) field which is pulsed on for times as short as a few microseconds. The response of the entire spin system is picked up in the normal manner, amplified, and detected in the spectrometer system. The resultant signal is digitized in an analog-to-digital converter (ADC) and stored in a computer. Subsequent "transients" are taken in a similar manner and are coherently added to the previously stored data for signal-to-noise improvement. After completion of the data accumulation and the Fourier transformation, the frequency domain data may be plotted in the normal manner.

There are a few simple relationships which are fundamental to an understanding of FT spectroscopy. First, to unambiguously assign a frequency of "f" Hz, the ADC must perform conversions at a rate of at least "2f" conversions/sec. Hence, a spectral width of 1000 Hz (the transmitter frequency is defined as 0 Hz) requires an ADC rate of 2000 conversions/sec. Each conversion will result in a time-indexed sampling of the transient response which must be added to the appropriately stored samplings in the computer memory. Therefore, the length of time spent acquiring data will determine the number of conversions and hence the number of memory channels used for data storage. Of course, the normal case is that of a limited memory capacity, typically 8192 channels or "words". Hence, for a full data table, 8192 = (ADC rate) × (acquisition time); 8192 = 2 × SW × AT.

Another fundamental relationship is that the attainable resolution is determined by the length of time recording the response of the spin system to the pulse (the free induction decay or FID), i.e., resolution ~ 1/AT (Figure 2). Of course, the limiting resolution attainable will be that corresponding to the natural linewidth $(\pi T_2)^{-1}$ or magnet homogeneity-dictated linewidth.

Characteristics of ^{13}C NMR Spectrum

In many respects, the ^{13}C chemical shift scale (Figure 3) closely follows that of the proton, even including the commonly accepted internal reference tetramethylsilane (TMS). Taking TMS as the zero of the scale, resonances to higher *frequency* (at constant field) are considered to have *positive* chemical shifts in ppm—the same as the proton "δ" scale. Aliphatic carbons are at higher shielding (lower frequencies) and therefore have the

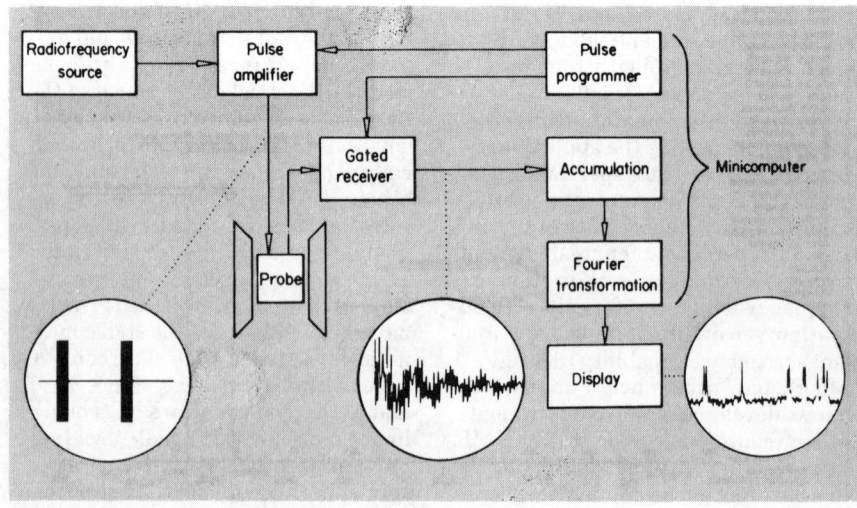

Figure 1. Schematic of FT spectrometer

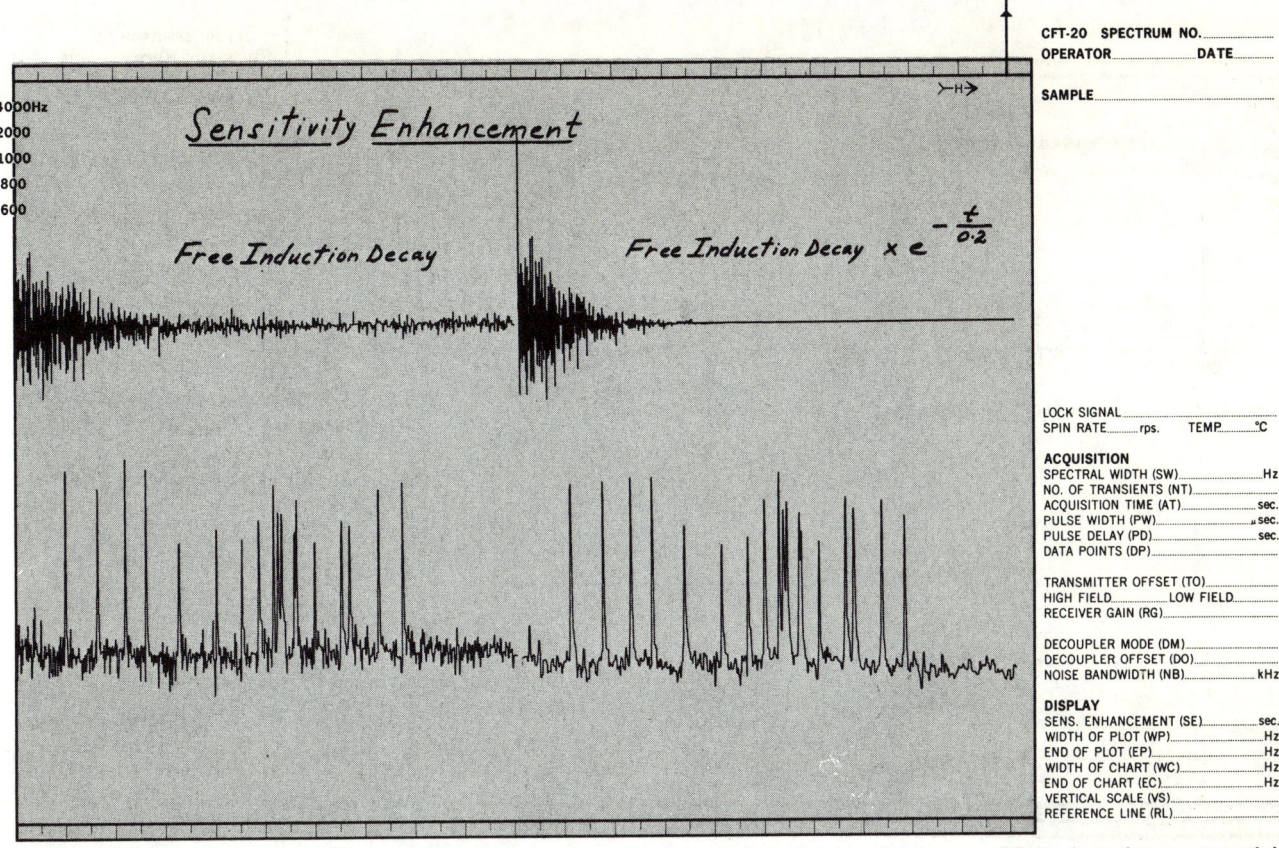

Figure 2. Signal-to-noise improvement obtained (at expense of some resolution) by multiplying raw FID by decaying exponential function $e^{t/(-SE)}$, thus favoring high S/N portion of FID over low S/N portion

smaller chemical shifts. With increasing substitution, and particularly heteroatom substitution, the resonances appear at greater chemical shifts. Olefinic and aromatic carbons occur at 90–170 ppm, whereas carbonyl carbons have chemical shifts at 160–220 ppm. Note that in ppm the carbon scale is greater than 20 times as wide as that of the proton. Since the linewidths are similar, this represents a true increase in dispersion.

Considerable success has been realized in empirical prediction of ^{13}C chemical shifts based on substituent parameters. Usually, this approach relies on model compounds and a minimal number of substituted compounds. For example, Grant and Paul (1) and later Lindeman and Adams (2) have systematically developed regressional analyses for linear and branched hydrocarbons which give excellent comparisons between calculated and experimental shifts. Thus, the chemical shift of a paraffinic carbon k can be expressed as:

$$\delta^k = A + \sum_i B_i n_{ik}$$

where A represents an appropriate reference compound, B_i the ith substituent effect, and n_{ik} the number of substituents of type i relative to carbon k. Typical substituent effects (appropriate for the reference methane) are 9.1 ppm for replacement of a proton by a carbon (α effect), 9.4 ppm for replacement of a β hydrogen, and -2.5 for replacement of a γ hydrogen. Further parameters are necessary for branching situations, but they are straightforward, and calculating shifts for an *arbitrary* branched or linear hydrocarbon is simply an arithmetic exercise. Similar parameter sets have been developed for cyclic hydrocarbons (3).

Replacement of a substituent carbon with a polar substituent leads to extremely large effects on the ^{13}C shifts.

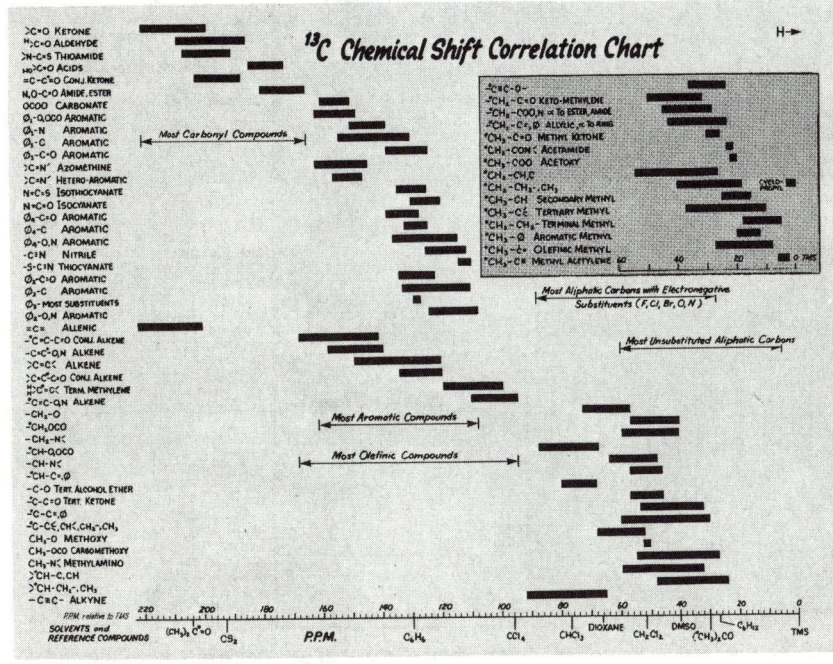

Figure 3. ^{13}C chemical shift correlations

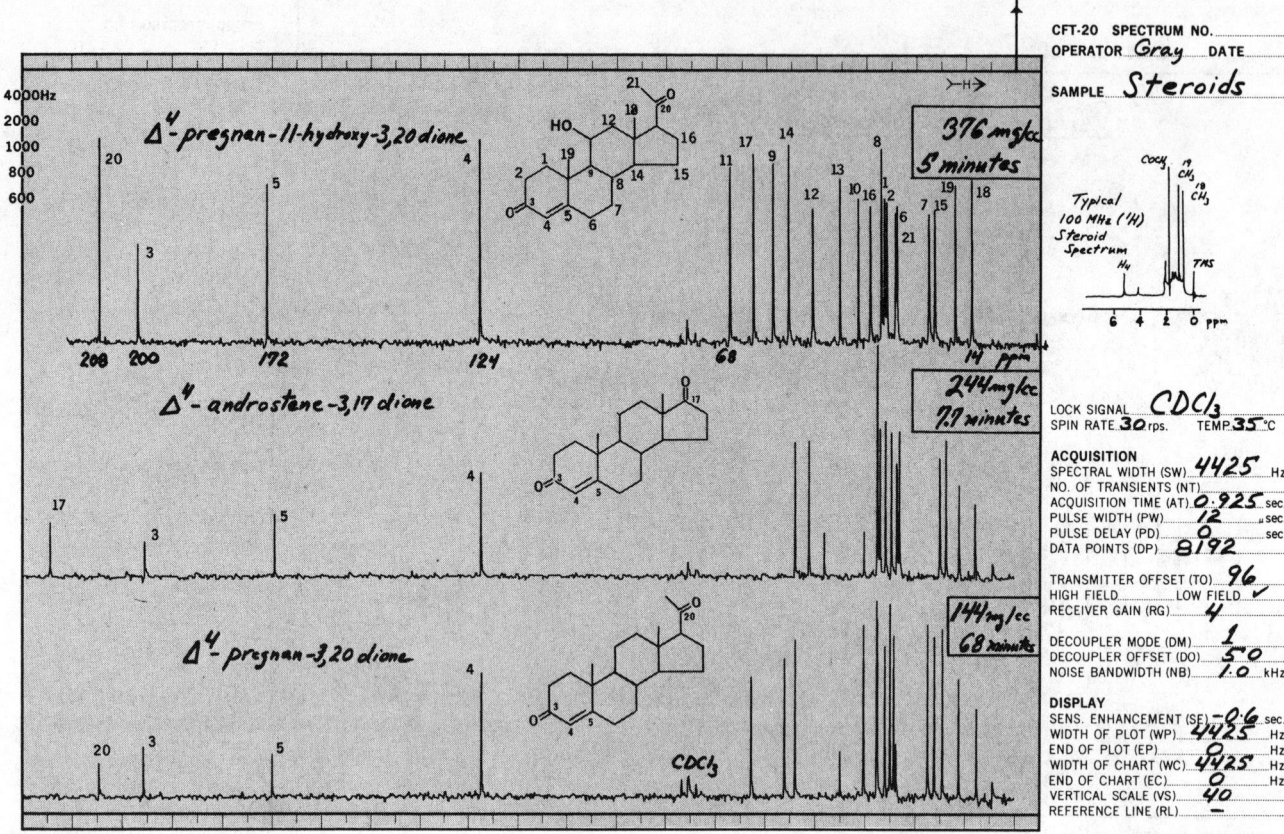

Figure 4. Closely related steroids showing sensitivity of ^{13}C shifts to chemical structure. Note only a few resonances may be assigned in any proton spectrum even at 100 MHz, and proton shift range covers 6 ppm whereas carbon shift ranges over 200 ppm

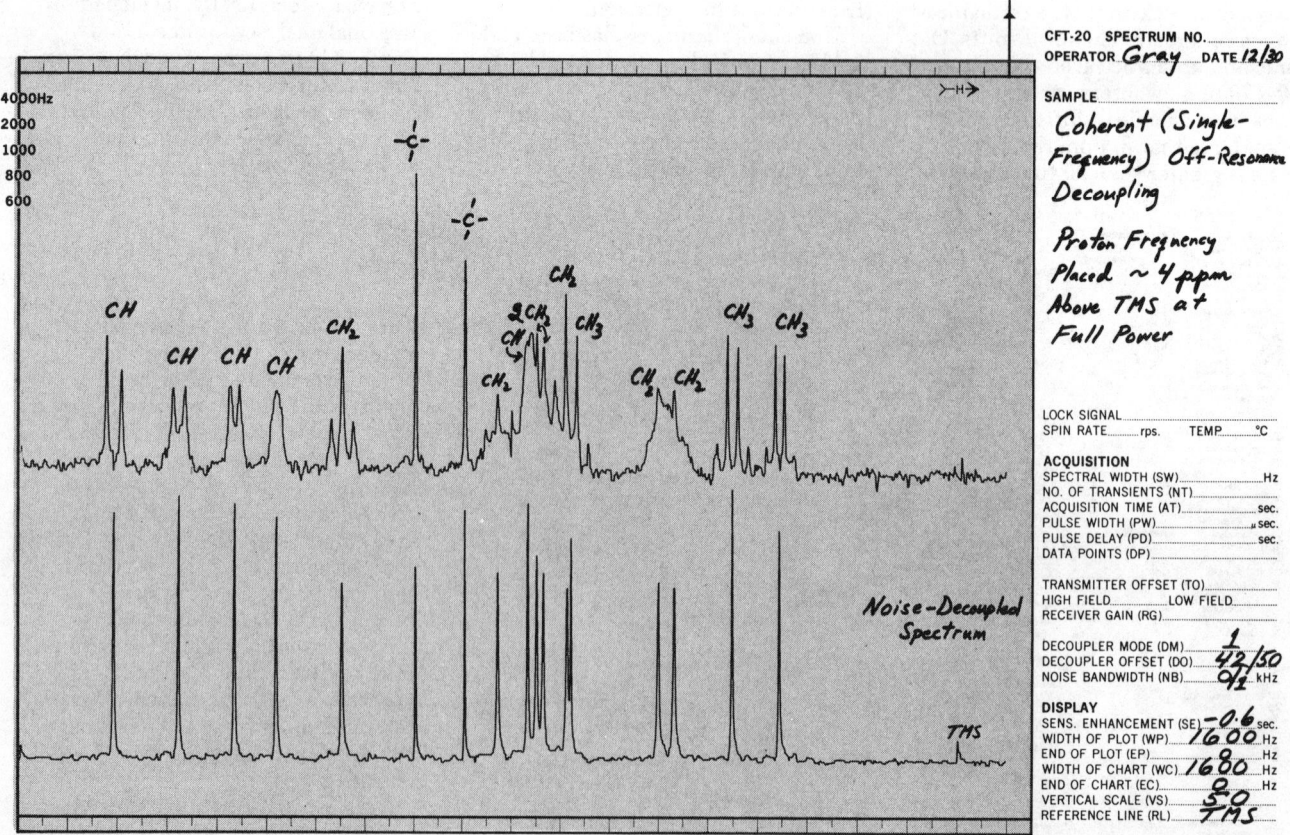

Figure 5. Off-resonance spectra allow determination of degree of protonation of carbon. Only changes necessary are moving decoupler frequency *out* of proton region and removal of noise bandwidth

Changes of ^{13}C Chemical Shifts upon Replacement of Methyl Group by Polar Substituent

Substituent	C-1	C-2	C-3
—OR	+45	−3	−1
—OH	+40	+1	−1
—OCOR	+43	−2	−1
—NH$_2$	+20	+2	−1
—Cl	+23	+2	−1
—F	+61	−1	−2
—COX	+15	−5	0
—COOR	+10	−1	−1
—COOH	+12	−3	−1
—CN	−2	−1	−1

R = alkyl, X = Cl or NH$_2$

Depending on substituents, carbonyl shifts range from 152.0 ppm for $(C_6H_5O)_2C=O$ to 215.8 ppm for di-t-butylketone. Aromatic and olefinic carbons exhibit similar large ranges of shifts and have also been systematically cataloged (4).

This inherent sensitivity to substituents is responsible for the power of ^{13}C NMR in structural studies (Figure 4). The shifts alone give significant information in this regard but are usually insufficient, except in the simplest cases, to completely assign the resonances and subsequently formulate a structure for an unknown compound. To attain these objectives, the chemist has several powerful diagnostic techniques:
- Coherent single-frequency off-resonance decoupling
- Gated decoupling
- Suppressed Overhauser spectra
- Selective proton decoupling
- $^{13}C-^1H$ shift cross-correlation
- Deuteration
- Specific labeling
- Spin-lattice relaxation time measurements.

Coherent Single-Frequency Off-Resonance Decoupling

When the ^1H decoupling power is insufficient and/or the ^1H frequency not exactly set at the resonance frequency of the proton(s) coupled to the carbon of interest, there will be some residual coupling, manifested as a line broadening or actual breaking up of the singlet into the *type* of multiplet characteristic for the number of protons on the carbon. Thus, a methyl carbon will appear as a compressed quartet, a methylene as a triplet, and a methine as a doublet. The advantages of this are the characterization of the carbon as to degree of protonation, retention of the full nuclear Overhauser effect, and reduced splittings, making interpretation convenient. In most cases, useful "off-resonance" spectra can be obtained in about the same time as the noise-decoupled spectra (Figure 5).

Gated Decoupling

A *coupled* spectrum may be obtained simply by turning off the proton decoupler, although a significant sensitivity penalty is often paid since no decoupling is permitted. It is possible to gain back some of the Overhauser enhancement by turning on the decoupler for a period just prior to the monitoring pulse and acquisition (Figure 6). Since the pulse samples the state of magnetization existing immediately *prior* to the pulse, the coupled spectrum *and* Overhauser enhancement may be obtained by automatically turning off the decoupler before the pulse and on again after completion of acquisition. Couplings appear instantaneously so that a fully coupled spectrum is obtained. The reverse "gating" sequence accomplishes the "suppressed Overhauser spectrum".

Suppressed Overhauser Spectrum

In this mode the decoupler is turned *on* just prior to the pulse and off *after* acquisition. Any Overhauser enhancement built up during the acquisition does not affect the signal being recorded since it only contributes to the magnetization along the magnetic

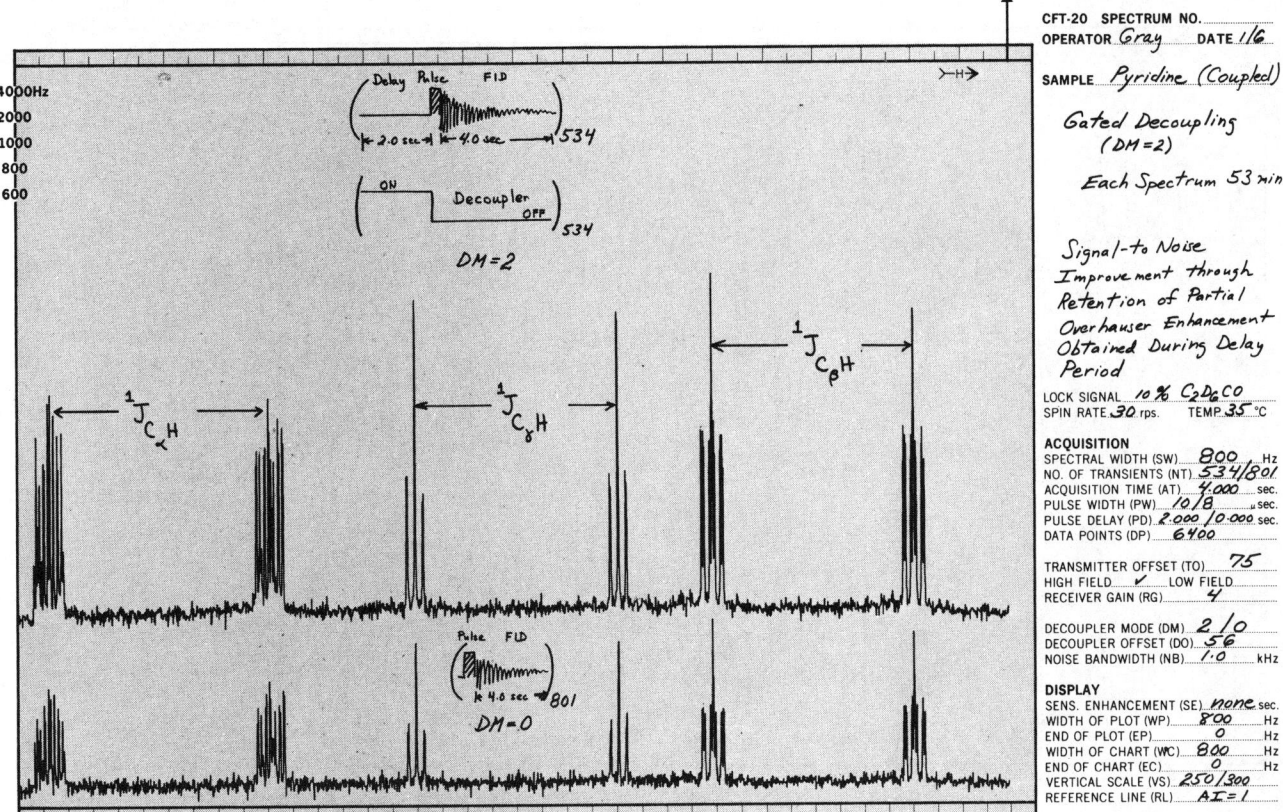

Figure 6. Greater signal-to-noise ratios obtained in given amount of time through use of gated decoupling

Insertion of pulse delay and setting DM = 2 automatically sets proper conditions. Note about a factor of two in S/N gained through use of gated decoupling, or same S/N as in lower trace could have been obtained in one-fourth of time. Obviously, even in very simple molecules, a very substantial S/N price paid to obtain coupled spectrum

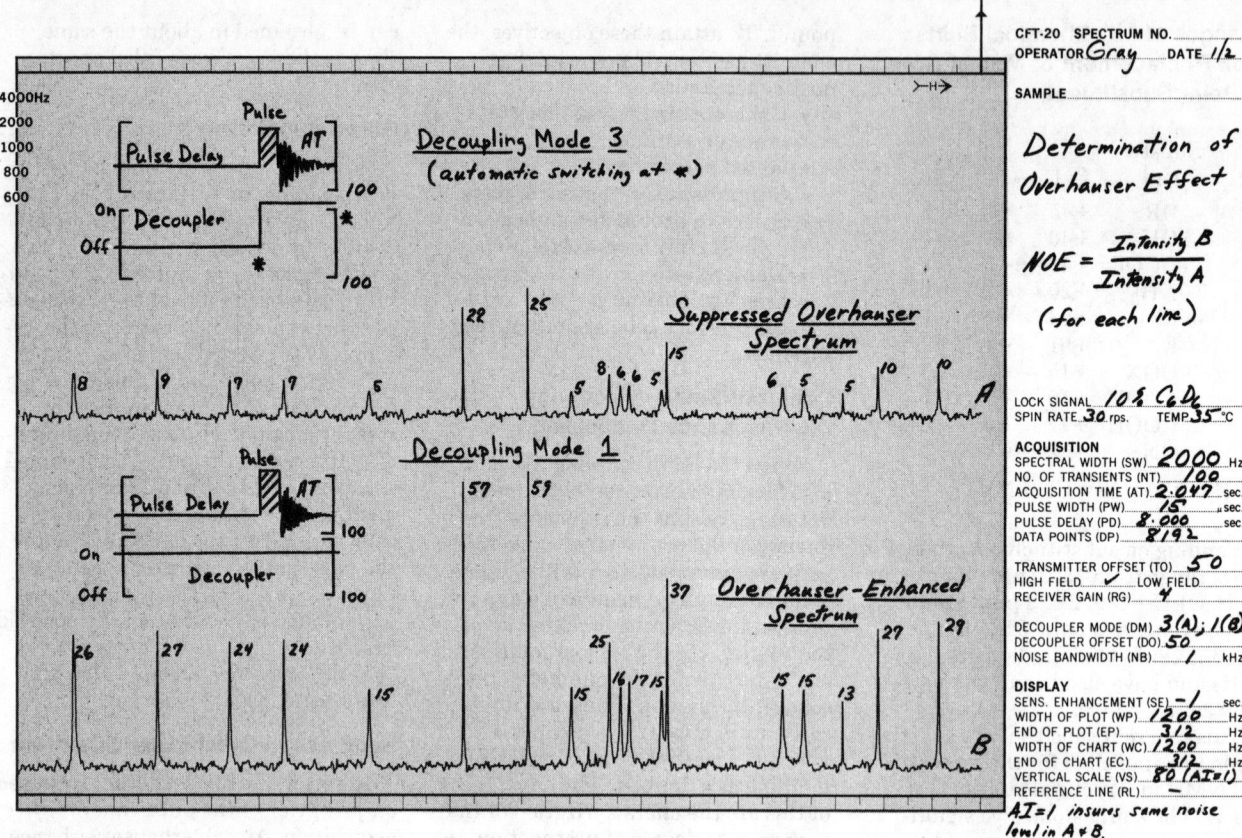

Figure 7. Only difference in two spectra is presence or absence of decoupling prior to pulse and concomitant nuclear Overhauser enhancement. Ratio of intensities or, better, areas, gives NOE "factor" directly. Decoupler gating automatic as specified by selecting DM parameter 1 or 3

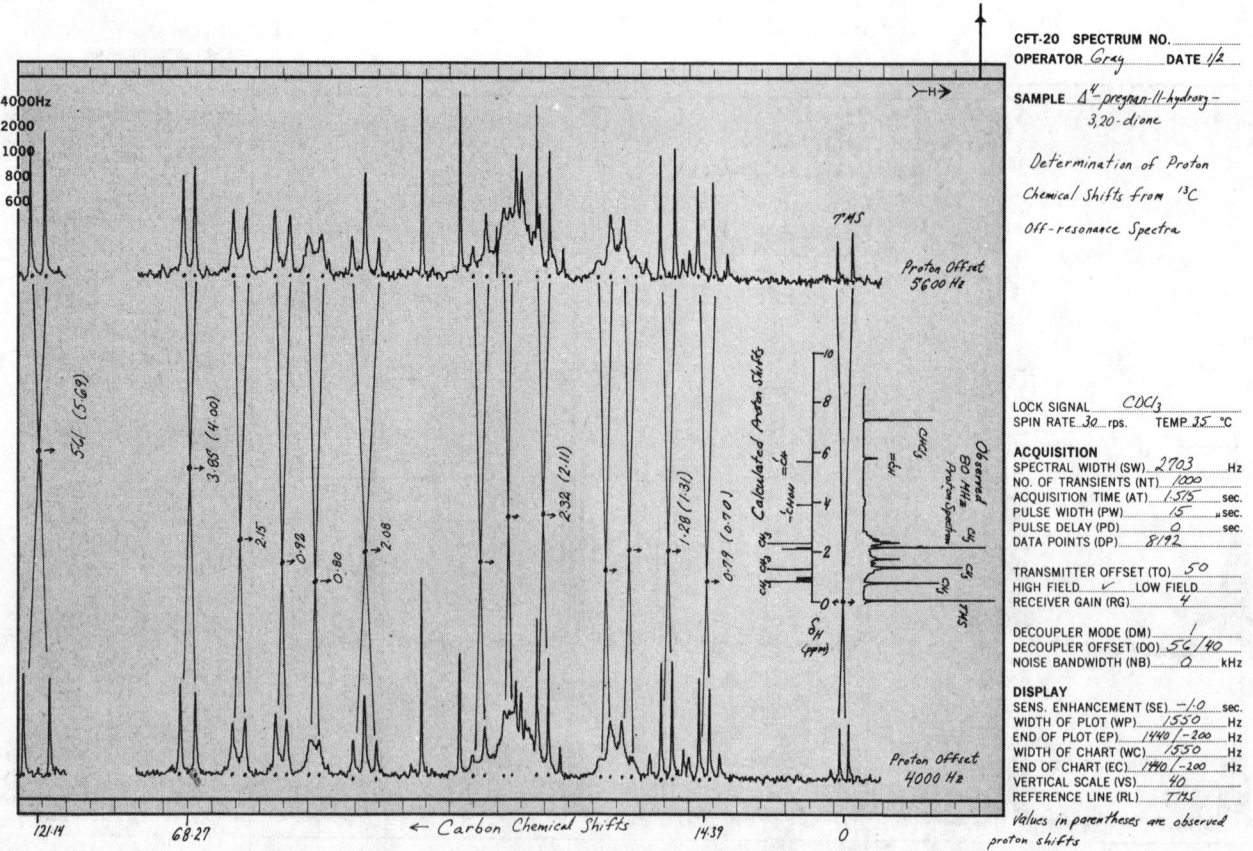

Figure 8. Method uses at least two off-resonance spectra (containing TMS) to establish proton shifts

For case of two as shown, baselines are positioned to provide convenient vertical proton frequency scale which can be converted subsequently to ppm. Two proton offsets bracket proton spectrum. Proton shift numbers taken from more accurate plots of residual couplings for spectra where proton offsets were 5800 (DO = 58), 5600, 5100, 4200, 4000, and 3800 Hz (DO = 38). Proton spectra of compound would only permit methyl, CHOH, and olefinic proton assignment

field (z) axis, while the receiver coil only picks up signals along one of the other orthogonal axes. Any subsequent monitoring pulse would measure signals with Overhauser enhancement (Figure 7). This may be avoided by making the interval between the end of acquisition and the next pulse at least 3-5 times as long as the acquisition time. If this delay is also ≥3-5 T_1^{max} (see below), each carbon will have completely returned to its equilibrium magnetization. The spectral integrations, therefore, are reflective of carbon concentrations and useful for quantitative analysis. Further, when this measurement is compared with data obtained under identical conditions except that the decoupler is left on continuously, the ratio of line intensities gives directly the nuclear Overhauser enhancement *factors* (minimum of 1.00 to maximum of 2.988). These factors are valuable in analyzing and using spin-lattice relaxation time data.

Selective Proton Decoupling

Further information may be obtained by relating one particular carbon to a proton (or protons) to which it is coupled, particularly if the proton spectrum gives structurally unambiguous information. Specific decoupling (single-frequency *on*-resonance) will result in a sharp singlet for any carbon directly bonded to protons resonating at that frequency. Certain complications can occur in those cases with overlapping $^{13}C-^1H$ satellite spectra (recall that it is the satellite peaks which must be irradiated, not those in the normal proton spectrum) or when the number of signals is too large for practicability. Both of these objections are circumvented by a simple graphical technique.

^{13}C, ^1H Chemical Shift Cross-Correlation

Selective decoupling is a time-consuming and tedious operation and therefore of limited use. A far faster and more informative technique utilizes two or more "off-resonance" spectra in a graphical manner to correlate individual carbons and protons (Figure 8). The residual $^{13}C-^1H$ couplings are plotted as a function of decoupler frequency, and points of intersection give exact decoupling frequencies. When an internal reference such as TMS is used, the decoupling frequencies can be placed on the chemical shift ppm scale, thus connecting the observed proton shifts with the observed carbon shifts. This is done *simultaneously* for all protonated carbons and hence is productive and useful as a routine diagnostic tool.

Deuteration

Replacement of hydrogen by deuterium is a standard and well-used technique in chemistry, and it has dramatic effects in the ^{13}C NMR spectrum. Deuterium has a spin of 1, is not affected by the proton decoupling, and produces a $^{13}C-^2H$ multiplet in the ^{13}C spectrum. If a carbon is completely deuterated, its spin-lattice relaxation time is greatly increased, so that its resonance is easily saturated and consequently its intensity in the normal rapidly pulsed mode is greatly diminished, perhaps to within the noise level. The absence of a resonance therefore indicates the deuterated carbon. Neighboring protonated carbons, however, still spin couple to the deuterium atoms, marking those carbons within two bonds of the originally labeled carbon. Sequence information is therefore an additional benefit of this technique.

Specific Labeling

Deuterium is just one example of specific labeling. This may alternately be done with both magnetically active *or inactive* nuclei. For example, labeling with ^{13}C *depleted* material would also give rise to one or more missing resonances when compared to the natural abundance ^{13}C spectrum. ^{15}N enrichments of ~90% are available commercially. Use of the ^{15}N will produce $^{13}C-^{15}N$ splittings of those resonances corresponding to carbons in proximity to nitrogen. Of course, ^{13}C enrichment is a classic avenue toward better signal-to-noise, but it can also be used to mark one type of carbon and follow its resonance unambiguously. This is extremely important for low concentration biochemical samples or in following metabolic or process development.

Spin-Lattice Relaxation (T_1)

The essentially instantaneous sampling of the magnetization state of the sample via a pulse allows following of time-dependent phenomena which were very difficult or impossible by traditional CW spectroscopy. One time-dependent process is the actual magnetization of the sample following placing the sample into the probe. In CW spectroscopy the relative slowness of the spectral determination made this time inconsequential. Our interest in this "relaxation" process (recall that the equilibrium populations in the spin levels are equal prior to placing the sample in the magnetic field; therefore, the sample is in an "excited" state) arises from the observation that difficult spectral lines in the NMR spectrum can and do have different relaxation times, differences that can be interpreted in terms of chemical and physical concepts.

Spin-lattice relaxation times can be determined experimentally using a two-pulse sequence as diagrammed in Figure 9. The 180° pulse inverts the magnetization to along the −z axis. At an arbitrary time later, the state of magnetization is sampled by a 90° pulse. For very short times between the 180° and 90° pulses, the resulting spectrum will appear inverted. Of course, as the interval time becomes very long, it is just equivalent to a single 90° sampling pulse, producing upright peaks. When a set of different experiments is performed at different interval times, a sequence of spectra is obtained in which the peaks start out negative and systematically go positive. The value of this process is the variability of rates of return to equilibrium for the various lines in the spectrum, a variability which can be tied to the chemistry of the molecule.

Although more specialized theory is necessary to explain T_1's of macromolecules tumbling very slowly in solution, there is a very simple form for the T_1 of a carbon in intermediate to small molecules (MW <10,000):

$$\frac{1}{T_1^{obs}} \cong \frac{1}{T_1^{dipolar}} = \sum_{\substack{all \\ H}} \frac{N_H}{r_{CH}^6} \cdot \tau_c$$

where N_H is the number of protons, r_{CH} the distance between the carbon and those N_H protons, and τ_c is the rotational correlation time for reorientation of the C—H vector (assumed isotropic here). Distance to nearby (r^{-6} dependence) protons and mobility of the C—H fragment are thus seen to be the controlling factors in determining the T_1 (within this dipolar mechanism) (Figure 9). There are other mechanisms active, particularly for very small molecules (MW <200) and in special situations. But for most organic molecules with 200 < MW < 10,000, this mechanism predominates.

Spin-lattice relaxation times are fast becoming another parameter with practical as well as theoretical significance. This follows from the inherent variability of these numbers for different lines in a spectrum. With the above formulation it is possible to rationalize differences in T_1's solely based on chemical structure and dynamics. After T_1's have been corrected for the number of protons, the resulting $N_H T_1$ values are directly proportional to *mobility* of the molecular fragment containing the C—H fragment. A graphical example of this is observed in branched polymers where side-chain carbons (of greater mobility) have T_1's much longer than backbone carbons. For example, poly (*n*-

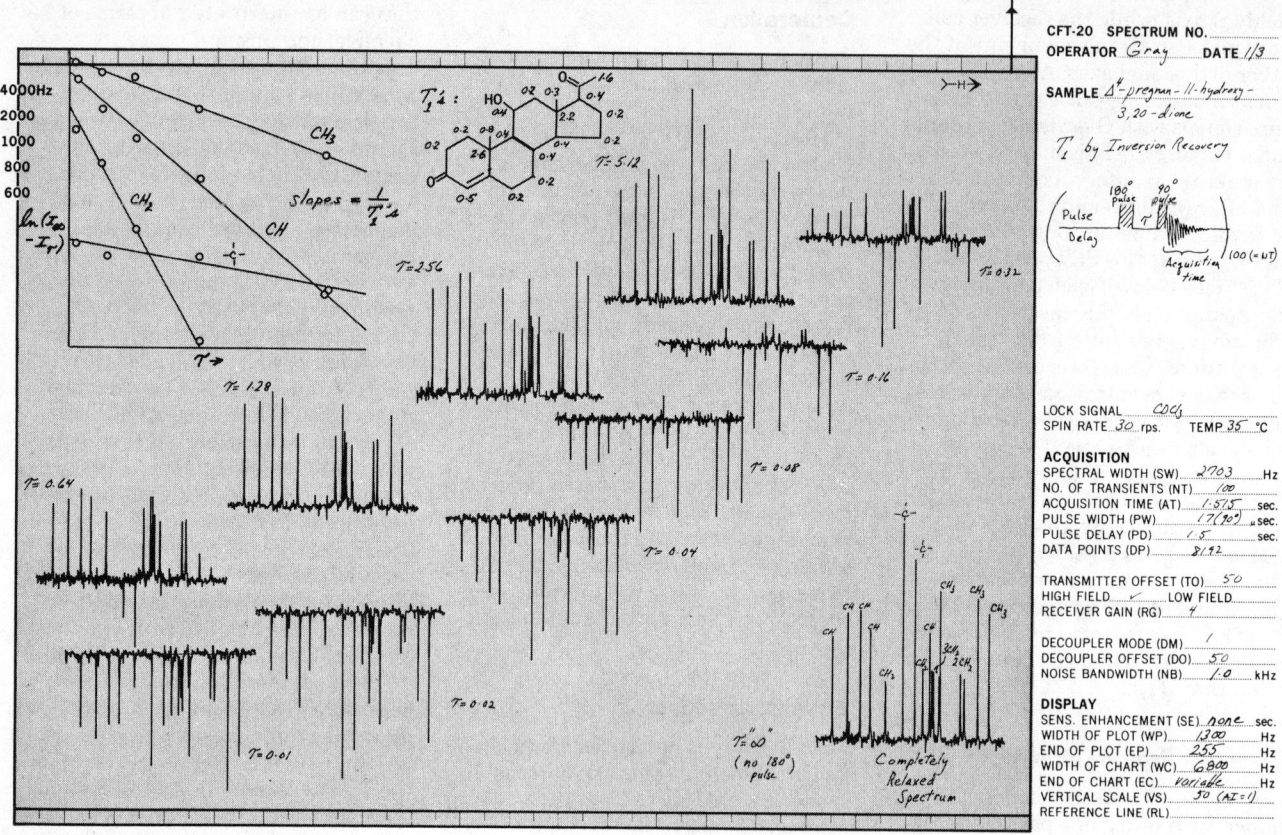

Figure 9. T_1's determined using several partially relaxed spectra

Pulse delay and acquisition time should be at least 3 T_1^{max}, a condition not met for quaternaries. Their T_1's will therefore be somewhat longer. Intensities for given τ value, I_τ, coupled with intensities of completely relaxed spectrum, I_∞, allow plotting (upper left) to obtain T_1's. Note T_1's directly related to number of protons for methylene and methine carbons. Methyl T_1's are longer than what might be expected because of their internal rotation, leading to shorter τ_c's. Intensities in completely relaxed spectrum are not proportional to populations since T_2's are also very short, leading to linewidths $CH_2 > CH > CH_3 > C$. Integrals directly related to population

butylmethacrylate) has T_1's (5) (in sec):

$$\begin{bmatrix} \text{CH}_3 & 0.031 \\ 0.6 & \\ -\text{C}-\text{CH}_2- & 0.075 \\ | & \\ \text{O}=\text{C} & 1.100 \\ | & \\ \text{CH}_2 & 0.120 \\ | & \\ \text{CH}_2 & 0.310 \\ | & \\ \text{CH}_2 & 0.760 \\ | & \\ \text{CH}_3 & 1.390 \end{bmatrix}_n$$

Although the backbone quaternary has low mobility like its neighboring CH_2, the distance to the nearest proton is considerably greater. Therefore, its T_1 is relatively long, as is that of the carbonyl carbon. The T_1's for the methylenes are directly comparable in terms of mobility. The tenfold increase in T_1 points out the substantial segmental motion in the butyl chain. The methyl carbons are interesting in that the side-chain methyl has a T_1 characteristic of a much smaller molecule, indicating a high degree of internal motion, whereas the backbone methyl has an unusually short T_1, appropriate to very slow motions. This dispersion of T_1 values bodes well for detailed analysis of motion in polymer chains. Of course, biopolymers will also have these same characteristics.

The utility of two-pulse sequences need not be restricted to full spin-lattice relaxation time measurements. If a reasonable T_1 prediction can be made for two lines in a spectrum too close to assign by chemical shifts where these T_1's are expected to differ substantially, a simple $[180° \ldots t \ldots 90° \text{ (FID)} \ldots \text{Delay}]_N$ sequence can result in one peak inverted (longer T_1) while the other is upright (shorter T_1), thus assigning the resonances. Even where no reasonable prediction can be made, this two-pulse sequence can provide helpful information.

Practical Considerations

Samples usually fall into two categories: sample limited or solubility limited. In the latter case, signal-to-noise ratios are obtained by using the largest tube, for example, 10 or 12 mm o.d. A typical gain of three to fourfold is realized for the same concentration in going from a 5- to a 10-mm tube. The sample-limited situation is best handled by dissolving the material in the least amount of solvent and running in a tube so that the sample height is no greater than 2–3 times the receiver coil height. Thus, 0.1-cc volumes are practical for 5-mm tubes, and 1 cc represents about the minimum for a 10-mm tube. In combining these figures, it is obvious that running a 0.1-cc sample in a 5-mm tube can result in a gain of about a factor of three in signal strength.

Fifty milligrams of a compound of MW 200 dissolved in 1 cc in a 10-mm tube should result in a usable spectrum in a few minutes. If 5 mg of the same compound were dissolved in 0.2 cc in a 5-mm tube and run in a 5-mm probe, it would produce a usable spectrum within less than 1 hr. It is obvious that ^{13}C spectra can be obtained on quantities of sample normally used for proton single-scan NMR analysis.

For those situations where sample quantity or solubility is no problem, ^{13}C NMR is useful for impurity analysis. One percent levels can be observed within a relatively short time. The satellite peaks corresponding to the $^{13}C-^{13}C$ coupling in those molecules having two ^{13}C's are 0.5% of the main peaks, and these serve as convenient internal standards.

Quantitative Analysis

It has been an accepted article of

ACENAPHTHENE

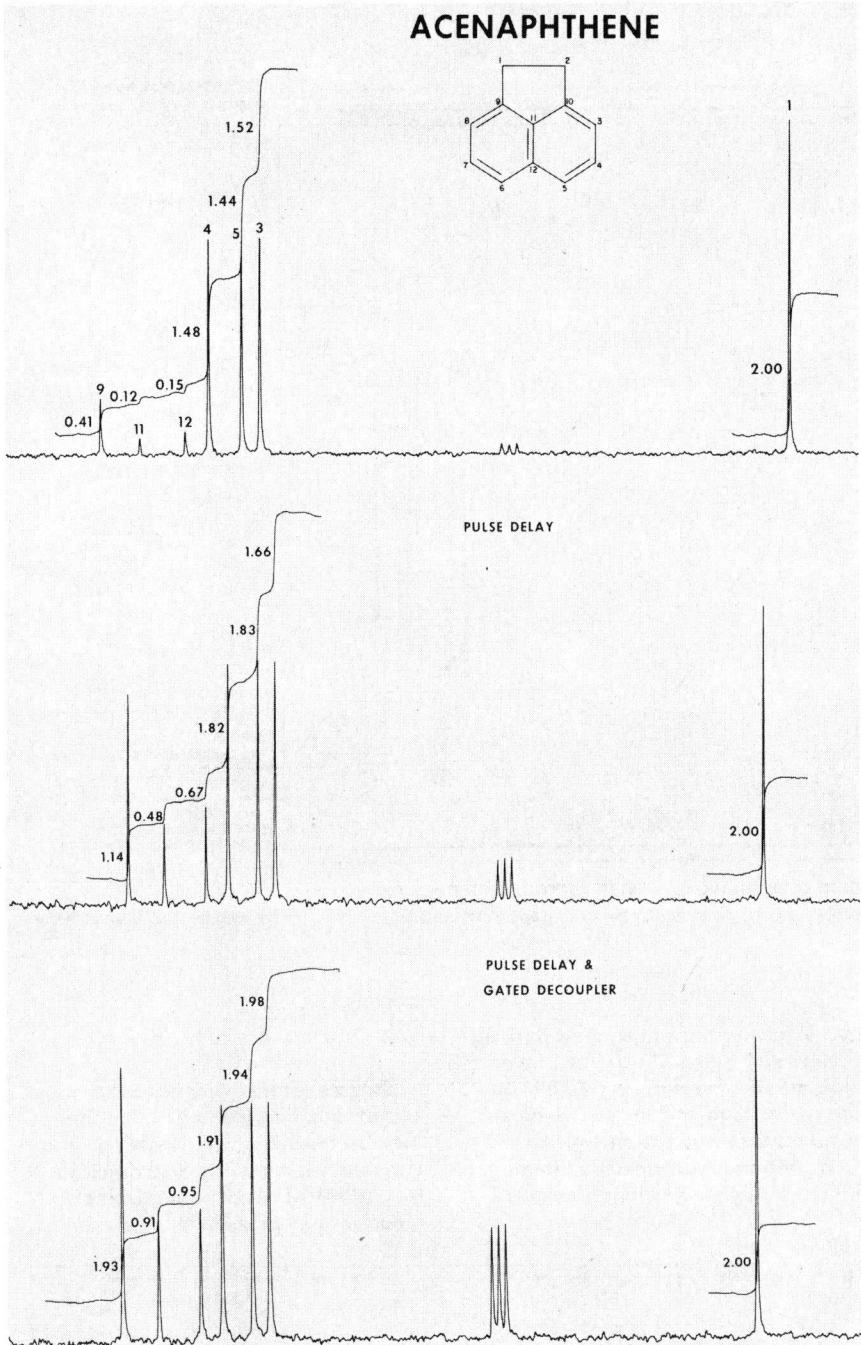

Figure 10. Quantitative analysis
Acenaphthene provides rigorous test for quantitative ^{13}C analysis since it has two quaternaries with very long T_1's. Top spectrum of 1.5 g/2-ml acenaphthene, 400-sec total time, in which relative areas are distorted. 400-sec pulse delay inserted between pulses for middle spectrum, and 16 transients recorded. This allows for T_1 relaxation, but inaccurate relative areas remained. Suppressed Overhauser spectrum at bottom successful in reproducing relative areas but required 150 transients to overcome loss of Overhauser effect

faith that ^{13}C NMR spectra are incapable of being used for quantitative analysis. As with most myths, this is incorrect. Variations of primarily T_1 and secondarily NOE do lead to inaccurate integrals mostly because of the rapid nature of the FT experiment. These inaccuracies can be removed (at the expense of sensitivity) in any system by performing the suppressed Overhauser experiment with delays between pulses of $\geq 3-5$ T_1^{max}, depending on the types of carbon of interest and the necessary accuracy of the experiment. If the observing pulse power leads to 90° pulse widths of >30 µsec, there may be some noticeable drop-off in the pulse strength across the spectrum, leading to some inaccuracies, but this is not much of a problem in modern spectrometers.

For samples in which paramagnetic doping is possible, chromium or iron acetylacetonate may be used to shorten T_1's and "level" the NOE's, but some care should be used to ascertain the procedure's accuracy by running a suppressed Overhauser calibration spectrum (Figures 10 and 11). Doping can reduce the analysis time enormously and make routine analysis practical.

The Spectrometer

Modern spectrometers have come a long way in making widespread the application of ^{13}C NMR. Until recently, the cost of spectrometers capable of this usually ranged from $120,000 up, depending on other capabilities. Today, smaller digital computer-equipped FT spectrometers are marketed starting around $50,000, not too far from what a fully-loaded A-60 sold for not too long ago! Although features and specifications vary, in general, this class provides for observation of ^{13}C at 15 or 20 MHz (^1H at 60 or 80 MHz) in 5-, 8-, or 10-mm tubes in the FT mode. The extent of computer involvement varies but can include computer control of observing frequency transmitter, decoupler frequency, pulse widths, gating operations, data acquisition display and plotting, receiver gain, and T_1 sequences, as well as others. These can be equipped with external memory devices and line printers for hard copy of parameters and chemical shifts. Magnetic tape cassette units are especially useful because of the low cost of the cassettes, their larger capacity (~20 8K data tables), and their ability to sequentially store data as generated from the spectrometer. It is now possible because of the computer control to instruct the spectrometer to perform a *series* of experiments unattended, perhaps overnight or over a weekend (the stability of the internally locked, self-shimming, phase-locked system insures that no drift in frequency or resolution occurs). For example, an 8-hr long-term time-averaging, an off-resonance, and T_1 experiment may all be executed automatically in an overnight run. Or a combination of noise decoupled and two separate off-resonance spectra could be run, giving ^{13}C shifts, ^1H shifts and a ^{13}C, ^1H cross-correlation. It is clear that productivity is significantly enhanced.

More powerful spectrometers are, of course, available, having better specifications and more capabilities including more nuclei, larger field strengths, higher sensitivity, more extensive data systems, and automatic T_1 calculation.

Along with the advancement of instrument performance has come a drastic improvement in the ease of operation. Certainly, the dedicated FT instruments are as easy, if not easier, to operate than the more familiar 60-MHz proton NMR spectrometers. Parameter input usually is via electronic means with alphanumeric oscilloscopes usually incorporated into the spectrometer. Inexperienced users are usually running their own spectra with only a minimum of instruction. The computers will generate, upon demand, a complete parameter list for

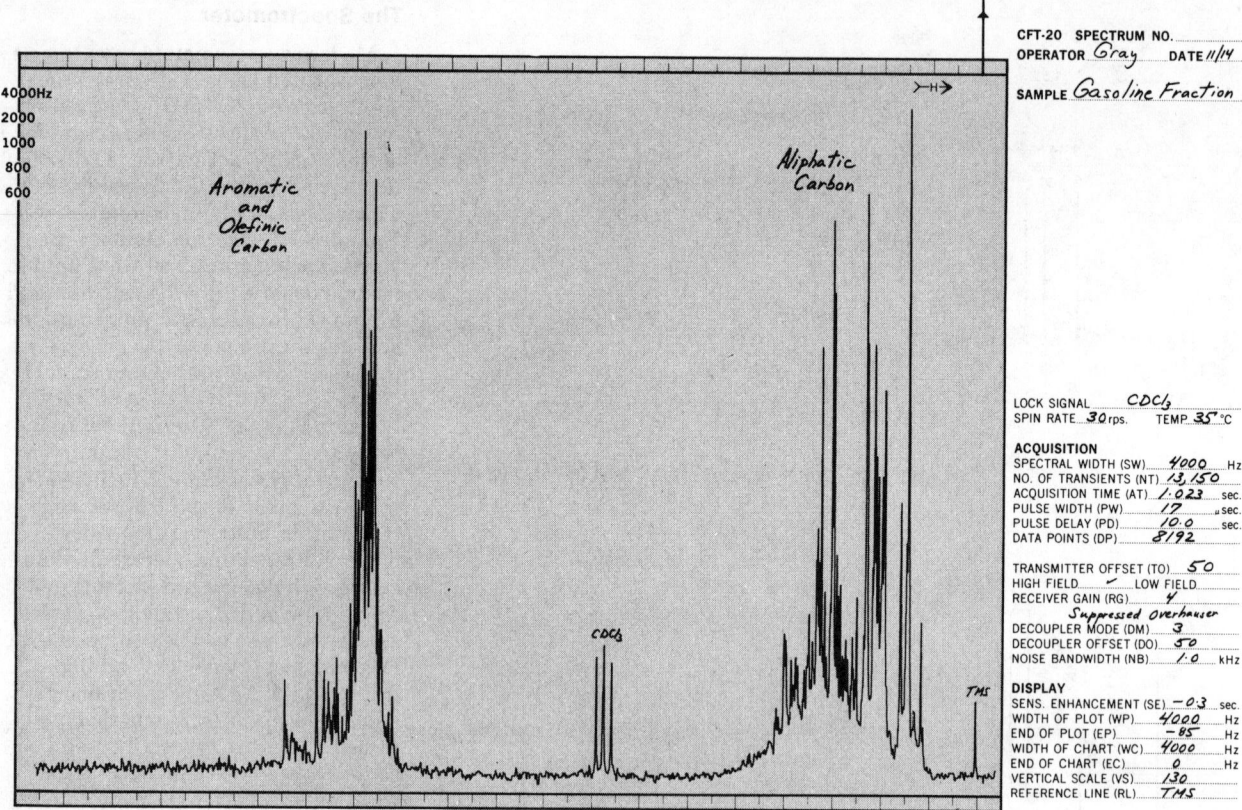

Figure 11. Spectra useful in determining percent of aromatic carbon in hydrocarbon mixtures
Spectrum run for quantitative analysis took longer than simple survey spectrum. Similar looking spectrum obtained in 1 hr in normal manner. Calculated difference in aromatic carbon content only about 3%

hard copy and a table of chemical shifts with respect to a designated line in the spectrum. Integrals may be displayed on the scope, and since they are digitally calculated in real time from the stored spectrum in the computer memory, the investigator is free to adjust phasing and amplitude, as well as to employ both linear and nonlinear baseline corrections to attain the most accurate integrations. Different portions of the stored spectrum may be displayed on the scope for detailed examination, plotting, and subsequent integral expansions.

The external memory device often greatly enhances the instrument capabilities. Usually, the device employs tape cassettes (digital grade) for both program and data storage. These cassettes are inexpensive and reusable. In a multiuser situation, each user can keep valuable data or programs fully protected. In certain very stable spectrometer configurations, it is possible to interrupt data averaging, dump to cassette, do some other sample, restore the original sample and data, and resume data averaging at the point of interruption. In this manner, very weak samples requiring long periods of averaging need not tie up an instrument during high-demand periods. This can be also useful in those situations when several experiments have been done unattended during an overnight period, and the raw data automatically taped. If no time is available the next morning for data output, the stored data on tape can be plotted at any future convenient time.

Examples of information provided by ^{13}C NMR are given in Figures 12–14.

The Future

Progress in instrument capabilities and design, like that which has made the observation of ^{13}C NMR now routine, will certainly proceed quickly toward a goal which is increasingly obvious: a widespread and universal

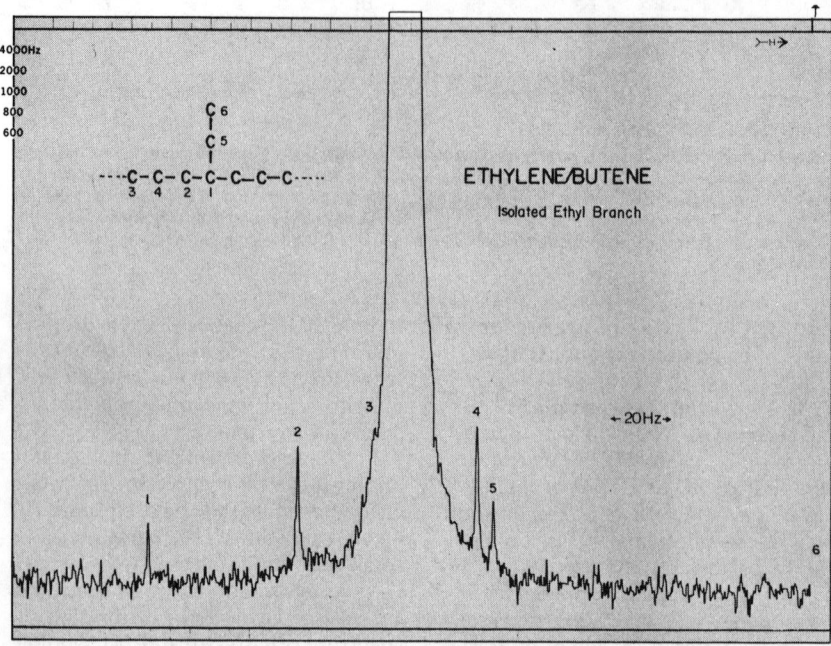

Figure 12. Branching in very high-purity polyethylene readily observed in its ^{13}C spectrum. Type of branching easily assigned from pattern of carbons near branch point

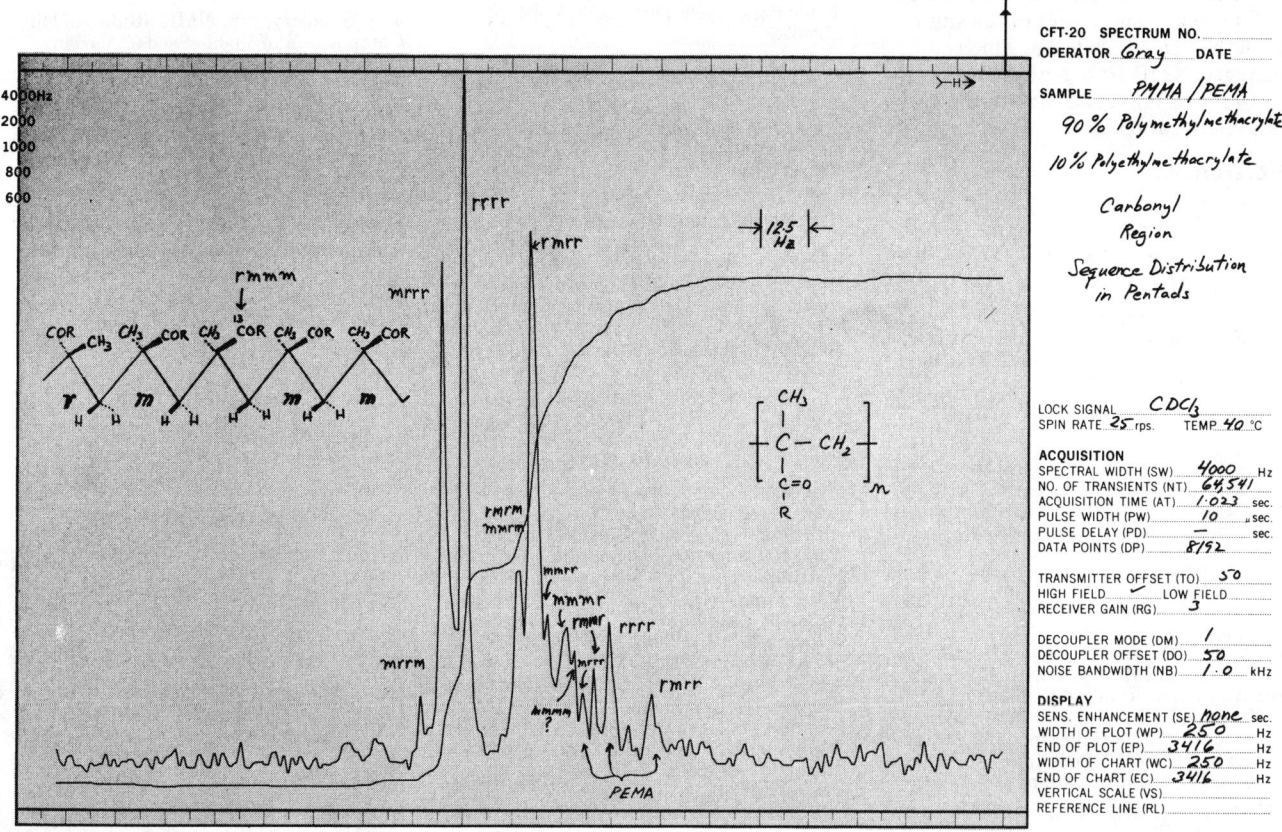

Figure 13. Carbonyl carbon of methacrylate polymer has high degree of sensitivity to sequence

Here are observed shifts resulting from change in conformation *six* bonds removed. Pentad distributions extremely important in testing of statistical methods as well as providing possible molecular indications for physical properties

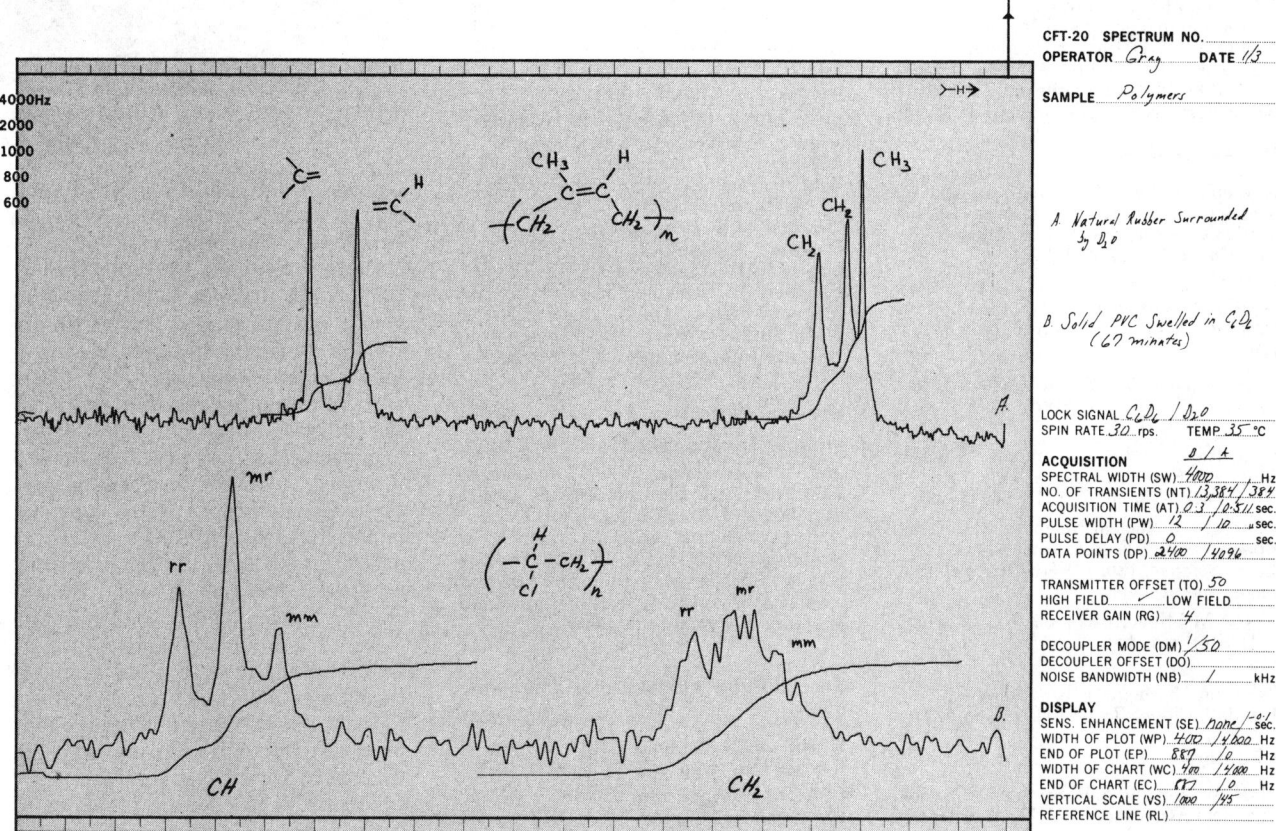

Figure 14. Solid polymers can give good ^{13}C spectra provided they are above their glass-transition temperatures

A. Natural rubber is classic example. Linewidths essentially natural linewidths, from which T_2's may be calculated. Trans configurations give markedly different spectra. B. PVC spectrum another example of solid polymer, here at fairly low temperature. PVC in solution gives narrower lines, in which *tetrad* sequences can be assigned

usage comparable to that now enjoyed by proton NMR. This explosive surge will be very reminiscent of the similar growth over a decade ago which changed NMR from a physicist's curiosity into a universal tool for the various chemical sciences.

References

(1) D. M. Grant and E. G. Paul, *J. Am. Chem. Soc.*, **86**, 2984 (1964).
(2) L. P. Lindeman and J. Q. Adams, *Anal. Chem.*, **43**, 1245 (1971).
(3) D. K. Dalling and D. M. Grant, *J. Am. Chem. Soc.*, **89**, 6612 (1967).
(4) For details see: J. B. Stothers, "Carbon-13 NMR Spectroscopy," Academic Press, New York, NY, 1972.
(5) G. C. Levy, *J. Am. Chem. Soc.*, **95**, 6117 (1973).

As entry points to both general and specific literature on applications of ^{13}C NMR and Fourier transform NMR, the references given below should be satisfactory.

General Texts

"Carbon-13 NMR Spectroscopy," J. B. Stothers, Academic Press, New York, 1972 (Vol 24 of "Organic Chemistry," A. T. Blomquist and H. Wasserman, Eds.) 559 pp. An extremely valuable collection of ^{13}C shifts and ^{13}C–X couplings with exhaustive coverage of the literature into 1970. It discusses experimental techniques, theoretical treatments, and applications.

"Carbon-13 Nuclear Magnetic Resonance for Organic Chemists," G. C. Levy and G. L. Nelson, Wiley-Interscience, New York, 1972, 222 pp. A good introduction to ^{13}C FT, problems (with solutions), 80 tables, and 450 references (good 1971 and early 1972 coverage). Good treatment of T_1 done by FT.

"Pulse and Fourier Transform NMR: Introduction to Theory and Methods," T. C. Farrar and E. D. Becker, Academic Press, New York, 1971, 115 pp. A good introduction to the "nuts and bolts" of FT NMR, particularly as it applies to ^{13}C FT.

Review Articles

E. Wenkert et al., "Carbon-13 Nuclear Magnetic Resonance Spectroscopy of Naturally Occurring Substances. Alkaloids," *Acc. Chem. Res.*, **7**, 46 (1974).

G. C. Levy, "Carbon-13 Spin Lattice Relaxation Studies and their Application to Organic Chemical Probems," *ibid.*, **6**, 161 (1973).

U. Sequin and A. I. Scott, "Carbon-13 as a Label in Biosynthetic Studies," *Science*, **186**, 101 (1974).

G. Koltwycz and R. U. Lemieux, "Nuclear Magnetic Resonance in Carbohydrate Chemistry," *Chem. Rev.*, **73**, 669 (1973).

"Carbon-13 Nuclear Magnetic Resonance Spectroscopy," F. A. L. Anet and G. C. Levy, *Science*, **180**, 141 (1973). An eight-page general review.

"^{13}C NMR Spectroscopy: A Brief Review," J. B. Stothers, *Appl. Spectrosc.*, **26**, 1 (1972).

"Carbon-13 Magnetic Resonance," E. W. Randall, *Chem. Br.*, 371 (1972), 8 pp.

Chapters

N. K. Wilson and J. B. Stothers, "Stereochemical Aspects of ^{13}C NMR Spectroscopy," in "Topics in Stereochemistry," E. L. Eliel and N. J. Allinger, Eds., Wiley, New York, 1974, Vol 8, Chap. 1, pp 96–126.

"Determination of Organic Structures," P. S. Pregosin and E. W. Randall (F. C. Nachod and J. J. Zuckermann, Eds.), Academic Press, New York, 1971, Vol 4, pp 263–322.

"Applications of ^{13}C Nuclear Magnetic Resonance in Biochemistry," G. A. Gray, *Crit. Rev. Biochem.*, **1**, 247 (1973), CRC Press, Cleveland, Ohio 44128. General review of applications of ^{13}C NMR in biochemistry up to early 1973. Titled references.

"Nuclear Magnetic Resonance," Vols 1–3, R. K. Harris, Senior Reporter, Specialist Periodical Reports of the Chemical Society (London), 1972, 1973, 1974. Valuable reviews of recent NMR literature; important for ^{13}C since so much of the total work done in ^{13}C has been in the last 2–3 years.

"High Resolution Nuclear Magnetic Resonance Spectroscopy," J. W. Emsley, J. Feeney, and L. H. Sutcliffe, Vol 2, Pergamon Press, New York, 1966, pp 982–1033. Review of early (pre-1965) literature.

"Nuclear Magnetic Resonance Spectroscopy of Nuclei Other Than Protons," T. Axenrod and G. A. Webb, Eds., Wiley, New York, 1974, containing chapters: 11, "^{13}C NMR Spectroscopy of Organo-Transition Metal Complexes," B. E. Mann; 12, "Assignment Techniques in ^{13}C NMR Spectroscopy," F. W. Wehrli; 13, "^{13}C–^{13}C Coupling Constants," G. E. Maciel; 14, "Uses and Misuses of ^{13}C–^{1}H Coupling Constants Between Directly Bonded Nuclei," V. M. S. Gil and C. F. G. C. Geraldes; 15, "^{13}C NMR Studies of Proline Peptides and Carbohydrates," O. Oster, E. Breitmier, and W. Voelter; and 16, "Applications of ^{13}C NMR Spectroscopy to the Solution of Structural Problems in Some Santonin Derivatives," P. S. Pregosin and E. W. Randall.

"Annual Review of NMR Spectroscopy," E. F. Mooney and P. H. Winson, Vol 2, E. Mooney, Ed., Academic Press, New York, 1969, pp 153–218. General review of 1964–1967 literature; Vol 5A, 1972, pp 557–630, D. G. Gillies and D. Shaw, "The Application of Fourier Transform to High Resolution Nuclear Magnetic Resonance Spectroscopy."

"Topics in Carbon-13 NMR Spectroscopy," Vol 1, G. C. Levy, Ed., Wiley-Interscience, New York, 1974, containing: "Theory of ^{13}C Chemical Shifts," R. Ditchfield and P. D. Ellis; "Substituent Effects on ^{13}C Chemical Shifts," G. E. Maciel; "Carbon-13 Nuclear Spin Relaxation," J. R. Lyerla, Jr., and G. C. Levy; "The Carbon-13 NMR Analysis of Synthetic High Polymers," J. Schaefer; "^{13}C NMR at High Magnetic Fields," F. A. L. Anet; and "^{13}C NMR Studies of Reaction Mechanism and Reactive Intermediates," J. B. Stothers.

Spectral Collection

"Carbon-13 NMR Spectra," L. F. Johnson and W. C. Jankowski, Wiley-Interscience, New York, 1972. Shows 500 actual spectra obtained on HA-100 or XL-100 FT. Uses coding system identical with previous Varian proton NMR catalogs to allow prediction of shift from molecular structure and reverse. It also has the shifts assigned and experimental conditions given.

Special Publications (Obtainable on Request)

H. Hill and R. Freeman, "Introduction to Fourier Transform NMR," Varian, 1970.

G. A. Gray, "Applications of ^{13}C NMR in Biochemistry," Varian, 1973. Collection of titled references covering a period up to March 1973.

F. W. Wehrli, "Some Applications of ^{13}C Fourier Transform NMR in Organophosphorus Chemistry," Varian (NMR-73-1), 1973.

J. N. Shoolery, "^{13}C NMR Studies of Oil Composition in Viable Seeds," Varian (NMR-73-3), 1973.

J. N. Shoolery and W. C. Jankowski, "Quantitative Aspects of ^{13}C NMR Spectroscopy," Varian (NMR-73-4), 1973. Comprehensive analysis of factors affecting quantitative use of ^{13}C NMR.

J. N. Shoolery, "The Use of ^{1}H and ^{13}C NMR Spectroscopy as Complementary Aids in Determination of Unknown Molecular Structures," Varian (NMR-73-5), 1973.

F. W. Wehrli, "Solvent Suppression Techniques in Pulsed NMR," Varian (NMR-74-1).

George A. Gray is senior NMR applications chemist at the Eastern Region Applications Laboratory of Varian Associates Instrument Division, Springfield, NJ. A 1967 graduate of the University of California at Davis with a PhD in chemistry, his research centered on determination of ^{13}C–^{13}C nuclear spin coupling constants within the ^{13}C NMR spectra of simple organic molecules. After a postdoctoral at Stanford University focusing on ion cyclotron resonance spectroscopy, he joined the faculty of the Oregon Graduate Center, Portland, OR, where he pursued interests in ^{13}C and ICR. Joining Varian in early 1972, his major research and publication interests have been in understanding the relationship of molecular structure, conformation and bonding of organophosphorus compounds as examined using ^{13}C NMR, as well as several areas of collaborative biochemical research. He is a member of AAAS and ACS and an officer of the North Jersey ACS NMR Topical Discussion Group and Section Executive Committee.

Atomic Absorption Spectroscopy — Stagnant or Pregnant?

Alan Walsh

Division of Chemical Physics
CSIRO
P.O. Box 160
Clayton, Vic., 3168, Australia

When Foil Miller invited me to participate in the Silver Anniversary Symposium on Great Moments in Analytical Chemistry, he suggested that I may care to indulge in some personal reminiscences and comments regarding the development and present status of atomic absorption spectroscopy. I hope the ones I have selected may illuminate, if not answer, the problem posed by the title of my address.

I realize that anyone who reminisces is usually so decrepit that his memory is totally unreliable. I shall, therefore, try to avoid too many errors of fact by restricting my comments largely to matters which are documented in reports of CSIRO, in correspondence, or in publications.

Presented at the Pittsburgh Conference on Analytical Chemistry and Applied Spectroscopy, Cleveland, Ohio, March 6, 1974. Silver Anniversary Symposium on Great Moments in Analytical Chemistry and Applied Spectroscopy.

My initial interest in atomic absorption spectroscopy was a result of two interacting experiences: one of the spectrochemical analysis of metals over the period 1939–46; the other of molecular spectroscopy over the period from 1946–52. The interaction occurred early in 1952, when I began to wonder why, as in my experience, molecular spectra were usually obtained in absorption and atomic spectra in emission. The result of this musing was quite astonishing: there appeared to be no good reasons for neglecting atomic absorption spectra; on the contrary, they appeared to offer many vital advantages over atomic emission spectra as far as spectrochemical analysis was concerned. There was the attraction that

- 2 -

The purpose of this report is to suggest a new technique for recording absorption spectra which offers many interesting possibilities. The method is basically simple and is illustrated in the diagram below.

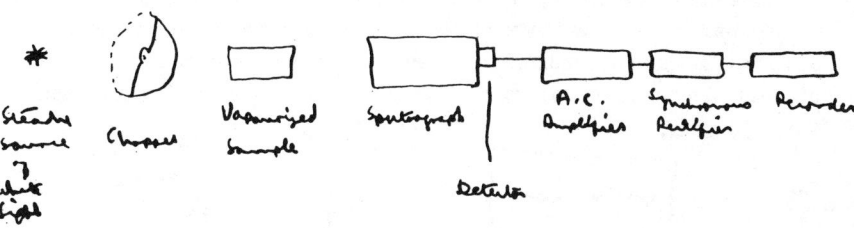

Assuming that the sample is vaporised by the usual methods, e.g flame, arc, or spark, then the emission spectrum is "removed" by means of the chopper principle. Thus, the absorption spectrum of the vapour is measured by passing through it white light which is chopped. The absorption spectrum and the emission spectrum are then scanned by a detector, the output from which is amplified by an amplifier tuned to the same frequency as the chopper. Thus the emission spectrum produces no output signal and only the absorption spectrum is recorded.

For analytical work it is proposed that the sample is dissolved, and then vaporised in a Lundegardh flame. Such flames have a low temperature (2000°K) compared to arcs or sparks (5000°K) and have the advantage that few atoms would be excited, the great majority being in the ground state. Thus absorption will be restricted to a small number

Figure 1. Extract from report for February–March 1952

absorption is, at least for atomic vapours produced thermally, virtually independent of the temperature of the atomic vapour and of excitation potential. In addition, atomic absorption methods offered the possibility of avoiding excitation interference, which at that time was thought by many to be responsible for some of the interelement interference experienced in emission spectroscopy when using an electrical discharge as light source. In addition, one could avoid problems due to self-absorption and self-reversal which often make it difficult to use the most sensitive lines in emission spectroscopy.

As far as possible experimental problems were concerned, I was particularly fortunate in one respect. For several years prior to these first thoughts on atomic absorption, I had been regularly using a commercial infrared spectrophotometer employing a modulated light source and synchronously tuned detection system. A feature of this arrangement is that any radiation emitted by the sample produces no signal at the output of the detection system. This experience had no doubt prevented the formation of any possible mental block associated with absorption measurements on luminous atomic vapours.

In an internal report for the period February–March 1952, I suggested that the same type of modulated system (Figure 1) should be considered for recording atomic absorption spectra. The following extracts from that report may be of interest:

"Assuming that the sample is vaporised by the usual methods, e.g., flame, arc, or spark, then the emission spectrum is 'removed' by means of the chopper principle. Thus the emission spectrum produces no output signal and only the absorption spectrum is recorded."

"For analytical work it is proposed that the sample is dissolved and then vaporised in a Lundegardh flame. Such flames have a low temperature (2000°K) compared to arcs and sparks (5000°K) and have the advantage that few atoms would be excited, the great majority being in the ground state. Thus absorption will be restricted to a small number of transitions and a simple spectrum would result. In addition, the method is expected to be sensitive since transitions will be mainly confined to those from the ground level to the first excited state."

At this stage I was thinking of electrical discharges, as well as flames, as a means of atomization. It will also be apparent that initially I had not appreciated the difficulties which would be involved in recording atomic absorption spectra when using a continuum source.

The next Bimonthly Report, for the period April–May 1952, includes the diagram shown in Figure 2 and describes our first experiment as follows:

"The sodium lamp was operated from 50 cycles/sec. and thus had an alternating output so that it was not necessary to use a chopper. The D lines from this lamp were isolated—but not resolved from each other—by means of a direct vision spectroscope and their intensities were measured by means of a photomultiplier tube, the output from which was recorded on a cathode ray oscillograph. Amplification of the signal was achieved by the A.C. amplifiers in the oscillograph. With the slit-width used the signal gave full-scale deflection on the oscillograph screen. A Meker flame was interposed between the sodium lamp and the entrance slit of the spectroscope. When a solution of sodium chloride was atomised into the air supply of the flame the signal at the oscillograph was reduced to zero. The principle of the method is therefore established."

In retrospect, such optimistic naivety is quite incredible.

This simple experiment gave me a great thrill, and I excitedly called in John Willis, who at that time was working on infrared spectroscopy and was later to make important contributions to the development of atomic absorption methods of chemical analysis. "Look," I said, "that's atomic absorption." "So what?" was his reply, which was the precursor of many similar disinterested reactions to our atomic absorption project over the next few years.

My report for June–July 1952 discusses the problems of recording atomic absorption spectra of flames with a continuum source and concludes that a resolution of about 0.02 Å would be required; this was well beyond the best spectrometer available in our laboratory at that time. The report concluded as follows:

"One of the main difficulties is due to the fact that the relations between

---

CHEMICAL PHYSICS SECTION

42nd Bimontly Report.   April–May, 1952.

C.P. 1/14. Atomic Absorption Spectra

  In the previous report the application of atomic absorption spectra to spectrochemical analysis was suggested. The possibilities of this approach have been explored and the results obtained to date are most encouraging. In the preliminary work the apparatus shown below was used.

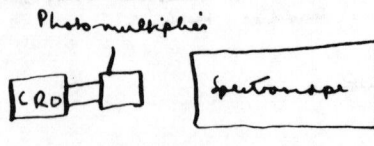 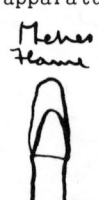 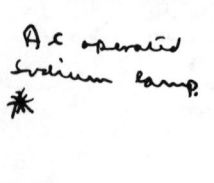

  The sodium lamp was operated from 50 cycles/sec. and thus had an alternating output so that it was not necessary to use a chopper. The D lines from this lamp were isolated–but not resolved from each other–by means of a direct vision spectroscope and their intensities were measured by means of a photo-multiplier tube, the output from which was recorded on a cathode ray oscillograph. Amplification of the signal was achieved by the A.C. amplifiers in the oscillograph. With the slit-width used the signal gave full-scale deflection on the oscillograph screen. A Meker flame was interposed between the sodium lamp and the entrance slit of the spectroscope. When a solution of sodium chloride was atomised into the air supply of the flame the signal at the oscillograph was reduced to zero. The principle of the method is therefore established. No attempt

Figure 2. Extract from report for April–May 1952

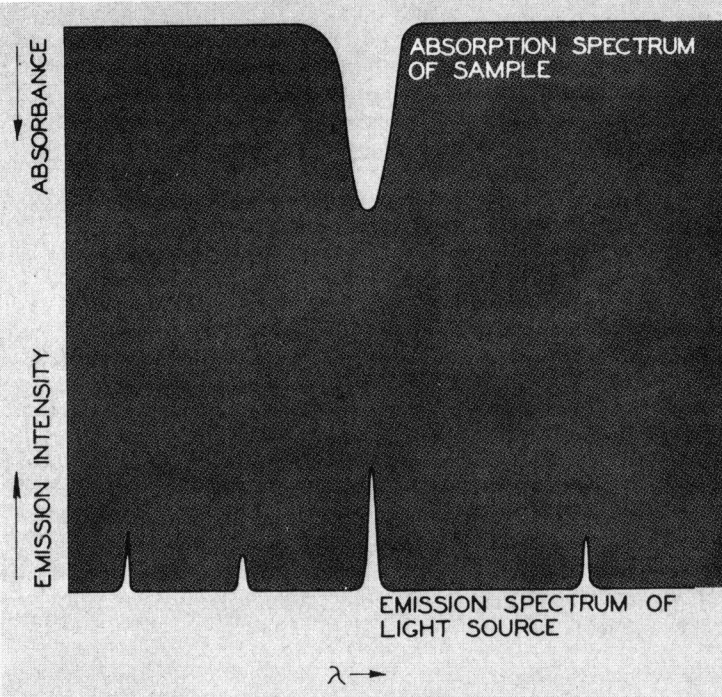

Figure 3. Schematic diagram of use of sharp line source to measure peak absorption

absorption and concentration depend on the resolution of the spectrograph, and on whether one measures peak absorption or total absorption as given by the area under the absorption/wavelength curve."

At this juncture the possibilities of measuring peak absorption were obviously coming into consideration.

There is then a gap of four months in my reports, owing to absence from the laboratory on sick leave for much of this period. The next report, for the period December 1952–January 1953, refers to the poor sensitivity obtained in the determination of copper by use of a continuum source and a monochromator obtained by placing a slit and detector on the focal wave of a Littrow spectrograph. The report states:

"It is thought that this (poor sensitivity) is due to the low resolution of the Littrow spectrograph and to the excessive amount of scattered light at low wavelengths. It is hoped to overcome this difficulty by using a hollow-cathode source (copper cathode) as source. This will emit sharp lines and a low resolution spectrometer will then be sufficient. The first attempts at producing hollow-cathode sources have not been successful."

This use of a sharp line source to measure peak absorption is illustrated schematically in Figure 3. In this case, the function of the monochromator is to isolate the required line for measurement from all other lines emitted by the source. The high resolution required for atomic absorption measurements is, in effect, provided by the sharp line source.

At this stage we had arrived at the principle of the technique which, in due course, became the generally accepted method of making the intensity measurements required in atomic absorption methods of chemical analysis.

These early experiments, carried out in collaboration with John P. Shelton, were originally confined to hollow-cathode lamps through which argon was flowed by a closed circulating system. We did not commence the development of sealed-off hollow-cathode lamps until December 1953–January 1954, when we first became aware of the work of Dieke and Crosswhite (1), which had been published in 1952.

The first person to express any interest in the application of the technique we had developed was John David, at the CSIRO Division of Plant Industry in Canberra; and on February 24th, 1953, he wrote to me a letter which began as follows:

"I understand from several sources that you have in mind a new technique of spectrochemical analysis involving the measurement of the absorption by the analysis line of an element in a vaporized sample rather than the measurement of its emission.

"I would very much appreciate any information which you may have and are prepared to give me regarding its application to analysis for traces of metallic and semi-metallic elements in plant ash, soil, mineral or similar samples."

My reply, dated 27th February 1953, was as follows:

"At the moment my work on atomic absorption spectra is still in the development stage and I cannot specify exactly what equipment will be necessary, but I think it will include the following items:-

(a) Monochromator having a resolution of 1 Å.
(b) Discharge tubes and gas-circulating system.
(c) Flame burner assembly.
(d) Photomultiplier plus associated power packs.
(e) Amplifiers having two homodyne rectifiers on the output side.
(f) Ratio-meter, potentiometer or ratio-recorder.
(g) One or two choppers.

"It is still too early to make any definite claims for the method, but it certainly offers most exciting possibilities and will, I believe, prove particularly valuable in your work, provided the sample can be taken into solution. It may prove possible to extend the method to solids.

"I should add that we are most anxious not to divulge any information to people overseas, so please regard this letter as confidential."

In view of the almost complete lack of interest in our work over the next few years, this request for secrecy was quite superfluous.

Later in 1953 I discussed possible commercial exploitation of our ideas with various instrument manufacturers in the United States and England, but the only person who viewed our work with enthusiasm was A. C. Menzies of Hilger and Watts Ltd., London; CSIRO arrived at a tentative exclusive licence agreement with that firm, based on the patent which we lodged in November 1953.

The next significant event was the first public exhibition of a working atomic absorption spectrophotometer (Figure 4). It was demonstrated in March 1954 in Melbourne University as part of an Exhibition of Scientific Instruments, arranged by the Australian Branch of the (British) Institute of Physics.

The apparent complexity of the instrument was due largely to its being of the double-beam type, which in our early experiments we regarded as essential because of the poor stability of many of our hollow-cathode lamps. The viewer was possibly further confused by the optical path being in opposite directions on the instrument and on the explanatory diagram! Whatever the reason, the instrument aroused no interest whatsoever during the three days it was on exhibition.

However, when Dr. Menzies visited Melbourne shortly afterward to assess its performance, he was sufficiently impressed for his firm to decide to produce, under licence to CSIRO, the first commercial atomic absorption spectrophotometer.

As soon as our final patent specification was lodged on October 21, 1954 (3), I submitted to *Spectrochimica Acta* my first paper (4), in which I discussed the factors governing the relationship between atomic absorption and atomic concentration, and the experimental problems involved in making atomic absorption measurements. The paper was published early in 1955, at about the same time as the paper by Alkemade and Milatz (5), who had independently arrived at the atomic absorption method. Neither paper created any great impact, and Alkemade and Milatz did not pursue their work further, possibly because they regarded this method merely as one for determining "all metals usually to be determined by flame photometry."

In 1956 and 1957 we published papers (6, 7) describing results obtained with our instrument, but these papers also created little interest. John Shelton wrote to me from London on March 5th, 1956, after giving a lecture on our work to the Institute of Physics, and reported that my first paper had given the impression that "the method was a scientific curiosity rather than a practical analytical method."

When I gave a series of lectures on the subject at the Louisiana State University Symposium in 1958, the net result was that only one person, Jim Robinson, was stimulated into activity. He became a "hot gospeller" and in due course played an important part in stirring up interest in the United States.

The surprising thing is that the appearance in 1958 of the papers by Allan (8) in New Zealand and David (9) in Canberra did not arouse any sizable impact, even though they described eminently successful applications of the technique.

On the commercial side, Hilger and Watts had produced an instrument which did not incorporate a modulated source and therefore could not fully exploit the technique. Other instrument manufacturers subsequently perpetrated the same error. By 1958 there was no sign of any instrument manufacturer willing to produce the type of instrument which we thought desirable. This was most curious since by that time there was some interest by other Australian laboratories in possible applications of this technique. It was at this stage that we decided to arrange for the production of appropriate equipment in Australia. The necessary items were manufactured by three small companies in Melbourne and then assembled by the user, according to our instructions. As it transpired, for the next few years the members of our research group were increasingly involved in supporting the commercial production in Australia of atomic absorption equipment. That a new type of Australian industry was eventually created was, of course, cause for much satisfaction, but it was inevitable that there was a substantial reduction in our research effort over a period of several years.

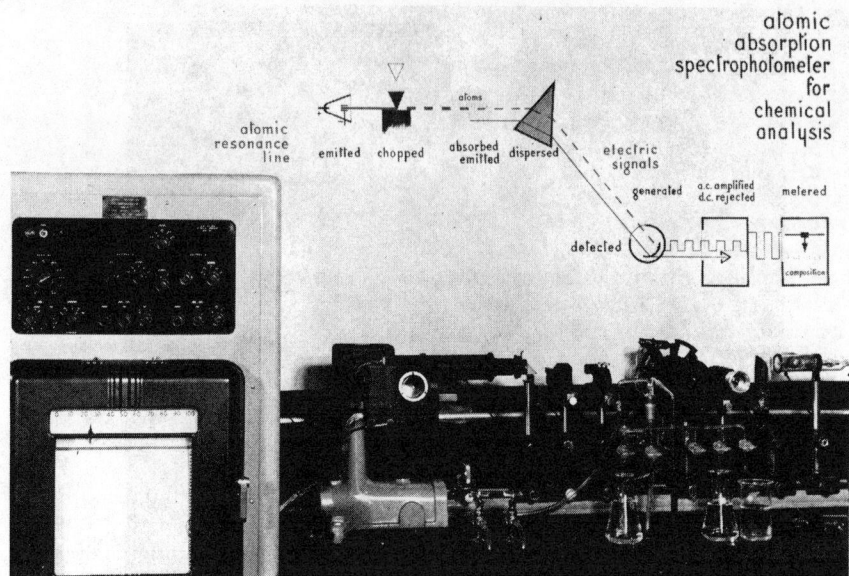

Figure 4. Photograph of atomic absorption spectrophotometer demonstrated at Institute of Physics Exhibition, Melbourne, March 1954

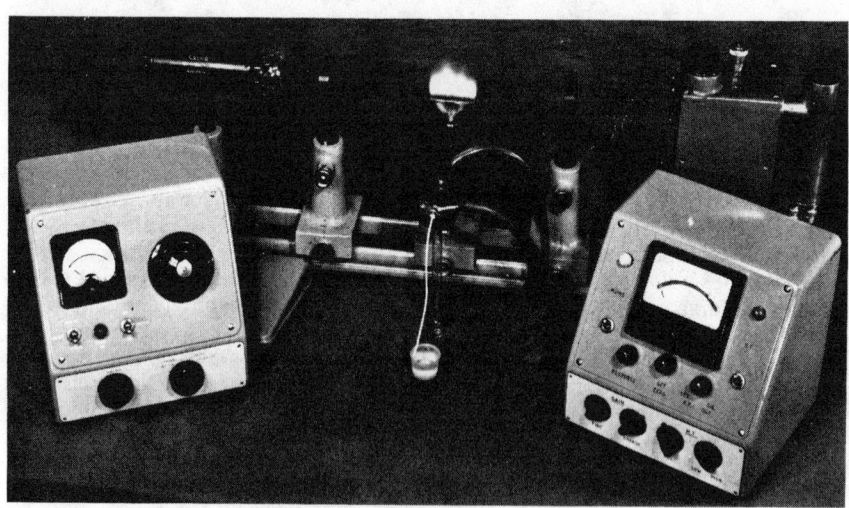

Figure 5. Photograph of simple atomic absorption spectrophotometer produced commercially in Australia

Figure 5 shows a typical instrument produced in this manner, the electronic units having been produced to our specifications by a firm in Melbourne called Techtron Appliances Pty. Ltd., which at that time had a total staff of five. During the period 1958–62, some 30 Australian laboratories were equipped by these "do it yourself" units.

While knowledge of the technique spread rapidly throughout Australian industry, there was one memorable exception. I recall the technical director of one of our biggest mining companies phoning CSIRO Head Office in the early 1960's and stating that he had just returned from South Africa where they were using a brand new instrument called an atomic absorption spectrophotometer. He wanted to know if there was anyone in CSIRO who knew anything about it. Our man in Head Office said he didn't know but he would make inquiries!

In this period there was also a slightly increased interest by manufacturers in other countries. For me, one of the "great moments" was in 1962 when I described to various members of staff of the Perkin-Elmer Corp. in Norwalk the impressive results which were being obtained by the laboratories in Australia which were by that time successfully using the technique. It was during these discussions that Chester Nimitz asked, rather tersely: "If this goddam technique is as useful as you say it is, why isn't it being used right here in the United States?" My reply, which my friends in Norwalk have never allowed me to forget, was to the effect that he would have to face up to the fact that, in many ways, the United States was an underdeveloped country! The Perkin-Elmer decision to embark on a large-scale project relating to the production of atomic absorption equipment was made shortly afterward. They were, in fact, guilty of overreacting, as witnessed by their subsequent claim that atomic absorption was "the greatest invention since the bed."

It was also in 1962 that Techtron decided to market a complete spectrophotometer. Initially, this incorporated an imported monochromator, Australian production of an appropriate monochromator being delayed until 1965. By that time the ruling engine constructed in the CSIRO Division of Chemical Physics was in full operation and has since supplied all the master gratings required by Techtron. Other firms also were becoming increasingly interested, and from that time onward I do not think the technique has suffered any major setback. In 1965 its future was virtually secured by the development (downunder) of the nitrous-oxide flame which made the technique applicable to more than 65 elements. I might add that the development of this flame by workers in atomic absorption spectroscopy was most altruistic, since it put back on its feet emission flame photometry which, at that stage, appeared to be dying rapidly.

The recent remarkable growth in the number of atomic absorption spectrophotometers produced is shown in Figure 6 and is indicative of the increasingly widespread acceptance of atomic absorption methods. In view of such growth, there would appear to be little justification for describing the subject as having been stagnant at any stage during the past decade. But the growth in the number of applications has resulted mainly from fairly straightforward extensions of a technique which originated more than 20 years ago. I believe it is no exaggeration to state that more than 99.99% of all analyses are still carried out by that original technique. In this respect the subject has been stagnant over an extended peri-

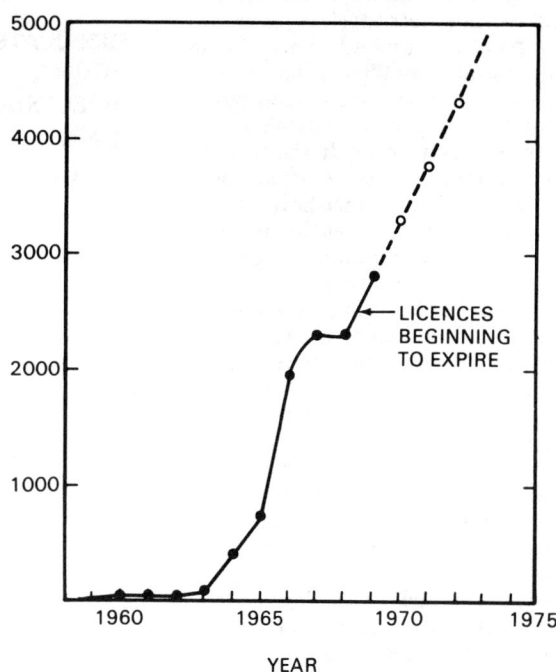

Figure 6. World sales of atomic absorption spectrophotometers

od. The situation is particularly depressing in view of the well-known limitations of existing flame methods, of which the most serious was described as follows in our first paper (3) describing our spectrophotometer:

"By far the most serious difficulty in the absorption method is due to the difficulty in atomizing various elements. This problem of complete atomization of the sample seems to us to be the outstanding problem at the present time."

As Allan (10) stated in his review of the subject in 1962, "It still is."

In a review (11) I presented in that same year, I pointed out that when a flame is used some elements are only partially atomized, "thus resulting in loss in sensitivity and the possibility of chemical interference due to variations of the degree of atomization of one element with the concentration of other elements, radicals or compounds in the solution . . . . This type of interference is present to the same extent in emission and absorption methods and is responsible for serious limitations in flame methods."

Several years later, exactly the same statement was being made, as if it represented the disclosure of a new fundamental truth regarding flame spectroscopy.

It is not my intention to decry the use of a flame as an atomizer, since it is unlikely that any other method will match its enormous range of application, its speed of operation, or its remarkable convenience. Nevertheless, I remain convinced that the best means of extending the range of application of atomic absorption methods of analysis is by developing new methods of atomization.

In some respects I imagine the above remarks present a gloomy picture of the present state of atomic absorption spectroscopy. Fortunately, there is a happier way of assessing the situation, and I want to mention briefly some aspects which lead me to think that this stagnation is more illusory than real. In the first place, so-called flameless methods of atomization, based on developments of the L'vov furnace, have in the last two years become of rapidly increasing importance, and their full potentialities have by no means been fully exploited. They now provide a remarkable ability to analyse extremely small amounts of material, and their importance, particularly in biochemical and clinical work, is already apparent. The technique is still in its infancy, and we can expect it to develop rapidly over the next few years. Of particular interest to me was the paper presented at the Toronto Conference last year in which Segar and Gonzalez (13) described, I believe for the first time, the coupling of a gas chromatograph to the graphite cuvette of an atomic absorption spectrophotometer. This combination may well make even more ubiquitous the remarkable techniques of chromatography.

The recent development of improved methods for the operation of electrodeless lamps seems likely to produce a new interest in the flame fluorescence methods since, as was first pointed out by Alkemade (14) and demonstrated by Winefordner and Vickers (15), they have striking potential advantages over atomic absorption spectroscopy. In this respect, it has always surprised me that workers in flame fluorescence have, in general, failed to exploit the fact that the fluorescence phenomenon provides its own monochromator. In the arrangement shown in Figure 7, for example, if the illuminating source is "pure," the only need to have any wavelength selection is to avoid excessive noise owing to radiation from the atomic vapour.

I would also like to refer to our recent work on the development of atomic absorption and atomic fluorescence methods for the direct analysis of solids (16, 17). The sample is made the cathode of a low-pressure discharge, and the atomic vapour is produced by cathodic sputtering. The encouraging results obtained in the analysis of low-alloy steels have recently been published (17). In that paper we reported the difficulties encountered when the method was applied to the analysis of aluminum- and zinc-base alloys. These difficulties have now been overcome, and it seems that the sputtering technique will be applicable to the analysis of a wide range of metals and alloys. It remains to compare the performance and range of these methods with those of other analytical techniques.

In the meantime, I would not expect the scientific instrument manufacturers to be greatly interested in the simple sputtering cell shown in Figure 8. I would, however, like to think that some of them are musing on possible ways of embellishment to ensure that any commercial version will have an impressive price tag.

We are now exploring the extension of our work to the vacuum ultraviolet for the determination of carbon, phosphorus, and sulfur. We also propose to extend our experiments to the analysis of powders and solutions. We are increasingly conscious of the important advantages of cathodic sputtering over thermal methods of atomization in isotopic analysis. Finally, I believe our recent work on cathodic sputtering takes us one step nearer the goal I discussed in my first paper on atomic absorption spectroscopy, the development of absolute methods of spectrochemical analysis.

It would appear, therefore, that the subject has not really been stagnant, but merely pregnant, and has now given birth to new offspring on which I trust Bunsen and Kirchhoff will look with approval and regard as worthy descendants of their original brainchild.

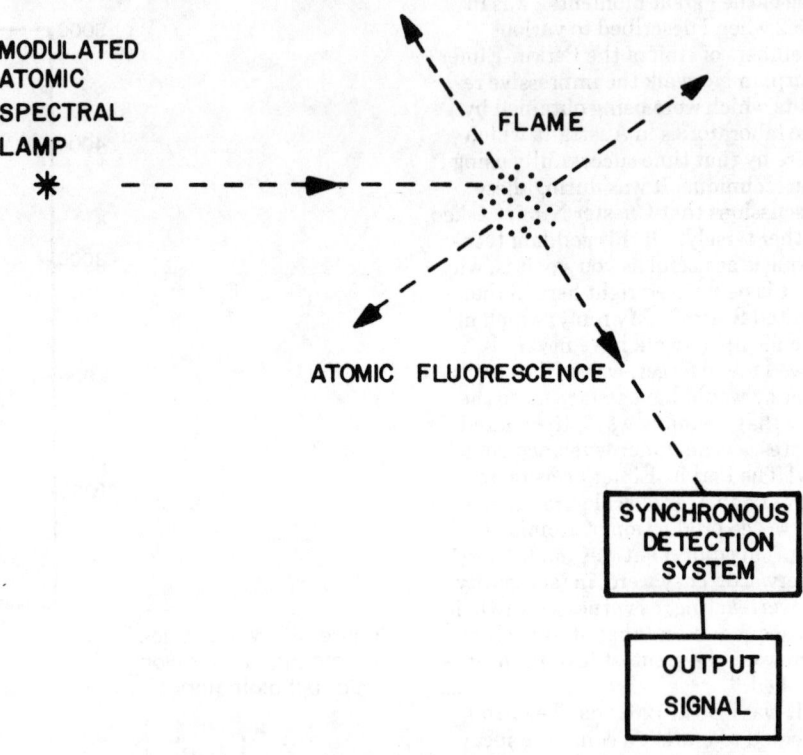

Figure 7. Schematic diagram of nondispersive atomic flame fluorescence spectrophotometer

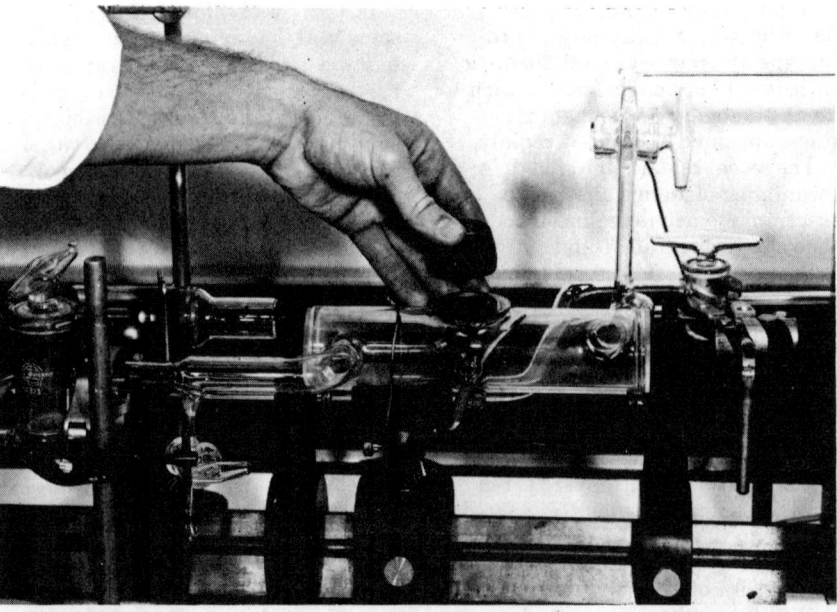

Figure 8. Photograph of atomic absorption spectrophotometer incorporating sputtering chamber for atomization of solid samples

References

(1) G. Dieke and H. M. Crosswhite, *J. Opt. Soc. Amer.*, **42**, 433 (1952).
(2) Australian Patent Application 23,041/53 (Nov. 17, 1953).
(3) Australian Patent Specification 163,586 (Oct. 21, 1954).
(4) A. Walsh, *Spectrochim. Acta*, **7**, 108 (1955); Erratum, *ibid.*, p 252.
(5) C. T. J. Alkemade and J. M. W. Milatz, *Appl. Sci. Res.*, **B4**, 289 (1955).
(6) J. P. Shelton and A. Walsh, Proc. XVth Congress IUPAC, **2**, IV-50, Lisbon, 1956.
(7) B. J. Russell, J. P. Shelton, and A. Walsh, *Spectrochim. Acta*, **8**, 317 (1957).
(8) J. E. Allan, *Analyst*, **83**, 433 (1958).
(9) D. J. David, *ibid.*, p 536.
(10) J. E. Allan, *Spectrochim. Acta*, **18**, 605 (1962).
(11) A. Walsh, Proc. Xth Colloquium Spectroscopicum Internationale, p 127, Spartan Books, Washington, 1962.
(12) B. V. L'vov, *Spectrochim. Acta*, **17**, 761 (1961).
(13) D. A. Segar and J. G. Gonzalez, paper presented at Fourth International Conference on Atomic Spectroscopy, Toronto, Canada, 1973.
(14) C. T. J. Alkemade, Proc. Xth Colloquium Spectroscopicum Internationale, p 143, Spartan Books, Washington, 1962.
(15) J. D. Winefordner and T. J. Vickers, *Anal. Chem.*, **36**, 161 (1964).
(16) B. M. Gatehouse and A. Walsh, *Spectrochim. Acta*, **16**, 602 (1960).
(17) D. S. Gough, P. Hannaford, and A. Walsh, *ibid.*, **B28**, 197 (1973).

Alan Walsh is assistant chief of the Division of Chemical Physics, Commonwealth Scientific and Industrial Research Organization (CSIRO), Melbourne, Australia. He received his DSc from Manchester University in England. He is considered the "father" of atomic absorption spectroscopy and exhibited a complete apparatus for the technique in March 1954. His now classic paper appeared in *Spectrochimica Acta* in 1955. He holds basic patents in Australia, the U.S., and other countries on atomic absorption spectroscopy, multiple monochromators, and improvements in grating monochromators. One of his major contributions was the development of intense, stable, and inexpensive hollow-cathode discharge lamps. He has published over 60 papers in atomic, infrared, and Raman spectroscopy. Dr. Walsh is currently interested in what he considers the ultimate goal of spectrochemical analysis, the development of absolute methods. This interest has resulted in the development of methods of atomization by using cathodic sputtering; and these, in turn, have led to the development of resonance detection and selective modulation techniques. More recently, he and his colleagues have adapted these techniques to the development of atomic absorption and atomic fluorescence methods for the direct analysis of solid samples.

Multielement Flame Spectroscopy

The important role of small amounts of elements in physical, chemical, and biological systems has emerged as methods of analysis have increased in sensitivity. Much of this progress may be attributed to the demands of materials science and biological research and to the availability of modern instrumentation.

The success achieved in trace element analysis has resulted from the contribution of a wide variety of instrumental techniques involving widely different principles and capabilities. In terms of the amount of information obtained in a given analysis, these trace methods can be conveniently classified into single and multielement methods. A single-element method is optimized to determine a given element with high accuracy and precision. Multielement methods are particularly valuable for survey analyses where simultaneous information on a large number of elements is desired. Obviously, for comprehensive coverage a compromise in operating conditions is often necessary to encompass the different behavior of some elements.

The tremendous popularity of the various flame techniques, i.e., atomic emission (AE), atomic absorption (AA), and atomic fluorescence (AF), for the solution of a wide variety of trace analytical problems can be related to the speed, high precision and accuracy, simplicity, reasonable sensitivity, and relatively low cost of these techniques. Flame spectrometry, however, has developed primarily as a single-element method. This paper will discuss the multielement potential of each of the three flame techniques and the problems which must be overcome before multielement flame spectrometry becomes a practical reality. In addition, the various approaches currently under study to achieve multielement capability in flame spectrometry will be reviewed.

Flame Cells

Table I shows a comparison of the desirable characteristics of flames for multielement emission, absorption, and fluorescence.

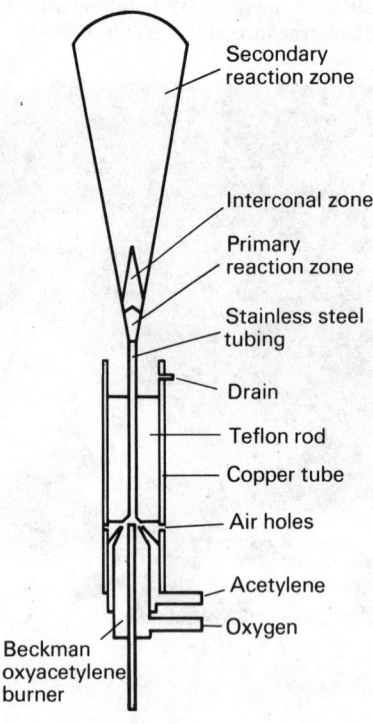

Figure 1. Fuel-rich oxyacetylene flame source for multielement atomic emission

Atomic Emission. With conventional single-element flame emission, 44 elements can be determined with flame sources at concentrations of less than 1 μg/ml (1). This does not mean that 44 elements can be simultaneously determined at these concentrations by flame emission. These detection limits are obtained by optimization of conditions for each element according to its spectrochemical properties.

These spectrochemical properties include: the excitation potential of the analytical line; the ionization potential of the element; and the free-atom population of the element for a given flame in the zone sampled by the spectrometer. Since these properties vary from element to element and since the flame is not uniform, single-element determinations are conventionally carried out by optimization of the flame zone sampled by the spectrometer, i.e., height, and by altering the chemical and physical environment by optimization of the fuel-to-oxidant ratio. In simultaneous multielement analysis, individual optimization is not possible, and a compromise must be reached. The ideal flame cell for simultaneous multielement flame emission analysis should provide optimum excitation conditions for a wide variety of elements in as small a geometrical region as possible so that this entire region can be viewed by the spectrometer at one time.

Boumans and DeBoer (2) studied the premixed nitrous oxide–acetylene flame as a potential excitation source for simultaneous multielement analysis. They evaluated this flame with a variety of burners and concluded that shielding with either nitrogen or argon is necessary for simultaneous multielement determinations. Shielding produces a large temperature gradient in this flame (3), and Boumans and DeBoer conclude that this provides the required wide range of excitation conditions over a small flame volume.

Another flame which has not been studied in terms of simultaneous multielement analysis is the fuel-rich oxyacetylene flame (Figure 1). Fassel and coworkers ($4-6$) showed that this flame is capable of exciting 43 elements at less than 1 μg/ml concentrations by conventional single-element flame emission. The region of optimum emission is the interconal zone. In a fuel-rich oxyacetylene flame, this region provides a reducing atmosphere necessary for the production of a large free-atom population of analyte, particularly for those elements that have a tendency to form refractory oxides. Further work is necessary to determine the actual potential of this flame for simultaneous multielement analysis.

Atomic Fluorescence. In atomic fluorescence the flame functions only to provide a means of producing free atoms from the analyte. In addition to being efficient in producing a large free-atom population, the flame used for atomic fluorescence should ideally

**Kenneth W. Busch and
George H. Morrison**

Department of Chemistry
Cornell University
Ithaca, N.Y. 14850

Multielement flame spectrometric methods are easily amenable to automation and have many potential applications to samples in the clinical, metallurgical, and environmental fields

have a low spectral radiance, i.e., background. Whereas the influence of flame background on the detection limit in atomic fluorescence is most severe with an unmodulated source and dc detection, the presence of intense flame background is detrimental even for systems employing modulation. Although the unmodulated flame background is not amplified directly with ac detection, its presence results in noise at the output of the amplifier. This reduces the signal-to-noise ratio and adversely affects the resulting detection limits. The ideal flame cell for simultaneous multielement analysis by atomic fluorescence would be a low background flame which is capable of producing a large free-atom population for a wide variety of elements in a small flame region. In addition, the flame gases should have a low quenching cross section.

A low background flame commonly employed for single-element determinations by atomic fluorescence is the turbulent hydrogen/argon/entrained-air flame (7). This flame is actually a hydrogen-air flame where the argon serves as an aspirating gas which has a low quenching cross section. Although this flame has been used in atomic fluorescence studies involving single-element determinations of volatile elements, the low flame temperature and nonreducing environment are likely to lead to severe interferences (8) in multielement determinations.

A promising flame for simultaneous multielement atomic fluorescence is the separated nitrous oxide-acetylene flame reported by Kirkbright and West (9) (Figure 2). Flame separation allows observation of the reducing interconal zone of the flame without interference from the radiation emitted by the secondary reaction zone. Thus, the observed background radiation for the hottest flame region, i.e., just above the primary reaction zone, is lower in certain wavelength regions than if a conventional unseparated nitrous oxide-acetylene flame were used. The suitability of this flame will depend on its ability to simultaneously produce large free-atom pop-

ulations for a wide variety of elements.

Finally, Slevin et al. (10) described the use of an air-acetylene flame in conjunction with a specialized Méker burner for use in atomic fluorescence. This burner provides an outer shielding flame which surrounds an inner flame into which analyte is introduced. The outer flame causes a more uniform temperature gradient across the inner flame, resulting in more uniform atom distributions compared with a standard Méker burner. It was concluded that the sheathed Méker burner is superior to the standard Méker burner for atomic fluorescence. An air-acetylene flame was used because of its relatively low spectral radiance. The ability of this flame and burner combination to efficiently produce a large free-atom

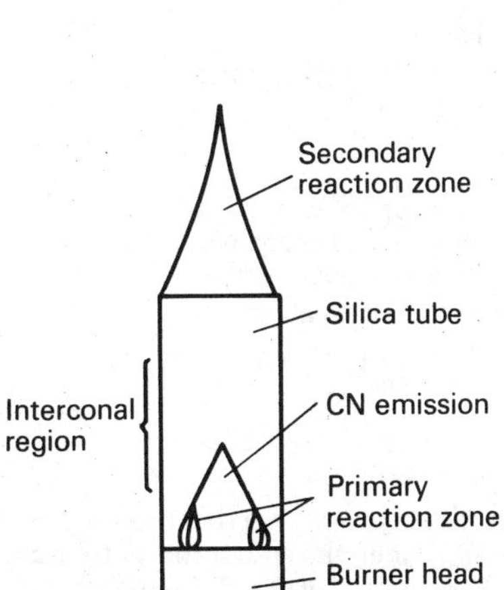

Figure 2. Separated nitrous oxide-acetylene flame source with circular slot burner for atomic fluorescence. Flame separation can also be achieved without silica tube by inert gas shielding of flame for atomic emission

population for a wide variety of elements needs to be evaluated before its suitability for multielement analysis is demonstrated.

Atomic Absorption. The flame requirements for simultaneous multielement flame atomic absorption are the production of a large free-atom population for a wide variety of elements in a small flame region. The shielded nitrous oxide-acetylene flame described by Boumans and De-Boer (2) seems most promising in this respect.

Primary Sources for AA and AF

Table II shows a comparison of primary light sources for multielement absorption and fluorescence.

Atomic Absorption. Flame absorption is conventionally carried out by use of the principle originally discussed by Walsh (11). This entails the determination of the absorption at the line center by using a narrow-line source emitting the given resonance line of the element, whose emission line profile is less than the absorption line profile of the analyte in the flame.

Although single-element hollow cathode discharge tubes are readily available for conventional single-element atomic absorption, their use in multielement atomic absorption is limited by the optical requirements of the atomic absorption experiment. Thus, some form of optical arrangement must be found to combine the separate radiation beams produced by each lamp into a single beam prior to passage through the flame.

One approach to this problem was described by Mavrodineanu and Hughes (12) who used the principle of reverse optics to produce a single beam from several hollow cathode lamps. The lamps were arranged along the focal plane of a grating spectrograph at positions corresponding to the wavelength of the desired resonance line. Radiation from each lamp after striking the grating was recombined into a single beam which emerged from the entrance slit and was subsequently passed through the flame. Their arrangement is shown schematically in Figure 3.

Table I. Flame Characteristics Desirable for Multielement Flame Spectrometry

Technique	Characteristics	Potential flames
Atomic emission	Good atomization efficiency for wide variety of elements	Shielded N_2O–C_2H_2
	Long residence time in optical path	Fuel-rich oxyacetylene
	High temperature for good excitation efficiency	
	Large variation in excitation conditions over small flame region to allow simultaneous excitation and observation of many elements	
Atomic absorption	Good atomization efficiency for wide variety of elements	N_2O–C_2H_2
	Long residence time in optical path	
Atomic fluorescence	Good atomization efficiency for wide variety of elements	Separated N_2O–C_2H_2
	Long residence time in optical path	Flame sheathed air–acetylene
	Low background spectral radiance	
	Low concentration of quenchers	
	Low scattering of exciting radiation	

Table II. Primary Light Sources for Multielement Flame Spectrometry

Technique	Approach	Limitations
Atomic absorption	Single-element hollow cathode lamps	Complicated optical arrangement required
	Multielement hollow cathode lamps	Number of elements which can be incorporated in given lamp limited
	Tandem hollow cathode lamps	Limited to three lamps by light losses
	Time-resolved spark	Less sensitive & less reproducible than hollow cathode lamp
	Continuum	Low spectral radiance at short wavelengths & less sensitive than hollow cathode lamp
Atomic fluorescence	Multielement electrodeless discharge tubes	Number of elements which can be incorporated into given lamp limited
	Banks of single-element electrodeless discharge tubes	Multiple microwave power supplies required
	Banks of pulsed hollow cathode lamps	No serious limitations except system may become unwieldy for large number of lamps
	Tunable laser	Very expensive at present
	Continuum	Low spectral radiance at shorter wavelengths

An alternative approach to the primary source problem of simultaneous multielement analysis by atomic absorption is the production of a light source which simultaneously radiates resonance lines of a variety of elements, i.e., multielement light sources. Massman (13) studied the factors involved in the production of multielement hollow cathode lamps. Concentric rings of metals retained in a copper or steel sheath have produced lamps with up to four elements. Since the sputtering rates and melting points of the elements are parameters affecting the production of multielement hollow cathode lamps, it seems unlikely that this approach will yield lamps having a large enough number of elements for true multielement analysis.

Strasheim and Butler (14, 15) made specially designed hollow cathode lamps which allow tandem mounting (Figure 4). This system, however, is limited to three lamps by light losses. If each lamp were a four-element multielement lamp, a maximum of twelve elements could be done simultaneously with a tandem arrangement.

Strasheim and Human (16) described the use of a time-resolved spark as a primary light source for multielement atomic absorption. A rotating graphite disk electrode was used to introduce solutions containing copper, zinc, calcium, and magnesium into the spark gap. Results indicate, however, that the hollow cathode provides higher sensitivity and better precision. Further work will be required to determine whether this system is capable of functioning with more than four elements.

A final approach to the problem of light sources for simultaneous multielement atomic absorption is the use of the continuum source. Various workers (17–20) demonstrated that a continuum source can be used for atomic absorption in conjunction with a monochromator having sufficient resolving power. In general, however, the detection limits are poorer with a continuum source as compared to a hollow cathode. Most workers have used a 150-W xenon arc lamp for these experiments. An additional problem with this source is the low spectral radiance produced in the far ultraviolet, i.e., less than 250 nm. Nevertheless, a continuum source offers one possibility for simultaneous multielement atomic absorption.

Atomic Fluorescence. In atomic fluorescence the narrowness of the emission line of the source is less important (except from the point of view of scatter) since the detector does not view the source directly. What is important, however, is that the source be unreversed and have a high radiance over the absorption line width.

Although sources possessing a high radiance are more difficult to produce, the optical arrangement of the atomic fluorescence experiment is more suited to multielement analysis. Specifically, banks of lamps may be arranged around the flame without the necessity for complicated optical arrangements or multielement lamps.

The most common light source for atomic fluorescence in the past has been the electrodeless discharge tube. Although some work was done to produce multielement electrodeless discharge tubes (21–25), the number of elements is limited. Using a bank of single-element electrodeless discharge tubes would be prohibitively expensive unless one microwave power supply could be used to run them all.

The most promising prospect for

Table III. Detection Systems Used for Multielement Flame Spectrometry

Approach	Advantages	Disadvantages
Temporal		
Rotating filters	Low cost	Wide bandpass may lead to interferences
	Large solid angle collected by detector	Lack of versatility in programming for different spectral lines
Scanning monochromator	Easily programmed for different spectral lines	Some form of wavelength control necessary
	Narrow bandpass reduces spectral interferences	Scan speed limited by electronic response time
	Wide wavelength range	Small measurement time for any given resolution element
		Adversely affected by source drift
Sequentially programmed monochromator	Easily programmed for different spectral lines	Some form of wavelength control necessary
	Narrow bandpass reduces spectral interferences	Adversely affected by source drift
	Permits optimization of flame & source for each element	
	Wide wavelength range	
Image-dissecting photomultiplier	Electronic scanning; no moving parts	Scan speed limited by electronic response time
	High scan speed possible	Small measurement time for any given resolution element
		Adversely affected by source drift
		Wavelength range limited by monochromator dispersion
		Compromise between wavelength range & resolution
		High cost
Spatial		
Direct-reading spectrometers	Simultaneous integrated measurement reduces influence of source drift	High cost
	Wide wavelength coverage	Measurement of line & background difficult
		Not easily programmed for different spectral lines
		Alignment critical
		Limited number of wavelengths can be monitored
Vidicon detectors	Simultaneous integration of large numbers of wavelengths possible	High cost
	Background subtraction possible	Wavelength range limited by monochromator dispersion
	Storage	Compromise between resolution & wavelength coverage
	Used in conjunction with conventional table-top monochromator without exit slit	Less sensitive than photomultiplier
	Easily changed to monitor different spectral lines	

light sources for multielement atomic fluorescence is the hollow cathode lamp. Operated in its conventional mode these lamps do not possess a high enough radiance for satisfactory atomic fluorescence. Recent work (26–28) showed that the radiance of these lamps may be increased sufficiently by operation in the pulsed mode. Power supplies capable of simultaneously operating banks of these lamps could easily be designed.

Multielement Detection Systems

In conventional flame spectrometry the radiation of analytical interest is present along with unwanted radiation from a variety of sources. To separate the desired analytical radiation from the unwanted radiation, a monochromator is used. The nature of the unwanted radiation differs in flame emission, absorption, and fluorescence; hence, the requirements placed on the monochromator differ. Nevertheless, use of a conventional monochromator restricts flame spectrometry to single-element determinations.

Table III classifies the various approaches which have been taken to achieve multielement detection. To measure intensities at different wavelengths, a multichannel device is necessary. Multichannel devices may be divided into two major classes: temporal multichannel devices and spatial multichannel devices. Temporal devices employ a single detector, where each channel is separated from the previous one in time. Spatial devices employ multiple detectors, where each channel is separated in space.

Temporal Multichannel Devices

Scanning Spectrometers. A single-channel spectrometer is one where, at any instant, light from just one resolution element is being detected and recorded. The most common example of a single-channel spectrometer is a monochromator, where an exit slit blocks all of the dispersed radiation except that within a given resolution element.

An obvious approach to using this type of system for temporal multichannel detection is to provide for some form of spectral scanning. A variety of spectral scanning systems were developed (29, 30) for various purposes. These systems are based on moving the detector if the spectrum is stationary (31) or scanning the spectrum past a fixed exit slit either by oscillating the dispersing element (32) or oscillating an auxiliary mirror (33).

There are several limitations to the use of spectral scanning systems for multielement analysis. One of the

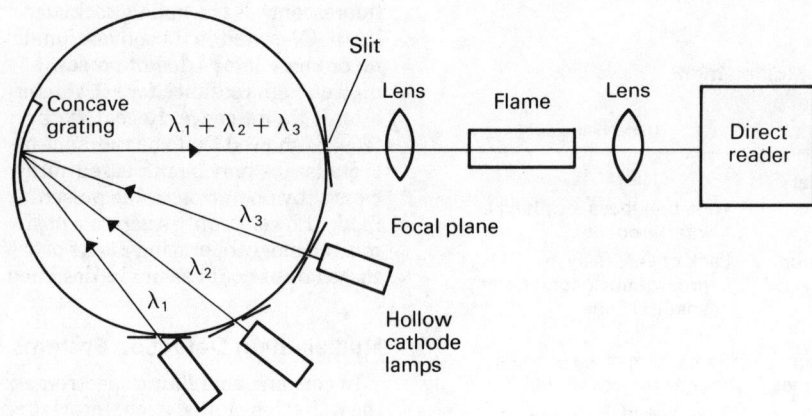

Figure 3. Optical arrangement for synthesis of multicomponent radiation beam from single-element hollow cathode lamps

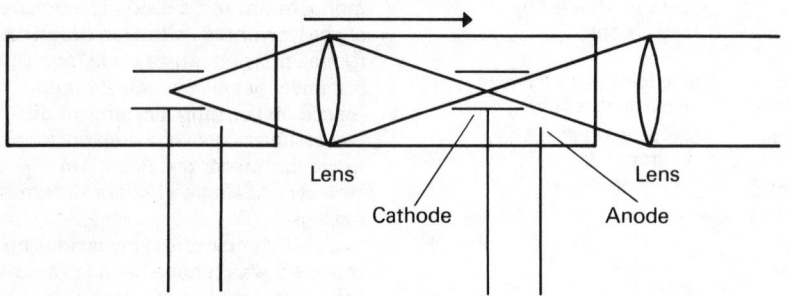

Figure 4. Tandem hollow cathode lamps for production of multicomponent radiation beam

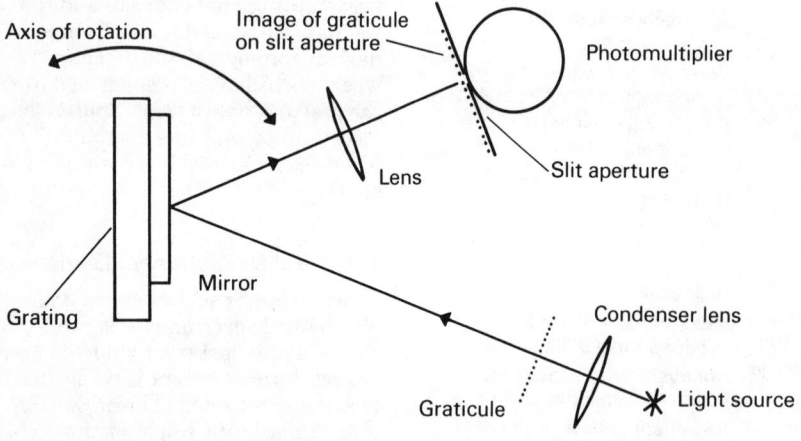

Figure 5. Electrooptical system for wavelength tracking for scanning monochromators

reasons for the development of simultaneous multielement techniques in analytical chemistry comes from the need to analyze limited amounts of sample for many elements. Under these conditions single-element techniques generally consume too much sample and time to permit the sequential determination of many elements. The first limitation of spectral scanning is that it is not truly simultaneous. Such systems approach simultaneity as the scan speed is increased. However, as the scan speed is increased, the time constant of the electronics must be decreased. The maximum scan speed is then determined by the response time of the electronics. Secondly, the use of spectral scanning will be less precise because of the short sampling time for any given resolution element. Finally, spectral scanning will be adversely affected by long-term drift in the signal as the scan proceeds.

A vital factor in the use of a scanning system for multielement analysis is the development of a highly reproducible means for accurately determining the wavelength being passed by the exit slit of the spectrometer at any instant. Dawson et al. (*34*) describe a system which accomplishes this (Figure 5). With their instrument, spectral scanning is obtained by an oscillating diffraction grating. The wavelength setting is determined electrooptically by a mirror attached to the reverse side of the grating. The mirror is arranged to project an image of a finely ruled graticule on a slit. A photomultiplier placed behind the slit produces a reference pulse train as the image of the graticule passes across the slit. This reference pulse train is used to gate the analytical signal into the appropriate integrator. The system, which scans at a rate of 1800 nm/sec, was used to determine Na, K, Ca, and Mg in clinical samples with a precision of 3%. Sodium, potassium, and calcium were determined by flame emission, and magnesium by atomic absorption. An air-acetylene flame was used.

Malmstadt and Cordos (*35*) designed an extremely flexible, modular automated instrument for atomic fluorescence and atomic emission which is also adaptable for atomic absorption. This instrument, which operates on a rapid sequential basis, overcomes the limitations of conventional scanning systems described earlier. The heart of the system is the programmable monochromator (*36*). This monochromator, controlled by digital logic, can be made to slew between wavelengths of analytical interest and remain at a given analytical wavelength until the rest of the system has been optimized for that particular element and until sufficient signal has been sampled, before proceeding to the next wavelength. Eight elements can be determined in less than 1 min with this system by atomic fluorescence. An average time of 3 sec is required for wavelength setting and optimization of flame conditions and primary light source for a given element, whereas 1–2 sec is required for the measurement step. This system has been used to determine Zn, Cd, Fe, Mn, Mg, Cu, Cr, and Ca by atomic fluorescence in a variety of NBS alloys. It was also used to determine Cu, Zn, and Mg in urine and Zn, Fe, Mn, Mg, and Ca in NBS orchard leaves.

Scanning Detectors. An approach which has not yet been applied to multielement flame analysis involves the use of image-dissector photomultipliers (*37*). Figure 6 shows a schematic diagram of this device. The image dissector consists of a photocathode and an electron imaging section separated from a conventional dynode chain electron multiplier by a plate

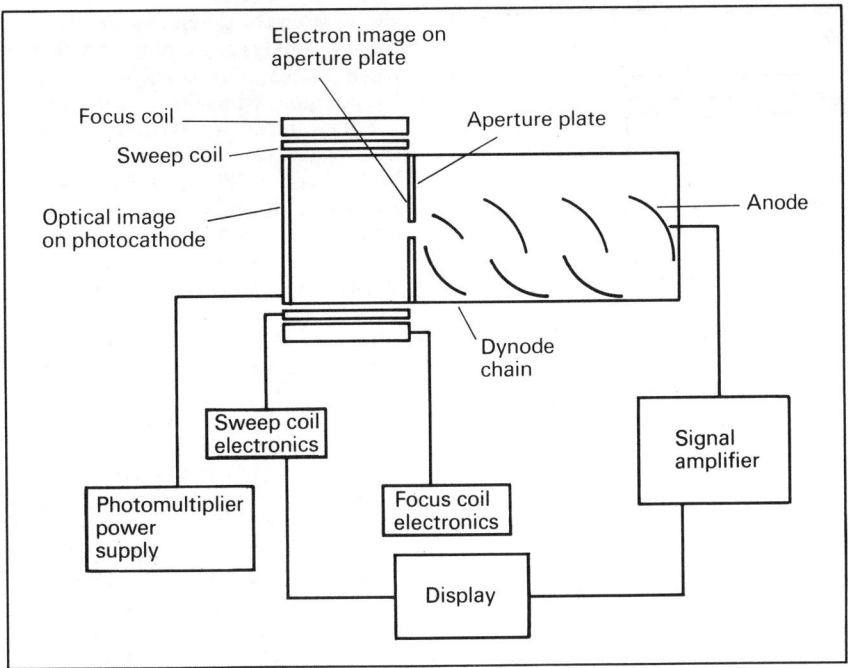

Figure 6. Image-dissector photomultiplier

with a slit aperture. The electron imaging section forms an electron image on the aperture plate from the light image incident on the photocathode. This electron image is swept across the face of the plate by electromagnetic sweep coils surrounding the tube. Only the portion of the electron image which passes through the slit in the aperture plate can reach the first dynode and be amplified by the dynode chain. Thus, instead of mechanical scanning, the image dissector utilizes electronic scanning.

Harber and Sonnek (38) described an electronic scanning spectrometer using an image-dissector photomultiplier. The spectrometer utilizes a 12.7-cm Czerny-Turner mount, f/4.5, with a reciprocal linear dispersion of 131 Å/mm. A spectral range of 250 nm was covered at a scan rate of either 100 or 1000 scans/sec. Such a system could profitably be used in conjunction with signal averaging techniques.

Filter Photometers. Another approach to temporal multichannel measurement for multielement analysis employs a filter wheel instead of a scanning monochromator. Mitchell and Johansson (26) described a system for atomic fluorescence. Four hollow cathode lamps are appropriately focused on the flame. A filter wheel is placed between the flame and a single photomultiplier. Four interference filters are sequentially rotated into position in front of the photomultiplier by a synchronous motor operating at 1 Hz. Trigger pulses derived from the rotation of the wheel by a magnet-coil device control the electronic system. The photomultiplier signal is amplified and appropriately gated to one of four integrators, prior to readout with a digital volt meter. The instrument was used for the determination of Ag, Cu, Fe, and Mg by use of an air-hydrogen flame.

Dagnall et al. (39) used a Technicon Corp. prototype (AFS-6) multielement atomic fluorescence filter photometer based on the design originally proposed by Mitchell and Johansson (26) to determine Cu, Fe, Mg, Mn, Ni, and Zn in aluminum alloys. The accuracy of the results compared with the published values of the British Chemical Standards was good, and the precision generally in the 1–5% range.

Dagnall et al. (40) also used this same instrument in the analysis of soil extracts for Ca, Cu, Mg, Mn, and Zn. The analysis must be carried out at two dilutions to ensure that each element is determined in a linear region of the calibration curve.

Spatial Multichannel Devices

A spatial multichannel device is one where, at any instant, light from several resolution elements is being detected by multiple detectors and recorded. Such systems are potentially more efficient than single-channel systems.

Multiple Slit–Multiple Detector Systems. This approach is exemplified in the familiar direct-reading spectrometer, where selected spectral lines are isolated from the dispersed radiation by exit slits positioned along the focal plane of the spectrometer and detected by photomultipliers positioned behind each slit.

The most obvious problem associated with direct-reading spectrometers is the positioning and alignment of the exit slits at the proper point on the focal plane. Even with the most careful alignment of the exit slits, environmental factors such as changes in the temperature of the spectrometer and the refractive index of the air are apt to result in misalignment. In addition, because of the problems associated with slit alignment, the spectrometer is not conveniently changed to monitor new wavelengths, i.e., new elements.

Multichannel direct readers are severely limited in the number and location of the wavelengths at which intensities can be measured simultaneously. The system becomes unwieldy and expensive if the number of separate detectors increases beyond 10 or 20, whereas the measurement of line and background intensities is impossible. Scanning methods employing vibrating slits or quartz plates (41–46) have been used for background correction. Laqua and co-workers (47, 48) used glass fiber optics to simultaneously measure adjacent wavelengths. The use of glass restricts this approach to the visible region of the spectrum.

Finally, the optical speed of direct-reading spectrometers is generally slow owing to the long focal length required to obtain the necessary high dispersion to allow detector placement. This adversely affects the use of these systems in conjunction with flames, which are relatively low-intensity sources compared with arcs and sparks.

To alleviate some of the limitations of direct-reading spectrometry, Golightly et al. (49) described the use of image-dissecting photomultiplier tubes. Since these tubes have an internal slit, the need for spectrometer exit slits is eliminated. This factor, combined with the scanning nature of the tube, alleviates the critical alignment problems conventionally associated with direct readers. As a result of reduced alignment problems, the instrument may be more conveniently programmed for different sets of spectral lines. The use of these tubes also permits the determination of background and dark current contributions to measured line intensities. Unfortunately, these tubes and their ancillary electronics are, at present, too costly to permit utilization of banks of these detectors in direct-reading spectrometers.

The direct-reader approach was used by Vallee and Margoshes (50) in 1956 to determine simultaneously Na, K, Ca, and Sr by flame emission; however, since that time little has appeared in the literature using this approach (12, 15, 16).

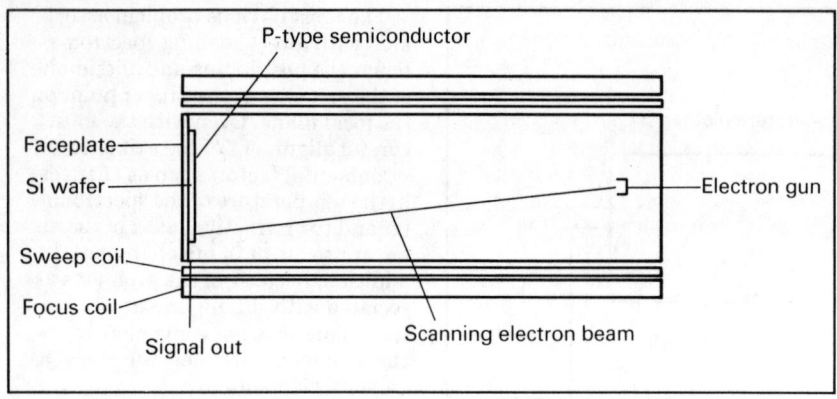

Figure 7. Silicon diode vidicon tube detector for multielement analysis

Multichannel Detectors. The first and perhaps most widely used multichannel detector has been the photographic plate. Among the limitations of photographic detection are low efficiency (<1%), limited dynamic range, and nonlinearity, but the most serious has been the length of time involved in acquiring and processing the data. Among the potential multichannel detectors proposed as alternatives to photographic detection have been television camera tubes (51) and mosaics of either photodiodes, phototransistors, or photoresistors (52). The advantages of multichannel detection include the measurement of intensities at closely spaced wavelengths, i.e., line and background, and a substantial increase in the number of wavelengths which can be monitored.

Boumans and Brouwer (52) studied arrays of photodiodes and phototransistors as multichannel detectors. Each array consisted of 20 equally spaced semiconductor photoelectronic devices, each 3-mm long and 25-μm wide. The space between each element was 5 μm. The detector arrays were used at the focal plane of a 1-m Czerny-Turner spectrometer. The response of phototransistors was reported to be 200 times that of photodiodes. Two phototransistors were used for simultaneous dual-channel intensity measurements in flame emission spectrometry, i.e., the determination of small concentrations of Ba in the presence of excess Ca.

The most promising approach to multichannel detection at this time appears to be the use of television camera tubes. Astronomers have already pioneered the use of these tubes in photoelectric image detection (53). With regard to atomic spectroscopy, the vidicon detector is already being applied (54).

Figure 7 shows a schematic diagram of a silicon vidicon tube. At one end of the tube is an electron gun which emits a beam of electrons. Focusing coils surrounding the tube

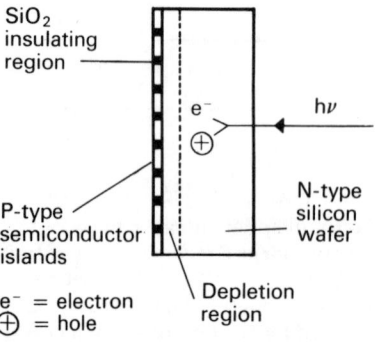

Figure 8. Target for silicon vidicon tube

focus the electron beam on the tube target. Deflection coils permit the electron beam to scan the target. The target (Figure 8) consists of an array of p-type semiconductor islands grown on an n-type silicon wafer to form a mosaic of photodiodes (55–57). The regions between the p-type islands are coated with silicon dioxide to shield the n-type silicon wafer from the electron beam.

During operation the electron beam scans the tube target, charging the p-type islands to the negative potential of the electron gun. Application of a positive voltage to the n-type substrate causes the diodes to be reversed biased. Because of the insulating properties of the depletion layer formed in reversed biased diodes, each diode acts as a storage capacitor. Photons absorbed by the n-type silicon wafer produce electron-hole pairs. The holes produced in this manner diffuse through the depletion layer into the neighboring p-type islands, discharging the capacitors in those regions exposed to light. This makes these p-islands relatively more positive than those where photon-induced holes have not been collected. No output signal results when the scanning electron beam scans over a region which has not lost its negative charge from the previous scan. An output signal is generated, however, when the scanning beam passes over those regions which have partially lost their initial negative charge either by leakage or photon-induced hole collection. The displacement current produced during the recharging of these partially discharged capacitors is amplified as the signal.

The advantages of this detector are a broad spectral range (200–1100 nm), a high quantum efficiency (0.8 max), and a linear response over a wide dynamic range.

One limitation of this type of detector is that it is less sensitive than a conventional photomultiplier. Silicon intensifier target tubes are available, however, which are more sensitive than the standard silicon target vidicons. These tubes differ from the standard silicon target vidicon by the addition of a photocathode stage prior to the silicon target. Photoelectrons ejected from the photoemissive surface are accelerated toward the n-type silicon wafer where they are focused as an electron image. Since electron-hole production is more efficient with the accelerated electrons than with photon absorption, the silicon intensifier target tube is more sensitive than a standard vidicon.

At present, a so-called optical multichannel analyzer (OMA) utilizing a silicon vidicon tube is commercially available (SSR Instruments Co.). The OMA has 500 channels across the 0.5-in. surface of the vidicon tube. The wavelength range covered across the face of the tube depends on the dispersion of the monochromator to which the tube is attached. The unit has a dual memory which allows background to be stored and subsequently subtracted from the data spectrum. Signal averaging can be accomplished by accumulating a preset number of scans in a given memory. By delaying the scanning electron beam, integration on the vidicon target is possible.

Multichannel Resonance Radiometers. A novel approach to multielement atomic absorption is that of Sullivan and Walsh (58). Emission from hollow cathode lamps passes through a flame in a conventional fashion, and absorption by the sample takes place. This absorbed beam is then passed through a resonance detector consisting of an atomic vapor of the element of interest produced by cathodic sputtering or indirect electrical heating of the metal. The atomic vapor absorbs resonance lines of the sample element. The absorbed energy is subsequently reemitted and is detected by viewing at right angles with a photomultiplier tube. A series of hollow cathode lamp–resonance detector combinations is arranged

Table IV. Comparison of Multielement Capability of Flame Techniques

Technique	Advantages	Limitations
Flame emission	Simple optical arrangement No primary light sources necessary Versatile—number of elements limited only by excitation ability of flame	Optimum excitation conditions vary for each element Less sensitive than AA or AF for elements with resonance lines less than 350 nm Requires high-temperature flame
Flame absorption	More sensitive than AE for elements with resonance lines less than 350 nm Dependent on atomization efficiency of flame but independent of excitation conditions	Lack of adequate multielement light sources Complicated optical arrangements required to produce multicomponent radiation beams Cost of lamps and power supply Versatility restricted by availability of light sources
Flame fluorescence	More sensitive than AE for elements with resonance lines less than 350 nm Dependent on atomization efficiency of flame but independent of excitation conditions	Lack of adequate multielement light sources Cost of lamps and power supply Versatility restricted by availability of light sources Requires low background flame with low concentration of quenchers and low scattering tendency

Table V. Summary of Multielement Flame Spectrometry

Approach	Elements	Method	Sample type	Ref.
Scanning monochromator	Na, K, Ca, Mg	AE, AA	Urine	34
Sequential programmable monochromator	Zn, Cd, Ni, Fe, Mn Mg, Cu, Cr, Ca	AF AF	Urine, alloys Orchard leaves	35
Rotating filters	Ag, Cu, Fe, Mg	AF	Synthetic solutions	26
	Cu, Fe, Mg, Mn, Ni, Zn	AF	Aluminum alloys	39
	Ca, Cu, Mg, Mn, Zn, K	AF, AE	Soil extracts	40
Direct-reading spectrometers	Na, K, Mg, Ca, Sr	AE	Synthetic samples	50
	K, Na, Ca, Cr, Ni, Mg	AE, AA	Synthetic samples	12

Table VI. Comparison of Multielement Flame Spectrometry with Other Multielement Techniques

Technique	No. of elements simultaneously determinable	Analysis time	Sample type	Precision, %
Spark source mass spectrometry	30–50	Photographic detection, slow Electronic detection, fast	Solids	5–25
Neutron activation analysis	30–45	Fast to slow depending on element	Solids or liquids	5–10
X-ray fluorescence	15–20	Fast	Solids or liquids	5
Emission spectrometry	20–30	Fast	Solids or liquids	25
Flame spectrometry	?	Fast	Liquids	5

around the flame to provide simultaneous multielement measurements.

Limitations to this approach include limited life of the resonance detector and variable performance. For many elements the maximum sensitivity will be considerably less than with a conventional atomic absorption spectrophotometer. Nevertheless, the potential of resonance detectors for low-cost multielement atomic absorption analyses is obvious.

Conclusions

Comparison of Flame Methods. Table IV shows a comparison of the potential multielement capability of the three flame techniques. Although multielement flame emission puts stringent requirements on the flame, it is still the most easily adapted to multielement analysis. The scope, i.e., the number and kinds of elements, is wide and limited only by the excitation ability of the flame. For certain elements, however, flame emission is not the most sensitive of the three techniques. It is generally accepted (59) for a 3000°K flame that absorption and fluorescence are more sensitive than emission for those elements whose resonance lines are less than 350 nm. Thus, for those situations calling for the widest possible scope at the highest possible sensitivity, hybrid instruments may be required. Since absorption appears to offer no advantages over fluorescence for multielement analysis (primarily owing to the optical arrangement required), these instruments will probably be hybrid emission–fluorescence instruments. Table V shows a summary of the elements done simultaneously by various approaches. Three out of eight approaches involved hybrid instruments.

Comparison of Detection Systems. Of the various approaches currently under study to provide multichannel detection (either temporal or spatial), the sequential programmable monochromator and the vidicon tube seem the most promising. The sequential programmable monochromator provides a wide wavelength coverage and allows optimization of the flame for each element. This optimization is a definite advantage in flame spectrometry.

The vidicon, on the other hand, requires a compromise between wavelength coverage and resolution, although the availability of wider vidicon surfaces in the future should ease this compromise. Generally speaking, since the vidicon is less sensitive than a photomultiplier, the tube is used in an integration mode. In principle, computer control will permit the readout of different channels at different integration times, varying the "exposure" times for selected wavelengths. This will permit the measurement of a strong line for a short integration time, while allowing weaker lines to accumulate for longer periods. In flame emission where fewer lines are emitted, the analyst's choice of lines of varying sensitivity is restricted. The ability to vary the integration time for different lines permits the analysis of samples for elements present in widely different amounts without dilution.

Comparison with Other Multielement Techniques. As mentioned earlier, a number of alternative multielement trace techniques have been used with considerable success. Table VI compares these techniques with regard to scope, analysis time, sample type, and precision. At the present time, relatively few applications of multielement flame spectroscopy to real problems have been reported as seen in Table V. In fact, almost all of the studies have been concerned with the behavior of a limited number of elements in a few systems. To properly assess the multielement analytical potential of flame spectroscopy will require additional research. Considerations of importance include speed of analysis, dynamic range and linearity, sensitivity, scope, sample type and size, simplicity of operation, and finally, cost.

Further Work. It is hoped that this review has served to highlight the previous efforts toward the development of multielement flame spectrometry and to emphasize that more research is needed in this area.

Although this review is confined to flame techniques, recent advances with flame-like plasmas (2, 60) showed that these high-temperature plasmas have the ability to excite a wide variety of elements with high sensitivity. Analyte solutions are aspirated into these plasmas with the same convenience as the flame. Further development of these nonflame emission sources may eventually lead to the displacement of the flame in emission analysis.

Other areas of future work include research on the development of new multielement primary light sources, including tunable lasers. More sensitive and comprehensive detectors along the lines of the vidicon tube are needed. In addition, hybrid AE and AF instruments are needed to extend the scope of multielement flame spectrometry.

Finally, multielement flame spectrometry has the potential for simple, rapid, inexpensive methods, which are easily amenable to automation. Analytical development of such methods will find application to a variety of samples, including clinical, metallurgical, and environmental samples.

Acknowledgment

The authors thank Marc Feldman for assistance in the literature search.

References

(1) E. E. Pickett and S. R. Koirtyohann, *Anal. Chem.*, **41** (14), 28A (1969).
(2) P. W. J. M. Boumans and F. J. DeBoer, *Spectrochim. Acta*, **27B**, 391 (1972).
(3) G. F. Kirkbright and S. Vetter, *ibid.*, **26B**, 505 (1971).
(4) V. A. Fassel and D. W. Golightly, *Anal. Chem.*, **39**, 466 (1967).
(5) A. P. D'Silva, R. N. Kniseley, and V. A. Fassel, *ibid.*, **36**, 1287 (1964).
(6) R. N. Kniseley, A. P. D'Silva, and V. A. Fassel, *ibid.*, **35**, 910 (1963).
(7) C. Veillon, J. M. Mansfield, M. L. Parsons, and J. D. Winefordner, *ibid.*, **38**, 204 (1966).
(8) R. Smith, C. M. Stafford, and J. D. Winefordner, *Anal. Chim. Acta*, **42**, 523 (1968).
(9) G. F. Kirkbright and T. S. West, *Appl. Opt.*, **7**, 1305 (1968).
(10) P. J. Slevin, V. I. Muscat, and T. J. Vickers, *Appl. Spectrosc.*, **26**, 296 (1972).
(11) A. Walsh, *Spectrochim. Acta*, **7**, 108 (1955).
(12) R. Mavrodineanu and R. C. Hughes, *Appl. Opt.*, **7**, 128 (1968).
(13) H. Massman, *Z. Instrum.*, **71**, 225 (1963).
(14) A. Strasheim and L. R. P. Butler, *Appl. Spectrosc.*, **16**, 109 (1962).
(15) L. R. P. Butler and A. Strasheim, *Spectrochim. Acta*, **21**, 1207 (1965).
(16) A. Strasheim and H. G. C. Human, *ibid.*, **23B**, 265 (1968).
(17) V. A. Fassel, V. G. Mossotti, W. E. L. Grossman, and R. N. Kniseley, *ibid.*, **22**, 347 (1966).
(18) J. H. Gibson, W. E. L. Grossman, and W. D. Cooke, *Appl. Spectrosc.*, **16**, 47 (1962).
(19) L. DeGalan, W. W. McGee, and J. D. Winefordner, *Anal. Chim. Acta*, **37**, 436 (1967).
(20) W. W. McGee and J. D. Winefordner, *ibid.*, p 429.
(21) G. B. Marshall and T. S. West, *ibid.*, **51**, 179 (1970).
(22) A. Fulton, K. C. Thompson, and T. S. West, *ibid.*, p 373.
(23) J. D. Norris and T. S. West, *ibid.*, **55**, 359 (1971).
(24) M. S. Cresser and T. S. West, *ibid.*, **51**, 530 (1970).
(25) B. M. Patel, R. F. Browner, and J. D. Winefordner, *Anal. Chem.*, **44**, 2272 (1972).
(26) D. G. Mitchell and A. Johansson, *Spectrochim. Acta*, **25B**, 175 (1970).
(27) D. G. Mitchell and A. Johansson, *ibid.*, **26B**, 677 (1971).
(28) E. Cordos and H. V. Malmstadt, *Anal. Chem.*, **45**, 27 (1973).
(29) A. M. Harris and J. H. Jackson, *J. Phys. E: Sci. Instrum.*, **3**, 374 (1970).
(30) H. A. Kruegle and S. A. Dolin, *Appl. Opt.*, **8**, 2107 (1969).
(31) G. H. Dieke and H. M. Crosswhite, *J. Opt. Soc. Amer.*, **35**, 471 (1945).
(32) J. L. Dye and L. H. Feldman, *Rev. Sci. Instrum.*, **37**, 154 (1966).
(33) R. K. Brehm and V. A. Fassel, *J. Opt. Soc. Amer.*, **43**, 886 (1953).
(34) J. B. Dawson, D. J. Ellis, and R. Milner, *Spectrochim. Acta*, **23B**, 695 (1968).
(35) H. V. Malmstadt and E. Cordos, *Amer. Lab.*, **4**, 35 (1972).
(36) E. Cordos and H. V. Malmstadt, *Anal. Chem.*, **45**, 425 (1973).
(37) P. T. Farnsworth, *J. Franklin Inst.*, **218**, 411 (1934).
(38) R. A. Harber and G. E. Sonnek, *Appl. Opt.*, **5**, 1039 (1966).
(39) R. M. Dagnall, G. F. Kirkbright, T. S. West, and R. Wood, *Analyst*, **97**, 245 (1972).
(40) R. M. Dagnall, G. F. Kirkbright, T. S. West, and R. Wood, *Anal. Chem.*, **43**, 1765 (1971).
(41) V. V. Nalimov, V. V. Nedler, and N. A. Arakel'jan, *Ind. Lab.*, **28**, 342 (1962).
(42) Yu. I. Belyaev, L. M. Ivancov, B. I. Kostin, and V. V. Semev, "Proc. 12th Coll. Spectr. Intern.," Exeter 1965, p 606, Adam Hilger, 1966.
(43) V. I. Maslov, *J. Appl. Spectrosc., USSR*, **2**, 131 (1965).
(44) I. Balslev, *Phys. Rev.*, **143**, 636 (1966).
(45) F. R. Stauffer and H. Sukai, *Appl. Opt.*, **7**, 61 (1968).
(46) W. Snelleman, T. C. Rains, K. W. Yee, H. D. Cook, and O. Menis, *Anal. Chem.*, **42**, 394 (1970).
(47) U. Haisch, K. Laqua, and W. D. Hagenah, *Spectrochim. Acta*, **26B**, 651 (1971).
(48) W. D. Hagenah, U. Haisch, and K. Laqua, "Proc. 14th Coll. Spectr. Intern.," Debrecen 1967, p 706, Adam Hilger, 1968.
(49) D. W. Golightly, R. N. Kniseley, and V. A. Fassel, *Spectrochim. Acta*, **25B**, 451 (1970).
(50) B. L. Vallee and M. Margoshes, *Anal. Chem.*, **28**, 1975 (1956).
(51) M. Margoshes, *Spectrochim. Acta*, **25B**, 113 (1970).
(52) P. W. J. M. Boumans and G. Brouwer, *ibid.*, **27B**, 247 (1972).
(53) J. D. McGee, "Astronomical Techniques," p 302, W. A. Hiltner, Ed., Univ. Chicago Press, Chicago, Ill., 1962.
(54) G. H. Morrison, paper presented at Pittsburgh Conference on Analytical Chemistry and Applied Spectroscopy, March 1973.
(55) P. H. Wendland, *IEEE Trans. Electron Devices*, **ED-14**, 285 (1967).
(56) M. H. Crowell and E. F. Labuda, *Bell Syst. Tech. J.*, **48**, 1481 (1969).
(57) S. M. Blumenfeld, G. W. Ellis, R. W. Redington, and R. H. Wilson, *IEEE Trans. Electron Devices*, **ED-18**, 1036 (1971).
(58) J. V. Sullivan and A. Walsh, *Appl. Opt.*, **7**, 1271 (1968).
(59) J. D. Winefordner and R. C. Elser, *Anal. Chem.*, **43** (4), 25A (1971).
(60) R. H. Wendt and V. A. Fassel, *ibid.*, **37**, 920 (1965).

Work supported by the National Institutes of Health Grant No. GM 19905-01.

Kenneth W. Busch received his BS in chemistry from Florida Atlantic University in 1966. In 1971 he received his PhD in analytical chemistry from Florida State University, where his research advisor was T. J. Vickers. From September 1971 to June 1972, he taught general chemistry at Florida State University as a postdoctoral teaching fellow. He is currently a postdoctoral fellow at Cornell University, where he is working with George H. Morrison on the development of multielement methods of analysis by flame spectrometry. In addition, his research interests include the development of spectroscopic instrumentation for analysis and the theory of spectrochemical excitation in flames and plasmas. Dr. Busch is a member of the American Chemical Society, the Society for Applied Spectroscopy, and a Fellow of the Chemical Society (London). He is also a member of Sigma Xi.

George H. Morrison, professor of chemistry and director of the Materials Science Center Analytical Facility at Cornell University, is one of the leaders in the field of trace analysis and materials characterization. His current interests include atomic spectroscopy, mass spectroscopy, and neutron activation analysis, and he is a principal investigator in the Lunar Analysis Program of NASA. He is coauthor with Henry Freiser of the book, "Solvent Extraction in Analytical Chemistry," editor of the book, "Trace Analysis: Physical Methods," and author or coauthor of more than 110 publications. He received the 1971 American Chemical Society Award in Analytical Chemistry and is Chairman-Elect of the Division of Analytical Chemistry of ACS.

Atomic Fluorescence Spectrometry

Although more research work is needed before atomic fluorescence is used a great deal for real samples, this sensitive, selective, and versatile method will undoubtedly become increasingly important in trace analysis

ATOMIC FLUORESCENCE SPECTROMETRY is a rather new technique of trace element analysis. In this method, a sample solution is atomized in a flame or nonflame cell, the resulting atoms are illuminated with a light source, and a fraction of the atomic fluorescence, resulting when a portion of these excited atoms undergo radiational deactivation and emit radiation toward the detection device, is measured.

The early nonanalytical work on atomic fluorescence of metal vapors in different atmospheres in quartz containers is discussed by Mitchell and Zemansky (1) and by Pringsheim (2). The first reference to atomic fluorescence of metal vapors in flames was reported in 1924 by Nichols and Howes (3). In 1927, Badger (4) wrote a classic paper on the atomic fluorescence of several elements in unseparated and in separated flames. In 1962, Alkemade (5) suggested the use of atomic fluorescence flame spectrometry as an analytical technique. In 1964, Winefordner and Vickers (6) wrote their first paper on atomic fluorescence flame spectrometry as an analytical method. Since then, numerous manuscripts (7) concerning the basis, the instrumentation, and the possible uses and advantages of atomic fluorescence flame spectrometry for analytical and physical studies have been published. Several comprehensive reviews (8–12) of atomic fluorescence flame spectrometry have been written, and the reader is referred to them for a more detailed discussion. In this paper, the authors will at-

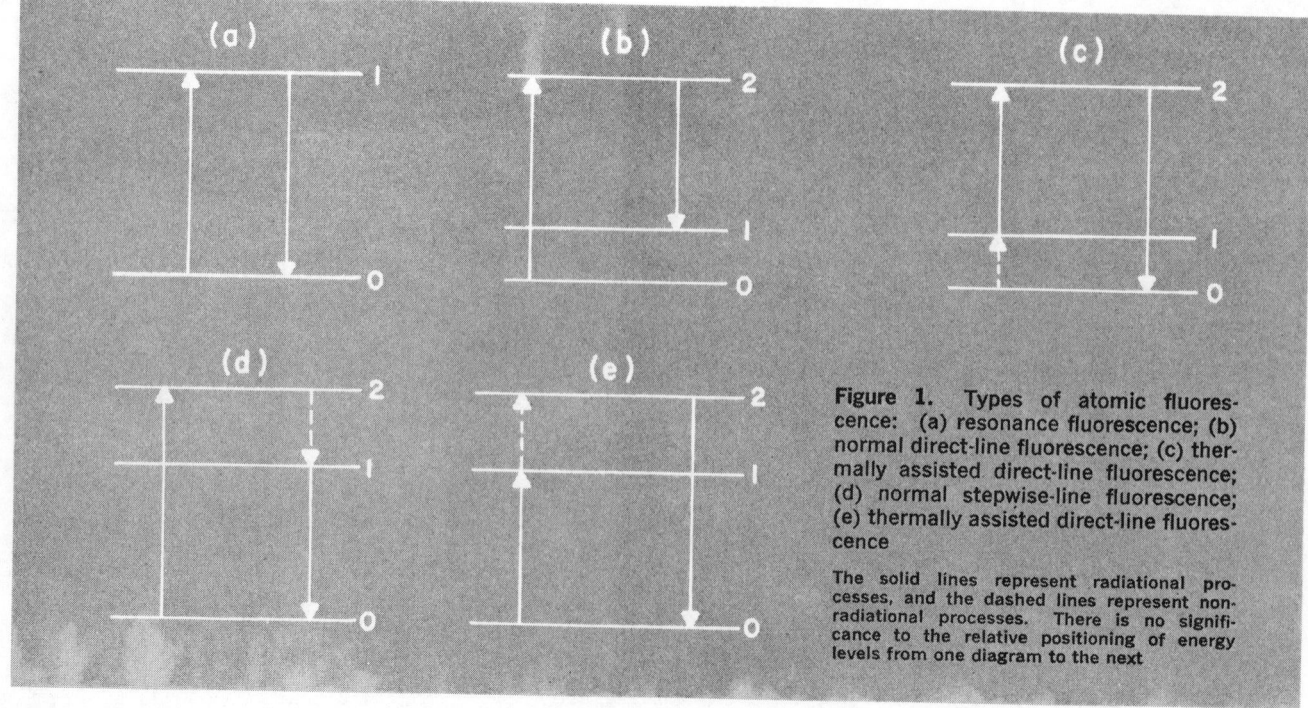

Figure 1. Types of atomic fluorescence: (a) resonance fluorescence; (b) normal direct-line fluorescence; (c) thermally assisted direct-line fluorescence; (d) normal stepwise-line fluorescence; (e) thermally assisted direct-line fluorescence

The solid lines represent radiational processes, and the dashed lines represent nonradiational processes. There is no significance to the relative positioning of energy levels from one diagram to the next

J. D. WINEFORDNER
R. C. ELSER
Department of Chemistry
University of Florida, Gainesville, Fla. 32601

tempt to review briefly, the fundamental aspects, the instrumental requirements, and the analytical uses of atomic fluorescence spectrometry.

Theoretical Considerations
(5, 6, 8, 11–17)

Mechanism of Atomic Fluorescence. In Figure 1, the types of atomic fluorescence processes are indicated. Resonance fluorescence has been most used for analytical studies. Both types of direct-line fluorescence have been of some use for analytical studies, whereas both types of stepwise atomic fluorescence have been of little analytical use. Energy transfer atomic fluorescence—i.e., sensitized atomic fluorescence—was not of any analytical use. In energy transfer atomic fluorescence, the analyte atoms are excited via donors that were excited by a light source; in flames, energy transfer atomic fluorescence is of no analytical use because of collisional deactivation of the donors and because of relatively low concentration of the donor.

Radiance of Atomic Fluorescence (8, 13, 14). Figure 2 gives a pictorial derivation of the radiance (most workers use the word "intensity" to signify radiance) of atomic fluorescence, B_F (ergs of fluorescence per second per unit area of the cell per unit solid angle). By evaluating the integral in E_A (see Figure 2) and f_s and by simplifying, the expressions in Table I result. These expressions were obtained by assuming the sample cell, which has the shape of a parallelepiped (absorption path length, ℓ, fluorescence path length, L, and height, ℓ'), is completely illuminated and the

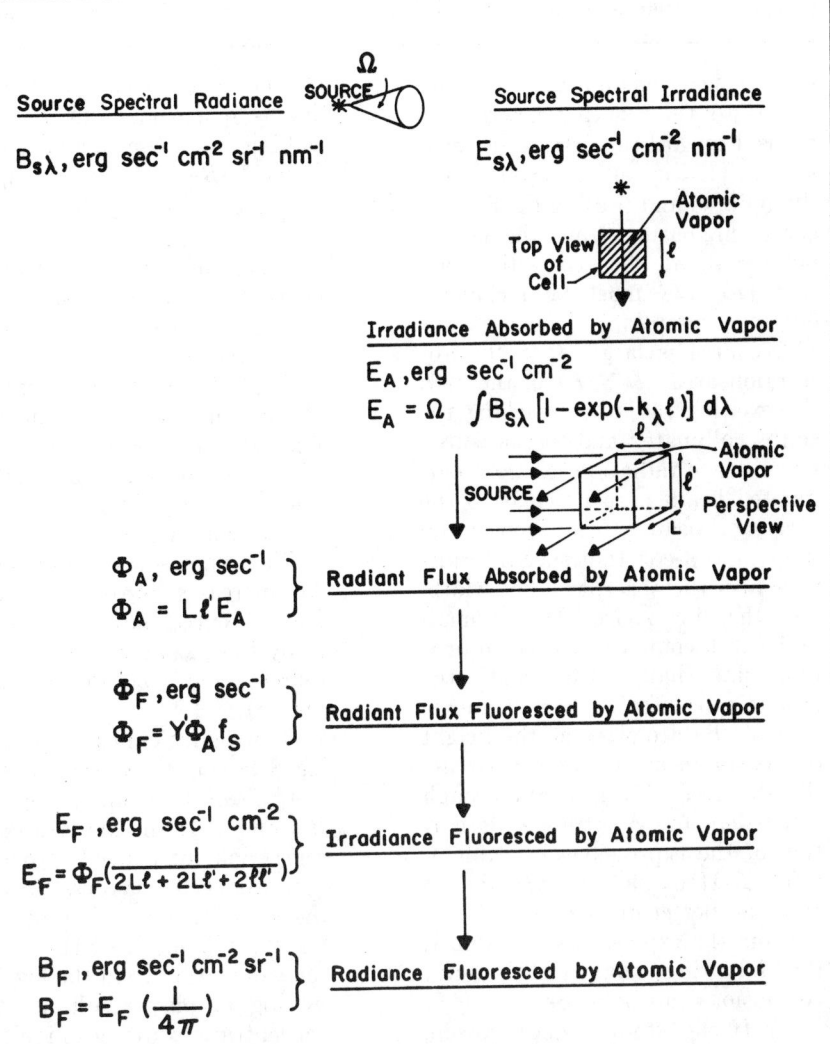

Figure 2. Pictorial derivation of fluorescence radiance for a dilute atomic gas

To obtain an expression explicit in terms of atomic concentration of analyte, it is necessary to evaluate E_A for the specific case—e.g., a specific analyte absorber at a specified concentration being excited by a specific source (continuum, line, or intermediate source). At high concentrations of analyte atomic gas, the fluorescence expressions must also be corrected for the reabsorption of fluorescence (the f_s factor) prior to emergence from the cell. The definitions of terms are: Ω = solid angle of exciting source collected and incident upon the analyte atomic vapor; κ_λ = atomic absorption coefficient for the analyte atomic vapor at wavelength, λ; Y' = fluorescence power yield for analyte atoms—i.e., erg sec^{-1} of fluorescence per erg sec^{-1} of absorption; λ = wavelength of absorption. All other terms including cell dimensions are defined in the figure; and f_s is a factor to account for reabsorption (self-absorption) of atomic fluorescence by analyte atoms within the illuminated portion of the cell

Table I. Atomic Fluorescence Radiance (erg sec^{-1}cm^{-2}sr^{-1}) Expressions[a]

	Low n_o	High n_o
Continuum Excitation Source	$B_F = C_1 \kappa_{\ell u} n_o X_\ell f_{\ell u} \Delta\lambda_D B_{C\lambda_{\ell u}} Y' \left(\dfrac{\Omega}{4\pi}\right)\left(\dfrac{L\ell\ell'}{A_s}\right)$	$B_F = 2B_{C\lambda_{\ell u}}\Delta\lambda_D a \ell' Y' \left(\dfrac{\Omega}{4\pi}\right)\sqrt{\dfrac{C_2 L \ell}{\sqrt{\pi}}}\left(\dfrac{1}{A_s}\right)$
Line Excitation Source	$B_F = \kappa_{\ell u} n_o X_\ell f_{\ell u} \delta_{\ell u} B_S Y' \left(\dfrac{\Omega}{4\pi}\right)\left(\dfrac{L\ell\ell'}{A_s}\right)$	$B_F = 2B_S \ell' Y' \left(\dfrac{\Omega}{4\pi}\right)\sqrt{\dfrac{a_F L}{\sqrt{\pi}\, \kappa_F n_o X_F f_F}}\left(\dfrac{1}{A_s}\right)$

[a] $C_1 = \sqrt{\pi}/2\sqrt{\ln 2}$, no units; $C_2 = \sqrt{\pi}/\ln 2$, no units; $\kappa_{\ell u}$ = modified atomic absorption coefficient for pure Doppler broadening = $X\lambda^2_{\ell u}/C_1\sqrt{\pi}\, c\Delta\lambda_D$; $X = \pi e^2/mc$, cm^{2}sec^{-1}, m and e = mass, in grams, and charge in esu, of the electron; n_o = concentration of atoms in the ground state, cm^{-3}; X_ℓ = fraction of analyte atoms in the lower state involved in the absorption transition, no units; $f_{\ell u}$ = absorption oscillator strength for transition $\ell \to u$, where ℓ is the lower state and u is the upper state, no units; c = speed of light, cm sec^{-1}; $\Delta\lambda_D$ = Doppler half-width of absorption line, in nm; $\lambda_{\ell u}$ = absorption line peak for transition $\ell \to u$, cm; $B_{C\lambda_{\ell u}}$ = spectral radiance for continuum source, erg sec^{-1}cm^{-2}sr^{-1}nm^{-1}; B_S = radiance of line source, erg sec^{-1}cm^{-2}sr^{-1}; Ω = solid angle of exciting radiation collected and impinging upon flame, in sr; L, ℓ, ℓ' = absorption path length, in cm, fluorescence path length, in cm, and atomizer height, in cm, respectively; A_s = total atomizer cell surface area, cm^2 (for parallelepiped cell, $A_s = 2L\ell + 2L\ell' + 2\ell\ell'$); Y' = fluorescence power yield, erg fluoresced sec^{-1} per erg absorbed sec^{-1}; $\delta_{\ell u}$ = factor to account for finite half-width of line source compared to absorption line, no units; a_F = damping constant for absorption of fluorescence; κ_F = modified absorption coefficient (same as $\kappa_{\ell u}$) but for reabsorption of fluorescence; X_F = fraction of atoms in lower state involved in reabsorption of fluorescence; and f_F = absorption oscillator strength for reabsorption of fluorescence.

fluorescence emanating at right angles to the exciting beam from the cell toward the detector is measured. [If the cells is incompletely illuminated and if all of the fluorescence emanating from the cell is not measured, then correction factors (13, 14) must be included.] Other assumptions include: the source area is larger than the absorption area ($L \times \ell'$) of the cell; the exciting radiation reaching the cell is collimated and the measured fluorescence radiation is also a collimated beam perpendicular to the exciting beam; the continuum source has a constant spectral radiance over the absorption line width, and the line source has a small half-width compared to the absorption line half-width; and, the atomic concentration and temperature of the atomizer at the height of measurement are constant across the atomizer. For a more thorough discussion and a complete derivation of the expressions in Table I, refer to Hooymayers (13) and to Winefordner et al. (8).

From the expressions in Table I, the following analytically useful conclusions can be made:

(i) If the atomic concentration, n_o, is low, and the cell is illuminated with the same radiant flux from the source of excitation, then the fluorescence radiance, B_F, is linearly related to n_o, whatever the source of excitation and whatever the cell shape.

(ii) The fluorescence radiance, B_F, is independent of n_o if n_o is high and a continuum source is used; B_F is dependent upon $1/\sqrt{n_o}$ if n_o is high and a line source is used.

(iii) Therefore, atomic fluorescence is primarily useful for trace analysis (low atomic concentrations).

(iv) The fluorescence radiance, B_F, depends directly upon the fluorescence power yield, Y', whatever the source of excitation, the cell shape, and the atomic concentration.

(v) The fluorescence radiance, B_F, increases linearly with effective source radiance (essentially $B_{C\lambda_{\ell u}}\Delta\lambda_D$ for a continuum source, and B_S for a line source).

(vi) The analytical curve plot of log S vs. log C, where S is the instrumental signal owing to B_F, and C is the analyte concentration introduced into the atomizer cell, has essentially the same shape as log B_F vs. log n_o (Figure 3). However, some deviation from the log B_F vs. log n_o may result at low concentrations owing to ionization of analyte atoms and at high concentrations owing to decreased sample introduction rate into the atomizer, reduced solute vaporization rate, and reduced aspirator yield, if one is utilized—e.g., in flame studies.

No equations accounting for incomplete illumination of the cell, incomplete measurement of atomic fluorescence over the cell area toward the detector, nonparallel exciting or fluorescing radiation, cell shapes other than rectangular parallelepipeds, sources intermediate between continuum and line, or the intersection region between high and low concentrations will be given here. The limiting expressions for the fluorescence radiance, B_F, will generally be quite similar to those given in Table I even for real analytical situations where the above inadequacies may be present. Alkemade (14) has given an excellent discussion and treatment of several of the above inadequacies. Also refer to Hooymayers (13) and Zeegers and Winefordner (15).

Atomization of Analyte. In atomic fluorescence spectrometry, the analyte must be atomized prior to radiational excitation. Atomization has been performed so far by flame and nonflame cells. These cells are discussed later in this review.

Predicted Limits of Detection (8). If the limit of detection is defined as that concentration (or amount) of analyte resulting in a specified signal-to-noise ratio (two, for instance), and if the instrumental signal [signal owing to fluorescence radiance, B_F, modified by the instrumental system (16)] and noise [primarily a result of flame flicker and shot noise (16, 17)] are evaluated for atomic emission flame spectrometry, ae, atomic absorption flame spectrometry, aa, and atomic fluorescence flame spectrometry, af, then comparison of relative limits of detection obtained with these three methods is possible. Assuming that the same resonance line of the same atom is measured by the same instrumental system and that the same flame atomizer is used for all three methods, then it can be shown (by using expressions for aa and ae similar to those in Table I) that ae should result in lower limits of detection than af and aa for atoms with resonance lines above approximately 4000 Å and af [and sometimes aa, especially for high temperature flames (8)] should result in lower limits of detection than ae for atoms with resonance lines below approximately 3000 Å. All methods should give similar results for atoms with resonance lines in the intermediate region of 3000 to 4000 Å (Figure 4).

With nonflame cells, af should result in even lower limits of detection than aa (assuming all other factors are equal) because the noise level should be reduced more in af than in aa. With nonflame cells, the background radiation and radiation flicker often can be made negligible compared to other noise sources. Therefore, in af, the major source of noise becomes shot noise which is often much less than flame flicker noise, whereas in aa, the major source of noise is source flicker noise with a contribution from shot noise (source flicker and shot noises are often greater than flame flicker noise, and so, nonflame cells should be more beneficial to af than aa, in terms of reducing limits of detection). Of course, the major advantage of nonflame cells is the possible use of small samples for analysis. Because of the vast difference in excitation energy available and in background in RF, microwave, and dc plasma discharges, it is impossible to compare aa and af with ae when using nonflame cells.

Predicted Interferences (8). In Table II, the predicted effect and relative extent of the major types of interferences in atomic flame spectrometric methods are given. The worst type of interference—physical, chemical—should be (and is) identical in all atomic flame methods. Scattering and band absorption interferences should be (and are) troublesome to some extent in aa and af, and the temperature variation interference should be (and is) a little worse in ae. Spectral interferences should be (and are) a little worse with ae, and aa and af with continuum sources, than with aa and af with line sources. Therefore, the total influence of interferences should be about the same for all atomic flame spectrometric methods as long as good instrumentation is utilized for all methods. Perhaps, if an overall relative magnitude of interferences was established, then atomic ad-

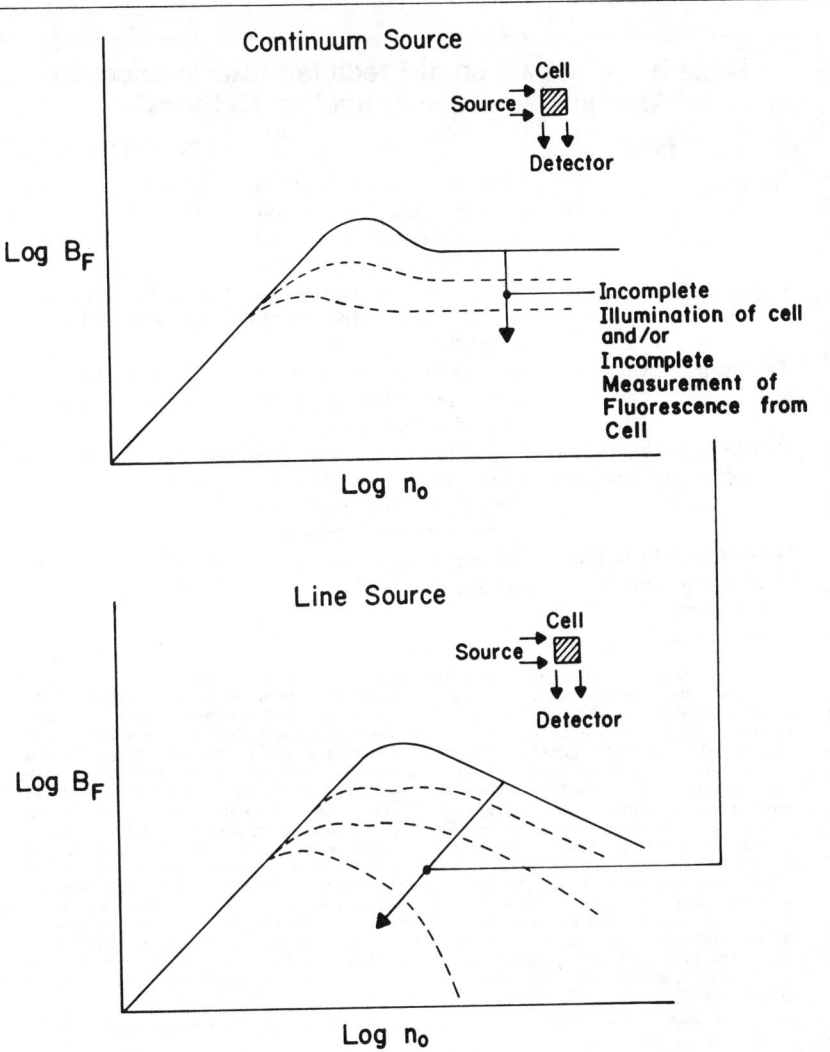

Figure 3. Hypothetical growth curves [log(B_F) vs. log (n_o), where B_F is the atomic fluorescence radiance and n_o is the concentration of analyte atoms in the ground state]

The dashed lines indicate the general influence of increasing the extent of incomplete illumination of the cell with exciting radiation and/or increasing the extent of incomplete measurement of atomic fluorescence from the cell

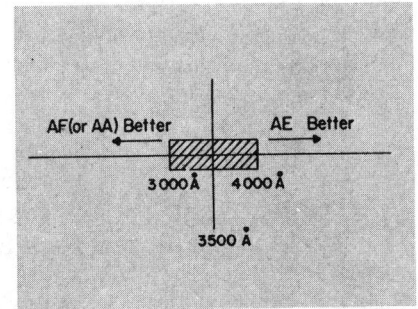

Figure 4. Theoretical prediction of wavelength range over which various atomic flame spectroscopic methods achieve lowest limits of detection

It is assumed that the same resonance line of any given element is measured by the three methods of atomic emission (ae), atomic absorption (aa), and atomic fluorescence (af), spectrometry. Also, it is assumed that the same flame cell and basically the same instrumental system is utilized for all three methods

Table II. Comparison of Predicted Interferences in Atomic Flame Spectrometric Methods[a]

Type	Effect	Relative Extent[b]
Spectral	Measurement of radiation characteristic of interferent rather than or as well as analyte radiation	aac, afc, ae > aal, afl[c]
Physical, chemical	Reduction in concentration of analyte atomic concentration[d]	Same for aac, aal, ae, afc, and afl
Temperature variation	Change in degree of excitation and change in degree of atomization	ae ≲ aac, aal, afc, and afl[e]
Scattering of radiation and band adsorption	Reduction in source radiance owing to nonatomic absorption and reduction in fluorescence radiance	aal, afl ≲ aac, afc > ae[f]
Quenching of excited atoms by species resulting from sample matrix	Reduction in fluorescence radiance	afc, afl ≲ aac, aal, ae[g]

[a] Ae = atomic emission flame spectrometry; aac = atomic absorption flame spectrometry with a continuum source; aal = atomic absorption flame spectrometry with a line source (this is the commonly used method, often called atomic absorption spectrometry or just aa); afc = atomic fluorescence flame spectrometry with a continuum source; afl = atomic fluorescence flame spectrometry with a line source (this is the commonly used method called atomic fluorescence spectroscopy or just af or afs). [b] > means specific interference is greater for methods listed to left of > than for those listed to right of >. A sign of ≲ means greater than or nearly the same. [c] The extent of spectral interference in aac, afc, and ae depends upon the spectral bandwidth and thus the dispersion of the optical system. The smaller the spectral bandwidth, the smaller the spectral interferences. [d] Interferences which reduce the atomic concentration of analyte in flames are the most severe type of interferences in flame spectrometric methods. [e] For most elements—i.e., those forming stable monoxides, monohydroxides, etc., in the flame gases—the effect of temperature change is essentially the same for all flame methods. [f] These interferences are not present in ae. Generally these interferences are not of great importance in either aa or af methods. [g] This interference is not present in aac, aal, or ae. However, it is also negligible in afc and afl.

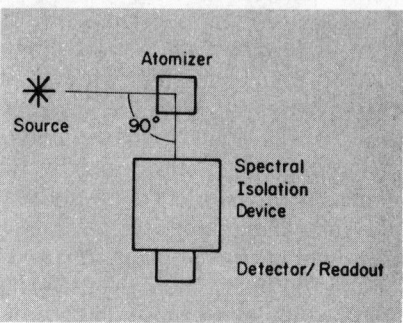

Figure 5. Schematic diagram of basic instrumental system for atomic fluorescence spectrometry

sorption and atomic fluorescence flame spectrometric methods with line sources should result in slightly lesser overall interferences than the other methods in Table II.

Interferences with nonflame cells will be similar to those predicted for flame cells but are difficult to compare in detail without specifying the exact nonflame cell type and method of sample introduction.

Atomic Fluorescence Instrumentation (18–92)

Basic System. The basic layout of an atomic fluorescence spectrometer is given in Figure 5. It can be thought of as an atomic absorption spectrometer bent at a 90° angle around the absorption cell axis. Looking at it another way, one sees that without the source, the setup is simply that of an atomic emission system. Basically, all atomic fluorescence systems have been built around this arrangement.

Sources. In Figure 6, the types of sources used in atomic fluorescence are schematically shown. As pointed out above, fluorescence response is a function of source radiance (intensity). Consequently, sources of high radiance are desirable. In Table III, the requirements for excitation sources for atomic fluorescence spectrometry are outlined. Both line and continuum sources have been used with success. Table IV compares line and continuum excitation sources. The choice of a line source or continuum source must be made on the basis of analytical requirements—i.e., number of different elements to be analyzed, interferences expected, and radiance required for reliable detection. The general types of line sources used in past atomic fluorescence studies are depicted in Figure 6a–f; these sources also have been used in many atomic absorption studies. The electrode-

Table III. Excitation Sources Requirements for Atomic Fluorescence Spectrometry

1. High radiance over the absorption line of analyte atoms
2. Long-(little drift) and short-(little flicker) term stability
3. Operation under either continuous or pulsed conditions
4. Simple tuning and focusing
5. Availability for all elements
6. Long lifetime
7. Safety of operation
8. Low cost

Table IV. Comparison of Line and Continuum Source of Excitation

Characteristic	Comments
Radiance over absorption line	Line generally higher
Spectral interferences	Line generally results in fewer
Wavelength adjustment	Line simpler
Multielement analysis	Single continuum source vs. many line sources
Stability	Continuum generally better
Tuning and focusing	Continuum generally simpler
Lifetime	Continuum generally longer

less discharge lamp (Figure 6c) has been the most-used line source in atomic fluorescence work because of its high radiance over most resonance lines. Electrodeless discharge lamps are operated in a microwave field which causes excitation of the atomic vapor contained in the discharge lamp. The field is directed on the discharge lamp by means of either antennae or waveguide cavities. Optimum performance and lifetime is a function of the power of the microwave field, the lamp temperature, and uniformity of the microwave radiation around the lamp. As a result, tuning of the lamp in the field is critical. Tuning requires primarily either movement of the lamp within the field or movement of the device directing the field. Consequently, difficulties arise in tuning the lamp properly and maintaining it in the correct optical position.

Xenon arc lamps of high spectral radiance are commonly available and offer the advantage of single lamp operation over multiple-line spectral lamps (one for each element). They suffer from the disadvantage of possessing low radiance in the uv below 2500 Å. They can be obtained either as "point" sources (Figure 6e) or as collimated beam sources (Figure 6f).

Metal vapor discharge lamps (Figure 6d) emit rather intense resonance radiation but are plagued with self-reversal problems. Hollow cathode lamps of the normal sealed variety (Figure 6a) used in atomic absorption studies are generally of insufficient radiance for atomic fluorescence work. Boosted-output hollow cathode lamps (Figue 6b) emit intense resonance radiation but are expensive and not commercially available for many elements.

Lamps which operate continuously at some arbitrary radiance can provide a greater peak radiance by pulsing them at higher current levels than is obtained at continuous operation. The only limitation is that the average pulsing current be about the same as the average continuous operating current. Overpowering at high current levels reduces the lifetime of the lamp. Optimally, detection in pulsed systems should be synchronous with source pulsing.

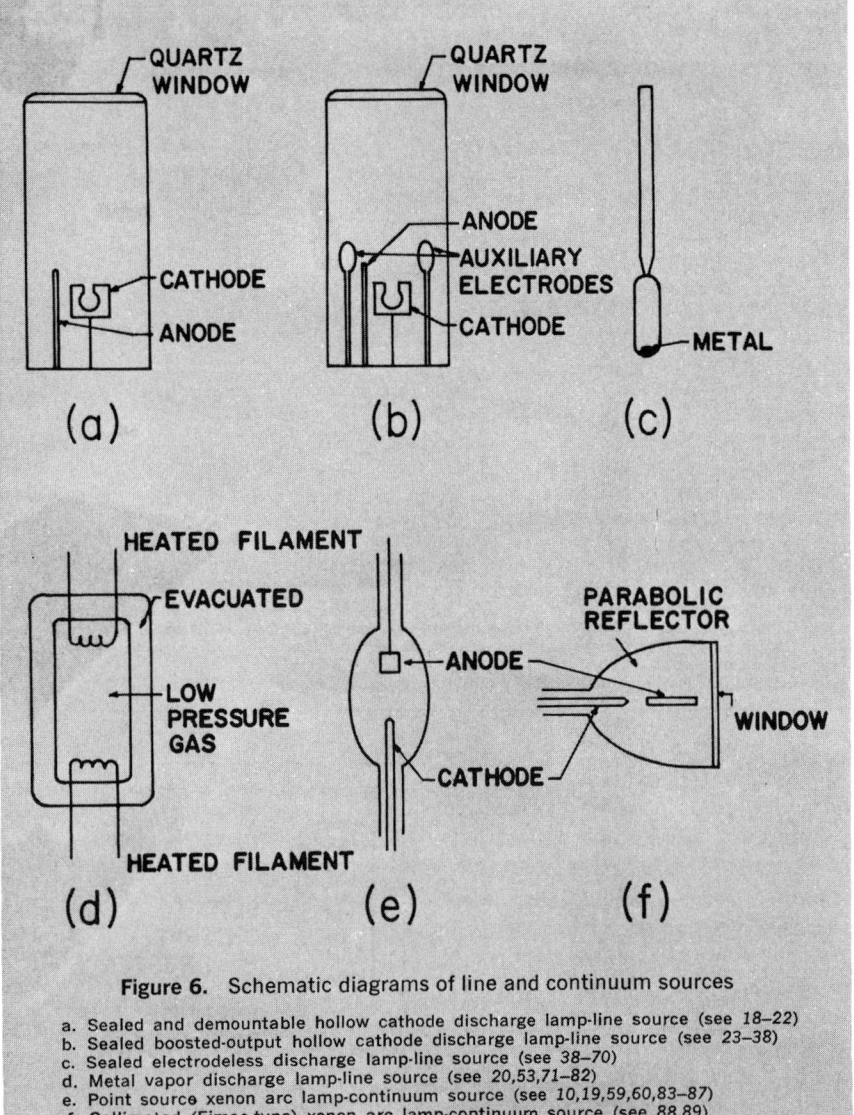

Figure 6. Schematic diagrams of line and continuum sources

a. Sealed and demountable hollow cathode discharge lamp-line source (see 18–22)
b. Sealed boosted-output hollow cathode discharge lamp-line source (see 23–38)
c. Sealed electrodeless discharge lamp-line source (see 38–70)
d. Metal vapor discharge lamp-line source (see 20,53,71–82)
e. Point source xenon arc lamp-continuum source (see 10,19,59,60,83–87)
f. Collimated (Eimac-type) xenon arc lamp-continuum source (see 88,89)

Atomizers. The requirements (Table V) of atomizers in atomic fluorescence spectrometry are essentially identical to those for atomic absorption and atomic emission spectrometry. The major difference lies in the third requirement of Table V; because nonradiational deactivation decreases the atomic fluorescence radiance regardless of how ideal the remainder of the system may be, radiational quenching processes should be minimized. This is accomplished by choice of the major cell gases. For example, the major quenchers in flames are CO, CO_2, and N_2; thus CO and CO_2 are avoided by using nonhydrocarbon flames such as oxyhydrogen or hydrogen diffusion flames. Also, N_2 can be minimized in flames by

Table V. Requirements of Atomizers Used in Atomic Fluorescence Spectrometry

1. Good atomization efficiency[a]
2. Low radiational background and background flicker
3. Low concentration of quenchers[b]
4. Long residence time of analyte atoms in optical path
5. Simplicity of operation
6. Low cost of initial purchase and operation

[a] The atomization efficiency means the efficiency of converting analyte within the sample matrix into analyte atoms in the atomizer. [b] Molecular species—*e.g.*, CO, CO_2, N_2, etc.—deactivate excited atoms *via* nonradiational means—*e.g.*, collisional processes.

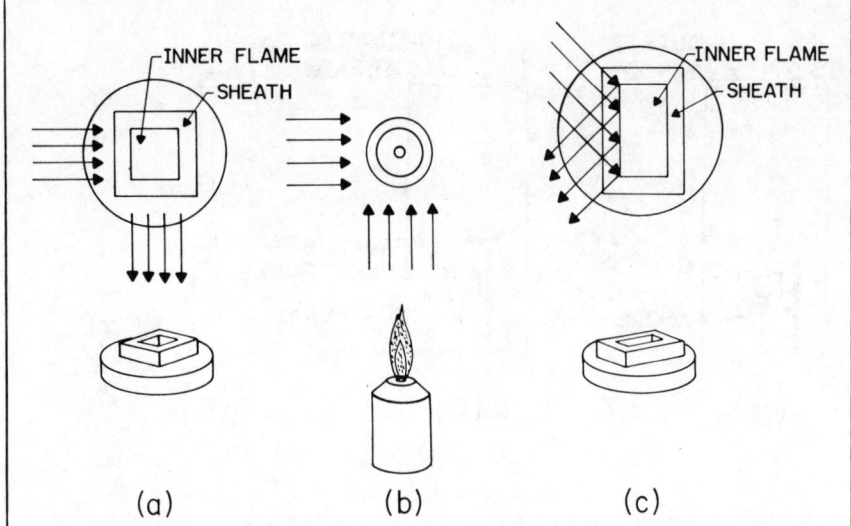

Figure 7. Schematic diagram of flame atomizer shapes (burner shapes) for atomic fluorescence flame spectrometry

 a. Rectangular flame—right-angle illumination—measurement (see 15,24,46,59,80,81,87)
 b. Round flame—right-angle illumination—measurement (see most references between 18 and 89 except 15,24,46,59,80,81,87,90)
 c. Rectangular flame—front-surface illumination—measurement (see 90)

ilar to the Massmann graphite tube furnace except the sample is introduced continuously with an efficient nebulizer. Metal tubes of platinum are simple to fabricate and are not porous compared to graphite tubes, but the upper temperature limit for platinum elements is only 1800°K compared to about 3000°K for graphite. Refractory metals can provide higher temperatures but are highly susceptible to air oxidation and become brittle after being heated. Tantalum tubes have been used allowing an increase in temperature of the atomizer to about 3000°K but oxidation problems, even in the presence of high flow rates of argon past the element, are appreciable at present. Sampling for the wire loop (platinum or tungsten, Figure 8d) is much more tedious and less precise than for any of the above non-

excluding air entrainment into the flame so far as possible by sheathing the analytical flame with either a burning sheath or an inert gas sheath. Nonflame cells are also sheathed with argon.

Some shapes of flames (burner-atomizers) used in atomic fluorescence spectrometry are shown in Figure 7. Generally, laminar-premixed flames with or without either flame or inert gas sheaths are supported on the square (Figure 7a) or rectangular (Figure 7c) burners. Turbulent or turbulent-premixed (39, 40) flames without flame or inert gas sheaths are usually supported on round burners (Figure 7b).

Nonflame atomizers are shown in Figure 8. The Massmann furnace (32) in Figure 8a and the West filament (30, 31, 67) in Figure 8b were used for both atomic fluorescence and atomic absorption spectrometric studies. All nonflame cells are electrically heated. An argon atmosphere is also maintained in most nonflame cells by either passing a stream of argon through them or around them. However, the graphite filament and the metal loop (Figure 8d) can be sheathed either by an argon stream or by a hydrogen diffusion flame. The metal tube furnace (Figure 8c) of Shull and Winefordner (91) is sim-

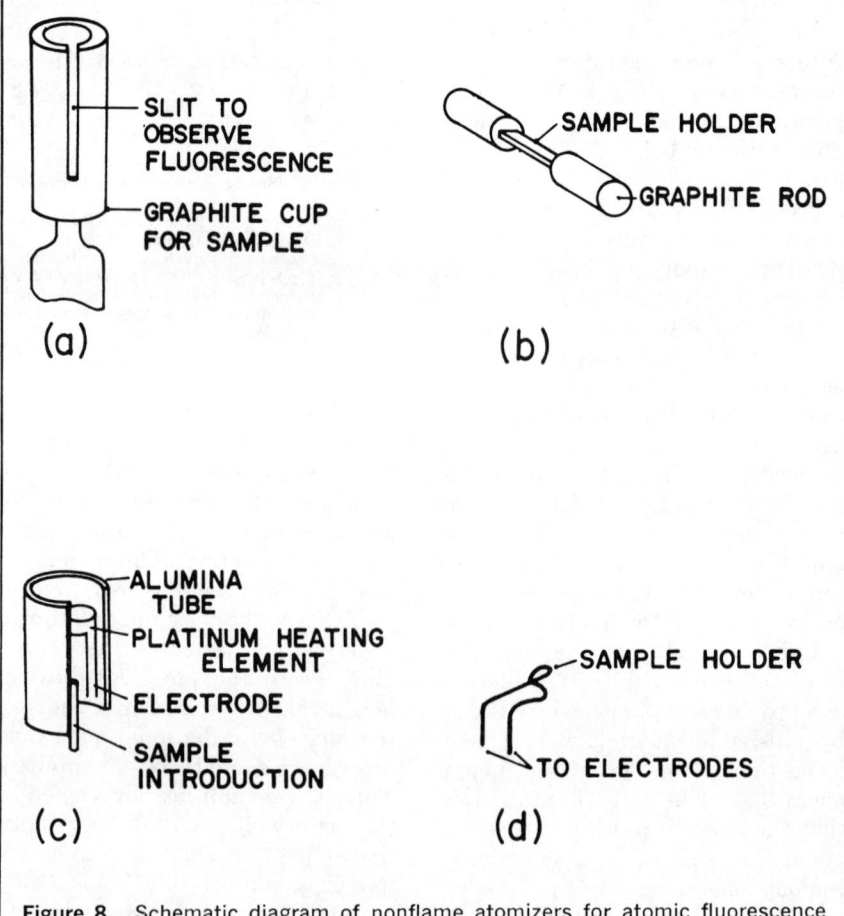

Figure 8. Schematic diagram of nonflame atomizers for atomic fluorescence spectrometry

 a. Massmann graphite furnace (see 32)
 b. West graphite filament (see 30,31,38,62)
 c. Shull metal tube furnace (see 91)
 d. Bratzel metal loop (see 63,64)

flame atomizers. The sample is either applied to the loop by dipping the loop into the solution to be analyzed, or is applied by means of a hypodermic syringe. For all of the nonflame cells except the metal tube and the metal loop, discrete sampling of solutions *via* a hypodermic syringe is utilized. Therefore, in systems employing atomizers of these types, the detector-readout system must respond reliably to the transient signal which results when the free atoms pass through the analytical detection zone of the optical system. Because of the requirement of short time constants in the detection system, the frequency response bandwidth of the measurement system is increased, and background noise may become a problem.

Entrance Optics. In Figure 9 schematic diagrams are given for various arrangements of entrance optics which have been reported for atomic fluorescence spectrometry. All of these arrangements, except the one in Figure 9e, employ the typical 90° arrangement. The simplest one, used in most laboratory-constructed systems, is shown as Figure 9a. In this arrangement, the amount of light collected from both the source and the sample cell is limited by the *f*-number of the lenses employed. Unfortunately, the amount of radiation collected depends upon the size of the lenses (cost increases with size).

An arrangement which improves the efficiency of irradiating the sample cell with source radiation and improves the efficiency of collecting fluorescence radiation from the cell is illustrated in Figure 9b. An advantage of this arrangement over the one in Figure 9a is that lower-cost, more readily available mirrors rather than lenses are used. Also mirrors are better than lenses in transferring images—*e.g.* fewer aberrations. In addition, since radiation from the cell and source is isotropic, placement of mirrors behind the source and the cell increases the amount of radiation collected from each—more than lenses having the same effective *f*-number and being arranged as in Figure 9a. The radiance of fluorescence is directly proportional to the radiant flux of the source incident upon the sample cell. Therefore, the arrangement in Figure 9b—compared to the one in Figure 9a—markedly improves the fluorescence response.

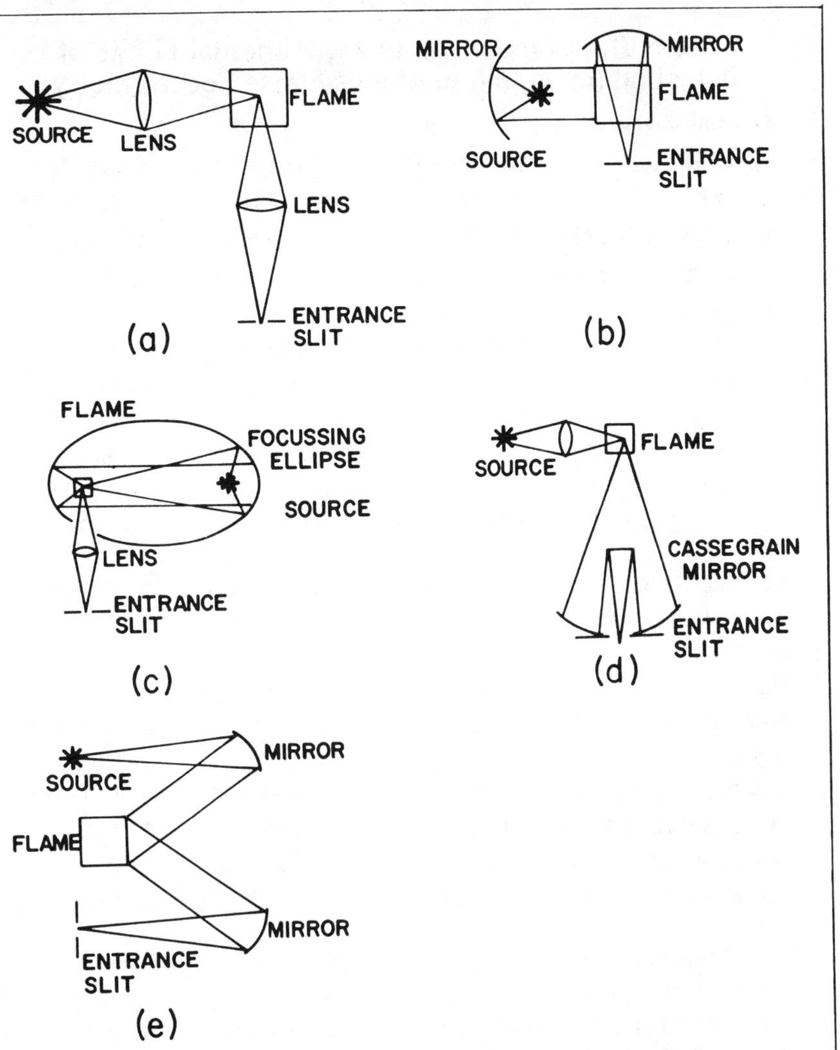

The focusing ellipse illustrated in Figure 9c provides an even more efficient means of collecting source radiation. By placing the source at one focus and the sample cell at the second focus, all of the source radiation in the plane of the ellipse is focused onto the sample cell. The cell is viewed through a slit in the side of the ellipse. Although a large gain in incident radiant flux is obtained with this arrangement, care must be taken to avoid measurement of scattered source radiation.

Figure 9. Schematic diagram of entrance optics used in atomic fluorescence spectrometry

a. All lens system—right-angle illumination—measurement (see 74)
b. All mirror or mirror-lens combination—right-angle illumination—measurement (see 23, 26, 29, 58)
c. Elliptical mirror system—right-angle illumination—measurement (see 91)
d. Systems with Cassegranian mirror—right-angle illumination—measurement (see 92)
e. All mirror systems—front-surface illumination—measurement (see 90)

A Cassegranian system (Figure 9d) has been employed in the Technicon AFS 6 (*92*). This expensive mirror system has the great advantage of being capable of collecting a large, solid angle of fluorescence radiation.

Front-surface illumination of the sample cell is employed in the arrangement depicted in Figure 9e. Here, also, mirrors with small aperture ratios having great light-gathering power may be employed at relatively low cost. The principal advantage to using front-surface illumination is that source radiation scatter is less.

Both dispersive and nondispersive spectral isolation devices have

Table VI. Comparison of Experimental Limits of Detection (in µg/ml) in Atomic Flame Spectrometry[a]

Element–Wavelength (Å)[b]	aal[c]	afl[c]	ae[c]
Ag–3281	0.0005(93)	0.0001*(57)	0.02(94)
Al–3962	0.04(103)	0.1(69)	0.005*(95)
As–1937,1937,2350	0.1 (93)	0.1*(45)	50(96)
Au–2428,2676,2676	0.01*(103)	0.005*(33)	4(96)
Be–2349	0.002*(93,103)	0.01(40,54)	0.1(97)
Bi–2231	0.05(93,103)	0.005*(59)	2(98)
Ca–4227	0.0005(103)	0.02(57)	0.0001*(94)
Cd–2288,2288,3261	0.0006(103)	0.000001(57)	2(94)
Co–2407,2407,3454	0.005*(93,103)	0.005*(22,52)	0.05(94)
Cr–3579,3579,4254	0.005*(93)	0.05(45)	0.005*(94)
Cu–3247,3247,3274	0.003*(103)	0.001*(18,24,37,74)	0.01(94)
Fe–2483,2483,3720	0.005*(93,103)	0.008*(45)	0.05(94)
Ga–2874,4172,4172	0.07(93)	0.01(59)	0.01*(94)
Ge–2652	0.1*(103)	0.1*(66)	0.5(94)
Hg–2537	0.2(103)	0.0002*(80)	40(96)
In–3039,4511,4511	0.05(93,103)	0.1(57)	0.005*(95)
Mg–2852	0.0003*(93,103)	0.001*(22,30)	0.005(94)
Mn–2795,2795,4031	0.002*(93)	0.006*(57)	0.005*(94)
Mo–3133,3133,3903	0.03*(93)	0.5(69)	0.1(94)
Ni–2320,2320,3415	0.005*(93)	0.003*(25)	0.6(96)
Pb–2833,4058,4058	0.01*(93)	0.01*(24,29,58)	0.2(94)
Pd–2746,3405,3635	0.02*(103)	0.04*(34)	0.05*(94)
Rh–3435,3692,3692	0.03*(93)	3(19)	0.3(96)
Sb–2175,2311,2598	0.07*(103)	0.05(48)	20(96)
Si–2516,2040,2516	0.1*(93)	0.6*(44)	5(96)
Se–1960,1960,1960	0.1*(93)	0.04(59)	N[d]
Sn–2246,3034,2840	0.03(93,103)	0.05(42)	0.3(94)
Sr–4607	0.004(103)	0.03(57)	0.0002*(94)
Te–2143,2143,2383	0.1(93,103)	0.005*(59)	200(96)
Tl–2768,3776,3776	0.02*(103)	0.008*(57)	0.02(94)
V–3184,3184,4379	0.02*(93)	0.07(69)	0.01*(94)
Zn–2138	0.002(93,103)	0.00002*(57,80)	50(96)

[a] Limit of detection is usually defined as that concentration resulting in a signal-to-noise ratio of two. All limits of detection are for elements atomized in "laminar" or "turbulent" flames produced using commercially available burners—e.g., limits of detection obtained by increasing the residence time of atoms in a flame by introducing the flame into a long quartz tube in aal are not listed in the above table. Only limits of detection are given for those elements which have been studied by all three methods and for those elements for which reliable limits of detection are listed. Limits of detection for aal are for the PE 303 or the Techtron AA 5. The limits of detection for the PE 303 were taken from the article by Kahn (93). More recent values are undoubtedly available for the PE 503 atomic absorption spectrometer but were not available to the authors. The limits of detection for the Techtron AA 5 were taken from recent literature (103). Also, lower limits of detection in atomic absorption spectrometry are available in the literature for some elements, but it was considered more reliable to take all aal values from the above two sources. It would also be more reliable to take all afl and ae values from one source, but no such tabulation is currently available. [b] If three wavelengths are listed, then the three limits of detection correspond to wavelengths used for aal, afl, and ae, respectively. If just one wavelength is listed, then the same spectral line was measured for all three methods. [c] aal = atomic absorption flame spectrometry with a line source; afl = atomic fluorescence flame spectrometry with a line source; ae = atomic emission flame spectrometry. [d] Not detectable by ae. * Asterisk superscript indicates that the listed value is threefold or more lower than other limits of detection for the same element. If an asterisk superscript is on two (or three) limits of detection for the same element, then those two (or three) methods give essentially the same result. A threefold range is used here because of the differences in defining limits of detection by some authors and to account for differences in imagination when measuring low signal-to-noise ratios.

been employed. Generally, a grating monochromator of fairly large aperture is used. Several investigators (77, 80, 90) have used a combination of interference filters and solar-blind multiplier phototubes. The choice as to spectral isolation device must be made on the basis of possible interferences which may be present. Nondispersive devices offer a definite cost advantage.

Analytical Studies (18–92)

Limits of Detection. In Table VI, a comparison is given of the best (state-of-the-art) experimental concentrational (in micrograms per milliliter) limits of detection for several elements measured by atomic absorption flame spectrometry with a line source (aal), atomic fluorescence flame spectrometry with a line source (afl), and atomic emission flame spectrometry (ae). The asterisk superscripts denote the methods giving the lowest limits of detection [threefold or lower than the other method(s)]. As seen from the data in Table VI, the results compare favorably with those predicted in Figure 4.

In Table VII, a comparison is also given of the best (state-of-the-art) absolute (in nanograms) limits of detection for atomic spectrometric methods with both flame and nonflame cells. Nonflame cells are useful for smaller amounts of analyte than flame cells. Also, with most nonflame cells, smaller sample sizes—e.g., one µg—are necessary. The best available absolute limits of detection for atomic absorption spectrometry with a line source and a graphite cell (gaal) and atomic fluorescence with a line source and either a graphite cell (gafl) or a metal loop (mafl) compare favorably with the best absolute limits of detection obtained with more exotic and expensive methods such as neutron activation and spark source mass spectrometry (102).

Interferences. Studies performed by West's group at Imperial College and Winefordner's group at the University of Florida indicate that the predicted degree of interferences listed in Table II and discussed previously is experimentally valid.

Comparison of Atomic Fluorescence Spectrometry with Atomic

Table VII. Absolute Limits of Detection (in ng) for Atomic Spectrometric Methods[a]

Element	aal[b,c]	afl[b,c]	ae[b,c]	gaal[c]	gafl[c]	mafl[c](63)	rfae[c](101)
Ag	0.1	0.02	4	0.0005*(99)	0.005*(32, 38)	0.002	
As	8	20	10,000	0.2*(100)	0.5*(38)		20
Au	2	1	800	0.07(99)	0.01*(38)		
Be	0.4	2	200	0.003*(99, 100)	0.03(38)	4	
Bi	10	1	400	0.02*(99)	0.01*(31)		
Ca	0.1	4	0.02	0.003(100)	0.0001*(38)	0.2	
Cd	0.1	0.0002	400	0.00006*(99)	0.00003*(38)	0.00002*	6
Cu	0.6	0.2	2	0.006*(99)	0.005*(38)		
Fe	1	2	10	0.02*(99)	0.01*(38)		1
Ga	14	2	2	0.02*(99)	0.05*(31)	20	
Hg	40	0.04	8,000	0.4(99)		0.02*	
Mg	0.06	0.2	1	0.003(99)	0.0000001*(30)	4	
Pb	2	2	40	0.02*(100)	0.01*(31)	4	2
Tl	4	2	4	0.002*(99)	0.05(31)	2	
Zn	0.4	0.004	10,000	0.001(99)	0.00005*(31, 32)	0.02	2

[a] Limit of detection is usually defined as that amount (in nanograms) resulting in a signal-to-noise ratio of two. Only limits of detection are given for elements measured by atomic absorption and atomic fluorescence flame and nonflame spectrometry. No wavelengths are given because they often differ for the various techniques. Exact experimental conditions can be obtained by consulting the literature references. [b] Values are calculated using concentrations in Table VI and assuming a 0.2-ml sample is used. [c] aal = atomic absorption flame spectrometry with a line source; afl = atomic fluorescence flame spectrometry with a line source; gaal = graphite cell atomic absorption spectrometry with a line source; gafl = graphite cell atomic fluorescence spectrometry with a line source; mafl = metal loop atomic fluorescence spectrometry with a line source; rfae = RF excitation atomic emission spectrometry (assuming a 0.2-ml sample is nebulized). Only results obtained by Dickinson and Fassel (101) are listed here because they are generally lower than measured by other workers using RF excitation. * Asterisk indicates the method (or methods) giving threefold or lower limits of detection than the other methods.

Absorption and Atomic Emission Spectrometry. The three methods with respect to limits of detection and interferences were compared above. The precision of measurements for a given sample depends primarily upon the sampling method, the source stability, and the instrumental system, but should be about the same for all three methods—e.g., the percent relative standard deviation should be about 1%. The accuracy of measurement depends primarily upon the analyte type, the sample matrix, and the chemical steps involved in the sample preparation. The instrumentation used in atomic emission flame spectrometry is simpler (less entrance optics and no external source needed) than the instrumentation required for atomic fluorescence or atomic absorption spectrometry. Atomic fluorescence spectrometry should be the simplest and best method to utilize for multielement analysis with multiplex spectrometers—e.g., source alignment is difficult in atomic absorption spectrometry and background radiation and flicker are limiting in atomic emission spectrometry.

Recently, Technicon Corp. announced the development of an atomic fluorescence spectrometer for multielement analysis (22). This instrument involves the use of pulsed hollow cathode lamps, a rotating interference filter wheel, a flame cell, and logic circuitry to measure the fluorescence of each metal when the proper interference filter is in place. Six elements per sample and 100 samples per hr can be measured with this unique and exciting new instrument.

Applications of Atomic Fluorescence Spectrometry. So far, few applications of atomic fluorescence spectrometry to real samples have resulted. Smith et al. (56) determined trace wear metals in jet engine lubricating oils. Sychra and Matousek (35) determined Ni in gas oils and petroleum distillates. Cotton and Jenkins (87) determined low concentrations of Cu, Fe, and Pb in hydrocarbon fuels. Amos et al. (38) determined lead in blood and urine by a unique approach involving measurement of direct-line fluorescence.

Most workers have determined metals in synthetic solutions with added interferences rather than in real samples. Finally, the new multielement Technicon atomic fluorescence instrument should be of particular use in the metallurgical industry. Certainly, more research involving the use of atomic fluorescence spectrometry for analysis of real samples is necessary before there will be a widespread use of this sensitive, selective, and versatile method of trace analysis.

References

(1) A. C. G. Mitchell and M. W. Zemansky, "Resonance Radiation and Excited Atoms," University Press, Cambridge, England, 1961.
(2) P. Pringsheim, "Fluorescence and Phosphorescence," Interscience, New York, N. Y., 1949.
(3) E. L. Nichols and H. L. Howes, *Phys. Rev.* 23, 472 (1924).
(4) R. M. Badger, *Z. Phys.* 55, 56 (1929).
(5) C. Th. J. Alkemade in "Proceedings of the Xth Colloquium Spectroscopicum International," E. R. Lippincott and M. Margoshes, Eds., Spartan Books, Washington, D. C., 1963.
(6) J. D. Winefordner and T. J. Vickers, ANAL. CHEM. 36, 161 (1964).
(7) J. D. Winefordner and T. J. Vickers, ANAL. CHEM. 42, 207R (1970).
(8) J. D. Winefordner, V. Svoboda, and L. J. Cline, *CRC Crit. Rev., Anal. Chem.* 1, 233 (1970).

(9) W. J. Price in "Spectroscopy," D. R. Browning, Ed., McGraw-Hill, New York, N. Y., 1969.
(10) D. W. Ellis and D. R. Demers, *Advan. Chem. Ser.* **73**, 326 (1968).
(11) R. Smith in "Spectrochemical Methods of Analysis," J. D. Winefordner, Ed., John Wiley, New York, N. Y., in press.
(12) J. D. Winefordner and R. Smith in "Analytical Flame Spectrometry, Selected Topics," R. Mavrodineanu, Ed., Centrex, Eindhoven, The Netherlands, in press.
(13) H. P. Hooymayers, *Spectrochim. Acta* **23B**, 567 (1968).
(14) C. Th. J. Alkemade in "Proceedings of Atomic Absorption Conference," Sheffield, England, July 1969, Butterworths, London, England, in press.
(15) P. J. T. Zeegers and J. D. Winefordner, *Spectrochim. Acta*, in press.
(16) J. D. Winefordner, M. L. Parsons, J. M. Mansfield, and W. J. McCarthy, ANAL. CHEM. **39**, 436 (1967).
(17) M. L. Parsons, W. J. McCarthy, and J. D. Winefordner, *J. Chem. Ed.* **44**, 214 (1967).
(18) J. I. Dinnin, ANAL. CHEM. **39**, 1491 (1967).
(19) D. L. Manning and P. Heneage, *At. Absorption Newslett.* **7**, 80 (1968).
(20) N. Omenetto and G. Rossi, *Anal. Chim. Acta* **40**, 195 (1968).
(21) G. Rossi and N. Omenetto, *Talanta* **16**, 263 (1969).
(22) D. G. Mitchell and A. Johansson, *Spectrochim. Acta* **25B**, 175 (1970).
(23) D. N. Armentrout, ANAL. CHEM. **38**, 1237 (1966).
(24) D. L. Manning and P. Heneage, *At. Absorption Newslett.* **6**, 124 (1967).
(25) J. P. Matousek and V. Sychra, ANAL. CHEM. **41**, 518 (1969).
(26) J. W. Robinson and C. J. Hsu, *Anal. Chim. Acta* **43**, 109 (1968).
(27) T. S. West and X. K. Williams, ANAL. CHEM. **40**, 335 (1968).
(28) T. S. West and X. K. Williams, *Anal. Chim. Acta* **42**, 29 (1968).
(29) V. Sychra and J. P. Matousek, *Talanta* **17**, 363 (1970).
(30) T. S. West and X. K. Williams, *Anal. Chim. Acta* **45**, 27 (1969).
(31) R. G. Anderson, I. S. Maines, and T. S. West, *Anal. Chim. Acta* **51**, 355 (1970).
(32) H. Massmann, *Spectrochim. Acta* **23B**, 215 (1968).
(33) J. P. Matousek and V. Sychra, *Anal. Chim. Acta* **49**, 175 (1970).
(34) V. Sychra, P. J. Slevin, J. P. Matousek, and F. Bek, *Anal. Chim. Acta* **52**, 259 (1970).
(35) V. Sychra and J. P. Matousek, *Anal. Chim. Acta* **52**, 376 (1970).
(36) J. B. Headridge and J. Richardson, *Lab. Pract.* **19**, 372 (1970).
(37) R. Smith, R. C. Elser, and J. D. Winefordner, *Anal. Chim. Acta* **48**, 35 (1969).
(38) M. D. Amos, P. A. Bennett, K. G. Brodie, P. W. Y. Lung, and J. P. Matousek, presented at 21st Pittsburgh Conference in Analytical Chemistry and Applied Spectroscopy, Cleveland, Ohio, March 1970.
(39) M. P. Bratzel, R. M. Dagnall, and J. D. Winefordner, ANAL. CHEM. **41**, 713 (1969).
(40) M. P. Bratzel, R. M. Dagnall, and J. D. Winefordner, ANAL. CHEM. **41**, 1527 (1969).
(41) M. P. Bratzel and J. D. Winefordner, *Anal. Chim. Acta* **39**, 394 (1967).
(42) R. F. Browner, R. M. Dagnall, and T. S. West, *Anal. Chim. Acta* **46**, 207 (1969).
(43) R. F. Browner, R. M. Dagnall, and T. S. West, *Talanta* **16**, 75 (1969).
(44) R. M. Dagnall, G. F. Kirkbright, T. S. West, and R. Wood, *Anal. Chim. Acta* **47**, 407 (1969).
(45) R. M. Dagnall, M. R. G. Taylor, and T. S. West, *Spectrosc. Lett.* **1**, 397 (1968).
(46) R. M. Dagnall, K. C. Thompson, and T. S. West, *Anal. Chim. Acta* **41**, 551 (1968).
(47) R. M. Dagnall, K. C. Thompson, and T. S. West, *Talanta* **14**, 557 (1967).
(48) R. M. Dagnall, K. C. Thompson, and T. S. West, *Talanta* **14**, 1151 (1967).
(49) R. M. Dagnall, K. C. Thompson, and T. S. West, *Talanta* **14**, 1467 (1967).
(50) R. M. Dagnall, K. C. Thompson, and T. S. West, *Talanta* **15**, 677 (1968).
(51) R. M. Dagnall and T. S. West, *Appl. Opt.* **7**, 1287 (1968).
(52) B. Fleet, K. V. Liberty, and T. S. West, *Anal. Chim. Acta* **45**, 205 (1969).
(53) G. F. Goodfellow, *Anal. Chim. Acta* **36**, 132 (1966).
(54) D. N. Hingle, G. F. Kirkbright, and T. S. West, *Analyst* **93**, 522 (1968).
(55) J. M. Mansfield, M. P. Bratzel, H. O. Norgordon, D. N. Knapp, K. E. Zacha, and J. D. Winefordner, *Spectrochim. Acta* **23B**, 389 (1968).
(56) R. Smith, C. M. Stafford, and J. D. Winefordner, *Can. Spectrosc.* **14**, 2 (1969).
(57) K. E. Zacha, M. P. Bratzel, J. D. Winefordner, and J. M. Mansfield, ANAL. CHEM. **40**, 1733 (1968).
(58) R. F. Browner, R. M. Dagnall, and T. S. West, *Anal. Chim. Acta* **50**, 375 (1970).
(59) A. Hell and S. Ricchio, presented at 21st Pittsburgh Conference on Analytical Chemistry and Applied Spectroscopy, Cleveland, Ohio, March 1970.
(60) M. S. Cresser and T. S. West, *Anal. Chim. Acta* **51**, 530 (1970).
(61) A. Fulton, K. C. Thompson, and T. S. West, *Anal. Chim. Acta* **51**, 373 (1970).
(62) J. F. Alder and T. S. West, *Anal. Chim. Acta* **51**, 365 (1970).
(63) M. P. Bratzel, R. M. Dagnall, and J. D. Winefordner, *Anal. Chim. Acta* **48**, 197 (1969).
(64) M. P. Bratzel, R. M. Dagnall, and J. D. Winefordner, *Appl. Spectrosc.* **24**, 518 (1970).
(65) G. B. Marshall and T. S. West, *Anal. Chim. Acta* **51**, 179 (1970).
(66) R. M. Dagnall, G. F. Kirkbright, T. S. West, and R. Wood, *Analyst* **95**, 425 (1970).
(67) P. C. Wildy and K. C. Thompson, *Analyst* **95**, 562 (1970).
(68) M. P. Bratzel and J. D. Winefordner, *Anal. Lett.* **1**, 43 (1967).
(69) R. M. Dagnall, G. F. Kirkbright, T. S. West, and R. Wood, ANAL. CHEM. **42**, 1029 (1970).
(70) K. C. Thompson and P. C. Wildy, *Analyst* **95**, 776 (1970).
(71) R. M. Dagnall, T. S. West, and P. Young, *Talanta* **13**, 803 (1966).
(72) R. S. Hobbs, G. F. Kirkbright, M. Sargent, and T. S. West, *Talanta* **15**, 997 (1968).
(73) D. R. Jenkins, *Spectrochim. Acta* **23B**, 167 (1967).
(74) J. M. Mansfield, J. D. Winefordner, and C. Veillon, ANAL. CHEM. **37**, 1049 (1965).
(75) N. Omenetto and G. Rossi, *Spectrochim. Acta* **24B**, 95 (1969).
(76) T. J. Vickers and S. P. Merrick, *Talanta* **15**, 873 (1968).
(77) T. J. Vickers and R. M. Vaught, ANAL. CHEM. **41**, 1477 (1969).
(78) J. D. Winefordner and R. A. Staab, ANAL. CHEM. **36**, 165 (1964).
(79) J. D. Winefordner and R. A. Staab, ANAL. CHEM. **36**, 1367 (1964).
(80) P. D. Warr, *Talanta* **17**, 543 (1970).
(81) D. R. Jenkins, *Spectrochim. Acta* **25B**, 47 (1970).
(82) N. Omenetto and G. Rossi, *Spectrochim. Acta* **25B**, 297 (1970).
(83) R. M. Dagnall, K. C. Thompson, and T. S. West, *Anal. Chim. Acta* **36**, 269 (1966).
(84) D. W. Ellis and D. R. Demers, ANAL. CHEM. **38**, 1943 (1966).
(85) C. Veillon, J. M. Mansfield, M. L. Parsons, and J. D. Winefordner, ANAL. CHEM. **38**, 204 (1966).
(86) M. Cresser and T. S. West, *Spectrochim. Acta* **25B**, 61 (1970).
(87) D. H. Cotton and D. R. Jenkins, *Spectrochim. Acta* **25B**, 283 (1970).
(88) M. P. Bratzel, R. M. Dagnall, and J. D. Winefordner, *Anal. Chim. Acta* **52**, 157 (1970).
(89) R. A. Miller and J. D. Winefordner, submitted for publication in *Appl. Spectrosc.*
(90) R. C. Elser and J. D. Winefordner, *Appl. Spectrosc.*, in press.
(91) M. Shull and J. D. Winefordner, unpublished results, University of Florida, Gainesville, Fla., 1970.
(92) D. G. Mitchell, presented at 1970 Technicon International Congress, New York, N. Y., November 1970.
(93) H. Kahn, *Advan. Chem. Ser.* **73**, 183 (1968).
(94) E. E. Pickett and S. R. Koirtyohann, *Spectrochim. Acta* **23B**, 235 (1968).
(95) E. E. Pickett and S. R. Koirtyohann, *Spectrochim. Acta* **24B**, 325 (1969).
(96) V. A. Fassel and D. W. Golightly, ANAL. CHEM. **39**, 446 (1967).
(97) D. N. Hingle, G. F. Kirkbright, and T. S. West, *Analyst* **93**, 522 (1968).
(98) R. S. Hobbs, G. F. Kirkbright, and T. S. West, *Analyst* **94**, 554 (1969).
(99) B. V. L'vov, *Spectrochim. Acta* **24B**, 53 (1969).
(100) D. C. Manning and F. Fernandez, *At. Absorption Newslett.* **9**, 65 (1970).
(101) G. W. Dickinson and V. A. Fassel, ANAL. CHEM. **41**, 1021 (1969).
(102) G. H. Morrison and R. K. Skogerboe in "Trace Analysis, Physical Methods," G. H. Morrison, Ed., Interscience, New York, N. Y., 1965.
(103) Detection Limits For Model AA 5 Atomic Absorption Spectrophotometry, Varian Techtron, Walnut Creek, Calif.

This work supported by AFOSR(SRC)-OAR-USAF, AF-AFOSR-70-1880 B.

James D. Winefordner received his B.S., M.S., and Ph.D. degrees in chemistry from the University of Illinois in 1954, 1955, and 1958, respectively. His research advisor was Professor H. V. Malmstadt. In 1967, after eight years on the Department of Chemistry staff, he was promoted to Full Professor at the University of Florida. He is currently Chairman of the Analytical Division of that department. Dr. Winefordner's research interests include: atomic and molecular emission, absorption, and fluorescence in flames and other hot gases; molecular fluorescence and phosphorescence of species in the condensed phase; development of sensitive, selective, accurate methods of trace analysis of metals and molecules in materials based on the above spectroscopic methods; development of sensitive, selective gas and liquid chromatographic detectors; and development of spectroscopic instrumentation for analysis. He has published more than 125 scientific papers and chapters on the above topics and given over 75 invited talks and seminars at international and national conferences and symposia and at universities, colleges, and industries. Since he has been at the University of Florida, 24 of his graduate students received Ph.D. degrees and 9 more received M.S. degrees. He has also had 11 postdoctorate fellows work with him. He is a member of the American Chemical Society, Phi Lambda Phi, Alpha Chi Sigma, and the American Association for the Advancement of Science. He is a member of the Advisory Boards of ANALYTICAL CHEMISTRY, Spectrochimica Acta B, and Chemical Instrumentation and is an associate member of the International Union of Pure and Applied Chemistry on the Spectrochemical and Other Optical Procedures for Analysis.

Robert C. Elser received his B.S. degree in chemistry from Lehigh University in Bethlehem, Pa. in 1963. From 1963 to 1964, he was employed as the assistant biochemist at the Allentown Hospital in Allentown, Pa. Between 1964 and 1967, he supervised the clinical chemistry laboratories of the Reading Hospital in West Reading, Pa. Since September 1967, Mr. Elser has been a graduate student at the University of Florida in Gainesville, Fla. His research director is James D. Winefordner. Currently he is working on the dissertation for his Ph.D. degree. The topic involves the use of continuum light sources in atomic absorption. His research interests include atomic emission, absorption and fluorescence in flames; the application of sophisticated analytical techniques to clinical chemical problems; and automation and computerization in clinical chemistry and toxicology. He is the coauthor of five scientific papers related to the above topics. Upon receiving his Ph.D. degree in June 1971, he will become affiliated with the York Hospital in York, Pa. where he will direct the clinical chemistry laboratories. He is a member of the American Chemical Society and the American Association of Clinical Chemists.

Alan G. Marshall and Melvin B. Comisarow
Department of Chemistry
University of British Columbia
Vancouver, Canada V6T 1W5

Fourier and Hadamard Transform Methods in Spectroscopy

To understand why Hadamard and Fourier methods have proved so valuable in spectroscopy, it is first necessary to recognize the disadvantages of the conventional way of doing spectroscopy, in which the spectrum is obtained by scanning across the spectroscopic region of interest with a narrow observation window. Surprisingly, the basic explanation may be expressed very simply, by analogy to the best way to use an ordinary double-pan balance.

Weights on a Balance

Consider the common problem of determining the weights of four unknown objects, by use of the schematic balance shown in Figure 1. Conventionally, we would solve the problem by weighing the unknowns one at a time in (say) the left pan, by putting the appropriate (known) weights on the right, as shown schematically below:

Measure-	Unknown			
ment	#1	#2	#3	#4
#1	1	0	0	0
#2	0	1	0	0
#3	0	0	1	0
#4	0	0	0	1
No. of weighings	1	1	1	1

The obvious advantage of this procedure is that the measurements yield each unknown weight directly; therefore, no data reduction is required. However, the disadvantage is that each unknown has been weighed only once.

In any experimental measurement characterized by a certain level of imprecision or random noise, it is desirable to repeat the measurement many times to obtain a more accurate result. The signal (in this case, the weight of a given unknown) will accumulate as the number of weighings, N. But if the noise is random its magnitude may be treated as a random walk about zero (the average noise level), and the average absolute distance away from zero

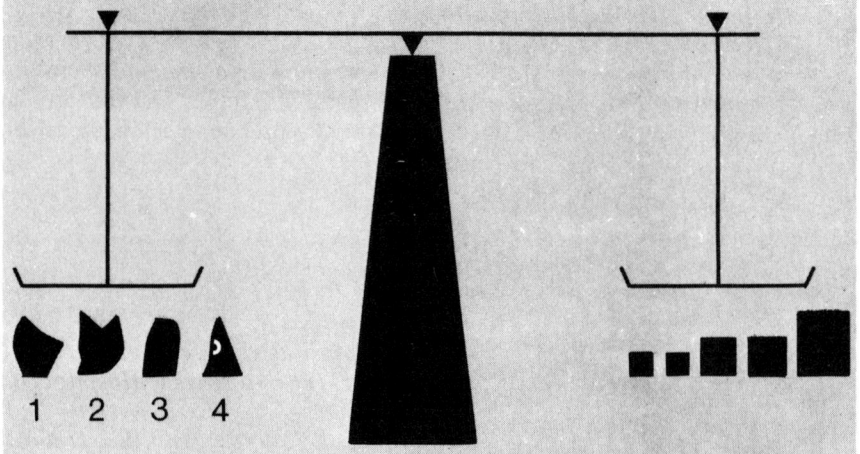

Figure 1. Schematic diagram of double-pan balance, set of standard weights, and four unknown weights

after N steps of a random walk (more precisely, the root-mean-square distance) is proportional to $\sqrt{N}$ (1). Thus, the true measure of precision of the repeated measurement, the signal-to-noise ratio, is proportional to $(N/\sqrt{N})$ or just $\sqrt{N}$.

Let us return to the balance problem. Suppose that two unknown weights are placed on the left pan at once and that four linearly independent arrangements of unknowns are weighed, two at a time:

Measure-	Unknown			
ment	#1	#2	#3	#4
#1	1	1	0	0
#2	0	1	1	0
#3	0	0	1	1
#4	1	0	0	1
No. of weighings	2	2	2	2

This time, the four unknown weights are related to the four observed weights by four linear algebraic equations, which may then be solved to yield the desired unknown weights. However, since each unknown has now been weighed twice, the precision (signal-to-noise ratio) of each calculated weight is now better by $\sqrt{2}$ than for the original conventional experiment.

Since the same total number of weighings (four) are required, the total time required for the new experiment is also the same. For an arbitrary number, N, of unknown weights, the general improvement in signal-to-noise ratio for this encoding-decoding scheme is $\sqrt{N/2}$ for an experiment that takes no longer to carry out than N conventional single-object weighings. [The precision of the calculated weight will be better only if the average error in a particular weighing depends upon the balance and is independent of the magnitude of the measured weight. This general condition is called "detector-limited" noise and is distinguished from the "source-limited" noise situation in which the rms noise is proportional to the square root of the signal magnitude. If the noise is source limited, the above encoding-decoding method will *not* improve the precision of the calculated weights over those determined in a simpler one-at-a-time weighing procedure (2). Spectroscopic examples in which the noise is detector limited include infrared, microwave, nuclear magnetic resonance, and ion cyclotron resonance experiments. Examples in which the noise is source limited include optical (VIS–UV) and charged-particle (photoelectron, ESCA, electron impact) spectroscopy.] Although the experimental details of the spectroscopy experiment are completely different from those of the balance experiment, the preceding encoding-decoding scheme still applies and forms the basis for Hadamard transform spectroscopy, as will shortly be evident.

By logical extension of the preceding argument, one might think of putting unknown weights on both sides of the balance rather than just on one side, while keeping track of the (known) weight required to balance any particular arrangement of unknowns:

Measurement	Unknowns in	
	Left pan	Right pan
#1	#1	#2, #3, #4
#2	#1, #2	#3, #4
#3	#1, #3	#2, #4
#4	#1, #4	#2, #3

(Each unknown weighed four times)

Again, it is possible to extract the four desired individual unknown weights by straightforward solution of four coupled linear algebraic equations. Since each unknown has now been weighed four times, the signal-to-noise ratio for each calculated weight is improved by $\sqrt{4} = 2$ times over conventional one-at-a-time weighing. For an arbitrary number of unknowns, N, it follows that the general improvement in signal-to-noise ratio will be a full $\sqrt{N}$. This improvement also applies to the (experimentally different) Fourier transform spectroscopy experiment, as discussed later in this paper.

Hadamard and Fourier methods, then, provide a means for improving the precision (signal-to-noise ratio) in a weighing experiment by a factor of $\sqrt{N/2}$ or $\sqrt{N}$, respectively, but require the *same* total time for measurement. This improvement is known as the Fellgett advantage (3); all that remains is to show that we ordinarily perform spectroscopy as inefficiently as we ordinarily use a double-pan balance, and we then explore the available means for exploiting the potential Fellgett advantage in the situation. [The Fellgett advantage can be realized only if the noise is detector limited, not when the noise is source limited (2).]

Direct Multichannel Spectrometers

Figure 2 is a highly schematic diagram of a generalized spectrometer. The dispersive element might be a prism or grating (infrared, optical), for example; the slit might be a band-pass filter for a low-frequency (microwave, nuclear magnetic resonance) case. The slit width is chosen sufficiently narrow

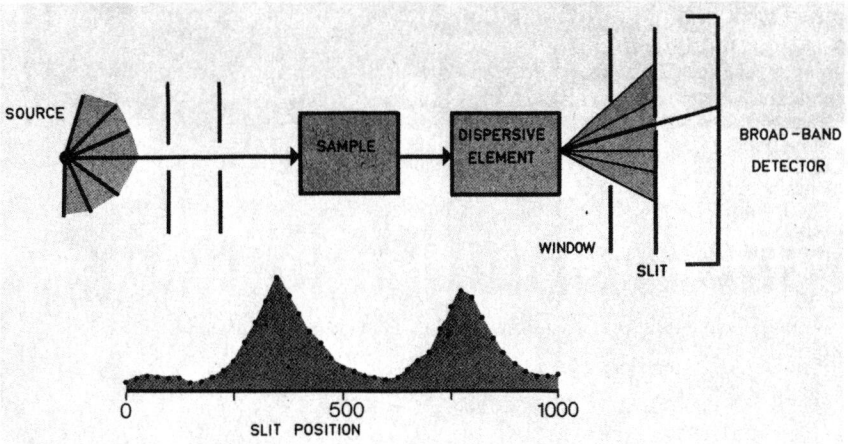

Figure 2. Top: schematic diagram of single-slit scanning absorption spectrometer. Single-slit scanning emission spectrometer lacks only broad-band source. Bottom: detector readings from number of individual slit positions

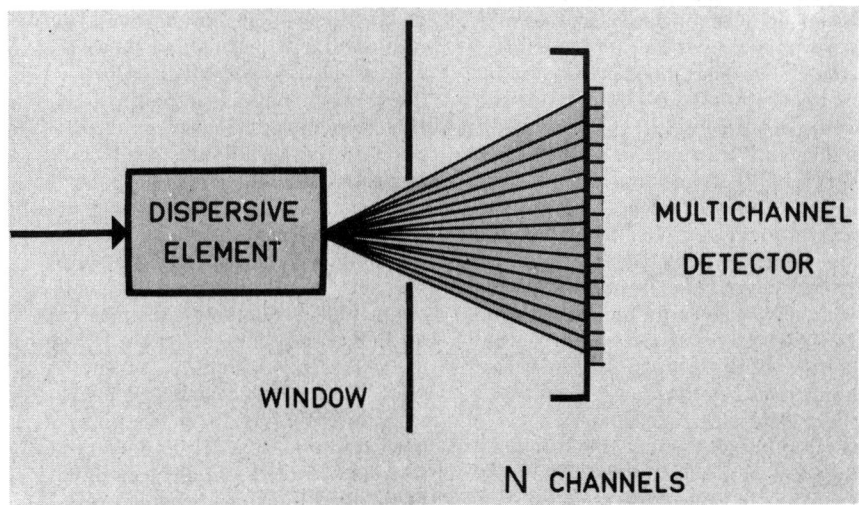

Figure 3. Schematic diagram of detector section of direct multichannel spectrometer composed of many separate single-channel detectors

that when detector readings are collected from a number of individual slit positions (bottom of Figure 2), there is sufficient resolution to distinguish spectral features of interest. The most important feature of such a spectrometer is that its detection of an absorption spectrum requires a procedure formally identical to the one-at-a-time method of determining the weights (spectral intensities) of N different unknown objects (spectral slit positions). It would thus be desirable to open up the slit aperture to the full width of the desired spectral window by using N separate single-channel detectors as shown in Figure 3. Since *all* slit positions are now monitored at once, rather than just one at a time, the spectrometer of Figure 3 offers (in principle) an improvement of the full $\sqrt{N}$ advantage in signal-to-noise ratio, compared to the result of a single complete spectral scan requiring the same total time by the spectrometer of Figure 2. Alternatively, it would be possible to obtain a spectrum having the *same* signal-to-noise ratio in $(1/N)$ the time required to scan the N individual slit positions one at a time.

Because of the conceptual simplicity of the spectrometer of Figure 3, it is logical to investigate its feasibility. It is desirable to be able to resolve spectral detail as narrow as the width of a typical spectral absorption line; therefore, the minimum number of channels that will be required in a multichannel spectrometer is simply the width of the entire spectral range of interest, divided by the width of a single spectral line. The resultant necessary number of channels for various forms of spectroscopy is shown in Table I.

From Table I, it would appear that electronic (VIS–UV) spectroscopy is the least likely candidate for success with a direct multichannel spectrometer; but in fact, multichannel detection of optical–UV radiation is readily accomplished photographically. The

Table I. Minimum Number of Channels Required for Various Types of Direct Multichannel Spectrometers

Type of spectroscopy	Largest usual frequency	Typical spectral range	Width of 1 line	Min no. of channels[a]
Mössbauer	6×10^{18} Hz	10^8 Hz	10^7 Hz	10
ESCA	3.5×10^{17}	10^{17}	10^{14}	1,000
Photoelectron	5×10^{15}	3×10^{15}	10^{12}	3,000
Electronic	1.5×10^{15}	1.2×10^{15}	10^9	1,250,000
Vibrational	2×10^{14}	1.5×10^{14}	3×10^9	50,000
Rotational	4×10^{10}	3×10^{10}	10^4	300,000
^{13}C NMR	8×10^7	2×10^4	0.5	40,000
ICR	2×10^6	2×10^6	10^2	20,000

[a] Number of channels is obtained by dividing the typical spectral range by the width of one line.

resolution of a fine grain photographic plate is sufficient to provide for the huge number of required channels, since the desired spectrum may be dispersed over the necessary distance (a few meters) without undue effort.

In ESCA (electron spectroscopy for chemical analysis) (4) and photoelectron spectroscopy (5), electrons are dislodged from atoms or molecules by X-ray or UV radiation, respectively, and the electrons released have a translational energy which depends on the energy of the bound state occupied by that electron in the original atom or molecule. By scanning the energy of the observed dislodged electrons, the energies of the original molecular electronic states can be determined. By passing the electrons between two charged parallel plates, the dislodged electrons may be dispersed in space, according to their velocity, to achieve the arrangement shown in Figure 3. This multichannel electron detection scheme has recently become feasible with the advent of the vidicon detector (6), in which an arriving electron strikes a fluorescent screen on the face of a television camera. Since electrons of different velocity can be dispersed to strike different regions of the screen, their arrival will be recorded independently by different elements of the television camera grid. Because of the small required number of detector channels (see Table I), the Figure 3 spectrometer is thus now feasible for ESCA and photoelectron spectroscopy.

With the other forms of spectroscopy listed in Table I, direct multichannel methods are less attractive. For microwave (rotational) spectroscopy, for example, there is no broad-band radiation source available: a blackbody radiation source, such as employed for other radiation energies (xenon or hydrogen discharge for UV, hot tungsten wire for visible, globar for near infrared and infrared, mercury vapor for far infrared) would have to be operated at an unreasonably high temperature to obtain sufficient radiation flux for use as a radiation source. It would be conceivable to construct an array of individual (narrow-band) microwave transmitters (about $5,000 each) as the "broad-band" radiation source, but Table I shows that the cost would be excessive (10^9 ...!). For infrared spectroscopy, on the other hand, the necessary broad-band source is available, but it would be necessary to disperse the spectrum over many meters to be able to resolve the desired spectral detail with existing individual (thermopile) detectors of about 1-mm width each. At a cost of about $200 per detector, the total cost again becomes unmanageable. (Photographic detection does not extend beyond about 12,000 Å and is thus unavailable.) Finally, for nuclear magnetic resonance (NMR) and ion cyclotron resonance (ICR) spectroscopy, broad-band sources are again available, but the cost of an array of tens of thousands of individual narrow-band mixer-filter detectors (see below) is again unreasonably high.

For infrared, microwave, and radiofrequency spectrometers, then, the direct multichannel approach is just not feasible, either geometrically or financially. We will now consider two recent and valuable indirect approaches: Hadamard transform spectroscopy and Fourier transform spectroscopy.

Hadamard Transform Encoding-Decoding ("Multiplex") Spectrometers

Figure 4 shows the instrumental modification which allows for use of the Hadamard scheme: by use of the original (inexpensive) broad-band detector of the spectrometer of Figure 2, a mask is interposed between the desired spectroscopic "window" and the detector. The mask is constructed so that its smallest opening is the same as the (narrow) slit width of the conventional spectrometer (Figure 2) but with approximately *half* the total possible slit positions open. The pattern of open and shut slits is random.

Let the spectrum of transmitted intensities (bottom of Figure 2) be represented by spectral elements: $x_1, x_2, \ldots, x_N$. When the mask in Figure 4 is in position, the detector total response, y, is composed of a sum of all the desired spectral elements, each weighted by a factor, a_n, of either zero or one, depending on whether that particular slit was shut or open, respectively:

$$y = a_1 x_1 + a_2 x_2 + \ldots + a_N x_N, \quad a_n = 0 \text{ or } 1 \quad (1)$$

In other words, the detector has provided one observable (y), expressed in terms of N unknowns (x_1 to x_N), according to the "code" (a_1 to a_N) of Equation 1. This situation is clearly parallel to that of putting half the un-

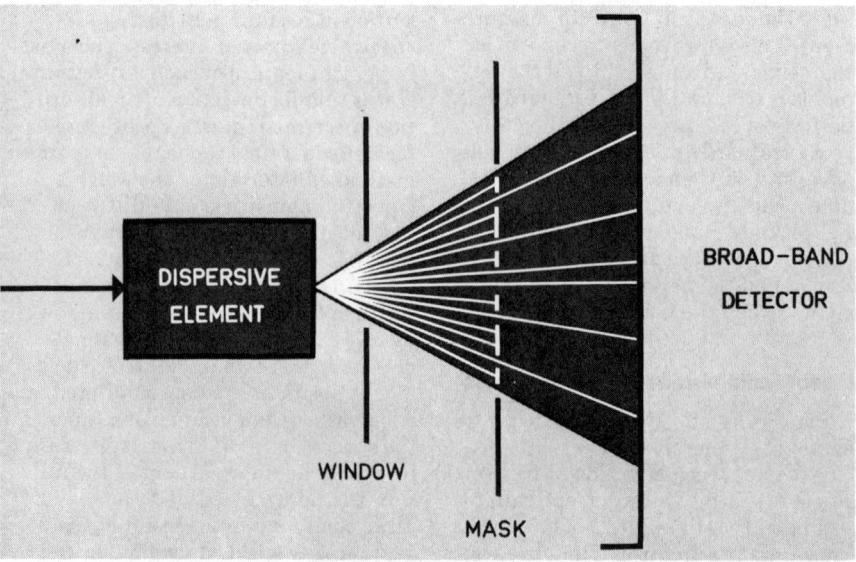

Figure 4. Disperser, mask, and (single, broad-band) detector of Hadamard spectrometer

Figure 5. Functional equivalence of four separate masks (N possible slits each) and movable single mask [(2 N − 1) possible slits]

known weights on the left pan of the balance, as already discussed. To recover the desired spectrum, x_1 to x_N, the next step is to remove the first mask and introduce a second mask, again with a random arrangement of open and shut slits with approximately half the slits open, so that this second slit arrangement is linearly independent from the first. Proceeding in this way, one readily obtains N observables (the total transmitted intensity through each of N different masks, y_1 to y_N), expressed in terms of N unknowns (the spectrum, x_1 to x_N), according to a "code" in which all the coefficients are either zero or one, and roughly half the coefficients in any one row are zero:

$$y_1 = a_{11}x_1 + a_{12}x_2 + \ldots + a_{1N}x_N$$
$$y_2 = a_{21}x_1 + a_{22}x_2 + \ldots + a_{2N}x_N$$
$$\vdots$$
$$y_N = a_{N1}x_1 + a_{N2}x_2 + \ldots + a_{NN}x_N$$
(2)

A particularly convenient method for providing the N linearly independent slit arrangements is shown in Figure 5. Instead of N separate masks, it suffices to construct just a single mask consisting of $(2N - 1)$ potentially open slit positions: then by opening the window over just the first N positions, one constructs the first slit arrangement at the left of Figure 5; by opening the window over the second through $(N + 1)$ positions, one creates the second slit arrangement at the left of the figure, etc. Mechanically, the change from one slit arrangement to another simply consists of translating the linear $(2N - 1)$-slit mask across the window by one position per change.

On both computational and mechanical grounds, the Hadamard approach of Figures 4 and 5 conveniently achieves an improvement of a factor of $\sqrt{N/2}$ in signal-to-noise ratio over the conventional one-slit-at-a-time scanning spectrometer of Figure 2, because half the N possible slits are open during each measurement, rather than just one, while both experiments require the *same* length of time for execution of N separate intensity measurements. [In the procedure of Figure 5, each successive slit arrangement or mask differs from the immediately preceding one by cyclic permutation of the slit pattern. Successful solution of a given set of simultaneous linear algebraic equations, such as Equations 2, by a digital computer requires that the matrix of the coefficients be "well conditioned" (7). Fortunately, the matrix of coefficients generated by the scheme of Figure 5 can readily be transformed to such a well-conditioned form, and Equations 2 can thus be solved accurately.] Alternatively, the Hadamard spectrometer can provide the same signal-to-noise ratio in a factor of $(2/N)$ as much time as would be required by the conventional single-slit spectrometer.

Hadamard encoding-decoding methods have been applied most prominently in infrared spectroscopy (8, 9).

Incoherent and Coherent Spectrometers

To proceed to Fourier methods, it is important to understand that the spectrometers discussed up to now (Figures 2–5) can operate with an incoherent radiation source; that is, there is no necessary common phase relationship (see below) between the various radiation components issuing from the source. For such incoherent source spectrometers, Hadamard mask techniques provide a means for effectively opening up the slit width without sacrificing resolution (Figure 6). There are, however, some major advantages (see below) in using a scanning spectrometer having a coherent source; the coherent source makes possible another type of encoding-decoding scheme (Fourier) for opening the spectral window while preserving resolution (Figure 6). Finally, one reason that both Hadamard (incoherent source) and Fourier (coherent source) methods can be applied to infrared spectroscopy is that the Michelson interferometer can be thought of as a device which effectively converts incoherent to coherent radiation in the present context.

A coherent radiation source and coherent detector in a spectrometer provide two important advantages to the spectroscopist. First, since the frequency of the coherent radiation source is easily determined to very high accuracy by use of an electronic counter, the line positions in a spectrum may be determined very accurately, simply by measuring the frequency of the source as it is (slowly) scanned over the spectral window.

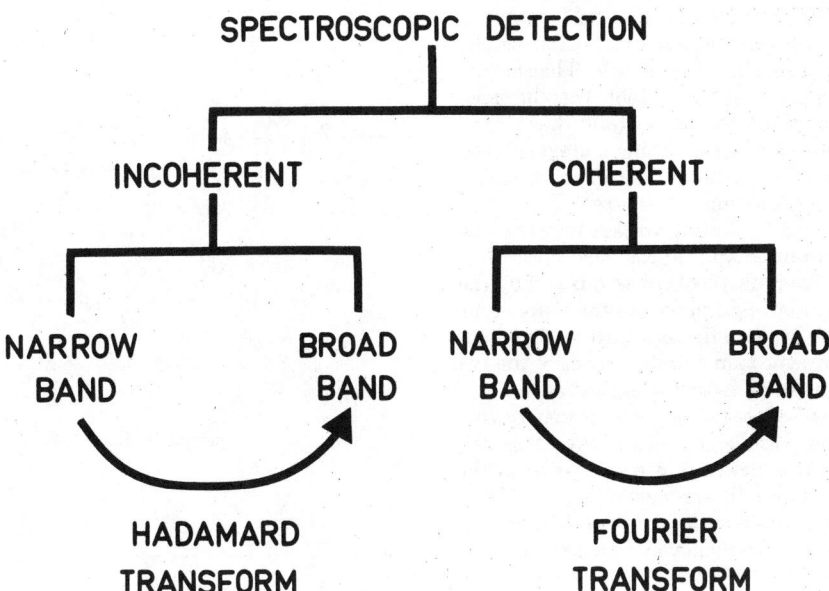

Figure 6. Spectroscopy classified by radiation source and spectrometer bandwidth

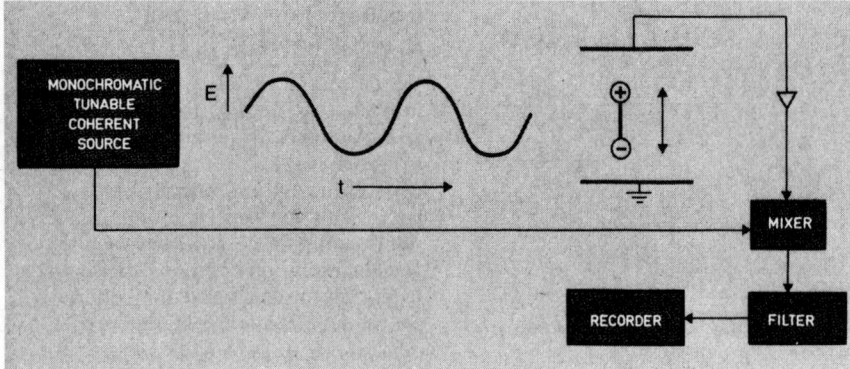

Figure 7. Hypothetical infrared laser source spectrometer

Fourier Transform Spectroscopy

Fourier transform methods at first seem strange to our intuition, because we are prejudiced by our eyes and ears to analyze our surroundings in the frequency domain—we judge light by its color and sound by its pitch. It is, however, equally useful to analyze observations in the time domain (Figure 9). The upper left diagram of Figure 9 shows a simple DC (zero frequency) pulse, which is turned on at time zero, and turned off at time, T. Intuition would suggest that the frequency representation of such a pulse should consist simply of a signal at zero frequency, but the actual frequency representation consists of a signal which is spread over a range of frequencies near zero. By using a shorter pulse (middle of Figure 9), the frequency representation is spread over an even wider range, and in the limit that the DC pulse is made infinitely narrow (bottom of Figure 9), the frequency representation is a completely flat spectrum. [The mathematical correspondence between the time domain and frequency domain diagrams of Figures 9 and 10 is called a (one-sided) Fourier transformation (14).]

The diagrams in Figure 9 suggest that the broad-band frequency excitation required for a multichannel or multiplex spectrometer can be generated by use of a sufficiently narrow electromagnetic radiation pulse. (If the pulse consists of an AC rather than a DC waveform, then the pictures of Figure 9 still apply, except that the frequency representation is now centered at the AC frequency rather than at zero frequency—see top trace of Figure 10.)

For NMR, for example, Table I indicates that an excitation bandwidth of about 10 kHz is required—Figure 10 (top trace) indicates that such an

Second, coherent radiation permits the implementation of electronic filtering techniques which can make the spectrometer resolution arbitrarily high. Thus, the spectral lineshape determined by a coherent radiation spectrometer can be made characteristic of the sample, by making the instrumental broadening arbitrarily small.

The basic operation of a coherent source spectrometer is shown in Figure 7, consisting of a hypothetical infrared laser source spectrometer for use in vibrational spectroscopy. The radiation issuing from the source consists of a plane-polarized electric field whose magnitude varies sinusoidally with time. Upon encountering an electric dipole (i.e., a polar molecule), the electric field will force the dipole to oscillate at the frequency of the radiation, and the amplitude of that dipolar oscillation will be greatest when the electric field oscillation frequency is the same as ("in resonance with") the "natural" vibration frequency of the dipole. If the source radiation is coherent, then all the dipoles in a given region of space will oscillate together, forming a macroscopic oscillating electric dipole in the sample. That macroscopic oscillating dipole then induces an oscillating charge (and thus a corresponding oscillating voltage) on the parallel plates of the capacitor enclosing the sample in Figure 7. That induced oscillating voltage may then be amplified and (in the most important step) multiplied (in a "mixer") by the oscillating signal from the source and the product decomposed electronically into the sum and difference of the two sine wave frequencies, just as the product of two sine waves may be decomposed algebraically (by a trigonometric identity) into sine waves of the sum and difference frequencies. The low-pass filter rejects the (higher) "sum" frequency and passes the (lower) "difference" frequency which is then recorded. The above mixing process effectively extracts a small spectral segment which is centered at the source frequency and whose width is determined by the bandwidth of the low-pass filter. For example, the transient ion cyclotron resonance signal in Figure 11 (see below) was obtained by just this sort of mixing and filtering procedure.

In more familiar language, this sort of spectrometer provides a slit position which is determined by the frequency of the source and a slit width which is determined by the bandwidth of the electrical low-pass filter and which may therefore be made arbitrarily wide or narrow without any mechanical adjustment of the spectrometer geometry. Spectrometers in which a macroscopic change in a physical property of the sample is induced by radiation from a coherent source, and that macroscopic change is detected electronically, in the manner described above, have long been employed in NMR spectroscopy (10) and ion cyclotron resonance (Figure 8) spectroscopy (11) and have recently been introduced in microwave (12) and infrared (13) spectroscopy.

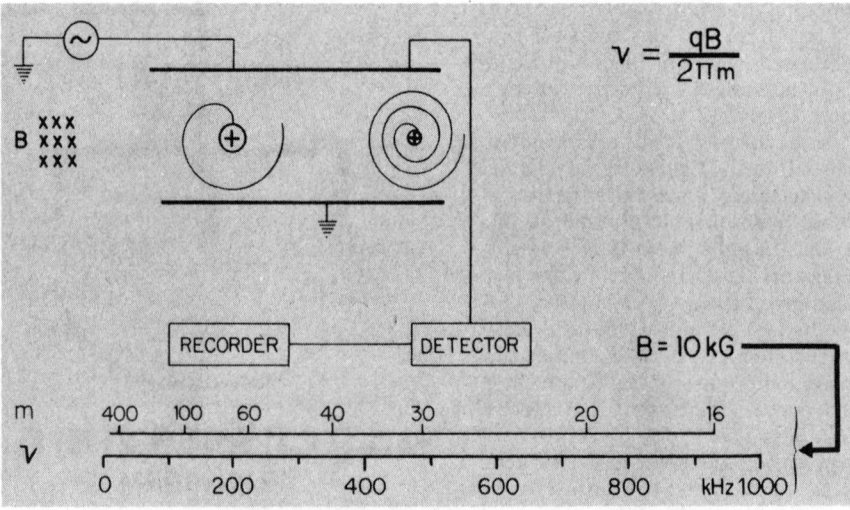

Figure 8. Schematic diagram of ion cyclotron resonance (ICR) spectrometer

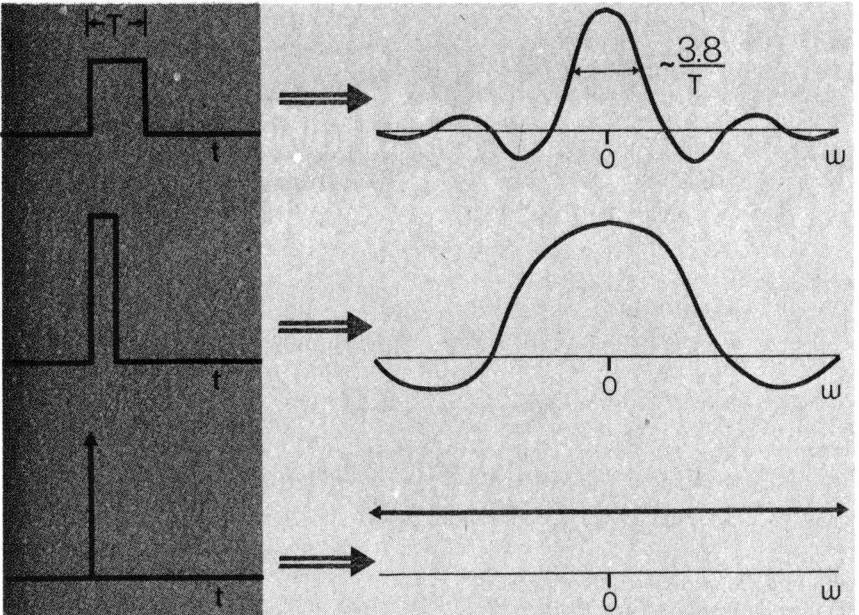

Figure 9. Time domain (left) and frequency domain (right) representations of DC pulses of three different durations

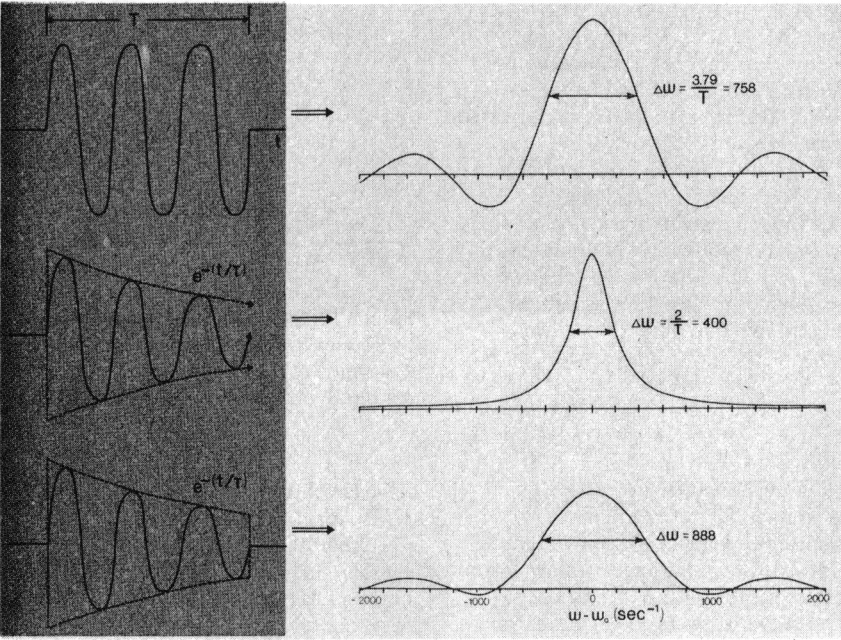

Figure 10. Frequency representations (right) of three types of time domain (left) spectrometer signals

excitation may be produced simply by applying a radiofrequency pulse whose duration is of the order of 10 μsec. As another example, electron impact spectroscopy (15) is based on the rapid passage of an electron past a molecule. This passing electron produces a very short, sharp pulse of electric field at the molecule and thus acts as a very broad-band, nearly flat source of irradiation. In this case, the frequency bandwidth is sufficient to excite the same sort of transitions as are more conventionally studied in photoelectron and ESCA spectroscopy.

When a given single oscillator is subjected to irradiation at its resonant frequency, the amplitude of oscillation will increase. If the irradiating excitation is then removed, the oscillation will persist with an amplitude which decreases (usually exponentially) with time, as shown for three convenient limiting situations at the left of Figure 10. If the oscillation is *not* appreciably reduced during the acquisition time, T (top trace of Figure 10), then the corresponding frequency representation has a functional form which resembles the amplitude of (Fraunhofer) diffraction by a slit (16). If, on the other hand, the oscillation is observed for several lifetimes of its decay (middle trace of Figure 10), the spectral representation approaches the familiar Lorentzian line shape encountered in many forms of spectroscopy. Finally, the bottom trace of Figure 10 illustrates the intermediate case in which the acquisition time, T, is of the order of the decay lifetime, τ. The irreversible decay of the oscillation is due to radiative damping ("spontaneous emission") (17) and to interactions of the sample (nucleus, ion, molecule) with its surroundings, where the interaction may be neutral-neutral collisions (microwave, infrared, optical); ion-molecule collisions (ion cyclotron resonance); rotational diffusion (nuclear magnetic resonance, electron spin resonance); or depletion of the excited species owing to chemical reaction.

The multichannel advantage of the Fourier approach can now be understood. Suppose that the time domain response, $y(t)$, is sampled at N equally spaced intervals during a total acquisition time, T. Each of these sampled time domain points, $y(t_n)$, $n = 1$ to N, is then a linear combination of all the discrete frequency domain spectral points, $x(\omega_m)$, according to:

$$y(t_1) = a_{11}x(\omega_1) + a_{12}x(\omega_2) + \ldots + a_{1N}x(\omega_N)$$
$$y(t_2) = a_{21}x(\omega_1) + a_{22}x(\omega_2) + \ldots + a_{2N}x(\omega_N)$$
$$\vdots$$
$$y(t_N) = a_{N1}x(\omega_1) + a_{N2}x(\omega_2) + \ldots + a_{NN}x(\omega_N) \quad (3)$$

in which

$$a_{nm} = \exp[2\pi i m t_n / T] \quad (4)$$

or just

$$a_{nm} = \exp[2\pi i m n / N] \quad (5)$$

Equations 3 should be compared to Equations 2: since there are now N independent observed sampled time domain points, $y(t_1)$ to $y(t_N)$, each expressed in terms of all N discrete frequency domain points, $x(\omega_1)$ to $x(\omega_N)$, it is again possible to "decode" the observed data to obtain the desired spectrum. The decoding procedure, called a discrete Fourier transformation, may be calculated rapidly by a digital computer (18). In contrast to the Hadamard technique, in which half the possible spectrum is detected in any given observation (i.e., half the a_{nm} in any one row of Equations 2 are zero), the magnitude of each a_{nm} in the Fourier experiment of Equations 3 is unity:

$$|a_{nm}| = |\exp[2\pi i n m / N]| = 1 \quad (6)$$

so that in the Fourier experiment, it is as if all the possible slits are open.

By the arguments previously used for the double-pan balance example, it is now clear that detection of the time domain response, followed by Fourier transformation to obtain the frequency domain response, provides a frequency spectrum exhibiting either signal-to-noise improvement of a factor of $\sqrt{N}$ in the *same* total observation period; or a spectrum having the *same* signal-to-noise ratio in a factor of ($1/N$) as much time as required by a conventional spectrometer which scans the spectrum slowly with a narrow-bandwidth detector. [For the unique case of Fourier transform infrared spectrometers based on the Michaelson interferometer, the spectrum is obtained by discrete Fourier transformation of the (spatially dispersed) sampled interferogram (*19*). Since half the spectral intensity is necessarily lost at the half-silvered mirror of the interferometer, a Fourier transform infrared spectrometer provides only half the full (factor of N) Fellgett time advantage.]

The final consideration in interpreting frequency spectra obtained by Fourier transformation of a time domain response is a comparison to the spectra obtained by a conventional slow-sweep spectrometer. Under the very generally valid condition that the system response be linear (i.e., proportional to the magnitude of the irradiating excitation), the slow-sweep and Fourier transform spectra are identical in the limit of long acquisition time.

Example: Fourier Transform Ion Cyclotron Resonance Spectroscopy

Figure 8 shows the essential components of an ion cyclotron resonance (ICR) spectrometer. The operation of this spectrometer can be understood by direct analogy to the hypothetical infrared spectrometer shown in Figure 7. A moving ion of mass, m, and charge, q, in a magnetic field, B, is constrained to circular motion at a "cyclotron" frequency,

$$\nu = qB/2\pi m, \text{ [mks units; } \nu \text{ in Hz]}$$

If such ions are placed between two parallel plates, as in Figure 8, and irradiated with a coherent (circularly polarized) electric field whose frequency is close to the ion cyclotron frequency, the resultant ion motion can be shown to become spatially coherent as the ions absorb energy from the irradiation by increasing the radii of their cyclotron orbits. Once the ions are all moving essentially together, their composite cyclotron motion will induce a macroscopic voltage in the surrounding plates, and that voltage signal may be amplified and recorded as shown in Figure 8. The main conceptual difference between the ion cyclotron and vibrational cases is that the system motion is circular rather than linear. For a magnetic field of 10 kG (1 tesla in mks units), ion cyclotron frequencies for singly charged ions of mass 16 to mass 400 fall between about 35 kHz and 1 MHz, as indicated at the bottom of Figure 8.

It is now apparent that an ICR spectrometer provides a signal whenever ions of a given mass have an ion cyclotron frequency which matches that of the irradiation. In other words, the device can function as a mass spectrometer to detect ions over a range of charge-to-mass ratios, by irradiating the ions with an oscillating electric field whose frequency is scanned over the range required by: $\nu = qB/2\pi m$.

Figure 11 shows the ion cyclotron resonance (ICR) time domain response following excitation of a bandwidth sufficient to excite ion cyclotron resonance for both N_2^+ and CO^+. Since the masses (and thus the cyclotron frequencies) of N_2^+ and CO^+ differ slightly, the time domain response of Figure 11 is a superposition of two decaying sine waves, whose beat pattern is evident in the figure. Fourier transformation of the time domain data of Figure 11 yields the ICR frequency spectrum shown in Figure 12. Based on these and other (*20–23*) prototype experiments, Fourier techniques promise to reduce by a factor of 1,000 the time required to obtain an ICR mass spectrum.

Conclusions

The value of any instrumental improvement must be gauged on the basis of its impact in making possible new experiments for experts and better routine measurements for nonexperts. On this basis, Fourier methods

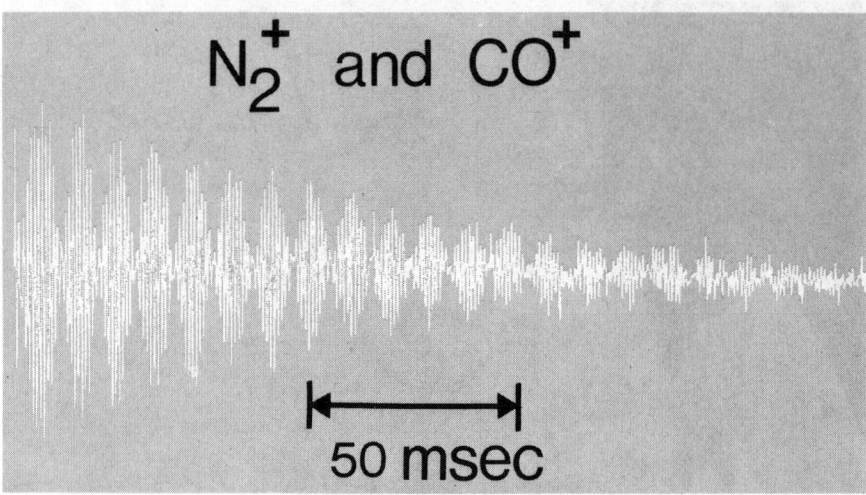

Figure 11. Transient ICR signal from sample of N_2^+ and CO^+

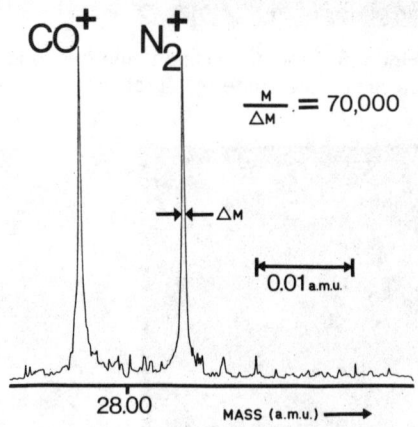

Figure 12. ICR mass spectrum obtained by Fourier transformation of transient ICR signal of Figure 11. Data for this spectrum obtained in 205 msec

have revolutionized infrared and NMR spectroscopy, by making it possible to obtain spectra of very weak signals, such as infrared spectra of planets (*24*) and carbon-13 NMR spectra of large organic molecules (*25*). Before 1965 (the advent of Fourier data reduction in NMR), for example, carbon-13 NMR spectra were obtainable only with great difficulty. Today (1975), most major chemistry departments use carbon-13 NMR spectra routinely in structural and kinetic analysis because it now requires only a few minutes (rather than several hours prior to Fourier methods) to obtain a carbon-13 NMR spectrum.

Based on the substantial proven advantages of Fourier data reduction in infrared (*26*) and NMR (*27*) spectroscopy, the recent application of Fourier techniques to electrochemical (*28*), microwave (*12*), ion cyclotron resonance (*20–23*), dielectric (*29*), and solid-state NMR (*30*) phenomena promises to make available to practicing chemists a broad new range of experiments not previously accessible. Ordinarily, the details of experimental measurement, although crucial to those working in the field, are relatively uninteresting to chemists in general. In this article, we hope to have shown that by taking the time to see how a double-pan balance should best be used, it is possible to encompass a wide spectrum of spectroscopic applications of direct chemical interest.

References

(1) D. F. Eggers, Jr., N. W. Gregory, G. D. Halsey, Jr., and B. S. Rabinovitch, "Physical Chemistry," p 392, Wiley, New York, NY, 1964; W. J. Moore, "Physical Chemistry," 3rd ed., p 232, Prentice-Hall, Englewood Cliffs, NJ, 1962; N. Davidson, "Statistical Mechanics," p 283, McGraw-Hill, New York, NY, 1962.
(2) L. Mertz, "Transformations in Optics," p 9, Wiley, New York, NY, 1965; R. J. Bell, "Introductory Fourier Transform Spectroscopy," p 23, Academic Press, New York, NY, 1972.
(3) P. Fellgett, *J. Phys. Radium*, **19**, 187 (1958).
(4) K. Siegbahn, C. Nordling, A. Fahlman, R. Norderg, K. Hamrin, J. Hedman, G. Johansson, T. Bergmark, S. Karlson, I. Lindgren, and B. Lindberg, "ESCA: Atomic, Molecular, and Solid State Structure Studied by Means of Electron Spectroscopy," Almqvist and Wiksell, Uppsala, Sweden, 1967; K. Siegbahn, C. Nordling, G. Johansson, J. Hedman, P. F. Heden, K. Hamrin, U. Gelius, T. Bergmark, L. O. Werme, R. Manne, and Y. Baer, "ESCA: Applied to Free Molecules," North Holland, Amsterdam, The Netherlands, 1969.
(5) D. W. Turner, A. D. Baker, C. Baker, and C. R. Brundle, "High Resolution Molecular Photoelectron Spectroscopy," Wiley, New York, NY, 1970.
(6) SSR Instruments Co., 1001 Colorado Ave., Santa Monica, CA 90404, for example.
(7) G. Forsythe and C. B. Moler, "Computer Solution of Linear Algebraic Systems," Prentice-Hall, Englewood Cliffs, NJ, 1967.
(8) E. D. Nelson and M. L. Fredman, *J. Opt. Soc. Am.*, **60**, 1664 (1970).
(9) J. A. Decker, Jr., *Anal. Chem.*, **44**, 127A (1972).
(10) J. A. Pople, W. G. Schneider, and H. J. Bernstein, "High-Resolution Nuclear Magnetic Resonance," Chap. 4, McGraw-Hill, New York, NY 1959.
(11) J. D. Baldeschwieler, *Science*, **159**, 263 (1968); J. L. Beauchamp, *Ann. Rev. Phys. Chem.*, **22**, 527 (1971); J. H. Futrell, in "Dynamic Mass Spectrometry," D. Price, Ed., Vol 2, Heyden and Son, New York, NY, 1971; J. M. S. Henis in "Ion-Molecule Reactions," J. L. Franklin, Ed., Vol 2, Plenum, New York, NY, 1972; G. A. Gray, *Adv. Chem. Phys.*, **19**, 141 (1971); C. J. Drewery, G. C. Goode, and K. R. Jennings, in "MTP International Review of Science, Mass Spectroscopy, Physical Chemistry," Series One, A. D. Buckingham and A. Maccoll, Eds., Vol 5, p 183, Butterworths, London, England, 1972; J. I. Brauman and L. K. Blair, in "Determination of Organic Structures by Physical Methods," F. C. Nachod and J. J. Zuckerman, Eds., Vol 5, p 152, Academic Press, New York, NY, 1973.
(12) J. C. McGurk, T. G. Schmalz, and W. H. Flygare, *J. Chem. Phys.*, **60**, 4181 (1974).
(13) R. G. Brewer and R. L. Shoemaker, *Phys. Rev.*, **A6**, 2001 (1972); for a very recent example of the application of infrared coherent detection methods which yield the advantages mentioned above (high accuracy in the measurement of line positions and very high instrumental resolution), see M. Mumma, T. Kostiuk, S. Cohen, D. Buhl, and P. C. von Thuna, *Nature*, **253**, 514 (1975).
(14) R. Bracewell, "The Fourier Transform and Its Applications," p 360, McGraw-Hill, New York, NY, 1965.
(15) C. E. Brion, in "MTP International Review of Science, Mass Spectroscopy, Physical Chemistry," Series One, A. D. Buckingham and A. Maccoll, Eds., Vol 5, Chap. 3, Butterworths, London, England, 1972.
(16) W. H. Furry, E. M. Purcell, and J. C. Street, "Physics for Science and Engineering Students," pp 498–502, McGraw-Hill, New York, NY, 1960.
(17) J. C. Davis, Jr., "Advanced Physical Chemistry: Molecules, Structure, and Spectra," pp 252–54, Ronald, New York, NY, 1965; L. Pauling and E. B. Wilson, Jr., "Introduction to Quantum Mechanics," pp 299–301, McGraw-Hill, New York, NY, 1935.
(18) J. W. Cooley and J. W. Tukey, *Math. Comp.*, **19**, 297 (1965).
(19) M. J. D. Low, *J. Chem. Educ.*, **47**, A163, A255, A349, A415 (1970).
(20) M. B. Comisarow and A. G. Marshall, *Chem. Phys. Lett.*, **25**, 282 (1974).
(21) M. B. Comisarow and A. G. Marshall, *ibid.*, **26**, 489 (1974).
(22) M. B. Comisarow and A. G. Marshall, *Can. J. Chem.*, **52**, 1997 (1974).
(23) M. B. Comisarow and A. G. Marshall, *J. Chem. Phys.*, **62**, 293 (1975).
(24) P. Connes, *Annu. Rev. Astron. Astrophys.*, **8**, 209 (1970); J. P. Maillard, "IAU Highlights of Astronomy 1973," Contopoulos et al., Eds., Reidel, Dordrecht, Holland, 1974.
(25) J. B. Stothers, "Carbon-13 NMR Spectroscopy," Academic Press, New York, NY, 1972; G. C. Levy and G. L. Nelson, "Carbon-13 Nuclear Magnetic Resonance for Organic Chemists," Wiley, New York, NY, 1972; L. F. Johnson and W. C. Jankowski, "Carbon-13 NMR Spectra," Wiley-Interscience, New York, NY, 1972.
(26) G. Horlick and H. V. Malmstadt, *Anal. Chem.*, **42**, 1361 (1970); G. Horlick, *ibid.*, **43**, 61A (1971); E. G. Codding and G. Horlick, *Appl. Spectrosc.*, **27**, 85 (1973).
(27) R. R. Ernst, *Adv. Mag. Reson.*, **2**, 1 (1968).
(28) S. C. Creason, J. W. Hayes, and D. E. Smith, *Electroanal. Chem. Interfacial Electrochem.*, **47**, 9 (1973); S. C. Creason and D. E. Smith, *Anal. Chem.*, **45**, 2401 (1973) and references quoted therein.
(29) G. A. Brehm and W. H. Stockmayer, *J. Phys. Chem.*, **77**, 1348 (1973); R. H. Cole, *ibid.*, **78**, 1440 (1974).
(30) A. Pines, J. J. Chang, and R. G. Griffin, *J. Chem. Phys.*, **61**, 1021 (1974).

Work supported by grants (to A.G.M and M.B.C.) from the National Research Council of Canada and the Committee on Research, University of British Columbia.

Addendum

In this addendum, we simply point out that the particular 4 arrangements of weights in our Hadamard example are not linearly independent, since the determinant of the matrix of their coefficients vanishes. This condition is readily rectified by using the 4 arrangements corresponding to Eq. [1] below, and thus it is evident that the analogy between use of weights on a balance and Hadamard spectroscopy is indeed valid.

$$\begin{bmatrix} 1 & 1 & 0 & 0 \\ 1 & 0 & 0 & 1 \\ 1 & 0 & 1 & 0 \\ 0 & 1 & 0 & 1 \end{bmatrix} \begin{bmatrix} x_1 \\ x_2 \\ x_3 \\ x_4 \end{bmatrix} = \begin{bmatrix} y_1 \\ y_2 \\ y_3 \\ y_4 \end{bmatrix} \cdot \quad [1]$$

In fact, for any N, it is always possible to find N linearly independent arrangements of N weights, where weights are placed on only one pan of the balance.

In Hadamard spectroscopy, it is additionally desirable that the arrangements not only be linearly independent, but also that successive arrangements differ by cyclic permutation, in order that the N arrangements may be derived by translation of a mask of $(2N - 1)$ slits, as discussed for Figure 4 of Reference 1. It may be shown that by choosing $N = 2^m - 1$, $m = 2, 3, 4, \ldots$, it is always possible to construct the desired array. More generally, possible values for N include $N = 3, 7, 11, 15, 19, 23, 31, 35, 43, 47, 63, \ldots$, according to the conditions (2),

$N = 4n - 1, n = 1, 2, 3, \ldots,$

and [2]

either $N = 2^m - 1, m = 2, 3, 4, \ldots,$

or $N = p$, where p is a prime number,

or $N = p(p + 2)$, where p is a prime number.

As examples, the arrays for $N = 3$ and $N = 7$ are shown below:

```
           0 1 0 1 1 1 0
           1 0 1 1 1 0 0
1 0 1      0 1 1 1 0 0 1
0 1 1      1 1 1 0 0 1 0
1 1 0      1 1 0 0 1 0 1
N = 3      1 0 0 1 0 1 1
           0 0 1 0 1 1 1
               N = 7
```

Of course, the scheme shown in Figure 4 of Reference *1* should be illustrated for a value of N chosen from Eq. [2] above. As an unrelated typographical correction, the signal-to-noise improvement for Hadamard spectroscopy over use of a single-slit scanning spectrometer should appear as $(\sqrt{N}/2)$ rather than $(\sqrt{N/2})$ at appropriate places in the text.

References

(1) A. G. Marshall and M. B. Comisarow, *Anal. Chem.* **47**, 491A (1975).
(2) N. J. A. Sloane, T. Fine, P. G. Phillips, and M. Harwit, *Appl. Opt.* **8**, 2103 (1969).

ATOMIC SPECTROCHEMICAL MEASUREMENTS WITH A FOURIER TRANSFORM SPECTROMETER

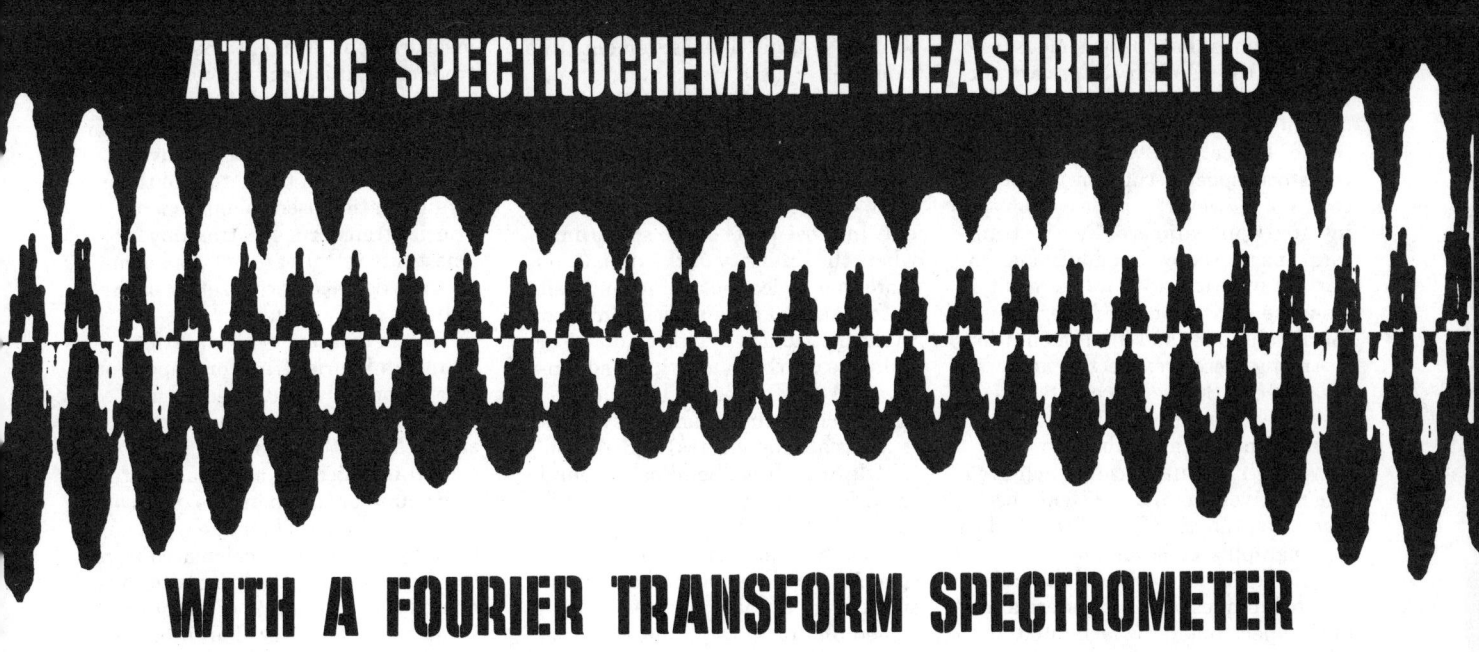

Gary Horlick and W. K. Yuen

Department of Chemistry
University of Alberta
Edmonton, Alta., Canada T6G 2E1

The development of effective simultaneous multielement analyses based on atomic spectrochemical methods is one of the major goals of a number of academic, government, and industrial laboratories. A key aspect of this development is the design of spectrochemical measurement systems capable of simultaneously measuring spectral information over a wide range of wavelengths. To date, the dominant technique for the measurement of spectra is the dispersive system based on the diffraction grating. The detection of the spectral information in a dispersive system has been accomplished in a number of ways. The two main ways are with a photographic plate or with a combination of an exit slit and a photomultiplier tube.

The photographic plate, although capable of recording thousands of lines in a single exposure as a permanent record, has a nonlinear response, limited dynamic range, and tedious readout. Thus, even though the exit slit–PM tube combination is limited to the measurement of one spectral line at a time, its wide linear dynamic range, sensitivity, and the fact that it transduces light intensity directly to an electronic signal have made it the detection system of choice for the majority of spectrochemical measurements. However, it is difficult to design truly effective and versatile dispersive PM tube-based systems for simultaneous spectrochemical measurements. The classic direct reader, in which an exit slit–PM tube is mounted at each point in the exit focal plane of a dispersive instrument at which measurements are to be made, is a reasonably powerful multichannel measurement system. But even at best, only a very small fraction of the spectral information available in the spectrum can be measured. This limitation is now being overcome by the utilization of electronic image sensors as detectors (1) which over a moderate wavelength range provide a continuous multichannel measurement system. Such devices, which include silicon vidicons (2–5), secondary electron conduction (SEC) vidicons (6), and silicon photodiode arrays (7), have recently been applied to a variety of multichannel spectrochemical measurements.

In addition to these multichannel systems, rapid scanning sequential systems have also proved useful for multielement analysis. Both a mechanical slew scan system based on a programmable monochromator (8) and what amounts to an electronic slew scan system based on an image dissector photomultiplier tube (9, 10) have been developed.

Another potential approach to the overall problem is to dispense with the dispersive system completely and use a multiplex technique. In a multiplex technique the spectral information is encoded so that a single detector can be used to simultaneously measure a wide wavelength range. The most common examples of multiplex techniques are Fourier transform spectroscopy and Hadamard transform spectroscopy (11). In Hadamard transform spectroscopy, the spectral information is encoded in a binary code based on Hadamard matrices. In Fourier transform spectroscopy, the spectral information is encoded in sine and cosine oscillations. This encoding is accomplished in the optical region of the electromagnetic spectrum by a Michelson interferometer. The encoded signal which is measured with a single detector is called an interferogram, and it is necessary to take the Fourier transform of the interferogram to decode it and obtain the desired spectrum.

Fourier transform spectroscopy has been extensively used for spectral measurements in the infrared regions, but so far it has found little application for atomic spectrochemical measurements in the UV–VIS spectral regions. One potential drawback to this application is that spectral measurements in the UV–VIS regions are normally signal noise limited, and in this type of situation the so-called multiplex or Fellgett's advantage of Fourier transform spectroscopy may not be realized. However, it is clear from the infrared applications of Fourier transform spectroscopy that additional important practical advantages result from the utilization of Fourier transform techniques and instrumentation.

Among these are: spectra can be measured with a very accurate and precise wave number axis which is predetermined (no calibration necessary), high resolution can be achieved in a relatively compact system, the resolution function is easily controlled and manipulated as an inherent step in data reduction by use of apodization techniques, and computerization of the spectrometer is facilitated. In addition, with proper utilization of aliasing, a Fourier transform spectrometer can be very versatile in simultaneously covering a wide range of wavelengths. This is one aspect that can be a major limitation of array-based multichannel measurement systems, such as vidicons and photodiode arrays, as a consequence of the finite length and detector element density limitations of these devices. In this article some considerations for using Fourier transform spectroscopy in the UV–VIS regions will be presented along with examples of atomic spectrochemical measurements made with a Fourier transform spectrometer.

Multiplex Advantage?

Much of the impetus for the development of infrared Fourier transform spectroscopy came from the promise of achieving the multiplex (Fellgett's) advantage. In contrast to a scanning spectrometer, a Fourier transform spectrometer using a Michelson interferometer may achieve, in the same measurement time, a signal-to-noise ratio that is superior by a factor of $(M/2)^{1/2}$, where M is the number of resolution elements being observed (11). The advantage is realized only in measurement situations that are detector noise limited, i.e., the increased light level falling on the detector must not increase the overall limiting noise of the system. Most UV–VIS spectral measurements utilize the photomultiplier tube detector and thus are signal noise limited, with the noise level increasing as the square root of the signal level, thereby eliminating any potential multiplex advantage. However, the situation is not quite this clear cut. Both the nature of the spectrum and the distribution of the noise in the spectrum can affect the existence of a multiplex advantage in signal (photon) noise-limited measurement situations.

As early as 1959, Kahn (12) discussed the signal-to-noise ratio of Fourier transform spectroscopy in the photon (shot) noise limit. He concluded that "the interferometer is better only in those parts of the spectrum where the intensity of the radiation is more than twice the average intensity in the range of frequencies admitted to the instrument." This result essentially means that a multiplex advantage will only be realized for simple spectra consisting of only a few lines (such as atomic spectra) and also only if little or no broadband background radiation is present. Essentially the same conclusion has recently been reached for Hadamard transform spectroscopy in a photon noise-limited situation (13). Filler (14) has also discussed photon–noise-limited Fourier spectroscopy. Again, it was concluded that a multiplex advantage will be present only for line emission spectra.

Another factor that must be considered in this discussion is the possibility of a so-called multiplex disadvantage. This arises primarily because the distribution of photon noise in a spectrum measured with a scanning instrument and one measured with a Fourier transform spectrometer is different. In the scanning case, the rms noise level is greatest where the signal is greatest (i.e., at the top of the spectral peaks). In the Fourier case, the noise tends to be spread out more or less uniformly throughout the entire spectrum. Thus, the signal-to-noise ratio of strong peaks should improve, but weak spectral lines may be obscured by the noise from strong lines that ends up distributed along the baseline of the spectrum. This problem is also present in Hadamard transform spectroscopy (13), and some experimental verification of this multiplex disadvantage has already been reported (15).

Thus, at this time no definitive conclusions can be reached about the existence and importance of multiplex advantages and/or disadvantages in photon noise-limited multiplex spectrochemical measurements. More experimental work is necessary to clear up the situation. What the above discussion does mean, however, is that the promise of any substantial multiplex advantage cannot be a driving force in extending Fourier transform techniques into the UV–VIS region; in fact, some disadvantages may exist with respect to signal-to-noise ratio in certain measurement situations.

Additional Considerations

Signal-to-noise ratio is not always the only and overriding consideration when carrying out a spectrochemical measurement. As mentioned in the introduction, it is clear from the impressive success and capabilities of the Fourier transform technique in the infrared that some important advantages result from the nature of the instrumentation used to implement Fourier transform spectroscopy. These advantages are not dependent on the existence or realization of the multiplex advantage.

The digitization of the interferogram in a Fourier transform spectrometer is normally controlled with a clock signal derived from a He–Ne reference laser. This provides a final spectrum which has a very accurate wave number (frequency) axis. In addition, the values along the wave number axis can be easily calculated from the wavelength of the laser line; thus, no external calibration of the spectrometer is necessary. This inherent accuracy of the wave number axis greatly facilitates intercomparison of digitized spectra for small spectral shifts and peak shape perturbations. In fact, a strong case can be made for the statement that a Fourier transform spectrometer is the best system for the measurement of digitized spectra. Perhaps the main instrumental contenders to this statement are dispersive-based systems utilizing electronic image sensors as the detection system, particularly echelle grating spectrometers coupled to area array electronic image sensors (6).

A Fourier transform spectrometer is capable of very high resolution while maintaining relatively high throughput and with a relatively compact system. A commercial interferometer (EOCOM Corp., 19722 Jamboree Blvd., Irvine, Calif. 92664) which can easily sit on a tabletop has an optical retardation of 16 cm (mirror movement of 8 cm) which provides a nominal resolution of 0.0625 cm^{-1}. This translates to 0.00225 nm at 600.0 nm and 0.00056 nm at 300.0 nm. Also, the form of the resolution function is easily manipulated as an inherent step in data reduction (16).

Considered in the context of capability, both the cost and basic simplicity of Fourier transform spectroscopy can be considered advantages. Certainly, the present cost of commercial Fourier transform spectrometers does not back up this statement, but without question, effective systems can be made considerably less expensively. A Fourier transform spectrometer constructed in our laboratory had a component cost (optical, electronic, and mechanical but not including machining costs and computer) of about $2500.

With respect to simplicity, the interferometer has only one moving part, the interferometer mirror. A simple air bearing, which costs about $30, provides an excellent moving mirror support, and most commercial mirror drive systems are based on this device. In another context the simplicity advantage was stated by Fellgett several years ago (17):

> It appears probable, moreover, that an outstanding advantage will eventually prove to be that proposed by Mertz, namely simplicity. If this assertion appears at first sight surprising, let us imagine that we have been doing interferometric multiplex spectrometry all our lives, and someone were to come along and propose the grating spectrometer for the first time. We should find that there were slits to cut off most of the light, only one element of the spectrum could be observed at one time, thousands of lines would have to be laid down on the grating with sub-wavelength accuracy, the resolution function would depend on the figure and phase shifts of the optical surfaces. Then there would be questions of overlapping orders, grating aberrations, blaze, etc. Would it really be surprising if such a proposal were rejected as too complicated?

Finally, with proper utilization of aliasing, a Fourier transform spectrometer can be very versatile in simultaneously covering a wide range of wavelengths while maintaining relatively high resolution. This aliasing advantage will be considered in some detail in the next section.

As a result of considering these additional points, we have utilized a repetitive-scanning laser-referenced Fourier transform spectrometer developed in our laboratory to carry out some preliminary atomic spectrochemical measurements in the visible region. A photograph of the interferometer is shown in Figure 1. The interferometer has a He–Ne reference laser to control digitization, a white light source to control time averaging, and an air bearing–speaker coil mirror drive system. The moving mirror velocity for these studies was about 2 mm/sec which results in a modulation frequency for the 632.8-nm He–Ne laser line of about 6000 Hz. A paper is in preparation that will describe the complete interferometer system. The computer system used to acquire the interferogram signals and transform them is a PDP 8/e with an OS/8 operating system (18).

Figure 1. Photograph of Fourier transform spectrometer

Table I. Spectral Regions Covered with 0.6328-μm Sampling Interval (Direct Sampling Rate)

Region	cm^{-1}	Å
1	0–7,901	∞–12,656
2	15,802–7,901	6,328–12,656
3	15,802–23,703	6,328–4,219
4	31,604–23,703	3,164–4,219

Table II. Spectral Regions Covered with 1.266-μm Sampling Interval ($\div$2 Sampling Rate)

Region	cm^{-1}	Å
1	0–3,950	∞–2.5 μ
2	7,901–3,950	1.2–2.5 μ
3	7,901–11,851	12,656–8,438
4	15,802–11,851	6,328–8,438
5	15,802–19,752	6,328–5,062
6	23,703–19,752	4,219–5,062
7	23,703–27,653	4,219–3,616
8	31,605–27,653	3,164–3,616

Table III. Spectral Regions Covered with 2.532-μm Sampling Interval ($\div$4 Sampling Rate)

Region	cm^{-1}	Å
1	0–1,975	∞–5 μ
⋮	⋮	⋮
6	11,851–9,876	8,438–10,125
7	11,851–13,826	8,438–7,233
8	15,802–13,826	6,328–7,233
9	15,802–17,777	6,328–5,625
⋮	⋮	⋮
15	27,653–29,628	3,616–3,375
16	31,605–29,628	3,164–3,375

Table IV. Major Emission Lines of Cs, Rb, K, and Li in Near IR

	Region (Table III)	Å	cm^{-1}
Cs	6	8943	11,181
	6	8521	11,735
Rb	7	7947	12,583
	7	7800	12,820
K	7	7698	12,990
	7	7664	13,048
Li	8	6707	14,909

Aliasing Question

One of the first major points to consider in utilizing Fourier transform spectroscopy in the UV-VIS region is the sampling rate (19). With the standard He-Ne reference laser, the basic sampling interval for the interferometer system is 0.6328 μm (one cycle of the laser line modulation). This means that the shortest wavelength of light that can be properly sampled without aliasing is 1.266 μm (7901 cm^{-1}). Clearly then, to work in the UV-VIS region, either the sampling rate must be increased or aliasing must be tolerated. The frequency of the reference laser modulation can be increased by use of optical techniques such as double passage of the laser through the interferometer or by electronic frequency multiplication using phase-locked loops (20). However, the high sampling rate necessary to sample the UV-VIS modulation frequencies without aliasing quickly results in a prohibitively large number of data points that must be digitized in the interferogram to achieve reasonable resolution. Thus, the alternative of aliasing the spectral information in the interferogram must be considered.

Aliasing (19, 20) refers to the undersampling of modulation frequencies in the interferogram. Normally, in Fourier transform spectroscopy, aliasing is avoided as the undersampled modulation frequencies show up as spurious spectral information (foldover). However, with line spectra, as commonly measured in atomic spectrometry, it is possible to use aliasing to advantage. As mentioned above, the basic sampling interval of the interferometer system is 0.6328 μm. The first four specific spectral regions (7901 cm^{-1} bandwidth) that can be covered with this sampling interval are listed in Table I. Regions 2-4 are aliased regions. Spectral information in all of these regions can be measured simultaneously as long as aliased spectral lines do not exactly overlap. With a sampling interval of 1.266 μm (÷2 sampling rate), these four regions become eight regions, each covering a bandwidth of 3950 cm^{-1} (see Table II). With a 2.532-μm sampling interval (÷4 sampling rate), 16 regions each with a bandwidth of 1975 cm^{-1} can be simultaneously covered. A partial list of the regions is presented in Table III. Thus, given a fixed number of data points that can be acquired and transformed, spectral lines of widely different wavelength can be simultaneously measured with significantly better resolution than could be achieved if aliasing were avoided. Flame emission spectra of Cs, Rb, K, and Li were measured with our Fourier transform spectrometer to illustrate this point.

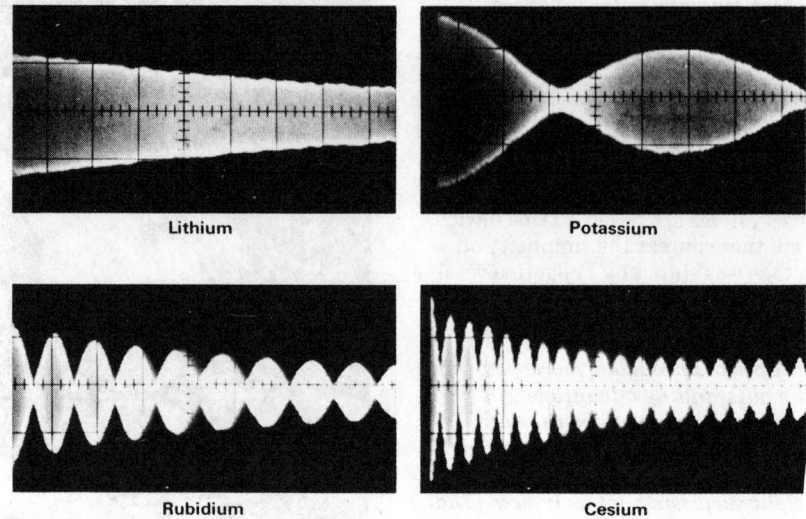

Figure 2. Analog interferograms of Li, K, Rb, and Cs

The emission lines of these elements in the near-IR region are listed in Table IV along with the region from Table III that they fall into. Solutions containing only one element each were aspirated into an air-C_2H_2 flame, and the emission was measured with the interferometer by use of a silicon photocell as the detector. The analog interferograms for the emission signal from each element are shown in Figure 2. The beat patterns in the K, Rb, and Cs interferograms clearly indicate that the spectra all consist of two lines, and the frequency of beating can be related directly back to the doublet separation, the potassium doublet having the least separation. These interferograms were digitized with the 2.532-μm sampling interval, and a total of 512 points was acquired. Therefore, the nominal resolution was

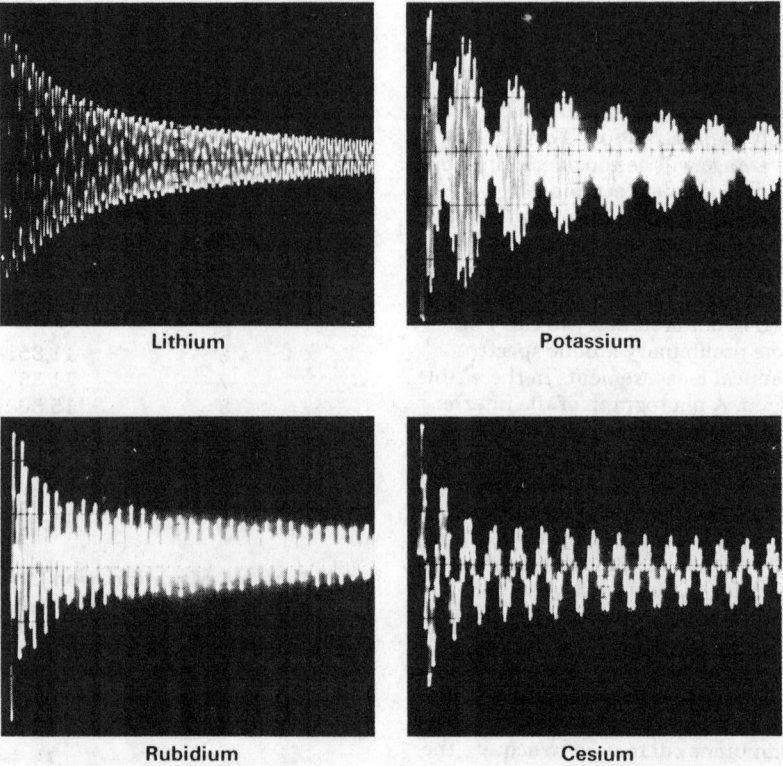

Figure 3. Digitized interferograms of Li, K, Rb, and Cs

about 8 cm^{-1}. The digitized interferograms are shown in Figure 3. Note the manner in which aliasing has affected the interferogram beat patterns. The spectra obtained by transforming these interferograms are shown in Figure 4. If aliasing were completely avoided, a bandwidth of 15,802 cm^{-1} would be necessary for these measurements, which would mean a sampling interval of 0.3164 μm. With the same number of points (512), resolution would be only about 80 cm^{-1}, or 4096 points would have to be digitized and transformed to achieve the same resolution. It is interesting to note that similar aliasing capability has recently been incorporated into a commercial Fourier analyzer to increase its frequency resolution without increasing the number of data points that must be processed (21).

None of the aliased lines in Figure 4 exactly overlaps; thus, these elements can be determined simultaneously. The analog interferogram measured when a solution containing all four elements was aspirated into the flame is shown in Figure 5 (A). An expanded (×5) oscilloscope trace of the first portion of the analog interferogram is shown in Figure 5 (B), and the complex beat pattern resulting from the modulation frequencies of the seven lines can easily be seen. Ten repetitive interferograms were digitized and time averaged with the same sampling interval (2.532 μm) as before. The resulting spectrum is shown in Figure 6. Although the measurement bandwidth is only 1975 cm^{-1}, spectral information over a range of 5926 cm^{-1} (regions 6–8 of Table III) has been measured while still maintaining the 8 cm^{-1} resolution. In this region of the spectrum, a 5926 cm^{-1} bandwidth represents 3797 Å. Also, the total measurement time for this spectrum was 3.3 sec (10 scans, 0.33-sec measurement time per scan). This spectrum is not of high quality and only represents preliminary results obtained primarily to illustrate the use of aliasing. However, it was measured in a relatively short period of time considering the effective wavelength range covered.

Conclusions

In this article we have tried to present some perspective on the potential application of Fourier transform spectroscopy to atomic spectrochemical measurements. In particular, the aliasing aspect of Fourier transform spectrochemical measurements can provide unique capability in the measurement of spectral data. Certainly more work and results are necessary to demonstrate the true overall capability of Fourier transform techniques for measurements in the UV–VIS region.

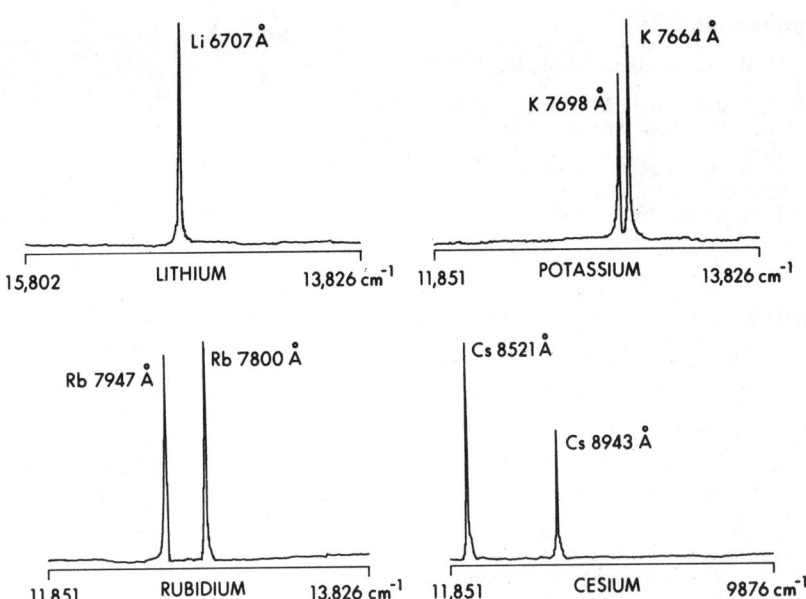

Figure 4. Flame emission spectra of Li, K, Rb, and Cs as measured with Fourier transform spectrometer

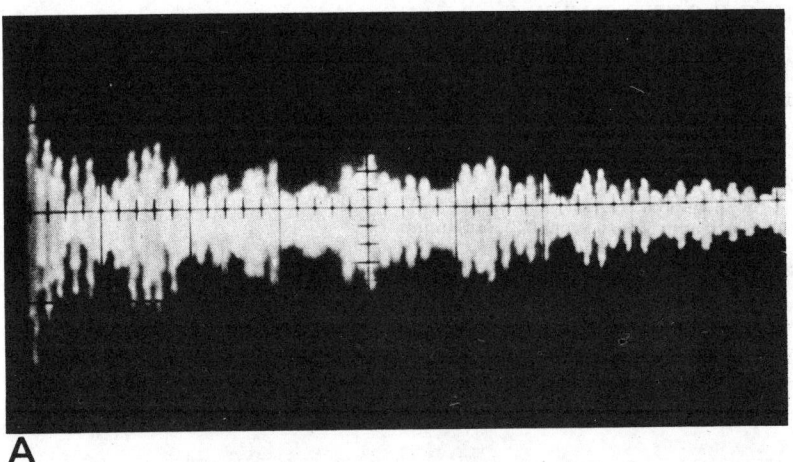

Figure 5. Analog interferogram for simultaneous flame emission determination of Li, K, Rb, and Cs (A); expanded scale (×5) oscilloscope trace of initial portion of interferogram (B)

References

(1) R. M. Barnes, *Anal. Chem.*, **46,** 150R (1974).
(2) K. W. Jackson, K. M. Aldous, and D. G. Mitchell, *Appl. Spectrosc.*, **28,** 569 (1974).
(3) K. W. Busch, N. G. Howell, and G. H. Morrison, *Anal. Chem.*, **46,** 2074 (1974).
(4) D. O. Knapp, N. Omenetto, L. P. Hart, F. W. Plankey, and J. D. Winefordner, *Anal. Chim. Acta*, **69,** 455 (1974).
(5) M. J. Milano and H. L. Pardue, *Anal. Chem.*, **47,** 25 (1975).
(6) D. L. Wood, A. B. Dargis, and D. L. Nash, Paper No. 310, Pittsburgh Conference, Cleveland, Ohio, March 3–7, 1975.
(7) G. Horlick and E. G. Codding, *Appl. Spectrosc.*, **29,** 167 (1975).
(8) E. Cordos and H. V. Malmstadt, *Anal. Chem.*, **45,** 425 (1973).
(9) A. Danielson, Paper No. 311, Pittsburgh Conference, Cleveland, Ohio, March 3–7, 1975.
(10) A. Danielson, P. Lindblom, and E. Soderman, *Chem. Scripta*, **6,** 5 (1974).
(11) A. G. Marshall and M. B. Comisarow, *Anal. Chem.*, **47,** 491A (1975).
(12) F. D. Kahn, *Astrophys. J.*, **129,** 518 (1959).
(13) N. M. Larson, R. Crosmun, and Y. Talmi, *Appl. Opt.*, **13,** 2662 (1974).
(14) A. S. Filler, *J. Opt. Soc. Am.*, **63,** 589 (1973).
(15) F. W. Plankey, T. H. Glenn, L. P. Hart, and J. D. Winefordner, *Anal. Chem.*, **46,** 1000 (1974).
(16) E. G. Codding and G. Horlick, *Appl. Spectrosc.*, **27,** 85 (1973).
(17) P. Fellgett, *J. Phys. (Paris)*, **28,** C2-165 (1967).
(18) G. Horlick, E. G. Codding, and S. T. Leung, *Appl. Spectrosc.*, **29,** 48 (1975).
(19) G. Horlick and H. V. Malmstadt, *Anal. Chem.*, **42,** 1361 (1970).
(20) H. V. Malmstadt, C. G. Enke, S. R. Crouch, and G. Horlick, "Optimization of Electronic Measurements," Benjamin, Menlo Park, Calif., 1974.
(21) H. W. McKinney, *Hewlett-Packard J.*, **26** (April), 20 (1975).

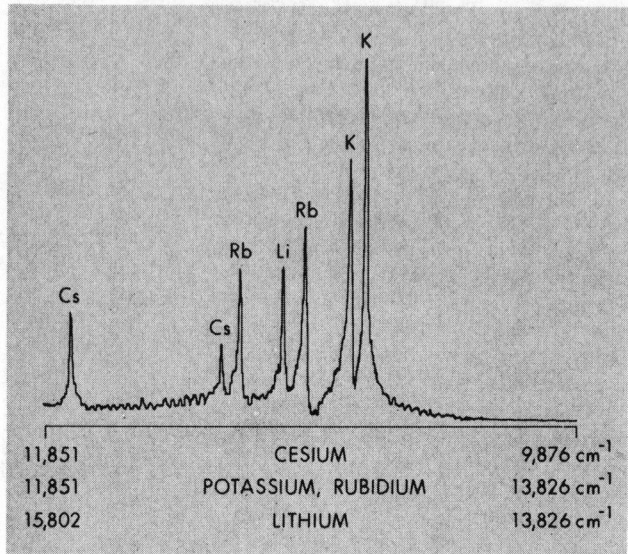

Figure 6. Flame emission spectrum of Li, K, Rb, and Cs as measured with Fourier transform spectrometer

X-Ray Energy Spectrometry

New applications are provided by the unique capabilities of X-ray energy spectrometry with the Si(Li) detector. Elemental analysis with this system is competitive with conventional X-ray fluorescence

David E. Porter and Rolf Woldseth

Kevex Corp.
898 Mahler Road
Burlingame, Calif. 94010

"On the spot" nondestructive elemental analysis is not here yet. The development of the Si(Li) detector X-ray energy spectrometer, however, is bringing it closer. X-ray spectrometry, intrinsically one of the simplest techniques for obtaining elemental analysis, gives information over a continuous range of elements and concentrations. The trace determination of lead, for example, cannot be done by neutron activation. It can be done by atomic absorption, but the sample must be diluted to a given concentration range. Both of these techniques are destructive, involving the loss of the sample, whereas X-ray spectroscopy does not. (For general information on X-ray spectroscopy, see References *1-3*.)

Background and Instrumentation

The means of transferring information in X-ray spectroscopy is the characteristic X-rays induced in a specimen by bombarding it with sufficiently energetic radiation. The inspection by radiation takes place without disturbing the specimen. This permits both the source of excitation and the detector system to be located physically apart from the specimen (this distance is limited by the need to attain a practical information rate). Conceptually, X-ray spectroscopy can be thought of as "chemical radar," an important feature for many process control types of applications (*4*).

An analysis of the characteristic X-rays results in a spectrum containing relatively few peaks that may be easily interpreted in both a qualitative and quantitative sense. The data, in most instances, are insensitive to structural and chemical environment (*5*).

Excitation System. The emission of characteristic radiation can be achieved, in principle, through any interaction that excites or removes an electron from its normal place in the atom. Subsequent rearrangement of electrons within the atom brings it back to its stable configuration, causing emission of electromagnetic radiation. Popular means of excitation are: X-rays or gamma rays from radioactive sources, X-rays from a tube, or charged particles such as electrons, protons, or alpha particles.

Specimen Preparation. Specimen preparation may be nonexistent for "on-site" qualitative analysis. On the other hand, for routine quantitative analysis it may involve surface preparation of a specimen that has been sized to fit within the confines of the spectrometer. In any case, the need for sample preparation is frequently minimal. It is feasible to analyze all phases of matter, provided the specimen is presented to the spectrometer in a reproducible way.

X-ray Analysis System. The spectral distribution of characteristic X-rays may be ascertained by diffraction with a crystal of known "d" spacing ("conventional" wavelength-dispersive X-ray spectroscopy) or by direct energy discrimination (X-ray energy spectroscopy).

Conventional Wavelength-Dispersive Spectrometry. The system used in conventional wavelength-dispersive spectrometry is shown schematically in Figure 1. It generally consists of a high-wattage X-ray tube with either a tungsten or chromium target, the tungsten tube being used for higher energy lines and the chromium tube for the lower energies. Specimens may be placed either singly or automatically in the path of the tube radiation. The characteristic X-rays induced in the sample are passed through a collimator on the periphery of the goniometer circle, diffracted by a suitable analyzing crystal and detected by a scintillation or proportional counter. The counter is located on the goniometer circle at an angle satisfying the Bragg condition. The wavelength-dispersive spectrometer has an overall low efficiency owing to several intensity losses through the restriction of solid angle, the low "reflectivity" of the analyzing crystal, and in some cases, detector losses. In applications where proportional counters are used, this technique suffers from low-detection efficiency at the higher energies.

Preparation and use of the conventional wavelength-dispersive system are complicated by the need to establish a number of operating conditions. For most routine analysis, there are well-established combinations of operating conditions. The choice generally resolves itself to finding an analyzing crystal that will diffract the characteristic X-rays of interest to a region on the Bragg circle where they may be resolved.

Figure 2 shows the effective resolution of some of the more popular analytical crystals, along with that of direct energy discrimination for the flow proportional counter and the lithium-drifted silicon X-ray Si(Li) energy spectrometer (6). A casual look at these curves points out that the conventional technique has significantly better resolution at energies below about 11 keV (atomic numbers below arsenic with the K lines, and lead with the L lines). One of the major advantages of the conventional technique over energy spectrometry lies in the superior resolution at the lower energies. In a practical sense, this means there are fewer instances of peak overlap in the lower energy region of the spectrum using the conventional method. At the high-energy end of the spectrum, the reverse is true, and the energy spectrometer has an advantage.

X-ray Energy Spectroscopy. The X-ray energy spectrometer is shown schematically in Figure 3. It consists of a means of excitation (be it charged particle or photon), the sample, and the detector ranged in a fixed geometry. The overall efficiency of this system is much better than the conventional technique because of fewer, less restrictive solid angle losses, the lack of diffraction losses, and near 100% detection efficiency over a wide energy range.

Elemental Range. For practical considerations, the analysis that may be performed by either system is limited on the low end by absorption. For the conventional technique the use of an ultrathin formvar window on a proportional counter allows analyses down to and including carbon (7). The current standard formulation of the X-ray energy spectrometer imposes a limit at approximately fluorine, owing to entrance window effects. Windowless Si(Li) detectors have been used to measure spectral peaks down to and including carbon (8–10). The need of keeping the detector free of contamination prohibits the use of windowless detectors in all but the most controlled conditions. At higher energies the silicon detector loses efficiency but is still useful where excitation of sufficient intensity is available. The conventional spectrometer suffers a greater performance loss at these energies.

Spectral Scanning. In the conventional technique the goniometer must be scanned from some predetermined low angle (high-energy limit) to another set limit at the high-angle end. The scanning may be done continuously or by stepping through the range. The latter is preferred owing to its better accuracy. A strip-chart recorder is generally synchronized to the scanning motor for continuous documentation of the intensity. Digital information is either obtained off the strip-chart record or from a teletype printout of each point with its angle and intensity. The analyst then can look up the energies corresponding to the peaks found in a table compiled for each of the popular analyzing crystals.

Although X-ray spectra, in general, are simple, they are complicated with the conventional technique by the possibility of higher order reflections. These are of lower intensity and for most purposes can be verified or eliminated from the first-order reflections. The rather cumbersome mechanical scanning of the spectrum is one of the serious drawbacks of the conventional technique, requiring extensive careful alignment and standardization procedures. To obtain a complete spectral scan, one must normalize the data from one analyzing crystal to another so that a representative intensity plot may result. (It is difficult to deal with absolute intensities since there are so many variables involved.) Although the complexity of the system and the data format may be accepted in the laboratory, this is one of the major drawbacks of the conventional X-ray spectroscopy as applied to its more widespread analytical uses.

An electronic scanning, simultaneous display of all elements is available in real time with X-ray energy spectroscopy. All the incoming signals are processed and sorted according to their energies; there is only the matter of choosing a range in which to look, by setting the electronic gain of the system. The memory of the multichannel analyzer may be continuously monitored by a display oscilloscope or television, allowing the operator to inspect the entire analytical range at a single glance.

The data may be recorded by photographing the display screen, in the form of an X-Y plot, or in digital form as a teletype readout of the

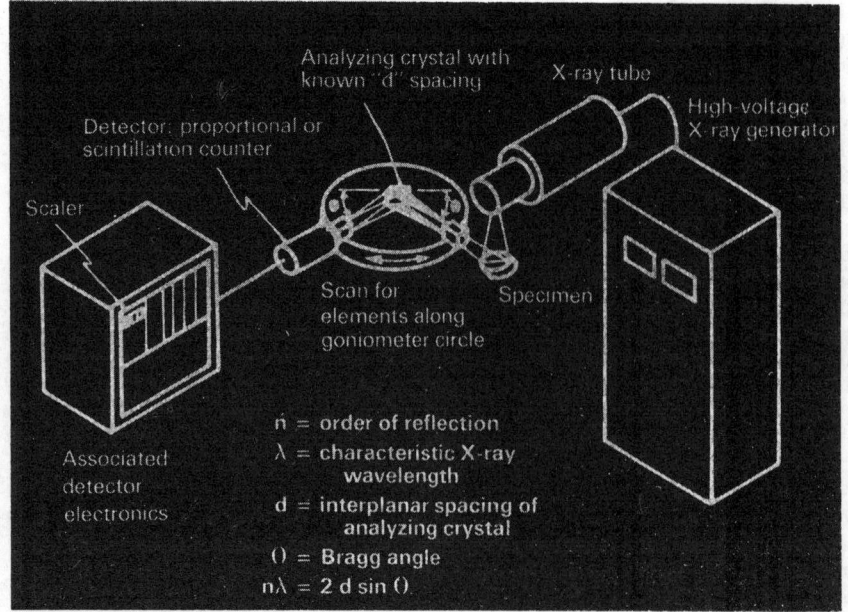

Figure 1. Schematic of wavelength-dispersive X-ray spectrometer

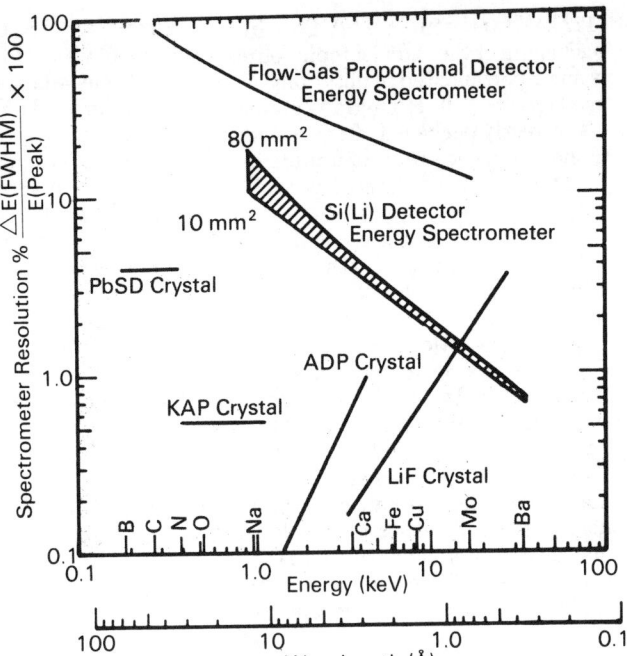

Figure 2. Comparison of wavelength and energy spectrometers in terms of energy resolution [adapted from Fitzgerald and Gantzel (6)]

an advantage because all data are accumulated simultaneously. Added benefits are the ease of on-line control and automatic specimen handling, items which are difficult at best with the conventional technique.

Applications: X-ray Energy Spectrometry

Choice of Photon Source. General analytical requirements are met by the use of photon excitation (14). The particular choice is dictated by the physical size limitations imposed on the analytical system and the quality of data required. Characteristic X-ray production is most efficient when the excitation energy is just above the absorption edge of the particular element of interest.

A second consideration in optimizing the excitation of characteristic X-rays is the need to minimize the spectrum background (15, 16). Background may be due to several effects. One of the major effects is controlled by the energy profile used for excitation because in most instances a significant part of this energy is "backscattered" by the sample. This backscattered energy is recorded along with the characteristic X-rays. The ideal excitation is a monochromatic source of X-rays having an energy just above the absorption edge of the elements of interest (17, 18). This type of source also allows the possible selective excitation of an element with respect to one of higher energy (important when the high-energy element predominates over the lower energy element of interest).

Radioisotopic sources are light in weight, of small size, have high stability, and require no utilities. They offer a limited selection of practical energies as exemplified by the following commonly employed isotopes: iron-55 (5.9 keV), cadmium-109 (22.1 keV), americium-241 (59.6 keV for the main line plus a more complex emission spectrum), and cobalt-57

memory. Since data are stored in the memory in a compact digital form, they are readily accessible for data reduction via a hard-wired data processor or the on-line type of minicomputer. The former, although limited in scope, offers many data reduction routines commonly used by the analyst at a maximum of speed and operator convenience. Such operations as peak integration, background subtraction, energy determination of the peak, and peak subtraction are among those operations available. Through the use of the minicomputer with the proper programming, essentially a total analysis is feasible, including matrix corrections and comparison to standards (11–13). All of these features are commonly available with the energy spectrometer but are generally only available on the high-cost conventional systems.

Data Acquisition Rate. In terms of the count rate capability, the conventional method has a distinct advantage on an element by element basis, since one is looking at only a single narrow energy range, geometrically disregarding the remainder of the spectrum. This is coupled with the ability of a proportional counter to handle extremely high count rates (of the order of 100,000 counts or more per second). By contrast, the energy-dispersive spectrometer must handle all of the incoming pulses electronically. The pulses are of necessity of long duration, limiting the useful data acquisition rate to a range up to 50,000 counts/sec for the total spectrum. A single element may be limited to a few hundred to a few thousand counts/sec.

For qualitative scans and for those quantitative scans where there are more than a few elements involved, the energy-dispersive technique has

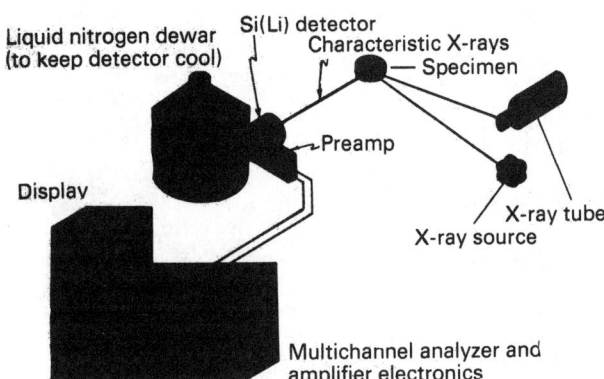

Figure 3. Schematic of X-ray energy spectrometer

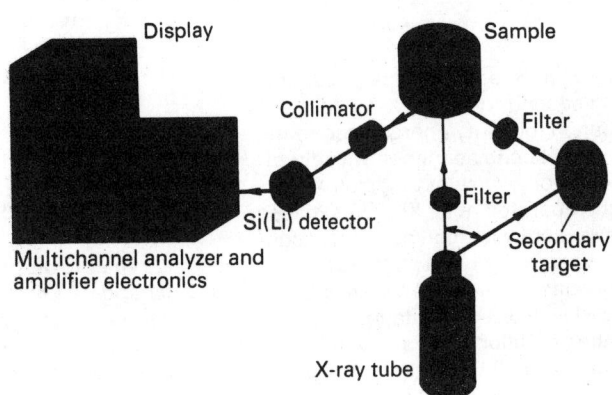

Figure 4. Schematic of secondary target excited X-ray energy spectrometer

69

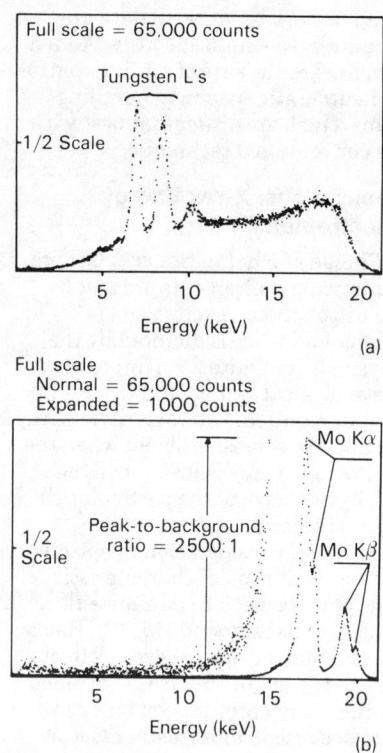

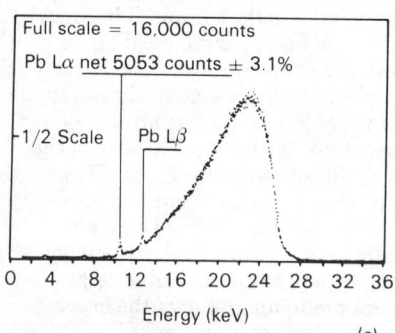

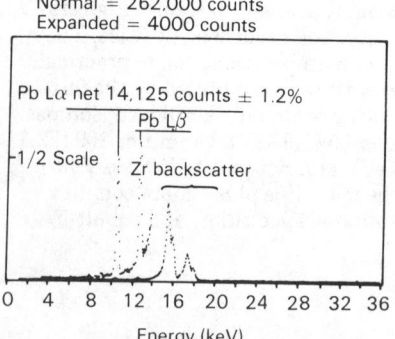

Figure 5. Spectra comparing direct and secondary target X-ray tube excitation for X-ray energy spectrometer: (a) Lucite specimen with direct excitation by use of tungsten X-ray tube operated at 20 kV; (b) same as (a) except with molybdenum secondary target; (c) 100 ppm lead in aqueous solution specimen as analyzed with direct excitation at optimum condition of 25 kV for total count rate 10,000 counts/sec and counting time of 300 sec; (d) same as (c) except with zirconium secondary target

(122 keV for the main line). Care in packaging of the radioisotopic sources is an important step in obtaining good results (19). High-grade materials correctly packaged offer the user a monochromatic source of limited intensity with an extremely low background. The selection of energies and the maximum intensities available with radioisotopes limit the overall efficiency of characteristic X-ray production.

X-ray tubes offer greater analytical flexibility at a cost of more complexity. Excitation by the characteristic tube lines may be accomplished by having several different target tubes. The continuum or bremsstrahlung radiation from high atomic number elements (generally tungsten or gold) may also be used for excitation, the accelerating voltage and spectral profile being changed to correspond to the analytical requirements. Both of these means of excitation give high background. Characteristic tube excitation can be used in conjunction with a filter of the target material to greatly reduce this background.

An alternative technique is the use of an X-ray spectroscopy tube to fluoresce a selectable secondary target (18, 21). This system is shown schematically in Figure 4. The bremsstrahlung radiation from a high atomic number element, such as tungsten or gold, is used to fluoresce characteristic lines of a variety of selectable secondary targets (such as titanium, silver, and barium). These characteristic lines are used, in turn, to fluoresce the sample. The secondary target reduces the background from the tube to the level of radiation backscattered from the secondary target material used. This low level can still be further reduced by selective filtration, which for all practical purposes eliminates the remaining contaminants.

The low X-ray conversion efficiency of this technique requires the use of a high-powered X-ray spectroscopy tube to provide practical data rates. The secondary target source offers superior performance to other excitation techniques and the desired flexibility, (18, 21). Figure 5 shows the difference between direct excitation with a tungsten target tube and excitation via a molybdenum secondary target. The secondary target technique is much more efficient (Figure 5, c and d).

Typical Analysis. *Analysis of Photographic Film.* For a densitometric measurement of photographic film, one analyst made up a lead mask to fix the region of exposure on the film, the mask being mounted on top of an americium-241 radioisotope source above the detector of an energy spectrometer. The typical spectrum (Figure 6) was obtained with a multichannel analyzer, whereas the actual system used for routine analysis consisted of a pair of single-channel analyzers. The energy window of one set is for the silver $K\alpha$ peak and the other for the background. This rudimentary system is capable of handling a great number of similar analytical problems involving the straightforward attainment of routine data.

Analysis for Chlorine in Rubber. This is basically the same analytical approach as described above except

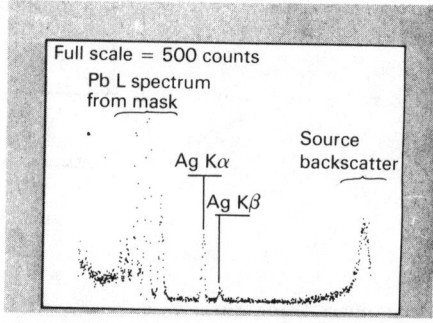

Figure 6. Spectrum of photographic film showing silver. Analysis conducted with 5 mCi of Am-241 and 300 mm² detector. 200-sec analysis time

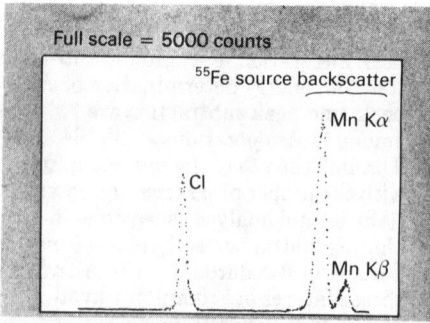

Figure 7. Spectrum of rubber showing chlorine. Analysis conducted with 25 mCi Fe-55 and 80 mm² detector. 100-sec analysis time

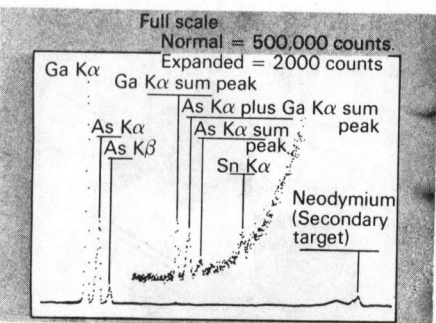

Figure 8. Spectrum of gallium–arsenide specimen showing tin dopant (level of the order of 1 ppm). Analysis conducted with neodymium secondary target. 1000-sec analysis time

that an iron-55 source is used in this instance. The analysis for chlorine in rubber (see Figure 7 for a typical spectrum) is another example in which a simple X-ray system is used in quality control. The system may be run by an untrained technician who "places the specimen" on top of the source and reads the resulting chlorine concentration by the number of counts in a single-channel analyzer window.

These two illustrative analyses could feasibly be run with a conventional wavelength spectrometer if it were modified to accept the various size samples. The resulting system cost and complexity, however, would generally preclude its use in everyday production situations.

Trace Analysis in Quality Control. The useful properties of many semiconductor materials are strongly dependent on the level of doping. There is a definite need for direct, nondestructive analysis of these materials; but the levels of doping can be extremely low, which complicates the analytical problem. The analysis can, in some cases, be performed by the X-ray energy spectrometer by use of the secondary target technique. Typical of this type of analysis is the spectrum in Figure 8, where the doping level for tin in gallium-arsenide is of the order of 1 ppm.

The extremely good analysis efficiency in this semiconductor material is most likely due to an optimum situation. There exists a moderately high atomic number matrix that tends to minimize the contribution of the backscatter peak with respect to the useful information of the spectrum. Also, the matrix is not too high in atomic number to cause severe absorption of the characteristic X-rays of interest. This condition, coupled with a relatively higher atomic number for the element of interest, produces a high overall fluorescence efficiency.

Duplex Excitation. A unique application of the secondary target technique is found in obtaining the ratio of silicon to germanium to within 0.1% in a single run. Statistically, this is most easily achieved if both of the elements produce equal count rates. If a single excitation energy is used, it must be above the germanium edge. The silicon X-ray production efficiency is lower than for germanium, and in this case, with the excitation being far away in energy from the silicon absorption edge, the production would be extremely low. A possible solution is to choose a secondary target energy that will strongly excite the silicon line and to use it without a filter. The higher energies from a tungsten target tube can then

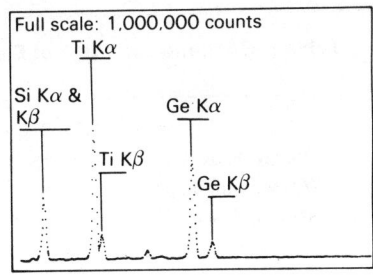

Figure 9. Spectrum of silicon-germanium specimen by use of titanium secondary target without filtration. 1000-sec analysis time

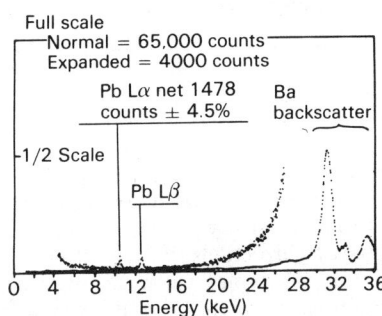

Figure 10. Spectrum of 100 ppm lead in aqueous solution specimen taken with same conditions as in Figure 5, c and d, except with barium secondary target

be backscattered for excitation of the germanium. The amount of backscatter increases as the atomic number of the secondary target decreases, and there is an optimum secondary target for this analysis. Figure 9 shows the results by use of a titanium secondary target representing the optimum condition. By use of the high count rate capable with this kind of system, the desired ratio was obtained in approximately 15 min.

Thick Vs. Thin Samples. The preceding examples of analysis, with the exception of silver in photographic film, have been conducted on thick samples. For our purposes, thick samples will infer that the incident or exciting radiation is either absorbed in the sample or backscattered from it. In a more general sense, there are three major considerations attributable to the thickness of the sample: self-absorption (more correctly, the specimen absorption) of the characteristic X-rays ("thick" sample consideration); scatter mass of the specimen which is directly proportional to its thickness ("thin" sample consideration); and susceptibility to system fluorescence that may occur when the excitation energy passes through the specimen and fluoresces material in the specimen chamber. This, in turn,

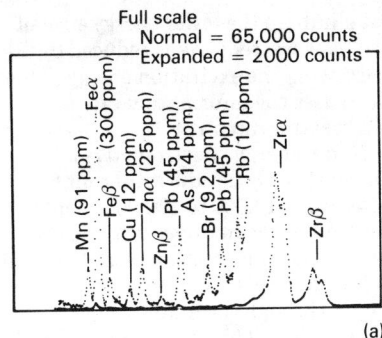

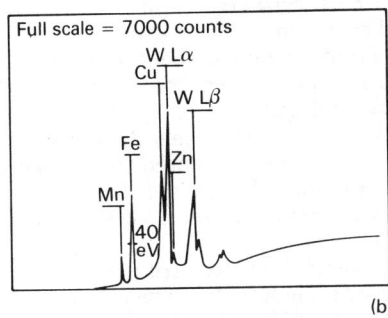

Figure 11. Spectrum of orchard leaves: (a) zirconium secondary target and analysis time of 1000 sec. Elemental concentrations, as reported by NBS, given after each element. (b) Same as (a), by use of wavelength spectrometer with LiF analyzing crystal

can come back through the specimen and be registered along with the characteristic X-rays ("thin" sample consideration).

In many problems of trace analysis the scatter mass limits the performance of the X-ray energy spectrometer in thick samples. The backscatter contribution to the spectrum increases with decreasing atomic number matrixes; we find that the inelastic or Compton contribution to the backscatter is predominant. At increasingly higher excitation energies, the spread between the Compton peak and the elastic peak becomes increasingly greater. For instance, Figure 10 shows the spectrum of a sample run under the same conditions as in Figure 5, c and d, by use of a barium secondary target. Compare the analysis for lead in aqueous solution to that for tin in gallium-arsenide (Figure 8) at two orders of magnitude lower concentration. The backscatter peak for low atomic number matrixes becomes the main contribution to the total system count rate. The ratio of the desired information rate to the backscatter rate is much more favorable in the analysis for tin in gallium-arsenide as opposed to lead in aqueous solution. The problem of having the peak of interest climb up the lower energy edge of the backscatter peak is reduced by increasing the excitation energy. This decreases the information rate to backscatter rate ratio.

X-ray energy spectrometers can currently only handle total count rates up to 10,000 to 100,000 counts/sec, imposing an upper limit on the obtainable information rate. A partial solution to this dilemma is sometimes found through the employment of thin samples (17). As we make the samples thinner, the information rate to backscatter rate ratio increases. There is an added benefit to the use of thin samples, in that the matrix correction problem becomes negligible. It is possible to make standards that will serve for general quantitative analysis; the procedure is particularly simplified through the use of a monochromatic excitation source (17).

Analysis of Orchard Leaves. A good example of the typical thin sample analysis is related to the field of environmental technology in which orchard leaves are commonly used to establish detection limits and sensitivities. This application shows the X-ray energy spectrometer's ability to conduct rapid survey analysis. The use of thin samples reduces the analytical problem to the simplest form. Figure 11 shows a typical spectrum for orchard leaves run by X-ray energy spectroscopy and by wavelength spectroscopy. By comparing with previous results, we see the advantages of using thin samples in improving the information rate to backscatter rate ratio. The comparison of the two techniques for the same sample is, however, quite unfair. The X-ray energy spectrometer was designed so that there was a "free" region behind the specimen to allow the incident radiation to escape. The crystal spectrometer, on the other hand, was designed primarily for the use of thick specimens where there was only a limited region behind the specimen. The multiple backscatter of the source X-rays in this confined area increased the background level above that which could possibly be obtained with this kind of sample.

The need for lowering the detection limit favors some sample preparation and effective preconcentration. Freeze drying, ashing, ion-exchange techniques, and filter retention are methods all commonly used for X-ray analysis to remove unwanted scatter mass from the sample. Though this adds a complicating step in the analysis procedure, it also extends the capability of the technique (22).

Quantitative Analysis. The use of thin samples along with monochromatic excitation offers an easy solution to the problem of providing quantitative data (17). A simple linear regression equation may be used to relate intensity to composition (matrix corrections are neglected). For thick samples, the mathematical equations to relate intensity to composition are more complicated. The matrix effects of absorption and secondary fluorescence enhancement cause nonlinear relationships between the peak intensity and concentration (5). The commonly used regression equation is

$$R_i = W_i / \sum_m A_{im} W_m$$

where R_i is the relative intensity of the element i, W is the weight fraction, and A_{im} is the interaction or correction coefficient accounting for the effect of the presence of element m on element i. Interelemental effects are taken into account by solving sets of simultaneous equations derived from measurements made on a fixed set of standards. The resulting solution is valid for a finite range of composition and is generally unique to a particular type of sample. The analysis of stainless steels by use of interelemental corrections is given in Table I (13); for comparison, results are also reported for wavelength-dispersive and NBS analysis of these samples.

An alternative to the empirical approach is the so-called fundamental parameter method (5) which requires the analytical description of the spectral distribution of the excitation source, the mass absorption coefficients, and fluorescence yields. The computed solutions of the resulting analytical equations, written for both the primary and secondary excited fluorescent X-rays, are accomplished by iteration. Iteration steps are repeated until the calculated and measured intensities agree. The main limitations in this method are in the uncertainty of mass absorption coefficients and fluorescent yields. The availability of new and better data for these parameters is gradually removing this limitation.

Charged Particle Excited X-ray Energy Fluorescence Spectrometry. Accelerator-produced particles, chiefly alpha and protons, have been used successfully for the analysis of thin samples in environmental technology (14, 23, 24). The large X-ray production available with charged particle excitation makes this technique attractive, particularly for the low atomic number elements. Though an interesting application of X-ray energy spectroscopy, this does not appear to be a practical solution to general analysis problems (14).

Microanalysis. The adaptation of the X-ray energy spectrometer to the scanning electron microscope, transmission electron microscope, and electron microprobe is much more favorably received than its cousin, the conventional crystal spectrometer (25). These instruments basically are used in a dynamic situation where rapid interpretation is necessary for the user to go on to the next step in an investigation. The ease of interpretation of X-ray energy analysis is largely responsible for the popularity of this technique. It also has a fundamental advantage over the conventional technique in that the exact sample position need not be known for correct element identification. Quantitative results are possible with these instruments, making the combination an extremely powerful tool in both research and quality control.

Acknowledgment

The authors thank Erwin Strahl, Kaiser Center for Technology, for supplying current information on wavelength-dispersive spectrometry.

Table I. Concentration (%) of Elements in Stainless Steels (13)

	Cr	Mn	Ni	Nb	Mo
NBS 846 (AISI 321 Cr 18, Ni 9)					
Energy dispersive	18.35	0.453	8.89	0.597	0.425
Wavelength dispersive	18.78	0.464	9.20	0.594	0.428
NBS analysis	18.35	0.53	9.11	0.60	0.43
NBS 849 (Cr 5.5, Ni 6.5)					
Energy dispersive	5.44	1.56	6.75	0.302	0.150
Wavelength dispersive	5.46	1.60	6.74	0.308	0.149
NBS analysis	5.48	1.63	6.62	0.31	0.15
NBS 850 (Cr 3, Ni 25)					
Energy dispersive	3.00	0.246	24.92	0.062	0.083
Wavelength dispersive	3.09	0.103	24.80	0.049	0.072
NBS analysis	2.99	...	24.8	0.05	...

References

(1) E. P. Bertin, "Principles and Practices of X-Ray Spectrometric Analysis," Plenum Press, New York, N.Y., 1971.
(2) R. O. Muller, "Spectrochemical Analysis by X-Ray Fluorescence," Plenum Press, New York, N.Y., 1972.
(3) L. S. Birks, "X-Ray Spectrochemical Analysis," Interscience, New York, N.Y., 1959.
(4) R. J. Gehrke, M. S. Cole, and W. A. Ryder, "Advances in X-Ray Analysis," Vol 15, p 276, K. F. J. Heinrich, Ed., Plenum Press, New York, N.Y., 1972.
(5) L. S. Birks, *Appl. Spectrosc.*, **23**, 303 (1969).
(6) R. Fitzgerald and P. Gantzel, Energy Dispersion X-Ray Analysis, p 3, ASTM STP 485, Philadelphia, Pa., 1971.
(7) E. P. Bertin, "Principles and Practices of X-Ray Spectrometric Analysis," p 193, Plenum Press, New York, N.Y., 1971.
(8) J. M. Jaklevic and F. S. Goulding, *IEEE Trans. Nucl. Sci.*, **NS-18**, 187 (1971).
(9) R. G. Musket and W. Bauer, *J. Appl Phys.*, **43**, 4786 (1972).
(10) R. G. Musket and W. Bauer, *Nucl. Instrum. Methods,* in press (1973).
(11) R. L. Heath, R. S. Frankel, R. J. Gehrke, and J. Barstow, *Proc., Instrum. Soc. Amer.*, Analysis Instrumentation Symposium, Houston, Tex., 1971; Vol 9; Instrument Society of America, Pittsburgh, Pa.
(12) J. R. Rhodes, C. B. Hunter, D. L. Kellogg, R. D. Sieberg, and T. Furuta, "Advances in X-Ray Analysis," Vol 14, p 127, C. S. Barrett, J. B. Newkirk, and C. Ruud, Eds., Plenum Press, New York, N.Y., 1971.
(13) W. G. Wood, *Amer. Lab.*, **4**, 27 (1972).
(14) J. A. Cooper, *Nucl. Instrum. Methods,* in press (1973).
(15) R. Woldseth, *Amer. Lab.*, **4**, 3 (1972).
(16) F. S. Goulding, J. M. Jaklevic, B. V. Jarrett, and D. A. Landis, "Advances in X-Ray Analysis," Vol 15, p 470, K. F. J. Heinrich, Ed., Plenum Press, New York, N.Y., 1972.
(17) R. D. Giauque and J. M. Jaklevic, *ibid.*, p 164.
(18) J. M. Jaklevic, R. D. Giauque, D. F. Malone, and W. D. Searles, *ibid.*, p 266.
(19) R. D. Giauque, *Anal. Chem.*, **40**, 2075 (1968).
(20) J. R. Rhodes, Energy Dispersion X-Ray Analysis, p 243, ASTM STP 485, Philadelphia, Pa., 1971.
(21) D. E. Porter, *X-Ray Spectrom.*, in press (1973).
(22) J. M. McCall, D. E. Leyden, and C. W. Blount, *Anal. Chem.*, **43**, 1324 (1971).
(23) R. L. Watson, J. R. Sjurseth, and R. W. Howard, *Nucl. Instrum. Methods,* **93**, 69 (1971).
(24) T. B. Johansson, R. Akelesson, and S. A. E. Johansson, "Advances in X-Ray Analysis," Vol 15, p 373, K. F. J. Heinrich, Ed., Plenum Press, New York, N.Y., 1972.
(25) D. R. Beaman and J. A. Isasi, *Mater. Res. Stand.*, **11**, 8 (1971).

David E. Porter is manager of the Electron Tube Subsystems Division at Kevex Corp. in Burlingame, Calif. He received his BS in metallurgy at the University of Arizona, Tucson (1963), and his MS and PhD in metallurgy at the University of California, Berkeley, under E. R. Parker (1966, 1969). He has conducted research and written several papers in the fields of solid state reaction kinetics, strengthening mechanisms in metals and fracture mechanics. Before coming to Kevex in 1970, Dr. Porter was a staff research metallurgist for Kaiser Aluminum and Chemical Corp., where he developed microstructure-fracture toughness relationships for aluminum alloys. During this work he became interested in X-ray energy spectrometry as a tool for the metallurgist. Pursuing these interests, he joined Kevex where he has had wide experience in X-ray energy applications. He has developed special diffractometers and spectrometers for X-ray energy applications.

Rolf Woldseth is director of the Analytical Applications Laboratory of Kevex Corp. He received his MS in engineering physics at the Technical University of Norway in 1955 and his PhD in physics at Washington University, St. Louis, Mo., in 1965. Before joining Kevex Corp. in July of 1970, Dr. Woldseth accumulated a broad exposure in the fields of atomic and nuclear physics and associated instrumentation as a member of the physics faculties at Wake Forest University, Winston-Salem, N.C., Rensselaer Polytechnic Institute, Troy, N.Y., and Washington University, St. Louis, Mo., and as a research physicist at the Institute for Atomic Energy (Norway).

X-Ray Photoelectron Spectroscopy

X-ray photoelectron spectroscopy is a powerful new analytical technique currently experiencing widespread growth in the scientific community. The technique is commonly called ESCA for electron spectroscopy for chemical analysis (1). ESCA is concerned with the measurement of core-electron binding energies. A molecule or atom is bombarded with an X-ray of sufficient energy to eject an electron. All electrons whose binding energies are less than the energy of the exciting X-ray are ejected. The kinetic energies of these photoelectrons are then measured by an electron analyzer. The core-electron binding energies (E_b) relative to the Fermi level can then be computed via the relationship

$$E_b = E_{\text{X-ray}} - E_k - \phi$$

where $E_{\text{X-ray}}$ is the energy of the exciting X-ray, and ϕ is the spectrometer work function, a constant for a given analyzer.

Chemical Shift

The utility of ESCA for the chemist is the result of "chemical shifts" that are observed in electron binding energies. The binding energies of core-electrons are affected by the valence electrons and therefore by the chemical environment of the atom. The attraction of the nucleus for a core-electron is somewhat diminished by the presence of the outer electrons. If one of these valence electrons is removed, the amount of shielding is diminished, and the effective nuclear charge experienced by the core-electron increases, thereby increasing the electron binding energy. Therefore, in a simple sense the shifts of the photoelectron lines in an ESCA spectrum reflect the increase in binding energy as the oxidation state of the atom becomes more positive. In general, any parameter (i.e., oxidation state, ligand electronegativity, coordination) that affects the electron density about the atom is expected to result in a chemical shift in electron binding energy.

The classic example of chemical shifts in electron binding energy is the C(1s) electron spectrum of ethyltrifluoroacetate shown in Figure 1 (1). The ethyltrifluoroacetate molecule contains four carbon atoms, each lo-

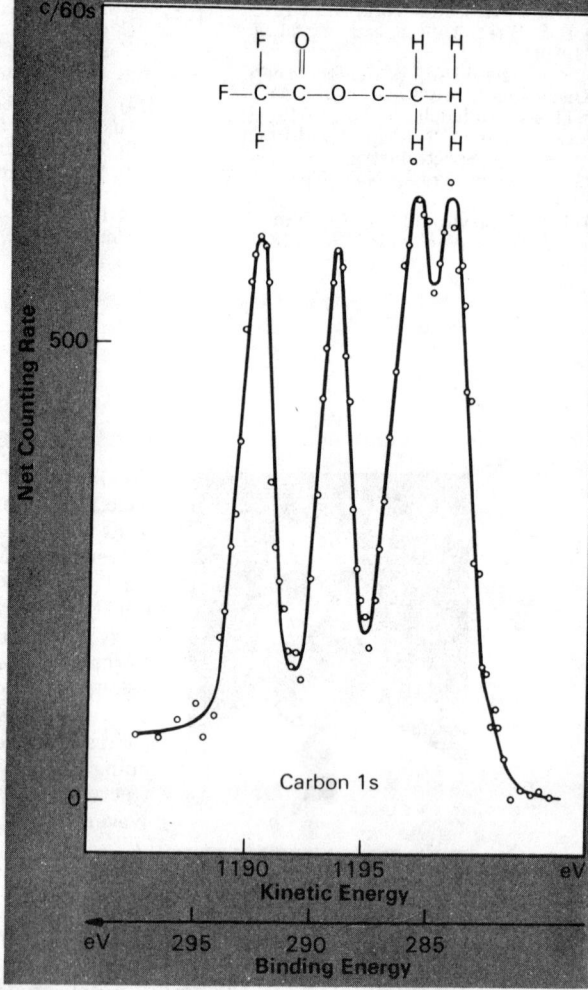

Figure 1. Carbon 1s photoelectron spectrum for ethyltrifluoroacetate [from Siegbahn et al. (1)]

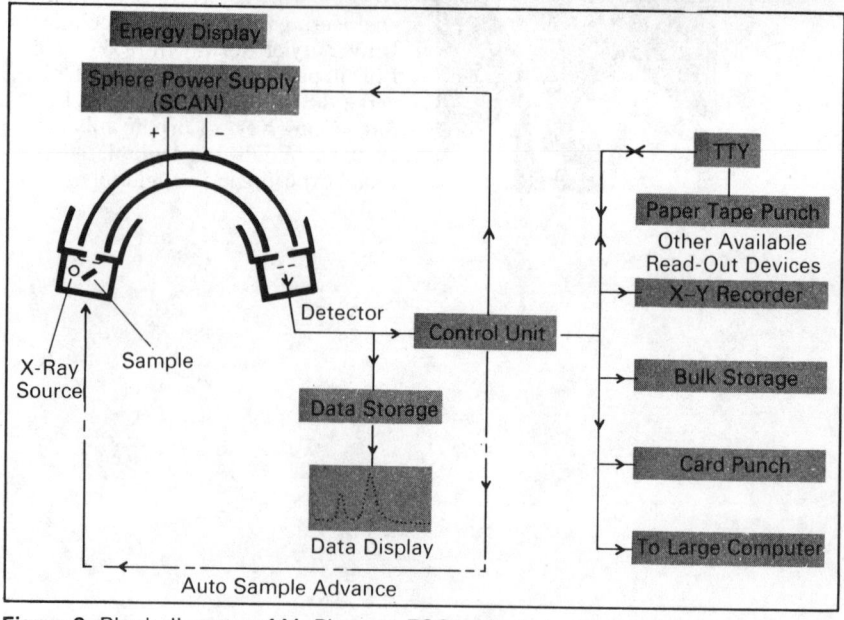

Figure 2. Block diagram of McPherson ESCA 36 photoelectron spectrometer

The availability of ESCA instrumentation and the qualitative and quantitative nature of the data produced promise much for future applications in chemical research and analytical problem solving

William E. Swartz, Jr.
Department of Chemistry, University of South Florida, Tampa, Fla. 33620

cated in a different chemical environment. This is shown by the C(1s) ESCA spectrum which contains four distinct C(1s) photoelectron lines. The trifluorocarbon yields the photoelectron line at highest binding energy since the fluorine atoms withdraw electron density from the carbon atom most efficiently. This results in the highest C(1s) binding energy, since the effective nuclear charge experienced by the remaining electrons is largest. In this instance, the relative positions of the photoelectron lines reflect the relative electronegativity of the various substituents.

Aside from the qualitative information one can obtain from an ESCA spectrum as exemplified above, ESCA also has quantitative possibilities. This arises from the fact that the intensity of a photoelectron line is dependent on the number of a particular type of atom present in the sample molecule. In the case of ethyltrifluoroacetate, the ratios of the carbon atoms are 1:1:1:1. This is evidenced by the 1:1:1:1 ratio of the four C(1s) photoelectron lines in the ESCA spectrum.

A major portion of the strength of ESCA as an analytical tool lies in the fact that chemical shifts can be observed for every element in the periodic chart except for hydrogen. Obviously, magnitudes of chemical shifts will vary from element to element, and the sensitivity for a particular element will vary with the photoelectric cross section. Thus, ESCA studies on some elements will be more valuable and easier to perform than on others. As of this writing, ESCA chemical shifts have been reported for approximately 84 of the elements. In this sense, ESCA is more valuable than NMR in that it is considerably more versatile from the perspective of elemental sensitivity. Coupling this elemental sensitivity with the qualitative and quantitative nature of ESCA data, one is able to understand the phenomenal growth of ESCA as an analytical technique.

Instrumentation

Figure 2 is a block diagram of a typical ESCA spectrometer, the McPherson ESCA 36 Photoelectron spectrometer. The basic components of the system are an X-ray source, a sample, an electron monochromator, a detector, and a scan and readout system. Since it is necessary to ensure that the mean free path of the photoelectron is large enough to allow it to traverse the distance from the sample to the detector without suffering energy loss, ESCA is a vacuum technique with a maximum operating pressure of approximately 5×10^{-6} torr.

The X-ray tube employed is of conventional design in that it consists of a heated cathode at high negative potential and a water-cooled anode maintained at ground potential. Since the major contribution to the width (FWHH) of a photoelectron line is the inherent width of the X-ray line, "soft" X-rays such as $Mg(K\alpha)$ and $Al(K\alpha)$ with a half-width of approximately 1 eV are usually employed for photoelectron excitation. With such "soft" X-rays, the FWHH of the photoelectron line is minimized. It is possible to further reduce the width of the exciting X-ray and thus the photoelectron line by using a crystal disperser to provide perfectly monochromatic X-radiation. Such a scheme was first described by Siegbahn et al. (1) and has been employed by Hewlett-Packard in their photoelectron spectrometer.

Since ESCA spectra can be obtained on solids, liquids, and gases, the physical form of the sample is not important. However, the technique is a vacuum technique; therefore, low vapor pressure solids are most easily run. A solid sample need only be placed on a probe that is appropriately positioned relative to the X-ray beam and the spectrometer slit. Liquids cannot be run per se but must be condensed onto a cryogenic probe and run in the condensed phase. Alternatively, liquids can be vaporized and run in the gaseous state. To obtain spectra on a gaseous sample, the spectrometer must be equipped with a differential pumping system to prevent the pressure in the analyzer from rising above 10^{-5} torr. Most commercial spectrometers have accessories that permit spectra to be obtained on liquid and gaseous samples.

Although the exciting X-rays can penetrate deeply into the solid sample, not all of the ejected electrons will escape for subsequent analysis. Those electrons ejected at considerable depths will lose all kinetic energy through internal collisions. The escape depth of the $Au(4f_{7/2})$ electrons has been determined to be on the order of 10–20 Å (2, 3). Therefore, ESCA is a surface technique. This places no restriction on the surface chemist. However, if bulk properties are to be measured, the surface must be a valid representation of the bulk.

Since ESCA samples only the first few angstroms of a solid sample, contamination and oxidation of the surface must be minimized. Oxidation can be a particularly serious problem if one is studying metals. Contamination by hydrocarbons during sample preparation or by deposition of residual vapors inside the sample chamber is routinely observed on most samples. Ion sputtering techniques are a convenient method for cleaning metallic samples (4). However, sputtering of powdered samples must be performed with care. By sputtering a Fe_2O_3 sample with Ar^+ ions, Yin et al. were able to follow the reduction of Fe_2O_3 to elemental iron (5). Most commercial ESCA instruments have sample preparation chambers and accessories that enable ion sputtering.

The electron monochromators used in ESCA instrumentation employ the double-focusing principle first described and developed by Siegbahn et al. (1). These monochromators employ either a magnetic or electrostatic field to energy sort the electrons. As the field is varied, electrons of appropriate kinetic energy are focused at the detector.

Since electron trajectories are influenced by magnetic fields, it is necessary to have the trajectory of the electron determined solely by the field of the monochromator. This means that the earth's magnetic field and any other stray fields in the vicinity of the analyzer must be reduced to effectively zero. Since paramagnetic shielding would perturb the field of a magnetic monochromator, it is necessary to use a Helmholtz coil system to cancel the extraneous field about a magnetic spectrometer. Such a coil system is generally very large and usually requires an additional feedback system to compensate for small fluctuations in the field. On the

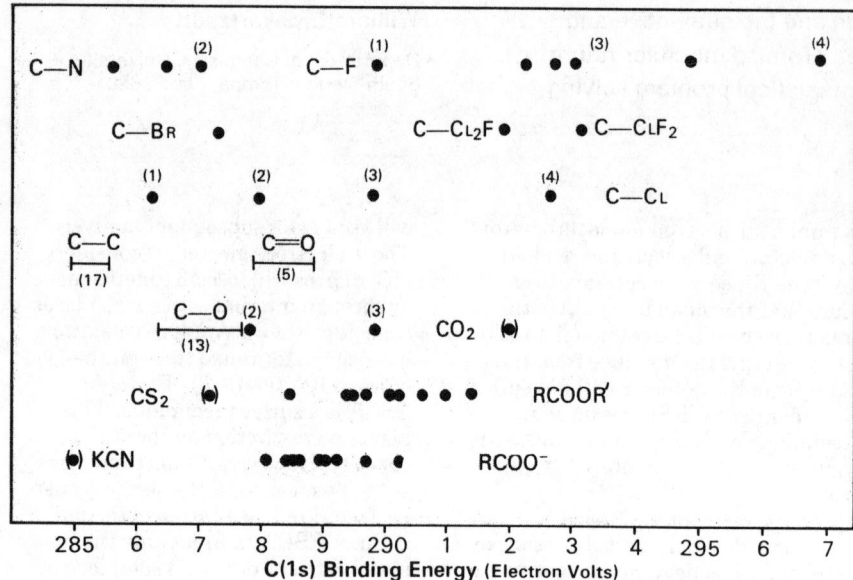

Figure 3. Correlation chart for C(1s) electron binding energies and functional groups. |———| = Range of observed energies for given bond where ( ) = number of compounds used for range above. (●) = Energies observed for single compounds. ● = Individual binding energies where ( ) = number of atoms attached to carbon. Particular functional group, compound or bond, is shown beside range or individual points [from Hercules and Hercules (9)]

other hand, paramagnetic shielding material can be used to cancel stray fields in the vicinity of an electrostatic monochromator. Such shielding is usually incorporated as a part of the analyzer housing. Thus, it is extremely easier to shield an electrostatic monochromator from stray fields; therefore, all commercial ESCA spectrometers employ an electrostatic monochromator.

Any detector capable of counting electrons can be employed in an ESCA spectrometer. In "state of the art" instrumentation, both continuous channel and discrete dynode electron multipliers are used. The continuous channel multiplier counts electrons with high efficiency to very low energies. Compared to discrete dynode multipliers, the continuous channel types are more stable to atmospheric and other vapor; therefore, they are considerably easier to handle without fear of deactivation.

Scan and readout systems used in ESCA instrumentation are either of a continuous or incremental type. In the continuous mode the focusing field is increased continuously as a function of time as the signal from the detector is simultaneously monitored by a rate meter. The focusing field and output from the rate meter are synchronized to allow accurate recording of spectra. In the continuous mode the energy region of interest can be scanned only once. This is somewhat of a handicap since no signal averaging can be performed. Signal averaging becomes necessary when the signal is weak or the quantity of sample material is small.

The incremental scanning mode increases the field in a series of small steps, counting the signal during each increment. When the counting rate at each increment is plotted as a function of focusing field, a spectrum is produced. Instrumentation using the incremental scanning mode uses either a multichannel analyzer or a small dedicated computer to accumulate the data. In such systems, signal averaging can be achieved by performing repetitive scans over the energy region of interest. The counting rate in each increment is then added to the preceding one. When the system contains a dedicated computer, both control of the energy scan and data acquisition are under its control. Commercial systems are usually designed so that several energy regions can be scanned sequentially, with the computer storing the data until they are retrieved by the operator.

Chemical Applications

Structure Determination. It is evident from the C(1s) electron spectrum of ethyltrifluoroacetate (Figure 1) that ESCA has a certain potential as a structural technique. As discussed above, it is apparent from the spectrum that there are four different carbon atoms in the molecule and that their relative abundances are 1:1:1:1. This potential has been compared to other powerful structural techniques such as infrared and nuclear magnetic resonance, in that empirical correlations between binding energy and previously compiled data libraries can be employed to determine the nature of the atom yielding a particular photoelectron line.

Considerable work has been reported in which structural correlations are attempted for elements of wide interest such as carbon, nitrogen, sulfur, and phosphorus. This work will be highlighted here in consideration of ESCA as a structural technique for the practicing chemist.

Carbon. The earliest work on carbon was summarized by Siegbahn et al. (1). The ethyltrifluoroacetate spectrum discussed in detail above is an integral part of these results. In their early work, Siegbahn and coworkers studied the C(1s) photoelectron spectra of some carboxylic acids and benzene carboxylic acids. The data indicated that a C(1s) electron line appears for each type of carbon atom present, except for the methyl and methylene carbons, which have identical binding energies. In addition, the intensity of the line owing to the carbonyl carbon is proportional to its compositional ratio and is shifted to higher binding energy owing to the electronegative oxygen.

The study of carbonaceous compounds was extended to σ-bonding electrons by Hamrin et al. (6). Gas-phase studies of methane and ethane yielded structural information for the bonding electrons which compared favorably to that of Turner et al. (7). Turner's data were obtained by use of the HeI resonance line at 21.21 eV for photoelectron excitation.

Figure 3 is an empirical correlation chart for the C(1s) binding energies that has been constructed with the data of Gelius et al. (8). Such a correlation chart may be of use to the analytical chemist. Gelius and coworkers studied an extensive series of organic molecules which contained sp^3, sp^2, and aromatic carbon. In order that the chemical environment of the carbon atom in question could be expanded, compounds containing elements such as oxygen, nitrogen, and halogens were studied. From Figure 3, the binding energies of carbon bonded to another carbon remain fairly constant at approximately 285 eV. In 13 compounds in which the carbon is bound to oxygen, the binding energy range is less than 2 eV. The carbonyl C(1s) electrons are shifted to slightly higher binding energies than the C—O carbons, and the range is also fairly constant. The halogenated carbons exhibit binding energies in the order expected on the basis of the electronegativity of the halogens (X), i.e., $CX < CX_2 < CX_3 < CX_4$. The C(1s) data for esters and acids cover a much wider range of binding energies than the other carbons, which indicates that some care must be taken when using empirical

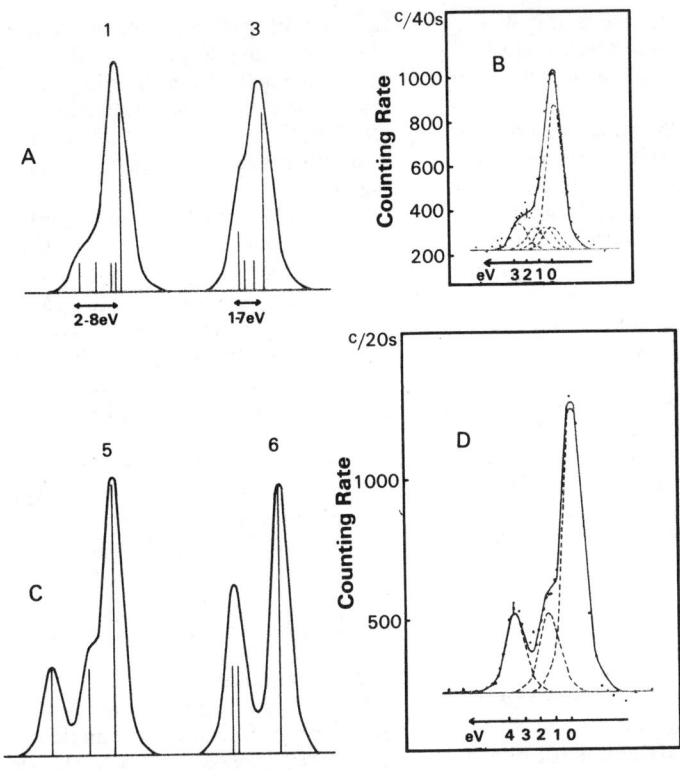

Figure 4. Use of ESCA as structural technique by comparing observed and predicted spectra. A: Predicted C(1s) spectra for structures 1 and 3 (see text). B: Observed C(1s) spectrum for material thought to be 1 or 3. C: Predicted C(1s) spectra for structures 5 and 6 (see text). D: Observed C(1s) spectrum for material thought to be 5 or 6 [from Hedman et al. (10)]

correlation charts for determination of structure.

Hedman et al. (10) used ESCA to determine the structure of a product resulting from the reaction of 2,4-pentanedione with H_2S in concentrated HCl. The following possible structures were postulated:

Nuclear magnetic resonance and mass spectrometry indicated 1 and 3 to be the most likely structures, whereas 2 was eliminated by UV and mass spectrometry. It was possible to predict the C(1s) ESCA spectra for structures 1 and 3 based on studies of model compounds in which carbon was bound to sulfur and oxygen. These predictions are shown in Figure 4A for structures 1 and 3 on the left and right, respectively. The C(1s) spectrum for the isolated material is shown in Figure 4B. The observed spectrum corresponds closely to that predicted for structure 1, which represents positive evidence that 1 is the major component in the fraction.

Gelius et al. (8) also studied the addition reaction of α-angelica lactone (structure 4) and were able to distinguish between two possible structures for the product:

The C(1s) spectra of the indicated carbons (*) form the basis for the distinction between structures 5 and 6. By use of previous C(1s) data (8), group shifts were used to predict the electron spectra for both possibilities based on the relative abundance of the particular carbon atoms in the compound. The predicted spectra for structures 5 and 6 are shown on the left and right, respectively, in Figure 4C. Figure 4D is the experimental C(1s) spectrum for the reaction product. The similarity of the experimental spectrum and that predicted for structure 5 clearly indicates that structure 5 is the correct structure of the reaction product.

Nitrogen. The N(1s) binding energies were reported for a large number of nitrogen compounds (1, 11). Siegbahn and coworkers (1) found that the N(1s) binding energies correlate well with the charge on the nitrogen atom calculated via Pauling's concept of partial ionic character. Results similar to those for carbon are found in that the nitrogens bearing the most positive charge have the highest 1s binding energies (NO_3^- at 407.2 eV), whereas the lowest N(1s) binding energies are found for the most negative nitrogens [$(\bar{N}=\overset{+}{N}=\bar{N})^-$ at 399.2 eV or $HN\text{-}COO^-Na^+$ at 397.3 eV]. Spectra for compounds containing several equivalent nitrogens contain only one N(1s) electron line. If two or more inequivalent nitrogens are present, multiple N(1s) lines are found in the ESCA spectrum.

From the N(1s) data reported by Nordberg et al. (11) and Siegbahn et al. (1), it is possible to construct a correlation chart for nitrogen compounds similar to that for carbon compounds (12). As illustrated in Figure 5, nitrogens within a given functional group have similar N(1s) binding energies. With the exception of the nitro group, the correlations in this chart were established with a limited number of compounds. However, it is possible to get a general idea of the N(1s) binding energy to be expected for a given group. For instance, the ammonium group lies within a range of a little more than 1 eV, 401–402 eV, whereas the nitro group lies within an even smaller range at higher binding energy.

Much more N(1s) binding energy data must be accumulated before this correlation chart can be used with the same degree of confidence as the Colthup chart for structure determinations by use of IR spectroscopy. This is clearly evident from the study of Jack and Hercules (13) which expanded the N(1s) data for quaternary nitrogen. The N(1s) binding energies were measured for more than 50 quaternary nitrogen compounds. The compounds studied included tetra-alkylammonium and mono-, bi-, and tricyclic aromatic quaternary nitrogens. The anions in this series of compounds covered a wide range: F^-, Cl^-, PF_6^-, I^-, Br^-, BH_4^-, ClO_4^-, BF_4^-, p-tosyl$^-$, and Br_3^-. The range of N(1s) binding energies was greater than 5 eV or more than twice the range indicated in the correlation chart in Figure 5. The anion obviously plays a major role in determining the binding energy of the quaternary nitrogen atom. If this 5-eV spread is compared to the 12-eV spread for all of the nitrogen compounds, great care must be taken in using single-atom correlations for determining structure. It is suggested that greater success might be expected if a multiple-atom approach is used (13).

Phosphorus. Phosphorus is another element which has been extensively studied. In a study of more than 50 phosphorus compounds, Pelavin et al. (14) found the total spread of P(2p) binding energies to be 8 eV. Since this total spread is smaller than that observed for N(1s) binding energies (12 eV), single-atom correlations would be more difficult than for nitrogen.

There are instances where ESCA can provide data about complex formation which were previously difficult or impossible to obtain. Transition metal complexes are thought to involve simple σ-type bonds in which electron pairs are shared as well as bonds formed by back-donation of electrons from the metal to empty ligand orbitals. Such transfer or donation of electrons will alter the electron density about both the metal and the ligand. Chemical shifts in ESCA spectra occur as a result of the dependence of core-electron binding energies on the electron density about the atom in question. Therefore, ESCA is a convenient technique for studying back-donation phenomena in transition metal complexes.

Blackburn et al. (15) studied back-donation in a series of complexes containing triphenylphosphine, ϕ_3P, as the ligand. When the P(2p) binding energies of the complexed and free ϕ_3P were compared, the binding energies in the complexes were either equal to or very slightly less than those in free ϕ_3P. Such behavior can be explained by π back-donation of electrons from the metal to the phosphorus. The resulting shift in electron density nearly cancels that caused by the σ-bonding which involves a shift of electrons from phosphorus to the metal. Since the P(2p) binding energies in the complexes are slightly smaller, the π back-donation occurs to a slightly greater extent than the σ-donation.

The study of the bis-(triphenylphosphine)iminium cation, $\{[(C_6H_5)_2P]_3\}^+$ (abbreviated PPN), by Swartz et al. (16) illustrates the use of ESCA in structural investigations. The N(1s) and P(2p) binding energies were measured for a series of simple salts, (PPN)X, where X was F^-, Cl^-, Br^-, I^-, SCN^-, N_3^-, and OCN^-.

Since ^{31}P NMR (17) and ESCA indicate the presence of only one type of phosphorus atom, any unsymmetrical structure for the PPN cation such as

$$[(C_6H_5)_3P = \overset{+}{N} - P(C_6H_5)_3]^+$$

can be eliminated.

X-ray crystallographic data (18) indicate the P—N—P bond angle to be 137–142°, decreasing the likelihood of formal phosphorus nitrogen double bonds. Proton NMR studies suggest the presence of some degree of electron delocalization over the entire cation (18).

The P(2p) binding energies for the PPN salts are very similar to those measured for quaternary phosphonium salts in which the phosphorus atom bears a large fractional positive charge. The similarity of the P(2p) binding energies in the two classes of compounds implies that the PPN phosphorus atoms are also largely positive.

The N(1s) binding energy in the PPN salts is approximately constant and equal to that of the negative nitrogens in the azide ion (1). Thus, the PPN nitrogen must have a large degree of negative charge. These data indicate that the structure of the PPN cation is:

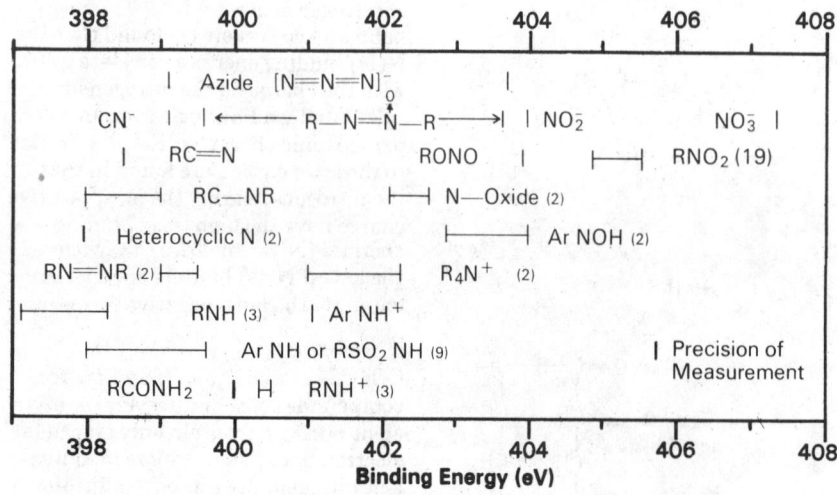

Figure 5. Correlation chart for N(1s) electron binding energies and organic functional groups. ⊢——⊣ = Range of observed energies. Number of compounds used in correlation is in parentheses [from Hercules (12)]

Sulfur. More ESCA data have been reported for sulfur than for any other element. The first papers reporting ESCA chemical shifts and their possible use in chemistry concerned sulfur. The early data on sulfur were summarized by Siegbahn et al. (1) and resulted in the first correlation chart of binding energies and functional group (12). Groups such as ethers, disulfides, sulfoxides, sulfones, and sulfonates showed narrow ranges of S(2p) binding energies.

The most extensive study of sulfur compounds was reported by Lindberg et al. (19). The S(2p) binding energies were determined for 136 sulfur-containing compounds. A correlation chart resulting from this work is shown in Figure 6. The range of S(2p) binding energies is approximately 10 eV, making possible the use of the chemical shift for structural studies in sulfur chemistry.

A qualitative correlation between the classical oxidation number of sulfur and its 2p binding energy relative to free sulfur is observed in that sulfides have the lowest S(2p) binding energies, whereas sulfates have the highest. A given type of compound has a range of S(2p) binding energy that is usually less than 1 eV.

Several additional aspects were also discussed by Lindberg et al. (19). A reasonable correlation is observed between experimental S(2p) binding energy and calculated charge. This linear correlation indicates the possibility of using shifts to characterize molecules. A study of a homologous series of nitrobenzenes attached to various sulfur-containing groups indicates that secondary substituent effects are usually insignificant. The S(2p) binding energies for atoms that are indirectly exposed to the strong electron-withdrawing power of the nitro group show no significant deviation from the linear plot obtained for the general correlation.

Other Elements. Although ESCA studies have been performed on most of the other elements, they have not been as extensive as those involving carbon, nitrogen, sulfur, and phosphorus. Several of these investigations will be briefly discussed below.

The Si(2p) binding energies were reported for 16 silicon compounds (20). Both inorganic and organosilicon compounds were studied. If elemental silicon and silicon bonded to fluorine, which exhibit extreme values, were excluded, the total range in Si(2p) binding energy was 3 eV. Such a narrow range of chemical shifts implies that ESCA probably will not be as useful a structural tool for silicon as it is for C, N, and S. Perhaps additional data are necessary.

The B(1s) binding energies were reported for 25 boron compounds (21). A binding energy range of approximately 8 eV was found for this series containing both organic and inorganic boron compounds. Again, eliminating the extremes ($NaBF_4$ at 195.1 eV and B_4C at 186.7 eV) implies that ESCA may not be extremely useful for structural analysis of boron compounds. For example, boric acid and the organoborates exhibit only slightly more than a 1-eV range of B(1s) binding energy.

Hendrickson et al. (21) found the Cr(3p) binding energies to range from 48.7 eV for $K_2Cr_2O_7$ to 43.2 eV for elemental chromium in a study of 16 compounds. A qualitative correlation between binding energy and the oxidation state of chromium was observed. The Cr(3p) binding energies in $K_3[Cr(CN)_6]$ and $K_3[Cr(CN)_5(NO)]$ were equivalent, indicating the latter to be a compound containing Cr(III) rather than Cr(II) as originally believed.

Matienzo et al. (22) demonstrated the ability of ESCA to differentiate between paramagnetic and diamagnetic isomers. The $[n\text{-}C_3H_7P(C_6H_5)_2]NiBr_2$ molecule crystallizes in either a square planar or tetrahedral geometry. The square planar form is diamagnetic, whereas the tetrahedral form is paramagnetic. Paramagnetic molecules exhibit satellite lines in their photoelectron spectra which are associated with valence electron excitations (e.g., 3d → 4s) owing to the perturbation of the central potential upon photoionization of an inner shell electron (23–25).

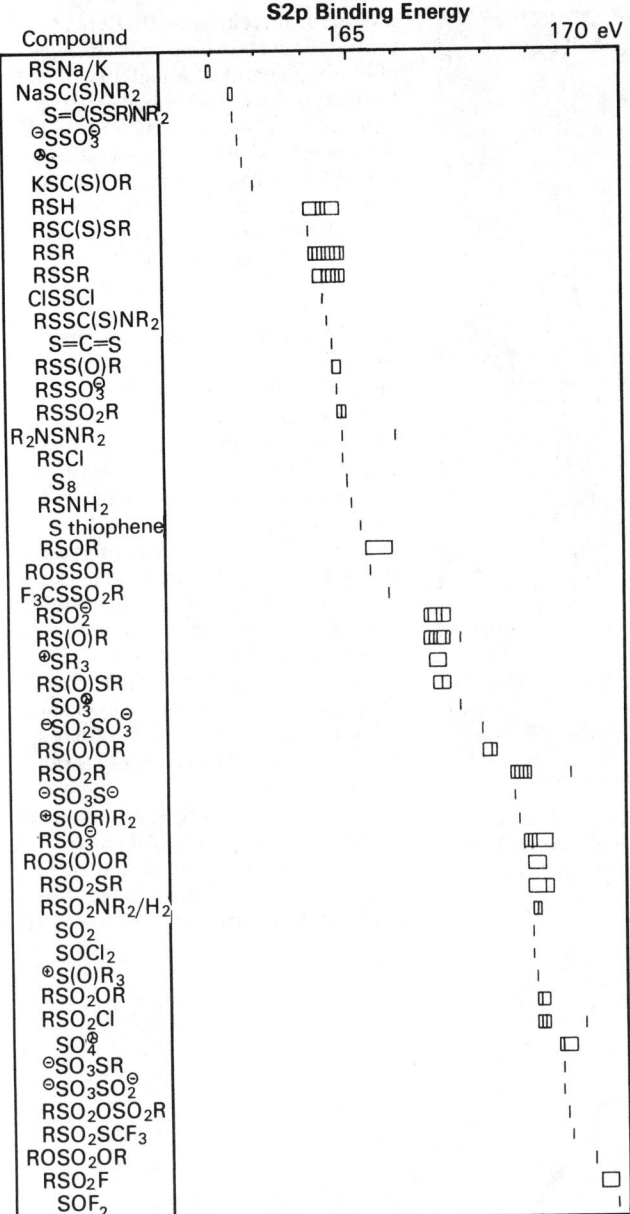

Figure 6. Correlation between S(2p) electron binding energies and chemical structure [from Lindberg et al. (19)]

These "shake-up" satellites occur in paramagnetic nickel compounds whenever the emission of a 2p photoelectron takes place simultaneously with the excitation of a valence electron. The satellite photoelectrons are then emitted with kinetic energies lowered by the amounts of the allowed "shake-up" transitions. "Shake-up" phenomena cannot occur in diamagnetic molecules; therefore, no satellite lines appear in their photoelectron spectra.

Figure 7 shows the Ni(2p) photoelectron spectra obtained for the diamagnetic square planar (A') and the paramagnetic tetrahedral form (B') of $[n\text{-}C_3H_7P(C_6H_5)_2]NiBr_2$. The presence of "shake-up" satellites in the spectrum of the tetrahedral form and the absence of them in that of the square planar form enable one to easily differentiate one isomer from the other.

To summarize the use of ESCA as

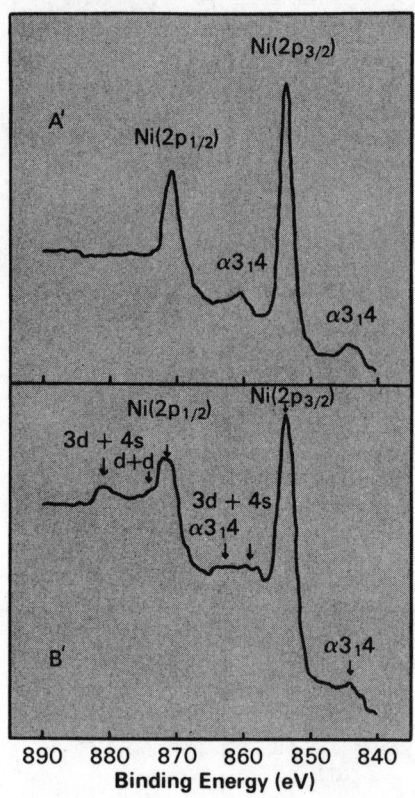

Figure 7. Ni(2p) photoelectron spectra of (A′) square planar and (B′) tetrahedral [n-C$_3$H$_7$P(C$_6$H$_5$)$_2$]$_2$NiBr$_2$ [from Matienzo et al. (22)]

a structural technique, it offers potential for certain elements. Ranges of binding energy of about 10 eV have been reported for carbon, nitrogen, and sulfur. These data indicate a definite potential of ESCA as a structural tool for these elements. However, considerably smaller ranges have been observed for phosphorus, silicon, and boron. Therefore, use of ESCA for structural studies of these elements is of more limited value. In general, certain elements in a functional group seem to have narrowly defined limits of binding energy, but others seem to vary widely. Therefore, it is unlikely that ESCA single-element correlation charts, similar to Colthup charts in the infrared, will have practical utility. Perhaps the most significant structural information will be obtained by taking a multielement approach. This would require measurement of the binding energies of all the atoms in a functional group and those adjacent to the group. However, more experimental data are necessary before such correlations can be accurately established.

Quantitative Analysis. As indicated, ESCA also has potential as a quantitative technique in that the intensity of a photoelectron line is proportional to not only the photoelectric cross section of a particular element, but also to the number of atoms of that particular element that are present in the sample. In some early work, Siegbahn et al. (1) first demonstrated the quantitative aspect of ESCA by analyzing some organic compounds for carbon, chlorine, and sulfur. The analyses were accurate to within 5–10%. The ability of ESCA to quantitatively analyze mixtures of MoO$_2$ and MoO$_3$ was recently demonstrated (26). Before the advent of ESCA, no instrumental technique existed that was capable of performing this analysis. By measuring the intensities of photoelectron lines in spectra obtained from known mixtures of MoO$_2$ and MoO$_3$ at binding energies that corresponded to each oxide, a calibration curve was constructed (Figure 8). Even though the surface of MoO$_2$ was significantly contaminated with MoO$_3$, a linear calibration curve resulted. A least-squares analysis of the curve indicated the error to be ±2% MoO$_2$. A series of analyses performed on synthetic unknowns yielded results consistent with this accuracy. The authors also indicate that PbO–PbO$_2$, Cr$_2$O$_3$–CrO$_3$, and As$_2$O$_3$–As$_2$O$_5$ mixtures could also be measured quantitatively.

Quantitative estimates of the total protein content of various grains can be made by measuring the intensities of the nitrogen and sulfur peaks (27).

Surface Analysis. ESCA perhaps has its greatest potential utility as a technique for the study of surfaces. Some exploratory studies in catalyst research were reported by Delgass and coworkers (28). The authors observed different nitrogen-containing molecules on high surface area zeolites and were able to use ESCA to qualitatively measure the effect of dispersion on the oxidizability of platinum metal supported on SiO$_2$ and on unsupported platinum. Spectra are also reported for Eu^{+3}-exchanged zeolites and supported Cu and Ni that give evidence of in situ reduction after treatment with hydrogen at elevated temperatures. Chemical changes in an oxide surface after use as a catalyst have also been observed.

In a study of copper oxide supported on alumina, Wolberg et al. (29) demonstrated the feasibility of using ESCA to characterize the structure of supported surface phases. In another study, ESCA spectra were obtained for calcined and uncalcined Mo–Al$_2$O$_3$ systems (30). After calcining, the Mo(3d$_{3/2}$–3d$_{5/2}$) doublet was no longer resolved owing to an interaction between the Mo and the alumina.

Kim et al. (31) studied the chemisorption of oxygen on a platinum sur-

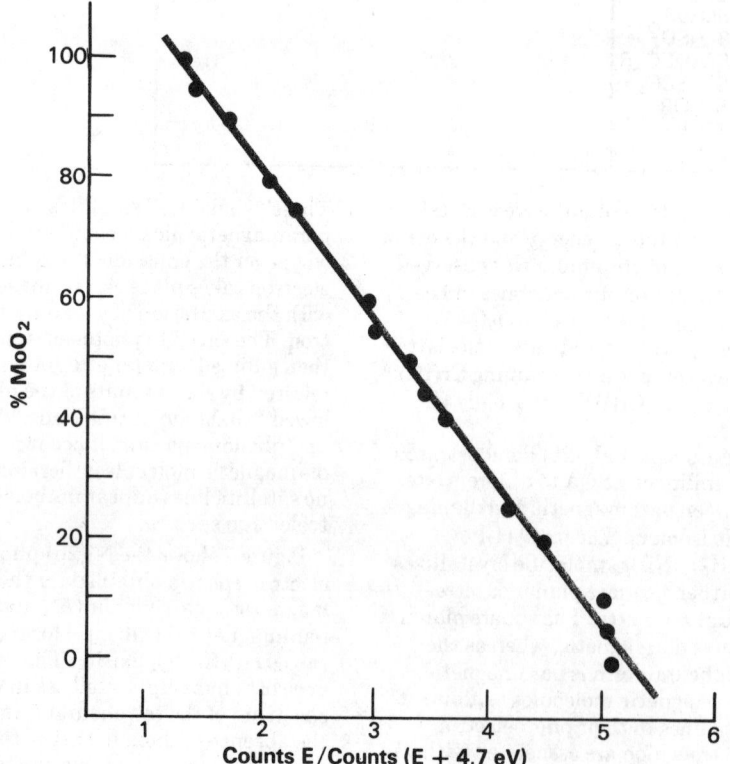

Figure 8. Calibration curve for quantitative analysis of mixtures of MoO$_2$ and MoO$_3$ [from Swartz and Hercules (26)]

face. Platinum oxide surfaces prepared by reaction with molecular oxygen or electrochemical oxidation were compared with bulk samples of various platinum oxides to identify the oxides formed in the two processes.

ESCA studies of oxygen adsorption on nickel were reported (32). The Ni(2p) spectrum of an oxidized nickel foil was characteristic of a mixture of nickel oxide and metallic nickel. The O(1s) spectrum exhibited two oxygen lines. After the temperature was raised to 300°C, the oxygen signal decreased in intensity, and the low binding energy line disappeared. In addition, no oxide structure appeared in the Ni(2p) spectrum. After exposure to the atmosphere, one intense oxygen signal appeared owing to adsorbed oxygen. These results indicate the potential for studying surface adsorption phenomena.

Polymers. Riggs and Fedchenko (33) reported an ESCA study of a series of polymers ranging from polyethylene to polytetrafluoroethylene. The stepwise substitution of one fluorine per monomer unit into the polymer was studied to achieve some understanding of the chemical shift effect in the system. The study found that the C(1s) binding energies fall into discrete regions depending on the type of carbon observed. The $—CH_2-$ carbons have binding energies which do not overlap with those corresponding to $—CHF-$carbons; these, in turn, are at lower energies than those of the $—CF_2-$carbons. Thus, by measuring the C(1s) binding energies, one can unequivocally identify the types of carbon present and, therefore, the structure of the fluoropolymer.

Pollution. Novakov and coworkers (34) described the application of ESCA to the determination of lead and the chemical states of sulfur and nitrogen in smog particles as a function of particle size and time of day in an attempt to elucidate the dynamics of atmosphere and aerosol interactions. They found the nitrogen to be present mainly as NO_3^- and NH_4^+ with significant amounts of amino- and pyridino-nitrogen. The sulfur was present in the S^{+4} and S^{+6} states. A similar study was reported by Araktingi et al. (35) in which ESCA is employed to determine the exact nature of the lead present in particulate pollutants.

Hulett and Carlson (36) compared the As(3d) binding energy measured in a soil sample with those in bulk arsenic compounds. They then estimated the charge state of the arsenic in the soil sample. These data demonstrate the potential of ESCA for studying herbicides and pesticides dispersed throughout our environment.

Summary

X-ray photoelectron spectroscopy is an exciting new analytical technique with both qualitative and quantitative applications to all of the elements except hydrogen. Its rapid growth is evidenced by the birth of symposia concerned solely with ESCA and by the somewhat phenomenal growth of the literature in the field. This rapid growth can be tied directly to the availability of ESCA instrumentation. "State of the art" instrumentation enables the nondestructive analysis of solid, "liquid," and gaseous samples. Although the resolving power is restrictive at times, the development of X-ray monochromators for use with ESCA spectrometers has already begun to remedy this situation.

The applications discussed in this report attest to the wide applicability of ESCA to chemical problems. The number of applications will continue to expand as the number of ESCA spectroscopists increases. At the present time, there may be more data awaiting publication than have thus far appeared in the literature.

References

(1) K. Siegbahn et al., ESCA, "Atomic, Molecular and Solid State Structure Studied by Means of Electron Spectroscopy," Almqvist and Wiksells, Uppsala, Sweden, 1967.
(2) R. G. Steinhardt, J. Hudis, and M. L. Perlman, *Phys. Rev. B*, **5**, 1016 (1972).
(3) Y. Baer, P. F. Heden, J. Hedman, M. Klasson, and C. Nordling, *Solid State Commun.*, **8**, 1479 (1970).
(4) S. Hufner, R. L. Cohen, and G. K. Wertheim, *Phys. Scr.*, **5**, 91 (1972).
(5) L. I. Yin, S. Ghose, and I. Adler, *Appl. Spectros.*, **26**, 355 (1972).
(6) K. Hamrin, G. Johansson, U. Gelius, A. Fahlman, C. Nordling, and K. Siegbahn, *Chem. Phys. Lett.*, **1**, 613 (1968).
(7) D. W. Turner, C. Baker, A. D. Baker, and C. R. Brundle, "Molecular Photoelectron Spectroscopy," Wiley-Interscience, London, England, 1970.
(8) U. Gelius, P. F. Heden, J. Hedman, B. J. Lindberg, R. Nordberg, C. Nordling, and K. Siegbahn, *Phys. Scr.*, **2**, 70 (1970).
(9) S. H. Hercules and D. M. Hercules, *Rec. Chem. Progr.*, **32**, 183 (1971).
(10) J. Hedman, P. F. Heden, R. Nordberg, C. Nordling, and B. J. Lindberg, *Spectrochim. Acta*, **26A**, 761 (1970).
(11) R. Nordberg, R. G. Albridge, T. Bergmark, U. Ericson, J. Hedman, C. Nordling, K. Siegbahn, and B. J. Lindberg, *Ark. Kemi*, **28**, 257 (1968).
(12) D. M. Hercules, *Anal. Chem.*, **42** (1), 20A (1970).
(13) J. J. Jack and D. M. Hercules, *ibid.*, **43**, 729 (1971).
(14) M. Pelavin, D. N. Hendrickson, J. M. Hollander, and W. L. Jolly, *J. Phys. Chem.*, **74**, 1116 (1970).
(15) J. R. Blackburn, R. Nordberg, F. Stevie, R. G. Albridge, and M. M. Jones, *Inorg. Chem.*, **9**, 2374 (1970).
(16) W. E. Swartz, Jr., J. K. Ruff, and D. M. Hercules, *J. Amer. Chem. Soc.*, **94**, 5227 (1972).
(17) R. Appel and A. Hauss, *Z. Anorg. Allgem. Chem.*, **311**, 290 (1961).
(18) L. B. Handy, J. K. Ruff, and L. F. Dahl, *J. Amer. Chem. Soc.*, **92**, 7312, 7327 (1970).
(19) B. J. Lindberg, K. Hamrin, G. Johansson, U. Gelius, A. Fahlman, C. Nordling, and K. Siegbahn, *Phys. Scr.*, **1**, 286 (1970).
(20) R. Nordberg, H. Brecht, R. G. Albridge, A. Fahlman, and J. R. Van Wazer, *Inorg. Chem.*, **9**, 2469 (1970).
(21) D. N. Hendrickson, J. M. Hollander, and W. L. Jolly, *Inorg. Chem.*, **9**, 612 (1970).
(22) L. J. Matienzo, W. E. Swartz, Jr., and S. O. Grim, *Inorg. Nucl. Chem. Lett.*, **8**, 1085 (1972).
(23) T. A. Carlson and M. O. Krause, *Phys. Rev.*, **140**, A1057 (1965); M. O. Krause, T. A. Carlson, and R. D. Dismukes, *ibid.*, **170**, 37 (1968).
(24) T. Novakov, *Phys. Rev. B*, **3**, 2693 (1971).
(25) G. K. Wertheim and A. Rosencwaig, *Phys. Rev. Lett.*, **26**, 1179 (1971).
(26) W. E. Swartz, Jr., and D. M. Hercules, *Anal. Chem.*, **43**, 1774 (1971).
(27) M. Klein and L. Kramer, Impr. Plant Protein Nucl. Tech., Proc. Symp., 242, 1971.
(28) W. N. Delgass, T. R. Hughes, and C. S. Fadley, *Catal. Rev.*, **4**, 179 (1970).
(29) A. Wolberg, J. Olgivie, and J. Roth, *J. Catal.*, **19**, 86 (1970).
(30) A. Miller, W. Atkinson, M. Barber, and P. Swift, *ibid.*, **22**, 140 (1971).
(31) K. S. Kim, N. Winograd, and R. E. Davis, *J. Amer. Chem. Soc.*, **93**, 6296 (1971).
(32) G. Scheon and S. T. Lundin, *J. Electron. Spectrosc.*, **1**, 105 (1973/73).
(33) W. M. Riggs and R. P. Fedchenko, *Amer. Lab.*, **4**, 65 (1972).
(34) T. Novakov, P. K. Mueller, A. E. Alcocer, and J. W. Otvos, *J. Colloid Interface Sci.*, **39**, 225 (1972).
(35) Y. E. Araktingi, N. S. Bhacca, W. G. Proctor, and J. W. Robinson, *Spectrosc. Lett.*, **4**, 365 (1971).
(36) L. Hulett and T. A. Carlson, *Appl. Spectrosc.*, **25**, 33 (1971).

William E. Swartz, Jr., received a BS degree in chemistry from Juniata College in 1966. He earned a PhD from the Massachusetts Institute of Technology in analytical chemistry in 1971 where he studied with D. M. Hercules. After postdoctoral study with E. R. Lippincott at the University of Maryland, he moved to the University of South Florida as assistant professor in September 1972. Dr. Swartz is a member of the ACS, Society for Applied Spectroscopy, and Sigma Xi. His research interests include the application of X-ray photoelectron spectroscopy to heterogeneous catalysis, chemisorption phenomena, minerals, and inorganic complexes.

James R. Allkins
15485 One Oak Lane
Monte Sereno, Calif. 95030

Tunable Lasers in Analytical Spectroscopy

It has long been the dream of many spectroscopists to have available a high-intensity source of monochromatic radiation that could be tuned to any desired wavelength at will. Such a source is the tunable laser.

It is not surprising, therefore, that in the nine short years since Sorokin and Lankard first observed stimulated coherent emission from a laser-pumped organic dye (1) and Schäfer et al. (2) realized the tunability of such a source, the tunable dye laser has experienced an extraordinary growth in its development (3–7). At present, owing in part to anticipated applications and in part to strong commercial competition, continuously tunable dye lasers are available at a cost that is well within the budget of most industrial laboratories and with an ease of operation that is comparable to that of fixed frequency gas lasers.

Before becoming too engrossed, however, it should be recognized that the tunable dye laser is not the immediate answer to all problems, and despite its tremendous potential, it still has certain drawbacks. The first is that the effective tuning range is restricted to the visible region of the electromagnetic spectrum (typically 750–350 nm). Tunable ultraviolet radiation can be and is obtained by frequency doubling the visible radiation inside the laser cavity with a suitable nonlinear crystal (8), but even then the upper limit of the tuning range is usually confined to around 260 nm. Secondly, to cover the entire visible region entails the use of many different dyes; therefore, various dye solutions have to be switched in and out of the pumping area as each region is scanned.

Tunable sources of coherent radiation other than dye lasers are available for other regions of the electromagnetic spectrum (9–12), particularly the infrared. Semiconductor diode lasers and spin-flip Raman lasers emit in the "information-rich" infrared fingerprint region (4000–400 cm^{-1}). Unfortunately, they too have their problems such as requiring cryogenic cooling and high-power magnetic fields to operate efficiently. They also cover a rather limited tuning range at any one time. However, their potential applications are so far reaching that these restrictions should be overcome in the very near future. Other tunable laser sources such as parametric oscillators and frequency mixing devices have, so far, been confined to the laboratory bench since they require a very stable temperature environment and very fine positioning controls. They also should achieve commercial viability once potential applications have been developed.

In light of the above preamble, it is not surprising that most applications of interest to chemists with tunable laser sources have been performed with the tunable dye laser. The major part of this article will, therefore, be concerned with the impact of tunable dye lasers on analytical spectroscopy. In particular, applications which have taken full advantage of the laser's unique properties of high intensity, narrow spectral bandwidth, coherence, small beam diameter, and small divergence directionality will be discussed. The potential impact of other tunable sources will serve as a conclusion.

Tunable Dye Laser

A simplified version of a typical dye laser is shown in Figure 1. Both the etalon for frequency narrowing and the nonlinear crystal for intracavity frequency doubling can be inserted in the laser cavity if and when required. To cover most of the near UV–VIS tuning range requires that a series of dyes in solution flow sequentially through the dye cell, since each dye only emits effectively over a limited range of some 20–50 nm. Alternatively, different dye cells containing different dyes can be swung in and out of the pumping cavity as required. Typical tuning elements and pumping sources are listed in Table I.

Dyes found most effective for stimulated emission are the coumarins, rhodamines, cresyl violet, and fluorescein. Factors that could reduce the efficiency of the lasing process such as nonradiative transfers or dimer formation by the dye are minimized by the addition of quenchers and deaggregating agents to the dye solution. A comprehensive discussion of various factors that affect the efficiency of dye lasers has been published recently by Drexhage (13).

Applications of Tunable Dye Lasers

Before becoming too deeply involved in the various applications of tunable dye lasers in analytical spec-

---

Table I. Typical Tuning Elements and Pumping Sources for Dye Laser Systems

(Flashlamp and N_2 laser pumping is usually by way of P_2, whereas other lasers pump axially along P_1; see Figure 1)

Pumping sources	Tuning elements
Flashlamp	Grating
N_2 laser	Prism/mirror assy
Freq. dbld. ruby laser	Birefringent filter
Freq. dbld. Nd: YAG	Tilted etalon[a]
Freq. dbld. Nd: glass	Wedged etalon
Another dye laser	Lyot filter[a]
CW argon laser	Acousto-optic filter
Xenon laser	Interferometer[a]

[a] Also used as frequency narrowing devices.

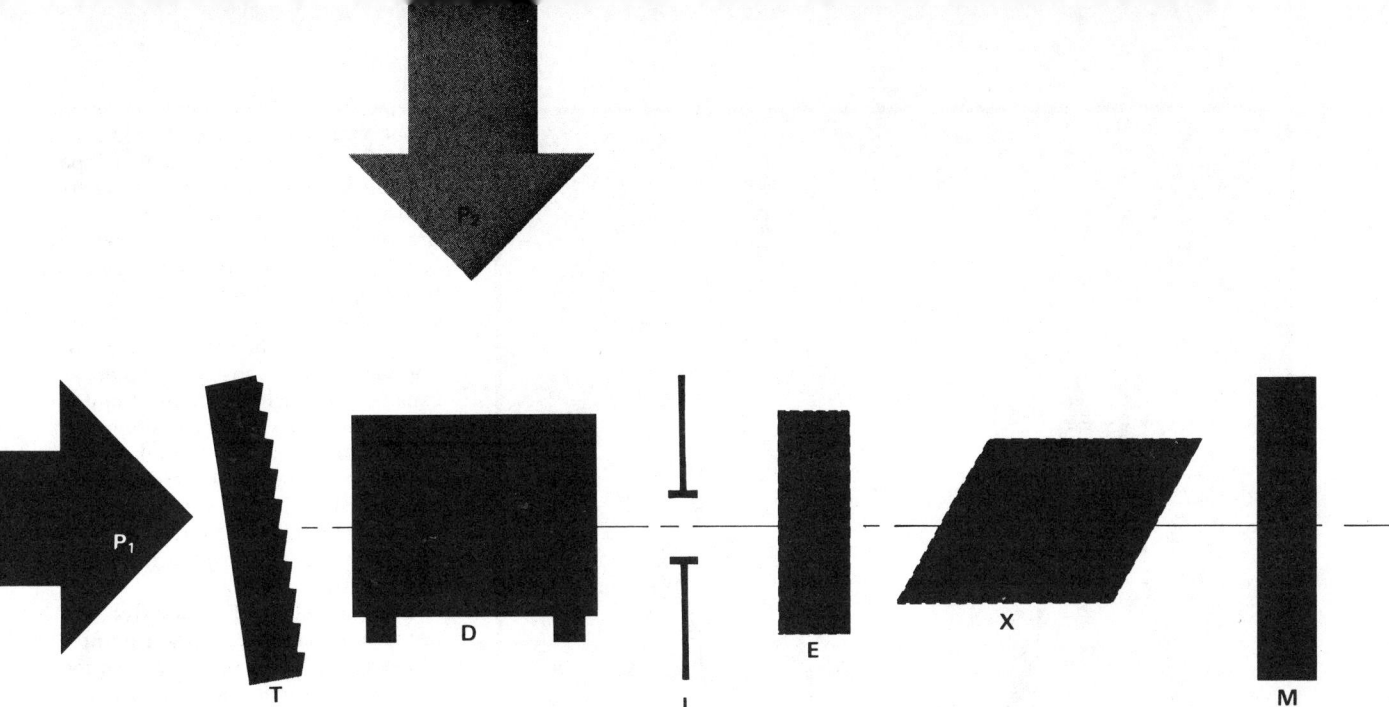

Figure 1. Schematic of typical dye laser
Solutions of dyes in dye cell D are pumped by external sources P_1 or P_2 to produce stimulated emission. Generated emission is wavelength tuned by tuning element T and frequency narrowed by optional etalon E. Nonlinear crystal X is inserted into cavity when tunable radiation is required in ultraviolet. A small fraction of tuned radiation is transmitted by output mirror M, the remainder being reflected for further optical amplification. An iris I is often inserted in cavity to control mode characteristics of intracavity beam

troscopy, it appears appropriate to define more fully those qualities of the laser that make it unique from other, more conventional, sources of radiation. The primary property of the laser is that all of the available laser radiation is confined in a physically very narrow, low divergence, directional beam that can be placed precisely at any point in a required area. Furthermore, the beam possesses spatial coherence which means that it can be focused to a very small spot at any point along its path with a concomitant increase in power density. Not only is the beam narrow in size (spatially), but it is also narrow in frequency spread (spectrally). This means that most of the radiation intensity is contained in a narrow band of frequencies around a central dominant mode. Furthermore, this bandwidth of frequencies can be made even narrower by the insertion of suitable etalons into the laser cavity. The narrow bandwidth is extremely useful for high-resolution spectroscopy since source broadening limitations are almost completely removed. Also, the loss of energy accompanying frequency narrowing of a standard source by filtering through very narrow slits of a standard monochromator is not encountered with the laser, since almost all of the available energy of the laser is channeled into the narrow band of frequencies selected. Thus, we have a narrow beam of high-intensity radiation concentrated in a narrow band of frequencies that can be directed to almost any small area in a system without any substantial loss of power in getting there.

Finally, there are the temporal properties of a laser, that is, the variations of output power with time. Tunable lasers fall into two broad classes, namely pulsed or CW (continuous wave). With pulsed dye lasers, i.e., those pumped by a pulsed excitation source, the available energy emerging from the laser is confined in a stream of short bursts or pulses. Pulse characteristics are generally determined by the temporal characteristics of the pumping source. With pulsed nitrogen laser pumped dye lasers, the pulse lengths tend to be short (nsec), whereas the peak powers of each pulse tend to be low. With flashlamp pumping the pulses tend to be longer (μsec), but the peak powers tend to be high. Flashlamp pumping, therefore, gives more energy per pulse, but nitrogen lasers have faster repetition rates (more pulses per sec). Pulsed lasers are used to great advantage in studies of time-related phenomena such as lifetimes, relaxation rates, reaction kinetics, or in studies involving nonlinear optical phenomena in which high peak powers are required. Pulsed lasers may be operated in either a single pulse mode or repetitively in which a stream of pulses is generated with a regular time interval between each pulse.

Continuous wave (CW) lasers are pumped by CW sources such as a CW argon ion laser. Although the power output of a CW laser cannot attain the peak power of a pulsed laser, the CW output is continuous and therefore more akin to the conventional sources of radiation used in most spectroscopic techniques. The average power output of a CW laser is usually higher than that of a pulsed laser, but, more importantly, the power stability of a CW source is usually much greater. The power output from a pulsed laser can vary considerably from pulse to pulse. This makes the CW laser especially useful in applications which demand a steady source of very stable power contained in a very narrow bandwidth of frequencies.

The following applications exemplify one or more areas where the properties of a tunable laser have proven superior, or at least equivalent, to those of previously used conventional sources. Advantages found with the laser will include greater sensitivity to improve limits of detection, greater resolution to reveal more information content, and better selectivity to aid in identification.

Absorption and Fluorescence Spectroscopy

Two of the more interesting properties of a tunable laser are its high spectral brightness and narrow frequency bandwidth. This implies that both high-sensitivity and high-resolution spectroscopy should be natural candidates for tunable laser applications. The sodium atom has received particular attention in this respect, since two of its primary atomic absorption lines (D_1 and D_2) lie close in wavelength to the emission peak of one of the more efficient lasing dyes, namely, rhodamine 6G.

To demonstrate the potential detec-

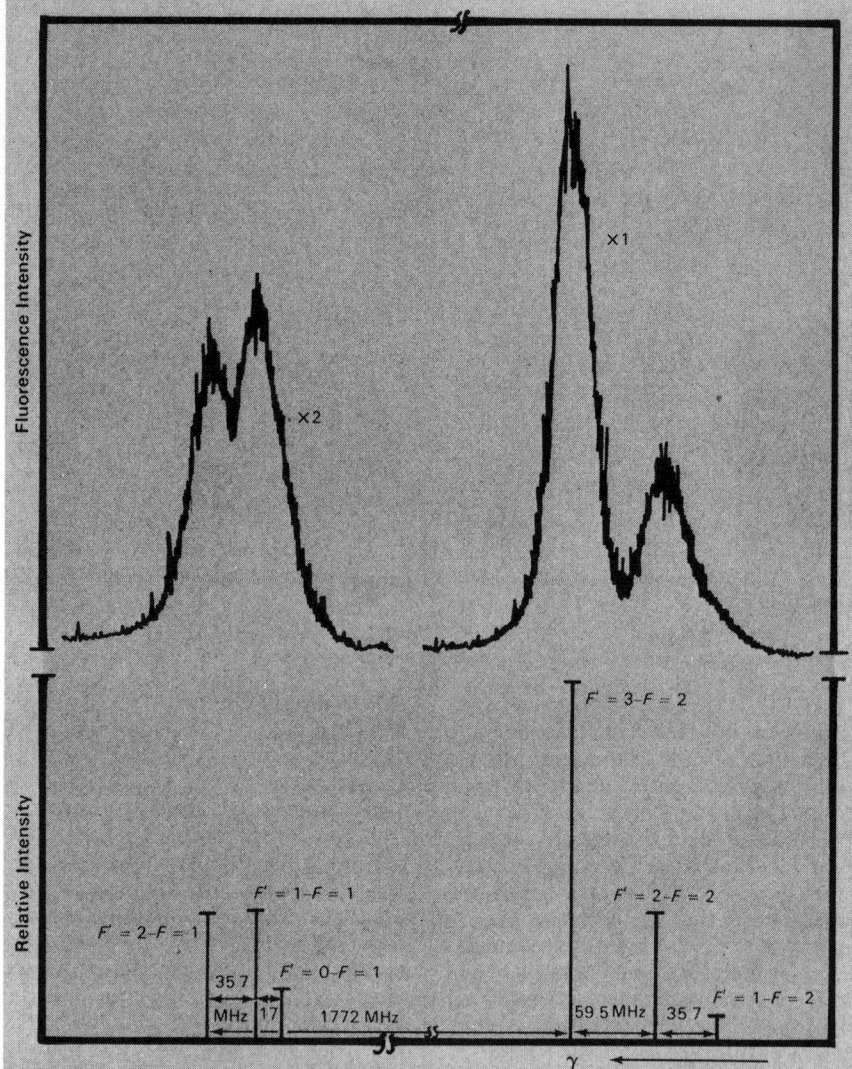

Figure 2. High-resolution spectrum showing hyperfine structure of sodium D_2 line recorded with CW dye laser

Lower part shows theoretical structure. F' and F denote quantum numbers of total angular momentum of hyperfine levels of $3\,^2P_{3/2}$ and $^2S_{1/2}$ levels, respectively
Figure taken from ref. 15 and reproduced with permission of Springer-Verlag, New York

tion sensitivity of a system involving a dye laser as its source, Jennings and Keller (14) observed the laser-excited fluorescence emission from a sample of sodium contained as a vapor in a simple heated cell. Using a frequency narrowed (~0.003 nm) CW dye laser with 50-mW output power at 589.6 nm as the source, they were able to detect as little as 0.016×10^{-15} gram (4.2×10^5 atoms) of sodium vapor. Detection was by simple visual observation of the fluorescence emission. Their system has since been modified to increase the sensitivity by a factor of 200 (5), which is an improvement in sensitivity of approximately 10^6 over sodium detection by vapor lamp excitation.

In the area of high-resolution spectroscopy, considerable interest has been generated in the past two years since this is one area in which the laser can show dramatic improvements over conventional techniques. As the linewidth of the laser can be frequency narrowed to a value of approximately 1/100th of that normally observed for a single atomic absorption line, hyperfine structure can be easily resolved with a laser source. To obtain high-resolution spectra requires a source with small bandwidth, excellent frequency stability, and good stable tunability. The frequency stability must be appreciably better than the desired resolution, and the laser must be able to tune smoothly without mode hopping. Figure 2 illustrates the high-resolution capabilities of a frequency stabilized CW dye laser as recorded by Hartig and Walther (15) for the sodium D_2 line. Demonstrated resolution was of the order of 3.5×10^{-5} nm (30 MHz) [this has since been improved to 10 MHz (6)], and the frequency of the laser was stabilized to ±1.5 MHz. Hyperfine structure is clearly resolved. To observe such hyperfine structure requires not only that the source be narrow, but also that line broadening of the sample be reduced to a minimum. The two primary causes of spectral line broadening in the visible are collision broadening and Doppler broadening. Collision broadening may be reduced by lowering the pressure in the sample cell. The high spectral brightness of the laser can more than compensate for the corresponding reduction in atom density. Doppler broadening, arising from axial motion of the sample atoms in the sampling beam, can be reduced by such techniques as saturation spectroscopy (16), by confining the sample in an atomic beam (15), or by the relatively new technique of two-photon absorption spectroscopy (17).

Using a special jet stream dye-flow system and a CW argon laser pump, Wu et al. (18) have recently recorded the hyperfine structure of I_2 with a resolution of one part in 10^9. The iodine sample was contained in a molecular beam, and the laser stabilized to six parts in 10^{13}.

On a more practical note, Gibson and Sandford have taken advantage of the sodium doublet:rhodamine 6G intensity match to remotely monitor sodium concentrations in the atmosphere. They were particularly interested in variances of the naturally occurring sodium layer found at an altitude of about 90 km. Results have been obtained for both day (19) and nighttime (20) concentrations. Other gaseous constituents of the atmosphere including NO_2, SO_2, and I_2 have also been detected remotely using tunable dye laser sources (21, 22), as have algae concentrations in the sea (23).

On the laboratory scale, one of the most widely reported uses of dye lasers in analytical spectroscopy has been in the field of laser-excited atomic fluorescence flame spectrometry (AFFS), a technique for analyzing trace elements (24–27). To detect small atom concentrations in a flame requires good signal-to-noise ratios for the atomic fluorescence emissions. To obtain strong fluorescence emission requires an excitation source of high spectral brightness that is also monochromatic. The source should also be stable when repetitively pulsed, have a high peak power output, and a small duty cycle (small ratio of on-to-off time). Such a source is the pulsed dye laser. The tunability of the dye laser gives the added advantage that one source may be used to study many different atoms in a sample, and a variety of atoms have been studied by this technique.

A typical system is shown in Figure 3, in which gated detection is employed to remove all noise except that generated by scattering of the laser ra-

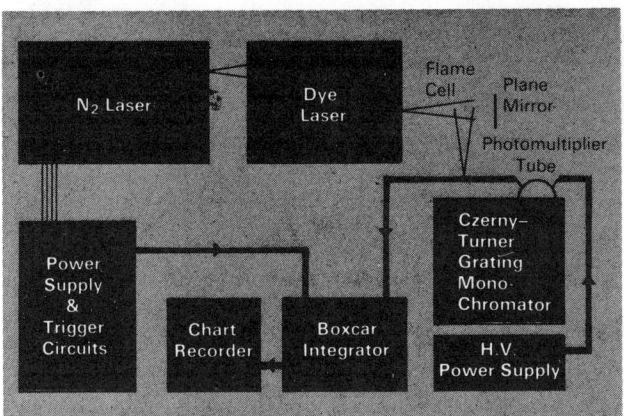

Figure 3. Block diagram of experimental system for laser-excited atomic fluorescence flame spectrometry
Figure taken from ref. 25

diation within the flame. Background scatter can be reduced even more by the use of suitable nebulizers or by recording fluorescence emissions from nonresonance transitions.

It would appear that with such an ideal source as the pulsed dye laser for AFFS, lower limits of detection and wider range, linear analytical curves would result from improved signal-to-noise ratios when compared with conventional trace analytical methods such as atomic absorption, atomic emission, or atomic flame fluorescence excited by conventional sources. Extensive results (28), however, show that at best the detection limits for laser-excited flame fluorescence are only slightly better than those of other atomic flame spectrometric techniques.

The reason for this anomaly has been traced in part to a saturation effect (29, 30). In this effect, the intensity of the laser irradiance reaches such a level that the atomic absorption transition becomes essentially saturated. Any further increase in laser irradiance simply results in increased background scattering. A limit is therefore set on the linear relationship between the fluorescence signal and the exciting laser radiation intensity. In more exact terms, the atomic fluorescence radiance assumes a nonlinear dependence on the exciting spectral radiance once the excitation radiance reaches a certain level. By irradiating only certain areas of the flame and by careful focusing of the laser beam to a power level approaching saturation, Omenetto and his coworkers (31) were able to extend the linear range of certain analytical curves by a factor of 1000. One other advantage observed when working near saturation is that the atomic fluorescence emission signal is virtually independent of source fluctuations and is also less affected by collision quenching. Further improvements in detection limits are expected when laser tuning ranges are finally extended further into the UV since many of the more sensitive absorption lines (i.e., those with high absorption oscillator strengths) are in the mid to far UV region. Initial results obtained in the ultraviolet region for Mg, Ni, and Pb have already been reported by Kuhl and Spitschan (32) using a frequency doubled dye laser as the excitation source.

Other recent applications involving tunable laser excitation of flame fluorescence have been in the detection and measurement of CH (33) and C_2 (34) radicals in flames. Flameless atomic fluorescence spectroscopy with laser excitation has also been demonstrated recently by Neumann and Kriese (35) using a frequency doubled flashlamp pumped dye laser. They were able to detect a low limit of 0.2 pg of Pb and extend the linear range of the Pb analytical curve by two orders of magnitude over that obtained with more conventional sources. Finally, the first report has recently appeared on the analytical applications of laser excited fluorimetry of molecules in the condensed phase (36).

Another technique that has received considerable attention in the past few years is laser enhanced absorption spectroscopy (37) or selective intracavity laser quenching, as it is sometimes called. In this technique, a sample is introduced into the cavity of a dye laser operating under broadband conditions, as shown in Figure 4. Broadband conditions are obtained by replacing the tuning element in the cavity by a broadband totally reflecting mirror. All wavelengths of the stimulated dye are then emitted simultaneously without spectral narrowing. If the sample has an absorption or absorptions at wavelengths within the broadband emission curve, losses will occur in the laser cavity at these wavelengths. The result is that the usually smooth emission profile from the dye laser now contains absorption bands definitive of the sample. A rather striking example is shown in Figure 5 in which a sample of NO_2 has been introduced into the cavity of a flashlamp pumped dye laser (38).

This method is particularly sensitive since it involves a selective spoiling of the gain inside a laser cavity. Single pass gain in a cavity is critical for lasing to occur, especially near threshold. A further advantage of the method is that the whole spectrum is obtained in one shot, as it were, since

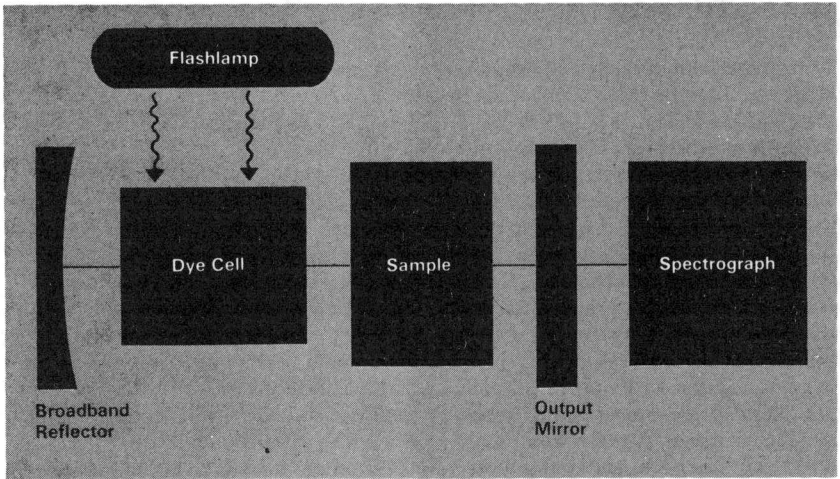

Figure 4. Dye laser system set up for broadband operation. Laser enhanced absorption spectrum of intracavity sample is recorded by spectrograph

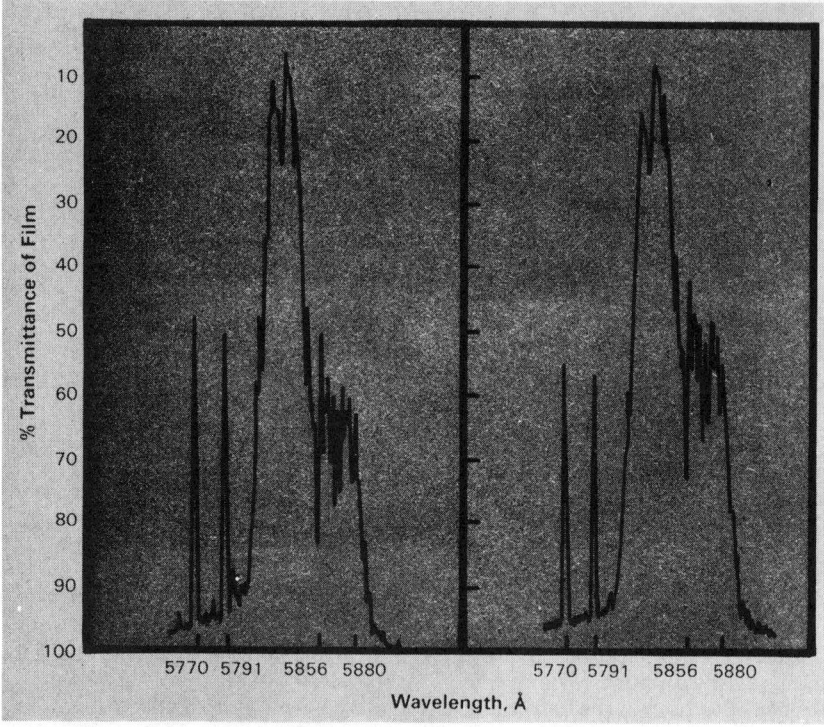

Figure 5. Intracavity absorption spectra of NO_2
Trace at right represents NO_2 pressure of 6 mm; trace at left, 7 mm. Mercury emission lines shown to left of each trace are included for wavelength calibration
Figure taken from ref. 38

one pulse of the flashlamp will produce one pulse of laser emission. Thus, high-resolution absorption spectra of short-lived species, such as transient free radicals, may be recorded (39). Since pulsed excitation is most often used for this technique, the absorption-emission spectrum is usually recorded photographically by means of a spectrograph. Photoelectric detection involving photodiode arrays has also been reported (40) and will probably become the method of choice, especially when high sensitivity is required. Most recent work has been directed toward developing both the quantitative and qualitative potential of intracavity absorption spectroscopy.

Raman Spectroscopy

One branch of spectroscopy that has benefited considerably from the discovery of the laser has been that of Raman spectroscopy (41, 42). It would seem only natural, therefore, that Raman spectroscopy should benefit even further from the development of the tunable dye laser. In a conventional laser-excited Raman system, the excitation source is fixed in frequency (for example, a CW Ar^+ or Kr^+ gas laser), and the scattered light from the sample analyzed with a scanning spectrometer. With the appearance of the tunable laser, one thought that immediately springs to mind is why not remove the rather costly spectrometer and instead tune the laser source? An investigation along these lines was performed by McNice in 1972 (43), but since then no reports appear to have been published. In McNice's experiment he used a stack of interference filters in place of the spectrometer and tuned the laser across the Raman region of interest. He successfully recorded Raman emission from oxygen and CO_2 but noted that the system sensitivity was limited by scattered light from the laser flashlamp. The old problem in Raman spectroscopy of scattered light rejection therefore prevails in tunable laser Raman spectroscopy. Further limitations included the restricted tuning range of the dye and the variance of laser output power with both wavelength and time. With future developments in stable output CW tunable lasers and advances in interference filter technology, further research in the area of no-spectrometer spectroscopy should be forthcoming.

Despite the tremendous impetus given to Raman spectroscopy by the laser, the most severe limitation is still the fact that the Raman effect is inherently very weak. Two techniques have been developed recently to specifically overcome this sensitivity problem by resonant enhancement of the Raman intensities. Both techniques depend heavily on the tunability of the excitation source.

The first technique is that of coherent anti-Stokes Raman spectroscopy (44) or CARS, as it is now known. Extensive theoretical descriptions of the process have already been discussed (45) and will not be repeated here. In essence, with reference to Figure 6, in the normal spontaneous vibrational Stokes-Raman effect, a molecule is excited to some virtual state by interaction with a source of excitation radiation of frequency ω_p. The molecule then returns instantaneously to the first excited vibrational energy level of the electronic ground state, emitting a photon of frequency ω_s. The difference $\omega_p - \omega_s$ is equal to a fundamental frequency of the molecule involved. For anti-Stokes emission, the initial energy level is the first excited vibrational level, and the transition terminates at the ground state, the emitted photon then having a frequency of ω_{as}. If an intense beam of coherent radiation of frequency ω_s such as can be produced by a tunable dye laser is now focused into the sample so that it crosses the original excitation pump beam ω_p at a small angle in the sample, then an intense, coherent beam of anti-Stokes Raman emission is generated from the crossing point with a frequency ω_{as}. The process is nonlinear and depends on the relationship $\omega_{as} = 2\omega_p - \omega_s$. The anti-Stokes beam is generated at an angle with respect to the pump beam and the dye laser beam and can therefore be spatially separated without a spectrometer. As ω_s is tuned in frequency across the normal Raman region, anti-Stokes emission will appear every time that the value of $\omega_p - \omega_s$ coincides with the frequency of a Raman active vibration.

Advantages of this process are numerous in that emission intensities are very high compared with the normal Raman process; fluorescence which may mask normal Raman emission is eliminated as the observed output is in the fluorescence-free anti-Stokes region; and no spectrometer is required. However, the efficiency of the process is quadratically dependent on the intensity of the pump beam and therefore requires the high peak powers available from pulsed lasers. Limitations of tuning range and cost are also involved, and the experimental conditions are still very much in the research stage. It is not expected that CARS will replace conventional Raman spectroscopy as an analytical technique but rather be used instead for very specific applications. Such an application has already been reported for the measurement of H_2 gas concentration, both in a flame and in a supersonic jet (46). It is felt, however, that the full potential of CARS for practical analytical spectroscopy has yet to be developed.

The second technique for Raman intensity enhancement is that of reso-

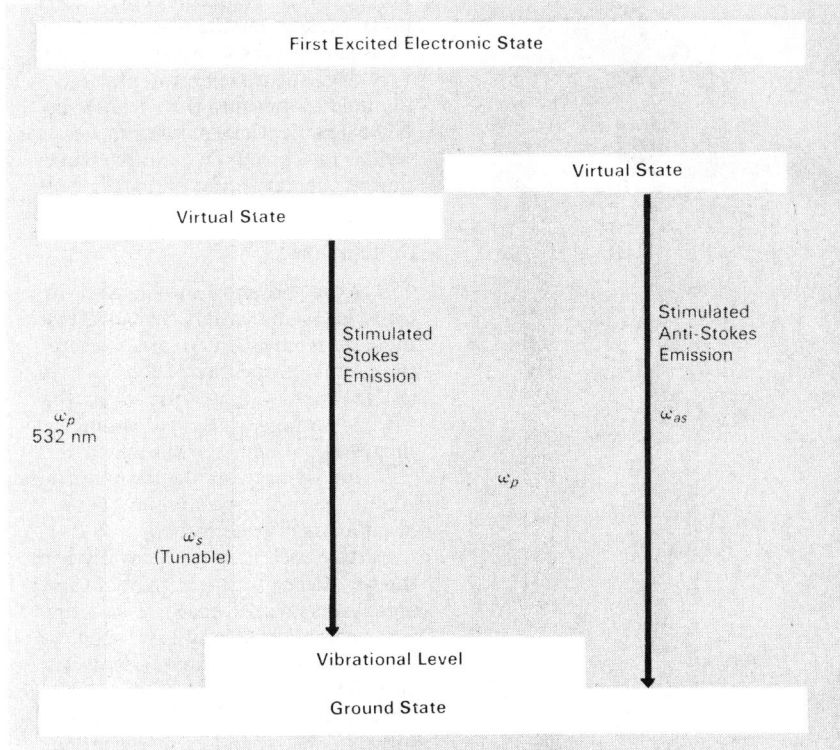

Figure 6. Energy level diagram for Raman process, adapted to show generation of stimulated anti-Stokes Raman emission

Figure taken from ref. 45 and reproduced with permission of the American Institute of Physics and the authors

nance Raman spectroscopy. If the frequency of the Raman excitation source is tuned to coincide with the frequency of an electronic absorption of the molecule being studied, then a resonance process is set up. The result of this resonance is that the Raman intensities of certain fundamental vibrations of the molecule are tremendously enhanced. Considerable research has already been performed in the investigation of this technique and in utilizing its advantages for determining spectroscopic constants of small molecules (47). It has also been found useful as an aid in assigning vibrational frequencies.

A more practical application of resonance Raman spectroscopy using tunable lasers has been in the study of biological molecules (48). If, in a complex molecule such as a protein, a group of atoms gives rise to an isolated electronic absorption band, then by tuning the excitation source into that isolated electronic absorption, the resonance Raman effect will enhance the intensities of only those Raman emissions associated with vibrations of the atoms of that group. Such a group of atoms may be a biological chromophore. Therefore, by the technique of resonance Raman, certain selective sites in a very complex molecule can be studied in isolation, as though they were independent units, without excessive interference from spectral lines arising from other sites in the molecule.

Using this technique, Spiro and his coworkers have carried out extensive studies of the heme proteins especially with regard to changes in structure and bonding arising from different spin and oxidation states (49). By tuning the laser excitation frequency into selective electronic absorption bands of the heme molecule, they have been able to study vibrations of the porphyrin rings in almost complete isolation. This particular application of tunable lasers is expected to promote a whole new approach to structure elucidation of complex biological systems.

Other Tunable Laser Sources

As mentioned earlier, other tunable laser sources, particularly those emitting in the infrared, have been fully developed. Only parametric oscillators appear to be available commercially at present. Various excellent reviews have been written on the characteristics, operation, and applications of tunable laser sources other than dye lasers (9–12, 50, 51); this report will, therefore, be restricted to recent applications pertaining to analytical spectroscopy.

For this restricted case (i.e., excluding such tunable sources as high-pressure gas lasers and frequency mixing generators), tunable lasers that operate in the infrared can be classified into three major types:

- Semiconductor diode (SD) lasers
- Spin-flip Raman (SFR) lasers
- Parametric oscillators (PO)

In each of the above, tuning is accomplished by varying either an applied magnetic field, the temperature of the device, or its position in an external pumping beam. At present, both SD and SFR lasers require cryogenic cooling which severely limits their application outside of the laboratory. SD and SFR lasers have limited continuous tuning ranges (of the order of cm^{-1}). However, they do operate with very narrow spectral bandwidths (10^{-6} cm^{-1}). They are, therefore, ideal for ultrahigh resolution spectroscopy, especially in the frequency region in which most fundamental molecular vibrations occur (4000–400 cm^{-1}: 2.5–25 μm). Parametric oscillators have a much wider tuning range, but the range is restricted mainly to the mid-infrared. Also, the spectral linewidths of PO's are broader than those of SD or SFR lasers, being typically 1–2 cm^{-1}, although this has been narrowed to 0.001 cm^{-1} in special cases. Work is progressing rapidly at present to extend the tuning range of PO's further into the infrared (52).

Most of the practical applications developed for SD, SFR lasers, and parametric oscillators have been in the area of atmospheric pollution studies (50). In the case of parametric oscillators, Henningsen et al. (53) have reported the remote detection of CO using a parametric tunable laser. By the technique of resonance absorption and by tuning the parametric laser over the frequency region of the rotational lines of the first vibrational overtone of CO at 2.3 μm, they have been able to measure CO concentrations over a distance of 107 meters.

Applications of tunable semiconductor diode lasers to the detection of air pollutants have been documented by Hinkley and Kelley (54). The unique property of the SD laser of condensing approximately 0.25 mW of power into a linewidth of around 10^{-6} cm^{-1} makes it a formidable source for trace pollution studies, both for point sampling and remote detection. The superior resolution enables many more infrared absorption lines to be detected than can normally be observed with a conventional spectrometer, and the superior power means that samples can be studied at such low pressures that collision broadening may be totally neglected. The ability to observe very narrow lines at very high resolution implies that the information content of any spectral region

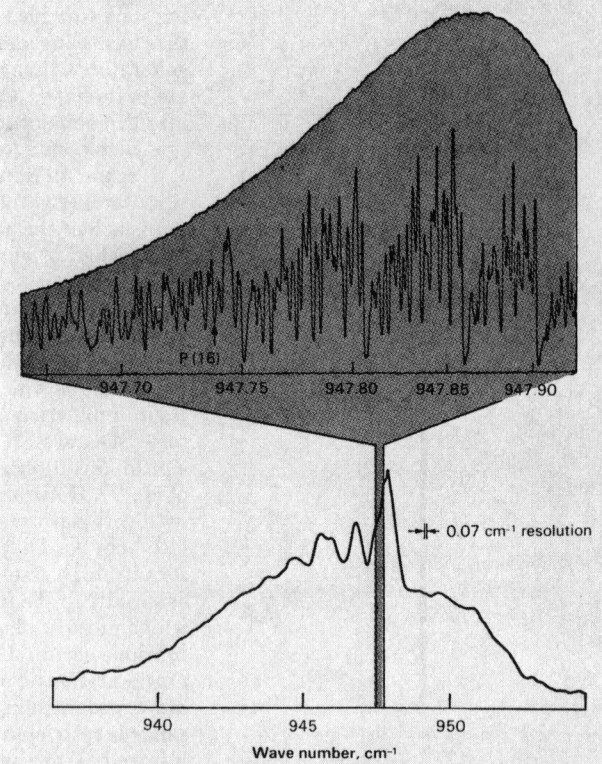

Figure 7. Infrared absorption spectrum of ν_3 band of SF_6

Lower trace taken with conventional grating spectrometer: sample pressure, 0.1 torr; cell length, 25 cm; resolution, 0.07 cm^{-1}. Upper trace taken with diode laser: sample pressure, 0.1 torr; cell length, 10 cm; resolution, 3 × 10^{-6} cm^{-1}. P (16) is CO_2 laser line used for frequency calibration of diode laser spectrum

Figure taken from ref. 54. Copyright 1971 by the American Association for the Advancement of Science. Reproduced with permission of the authors

containing infrared absorptions can be vastly increased. It then follows that many different pollutants may be specifically identified by diode laser spectroscopy, both qualitatively and quantitatively, even when all present together, providing at least some of their major absorption frequencies do not overlap. A recent example of the ability of diode laser spectroscopy to identify organic molecules by high-resolution analysis of the C—H stretching region has been given by Nill et al. (55).

The superior resolution capabilities of SD lasers over conventional infrared spectrometers (being somewhere of the order of 2000) is so strikingly illustrated in Figure 7 that it bears reproduction here, even though this figure has appeared in many recent reviews. With resolution capabilities such as those shown for just one small region of one band of SF_6 by laser excitation, it is fair to say that the whole field of infrared spectroscopy may well undergo a minor revolution in the near future once SD lasers become commercially available.

It is always very satisfying to be able to end any review with a state-of-the-art application that is also relevant to a concern of the day. It is fortuitous, therefore, that Patel et al. have recently published their initial findings on the spectroscopic measurement of NO and water vapor in the stratosphere (56). Their final results should prove conclusive in determining which of the many theoretical models for predicting the effect of SST's on the stratosphere is correct.

Spectroscopic measurements were taken using a spin-flip Raman laser as the source and a combination multipass cell and optoacoustic cell as the detector. The unit was carried beneath a balloon to a height of 28 km, and measurements were taken in situ both in the presence and absence of ultraviolet radiation from the sun (i.e., at day and at night). The detection limit of the system for NO was approximately 1.5 × 10^8 molecules cm^{-3} corresponding to a volumetric mixing ratio of approximately 0.2 parts per billion at an altitude of 28 km. No absorptions attributable to NO were found at night, implying that the maximum possible concentration at night was 1.5 × 10^8 molecules cm^{-3}. After the ultraviolet sunrise, distinct absorptions were observed which could be correlated with NO concentration up to 20 × 10^8 molecules cm^{-3} (3.8 ppb). Measurements of H_2O absorptions showed the stratosphere to be very dry. Future experiments are planned to measure the variance in NO concentrations as the sun sets, as well as measuring O_3 concentrations during the day–night cycle.

Conclusion

As expected, most applications of tunable lasers in analytical spectroscopy to date have been in areas where the unique properties of the laser can replace those of conventional sources with advantage. Extended tuning ranges at both ends of the spectrum (UV and IR) and stable power outputs should extend these advantages to other areas of spectroscopy.

Certain recent trends possibly point the way to the future impact of tunable lasers on spectroscopy. The first is the establishment of laser spectroscopy as a field in its own right as witnessed by the First International Conference (57) devoted solely to the subject. The second is the recent appearance of systems designed specifically to take advantage of the unique properties of tunable laser sources. These include enhanced sensitivity (58), ultrarapid tuning (59), and improved instrumentation (60). Finally, other novel techniques, similar perhaps to coherent anti-Stokes Raman spectroscopy, which are not possible with conventional sources may be discovered, thereby opening completely new avenues for exploration.

As a postscript, Fairbank et al. (61) have recently published their results on the resonance fluorescence detection of sodium using a CW laser and photoelectric detection. The minimum sodium vapor density detected was 10^2 atoms/cm^3 which translates into observation of $\frac{1}{15}$ resonant atoms, on the average, in the sampling region. They claim that it should be possible to measure many other atoms with the same sensitivity. It is applications such as these that point to the almost inevitable future interdependency of analytical spectroscopy and tunable laser sources.

Acknowledgment

It is a pleasure to acknowledge the assistance of Neil Sandow and Jack Wiley in bringing certain relevant literature to my attention.

References

(1) P. P. Sorokin and J. R. Lankard, *IBM J. Res. Dev.*, **10**, 162 (1966).
(2) F. P. Schäfer, W. Schmidt, and J. Volze, *Appl. Phys. Lett.*, **9**, 306 (1966).
(3) P. P. Sorokin, *Sci. Am.*, **220**, 30 (1969).
(4) J. P. Webb, *Anal. Chem.*, **44** (6), 30A (1972).

(5) R. A. Keller, *Chem. Technol.*, **3**, 626 (1973).
(6) H. Walther, *Phys. Scr.*, **9**, 297 (1974).
(7) F. P. Schäfer, Ed., "Dye Lasers," Springer-Verlag, New York, N.Y., 1973.
(8) R. W. Wallace, *Opt. Commun.*, **4**, 316 (1972).
(9) R. G. Smith, *Anal. Chem.*, **41** (10), 75A (1969).
(10) D. C. Hanna and R. C. Smith, *Sci. J.*, **5** (April), 53 (1969).
(11) J. Kuhl and W. Schmidt, *Appl. Phys.*, **3**, 251 (1974).
(12) J. K. Burdett and M. Poliakoff, *Chem. Soc. Rev.*, **3**, 293 (1974).
(13) K. H. Drexhage, *Laser Focus*, March, 35 (1973).
(14) D. A. Jennings and R. A. Keller, *J. Am. Chem. Soc.*, **94**, 9249 (1972).
(15) W. Hartig and H. Walther, *Appl. Phys.*, **1**, 171 (1973).
(16) T. W. Hänsch, I. S. Shahin, and A. L. Schawlow, *Phys. Rev. Lett.*, **27**, 707 (1971).
(17) F. Biraben, B. Cagnac, and G. Grynberg, *Phys. Lett. A*, **49A**, 71 (1974).
(18) F. Y. Wu, R. E. Grove, and S. Ezekiel, *Appl. Phys. Lett.*, **25**, 73 (1974).
(19) A. J. Gibson and M.C.W. Sandford, *Nature*, **239**, 509 (1972).
(20) A. J. Gibson and M.C.W. Sandford, *J. Atmos. Terr. Phys.*, **33**, 1675 (1971).
(21) R. M. Measures and G. Pilon, *Opto-electronics*, **4**, 141 (1972).
(22) K. W. Rothe, U. Brinkmann, and H. Walther, *Appl. Phys.*, **3**, 115; **4**, 181 (1974).
(23) H. H. Kim, *Appl. Opt.*, **12**, 1454 (1973).
(24) M. B. Denton and H. V. Malmstadt, *Appl. Phys. Lett.*, **18**, 485 (1971).
(25) L. M. Fraser and J. D. Winefordner, *Anal. Chem.*, **43**, 1693 (1971).
(26) J. Kuhl and G. Marowsky, *Opt. Commun.*, **4**, 125 (1971).
(27) L. M. Fraser and J. D. Winefordner, *Anal. Chem.*, **44**, 1444 (1972).
(28) N. Omenetto, L. M. Fraser, and J. D. Winefordner, *Appl. Spectrosc. Rev.*, **7**, 147 (1973).
(29) N. Omenetto, P. Benetti, L. P. Hart, J. D. Winefordner, and C.Th.J. Alkemade, *Spectrochim. Acta*, **28B**, 289 (1973).
(30) J. Kuhl, S. Neumann, and M. Kriese, *Z. Naturforsch.*, **28a**, 273 (1973).
(31) N. Omenetto, L. P. Hart, P. Benetti, and J. D. Winefordner, *Spectrochim. Acta*, **28B**, 301 (1973).
(32) J. Kuhl and H. Spitschan, *Opt. Commun.*, **7**, 256 (1973).
(33) R. H. Barnes, C. E. Moeller, J. F. Kircher, and C. M. Verber, *Appl. Opt.*, **12**, 2531 (1973).
(34) K. H. Becker, D. Haaks, and T. Tatarczyk, *Z. Naturforsch.*, **29a**, 829 (1974).
(35) S. Neumann and M. Kriese, *Spectrochim. Acta*, **29B**, 127 (1974).
(36) B. W. Smith, F. W. Plankey, N. Omenetto, L. P. Hart, and J. D. Winefordner, *ibid.*, **30A**, 1459 (1974).
(37) J. S. Shirk, *Res. Dev.*, Sept., 30 (1974).
(38) H. W. Latz, H. F. Wyles, and R. B. Green, *Anal. Chem.*, **45**, 2405 (1973).
(39) G. H. Atkinson, A. H. Laufer, and M. J. Kurylo, *J. Chem. Phys.*, **59**, 350 (1973).
(40) G. Horlick and E. G. Codding, *Anal. Chem.*, **46**, 133 (1974).
(41) S.P.D.S. Porto, *Ind. Res.*, May, 66 (1969).
(42) J. L. Koenig, *Res. Dev.*, June, 18 (1969).
(43) G. T. McNice, *Appl. Opt.*, **11**, 699 (1972).
(44) R. F. Begley, A. B. Harvey, R. L. Byer, and B. S. Hudson, *Am. Lab.*, Nov., 11 (1974).
(45) R. F. Begley, A. B. Harvey, and R. L. Byer, *Appl. Phys. Lett.*, **25**, 387 (1974).
(46) P. R. Regnier, F. Moya, and J.P.E. Taran, *AIAA J.*, **12**, 826 (1974).
(47) W. Kiefer, *Appl. Spectrosc.*, **28**, 115 (1974).
(48) T. G. Spiro, *Acc. Chem. Res.*, **7**, 339 (1974).
(49) T. C. Strekas and T. G. Spiro, *Biochim. Biophys. Acta*, **263**, 830; **278**, 188 (1972) and **351**, 237 (1974).
(50) E. D. Hinkley, *Opto-electronics*, **4**, 69 (1972).
(51) H. R. Schlossberg and P. L. Kelley, *Phys. Today*, **25**, July, 36 (1972).
(52) R. L. Byer, M. M. Choy, R. L. Herbst, D. S. Chemla, and R. S. Feigelson, *Appl. Phys. Lett.*, **24**, 65 (1974).
(53) T. Henningsen, M. Garbuny, and R. L. Byer, *ibid.*, p 242.
(54) E. D. Hinkley and P. L. Kelley, *Science*, **171**, 635 (1971).
(55) K. W. Nill, A. J. Strauss, and F. A. Blum, *Appl. Phys. Lett.*, **22**, 677 (1973).
(56) C.K.N. Patel, E. G. Burkhardt, and C. A. Lambert, *Science*, **184**, 1173 (1974).
(57) Proceedings of the First International Conference on Laser Spectroscopy, Vail, Colo., 1973, R. G. Brewer and A. Mooradian, Eds., Plenum Press, New York, N.Y., 1974.
(58) M. I. Bell and R. N. Tyte, *Appl. Opt.*, **13**, 1610 (1974).
(59) J. M. Telle and C. L. Tang, *Opt. Commun.*, **11**, 251 (1974).
(60) D. C. Harrington and H. V. Malmstadt, *Anal. Chem.*, **47**, 271 (1975).
(61) W. M. Fairbank, T. W. Hänsch, and A. L. Schawlow, *J. Opt. Soc. Am.*, **65**, 199 (1975).

James R. Allkins was born and educated in London, England. He obtained BS and PhD degrees from London University and then came to the United States as a postdoctoral fellow to work under Ellis Lippincott at the University of Maryland. After leaving Maryland he spent two years as manager of applications at Spex Industries and then joined Interactive Technology in California as director of applications research. His original interest in lasers was revived by a year at Chromatix where he became particularly interested in tunable lasers as spectroscopic sources.

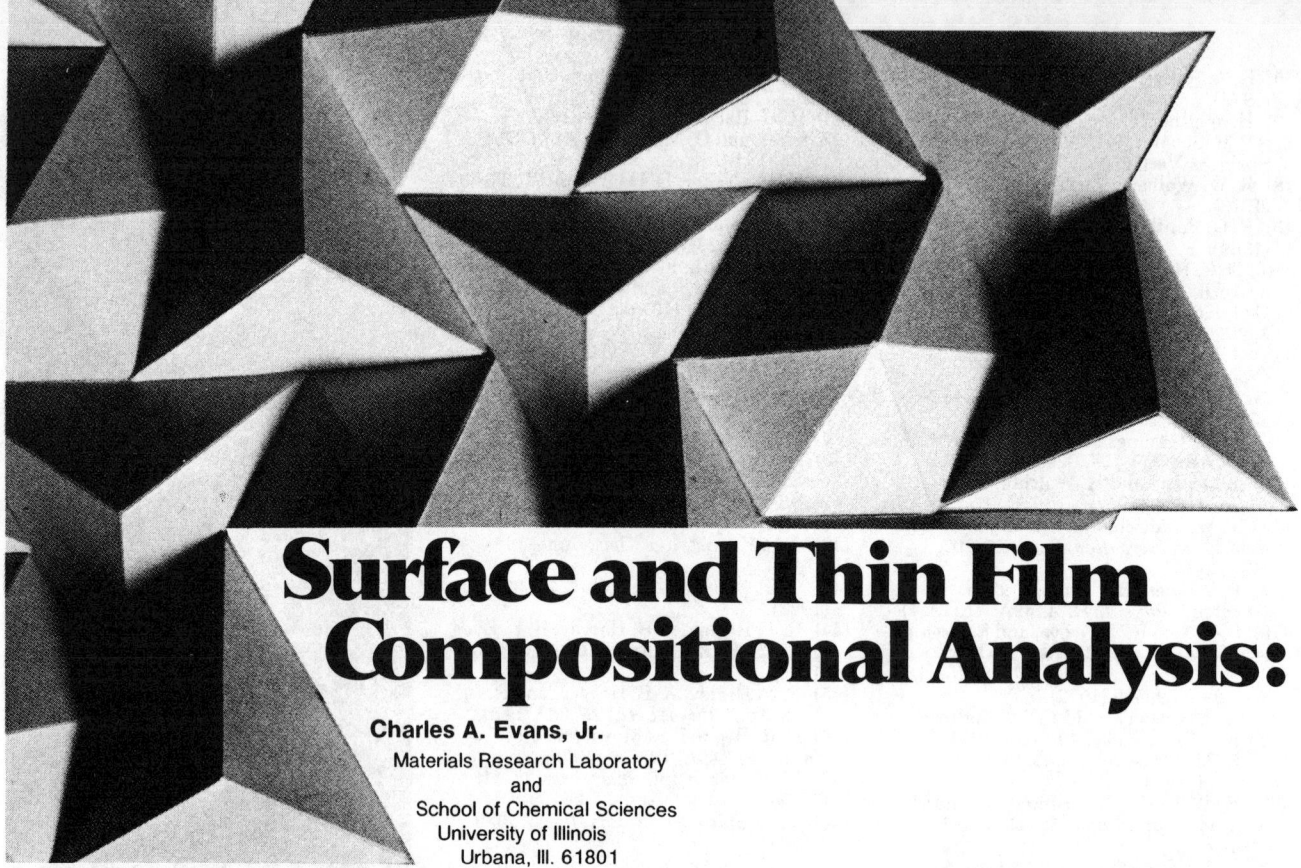

Surface and Thin Film Compositional Analysis:

Charles A. Evans, Jr.
Materials Research Laboratory
and
School of Chemical Sciences
University of Illinois
Urbana, Ill. 61801

The materials analyst today has available to him techniques for characterizing the three-dimensional elemental composition of a sample, some of which give a lateral spatial resolution on the order of 1 μm and all of which give a depth resolution of ~200 Å or less. The development of these powerful techniques has been stimulated by the move to thin-film technology in the semiconductor industry and by the growing technological importance of such phenomena as corrosion, embrittlement, and catalysis. In all these areas the composition and properties of surfaces determine many of the properties of materials. The earliest microanalytical technique was the electron microprobe, which made possible analyses with lateral (x, y) and depth (z) resolutions of about 1 μm. The 1960's saw the development of electron spectroscopies (ESCA, Auger) and secondary ion mass spectrometry (SIMS) which are sensitive to the outer 10–20 Å of a sample (i.e., have a shallow *escape* depth for the analytical signal). The coupling of these spectroscopies with sputtering (sputtering is, of course, intrinsic in the SIMS technique) allows the characterization of successive 10–20 Å layers as sputtering erodes the sample. Lateral (x, y) resolution is attained in the charged particle-induced spectroscopies by using a microfocused primary beam (Auger, ion microprobe) or by ion optically locating the origin of the secondary signal (ion microscope).

While the capability for x, y, and z microanalysis is important in many applications (integrated circuit technology is an obvious example), frequently samples are obtained which are laterally homogeneous and which require characterization only in the z-direction. Chemical analyses with z-resolution are referred to as *depth profiles*. (The investigation of thin-film systems is a case in point.) Techniques which are not microanalytical in the three-dimensional sense can then be employed. ESCA has been mentioned already. Low-energy ion scattering spectrometry (ISS) and MeV ion backscattering spectrometry (BS) are two others. In such techniques the exciting beam is usually unfocused owing to the lack of focusing systems (ESCA), signal loss associated with small beam size (ISS), or simply the expense of focusing systems (BS). As with the secondary ion and electron techniques, ISS exhibits a very shallow escape depth and thus must be combined with ion sputtering for depth profiling. Sputtering is not required in the use of MeV ion backscattering spectrometry.

The techniques introduced here have the common advantage of broad elemental sensitivity (H to U for SIMS, Li to U for the others). They differ widely in other aspects—ease of quantitation, cost, convenience, and sample throughput capability. This article will discuss the basic concepts of the more important techniques commonly used for surface and thin-film analysis and will compare them in their analytical features and capabilities. Those techniques which are commercially available will be emphasized. The INSTRUMENTATION article in this issue of ANALYTICAL CHEMISTRY will discuss the instrumental features of these surface analytical techniques and the commercially available instruments.

Ion Sputtering

As mentioned above, almost all techniques for the determination of concentration variations with depth require the use of ion sputtering. If an atomic particle with a kinetic energy in the range of 100 eV to 100 keV impinges on a material surface, energy and momentum transferred to the surface atoms cause some of them to be ejected from the surface. Continued particle bombardment will cause successive layers of atoms to be removed and the underlying layers to be exposed. In almost all applications of the sputtering technique, ions are used as the bombarding species since they are more easily accelerated, focused, and deflected than are neutrals.

Ion sputtering is generally accomplished either by ion beam or diode techniques. In the ion beam method the ions are produced in a separate ion source, extracted and accelerated to their final kinetic energy, and directed against the sample surface. In this

Description and Comparison of Techniques

manner the ion beam can be directed to a specific region of the sample. The sample is in a high-vacuum (<1 × 10^{-4} torr) region since the mean free path of the ion must be greater than the distance between the ion source and the sample. In diode sputtering the sample is immersed in a gas at a pressure of ~10^{-2}–10^{-1} torr. A dc or rf discharge is struck between the sample and another electrode, and sputtering occurs. The sample is sputtered over its entire surface and is actually part of the ion source.

The sputtering phenomenon is an integral part of depth profiling with most of the techniques to be discussed. As such, the analyst must be concerned about its effect on the quality of an analysis. Artifacts of the sputtering process such as nonuniform crater shapes, preferential removal of one element with respect to another, etc., can lead to erroneous conclusions from an analysis. A discussion of these effects is beyond the scope of this article, but the reader is forewarned that this process should be given serious consideration in its use with surface and thin-film characterization. The fundamentals of ion surface interactions are discussed by Kaminsky (1) and Carter and Colligan (2). The implications of artifacts encountered in the use of sputtering in surface analysis are discussed by Coburn and Kay (3), McHugh (4), and Honig (5).

Auger Electron Spectrometry (AES). Bombardment of a material by electrons in the 1–10-keV energy range causes a variety of secondary processes to occur. The two most important processes relevant to this discussion are the ejection of secondary electrons from atomic valence and conduction bands and ionization of the surface atoms. If an inner shell vacancy (K, L, or M) is created by a primary electron, an outer shell electron will fall into this hole. The recombination energy is carried away by a photon (as in the X-rays used for electron microprobe analyses) or another electron. The energy of the ejected or *Auger* electron is determined by the energy levels of the atom before and after ejection and is characteristic of the emitting atom. If this atom is sufficiently close to the sample surface (3–20 Å for 50–2000-eV electrons), there is a high probability that the Auger electron will not undergo any inelastic collisions and will escape the sample with its initial characteristic energy. Subsequent energy and intensity analysis of the Auger electrons will permit qualitative and quantitative analysis of the outer atomic layers of the sample. If the Auger electron does undergo an inelastic collision, as will most electrons emitted from deeper than a few atomic layers into the sample, it will lose part of its original energy and no longer be characteristic of the emitting atom. These inelastically scattered electrons and the other secondary electrons from the sample produce a large "continuum" background in the electron energy spectrum. The discrete Auger peaks are very small when compared to the large background and require special instrumentation techniques to resolve them. The differentiation and "lock-in" techniques used are described in the companion INSTRUMENTATION article.

Energy analysis of the Auger electrons permits qualitative and quantitative analyses to be performed on the volume of material bounded by the diameter of the primary electron beam (5–500 μm) and the Auger electron escape depth (3–20 Å). By combining Auger electron spectrometry with ion beam sputtering, depth profiling analysis may be performed. Lateral (x, y) resolution of the AES technique is determined by the diameter of the probing electron beam. Review articles by Morabito (6) and Chang (7) provide a more detailed discussion of AES and an entrance into the literature.

Energetic Ion Beam Techniques. There are three depth profiling techniques which employ high-energy (≥300 keV), monoenergetic ion beams as the probing species. These techniques depend on three principles. The first is that ions in this energy range can penetrate deep into the sample surface (>1000 Å) while causing little sputtering or damage. Second, the deceleration of these ions as they penetrate the sample results from the interaction of the ion with the electrons associated with the sam-

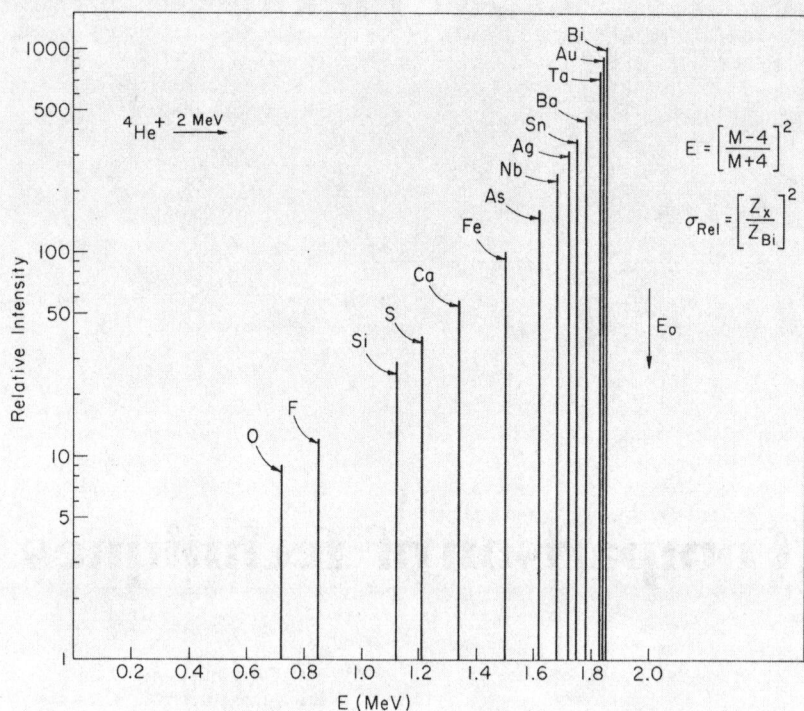

Figure 1. Relative number of backscattering events vs. atomic number for 2-MeV $^4\text{He}^+$

ple atoms. The magnitude of the interaction, called the electronic stopping power, is well known for each element; thus, the energy at any depth can be accurately predicted. The third principle is that the energetic probing ion can interact with a sample atom to produce one of three secondary processes: a Rutherford backscattering collision, an X-ray, or a nuclear reaction.

Backscattering Spectrometry (BS). When a surface is bombarded with H^+ or He^+ ions in the 300-keV to 3-MeV energy range, a small fraction ($\sim 10^{-6}$) of the incoming particles undergoes a Rutherford collision and is backscattered. The energy, E, of the ions scattered through an angle of 180° is related to the initial energy E_o by

$$E = \left[\frac{M-m}{M+m}\right]^2 E_o \quad (1)$$

where m is the mass of the scattered particle, and M the mass of the target atom. Thus, energy analysis of the primary ions backscattered from a sample surface provides a mass analysis of the surface atoms. The relative sensitivity of different surface atoms is determined by the differential cross section, σ, according to the relationship:

$$\sigma \propto \left[\frac{Z_1 Z_2}{E_o}\right]^2 \quad (2)$$

for 180° backscattering. Z_1 and Z_2 are the atomic numbers of the incident and target atoms. Figure 1 shows the mass scale and relative sensitivities for 2-MeV $^4\text{He}^+$ incident on several different surface atoms.

Because of the energy loss of the probing particle ($^4\text{He}^+$ is generally used) as it traverses successively deeper atomic layers, the E_o for scattering from an internal atom will be reduced according to the relationship:

$$\Delta E = \frac{dE}{dx} t \quad (3)$$

where t is the sample thickness traversed, and dE/dx, the stopping cross section or stopping power, is the energy loss per unit length traversed in the material under analysis. The stopping cross section can be calculated or is available in empirical tables (8). The backscattered ion energy is determined by the relationship:

$$E_{\text{final}} = E_o - E_{\text{inward}}^{\text{loss}} - E_{\text{backscattering}}^{\text{loss}} - E_{\text{outward}}^{\text{loss}} \quad (4)$$

The energy spectrum is a convolution of a mass scale (established by the backscattering process) and a depth scale (established by energy loss in the sample before and after the backscattering event). Through the use of tabulated physical constants, experimental techniques such as sample tilting, and good analytical reasoning, the analyst can obtain elemental atomic ratios and distribution of elements vs. depth without standards.

Nuclear Reactions (Also Called Nuclear Microanalysis). The production of nuclear reactions by the interaction of 0.5–2-MeV protons or deuterons with the low Z elements (H to F) can be used to determine the presence of these elements on surfaces and in thin films. These reactions occur over a narrow range of the incident particle energy. Since the incident particle loses energy as it penetrates into the sample, a given initial particle energy will result in a reaction occurring within a given, small depth range. Depth profiling is accomplished by monitoring the flux of the reaction product as the initial ion energy is varied. Thus, nuclear microanalysis establishes elemental depth distributions through "tuning" of the incident ion energy and knowledge of the stopping powers of the matrix under study.

Ion-Induced X-rays. Bombardment of a material by energetic ions (2-MeV H^+, for example) produces X-rays just as electron bombardment does in the electron microprobe. The large cross sections for X-ray production and lower intensity bremsstrahlung (background) X-radiation make this a sensitive method for detection of trace elements in thin films. At the present time, the lack of information on the physical processes and cross sections involved limits the usefulness of this technique. However, it is anticipated that research in these areas will result in the development of a sensitive surface and thin-film analytical technique. The most comprehensive discussion on the energetic ion techniques is provided in a book edited by Mayer and Ziegler (9). Most of the information in this book is contained in pages 1–406 of *Thin Solid Films,* Volume 19.

Electron Spectroscopy for Chemical Analysis (ESCA). The interaction of X-rays with a sample atom causes photoionization of the atom and ejection of an inner shell electron. The use of a discrete photon energy (Al K_α or Mg K_α, for example) and electron energy analysis permits the determination of the binding energy of the electron from the relationship:

$$E_{\text{photoelectron}} = E_{\text{photon}} - E_{\text{binding}} + \Delta E \quad (5)$$

where ΔE is a correction for the work function difference of the sample and the electron spectrometer. Since the photon energy (E_{photon}) is fixed, the binding energy (E_{binding}) is characteristic of the emitting atom, and the escape depth of the photoelectrons (20–1500 eV) is of the order of 3–20 Å, a surface analysis can be obtained from the electron energy spectrum. The binding energy of the electron is modified by the chemical surroundings of the emitting atom. The photoelectron energy can therefore be used to determine the bonding characteristics of (or the elements bonded to) the emitting atom. Hence, the name *Electron Spectroscopy for Chemical Analysis*.

Although it is not yet widely practiced, ESCA can be combined with ion sputtering to determine depth profiles. There are two concerns in combining ESCA and sputtering. The first is that the signal levels are lower in ESCA than in AES and data accumulation is slower, thereby requiring longer times to profile a certain sample thickness. Of more critical importance is the concern that sample damage by the sputtering process will distort or destroy the chemical bonding in the sample reducing ESCA/sputtering to an *elemental* rather than a *chemical* depth profiling technique (*10*).

Ion Scattering Spectrometry (ISS). As in high-energy ion backscattering spectrometry discussed above, the interaction of a low-energy ion (0.5–2 keV) and a sample atom depends on the masses of the target and incident atom and the scattering angle. For a scattering angle of 90° (the most commonly used):

$$E^m_{\text{scattered}} = E^m_{\text{initial}} \left[\frac{M-m}{M+m} \right] \quad (6)$$

where E^m is the energy of the incident ion before and after scattering, m is the mass of the incident ion, and M the mass of the scattering or target atom. Since these low-energy atoms penetrate to only a few atomic layers and an elastic scattering event occurs only for the outermost atomic layer, the spectrum provides qualitative and quantitative information only on this layer. The most commonly used probe ions are helium, neon, and argon. Ion scattering spectrometry uses several different ion species to maximize the elemental coverage and mass resolution, since the scattering process is blind to masses below the probe ion's mass and the ability to resolve atoms of similar mass degrades as the difference between the probing and target masses increases.

Depth profiling analysis results from the removal of surface layers (sputtering) by the probe ion during the ISS spectral analysis. The use of helium ions provides a very slow rate of removal, whereas neon and argon are used for higher sputtering rates. It is important to note that in ISS as well as any other technique using 0.1–100-keV ions as the probing or exciting species, the excitation and surface layer removal (sputtering) processes *cannot* be independently controlled as they can when electron or X-ray excitation is combined with ion sputtering. When higher-energy ions (>300 keV) are employed, as in backscattering spectrometry, the highly penetrating nature of the ions minimizes sputtering during the analysis, and the analytical step becomes independent of the material removal step. Honig (*11*) and Honig and Harrington (*12*) provide detailed discussions on the ISS technique as well as illustrative examples of its use.

Secondary Ion Mass Spectrometry (SIMS). In addition to the removal of neutral atoms and molecules, sputtering of a material with 1–30-keV ions produces ions of the material under bombardment. Although only a small fraction of the ejected species is ionized (0.1–10%), these secondary ions can be analyzed and detected by a mass spectrometer to provide a sensitive surface and depth profiling analysis. As with the AES, ESCA, and ISS techniques, the escape depth for the analytical signal (the secondary ions in this case) is of the order of 10–20 Å, thus permitting localization of the analysis to a shallow depth. Since sputtering of the surface atomic layers accompanies the production of the secondary ions, one or more elements can be depth profiled by monitoring the appropriate masses vs. time. Secondary ion images of lateral elemental distributions can be obtained either directly by the use of stigmatic secondary ion optics as in the *ion microscope* or by rastering a finely focused (1–3 μm) primary ion beam as in the *ion microprobe*. References *4*, *11*, and *13–17* provide detailed discussions of the SIMS technique and the ion production process.

Related Techniques. There are three additional techniques presently under development in research laboratories which promise to be valuable surface and thin-film profiling techniques.

Surface Composition by Analysis of Neutral and Ion Impact Radiation (SCANIIR). In addition to the sputtering of neutral and ionized sample atoms, energetic ion bombardment of a surface also produces photon emission in the 100–1000-nm region (*18*, *19*). By combining a primary ion source and ion transport optics with a light optical spectrometer and readout, this light emission can be characterized and used for surface and depth profiling analysis. As in SIMS, the production of the analytical signal by the ion beam is accompanied by sputtering of the surface layers; hence, a depth profiling analysis is obtainable. To date, work in the SCANIIR field has not used microfocused ion beams for high lateral resolution analysis, but there appears to be no fundamental limitation to this mode of operation. Although the detection limits demonstrated by SCANIIR to date appear to be an order or two in magnitude poorer than with SIMS, they seem to be better than can be attained by AES, BS, ISS, or ESCA. In addition, the instrumentation is less expensive and less complicated than in SIMS since a light, rather than a mass, spectrometer is used for signal detection.

Glow Discharge Mass Spectrometry (GDMS) and Glow Discharge Optical Spectrometry (GDOS). These two techniques which employ high-pressure (10^{-1}–10^{-3} torr) diode or discharge sputtering rather than ion beam sputtering are somewhat analogous to the SIMS and SCANIIR techniques. In GDMS and GDOS the sample is part of a diode sputtering system (either dc or rf). The sample atoms are sputtered into the high-pressure region immediately above the sample surface. In this region the atoms are ionized or excited to optical transitions by interaction with the metastable sputtering gas (generally argon). In GDMS the resultant ions exit the high-pressure region through an orifice into a higher vacuum (10^{-5}–10^{-6} torr) where they are mass analyzed generally by a quadrupole mass filter. Full mass scans permit qualitative and quantitative analysis (*20*, *21*). Repetitive monitoring of one or more masses as sputtering exposes successively deeper layers is used for depth profiling analysis. The GDOS technique analyzes the light emission from the excited atoms in the discharge for qualitative and quantitative analysis. Depth profiles are performed by monitoring the characteristic wavelengths of the elements of interest vs. time (*22*, *23*). The detection limits demonstrated by GDOS and GDMS are similar to those of SCANIIR, i.e., between those of SIMS and AES, ISS, or ESCA. Since diode or discharge sputtering cannot be localized as ion beam sputtering can be, GDMS and GDOS provide little lateral resolution capability.

An interesting relationship exists among the SIMS, SCANIIR, GDOS, and GDMS surface analytical methods. SIMS and SCANIIR employ *ion beam* excitation *at* the sample surface (with 10^{-5}–10^{-9} torr pressure in the sample region) with ion and photon detection, respectively. GDMS and GDOS monitor the same secondary processes, ions and photons, respectively, but with low-energy ion sputtering of the sample surface and excitation *away* from the sample surface by interactions in a high-pressure region.

Comparison of Analytical Features

Since the above surface analytical techniques utilize a wide diversity of excitation and secondary processes, there are major differences in the analytical characteristics of each. The

Table I. Elemental Sensitivity

Technique	Coverage	Specificity	Sensitivity variation	Detection limits (atomic fraction)
Auger electron spectrometry (AES)	Li–U	Good	Less than a factor of 10	10^{-3}
MeV ion backscattering spectrometry (BS)	Li–U (ω/2-MeV ^{4}He$^+$)	Lo Z – good; Hi Z – poor	Sensitivity increases with Z – Bi/O $\approx$ 100	10^{-1}–10^{-4}
Electron spectrometry for chemical analysis (ESCA)	Li–U	Good	Less than a factor of 10	10^{-2}–10^{-3}
Ion scattering spectrometry (ISS)	Li–U	Small $\frac{M}{m}$ – good; Large $\frac{M}{m}$ – poor	Sensitivity increases with Z – Bi/O $\approx$ 5	$\sim 10^{-2}$
Secondary ion mass spectrometry (SIMS)	H–U	Good (also provides isotopic detection)	Depends on ionization efficiency 10^4–10^5	10^{-4}–10^{-8}

"best" technique for each given analytical system will differ depending on the information or characterization desired. Moreover, it is becoming increasingly obvious that several techniques must be used to fully characterize a "sample". The analyst must be familiar with or even expert in the many available techniques if he is to provide a complete sample characterization. This section provides a brief overview of the analytical features of some of the techniques discussed above to give the analyst an idea where to turn when trying to meet the demands of a given analytical system. Only those techniques which are most widely used and more fully evaluated are compared. Therefore, nuclear microanalysis, ion-induced X-ray emission, SCANIIR, and the glow discharge optical and mass spectrometry techniques are omitted.

Elemental Sensitivity. The first question generally asked of a surface analytical technique is whether it can detect this element of interest at this estimated concentration in the presence of this or these matrix elements. Several aspects of elemental sensitivity dictate the answer to this question.

• Coverage—What elements in the periodic table can each technique detect based on the fundamental principles involved?

• Specificity—How well do the fundamental processes and instrumentation employed permit the detection of one element in the presence of another? In other words, is the technique subject to spectral interferences?

• Sensitivity Variations—Are the excitation and detection of the analytical signal uniform from element to element? Sensitivity variations from element to element complicate quantitative analysis and can make it difficult or impossible if the sensitivity variations are not predictable or vary owing to matrix changes, exciting conditions, etc.

• Detection Limits—What is the minimum concentration of an element that can be detected in the absence of a spectral interference? Elemental detection limits can be modified by variations in sensitivity and a specific spectral interference.

Table I provides a summary of these four aspects of elemental sensitivity for the five techniques to be discussed in this comparison section. The comments and values of Table I are as *generally* practiced and in an average situation. Specific analytical situations can improve or degrade any or all of the elemental sensitivity features of Table I. Some general trends, observations, and groupings can be made.

• All the techniques have the same general coverage, Li to U, with SIMS also being able to detect hydrogen.

• The two electron spectroscopy techniques (AES and ESCA) are not highly subject to spectral interferences. The variation in sensitivity from the most sensitive to the least sensitive element is about a factor of 10, and the detection limits are in the range of 0.1–1% atomic.

• The specificity or elemental resolution capabilities of the high- and low-energy ion backscattering techniques are highly dependent on the relative masses of the probing and sample atoms. In both techniques, mass resolution decreases dramatically as the target atom's mass greatly exceeds that of the probing or incident particle. The use of probing ions of higher mass improves mass perception with both the backscattering techniques and is commonly employed in the low-energy technique (ISS). Since radiation damage to the solid-state detector used in the BS technique occurs very rapidly (a few hours) when ions with a mass above ^{4}He$^+$ are used, the practitioner of MeV ion backscattering spectrometry rarely uses probing ions of higher mass. He must accept the mass resolution illustrated in Figure 1. The ISS detection limits vary by only a factor of five from oxygen to bismuth, but the average detection limit of the ISS technique is of the order of 1% atomic. The major contribution to this poor average detection limit is that greater than 99% of the incident ions undergo charge neutralization during scattering and cannot be detected by the electrostatic analyzer/electron multiplier spectrometer system. Ionic charge state is of no concern to the surface barrier detectors used for detection and energy analysis of MeV backscattered ions in BS; hence, this technique can generally realize better average detection limits than ISS.

• Intrinsically, the SIMS technique provides excellent elemental specificity as well as isotopic resolution. However, the presence of spectral interferences can prevent the detection of a specific element in a specific matrix (24). As with any analytical technique, the potential for spectral interferences in SIMS increases as the analysis requires better detection limits. At the 0.1% level (the ultimate detection limit of most of the techniques under discussion), SIMS spectral interferences do not occur very often, and the use of high-resolution mass spectrometry can reduce their occurrence (25). As the concentration of the sought-for element approaches the ultimate detection of the technique (sub-ppm on the average), the possibility for interferences increases. Thus, SIMS does not encounter significant interference problems until one is analyzing at levels well below the detection limits of most surface techniques. As can be seen from Table I, the SIMS technique exhibits large variations in sen-

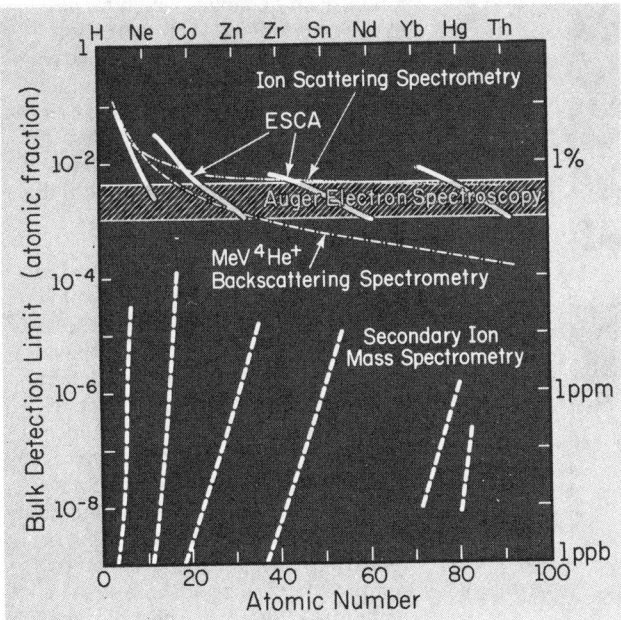

Figure 2. Bulk detection limits in interference-free situation vs. atomic number for AES, BS, ESCA, ISS, and SIMS

sitivity from element to element and consequently large variations in elemental detection limits. These variations result from the relative ion yields of the different elements under energetic ion bombardment.

Figure 2 graphically illustrates the relative detection limits of these five techniques and the extent of variations which can be encountered. This figure shows the bulk detection limits realizable under interference-free conditions for AES, ESCA, BS, ISS, and SIMS.

In-Depth Analysis. All of the five techniques being discussed provide a depth profiling capability. Backscattering analysis with MeV ^{4}He$^+$ ions samples all depths to about a micrometer simultaneously. The depth scale (i.e., the depth of a subsurface feature) is established by the energy loss process as the probing ^{4}He$^+$ ion traverses the sample atomic layers before and after the backscattering event. The energy loss per atomic layer traversed is well established in the published literature and by physical principles. Thus, depth scale assignments can be made without direct comparative standards. The parameters controlling depth assignment accuracy are the quality of the literature values (generally ±5%), energy straggling effects, and detector resolution. The depth resolutions attainable are of the order of 100–300 Å. However, this is a complex situation, and the reader is referred to the literature for more detailed discussions (9, 26).

All of the other techniques (AES, ISS, ESCA, and SIMS) combine a shallow escape depth for the analytical signal (3–20 Å) and ion beam sputtering for sequential removal of atomic layers to perform depth profiling analyses over 10–10,000 Å thick layers. Since all the techniques have essentially the same escape depth, the attainable depth resolution depends on the quality of the sputtering process and its calibration and control in each instrument. Since depth assignment accuracy is dependent on the conversion of sputtering time to material removed, the rate of material removal (Å/sec) must be calibrated for the sample matrix under study. This requires that the ion beam characteristics (ion species, energy, and angle of impact) and the sample characteristics (matrix atoms, chemistry, and structure) be identical in the standard and unknown to obtain depth assignment accuracies of ±5%. The need for an accurately known sputtering rate is the main limiting feature in depth profiles employing ion sputtering. If differential sputtering of the different elements or crystalline phases in the sample is not encountered and if all the analytical signal is taken from the same depth into the sample (i.e., crater edge effects are avoided), depth resolutions of 50–200 Å can be realized from these techniques.

Lateral (x, y) Analysis. For an analytical technique to provide 10-μm lateral microanalyses, some method must be used to localize the emission of the analytical signal. As generally practiced, only AES and SIMS can be considered microanalytical techniques. In the Auger microprobe and ion microprobe, this is accomplished by a microfocused electron or ion beam. Point-to-point analysis is accomplished by positioning this microfocused beam on the area of interest. Images of the lateral distribution of an element or elements is accomplished by synchronously rastering the probing beam and using an oscilloscope whose intensity (or z axis) is modulated by either the Auger transition or ion intensity of interest. The ion microscope has the same capabilities through the use of stigmatic secondary ion optics (see the accompanying INSTRUMENTATION article for details).

For the two techniques which can provide a localized analysis, there is an important interaction between the area analyzed and the detection limits which are obtained from this area. The detection limits of the AES technique are controlled by the number of atoms present in the volume bounded by the area bombarded and the electron escape depth, the number of primary electrons that can be delivered to the area of analysis (current density), signal-to-background, and instrumental parameters. The detection limits of SIMS are controlled by the number of atoms in the volume of material consumed (determined by the area bombarded and the depth of the sputtered crater), the efficiency with which these atoms are ionized, and the efficiency with which the mass spectrometer can extract, transport, and analyze these ions. Figure 3 illustrates this concept by plotting the concentration required for 10% statistics vs. the area bombarded by the primary electron beam (AES) or the primary ion beam (SIMS). Since this relationship is a complex function of many parameters, the relationships of Figure 3 were *calculated* for a given set of conditions. As can be seen, below ~30 μm^2 the attainable AES detection limits depend on beam diameter. This is due to the low bombarding absolute current at 10^{-7} A/μm^2 current density and the limitation of usable time constants in the lock-in amplifier. However, increasing the primary electron current and, hence, the number of Auger electrons by going above ~30 μm^2 does not improve the detection limits. At this point, fundamental and instrumental effects come into play and do not permit improved detection limits.

In the SIMS technique two factors are seen to control detection limits. First, the number of atoms available is determined by the area bombarded since the sampling depth is constrained to 50 Å, as might be the case in a depth profiling analysis. Second, the ionization efficiency of a given ele-

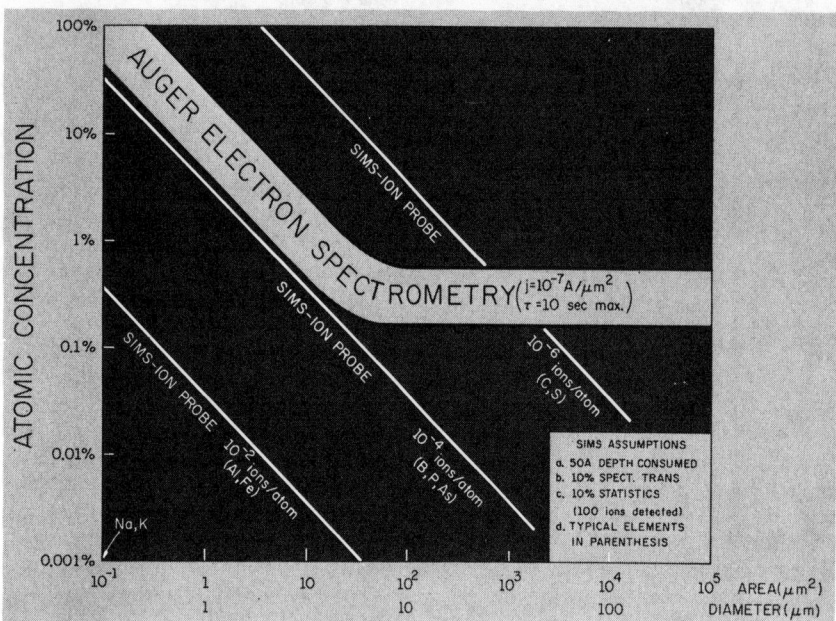

Figure 3. Detection limits attainable at 10% statistics vs. area of analysis with AES and SIMS where j is electron beam current density, τ is lock-in amplifier time constant for AES analysis, and SIMS conditions as noted

ment combines with the analytical area to determine that element's detection limit. Even though this figure illustrates the results of a general situation, two conclusions can be drawn. First, when using the SIMS technique, the analyst must be concerned with the volume of material available and elemental ionization efficiencies. To realize the best detection limits where the area is limited (localized x, y analysis), deep craters must be used; large areas should be analyzed when the depth per analytical point is constrained (depth profiling). Second, when the area and/or depth of analysis is constrained, AES will have better detection limits for some elements than will SIMS.

Sample Consumption and Alteration. Backscattering spectrometry with MeV ^{4}He$^+$ ions does not consume any sample during any part of the analysis. Sample alteration can manifest itself in radiation damage to the sample crystallinity upon extended exposure to the probing beam. In almost all applications, BS is a *nondestructive* technique.

Auger electron spectrometry and ESCA do not consume material during the actual analytical excitation process. However, some samples may be altered by the electron or photon bombardment used in the two techniques. Since depth profiling analyses with AES and ESCA require ion sputtering, these two techniques consume sample and are *destructive* techniques when operated in the depth profiling mode. The ability to separate the analytical excitation from the sputtering process is quite useful when the analyst wishes to fully characterize the surface or interfaces in a thin-film sample.

Sample consumption accompanies the analytical excitation to a greater or lesser degree in both the ISS and SIMS techniques. If low ion current densities and energies are employed, both techniques permit the analysis of a monolayer of sample while consuming only a fraction of that layer (*12, 15*). This mode is not widely used in SIMS, and the technique is generally classed as a *destructive* technique.

Matrix Effects. Because of the physical principles involved, BS exhibits no matrix effects, and relative spectral intensities are not altered by a change in sample chemistry or composition. Even though ISS also employs the Rutherford scattering process, variations in charge neutralization probabilities from sample to sample or from one region to another in a given sample can cause matrix effects.

Since changes in matrix atoms or sample chemistry influence the energy levels of an atom, both AES and ESCA exhibit matrix effects. In fact, matrix effects or "chemical shifts" form the basis for the whole ESCA technique. The reader is referred to Siegbahn et al. (*27*) for a detailed discussion of this point. Some matrix or chemical effects are observed in AES (*7*). These manifest themselves in spectral energy shifts or intensity changes for the particular Auger transitions. These effects are not generally a detriment in an analysis and can actually be used to an advantage.

Secondary ion mass spectrometry is rather susceptible to matrix effects. These matrix or chemical effects can be a problem since they can result in ion intensity changes which do not necessarily reflect a concentration change. Techniques such as reactive ion bombardment (*28*) or continuous exposure of the sample surface to a reactive gas ambient (the so-called "backfilling" technique) (*29*) have been developed to override and reduce the effects of sample chemistry changes.

Quantitation. One of the most important considerations when evaluating an analytical technique is the ability to quantitatively evaluate the data produced by that technique. Generally, this applies to the conversion of spectral intensities to atomic concentrations. This can be either conversion to absolute amounts of each atomic constituent or the relative amounts of two elements in the same sample or relative concentrations between two samples. In the use of these surface techniques, another aspect of quantitation becomes important: the quantitative assignment of a lateral or in-depth location to a particular element or elements. Since one of the principal uses of these techniques is to determine the location of an element, or elements, this aspect of quantitation can be as or more important than the assignment of a concentration to that element.

Generally speaking, quantitative assignments to lateral positions of sample constituents are no problem since instrumental design allows the easy assignment of image magnifications and distances. Depth scale determination in a depth profiling analysis is not generally as easy. For all the techniques employing ion sputtering for layer removal, the sputtering time must be converted to depth. Since sputtering rates depend on a multitude of instrumental parameters and sample composition and structure, the analyst must carefully determine the expected sputtering rates before he can assign an absolute value for the depth of a subsurface feature. If all that is needed are the relative depths of a feature between sample "A" and sample "B", good accuracies can be obtained if instrumental conditions are maintained and the two samples are similar in matrix composition and atomic structure.

Since the ion sample interaction processes involved in MeV ion backscattering spectrometry are well established, depth scale quantitation is somewhat more straightforward in this technique. The probing ion loses an incremental amount of energy for each atomic plane it passes (dE/dx). Thus, a knowledge of the energy lost before and after the scattering event (see Equation 4) and the sample density permits a direct conversion of the energy scale to a depth scale. The energy loss or stopping power parameters are known for most elements, and the assumption of the linear superposition of stopping powers permits their calculation for multielement regions [Bragg's rule (30)]. It is the need for an atomic density for a layer which keeps the conversion of energy loss to depth from being a foregone conclusion. However, the densities of the many different types of thin films encountered are known, or bulk densities can be assumed. With the exception of this generally minor problem, MeV ion backscattering provides the most readily quantitative depth scales of all the depth profiling techniques.

An evaluation of the relative ability of each technique to provide quantitative atomic concentrations is not as clear-cut with depth scale assignment. The BS technique permits quantitative determination of atomic ratios in each layer based solely on spectral intensities and tabulated values of backscattering cross sections. There is no ambiguity in this determination as there is in the depth scale assignments. Thus, in terms of quantitation of concentrations and depth scales, MeV ion backscattering qualifies as the most readily quantitated technique. Conversely, all of the other techniques (AES, ESCA, ISS, and SIMS) depend primarily on comparative standards to establish "relative sensitivity" or correction factors for concentration quantitation. As with depth scale calibration, the "standard" used may be either for absolute or relative conversion of intensity to concentration. To circumvent the need for a standard of element x in matrix y, researchers in each field are investigating other methods to simplify quantitation. If at all possible, one would like to have a correction procedure which depends on a knowledge of the mechanism of analytical signal production to correct intensities to concentrations without the need of an external standard.

• Efforts are being made in both the electron spectrometry techniques (AES and ESCA) to develop a mechanistic approach to quantitation. Morabito (31) uses a first-order approximation to the fundamental and instrumental processes governing Auger electron production and detection. Results from Morabito and selected examples taken in the authors' laboratory (32) have shown that this approach is successful when used carefully.

Quantitation of ESCA spectra incorporates two areas. It is generally assumed that the relative areas of the ESCA peaks for an element in two or more chemical states directly represent the relative number of atoms of that element in each chemical state. Quantitating the relative concentrations of two or more different elements is more complicated, and the use of a successful mechanistic approach is unknown to this author.

The practitioners of ISS generally assume the application of a backscattering coefficient correction (33) to the relative intensities from a given sample provides quantitative analysis. If future work validates this method, quantitation of ISS will be quite straightforward.

• As can be imagined, the large variations in secondary ion yields and matrix or chemical effects complicate quantitation of the SIMS techniques. The need for quantitation of SIMS intensities is great since its excellent sensitivities permit the detection of many more elements than are generally found by the other techniques. Thus, quantitation of many elements requires a very complete set of standards. Since the SIMS technique exhibits a linear relationship between intensity and concentration over many orders of magnitude, relative sensitivity factors can be established, using other techniques to calibrate at high concentrations for use at low concentrations.

Several different groups have proposed mechanistic approaches to SIMS quantitation. The most successful of these is the local thermal equilibrium (LTE) model of Andersen and Hinthorne (34). They have been able to provide corrected values to within a factor of two for almost all elements and to within 10–20% for many of these.

Chemical Information. Electron spectroscopy for chemical analysis (ESCA) is the only surface analytical technique which provides direct information on the chemical nature of the atoms in an unknown sample. The assignment of valence or bonding information is performed by comparison with published spectra of known compounds or by using one or more of the calculation methods which have been developed over the past few years (35).

As was mentioned above, there are some chemical shifts observed in the Auger electron spectra which can be used to determine sample chemistry. However, AES is not generally considered a chemical analytical technique. This difference in the amount of chemical information between the two electron spectroscopy techniques results from both the excitation and emissive modes of the two techniques. In the ESCA technique a discrete amount of energy is absorbed in the ionization step ($h\nu$), and the photoelectron carries away the energy difference between the photon and its binding energy. Ionization of the core shell by electron bombardment in the AES process does not always involve transfer of a discrete amount of the electron energy; thus, the first ejected electron carries no characteristic energy. Second, the final Auger electron's energy results from the differences between two energy levels, each of which may be chemically shifted by an unpredictable amount.

As with AES, some chemical information can be inferred from SIMS data. This can be accomplished by interpretation of chemical enhancement effects or from the molecular ion spectra produced by the ion bombardment process (15).

Neither of the ion backscattering techniques, ISS or BS, provides chemical association information.

Table II. Summary of Relative Advantages and Disadvantages of Five Common Surface Analytical Techniques

	Advantages	Disadvantages
AES	Sensitivity to low Z Minimal matrix effects Microanalysis Good all-around technique	Difficult to quantitate 0.1% detection limits
BS	Fast Quantitative in concentration and depth	Low sensitivity, especially for low Z Poor lateral resolution
ESCA	Chemical information	0.1% detection limits Poor lateral resolution Slow profiling
ISS	Outer monolayer analysis	Low sensitivity Poor lateral resolution Slow profiling
SIMS	ppm detection limits for many elements Isotopic resolution Fast Microanalysis	Quantitation and matrix effects

Speed of Analysis. When a high sample throughput is required, the time required for an analysis can be very important to the analyst. There are many different analytical situations which can be envisaged to compare the above techniques. For the sake of a comparative example, the following discussion pertains to the depth profiling analysis of four different elements in 10 different samples over a depth of 5000 Å (0.5 μm).

• Backscattering spectrometry would provide such an analysis faster than any of the other four techniques. Since vacuum requirements are minimal, the analysis could begin about 15 min after samples are loaded. Since all depths are sampled simultaneously, the data acquisition time depends on the counting statistics desired. The generally accepted time for spectrum acquisition is 10–15 min/sample. About 3 hr would be consumed in sample loading and data acquisition for the hypothetical 10 samples.

• A SIMS instrument equipped with an automated or computerized data system would not take much longer than the BS technique since sample loading would average 3–5 min/sample, and sputtering rates of 5–10 Å/sec (300–600 Å/min) are readily obtained for most matrices with most SIMS instruments. These estimates would result in about 4 hr consumed for analyzing the 10 hypothetical samples including data system set-up time.

AES would require somewhat more time since sample loading and pump-down is more time consuming, the spectrometer must be calibrated for each sample, and sputtering rates of 150–200 Å/min are the maximum generally available. Almost all of an 8-hr day would be required if a 1×10^{-8} torr vacuum were sufficient for the analysis.

• There are few or no examples of depth profiling of 0.5-μm thick films with either ISS or ESCA on which to base an estimate. With the fact that low sputtering rates (≤10 Å/min) must be used with both techniques to permit data acquisition kept in mind, at least one to two weeks would be required by both of these techniques.

The above estimates are clearly for only one of many analytical situations, but they do illustrate the relative times required for an analysis of several thin-film samples.

Conclusions

The above discussion is by necessity brief and generalized. It does not provide an exhaustive survey of the techniques or the literature on these surface analytical tools. To obtain such information the reader must delve into the details of the literature on each technique. However, several general concepts are summarized in Table II and should be used only as superficial guides. The presently available surface analytical techniques are not generally competitive. Each technique has its own particular advantages and features. There is sufficient overlap in these methods to sometimes provide the analyst with that all-important confirmation of an analysis by using more than one technique. As the demands placed on an analyst are increased, he will have to move away from dependence on one technique and bring several surface analytical methods to bear on the problem. Through this article, the author hopes to assist such an analyst in making the right choice. For other perspectives on the relative features of these surface and thin-film analytical techniques, see refs. 3, 5, and 26.

References

(1) M. Kaminsky, "Atomic and Ionic Impact Phenomena on Metal Surfaces," Academic Press, New York, N.Y., 1965.
(2) G. Carter and J. S. Colligan, "Ion Bombardment of Solids," American Elsevier, New York, N.Y., 1968.
(3) J. W. Coburn and E. Kay, CRC Crit. Rev. Solid State Sci., **4**, 561 (1974).
(4) J. A. McHugh, "Secondary Ion Mass Spectrometry," in "Methods and Phenomena, Methods of Surface Analysis," S. P. Wolsky and A. W. Czanderna, Eds., Elsevier, Amsterdam, in press, 1975.
(5) R. E. Honig, "Surface and Thin-Film Analysis of Semiconductor Materials," Thin Solid Films, in press, 1975.
(6) J. M. Morabito, Thin Solid Films **19**, 21 (1973).
(7) C. C. Chang, "Analytical Auger Electron Spectroscopy," in "Characterization of Solid Surfaces," P. F. Kane and G. B. Larrabee, Eds., Plenum Press, New York, N.Y., 1974.
(8) J. F. Ziegler and W. K. Chu, Thin Solid Films, **19**, 281 (1973); At. Nucl. Data Tables, **13**, 463 (1974).
(9) J. W. Mayer and J. F. Ziegler, "Ion Beam Surface Layer Analysis," Elsevier Sequoia, Lausanne, Switzerland, 1974.
(10) N. Winograd, Purdue University, Lafayette, Ind., private communication, 1975.
(11) R. E. Honig, "Analysis of Surfaces and Thin Films by Mass Spectrometry," in "Advances in Mass Spectrometry," Vol 6, p 337, A. R. West, Ed., Elsevier, Barking, England, 1974.
(12) R. E. Honig and W. L. Harrington, Thin Solid Films, **19**, 43 (1973).
(13) C. A. Evans, Jr., Anal. Chem., **44**, 67A (1972).
(14) C. A. Evans, Jr., Thin Solid Films, **19**, 11 (1973).
(15) A. Benninghoven, Surf. Sci., **35**, 427 (1973).
(16) H. W. Werner, ibid., **47**, 301 (1975).
(17) G. Slodzian, ibid., **48**, 161 (1975).
(18) C. W. White, D. L. Simms, and N. H. Tolk, Science, **177**, 481 (1972).
(19) N. H. Tolk, D. L. Simms, E. B. Foley, and C. W. White, Radiat. Eff., **18**, 221 (1973).
(20) H. Oechsner and W. Gerhard, Phys. Lett., **40A**, 211 (1972).
(21) J. W. Coburn, E. Taglauer, and E. Kay, J. Appl. Phys., **45**, 1779 (1974).
(22) J. E. Greene and J. M. Whelan, ibid., **44**, 2509 (1973).
(23) J. E. Greene and F. Sequeda-Orsorio, J. Vac. Sci. Technol., **10**, 1144 (1973).
(24) B. N. Colby and C. A. Evans, Jr., Appl. Spectrosc., **27**, 274 (1973).
(25) D. K. Bakale, B. N. Colby, and C. A. Evans, Jr., Anal. Chem., **47**, 1532 (1975).
(26) C. A. Evans, Jr., J. Vac. Sci. Technol., **12**, 144 (1975).
(27) K. Siegbahn, C. Vordling, A. Fahlman, R. Nordberg, K. Hamrin, J. Hedman, G. Johansson, T. Bergmark, S.-E. Karlsson, I. Lindgren, and B. Lindberg, "ESCA—Atomic, Molecular and Solid State Structure Studied by Means of Electron Spectroscopy," Almquist and Wiksells, Uppsala, Sweden, 1967.
(28) C. A. Andersen, H. J. Roden, and C. F. Robinson, J. Appl. Phys., **40**, 3419 (1969).
(29) G. Slodzian and J. F. Hennequin, CR Acad. Sci. B, **263B**, 1246 (1966).
(30) W. H. Bragg and R. Kleeman, Philos. Mag., **10**, S318 (1905).
(31) J. M. Morabito, Surf. Sci., **49**, 318 (1975).
(32) R. J. Blattner and C. A. Evans, Jr., unpublished results.
(33) F. W. Bingham, Tabulation of atomic scattering parameters calculated classically from a screened Coulomb potential, Sandia Research Report, SC-RR-66-506, TID-4500-Physics, 1966.
(34) C. A. Andersen and J. R. Hinthorne, Anal. Chem., **45**, 1421 (1973).
(35) "Electron Spectroscopy," D. A. Shirley, Ed., North-Holland, Amsterdam, The Netherlands, 1972.

Research supported in part by National Science Foundation Grants DMR 72-03026 and MPS 74-05745.

Charles A. Evans, Jr., is senior research chemist for the Materials Research Laboratory and associate professor of chemistry in the School of Chemical Sciences at the University of Illinois at Urbana-Champaign, Ill. After receiving BA (1964) and PhD (1968) degrees in chemistry from Cornell University, he joined Ledgement Laboratory of Kennecott Copper Corp. in Lexington, Mass., as analytical chemist/mass spectroscopist. Dr. Evans has been at the University of Illinois since 1970. His major areas of research include surface and thin-film characterization, ion microprobe mass spectrometry, and mass spectrometric ion sources. He has authored or coauthored over 30 invited and contributed publications on these and related topics with approximately five additional manuscripts in review.

Charles A. Evans, Jr.
Materials Research Laboratory
and
School of Chemical Sciences
University of Illinois
Urbana, Ill. 61801

Surface and Thin Film Analysis

The surface and thin-film techniques discussed in the companion REPORT article (1) require highly specialized instrumentation if the full capabilities of each technique are to be realized. Thus, to fully understand and appreciate the strengths and weaknesses of a technique, or more importantly a specific instrument, the instrumentation requirements of each must be understood. This article will discuss the general instrumental features of each of these surface and thin-film analytical techniques and the commercial instruments presently available to the scientific community. These requirements will be discussed as the technique is generally practiced and in light of the needs of someone evaluating and comparing the instruments commercially available.

General Considerations

There are two instrumental concepts and requirements which are common to all of the surface and thin-film analysis techniques: a high or ultrahigh vacuum and an ion source. Each of the techniques to be discussed places different demands on these two instrumentation components, depending on the analytical process involved and the requirements of the particular sample under analysis.

Vacuum. Since each of these techniques is used to analyze the surface of a sample, it is important to understand any interaction that surface might have with its surroundings. It is generally assumed that a *reactive* gaseous species will form one monolayer on a surface in 1 sec at a partial pressure of 1×10^{-6} torr of that species. Therefore, one must be concerned with the analytical implications of reactive species such as O_2, H_2O, N_2, CO, CO_2, and hydrocarbons when they are present in the residual vacuum. There are three regimes of vacuum requirements encountered in these surface analytical techniques.

Those techniques which have a shallow *escape depth* for the analytical signal (AES, ESCA, ISS, and SIMS) place the most stringent requirements on the cleanliness of the residual vacuum in the vicinity of the sample surface. In all of these techniques, the presence of a surface contaminant will attenuate the analytical

signal coming from the true sample and add the contaminant's characteristic spectrum to that of the real sample. Thus, vacua of *at least* 1×10^{-8} or 1×10^{-9} torr are required to maintain a sufficiently clean surface for analysis after an initial ion etch or system bake-out. In practice the partial pressure of reactive gases in all commercial AES, ESCA, and ISS instruments can be reduced to $1-3 \times 10^{-10}$ torr. Since sputtering accompanies the excitation process in SIMS, less stringent requirements are placed on a SIMS vacuum system. Hence, 1×10^{-8} torr can be tolerated except when very low sputtering rates are employed or when one wishes to analyze for elements which might contaminate from the residual vacuum (i.e., C, O, N, and H). In these instances vacua of 1×10^{-9} to 1×10^{-10} torr should be used. Although there is little information on the SCANIIR technique, its vacuum requirements would seem to parallel those of SIMS. Therefore, ultrahigh vacuum (UHV) should be considered for all these techniques.

In all but secondary ion mass spectrometry and SCANIIR, the shallow escape depth techniques use inert gas ion bombardment (He^+, Ne^+, or Ar^+) for ion sputtering of the surface. Since these inert gases do not generally react with the surfaces studied by AES, ESCA, and ISS, their presence at a high partial pressure near the specimen surface is not detrimental. It is not uncommon to introduce these gases at a high partial pressure (5×10^{-5}–5×10^{-6} torr) to feed the ion gun used for sputtering while maintaining a 1×10^{-10} torr partial pressure of the reactive gases.

The second vacuum regime is for the energetic ion beam techniques with a very large escape depth (backscattering spectrometry, nuclear microanalysis, and ion-induced X-rays). Since these techniques are not generally used for outer monolayer analysis and the analytical signal is unaffected by a monolayer surface contaminant, vacua of about 1×10^{-6} torr are generally employed. This pressure is sufficiently low to not scatter the exciting ion beam and is easily and inexpensively attainable with modern vacuum technology.

The third operating vacuum regime is that employed in the glow discharge techniques. In GDOS and GDMS the partial pressure of reactive gases is usually kept low ($<1 \times 10^{-6}$ torr) so that they will not influence the analysis. However, the actual pressure in the region of the sample surface is quite high, 1×10^{-3}–1×10^{-1} torr, to provide the inert gas ions for sputtering and the metastable atoms for excitation.

Ion Sources. All of the surface and thin-film analytical techniques require ions for either excitation or sputtering. Just as in the vacuum system, the exact ion source, focusing system, etc., used varies with the technique and application.

The AES, ESCA, and ISS instruments generally employ simple electron impact, low accelerating potential (0.5–3 kV), and low current density (100–500 µA/cm^2) ion sources to produce a large diameter (1–5 mm) ion beam. The beam is usually kept large and the analytical signal localized to a small area to ensure that the data come from a region of uniform depth.

Secondary ion mass spectrometers generally employ a more sophisticated ion source and focusing column. These range from a relatively simple plasma discharge source to the duoplasmatron with two stages of electrostatic lens demagnification used in ion microprobes. Since the ion beam used in SIMS fulfills the excitation as well as sputtering requirements of the technique, the ion source is a key part of the instrument and determines many of the instrumental capabilities.

The ion sources used in the energetic ion beam techniques are an integral part of the accelerators needed for these techniques. As in SIMS, the ion source/accelerator combination is a key part of the overall analytical capability of the final system. The reader is referred to Wilson and Brewer (2) for a discussion of accelerator technology.

Specific Instrumentation

Auger Electron Spectrometry (AES). The basic functions and corresponding required instrumentation for Auger electron spectrometry are:

Excitation. An electron beam of 3–10 keV in energy and up to 50 µA is employed for producing the Auger electrons from the sample surface to be characterized. Auger electron spectrometers are available as large-beam (>25 µm) depth profiling instruments or as Auger microprobes with minimum electron beam diameters of <5 µm. It is important to remember that the detection limits attainable by AES are related to the spectrometer detection system's integration time and the exciting electron current; thus, the reduction in beam currents which accompanies the decrease in probe diameter will result in poorer detection limits and/or increased analytical time. If an instrument is to be used as a high-sensitivity "depth profiler" and as a high lateral resolution microprobe, two electron guns may be required to realize optimum use of each operating mode.

Sample Mounting and Manipulation. The samples to be analyzed must be mounted in a UHV chamber. To obtain reasonable turnaround times at ultrahigh vacuum, a multisample carrousel (10–12 samples) or a vacuum lock arrangement must be available. Sample movement in x, y, and z is required for spectrometer energy calibration as well as for positioning the area to be analyzed within the area viewed by the spectrometer.

Spectrum Acquisition. The cylindrical mirror analyzer (CMA) (3) is the most commonly used dispersing element in AES with an electron multiplier for electron detection. The spectrometer voltage is modulated by a few volts, and a lock-in amplifier is used for signal amplification to differentiate the low-level Auger peaks from the large continuous background. Recently, pulse counting (with no spectrometer voltage modulation) techniques have been used for detecting the low Auger intensities produced by low-intensity exciting beams (4).

Ion Sputtering. To provide ion sputtering for surface cleaning and/or depth profiling, the sample chamber is generally backfilled with argon and an electron impact ion source used. Since the ion beam's current density controls the sputtering rate, high current densities (>100 mA/cm^2) may be desired when depth profiles over large distances (≥1 µm) are desired.

Other Capabilities. A variety of options or accessories are available with AES systems for specific applications. A multiplexer or computerized data system for recording the intensities of several peaks vs. time is a must if the instrument is to be used extensively for depth profiling. Sample heating, cooling, and/or fracturing facilities are required for certain metallurgical and annealing studies. Auger microanalysis and imaging require primary electron beam rastering and readout oscilloscopes for recording.

Table I summarizes some of the features and capabilities of the commercially available AES systems. As with all the instruments to be discussed, the tables are only summaries of information provided to this author (marked with an *) or information gleaned from commercial brochures and publications. This table includes information on both the depth profiling AES systems as well as the Auger microprobes. As can be seen, a wide variety of capabilities are provided by the available AES systems.

Energetic Ion Beam Techniques. Complete systems for the three high-energy ion techniques of backscattering spectrometry (BS), ion-induced X-rays, and nuclear microanalysis are not commercially available. However, the accelerators, vacuum system, sample chamber detectors, and readout devices are readily available and can be assembled in many laboratories.

Accelerators. Depending on the energies desired, ion implantation units (up to 300 keV) or Van de Graaff units (0.5–3 MeV) are available in many laboratories or can be purchased for $40,000–$100,000. The type of ions required depends on the particular technique to be used with ^{4}He$^+$ generally used in BS, protons and deuterons in nuclear microanalysis, and protons for ion-induced X-rays.

Sample Chamber and Manipulation. Generally, standard diffusion pump vacuum systems are employed in all three techniques. Sample manipulation requirements include a multiple sample holder to increase sample throughput and a goniometer if channeling measurements are to be made.

Detection Systems. Surface barrier detectors are used for detecting the backscattered ^{4}He$^+$ in the BS technique. This detector produces a pulse which is proportional to particle energy. The pulse is then processed by a multichannel analyzer to produce a spectrum of backscattered events vs. energy. Digital recorders or point plotters then provide hardcopy of the data.

The X-rays produced by ion bombardment are detected by a lithium-drifted silicon detector which produces a pulse proportional to each X-ray photon's energy. Multichannel analyzer processing of these pulses results in an X-ray spectrum which can be read out by a variety of methods.

The detector requirements of nuclear microanalysis are dependent on the particle produced by the nuclear reaction, i.e., ^{4}He$^+$, protons, or gamma rays. Multichannel analyzers are not required since counting data are taken for each setting of the probing beam energy. Simple electronics can be used with manual data recording.

Electron Spectroscopy for Chemical Analysis (ESCA). There are several commercially available ESCA systems. This discussion will be limited to X-ray photoelectron spectroscopy (XPS) since this is the technique most generally applicable to a complete surface characterization. The use of ultraviolet photoelectron spectroscopy (UPS) is limited by the energy of available UV sources to the study of loosely bound electrons (<50 eV); thus, it lacks the broad coverage of X-ray excitation.

The purpose of an ESCA analysis is to accurately measure the binding energy of an electron both for elemental characterization and for the measurement of chemical shifts resulting from the atomic environment of each element. The effect of each instrumental component on spectral resolution must be considered.

X-ray Source. The X-ray source used for photoelectron excitation is a key part of an ESCA spectrometer. The radiation from an X-ray source consists of the line or lines characteristic of the target element superimposed on a continuous background. The intensity of the X-ray background is much less than that of the characteristic line (ex., Al Kα) but will still contribute photoelectrons to the spectrum which will become part of the spectral background. In addition, the presence of more than one characteristic line in the X-ray spectrum (ex., Kα_1 and Kα_2 or Kα and Kβ) will produce a "shadow" spectrum which must be accounted for in the interpretation process and may cause spectral interferences.

Two different types of X-ray systems are used in commercial ESCA instrumentation. The first is the use of direct illumination of the sample by the output of the X-ray source. This "polychromatic" type of X-ray source is usually quite simple and permits the use of different target materials and hence different energies of photoelectron excitation. The other system incorporates an X-ray spectrometer in the source which disperses the X-radiation and provides monochromatic illumination of the sample surface. Although this reduces spectral interferences and background, only one X-ray target material can be used, and low X-ray intensities impinge on the sample surface. One manufacturer (AEI Scientific Apparatus) provides either or both types of sources in its instrument.

Sample Chamber and Manipulation. As with the other shallow escape depth techniques, ESCA requires UHV techniques to reduce surface contamination by residual vacuum species. Of all the surface sensitive techniques, ESCA systems are available with the widest variety of sample handling facilities. Accessories are generally available for the introduction of gases, liquids, and solids, for sample heating and cooling, for mounting of several samples, and for ion etching to clean a surface or depth profile an element or elements. In addition, some ESCA's can be fitted with an electron source to provide Auger electron spectrometry using the ESCA spectrometer.

Electron Spectrometer. The energy resolution of an ESCA system is not necessarily determined by the spectrometer resolution. Other effects such as those discussed above (i.e., the X-ray source and its line width) and the electron energy line width combine with the spectrometer resolution to produce the attainable spectral resolution. Two types of electrostatic spectrometers are used in ESCA systems—the double-pass cylindrical mirror analyzer (Physical Electronics) and spherical analyzers.

Data Acquisition and Handling. After dispersion by the electron spectrometer, the photoelectrons are detected by an electron multiplier of either the discrete or continuous dynode type. Analog intensity or pulse count vs. spectrometer energy pass provides the ESCA spectrum. The spectrum may be recorded with an x-y recorder, multichannel analyzer, or a computer-based data system. The latter usually provides software for data manipulation such as peak identification, spectral smoothing, and background subtraction. Table II summarizes the analytical instrumentation available for ESCA experiments.

Ion Scattering Spectrometry (ISS). There is only one commercially available ion scattering spectrometer (Model 520, 3M Corp., St. Paul, Minn.). An electron impact ionization source provides rare gas ions (He$^+$, Ne$^+$, or Ar$^+$) with a low initial kinetic energy spread. These ions are accelerated to 0.5–3 keV and directed toward the sample surface at 45°. The primary beam is about 1 mm in diameter with a current density of 10 μA/cm^2. Instrumentation to produce a 0.1-mm diameter beam is under development to provide better resolutions. This facility will also permit primary beam rastering to flatten the crater bottom and reduce crater edge effects.

The ISS instrumentation provides a sample chamber residual vacuum of 1 $\times$ 10^{-9} torr with ion and titanium sublimation pumping. A multiple sample holder is used to reduce sample turnaround time. Sample charging of insulating surfaces is alleviated with a charge neutralization filament.

Scattered ion energy at a 90° scattering angle is measured by a 127° electrostatic spectrometer situated at 45° to the sample surface. Ions are detected with a continuous dynode multiplier, pulse counting readout electronics, and a ratemeter for analog chart recording of the spectrum.

A secondary ion mass spectrometry (SIMS) attachment is available to complement and extend the ISS capability. This SIMS attachment uses a quadrupole mass filter for secondary ion spectrometry. The reader is referred to the next section for a discussion of specific SIMS instrumental requirements.

Secondary Ion Mass Spectrometry (SIMS). Instrumentation for SIMS ranges from simple discharge ion gun/quadrupole mass analyzers to sophisticated ion microprobes and ion microscopes. The nature of the primary ion source, sample chamber, and mass spectrometer used in any SIMS system controls its capability and price. The SIMS instrumentation can be divided into three general classes: ion microprobes which use a microfocused primary ion beam to provide lateral microanalysis with high spatial resolution (2–10 μm) as well as surface analysis and in-depth profiling; the ion microscope which employs a large primary beam combined with a stigmatic mass spectrometer to provide high lateral resolution ($\leq$1 μm) (surface and in-depth profiling analyses are also available with this instrumental concept); and the "big beam" ion probes which provide surface and depth profiling analyses with minimal lateral resolutions ($\geq$100 μm). The following is based on the above classifications for ease of discussions.

Primary Ion Source and Optics.
ION MICROPROBE MASS SPECTROMETERS. The duoplasmatron is the ion source of choice for ion microprobes. This source employs a plasma with both electrostatic and magnetic constriction (hence the term *duo*) to produce a high brightness source of inert or reactive gas ions. Ions are extracted from the source and accelerated to 15–25 keV. Two electrostatic lenses then provide demagnification of the duoplasmatron source image to produce an ion beam diameter of 2–10 μm. The primary beam spot is held stationary for a local analysis or rastered about the sample surface for secondary ion imaging or for producing a flat-bottomed crater for accurate depth profiling. Primary ion mass spectrometry may be employed to provide an elementally and isotopically pure primary beam. Although this primary ion mass analysis capability is analytically appealing, there are no comparative data yet available to demonstrate its overall utility or need.

ION MICROSCOPE. The ion microscope uses the duoplasmatron ion source and two electrostatic lenses to produce ion beams in the range of 20–250 μm diameter. Smaller diameters are not required since this instrumental concept depends on the secondary ion optics (i.e., a stigmatic focusing mass spectrometer) for control of the lateral resolution. A primary ion beam raster is also used in the ion microscope but only for crater geometry control.

LARGE-BEAM, SIMS INSTRUMENTS. There are about as many different ion sources used in this type of instruments as there are instruments available. Since the ion source in any ion probe is closely related to the instrument's ultimate use, manufacturers have used oscillating electron sources and electron impact sources as well as others to realize their specific aims. The advantages, limitations, and justifications for each source are beyond the scope of this article; thus, the reader is referred to the scientific literature and the manufacturers for details.

Sample Chamber and Manipulation. The quality and reliability of a SIMS analysis, microbeam or "big beam", are influenced by vacuum quality in the sample region. In gener-

Table I. Summary of Commercially Available Auger Electron Spectrometer Systems

Manufacturer, address, and model number	Primary electron column			Specimen chamber		Sputtering	
	Type of lenses	Primary energy keV	Spot size and current	Ultimate vacuum, torr	Sample introduction	Ion energy, keV	Current density, $\mu A/cm^2$
Etec Corp.* Hayward, Calif. SEM with Auger system attachment†	Magnetic	1–30	100 Å/300 μm 5 pA/10 μA	10^{-9}	Chamber venting with 8-sample carrousel	1–25 depending on option	125 or 250
JEOL Ltd.* Akishima, Tokyo, Japan Model JAMP-3 Auger scanning electron microscope	Magnetic		250 Å/1 mm 10 pA/50 μA	$<10^{-9}$	Airlock, one sample at a time	Up to 3	100
Physical Electronics* Eden Prairie, Minn. Model 540A thin-film analyzer	Electrostatic	5		10^{-10}	Chamber vent with 12-sample carrousel	Up to 5	200
Model 545 scanning Auger microprobe	Electrostatic	10	<5 μm/15 μm 50 nA/5 μA	10^{-10}	Chamber vent with 12-sample carrousel	Up to 5	200
Varian Associates* Palo Alto, Calif. Model 981-2000 high-resolution Auger spectrometer	Electrostatic	0–10	<5 μm/30 μm >100 nA/10 μA	10^{-10}	Chamber venting with 15-sample carrousel	Up to 3	250
Model 981-2000 with Auger microprobe options	Electrostatic	0–10	<5 μm/30 μm >100 nA/10 μA	10^{-10}	Chamber venting with 15-sample carrousel	Up to 3	250
V. G. Scientific Ltd.* East Grinstead, Sussex, England Model HBA 205 scanning Auger microscope	Magnetic (field emission gun)	25	100 Å/1 μm 5 nA/50 nA	2×10^{-10}	Airlock, one sample at a time	Up to 10	400
Model 850 CMA	Electrostatic	0.5–5	5 μm/5 mm 1 μA/1 mA	$<5 \times 10^{-11}$	Venting or airlock, 8-sample carrousel	0.5–10	100
Model SAE S2	Electrostatic	5	5 μm/5 mm 1 μA/1 mA	$<5 \times 10^{-11}$	Airlock, 4 samples at time	0.5–10	100

*Denotes information provided by the company to the author.

System	Sputtering rate for Si, Å/min	Electron spectrometer		Data acquisition			Basic system price	Other options
		Spectrometer type	Resolution transmission	Spectral display	Multiplexer?	Imaging modes		
	200	Cylindrical mirror analyzer (CMA)	0.5–1.0% / 10%	CRT, recorder	Yes, 6 channels	SEM, Auger, sample current, line scan	$100,000–$150,000	Sample introduction air-lock; Hot/cold stage; Tensile stage
	100	CMA	0.5% / 6%	CRT, recorder	Yes	SEM, Auger, sample current, line scan		Fracture stage; Hot/cold stage; X-ray detector; IC holder
	250	CMA	0.6% / 10%	CRT, recorder	Yes, 6 channels	No, optical microscope only	$60,000	Fracture stage; Cold stage
	250	CMA	0.6% / 10%	CRT, recorder	Yes, 6 channels	SEM, Auger, sample current, line scan	$100,000	Electron beam heating of sample
	250	CMA	0.25% / 8%	CRT, recorder	Yes, 8 channels	SEM and optical microscope	$68,000	Hot/cold stage; Field replaceable filament
	250	CMA	0.25% / 8%	CRT, recorder	Yes, 8 channels	SEM, Auger images and line	$75,000 (complete)	$7,000 as pkg to upgrade above to an Auger microprobe
	100	Hemi-CMA	Variable, 0.4–0.8% / >5%	CRT, recorder (optional data system)	Yes, 8 channels or with optional computer	SEM, Auger, sample current, line scan	$150,000	X-ray detector; 8–10 others; Data system
	160	CMA	<0.3% / 4%	CRT, recorder (optional data system)	Yes or with optional computer	SEM, sample current	$80,000	Auger imaging; Fracture stage; 6–8 others
	160	150° spherical sector in retarding mode	<25 MeV / 4%	CRT, recorder (optional data system)	Yes or with optional computer	SEM, sample current	$75,000	Fracture; ESCA option

Table II. Summary of Commercially Available ESCA Systems

Manufacturer and address	Exciting sources			Ultimate vacuum, torr	Sample chamber
	X-ray anode and power	UV lamp?	Electron impact?		Sample introduction and handling
AEI Scientific Apparatus Ltd.* Urmston, Manchester, England	Polychromatic: Mg and Al 600 W Monochromatic: Al Kα 600 W (optional)	Option He I He II	Option 0–10 μA at 1–10 keV	1×10^{-10}	Solids: 4-sample holder via airlock Liquids: cold stage Gases: with cell
E. I. du Pont* Monrovia, Calif.	Polychromatic: Mg and Al 350–600 W			2×10^{-9}	Solids: single-sample holder via airlock
Hewlett-Packard Corp. Palo Alto, Calif.	Monochromatic: Al Kα (600 W estimate)			$\sim 10^{-8}$	Solids: single-sample holder via airlock
McPherson Instruments Corp. Acton, Mass.	Polychromatic: Al, Mg, Cu 500 W	Option	Option	$< 10^{-10}$	Solids: 8-sample holder via chamber venting
Physical Electronics Industries* Eden Prairie, Minn.	Polychromatic: Al or Mg 600 W	Provisions for addition by user	Standard 50 μA at 5 keV	1×10^{-10}	Solids: 12-sample holder via chamber venting
V. G. Scientific Ltd.* East Grinstead, Sussex, England	Polychromatic: Al 1000 W Mg 600 W	Option He I He II	Option microfocus 1 μA–1 mA 5 μm–5 mm at 5 keV	5×10^{-11}	Solids: 4-sample holder via airlock Gases: via optional cooled sample holder

*Denotes information provided by the company to the author.

al, the particular pumping system required is independent of instrument type (microprobe, microscope, or "big beam") and depends on the application for the instrument and on the configuration of the ion source, sample chamber, and spectrometer used. The potential user must seriously consider each instrument in light of the analytical demands to be placed on that instrument.

Sample mounting and manipulation requirements for SIMS depend on the instrumental type and potential applications. The ion microprobe is the most demanding on sample manipulation since microanalyses generally require micromanipulation capabilities. In addition, high-quality light optics for sample viewing are also a critical part of a microprobe. The ion microscope does not require as sophisticated a sample stage as the microprobe owing to the instrumental and operating concepts involved. The large-beam ion probes usually have simple manipulation facilities since precision lateral control is not necessary. All of the SIMS instruments use either a multiple sample holder or a vacuum lock for maintaining a short turnaround time.

Mass Spectrometer. There are two basic types of mass spectrometers used for secondary ion mass analysis—magnetic deflection and quadrupole filters. Because of the large initial kinetic energy spread of the secondary ions, most SIMS instruments employ some form of energy filter, generally an electrostatic analyzer. All the ion microprobes employ a double-focusing instrument (i.e., electrostatic and magnetic analyzers) for mass analysis of the secondary ions. The Applied Research Laboratories (5) and Hitachi (6) instruments each use a specially designed mass spectrometer with mass resolutions of the order of 300 (10% valley). The AEI Scientific Apparatus IM-20 (7) uses the MS-7 spark source mass spectrometer and provides mass resolution in excess of 5000 (10% valley).

The Cameca Instruments' ion microscope employs the magnetic prism-electrostatic mirror-magnetic prism stigmatic optics of Castaing and Slodzian (8). The secondary ions are extracted by an electrostatic immersion lens and mass separated by the prism-mirror-prism optics. The electrostatic mirror provides an ion kinetic energy cutoff to reduce the effects of the large initial energy spread. This unique ion optical system extracts, transports, and mass separates the secondary ions while maintaining their spatial relationship. An image of any desired mass is produced simultaneously from all points in the bombardment area by the secondary ion optical system. By choice of the proper operating mode, a mass scan can be taken of a selected area, depth profiles obtained at a specific area, or the spatial distribution (ion image) of any element recorded. Recently, an electrostatic analyzer has been added to the ion microscope to permit high mass resolution (≥ 3000, 10% valley) in all operating modes except the imaging mode.

The large-beam instruments use the quadrupole mass filter with some form of energy analysis for secondary ion

Charge neutralization	Ion etching	Electron spectrometer	System control	Basic system price	Other options
Required only with monochromatic X-ray source (option)	Option 100 μA/cm² at 2 keV	180° hemispherical analyzer with retarding lens	Manual Optional MCA or computer	$110,000	Hot/cold stages Monochromator Gas cell Others
Optionally available	Option 30 μA/cm² at 2 keV	Nondispersive, series electrostatic filters	Manual Optional MCA	$55,700	Hot/cold stages Vacuum booster Evaporator Others
Optionally available	Option	180° hemispherical analyzer with dispersion compensation optics	Manual and with MCA Optional computer	$114,000	Hot/cold stages Evaporator
		180° hemispherical analyzer without retarding	Computer system		Custom sample chambers He cryo-pump for sample chamber
Optionally available	Standard 200 μA/cm² at 5 keV	Double-pass cylindrical mirror analyzer	Manual Optional MCA Provisions for computer addition	~$100,000	Hot/cold stage Fracture stage EB heating of sample
Optionally available	Option 1–100 μA at 0.5–10 keV	150° spherical sector with retarding	Manual Optional computer system	~$85,000	Evaporator AES electronics Hot/cold stage Quadrupole RGA

spectrometry. These energy analyzers range from cylindrical electrostatic analyzers to simple lenses and an aperture to pass ions of only a selected energy. The general aim of the large-beam instruments is to provide the excellent detection limits and other SIMS analytical features in a simple, low-cost depth profiling instrument. The quadrupole mass filter fulfills this need quite well. However, it exhibits reduced transmission at high mass and cannot provide the mass resolution of a double-focusing magnetic instrument.

Ion Detection and Data Handling. After mass analysis, the secondary ions are generally detected by electrical detection devices. The electrical detectors used include discrete and continuous dynode electron multipliers and the Daly-type detector (9). Analog measurement techniques are used for spectral recording, and digital pulse counting is used for accurate measurement of ion intensities. The AEI ion microprobe does have provisions for photographic plate recording of the mass spectrum in addition to the electrical detection mode. This mode can be quite useful since all masses are recorded simultaneously but is much less sensitive than the electron multiplier detection system.

The readout modes available with each instrument depend on the application and instrumental design concepts.

ION MICROPROBES. Because of their versatility and range of applications, the ion microprobes can be operated in several modes. For localized lateral microanalysis the primary beam is held stationary on the area to be analyzed. Survey analyses of this area are obtained by taking a mass spectrum in the analog recording mode as sample atoms are sputtered and ionized. Ion intensities are quantitatively determined by setting on the mass or masses of interest and digitally counting each for a preset time. Elemental distributions are provided in the secondary ion imaging mode by rastering the primary beam about the sample surface while monitoring the mass of interest. The intensity of that ion is used to modulate the intensity axis of an oscilloscope which is synchronously rastered with the primary beam. The relative number of secondary ions from different areas is thereby converted to regions of brightness and darkness on the oscilloscope screen. Depth profile measurements are performed by monitoring the mass of interest while rastering the primary beam over an area at least five or six beam diameters on a side. In this manner a flat bottom crater is produced so that proper gating of the detector counting electronics allows only those ions generated over the flat-bottomed area to be measured. This "electronic aperturing" technique rejects ions from the sloping crater walls and permits the assignment of a discrete depth to each data point. Tables III and IV summarize some of the instrumental features of the microanalytical and large-beam commercially available instruments.

ION MICROSCOPE. All of the different analytical modes of the ion micro-

Table III. Summary of Commercially Available Ion Microprobes and Ion Microscope

Manufacturer and address	Primary ion system			Sample Chamber		
	Primary ion mass analyzer	1° spot diam.	Raster 1° spot	Ultimate press, torr	Light optics	Sample introduction
ION MICROPROBES						
AEI Scientific Apparatus Ltd.* Urmston, Manchester, England	Optional electromagnet	<10–>300 μm (spec) 3 μm attained in actual use	Yes (optional)	2×10^{-8}	Yes, monocular	Chamber venting with 8-sample carrousel
Applied Research Laboratories Sunland, Calif.	Yes, electromagnet	2–300 μm (spec) <2 μm attained in actual use	Yes	10^{-7}–10^{-8}	Yes, binocular	Chamber venting with 4-sample carrousel
Hitachi Ltd.* Katsuta, Ibaraki, Japan	Optional Wien filter	2–1000 μm	Yes	2×10^{-7}	Yes, binocular	Chamber venting with 6-sample carrousel
ION MICROSCOPE						
Cameca Instruments 92 Courbevoie, France	No	15–300 μm	Yes	$<10^{-8}$	Yes, monocular	Airlock, one sample at a time

*Denotes information provided by the company to the author.

Table IV. Summary of Commercially Available Large-Beam SIMS Systems

Manufacturer and address	Primary ion system			Sample	
	1° source type	1° beam diam	1° beam rasterable	Ultimate press, torr	Sample introduction
Applied Research Labs* Sunland, Calif.	Duoplasmatron	70–100 μm	Yes, over 1 × 1 mm	$<10^{-8}$	Chamber venting and single-sample holder
Balzers GmbH Liechtenstein	Finkelstein plasma source	~1 mm		10^{-11}–10^{-10}	
Commonwealth Scientific* Alexandria, Va.	Magnetic, cold-cathode glow discharge	1–6 mm	No	3×10^{-7}	Chamber venting and 2-sample holder
Etec Corp.* Hayward, Calif. (accessory for scanning electron microscope)	High-density diode	0.1–1 cm	No	10^{-6}–10^{-9} depending on system configuration	Chamber venting or airlock
Extranuclear Laboratories* Pittsburgh, Pa. (in conjunction with Atomica, W. Germany)	"Telefocus gun"	0.1–1 mm	Yes, over 1.5 × 1.5 mm	10^{-10}	Chamber venting with 8-sample holder
3M Corp. St. Paul, Minn. (accessory for ion scattering spectrometer)	Electron impact	~1 mm	Optionally available	10^{-9} (estimate)	Chamber venting and 4-sample carrousel
V. G. Scientific Ltd.* East Grinstead, Sussex, England		0.5–5 mm	Yes, over 1 × 1 cm	5×10^{-11}	Airlock with 8-sample holder

*Denotes information provided by the company to the author.

Mass spectrometer						
Estd transmission at low resolution	Max. mass resolution	Detection	System control	Crater edge rejection	Basic price	Other options
5–10% at 250 RP; $1 \times 10^{+6}$ $^{56}Fe^+$/nA of primary O^+ on pure Fe	>5000 (10% valley) >10,000 (FWHM)	Photographic; Electrical (optional)	Manual	Optional electronic aperture	$276,000	System is modular with several options
10%; $2 \times 10^{+6}$ $^{56}Fe^+$/nA	600 (10% valley)	Electrical	Manual; Optional MCA and computer	Optional electronic aperture	$260,000 (estimate)	
~1% at 300 RP; $2 \times 10^{+5}$ $^{56}Fe^+$/nA	>300 (10% valley)	Electrical	Manual; Optional MCA and computer	Stop-scan method in manual control; Electronic aperture with computer	$150,000 in Japan	
5–10% at 200 RP; 10^{+6} $^{56}Fe^+$/nA	1000 (10% valley) See options	Electrical for all modes except images recorded photographically	Manual; Optional MCA and computer	Selected area aperture	$250,000	Electrostatic analyzer for resolution >3000 (10% valley)

chamber		Mass spectrometer					
Charge neutralization	Light optics	Mass spectrometer	Energy filter	Crater edge rejection	System operation	Basic price	Other options
Provided by negative 1° ion bombardment	Optional	Quadrupole	Yes, 160° spherical electric sector	Electronic aperture	Manual; Optional multiplexer or computer system	$100,000	Cryopanel in sample chamber; x,y rotational sample stage
		Quadrupole		Large beam	Manual		Auger spectrometer on same vacuum system
e^- flooding by hot filament	No	Quadrupole	Yes	Large beam and selected area aperture	Manual	$35,000	Ion counting
Not available	No, sample viewed by SEM	Quadrupole	Yes	Aperturing	Manual		
e^- flooding by hot filament	No, but can be seen by unaided eye	Quadrupole	Yes, parallel plate analyzer	Electronic aperture	Manual; Optional MCA and computer system	$95,000	
e^- flooding by hot filament	No	Quadrupole	Yes	Optional electronic aperture	Manual		
e^- flooding by hot filament	No, but can be seen by unaided eye	Quadrupole	Yes	Large beam and selected area aperture	Manual; Optional computer system	$85,000	Available as part of multi-technique instrument

probe are attainable with the ion microscope. As can be seen from the above discussions, each type of analysis is obtained in a significantly different manner in the microprobe and the microscope, owing to the dramatically different instrumental concepts employed.

BIG-BEAM ION PROBES. Since the complex instrumentation required for high spatial resolution microanalysis and ion imaging is not involved, the big-beam SIMS instrumentation is much simpler and easier to use. Surface analyses are performed by low current density primary ion beams while the desired spectra are acquired Depth profiling usually employs high-density primary beams (hence more rapid sputtering) for routine analysis over depths of several thousand angstroms. Several masses can be monitored if rapid peak switching facilities are available.

To realize efficient use of the copious and rapidly generated data, a SIMS instrument requires either a hardwired or computerized data system. This is especially true for depth profiling. In this mode, ions are produced from all elements in the sample volume and that volume is consumed and lost for future analysis. Thus, the ability to successively monitor as many masses as possible during a given profile permits more efficient material consumption and ion generation.

As can be seen by the above discussion, the instrumentation used for SIMS is quite varied and ranges from rather simple systems to the highly sophisticated ion microprobes and microscopes. The needs of a laboratory must be kept in mind when evaluating the overall technique as well as a particular instrument.

Future Instrumentation

In addition to the continued evolution of the present instrumental concepts, new instrumental concepts are being developed. General Ionex of Ipswich, Mass., will have a SCANIIR-type instrument available in mid-1975. As the development of the glow discharge techniques progresses and they show reasonable promise, commercialization of the appropriate instrumentation seems quite likely.

One of the more analytically important instrumentation developments is presently taking place. This is the combining of two or more analytical techniques in the same analytical system to permit rapid use of the complementary nature of the different surface and microanalytical techniques. Several manufacturers are or will soon be offering combined scanning electron microscopes and scanning Auger microprobes (Physical Electronics, Varian Associates, Vacuum Generators, and Etec Corp.). Auger electron spectrometry and ESCA have been combined in a single instrument by Physical Electronics and AEI Scientific Apparatus, albeit with significantly different approaches. As the capabilities and limitations of each technique are delineated and as more analytically demanding materials analysis problems are attacked, other combined systems will almost certainly be developed.

References

(1) C. A. Evans, Jr., *Anal. Chem.*, **47**, 818A (1975).
(2) R. G. Wilson and G. R. Brewer, "Ion Beams with Applications to Ion Implantation," Wiley, New York, N.Y., 1973.
(3) P. W. Palmberg, G. K. Bohn, and J. C. Tracy, *Appl. Phys. Lett.*, **15**, 254 (1969).
(4) H. Weinberg, Caltech, Pasadena, Calif., private communication, 1974.
(5) H. Liebl, *J. Appl. Phys.*, **38**, 5277 (1967).
(6) H. Tamura, T. Kondo, and H. Doe, "Advances in Mass Spectrometry," Vol V, p 441, A. Quayle, Ed., Institute of Petroleum, London, England, 1971.
(7) A. E. Banner and B. P. Stimpson, *Vacuum*, in press (1975).
(8) R. Castaing and G. Slodzian, *J. Microsc.*, **1**, 395 (1962).
(9) N. R. Daly, *Rev. Sci. Instrum.*, **31**, 264 (1960).

Research supported in part by the National Science Foundation Grants DMR 72-03026 and MPS 74-05745.

The Renaissance in Polarographic and Voltammetric Analysis

JUD B. FLATO

Chemical Instrument Group, Princeton Applied Research Corp.
P.O. Box 2565, Princeton, N.J. 08540

As more and more analysts adopt modern polarographic and voltammetric techniques, literature data will become more complete, and the use of such methods as differential pulse polarography, alternating current polarography, fast linear sweep voltammetry, direct anodic stripping voltammetry, and differential pulsed anodic stripping voltammetry for fingerprint purposes and analytical applications will increase

Since Heyrovsky discovered that the current flowing between a counter electrode and a dropping mercury electrode (1), at a particular potential, was related to the concentration of one of the species present in the solution through which the current was flowing, polarography has been used as an analytical technique. In 1927 when the initial work was performed, instrumental methods were a rarity, and the development of this technique, despite difficulties in using photographic recorders and mechanical galvanometers, provided a breakthrough in the analytical laboratory. For the first time, a method other than time-consuming volumetric and gravimetric techniques was available for the determination of metals in solution. For the first time, solutions below millimolar level could be analyzed without using extremely elaborate techniques.

Through the second quarter of this century, the growth of the technique was marked, and polarographic analysis found its way into many laboratories. In fact, the contribution of this technique to the advancement of analysis was considered sufficiently important to warrant the awarding of the Nobel prize for chemistry to Heyrovsky in the late 1950's.

The basic dc polarographic technique, however, suffered from a number of "defects" which made it less than ideal for routine analytical purposes and made the results obtained somewhat difficult to interpret. The idealized waveforms expected from dc polarography are quite easy to interpret, but the actual waveforms obtained from all but the best cases are considerably more difficult to use. The "Faradaic" current produced by the reduction or oxidation of the species of interest is only one of a number of currents flowing through the system, so that very dilute solutions, where the contribution of this Faradaic current to the overall signal is small, yield dc polarograms lacking in useful information. Problems such as these dampened enthusiasm for the technique somewhat and diminished its growth.

In addition, it was quite clear from the onset of polarographic investigations that many applications to organic chemistry existed for the technique. However, early two-electrode equipment could not cope with the higher resistance of nonaqueous solvent systems, so that the applications were additionally limited to systems which could be made up in water or water–alcohol systems.

Beginning in the mid-1950's when a degree of electronic sophistication entered the chemistry laboratory and, still more, with the advent of low-cost operational amplifiers in the late 50's and early 60's, modifications of the basic polarographic technique aimed at overcoming the various problems associated with it began to meet with success. Various workers applied waveforms different from and usually more complex than the simply varying dc potential normally used in classical polarography and applied various modes of signal processing to the measured currents obtained. Each of these efforts was aimed at overcoming one or more specific problems, and little by little, ways of overcoming each of these problems were developed. Investigations of such techniques as square wave polarography, pulse polarography, differential pulse polarography, and ac polarography began in various locations, and results began to appear in the literature.

In fact, the decade from 1955 to 1965 might be characterized as the one single period during which the greatest advancement in the technical aspects of polarography took place, while, simultaneously, the greatest decline in the practical everyday usage of these techniques occurred. Great progress was being made in the research laboratory in solving the basic problems associated with dc polarography and in applying the newer polarographic techniques to organic systems, but commercial instrumentation capable of exploiting this progress was either unavailable or extremely expensive. Thus, analysts who were unwilling or unable to build their own equipment were forced to rely upon primitive dc instrumentation. Then, when other analytical methods which were either less prone to difficulty or more profitable for the instrument manufacturers began to be aggressively promoted and when commercial instrumentation for these

Instrumentation

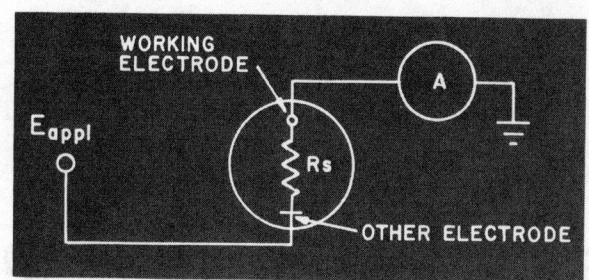

Figure 1. Simple two-electrode cell showing resistance

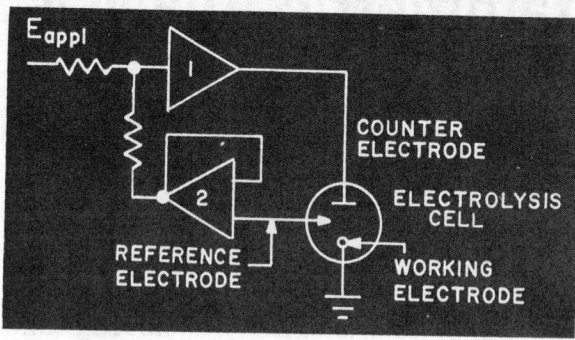

Figure 2. Potentiostatic three-electrode system

methods began to become readily available, the polarograph was shoved to the back of the bench; the analyses previously performed on it were transferred to the atomic-absorption spectrometer, the flame-emission spectrometer, the gas chromatograph, etc.

New Techniques

In classical dc polarography, it is assumed that the measured current is "diffusion-limited." This simply means that diffusion is the only mode of transport by which the electroactive particles whose reaction will produce the current can reach the electrode and that when the applied potential is noticeably more negative than the half-wave potential, any particle which arrives at the electrode surface by diffusion immediately and essentially instantaneously undergoes an electron transfer reaction. In such a process, the current that flows is determined by the diffusion rate of the material in question, which is determined by its concentration, its diffusion coefficient, and by the electrode area. One can derive a theoretical model for the diffusion-limited process by making assumptions about the area and shape of the electrodes, and in pure, relatively concentrated solutions, the results can be made to agree with experiment to a good approximation.

Unfortunately, the actual current that one measures contains contributions from a number of other current sources in addition to this diffusion-controlled Faradaic current. Such sources include the current necessary to charge the capacitance of the electrical double layer at the electrode surface, currents produced by the reduction of other species in the solution, and background or noise currents from either the electrode system or the instrument. The problem of distinguishing the current of interest from the various other signals is the one which each of the "modern" voltammetric techniques seeks to solve. In the remainder of this article we shall discuss the way in which these techniques can provide such information and indicate some of the applications which have developed in the past five years for these techniques. It is only with the advent of low-cost modern polarographic instrumentation that these techniques can be routinely applied, and the renaissance in polarographic analysis which forms the subject of this article can be directly attributed to the availability of such instrumentation.

Potentiostatic Control

One key characteristic common to all modern polarographic instrumentation is "potentiostatic" (2) control of the working electrode potential. Classical polarographs did not possess such "potentiostatic" capabilities and were thus not usable in high-resistance solutions and organic solvents. They applied the potential across the entire cell, rather than across the working electrode-solution interface, and thus yielded data which were considerably in error if solution resistances and the resultant voltage drops through them made an appreciable contribution to the recorded data (Figure 1). Modern instruments incorporate a potentiostat which controls the potential right at the working electrode-solution interface, eliminating errors owing to solution resistance, and are thus usable in a much wider range of systems.

A potentiostat accomplishes this end by making use of a three-electrode system, as shown in Figure 2. Here, a reference electrode of constant potential, inserted in the system and positioned as closely as possible to the working electrode, is connected to the instrument through a circuit which draws essentially no current from it. There is thus no current flow between the tip of the reference electrode (or its connecting bridge) and the instrument and thus no voltage drop. The output of the circuitry within the instrument is then the voltage right at the tip of the reference electrode. The operational amplifier control loop will then apply sufficient compensating potential to the counter electrode to insure that the potential at the reference electrode tip is the desired one, even if solution resistance is sufficiently high as to cause appreciable voltage drops when currents flow through the solution.

This procedure is in contrast to earlier instruments using only two electrodes. They either employed a simple metallic counter electrode with, at best, mediocre potential reference abilities or used the reference electrode as the only other electrode in the system, so that current was forced to flow through it. Since reference electrodes are often sensitive to current flow and can undergo potential change owing to polarization under these circumstances, the use of such a system, even when the electrode is positioned closely to the working electrode, can introduce significant errors.

Modern Voltammetric Techniques

For the remainder of this article, we shall assume that all the instrumentation discussed possesses potentiostatic capabilities, and we shall limit our discussion to those techniques which have found the greatest degree of analytical application. Commercial instrumentation for all these techniques is available. The techniques to be discussed are as follows:

Differential pulse polarography
Ac polarography
Fast linear sweep voltammetry
Direct anodic stripping voltammetry
Differential pulsed anodic stripping voltammetry

All these techniques are united in their applicability to electrodes other than the standard dropping mercury electrode employed in most normal dc polarography, and all these techniques are alike in their ability to detect and

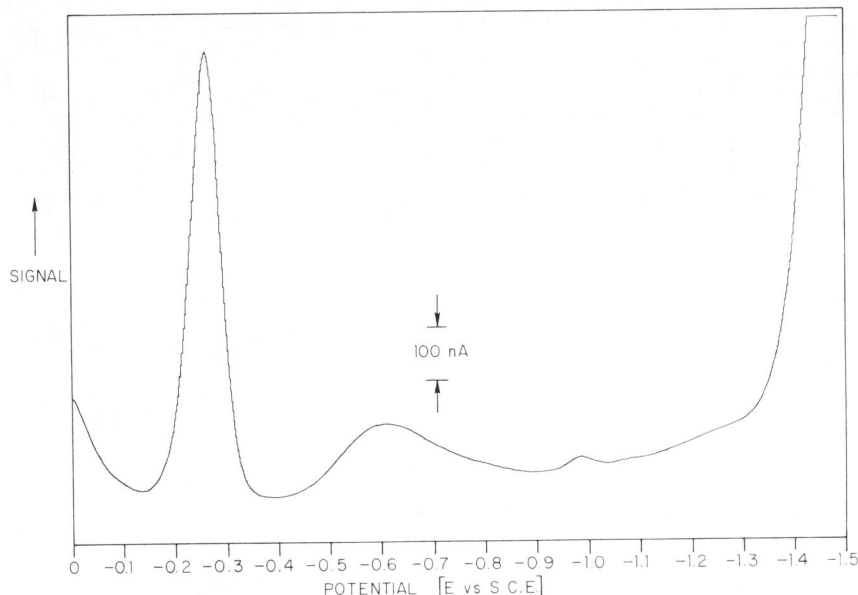

Figure 3a. Differential pulse polarogram. $1.3 \times 10^{-5}M$ chloramphenicol in $0.1M$ acetate buffer. PAR Model 174 polarographic analyzer, dropping mercury electrode, 50 mV pulse amplitude, 1-sec drop

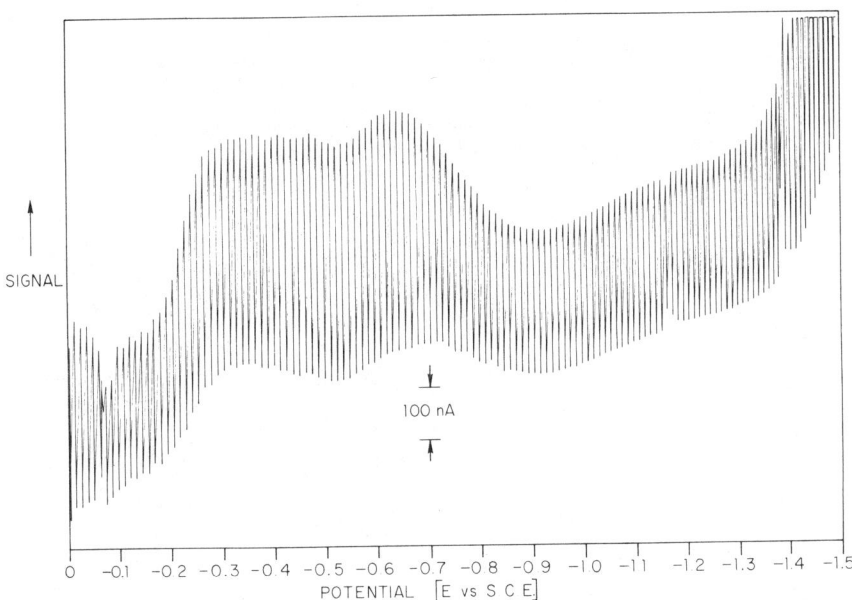

Figure 3b. Dc polarogram. Same solution and conditions

quantitate analytes at levels orders of magnitude below those normally associated with dc polarography. In addition, most make some attempt to separate capacitive and Faradaic currents, and all yield presentations where the signal arising from each individual analyte is in the form of a peak of some sort, rather than the step usually associated with dc polarography.

Differential Pulse Polarography. Figure 3a is the differential pulse polarogram of a $1.3 \times 10^{-5}M$ solution of the antibiotic chloramphenicol in a $0.1M$ acetate buffer supporting electrolyte. Figure 3b is a dc polarogram of the identical solution. Both curves were run at the same instrument sensitivity—$1 \mu A$ full scale. This concentration represents the approximate lower limit for the dc polarographic determination of the material. The wave is clearly discernible, but quantitation of the wave height would be difficult, and precise location of the half-wave potential almost impossible. In the pulse case, however, the clearly defined sharp peak allows precise measurement of peak height and exact location of peak potential.

In fact, differential pulse polarography can be used to determine antibiotics and many other materials at concentrations well below the ppm level. Figure 4a, for example, is the differential pulse polarogram of a 0.36 ppm tetracycline solution in a similar $0.1M$ acetate buffer solution. The curve, run at an instrument sensitivity of 200 nA full scale, gives a peak analytically useful. By contrast, however, Figure 4b is a dc polarogram of a 180 ppm solution of the same material—500 times more concentrated. The curve is too poorly defined to be of use.

The differential pulse polarographic technique, originally developed as an offshoot of square wave polarography in Britain during the 1950's (3), consists of superimposing a fixed-height potential pulse at regular intervals on the slowly varying potential associated with dc polarography (Figure 5). The pulse is repeated at intervals of perhaps 1 sec and is synchronized with the maximum growth of the mercury drop, if a dropping electrode is used.

Differential pulse polarography instrumentation then samples the current flowing into the working electrode twice during each operating interval, by use of electronic switching. The first current sample is taken just before the application of the potential pulse and is a sample of a current which is essentially equivalent to that which would be obtained in the normal dc polarographic case. Immediately after the conclusion of the sample-taking process, a sudden pulse of potential, usually between 5 and 100 mV, is applied to the electrode. The application of this sudden change in potential produces a concurrent sudden change in the current flowing, which comes from two primary sources. The first is the additional current which must flow to charge the double-layer capacitance of the electrode to the new applied potential, and this current decays exponentially at a rate governed by the magnitude of the capacitance and the series resistance of the system. Simultaneously, an additional current may flow if the applied potential has suddenly changed to a potential where the equilibrium between the reduced and oxidized forms of the electroactive species involved is shifted.

When both the potentials involved lie either before or after the rising portion of the polarographic wave, no change in the Faradaic current measured will be observed. However, when at least one of the two potentials is on the rising portion of the polarographic step, a significant change in the Faradaic current flowing takes place when the potential is suddenly stepped. A larger current is required than by the diffusion-controlled system, and this current is

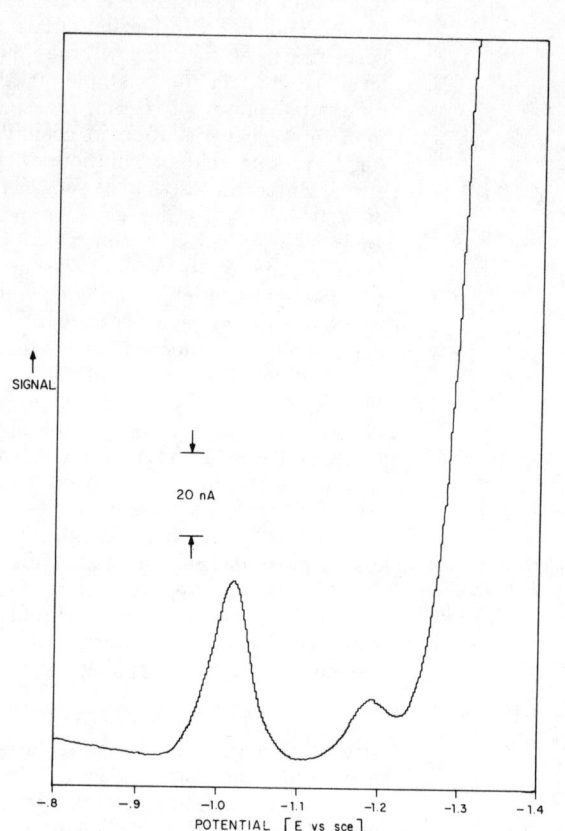

Figure 4a. Differential pulse polarogram. 0.36 ppm tetracycline·HCl in 0.1M acetate buffer, pH 4. PAR Model 174 polarographic analyzer, dropping mercury electrode, 50 mV pulse amplitude, 1-sec drop

Figure 4b. Dc polarogram. 180 ppm tetracycline·HCl in 0.1M acetate buffer, pH 4, similar conditions

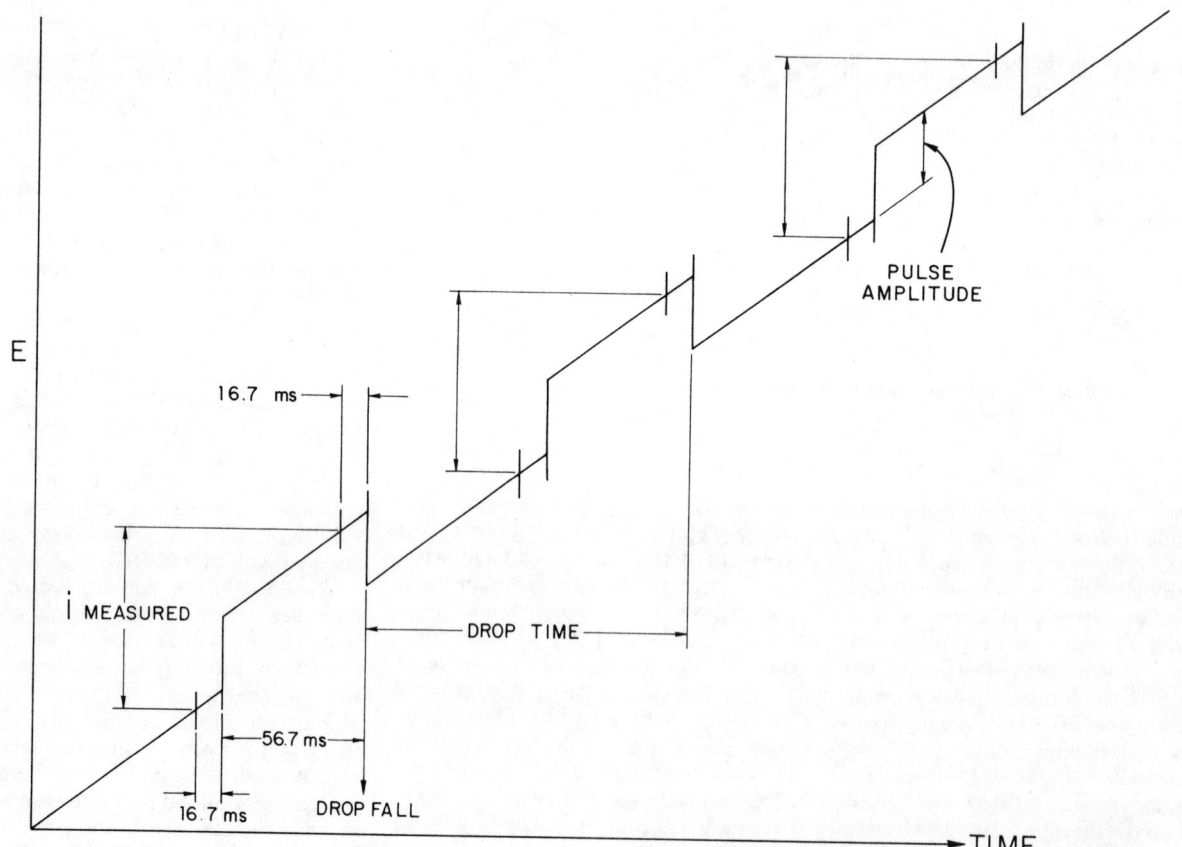

Figure 5. Potential excitation waveform used in differential pulse polarography and pulsed stripping voltammetry

further increased by the fact that static equilibrium conditions do not prevail because of the sudden change of potential.

The pulse potential is maintained for a period of time long enough to allow the capacitive current to decay to a low value, but during which the Faradaic current, although it also decays somewhat, still does not reach the diffusion-controlled level. At the end of this period, a current sample is again taken. The difference between these two samples, developed by applying the two signals stored in the memories to a differential amplifier, is then amplified and presented to the output of the system. This difference current-curve, which is proportional to the concentration of material, all the other usual polarographic parameters, and the amplitude of the pulse, has the appearance of a peak rather than the usual polarographic step. It is also not complicated by significantly sloping baselines, since the capacitive current contribution is minimized by the delay in the sampling process.

Further refinements in the data can be obtained by synchronizing the timing of the mercury drop with the power line frequency so that the entire sampling process takes place on equivalent portions of the power line sinusoid, and by making the width of the sampling gate equal to a known multiple of the power line frequency so that slight variations in the signal produced by the pickup of power line signals are minimized. These precautions are often necessary because reference electrodes and high-resistance organic solutions may act as antennas which pick up significant amounts of power line signal; this can cause a great deal of noise on the output unless it is removed.

The differential pulse technique provides solutions to just about all the problems which plague polarographers. The influence of capacitive currents is minimized by the pulsing and sampling process, and peaks are obtained rather than steps, so that resolution can be improved. Such problems as polarographic maxima, poorly defined waves, and severely sloping background baselines are all at least partially attacked by the differential pulse technique, and all yield to its effects to some degree.

Instrumentation for differential pulse polarography requires various timing and sampling circuits, low-drift analog memories, good differential amplifiers, etc. Until recently, available commercial instrumentation was quite old and expensive, used tubes with their attendant service problems, and was nonpotentiostatic. These factors precluded the rapid development of a body of applications information and served to prevent widespread adoption of the technique. However, with the recent advent of instruments employing low-cost, highly stable integrated circuit amplifiers and high-impedance junction field-effect transistors, differential pulse polarography has been brought within the reach of every laboratory and is being applied to many different analytical areas.

Subject only to limitations on the duration of the applied pulse, the technique is applicable to both reversible and irreversible reactions and yields peak shapes which closely approximate the theoretically predicted derivative of the dc waveform in most cases. It thus permits one to obtain the maximum possible resolution between closely spaced waves, while permitting examination of large and small signals during the same scan. In addition, when used with a modern potentiostatic instrument, solutions containing low concentrations of supporting electrolyte will yield curves as well-shaped as those obtained with solutions containing high concentrations of supporting electrolyte.

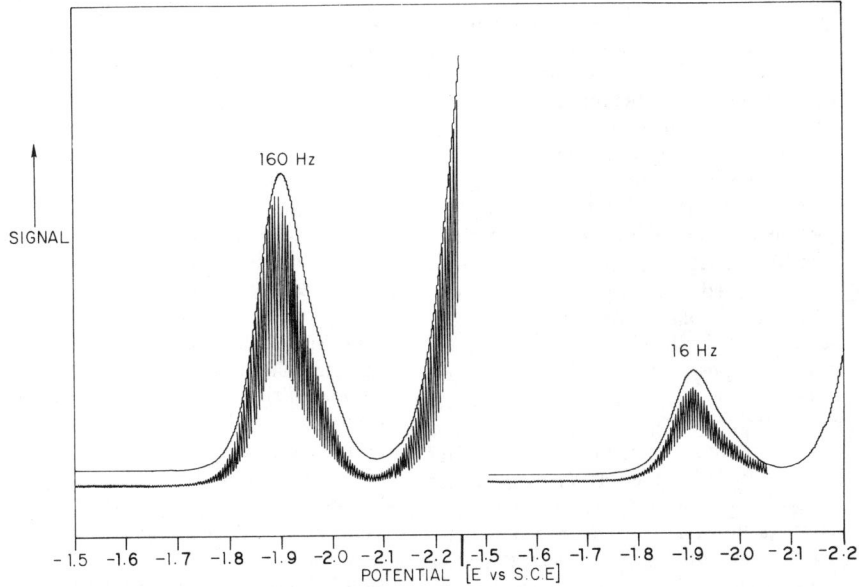

Figure 6a. Phase-sensitive ac polarograms. Hydrolysis products in acetonitrile. PAR Model 174 polarographic analyzer, Model 174/50 ac polarographic interface, and Model 124 lock-in amplifier. Dropping mercury electrode, 30 mV modulation, 1-sec drop

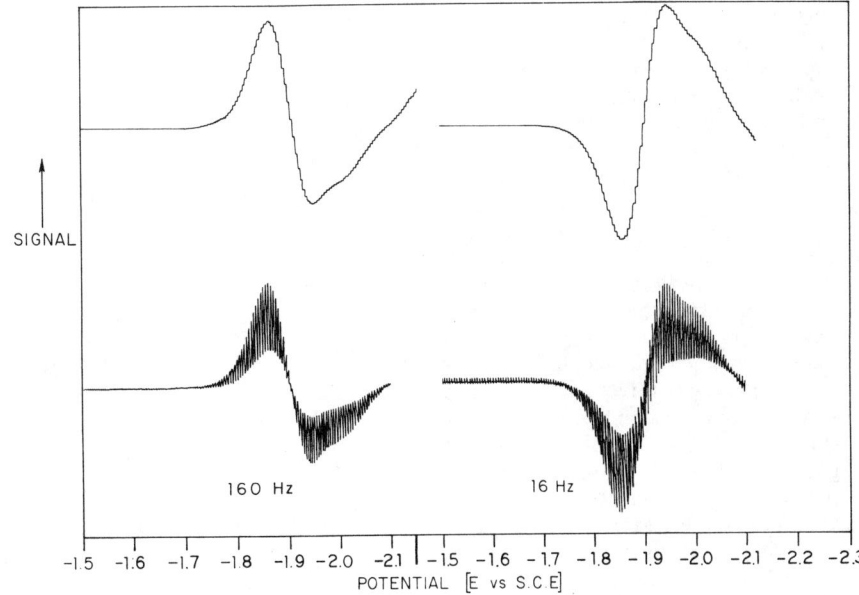

Figure 6b. Second-harmonic phase-sensitive ac polarograms. Similar conditions

Ac Polarography. Figure 6a is a phase-sensitive ac polarogram of hydrolysis products in acetonitrile containing $0.1M$ tetraethylammonium perchlorate as supporting electrolyte. The relatively smooth upper line is obtained by using drop-synchronized current sampling, with a single sample being taken just before the mercury drop is dislodged. Using this approach permits the analyst to obtain smooth curves from which the usual oscillations of drop growth are removed. The lower curve is the usual ac polarogram with the drop oscillations shown.

Figure 6b is the second harmonic, phase-sensitive ac polarogram of the same solution, where the system has been modulated at the same frequencies, but detection has been carried out at twice the modulation frequency. The second harmonic technique, by virtue of the fact that its signal crosses through zero at the half-wave potential, is useful for resolution problems, especially in complex mixtures. Each of the experimental curves has been run both in the full-signal mode and by using drop-synchronized current sampling.

In the ac polarographic technique (4), a small amplitude sinusoidal modulation is superimposed upon the slowly varying direct potential used in normal dc polarography. The instrumentation used employs circuitry which permits detection and presentation of only the alternating components of the total current flowing into the working electrode, and in the more sophisticated instruments, the signal presented is further selected by being limited to that signal which maintains a specific preselected phase relationship with the phase of the modulation of the applied potential. By looking only at the alternating portion of the current that flows and detecting its amplitude, we are in effect looking at the difference in current that flows between the minimum and maximum applied potentials during the modulation period. We thus get a peak waveform, rather than a dc step, and thus operate in an environment where the rate of arrival of electroactive material at the electrode surface is different than it would be in the dc case.

In differential pulse polarography, the signal arising from capacitive current is suppressed by sampling the flow of current after the capacitive contribution has decayed. In ac polarography the signal can be suppressed by using a phase-sensitive detector to measure only that portion of the alternating current which maintains a specific phase relationship with the applied potential. The capacitive current, just as would be the case in any other capacitor, will differ from the applied voltage by 90° in the ideal case, whereas the ideal reversible Faradaic current differs from the applied potential by 45°. Using a phase-sensitive detector thus permits one to measure only the Faradaic current or, as can be extremely important in studies of kinetics and adsorption, only the capacitive current.

It has been postulated that ac polarography can be employed successfully in some cases without the removal of dissolved oxygen from the system. This is because the reduction of oxygen on mercury is highly irreversible, and the signal from the oxygen reduction is thus quite small and does not interfere with the measurement of signals arising from more favorable reactions. This is a significant analytical advantage in cases where it is applicable. However, the electrochemistry of oxygen produces products which can in themselves severely affect the overall course of reactions within the system. When such effects are observed, outgassing is necessary even in ac polarography.

The peak position, height, and shape obtained in ac polarography are quite dependent on the specific electrode reaction taking place. The signals may be moved around, and their interrelationships changed significantly by varying the frequency of the applied ac modulation. This makes ac polarography, especially when phase-sensitive detection and second harmonic capabilities are available, particularly useful for analyzing complex systems where many different electroactive substances give signals. Some substances give well-defined ac waves whose phase characteristics correspond to ideal reversible behavior at a particular frequency but may completely change their behavior at higher frequencies. Others give no detectable ac wave under any conditions, so that analyses of mixtures in which they are present can be accomplished by combining ac determination of other components with pulse studies of these. The second harmonic technique can also be useful when resolution is the primary problem, since it permits resolution of much more closely spaced waves than other techniques. It has been applied to a large number of complex analytical problems, especially in the pharmaceutical field, for this reason.

Fast Linear Sweep Voltammetry. Figure 7 shows the polarograms which can be obtained with a variety of techniques from the well-known insecticide carbaryl ("Sevin"), which has been nitrosated to render it polarographically active. The polarographic determination of nitrosated carbaryl is an accepted technique for determining trace residues of this material. Curve A shows the normal dc polarogram of the material whose concentration in this solution is 6.4 ppm. No detectable wave is observed. Curve B shows the fast linear sweep voltammogram of the identical solution. The two curves were obtained on the identical system, at the same current sensitivity of $2~\mu A$ full scale, and on the same dropping mercury electrode. The difference, however, is that curve A was run at a scan rate of 10 mV/sec by use of many

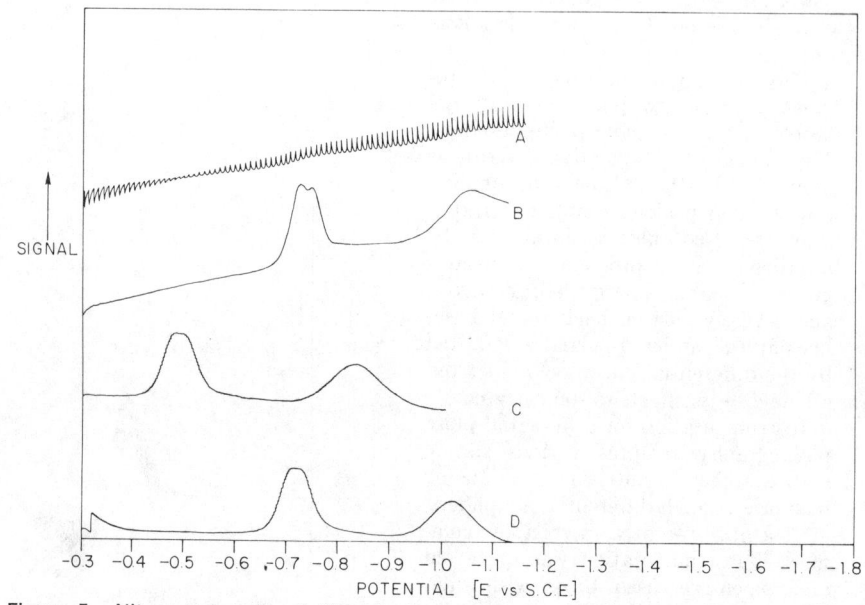

Figure 7. Nitrosated carbaryl ("Sevin"). 6.4 ppm. PAR Model 174 polarographic analyzer, Model 174/51 linear sweep accessory

A. Dc polarogram, $2~\mu A$ full scale, 1-sec drop, E vs. SCE
B. Fast linear sweep voltammogram, $2~\mu A$ full scale, 500 mV/sec sweep, E vs. SCE
C. Differential pulse polarogram, $10~\mu A$ full scale, 50 mV pulse, 1-sec drop, E vs. Pt
D. As in C, but E vs. SCE

drops, whereas curve B was obtained by applying a 500 mV/sec scan to a single drop. In curve B the wave is clearly discernible and easily quantitated. (For reference, curves C and D, which are differential pulse polarograms of the same system, are included.)

Clearly, the fast sweep technique provides a marked increase in sensitivity over dc polarography and offers the added advantage of giving rapid results. In the particular example shown, the differential pulse technique proved to be approximately five times more sensitive with the specific instrument settings used, but fast sweep voltammetry was much more rapid.

In this technique the normal slowly varying scan potential used in dc polarography is replaced by a fast sweep with no modulation of any sort (5). Potential change rates in dc polarography are usually of the order of 1–50 mV/sec, but fast linear sweep voltammetry is carried out at 100 mV/sec and higher. The technique was originally developed to obtain higher sensitivities than could be obtained by dc polarography without requiring the extremely sophisticated circuitry of the square wave or pulse polarographs. It succeeded in this endeavor because the application of a rapidly changing potential to an essentially stationary electrode produced a nonequilibrium condition around the working electrode.

In any electrochemical system, an equilibrated diffusion layer is established between the electrode and the bulk of the solution—a layer in which, once the reaction potentials have been reached, the concentration of the electroactive species of interest changes from the "bulk" concentration value at positions far removed from the electrode surface to essentially zero at the electrode surface, because any electroactive particle arriving at the surface instantly reacts.

When slow scans are employed, the slope of the concentration gradient within the diffusion layer is that which is dictated primarily by the rate of depletion of the electroactive species. This gradient varies from essentially zero, at potentials significantly more positive than the reduction potential, to a value governed by the concentration and diffusion coefficient of the substance in question, at potentials well past the reduction potential.

When the process in question uses rapid changes in potential, however, either because the scan rate is fast or because a pulse modulation of some sort is employed, the slope of the concentration gradient at any particular potential will be greater than in the slow-scan case, and the bulk concentration will be present closer to the electrode surface. Thus, the number of electroactive particles arriving at the surface per unit time will be greater, and larger signals will result.

When single drops are used, the typical concentration gradient discussed previously prevails at potentials more positive than the reduction potential. However, once this point has been passed, more and more material is used up, and the diffusion layer extends further and further into the solution. Unlike the case with dropping electrodes, this process is not periodically reversed by the stirring associated with drop fall, so that the decay of signal continues and peak-like readouts are obtained.

In addition, since all the data are obtained on a single drop, the annoying mercury drop oscillations superimposed on the usual polarographic waveform are not present.

Linear sweep voltammetry is usually performed on an instrument which either incorporates or is connected to an oscilloscope, since until recently, recorders with adequate response speeds were unavailable. The technique suffers from sloping baselines caused by the increase in both capacitive and Faradaic currents with drop size but still yields useful data, and dc differentiators can be used to overcome the baseline problems. A great deal of data has been published by using this method, and now that low-cost instrumentation which incorporates this capa-

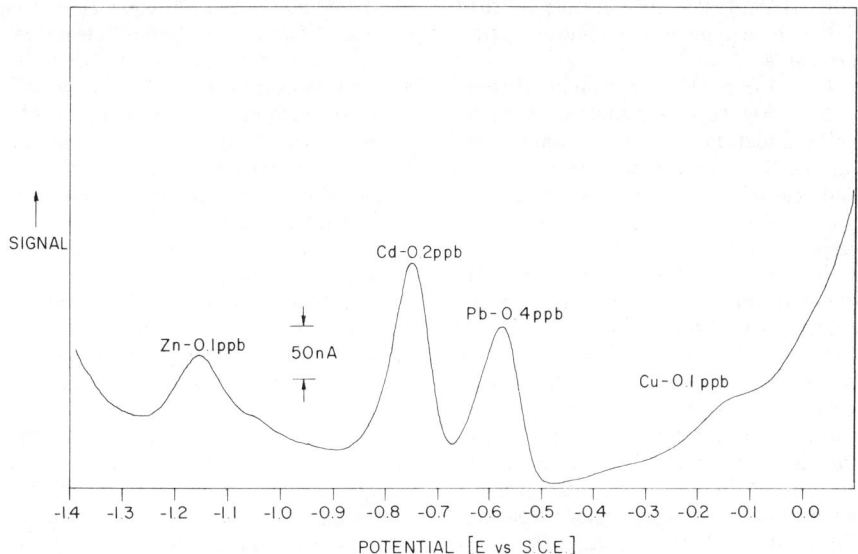

Figure 8. Differential pulse anodic stripping voltammetry. PAR Model 174 polarographic analyzer, Model 9319 wax-impregnated graphite electrode (mercury-plated). $2 \times 10^{-9} M$ Zn, Cd, Pb, and Cu

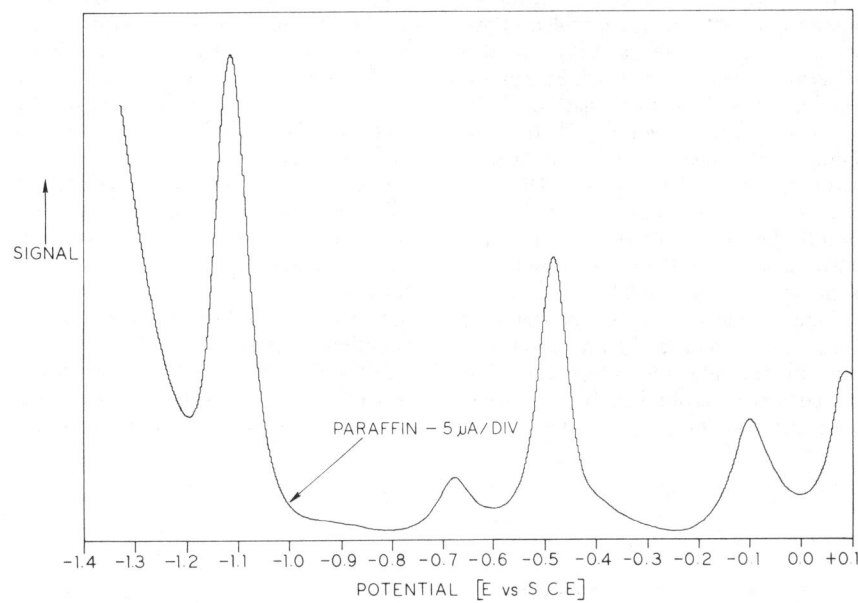

Figure 9. Differential pulsed stripping voltammetry. Equipment as in Figure 8, $2 \times 10^{-8} M$ Zn, Pb, approx. $0.8 \times 10^{-8} M$ Cd, Cu

bility is available, much of the published data can be applied to routine analytical problems.

Pulsed and Direct Stripping Voltammetry. Figure 8 is a pulsed stripping voltammogram of a solution containing $2 \times 10^{-9} M$ zinc, copper, lead, and cadmium (corresponding to 0.1 ppb, 0.1 ppb, 0.4 ppb, and 0.2 ppb, respectively). Clearly, the zinc, cadmium, and lead peaks are well-defined and easily quantitated. However, note that the copper peak is not as well-defined. This is primarily because an intermetallic compound forms between the copper and the zinc under the specific deposition conditions used for this experiment, and much of the copper is then not stripped properly when its oxidation potential is reached. For contrast, Figure 9 is a similar curve run on a solution approximately 10 times more concentrated, where intermetallic compound formation is not a problem, and the copper is easily seen.

Intermetallic compound formation should not be taken as a problem, however, in that, if the deposition in this experiment were carried out at, for example, -0.9 V vs. SCE instead of the -1.4 V which was employed, the zinc would not deposit, and the copper wave can easily be studied. The procedure for the analysis would then simply be to run it twice, once depositing at -1.4 V and a second time with deposition at -0.9 V.

Figure 10 is a direct anodic stripping voltammogram of the identical solution, run on the identical electrode. The pulsed curve was obtained by depositing, with stirring, for 3 min. After 5 min of stirred deposition, the dc case was stripped from the same electrode. Clearly, the data are there, but the allowed deposition time did not permit adequate stripping currents to be obtained. In addition, the severely sloping baseline obtained at these sensitivities makes the data somewhat more difficult to interpret.

Pulsed and direct anodic stripping voltammetry are techniques which are applicable almost exclusively to metals analysis (6–8). They are essentially offshoots of their polarographic counterparts which make use of the properties of a stationary mercury electrode to increase sensitivity, based on the premise that significant increases in signal strength can be obtained by concentrating the material of interest into an electrode before studying it. If a potential more negative than the half-wave potential is applied to a suitable electrode, the metal ions will be continually reduced at the electrode surface. As time passes, the concentration of the material of interest in the amalgam electrode will grow and, after a suitable interval, will be significantly higher than the concentration of the same material available for reduction in the body of the unknown solution. Application of a positive-going current will then cause the metals to oxidize back out of the mercury, and their higher concentration in the amalgam will give rise to much higher currents than would have been obtained in the reduction process.

It would seem at first glance that any electrode might be usable for this purpose, but severe complications arise, except in special cases when anything other than mercury is employed. This is because the potential at which the reactions take place is strongly dependent on the nature of the electrode surface, and this surface is changing significantly during the course of the experiment, as it becomes covered with reaction products. If mercury is used, however, metallic reaction products will dissolve in the electrode to form amalgams, and the electrode surface will thus only gradually be converted from pure mercury to amalgam. In addition, if more than one material is depositing on the electrode surface, intermetallic compounds may form among the various constituents, but this phenomenon will be minimized (though not eliminated) when the various metals are "diluted" with mercury.

In the basic dc anodic stripping process, a suitable mercury-containing electrode is initially maintained at a specific known potential, more negative than the reduction potential of the metal which is most difficult to reduce, for a known length of time. The material deposits into the electrode at a rate governed by its concentration and its rate of arrival at the electrode surface (either by diffusion or via stirring), and at the end of a known length of time, a known amount of material has deposited onto the electrode surface.

At the end of this process, a potential increasing in the positive direction is applied to the electrode, and the current obtained is measured as a function of potential. In the normal dc case, the most well-defined waves and the highest sensitivities are obtained when the rate of change of this potential is reasonably fast, so that peak waveforms are obtained. Baseline slopes are encountered but may either be compensated or ignored.

The sensitivity of this technique is theoretically limited only by the length of time over which one may keep the instrument and electrode system stable so that deposition may continue, the degree of reproducibility and control one can exercise over the deposition process, especially when stirring is employed, and the degree of one's patience. It is theoretically possible to analyze solutions well below $10^{-11} M$ concentrations by this technique through the use of long deposition times. However, instrument instabilities and electrode malfunctions may be a problem when experiments last many minutes, and more importantly, the excessive times involved can cause all of the analyte to disappear as it is adsorbed on the walls of the vessel. Sensitivities of the order of $10^{-9} M$ can be obtained, and the technique is often used for metals analysis at these levels.

The original electrode used for this technique, and the one most commonly

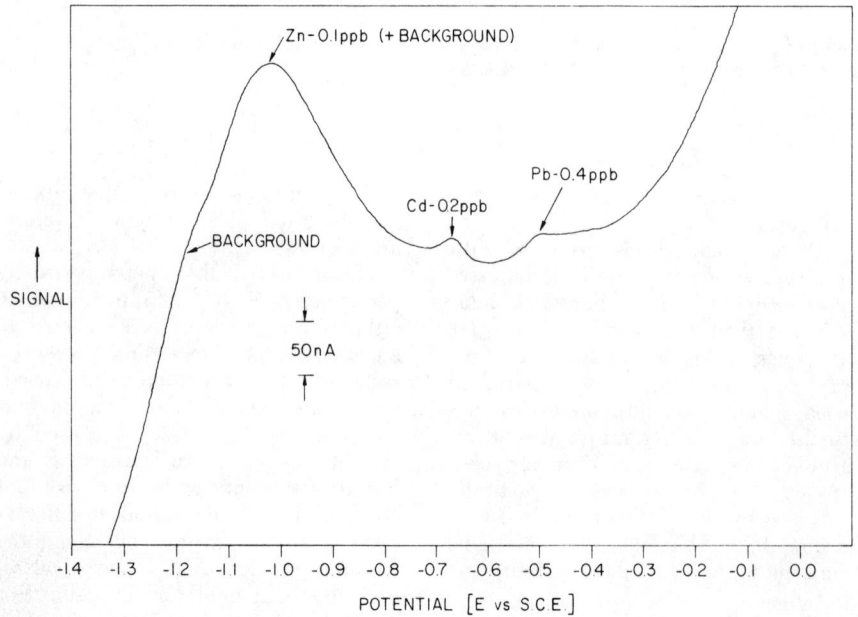

Figure 10. Direct anodic stripping voltammetry. Solution and equipment as in Figure 8

associated with it, is the hanging mercury drop electrode. Such an electrode in the modern syringe form is convenient and reproducible. It suffers from some problems, however, in that it is difficult to maintain a stable drop for times of one-half hour and in that diffusion of the species of interest into the electrode takes place to a significant degree during the course of the deposition, so that not all the material which has been reduced is available for oxidation when the scan is applied.

Recent efforts have aimed at developing mercury film electrodes of various types, which can overcome these problems because they are mechanically stable and employ thin films, so that diffusion into the body of the electrode is not a problem. Various degrees of success have been reported, and such electrodes as mercury-plated wax-impregnated graphite, mercury-plated pyrolytic graphite, and mercury-plated glassy carbon have all been employed for the purpose. One must be careful, however, in using thin-film electrodes, in that the time scales of the dc stripping technique involve massive depletion of the solution, when extremely dilute solutions are being investigated, and exhaustive stripping of the mercury film during the detection process. This means that reproducibilities will be subject to modification by factors which can affect such massive, exhaustive processes, such as the degree of constancy of the stirring rate, the exact thickness of the mercury film, and the amount of exposure, if any, of the substrate.

Differential pulsed anodic stripping voltammetry differs from dc anodic stripping voltammetry only in that the oxidation process is studied through the use of the same pulse-modulated ramps discussed under differential pulse polarography. The deposition process is carried out in exactly the same way (except that commercial practice often involves leaving the pulsing on during the deposition process since pulses of 25 or 50 mV amplitude have no appreciable effect on the deposition rates).

In differential pulsed stripping voltammetry, all the considerations of the dc stripping voltammetry case apply. However, the far greater sensitivity and signal processing capabilities of the pulse-modulated detection technique permit significantly higher instrument sensitivities to be used and thus allow either much shorter deposition times or much lower instrument gains. Under these circumstances, deposition times are kept to a few minutes so that electrode instabilities are less of a problem, and diffusion into the body of the electrode is less important. Similarly, the differential pulse scan rate of 5 mV/sec or so is such that the system remains in equilibrium with the electrode, except for the effects of the pulse potential change, so that greater reproducibilities may be obtained.

By use of the pulsed stripping technique on both hanging drop and film electrodes, sensitivities similar to those obtainable by dc stripping voltammetry can be obtained with much shorter experiment times. The same limitations which apply to the dc stripping voltammetry technique become the limits of sensitivity; that is, electrode stability, instrument stability, and loss of analyte to the vessels. However, the shorter times involved mean that the limits of sensitivity can be pressed perhaps one or two orders of magnitude further by extending deposition times. In addition, since the signal-processing circuitry of the pulse modes permits the use of higher instrument gains, sensitivities may be further improved.

Prognosis

Each of the above techniques provides a significant improvement in such key parameters as resolution and sensitivity over dc polarography. Many analytical problems which could never have been attacked by the dc technique fall within the possible operating realm of one or more of the advanced techniques, and many analytes which gave essentially useless signals in the dc case now yield useful data.

In the metal analysis area for which polarography was originally developed, the pulsed stripping technique is one of the most sensitive techniques available and certainly offers the most economical way of obtaining these sensitivities. The equipment is simple and easy to operate, and operator skill and knowledge are not required.

In the area of organic analysis, the specificity of voltammetric techniques permits these techniques to be applied to the analysis of relatively complex systems and to the unequivocal identification of complex substances. In any particular supporting electrolyte–solvent combination, only certain organic functional groups will exhibit electroactivity in a particular potential region, and fairly small changes in the nature of the molecule can provide significant shifts in the reduction potential. In addition, even in a relatively complex molecule, only a few peaks will be obtained since only a few of the functional groups will be electroactive. As a result, the worker using organic polarographic techniques for analysis or identification can usually identify the electroactive species with relative ease by comparison with published data. If a sample of pure substance is available, unequivocal identification can often be made by selective addition, but even if this is not the case, the simplicity of the recorded data makes interpretation much more simple.

As more and more analysts adopt modern polarographic and voltammetric techniques, more and more data will find its way into the literature, and the use of the techniques for fingerprint purposes and analytical applications will increase. Similarly, as more and more investigations of the electroactivity of compounds of interest take place, these techniques will become even more widespread, and their applicability will be accepted to an even higher degree. Thus, we see a technique, which is one of the oldest instrumental techniques available and which fell into disuse as the newer techniques came along, now returning to a place of prominence in the analytical laboratory, finding its way into new applications where it had previously not been tried, and owing to the availability of modern instrumentation, returning to its rightful lace in the laboratory.

References

(1) J. Heyrovsky, *Trans. Faraday Soc.*, **19,** 785 (1924).

(2) A. Hickling, *ibid.*, **38,** 27 (1942).

(3) G. C. Barker and A. W. Gardner, *Z. Anal. Chem.*, **173,** 79 (1960).

(4) B. Breyer and F. Gutmann, *Trans. Faraday Soc.*, **42,** 645 (1946).

(5) J. E. B. Randles, *ibid.*, **44,** 322 (1948).

(6) I. Shain and R. D. DeMars, *Anal. Chem.*, **29,** 1825 (1957).

(7) W. Kemula, Z. Galus, and Z. Kublik, *Bull. Acad. Pol. Sci., Ser. Sci. Chim.*, **7,** 723 (1959).

(8) G. D. Christian, *J. Electroanal. Chem.*, **23,** 1 (1969).

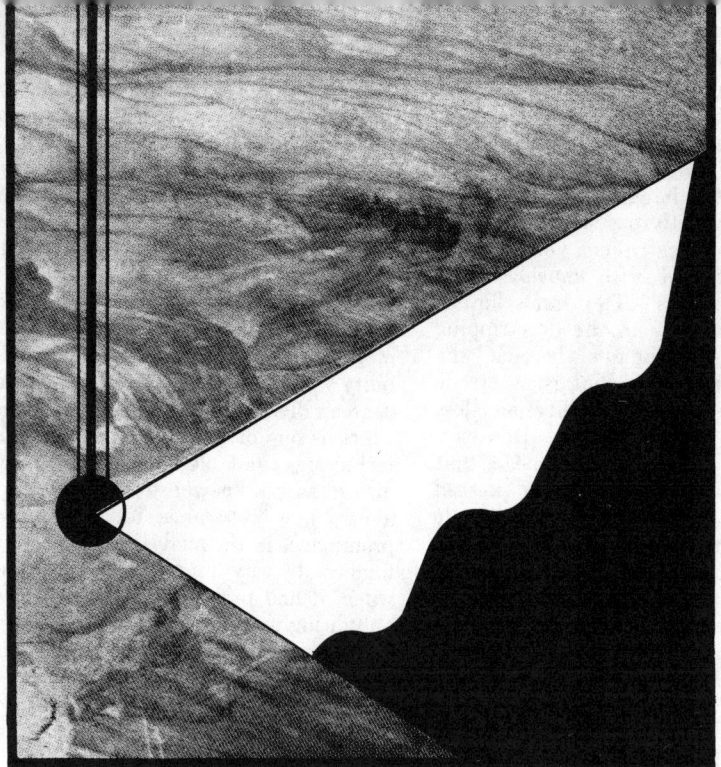

Anodic Stripping Voltammetry

T. R. Copeland
Department of Chemistry
Northeastern University
Boston, Mass. 02115

R. K. Skogerboe
Department of Chemistry
Colorado State University
Fort Collins, Colo. 80521

The use of anodic stripping voltammetry (ASV) has grown rapidly during the past few years. The increasing use of the technique is due to its ability to simultaneously determine several elements at concentration levels ranging down to the fractional parts per billion range with relatively inexpensive instrumentation. While ASV cannot be regarded as a new analytical technique, there have been several reasonably recent developments in instrumentation and methods associated with it; these have advanced the status of ASV considerably. The technique clearly offers capabilities for the solution of numerous difficult trace analysis problems. For this reason, it will surely become much more widely used. During this period of growth, it is certain that the technique will often be inappropriately used, misused, and abused, just as all relatively new techniques are. As a possible aid in avoiding these problems, the present report will: examine the fundamental aspects of the technique; discuss the important operational features of the instrumentation required; compare the more obvious differences in methodology; and summarize present and future applications.

Fundamentals

An ASV measurement involves two discrete steps: The analytical species is reduced (electrodeposited, plated) onto or into the working electrode and is then oxidized (stripped, electrolyzed) back into the electrolyte solution. For the deposition step, a suitable electrode is maintained at a potential cathodic of the reduction potential of the element(s) to be determined. The metals to be deposited arrive at the electrode surface at rates determined by their respective concentrations, the diffusion properties of the electrolyte solution, and the area of the electrode used. The deposition time is consequently carefully measured. The deposition results in preconcentration of the analyte(s) into a small (surface) volume. To strip this material from the electrode, its potential is systematically changed back in the direction required for oxidation, i.e., it is moved in the anodic direction. At the oxidation potential of each analytical species, the faradic current produced by its oxidation is measured.

The readout obtained is thus stripping current as a function of the electrode (oxidation) potential. The stripping current due to oxidation of each analyte is proportional to the concentration of that analyte on or in the electrode and, thus, in the analytical solution.

For any electrode material employed in the ASV system, the same general plating theory applies. An elementary evaluation of the theory governing the plating process affords an indication of the parameters which can be adjusted to maximize the plating efficiency, and hence the stripping current measured. For a single metal ion species (M^{+n}) being reduced at an electrode surface, it can be shown that the current flow (the deposition current at time t) is reasonably approximated by the Levich equation:

$$i(t)_{dep} = 0.62\, nFAD^{2/3}\omega^{1/2}\underline{\mu}^{-1/6}C(t) \quad (1)$$

where n = cation charge, F = the Faraday constant, A = the electrode surface area, D = diffusion coefficient, ω = the rate of electrode rotation or solution stirring, μ = the kinematic viscosity of the solution, and $C(t)$ = the ion concentration of deposition time t. The parameters in this equation which can be controlled are: $\mu, \omega, A,$ and t. The viscosity of aqueous solutions is typically variable only over a rather narrow range so its sixth root effect yields minimal benefits. The stirring or electrode rotation rate can be conveniently used to increase the deposition rate up to the point where

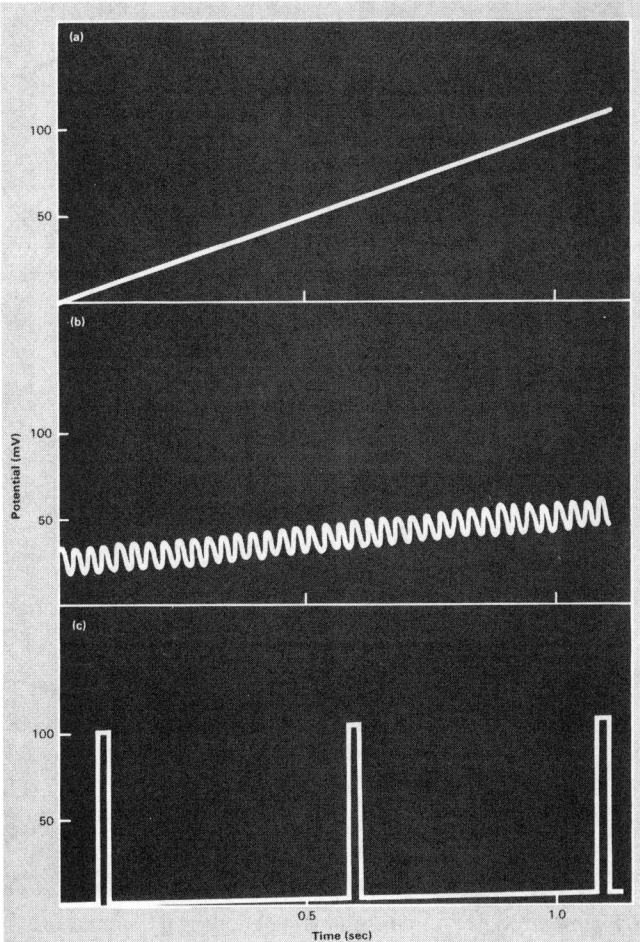

Figure 1. Potential-time waveforms used for anodic stripping
(a) Linear ramp stripping
(b) Ac stripping
(c) Differential pulse stripping

solution cavitation or the ability to maintain a drop, when hanging mercury drop electrodes are used, reaches critical conditions. Increases in the electrode area may also be used to maximize the amount of analyte plated while minimizing the deposition time commitment. These parameters are thus primary in the optimization of plating efficiency, i.e., the amount of element deposited per unit time.

Several different potential-time waveforms may be used to strip the deposited analyte from the electrode and obtain the quantitation parameter, the stripping current (i_p). By far the most common choice is a linear ramp of the potential [Figure 1 (a)]. This appears to be due to the general simplicity and wide availability of equipment having the linear ramp capability. For a hanging mercury drop electrode (HMDE), the stripping peak current is given by (1):

$$i_p = 2.72 \times 10^5 \; n^{3/2} A D^{1/2} C_E \nu \quad (2)$$

where C_E is the concentration of the analyte in the mercury drop, and ν is the ramp rate for the potential scan.

For mercury thin-film electrodes (MTFE), the theory developed by deVries and van Dalen (2) yields the following expression:

$$i_p = 1.1157 \times 10^6 \; n^2 A C_E L \nu \quad (3)$$

where L is the film thickness. The amount of analyte in a MTFE is given by (3):

$$Qm = nFAC_E L \quad (4)$$

Thus, Equation 3 reduces to

$$i_p = 11.6 \; n\nu Qm \quad (5)$$

This indicates that the peak current for thin films is determined by the amount of analyte present rather than its concentration in the film. On a practical basis, then, very thin films which are difficult to reproduce offer no sensitivity advantage over thick films which still exhibit thin-film behavior. DeVries and van Dalen (2) and Osteryoung and Christie (4) showed that films ranging in thickness from approximately 2–10,000 Å exhibit thin-film behavior, i.e., stripping peaks do not show distortion of any consequence.

The primary parameter for maximization of the stripping current is consequently the scan rate. Linear scan rates of typically 50 mv/sec or more are employed to improve the resolution of adjacent analyte peaks and provide larger peak currents. The rapid and continuous change in potential unfortunately generates a relatively large nonfaradic capacitative (or charging) current which generally constitutes a high background and severely degrades the signal-to-background and/or the signal-to-noise ratio of the response measured. The capacitative current may be approximated by:

$$i_c = AC_D \nu \quad (6)$$

where C_D is the differential capacity of the electrode double layer. In essence, the capacitative current is directly proportional to the higher scan rates indicated to be desirable by Equations 3 and 5. This imposes a practical limitation on the improvement that can be realized by increasing the scan rate. Consequently, all other potential-time waveforms used in the stripping process are employed to minimize the effect of the double layer charging current.

The waveform applied for ac stripping voltammetry (ACSV) is a linear potential ramp with a small amplitude sine wave superimposed on it as illustrated in Figure 1 (b). The equation for the stripping current has been given by Unterkofler and Shain (3). The major advantage of ACSV over the linear scan approach derives from the fact that the ac faradic stripping current precedes the applied ac potential by a 45° phase angle (⅛ cycle) while the capacitative current is at a maximum at a phase angle of 90°. Phase sensitive detection can thus be used to separate the faradic current from the nonfaradic component. This advantage is gained at the expense of requiring a more sophisticated waveform generator and phase sensitive detection. Without such detection other advantages still accrue from the use of ac stripping. The only faradic processes which generate ac faradic currents are those which are reversible within the time scale of the alternating potential imposed. Nonfaradic contributions from irreversible processes such as gas evolution may be consequently eliminated. Moreover, monitoring currents at harmonics of the ac voltage may also enhance signal-to-background ratios because the capacitative contributions are diminished (5). The effect of uncompensated resistance, i.e., resistance of the solution or the electrode, is to shift the phase angle of the ac signal. Such changes degrade the ability to resolve the two components of current and limit the utility of the method. The use of a HMDE of small area (low resistance) rather than a MTFE (high resistance) reduces the significance of this problem but prevents one from taking advantage of the high plating efficiencies associated with thin-film electrodes. The ac technique is generally reported to be 2–10 times more sensitive than the linear scan approach and provides better defined stripping peaks which improve the specificity of measurements for multielement analyses.

In differential pulse anodic stripping voltammetry (DPASV), relatively large amplitude (50–100 mV) pulses are periodically superimposed on a shallow linear potential ramp (~5 mV/sec) for short periods as shown in Figure 1 (c). The nonfaradic current induced by the application of the pulse decays more rapidly than the faradic current. This is governed in part by the rate at which the reoxidized metal can diffuse away from the electrode surface. The current is consequently measured just prior to the pulse application and again for the same period of time at the end of the pulse life. The difference in these two measurements is amplified and read out. The choice of the measurement periods permits a high level of discrimination against the capacitative current since the current measured near the end of the pulse is predominantly due to the faradic reaction. Figure 2 illustrates this.

Osteryoung and Christie (4) have shown that the peak current (i_p) is related to the amount of material in the MTFE (Qm) and the pulse width (t) by:

$$i_p = 0.138 \, Qm/t \quad (7)$$

Shorter pulses can thus be used to enhance the measured signal. An additional advantage of the method is illustrated conceptually in Figure 3. As the potential is pulsed through that at which stripping occurs, the analyte is oxidized from the electrode during the pulse but does not have time to diffuse away from the proximity of the electrode before the pulse ends. As the pulse is removed, the potential returns to a level at which reduction (plating) occurs. Consequently, a significant portion of the analyte stripped during the pulse is replated during the rest period between pulses as long as that period is cathodic of the reduction potential. As a result, the same analyte may make repetitive contributions to the current measured while it can contribute only once in a linear scan operation. Differential pulse stripping offers sensitivity enhancements of one to several orders of magnitude and appears to offer the highest signal-to-noise ratio of any stripping technique. This gain is achieved through the use of somewhat more complex instrumentation.

A modification of the ac method involves the superposition of a small amplitude square wave potential on the linear ramp. Again, the ac component of current is measured, typically with a gated or strobed detector system (6). By sampling the current just before the square wave changes polarity, the capacitive current will have

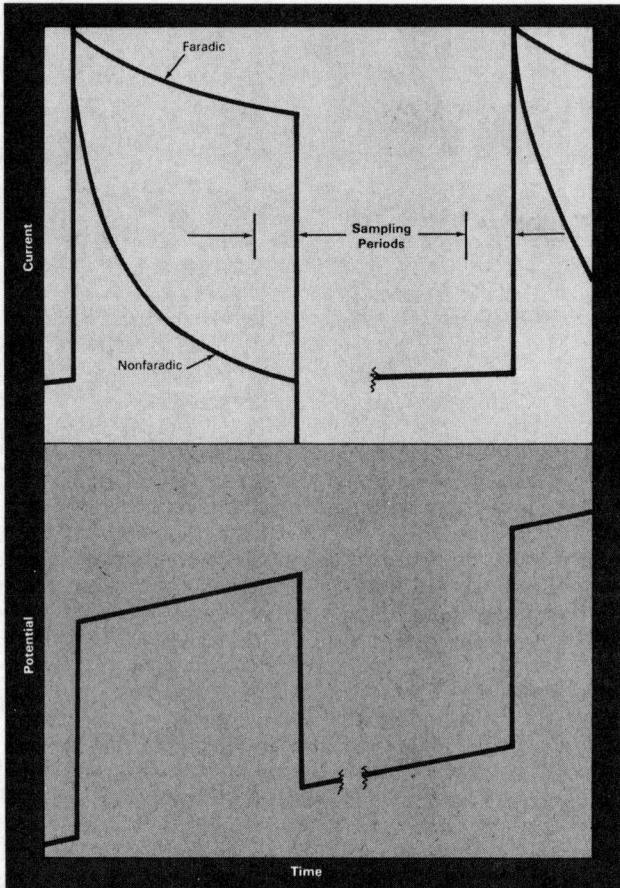

Figure 2. Choice of measurement periods in differential pulse stripping to reduce contributions of nonfaradic processes

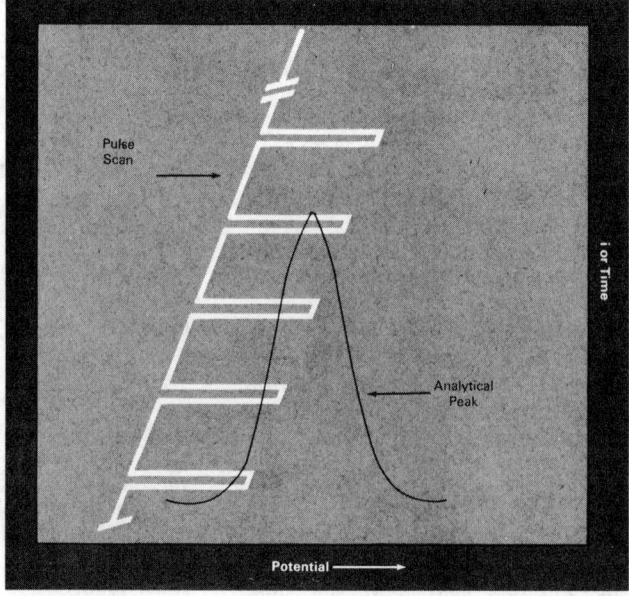

Figure 3. Differential pulse scan for which analyte stripped during pulse may be replated in pulse-off period

decayed substantially while the faradic current has undergone less decay. This technique for discriminating against the charging current does not benefit from the replating of analyte mentioned above and is inherently less sensitive than the large amplitude differential pulse technique.

A step function (staircase) waveform in conjunction with gated detection is another technique used to take advantage of the difference in the decay rates of the analytical and capacitive currents (7). The steps may be rather large so sensitivity greater than that obtained with the linear scan, ac, or square wave techniques may be realized. The replating phenomenon does not occur; therefore, this approach is less sensitive than DPASV.

The deposition step of the analysis and the parameters which affect it are the same for any of the techniques described above. The primary differences in technique are associated with the approach used during the stripping step. The preceding discussion has dealt with these differences in general terms. Certainly, specific analytical problems will arise for which the general comparisons made will not hold. In such instances, access to instrumentation offering a selection of stripping waveforms would be desirable.

Instrumentation

The basic instrumentation required for ASV is relatively inexpensive and readily available commercially. The simplest necessities include a three-electrode potentiostat and voltage ramp generator; current measuring circuitry; a cell with working, reference, and counter electrodes; and a recorder or other readout device. Instruments designed for dc, ac, or pulse polarographic measurements are generally quite adequate for stripping application. The currents to be measured in ASV are typically, however, two to four orders of magnitude greater than those generated in the corresponding polarographic techniques. Much of the noise rejection circuitry designed into commercial polarographs consequently becomes largely superfluous for ASV and, in some instances, can actually reduce ASV sensitivities.

The current measuring circuitry required for linear scan stripping is almost trivially simple. The DPASV approach requires reasonably complex circuitry to permit measurement of current at two time intervals per pulse and amplify the difference in these. Ideally, the gated sample and hold circuits used should have variable sampling widths and RC time constants. This provides the flexibility required to select conditions which minimize the capacitative current and maximize the signal-to-noise ratio. Those techniques relying on phase sensitive or harmonic detection impose the most complex circuitry requirements.

An ASV cell is simply a container with access for the three electrodes and a deoxygenation gas (nitrogen). Stirring may be with a Teflon-covered magnetic stirring bar. Somewhat more elaborate arrangements involve rotating the working electrode, using a flow through cell (8), or rotating the cell with stationary electrodes (9). For trace analysis, the cell and all other components in contact with the working solution should be of a material chosen to reduce contamination from leaching and loss due to adsorption. Teflon or acid-leached quartz and pyrex are in common use.

Almost any chemically inert material with a reasonable surface area satisfies the requirements for the counter (auxiliary) electrode. Platinum wire or foil is widely used. Silver–silver chloride or calomel electrodes are popular choices for the reference. A single overriding criterion in choosing any reference electrode for trace analysis is its leak rate. Contamination of the sample solution by leakage of impurities from the electrode or a salt bridge is a recurring problem. Glass frit isolation devices, for example, generally exhibit a higher leak rate and offer a large surface area which may enhance any analyte adsorption problems. Cracked glass bead, fiber wicks, and porous vycor plugs are isolation devices that generally have low leak rates.

A range of materials has found application as working electrodes for ASV. By far the most popular are those involving mercury. Solid electrodes of gold, silver, platinum, carbon, and bismuth have been used. The determination of those metals with oxidation potentials anodic of that of mercury (e.g., Au, Hg, and Ag) requires the use of solid electrodes. Some of the noble metals have a mutual tendency for the formation of intermetallic compounds (e.g., AuHg). To avoid this a carbon electrode is often used. All solid electrodes suffer the same disadvantage to some extent. During the deposition step, the surface of the electrode and the activity of the metal film being deposited are continuously changing. This is particularly apparent when several metals are to be codeposited. The problem is manifested by irreproducibility in the stripping peaks and often by multiple or split stripping peaks. The latter is presumably due to the analyte being stripped from surfaces composed of other analyte, other codeposited metals, or the working electrode material. Each possibility may occur at a different stripping potential. Experimental approaches have been devised to improve the reproducibility in such systems. One example involves deposition on the glassy carbon disk of a rotating ring disk electrode (10). Subsequently, the electrode ring was set at a cathodic potential while that of the disk was scanned anodically to strip the metal. The metal stripped was consequently deposited on the ring, and the redeposition or collection current was measured rather than the stripping current.

The use of any solid electrode (particularly carbon) requires that precise electrode cleaning, polishing, and pretreatment procedures be used to obtain reproducible results. The nature and the extent of these procedures depend on the material involved.

Hanging mercury drop electrodes are quite easily formed. The drop may be suspended from the tip of a platinum wire imbedded in a glass rod, or it may be formed at the tip of a microsyringe or capillary tube with a positive displacement micrometer calibrated to permit reproducible delivery of drops of known area. The HMDE offers simplicity, economy, and reproducibility when properly used. Its applicable potential range is limited on the anodic side by the mercury oxidation ($\sim$0V vs. the saturated calomel electrode, SCE) and by hydrogen evolution on the cathodic side ($\sim$1.8 V vs. SCE depending on pH). Two primary disadvantages are associated with the HMDE. First, they typically have a low surface area-to-volume ratio. The smaller area reduces plating efficiency while the larger volume causes a broadening of the stripping peaks which limits the ability to resolve adjacent peaks. This is due to the finite time required for the analyte to diffuse from the drop interior. If long deposition times are used, the analyte may diffuse up into the mercury column in the supporting capillary, causing further peak broadening. Second, only rather minimal solution stirring procedures can be used to avoid distortion or dislodging of the drop. The stirring and area limitations impose the requirement of longer deposition times as indicated in Equation 2. While this may not be a serious problem, it does generally mean that a longer time commitment per analysis is required when working at trace-to-ultratrace concentration levels.

These factors have been at least partially responsible for the development of mercury thin-film electrodes which offer the same potential range as the HMDE. The MTFE's offer a

Table I. Effect of Graphite Substrates and Wax Impregnation on Peak Currents[a,b]

Graphite type[c]	CT	AZ	AZ9	FX1	FX91	Vitreous (Beckman)
Apparent density, g/cm³	1.68	1.57	1.54	1.88	1.88	1.50
Scleroscopic hardness (unitless)	43	52	52	75	70	125
Peak current, μA (nonimpregnated)	0	5.8	1.6	16.8	22.0	27.0
Peak current, μA (wax impregnated)	4.2	8.5	5.0	16.5	20.0	...

[a] 4 ng/ml Pb solution; plating time, 2 min. [b] From *Anal. Chem.*, **45** (13), 2171–74 (Nov. 1973).
[c] Obtained from Poco Graphite, Inc., Decatur, Tex., unless otherwise noted.

large surface area-to-volume ratio and can be easily rotated or stirred at quite high rates. Rotation at 3600 rpm is not unusual. High plating efficiency and sensitivity are readily achieved under these conditions, and the stripping peaks are better defined because the diffusion time of the analyte from the film is minimal.

The ideal substrate for the MTFE must meet relatively few criteria—it must have reasonable electrical conductivity, be chemically inert to the mercury and to the analysis solution, and be electrochemically inert at potentials anodic of the most easily reduced of the elements of interest. Metals, although highly conductive, have several undesirable characteristics as possible substrates. Several have adherent oxide films (e.g., Pt, Ni), which inhibit uniform mercury deposition. Others have a low hydrogen overpotential and are slightly soluble in mercury (e.g., Ag, Ni, Pt). This results in a hydrogen overpotential for the film which decreases with time, making certain analyses highly irreproducible or impossible. If the substrate has a finite solubility in mercury, intermetallic compounds between substrate and analyte (e.g., AuCu, AuGa, AuIn, AuZn, AgCd, AgZn, etc.) may form, resulting in irreproducible stripping peaks (11, 12).

Carbon has a reasonably high hydrogen overpotential and electrical conductivity, is insoluble in mercury, and is chemically inert. However, many graphite types are porous enough to allow the analyte solution or the mercury film to creep into the substrate. This results in a constantly changing electrode area and irreproducible behavior. Wax impregnation of porous graphite or carbon prevents solution creep but can cause nonuniform mercury deposition if precautions are not taken to keep the polished surface free of wax films. Polishing is difficult even when wax impregnation is used.

The results of experiments designed to evaluate the utility of several types of graphite, having a range of physical properties, for ASV are summarized in Table I (13). These results demonstrate that wax impregnation is essential for the more porous (low density, low hardness) types of graphite. The primary effect of impregnation before deposition of the mercury film is the improvement in the surface area-to-volume ratio with its concomitant effect on plating efficiency. The results further show that the hard, nonporous (glassy or vitreous) types of carbon yield large and reproducible peaks without wax impregnation. Such types may be readily polished to a mirror-like finish so that the surface irregularities are small in comparison to the mercury film thickness, the surface-to-volume ratio is high, and a coherent film over the entire surface is readily obtained. The use of such types of graphite or carbon strikingly simplifies the preparation of thin-film electrodes. After polishing the carbon via exacting metallographic procedures, the thin film can be quite precisely electrodeposited on the surface.

As previously indicated, the use of very thin films results in erratic behavior, but films several thousand angstroms thick adhere to thin-film behavior (2–4). A larger electrode area results in the deposition of larger amounts of analyte per unit time (Equation 3), but for a given amount of material in the film, the currents measured are independent of film thickness for both linear and pulsed stripping procedures. Thus, inordinately thin films afford no increase in sensitivity over a thicker film of equivalent surface area. Moreover, the solubilities of certain metals in mercury or the formation constants of some intermetallic compounds may be exceeded due to the use of a mercury film having a small volume. The solubilities of several elements in mercury are given in Table II.

To illustrate the possible extent of mercury film saturation, the analytical solution concentrations required to reach saturation under fairly typical working conditions have been calculated and included in the table. These calculations were based on the use of a thin film ¼ in. in diameter by 2500 Å thick, plated for 2 min with rotation at 3600 rpm in an analytical solution volume of 25 ml. A 100% plating efficiency was assumed; therefore, the values given underestimate the actual concentrations necessary for saturation of the film. The results reveal, however, that film saturation may be readily attained for some elements, e.g., Cu, Ga, Pb, and Zn. The use of the method of standard additions as a means of quantitating the measurements may enhance the possibility of attaining saturation. The existence of such possibilities speaks for the use of thicker films or the alternative choice of plating conditions which reduce the plating efficiency and the sensitivity when necessary.

The occurrence of saturation even when the element attaining saturation is not of analytical interest can cause another problem. After saturation, continued plating results in the formation of solid metal on or in the mercury film, thereby changing the nature of the film, its effective area, and possibly other electrode properties, e.g., its hydrogen overpotential. Such changes result in irreproducibility and necessitate the modification of the analytical conditions used or the removal or dilution of the causative element.

All voltammetric techniques involve analyses in a supporting electrolyte. The electrolyte may be simply an acid medium, or it may be a buffer system, particularly when pH control is essential. Supporting electrolyte concentrations which are consistent from sample to sample are generally essential. Changes in the total electrolyte concentrations result in changes in the iR drop between the reference and working electrodes. Such fluctuations manifest changes in the stripping current per unit amount of analyte and in the positions of the stripping peaks; an increase in the iR drop shifts a peak. A second problem originates from the fact that the solution resistance and the double-layer capacitance of the working electrode define an RC time constant through which changes in potential must be effected. In highly resistive solutions, for example, this

Table II. Solubilities of Elements in Mercury and Calculated Concentrations Required for Saturation of Film

Element	Solubility in Hg, atomic %	Analyte concn for saturation, μg/ml[a]
Cd	10.0	17
Cu	0.006	0.006
Ga	3.6	2
In	70.0	74
Pb	1.3	4
Tl	43.0	6
Zn	5.83	6

[a] See text for conditions assumed in calculation.

makes it impossible to apply a sharp pulse or other rapid potential change to the electrode, thereby complicating the use of some of the more sophisticated stripping techniques. The effect is most evident for large area electrodes because the capacitance is proportional to the electrode area.

Supporting electrolyte concentrations of 0.05–0.5M are typical. At these relatively high concentrations, judicious care must be exercised in selecting the reagents used to avoid contaminant contributions from the electrolyte. Purification of all reagents by constant potential electrolysis or other methods is often required and is almost universally mandatory for analyses at the $10^{-9}M$ level. Almost any indifferent electrolyte can be used, but the choice often improves the reproducibility and the resolution of the stripping peaks. Zinc analyses, for example, are strongly affected by the pH of the analytical solution because hydrogen evolution contributes significantly to the background at the zinc stripping potential. The use of a buffer system such as acetate–acetic acid or citrate–citric acid as supporting electrolyte can reduce and stabilize the background.

ASV is primarily a trace analysis technique. The choice of instrumentation, the experimental approach, and the analytical conditions must always be made with the requirements of the analyses to be carried out in mind. If a diversity of elements in a variety of sample types defines the range of analytical problems to be encountered, it would be advisable to select equipment offering the versatility of two or more of the stripping modes described previously, e.g., linear scan and differential pulse.

Applications

The high analytical sensitivity of ASV has led to its application to a large number of analytical problems. While space does not permit a comprehensive discussion of the ASV literature, a number of selected publications can be used to indicate the general utility of the technique. These are summarized with other pertinent analytical information in Table III. Although it is considered possible to determine Ag, As, Au, Ba, Bi, Cd, Cu, Ga, Ge, Hg, In, K, Mn, Ni, Pb, Pt, Rh, Sb, Sn, Tl, and Zn by ASV techniques (37, 38), the vast majority of applications have dealt with those elements listed in the table. More comprehensive applications information covering work prior to 1966 has been given by Barendrecht (38).

The sensitivity of ASV and its dependence on the presence of species which can be readily dissociated during the deposition step have led to some interesting and possibly very fruitful applications. Labile metal ion concentrations in natural waters have been estimated by using short deposition periods (less than 5 min) (31, 32). The assumption required in this estimation is that "complex materials with slow solution dissociation kinetics do not contribute to the signal significantly" (31). Whether an appreciable amount of metal from these complex materials will be deposited during the plating period depends on the validity of this assumption for the materials in question. Following the initial measurement of the labile metal concentrations, the solutions were acidified, and the increases in metal ion concentrations observed were taken as measures of the "acid exchangeable" metal ion concentrations (31, 32). In another experiment designed to characterize copper-containing species in secondary sewage effluents, the effluent solution was passed through Sephadex columns to separate molecular-weight fractions (33). The ASV determination of copper in the eluted fractions indicated that it was primarily associated with two distinct molecular-weight ranges. Measurements of this type suggest that ASV can be used to solve other speciation problems.

When considering ASV as a viable technique for an analytical problem, the techniques which are commonly considered alternatives are atomic absorption (AA), atomic emission, and atomic fluorescence. Of these, atomic absorption is by far the most widely used, both with flame and nonflame atom reservoirs. Highly specific comparisons between ASV and AA are impossible outside the context of a given analytical problem. However, general

Table III. Summary of Selected ASV Applications

Element determined	Sample type(s)	Type of working electrode used[a]	Ref.	Comments
Ag	Natural water	WIG	14	Linear scan stripping (LSS)
	Rain & snow	WIG & PG	15	Two types of graphite electrodes compared
Au	Drugs & serum	CP	16	Rotating ring disk electrode as described above (9) used to improve reproducibility
Hg	Natural water	WIG	17	
Sn	Geological materials	HMDE	18, 19	Both linear scan and ac stripping techniques used
Zn, Cd, Pb	Seawater	HMDE	20	Stripping carried out with matched Hg drops, only one of which had been used for deposition, to cancel background by difference measurement techniques
Zn, Cd, Pb, Cu	Atmospheric particulates	HMDE	21	LSS
	Natural water	HMDE	22, 23	LSS
	Estuarine waters	HMDE	24	Ac stripping
Pb	Blood	HMDE	25	Small samples required
	Atmospheric particulates	HMDE	26	LSS
Zn, Cd, Pb, Cu	Atmospheric particulates	MTFE	27	LSS
Zn, Cd, Pb, Cu	Natural water	MTFE	28	Lability of metal complexes in water estimated
Pb	Blood	MTFE	29, 30	Large electrode areas and long deposition times used to analyze trace levels in acid digests
Pb, Cu, Cd	Natural water	MTFE	31, 32	Labile and acid exchangeable metal ion concentrations determined
Cu	Secondary sewage effluents	MTFE	33	Sephadex used to separate molecular-weight fractions and establish that Cu occurs in two distinct fractions
Pb, Cd, Cu	Natural water & reagents	MTFGC	34	Wax impregnation of electrode not required
Rh		MTFGC	35	$10^{-7}M$ Rh determined
Pb, Cd	Natural water, blood, urine	MTFGC	13	Small blood and urine samples analyzed by differential pulse techniques without prior acid digestion
Pb, Cd, Zn, Tl	Natural water, blood, & biological tissue	MTFGC	36	Gallium added to eliminate formation of interfering Cu–Zn intermetallic

[a] WIG = wax impregnated graphite; PG = pyrolytic graphite; CP = carbon paste; HMDE = hanging mercury drop electrodes; MTFE = mercury thin-film electrodes; MTFGC = mercury thin film on glassy carbon.

Table IV. Comparison of Detection Limits (ng/ml)

Element	ASV Differential[a] pulse	ASV Linear[b] scan	AA Flame[c]	AA Nonflame[d]
Bi	...	0.01	46.0	3.0
Cd	0.005	0.01	0.7	0.01
Cu	0.005	0.01	2.0	0.3
Ga	0.4	...	38.0	...
In	0.1	...	38.0	...
Pb	0.01	0.02	15.0	0.5
Rh	...	10.0	30.0	...
Sn	2.0 (ac)	...	30.0	0.1
Tl	0.01	0.04	13.0	1.0
Zn	0.04	0.04	1.0	0.008

[a] All values calculated from data and figures in ref. 39. All are 2-min plates with 3600-rpm electrode rotation. Sn entry from ref. 19. [b] All values are from ref. 29 except Rh which is from ref. 35. All values (except Rh) are for a 30-min plating period without electrode rotation. [c] All values from Varian Techtron literature. [d] Values are from refs. 40 and 41, calculated for a 10-μl sample.

analytical considerations afford some useful comparisons.

Although samples as small as 100 μl may be analyzed by ASV, typical samples are in the 1–50-ml range. Thus, sample sizes are approximately equal to those used for flame AA but are considerably larger than the volumes needed for nonflame atom reservoirs. It should be noted that for ASV, addition of supporting electrolyte is the only normal sample preparation requirement; thus, ASV may be essentially a nondestructive technique.

Acid digestion or ashing procedures for solid samples are equally applicable to both spectrometric and stripping techniques. Stripping analysis frequently requires the further addition of a supporting electrolyte to swamp out the effect of variable amounts of electrolyte naturally occurring in the samples.

Stripping techniques require a plating time which is typically 1–10 min. Differential pulse stripping is more sensitive than linear scan stripping and thus requires shorter plating times, but due to the stripping waveform, it requires approximately 3 min for a stripping scan (900 mV @ 5 mV/sec) as opposed to less than a minute for linear scan techniques. Thus, with typical analysis times of 10 min, ASV techniques are somewhat more time consuming than AA methods. Note that these plating times are typical for trace concentrations of $\sim 10^{-8}M$. The detection of smaller quantities would necessitate longer plating times or the use of more efficient plating procedures.

ASV affords multielement capabilities for approximately 4–6 elements with no additional plating time, equipment, or sample requirements. Quantitation is usually achieved by the standard additions procedure. In applications where multielement determinations are to be made, ASV requires less time for the analysis than sequential atomic absorption measurements. Simultaneous multielement AA analysis is difficult, requiring more complex equipment (multiple or multielement lamps, multiwavelength monochromator arrangements, etc.).

A complete and relatively versatile ASV system including cell and electrodes would generally cost ca. $3,000 to $4,000, about half the cost of the most widely used AA systems (excluding a nonflame atom reservoir accessory).

The detection limit comparisons given in Table IV are for elements representative of those which may be detected by either AA or ASV. The values given may be considered accurate to an order of magnitude for the conditions given and are indicative of the concentration ranges below which analyses become inordinately difficult. Clearly ASV offers greater sensitivity than flame AA techniques and is quite generally competitive with the nonflame systems. The primary advantage of AA consequently accrues from the wider range of elements that can be determined. Since the sample is not destroyed in an ASV measurement, the technique permits convenient cross-check analyses by independent analytical methods. Certainly, it is a technique that supplements and complements the analytical capabilities of any laboratory. Its utilization is expanding rapidly and will continue to do so.

References

(1) P. Delahay, "New Instrumental Methods in Electrochemistry," p 119, Interscience, New York, N.Y., 1954.
(2) W. T. deVries and E. van Dalen, *J. Electroanal. Chem.*, **14,** 315 (1967).
(3) W. L. Unterkofler and I. Shain, *Anal. Chem.*, **37,** 218 (1965).
(4) R. A. Osteryoung and J. H. Christie, *ibid.*, **46,** 351 (1974).
(5) H. Blutstein and A. M. Bond, *ibid.*, p 1531.
(6) G. C. Barker, *Anal. Chim. Acta*, **18,** 118 (1958).
(7) W. R. Matson, R. M. Griffin, and E. W. Zink, 1974 Pittsburgh Conference, paper #16, Cleveland, Ohio, 1974.
(8) W. R. Seitz, R. Jones, L. N. Klatt, and W. D. Mason, *Anal. Chem.*, **45,** 840 (1973).
(9) R. G. Clem, "MPI Applications Notes," Vol 8, p 1, McKee Pedersen Instruments, Danville, Ca., 1973.
(10) D. C. Johnson and R. E. Allen, *Talanta*, **20,** 305 (1973).
(11) W. Kemula, Z. Galus, and Z. Kublik, *Nature*, **182,** 1228 (1958).
(12) A. I. Zebreva and M. T. Kozlovskii, *Sovrem. Metody Anal.*, **1965,** p 214; *CA*, **65,** 15 (1965).
(13) T. R. Copeland, J. H. Christie, R. A. Osteryoung, and R. K. Skogerboe, *Anal. Chem.*, **45,** 2171 (1973).
(14) S. P. Perone, *ibid.*, **35,** 2091 (1963).
(15) U. Eisner and H. B. Mark, Jr., *J. Electroanal. Chem.*, **24,** 345 (1970).
(16) G. M. Schmid and G. W. Bolger, *Clin. Chem.*, **19,** 1002 (1973).
(17) S. P. Perone and W. J. Kretlow, *Anal. Chem.*, **37,** 968 (1965).
(18) A. M. Bond, *ibid.*, **42,** 1165 (1970).
(19) A. M. Bond, T. A. O'Donnell, and A. B. Waugh, *ibid.*, p 1168.
(20) A. Zirino and M. L. Healy, *Environ. Sci. Technol.*, **6,** 243 (1972).
(21) G. Colovos, G. S. Wilson, and J. Moyers, *Anal. Chim. Acta*, **64,** 457 (1973).
(22) I. Sinko and J. Dolezal, *J. Electroanal. Chem.*, **25,** 299 (1970).
(23) H. Siegerman and G. O'Dom, *Amer. Lab.*, **4** (6), 59 (1972).
(24) T. Rojahn, *Anal. Chim. Acta*, **62,** 438 (1972).
(25) L. Duic, S. Szechter, and S. Srinivasan, *J. Electroanal. Chem.*, **41,** 89 (1973).
(26) E. P. Parry and D. H. Hern, Joint Conference on Sensing of Environmental Pollutants, Paper 71-1119, Palo Alto, Calif, Nov. 8–11, 1971.
(27) W. R. Matson, PhD thesis, Massachusetts Institute of Technology, Cambridge, Mass., 1968.
(28) W. R. Matson, U.S. Clearinghouse Fed. Sci. Tech. Inform., AD666554, 1968.
(29) W. R. Matson, R. M. Griffin, and G. B. Schreiber, in "Trace Substances in Environmental Health," D. Hemphill, Ed., Vol 4, p 396, University of Missouri, Columbia, Mo., 1971.
(30) W. R. Matson, "Interface Newslett.," Jan. 10, 1974.
(31) W. R. Matson, H. E. Allen, and P. Rekshan, Amer. Chem. Soc., Div. Water, Air, Waste Chem., Gen. Pap., p 164, Minneapolis, Minn., Apr. 14–18, 1969.
(32) H. E. Allen, W. R. Matson, and K. H. Mancy, *J. Water Pollut. Contr. Fed.*, **42,** 573 (1970).
(33) M. E. Bender, W. R. Matson, and R. A. Jordan, *Environ. Sci. Technol.*, **4,** 520 (1970).
(34) T. M. Florence, *J. Electroanal. Chem.*, **27,** 273 (1970).
(35) G. N. Popov, V. V. Pnev, and M. S. Zakharov, *Zh. Anal. Khim.*, **27,** 2456 (1972).
(36) T. R. Copeland, R. A. Osteryoung, and R. K. Skogerboe, *Anal. Chem.*, in press.
(37) Princeton Applied Research Corp., Application Note AN-107.
(38) E. Barendrecht, in "Electroanalytical Chemistry," A. J. Bard, Ed., Vol II, Marcel Dekker, New York, N.Y., 1967.
(39) T. R. Copeland, PhD thesis, Colorado State University, Fort Collins, Colo., July 1973.
(40) M. D. Amos, *Amer. Lab.* (August 1970).
(41) T. K. Aidarov and L. G. Aleksondrova, *Zh. Anal. Khim.*, **28,** 998 (1973). [*J. Anal. Chem. USSR*, **28,** 886 (1973).]

Research of the authors referenced herein was supported by NSF Grant No. GI34813X.

It is not important for the scientist whether his own theory proves the right one in the end. Our experiments are not carried out to decide whether we are right, but to gain new knowledge. It is for knowledge sake that we plow and sow. It is not inglorious at all to have erred in theories and hypotheses. Our hypotheses are intended for the present rather than for the future. They are indispensable for us in the explanation of the secured facts, to enliven and to mobilize them and above all, to blaze a trail into unknown regions towards new discoveries.

R. Willstätter
Willard J. Gibbs Medal Address
Chicago, IL, September 14, 1933 (*24*)

Foundations of Modern Liquid Chromatography

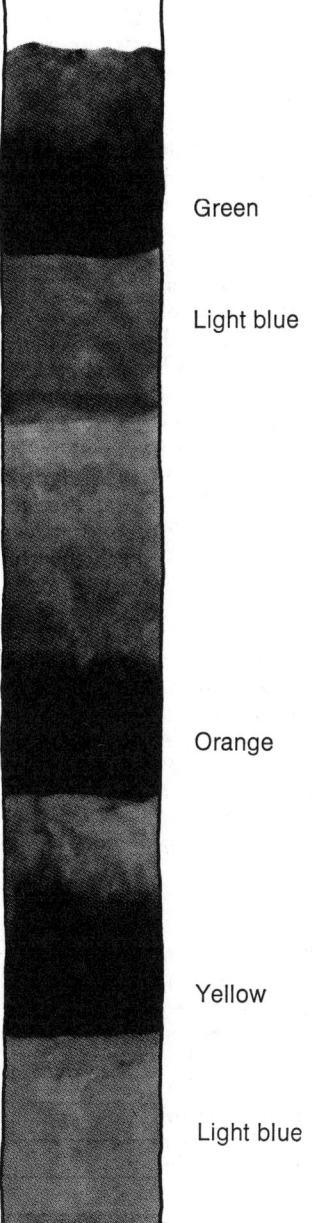

The design is based on a color drawing in the frontispiece of the book, "Chromatographic Absorption Analysis," by Harold H. Strain, published in 1942 by Interscience. Represented is the separation of carotenes by adsorption of petroleum ether extracts of leaves on a magnesia column. At the top of the column are xanthophylls and chlorophylls; in the center, beta-carotene; and bottom, alpha-carotene. Strain himself was a pioneer in the introduction of modern liquid chromatography in the U.S.

L. S. Ettre
Perkin-Elmer Corp.
Norwalk, CT 06851

C. Horvath
Department of Engineering and Applied Science
Yale University
New Haven, CT 06520

During the last few years, a number of articles have dealt with the development of the various chromatographic techniques and the work of some of the pioneers in this field (*1–9*). Today, we finally have a detailed picture of the life and activities of Tswett who, in 1903–1906, developed the technique we today call liquid column adsorption chromatography. Thanks to Martin's accounts (*10*), we have direct insight into the thinking and scientific motivation which led to the development of liquid-liquid partition chromatography, paper chromatography, and gas-liquid partition chromatography (GLPC). We also have a report on the pioneering works resulting in the development of gas adsorption chromatography (*11*) which actually preceded GLPC. All these, together with some of the older summaries (*12–16*), highlight major events yet unwritten in the history of chromatography. The picture is still very incomplete; many subjects not even touched await a chronicler.

In 1975 there is no need to stress the importance of chromatography; it has become the most widely used analytical method. Today, there is practically no laboratory in the whole world where some form of chromatography is not employed. The cursory observer might attribute the universal usage of the technique to the explosive development of gas chromatography in the last 20 years; however, this is not true. Twenty-eight years ago, well before the development of gas chromatography and just three years after the introduction of paper chromatography by Consden, Gordon, and Martin (*17*), Paul Karrer, the great organic chemist and cowinner of the 1937 Nobel prize in chemistry, already had given the following appraisal of chromatography in his plenary lecture at the 1947 Congress of the International Union of Pure and Applied Chemistry in London (*18*):

No other discovery has exerted as great an influence and widened the field of investigation of the organic chemist as much as Tswett's chromatographic adsorption analysis. Research in the field of vitamines, hormones, carotenoids and numerous other natural compounds could never have progressed so rapidly and achieved such great results if it had not been for this new method, which has also disclosed the enormous variety of closely related compounds in nature.

Karrer's positive appraisal of chromatography as *the* discovery which exerted the greatest influence on organic chemistry included a very peculiar remark: in 1947 he called chromatography—which after all was discovered by Tswett around 1906, i.e., 40 years earlier—"this new method". And indeed, this is one of the most intriguing questions in the history of chromatography: how could it be that for about 20 years, Tswett's technique remained stagnant, used only by a few researchers, and then, when it was brought back from oblivion by Kuhn, Winterstein, and Lederer (*19*), it became universally accepted within a few years? In other words, why was chromatography for Karrer (who received his PhD in 1911, just a few years after Tswett's basic papers) still a relatively "new" method in 1947?

Thus, the first question we want to investigate in this article is the reason for the 20-year-long stagnant period in the development and application of chromatography as an investigative method. We shall also deal with those who *did* utilize chromatography in their work during this period.

After the "reinvention" of chromatography in 1931, there was a fantastic explosion in its use, and at about the start of the Second World War, chromatography already was a generally accepted technique. Thus, we shall also briefly discuss this expansion, the establishment of the art of chromatography in the period following the first paper by Kuhn, Winterstein, and Lederer.

In the previous paragraph, we spoke about the "art of chromatography". The reason for this is simple: at that time chromatography had practically no theory at all. It was developed later, among others by Martin and Synge who, in their first paper on liquid-liquid partition chromatography (*20*), gave a model which could be used as the expression of column efficiency. The development of chromatographic theory received a great impetus from the results in gas chromatography and is still continuing. Now, it is a powerful tool which is applied in the field where it all began: liquid column chromatography. Therefore, the last subject we review is the early development of the theory of chromatography.

The subject of this paper is the accession of the art of liquid chromatography and the advent of chromatographic theory; these two represent the foundation of modern high-speed, high-performance liquid column chromatography. We conclude our discussion around 1955–1956, when the two started to converge and the evolution of liquid chromatography as used today commenced. It is certain that some of the readers will disagree with many of our conclusions, but this is natural: after all, we are dealing here with relatively recent events, and some of the scientists mentioned and their students are still alive.

Let us begin with the period after Tswett.

Period Following Tswett

It is often said that Tswett had only a few followers and his technique (liquid column adsorption chromatography) was almost forgotten until 1931. Strain, himself a pioneer in the introduction of modern liquid chromatography in the United States, debates this in a very detailed review article (*21*) and refers to the existence of "extensive use" of Tswett's method during this period. However, even he could list only a little over half-a-dozen references coming from about four laboratories; thus, we see no reason to change this general statement.

To investigate the reasons for this dormant period, we must look into the incredible development of chemistry in Germany and the interest of organic chemists in this period, up to about the second decade of the twentieth century.

In 1824 when Justus von Liebig (1803–1873) started to teach chemistry in Giessen, using a laboratory approach he had learned during his stay in Paris, he, in fact, created the basis of chemistry as a branch of science in Germany, resulting within two generations in the worldwide fame of German chemistry. [The laboratory-based teaching of chemistry had its origins at the College of Mining founded in 1735 in Hungary in the town of Selmeczbánya (today: Banská Stiavnica in Czechoslovakia). It was developed there by N. J. Jacquin (1727–1817), the Dutch-born scientist who was the first professor of chemistry at the college. This system was then later utilized at the Ecole Polytechnique founded in 1794 in Paris, and Liebig saw its advantages when staying in Paris for two years as a student at the Ecole and an assistant to Gay-Lussac. Since both authors of this article are graduates of the Technical University, Budapest, they are happy to emphasize this chain of development originated at the ancestor of their alma mater.]

In 1852 Liebig moved to the University of Munich. After his death his chair was taken in 1875 by Adolf von Baeyer (1835–1918) and then after Baeyer's retirement four decades later, by Richard Willstätter, Baeyer's former private assistant. With our 1975 thinking, we simply cannot comprehend the prestige and authority these scientists exercised during their lifetime. Liebig's lectures in Munich were routinely attended by members of the Bavarian royal family, and both Baeyer and Willstätter were bestowed the highest civilian honors. The immense achievements reached in this time are best summarized by Willstätter in his appraisal of Baeyer's work (*22*):

When Baeyer began to work in indigo, there were no dye factories in Germany... In the years followed, the chemical laboratories of German universities exerted an influence on the industry such as probably will never again be possible, in view of the tremendous growths of the undertakings. But at that time, the university teacher Baeyer could become a founder and promoter of the German dye industry... (He) created industries and contributed much to the economic welfare of his country.

We admire even more this rapid growth if we realize that in his first research in 1857, Baeyer was still trying to prove that there is a difference between methyl chloride prepared from methanol and hydrogen chloride, and chloromethane prepared from methane and chlorine; at that time the equivalency of the four hydrogen atoms in the methane molecule was still not clear. The structure of benzene was first described in 1865 by Baeyer's older friend, Kekulé (1828–1896), and within 15 years, Baeyer developed the total synthesis of indigo:

Adolf v. Baeyer founded modern German organic chemistry and industry. However, from our point of view, the scientist most important was Richard Willstätter, Baeyer's successor whose prestige and students dominated organic chemistry in Europe (and probably also in the United States) in the first four decades of this century. Since his life and activities interweave with the events discussed in this article, a brief summary is useful here.

Richard Willstätter was born in 1872 in Karlsruhe. He entered the Technical University at Munich in 1890; after graduation in 1895, he became a private assistant to Baeyer at the University of Munich. In 1905 he accepted the invitation of the Federal Technical University (ETH) in Zurich, Switzerland, and occupied the chair in chemistry until 1912. Most of his work on chlorophyll was carried out there; his book written in cooperation with A. Stoll (23) was published in 1913.

In 1912 Willstätter became associated with the Kaiser Wilhelm Institute of Chemistry in Berlin-Dahlem for only three years. In 1915 upon the retirement of A. v. Baeyer, he took over Baeyer's chair and institute. In 1924 when the appointment of Goldschmidt to professor and head of the Institute of Mineralogy was rejected because of his Jewish origin, Willstätter handed in his resignation as a matter of principle and remained firm in spite of many entreaties. He remained active in chemistry, and an assistant continued his experimental work; however, he never entered his old institute again, although he succeeded in getting Heinrich Wieland to be nominated as his successor. (Under Wieland's leadership, the institute later became actively involved in the development of chromatography. We shall deal later in detail with the pioneering work of Schwab in the extension of chromatography to inorganic analysis; it might also be interesting to mention that G. Hesse, who wrote one of the first monographs about chromatography, graduated in 1932 from Wieland's laboratory.)

After Hitler took over the German government, Willstätter started to have problems because of his Jewish origin, but for some time, he refused to leave the country. Finally, in March 1939 after grueling experiences, he crossed the border into Switzerland where he died in 1942.

In the period when Tswett published his chromatography papers and reported on his investigations about chlorophyll, Willstätter was the "pope" of this field; after all, his 1915 Nobel prize was awarded for "his results in the field of plant pigments particularly chlorophyll". Thus, it is natural that the opinion of Willstätter could have played an important role in the acceptance and application of Tswett's method, particularly since after all, Tswett was not an organic chemist but a botanist, and was an outsider in the closed society of European science and chemistry.

There is a controversy in the interpretation of Willstätter's opinion about Tswett. His autobiography (24)—one of the most moving personal documents with the beautiful subtitle, "About Work, Leisure and Friends"—contains the following statement:

The importance of the adsorption technique has already been proved by the botanist Micaël Tswett of Warsaw in his early research on chlorophylls, to which I have often drawn the attention.

This seems to show that Willstätter considered Tswett's results important. That is, however, not entirely true, at least not until the later part of his activities. In his famous book on chlorophyll (23), Tswett's work, "the separation of the natural pigment on an analytical scale by means of its frac-tional adsorption from its solution," is called "an odd way" (25a); another quotation from the same book summarizes Willstätter's two basic objections against chromatography: it seemed to be unsuitable for preparative work (isolation), and there was a possibility of chemical changes of the sample components during the adsorption process (25b):

The chromatographic method has as yet been used on a small scale only and it appears unsuitable for preparative work. How far Tswett succeeds, in the formation of the chromatogram and the isolation of the chlorophyll components from the same, in preventing the allomerization of chlorophyll is as yet unknown.

This negative opinion is also demonstrated by a later statement from the same book (25c) which refers to a controversy between Tswett (his book: 26) and a 1907 publication of Willstätter and Mieg (27) (Later it was proved that in this dispute, Tswett was right):

Tswett considers the xanthophyll of Willstätter and Mieg an isomorphous mixture of two or three xanthophylls... It is not unlikely that the assumption of the esteemed botanist, like many of his observations, is correct... It might also be possible, however, that xanthophyll has undergone a change in the chromatographic analysis as a result of oxidation, to which it is especially liable when in an adsorbed condition.

As pointed out by Dhéré (12), the best proof that in this period Willstätter did not consider Tswett's work really important is the fact that in the chapter written for Abderhalden's handbook, the definitive monograph of biological methodology of that period, in which he summarized the stand of knowledge on plant pigments (28), Willstätter does not even mention Tswett's work on chlorophyll.

E. Lederer (29), in retrospect, points out that Willstätter's negative opinion about the usefulness of chromatography in this field might have been based on the simple fact that his collaborators used an adsorbent which was really unsuitable for chlorophyll; after all, Tswett had previously observed that these pigments can decompose on "agressive" adsorbents and suggested therefore that sugar powder or inulin be used. Lederer continues by saying that this "apparently escaped the attention of Willstätter and Stoll". This, however, would be peculiar since in the chlorophyll book all three adsorbents, calcium carbonate, sugar, and inulin (which can be used to pack Tswett's columns) are specifically mentioned (25b).

Later, when working on enzymes, some chromatography was done [This is mentioned, e.g., in the footnote to page 1 of Turba's book (30)] in Willstätter's laboratory, and evidently, his opinion started to change. In his book on enzymes published in 1928 (31), Willstätter said:

Although his method was not suitable for preparative work, Tswett was still able, although, of course, only in analytical scale, to separate the mixture of leaf pigments to its components, with an adsorption procedure. (31a)

The chromatographic adsorption method which, in the hands of M. Tswett, lead to important results concerning chlorophyll and its derivatives, is not suitable for work in larger scale, i.e., for preparative purposes. (31b)

It is interesting to read these statements because, while still emphasizing his belief that chromatography is not suitable for preparative work, Willstätter seems to have indicated that on another scale, the technique could be important. We might add to this that in fact, Willstätter led Kuhn to realize the importance of chromatography by giving him the German translation of Tswett's magnum opus (*26*), prepared before 1913 especially for Willstätter.

The opinion of Willstätter, the overemphasis on preparative work, reflects a philosophy, the general attitude existing among organic chemists at that time. This is, again, a logical result of the development of organic chemistry.

At the time of Baeyer's youth, the key interest of organic chemists (and there were not too many of them) was to learn about reactions, to isolate pure substances, and to modify them with the help of relatively simple reactions. Again quoting Willstätter (*22*), this was

> ... *the period of simple organic compounds whose mighty stream was fed by the springs of vegetable and animal life and reinforced by the great tributary welling from coal tar. It was the period of materials which could be distilled and crystallized.*

Baeyer directed this interest to a new level: the synthesis of complex organic substances. (It is indicative of the general interest of this period that the title of Baeyer's inaugural academic address in 1878 was "The Chemical Synthesis".) This interest created a brand new industry which, in turn, made this type of work even more important. Willstätter then went one step forward: he started to investigate the complex pigments present in plants, without having direct industrial possibilities. However, at this time, the key-words were still isolation and purification: isolation from the accompanying material and purification from trace impurities—this was done at that time by extraction and crystallization. Tswett's method just did not fit this interest: there was no real interest to isolate all the compounds present—one only wanted to isolate a few key compounds, and there was really no desire or need to work with very small amounts. We are almost three decades away from the opinion expressed by Rosenthaler (*32*) commenting on one of Karrer's papers:

> *It would be a mistake to think that a preparation purified through crystallization must be purer than one prepared through chromatographic analysis. The chromatographic purification had been proved in every new investigation to be superior to that obtained through crystallization.*

Schwab, in his first paper on the use of chromatography in inorganic analysis published in 1937 (*78*), summarized very appropriately the changed emphasis which came only 25 years later:

> *After the highly gifted Russian botanist Tswett created the method and applied it successfully to plant extracts, it lie dormant for 25 years in the lap of the literature. Only after biochemistry, pressed by new problems, demanded methods for the reliable separation of small quantities of similar substances, could chromatography celebrate a rapid and brilliant resurrection, first at the laboratory in Heidelberg and then everywhere.*

The activities of the new generation, mostly Willstätter's students or scientists who got their initial impetus from him, brought forth the importance of total separation (i.e., separation of all components present), not just isolation of one or two major constituents, and the possibilities of carrying out this separation with small quantities. The rebirth of chromatography coincided with the great shift in chemical research from synthesis to analysis, a new way of thinking which has been dominant since then.

Early Utilization of Tswett's Technique in "Dormant Period"

Tswett's work was naturally not forgotten completely. It was utilized by a few, mostly botanists and biochemists, for the investigation of leaf pigments. The first was Dhéré (at Fribourg University in Switzerland) who started to apply Tswett's technique as early as 1911, first with Rogowski, a student from Poland, whose thesis is dated 1912 (*33*), and then with Vegezzi who graduated in 1916 (*34*). Dhéré's main interest was related to the studies of fluorescence and other optical phenomena in organisms, and he could be considered a pioneer in biophysics; his interest is reflected in the titles of two theses: Research on the Ultraviolet Absorption Spectra and on the Fluorescence Emission Spectra of Chlorophyllic Pigments (Rogowski) and Research on Certain Pigments Present in Invertebrates: Helicorubin, Hepatochlorophyll, Tetronerythrin (Vegezzi).

Without question, Dhéré was the first to grasp the importance of Tswett's method and put chromatography to work in the laboratory. He was also the first to publish (in 1943) a detailed biography of Tswett with an appraisal of his results and the influence of chromatography on the advancement of science (*12*). (In the preface of the second edition of his monograph published in 1938, Zechmeister mentions that he planned to introduce the new edition with a biography of Tswett; however, he could not find any reliable data on his life.)

A few other scientists such as Czapek (*35*), Coward (*36*), and Lipmaa (*37*) are also quoted by Zechmeister and Cholnoky in their monograph and by Strain and Sherma (*21*). These were, however, evidently isolated cases which had little impact.

During this period, the activities of an American scientist are most important; besides Dhéré, he is probably the only one who immediately understood the importance of Tswett and applied chromatography in his own work. He is Leroy Sheldon Palmer (1887–1944), a graduate of the University of Missouri. After graduation he worked at the Dairy Research Laboratory operated by the U.S. Department of Agriculture and the university. In 1919, following his mentor, C. H. Eckles (1875–1933), the head of the Department of Dairy Husbandry, he moved to the University of Minnesota where he served for almost 25 years as associate professor, professor, and head of the Division of Agricultural Biochemistry. The Palmer classroom building on the university campus in St. Paul is dedicated in his memory. [Palmer's life and activities have been discussed in detail by two of his former students (*38*).]

Palmer's first major interest was dairy chemistry which was later extended to nutrition problems. His doctoral thesis in 1913 was entitled "Carotin—the Principal Natural Yellow Pigment of Milk Fat" and was published in 1914 in five papers coauthored with Eckles (*39–43*). This work was followed by a number of papers (e.g., *44–46*) on yellow pigments in a wide variety of biological substances and animal products and their relationship with the growth, fecundity, and reproducibility of animals.

In his first publication (*39*), Palmer described Tswett's technique, column adsorption chromatography, which he extensively used in all his work. His magnum opus is the book published in 1922 as the first volume of the American Chemical Society Monograph Series (*47*). Besides giving a very thorough summary of his own work and the work of others in the field of carotenoids and related pigments, Palmer discusses in detail Tswett's work and technique. Moreover, he emphasizes that a chromatographic separation is not yet an absolute proof in identification since further examination of the fractions is needed, and that often a chromatographic fraction has to be rechromatographed to obtain high purity.

It is not clear whether Palmer read Tswett's book (published in Russian) (*26*) which Willstätter had in German translation; he quotes its title in French which would indicate that he took the reference from another publication. However, he certainly refers to the most important publi-

Leroy Sheldon Palmer

cations of Tswett which appeared in German and French journals.

We cannot judge the actual influence of Palmer on future researchers in chromatography although it is true that his book was the standard reference for many years. However, his direct connection to the next evolution step in chromatography is clear: Lederer, in his retrospective article (29), mentions that he learned about the existence of Tswett's method when reading Palmer's book. In this way, we arrived at the next period: the rebirth of chromatography.

Rebirth of Chromatography

In the previous section on the "dormant period", it may have appeared that we largely blamed Willstätter for the lack of recognition and universal acceptance of Tswett's work. However, we have only tried to place his influence in the proper context of the development of organic chemistry. It should be emphasized again that Willstätter was one of the most respected and loved leading scientists in the history of modern chemistry, and through his influence some of his students have also received the highest scientific recognitions. He implanted in them the need for better methods, the search for truth, and the love of the study of nature and its complex substances. Thus, it seems natural that when more refined separation methods were needed, they returned to the almost forgotten technique of Tswett; in fact, as we already discussed, Willstätter might have contributed to this through one of his closest former students: Richard Kuhn.

Richard Kuhn (1900–1967) was born in Austria; after spending one year at the University of Vienna, he continued his studies in Munich and graduated in 1922 from Willstätter's laboratory. For four years he remained there, and in 1926—when only 26 years old!—on the recommendation of Willstätter, he was appointed a professor at the Federal Technical University (ETH) in Zurich. In 1929 he moved to Heidelberg as the director of the Institute of Chemistry at the Kaiser Wilhelm Institute for Medical Research and as a professor at the University of Heidelberg. He brought with him from Zurich to Heidelberg Alfred Winterstein (1889–1960), a Swiss and the son of a well-known scientist. Winterstein studied at the ETH and in 1923 graduated under Professor Staudinger whom Kuhn succeeded in 1926.

We have an interesting account by Edgar Lederer on the events surrounding the "reinvention" of chromatography (29). Lederer was also born in Austria; after graduating in July 1930 at the University of Vienna, he arrived in September at Heidelberg to join the laboratory of Kuhn and Winterstein as a young assistant.

Both Kuhn and Winterstein were active in the field of natural pigments, carotenoids. When young Lederer joined them, his assigned project was also in this field: to test the

The adsorption properties of carotin and xanthophylls used throughout these experiments were an application of Tswett's interesting discovery that carotin when in anhydrous carbon bisulphide (or petroleum ether) solution is not adsorbed by pure, dry $CaCO_3$, while xanthophylls, under the same conditions, are adsorbed to a greater or less degree. Tswett has made a beautiful application of this discovery to a solution of carotin, xanthophylls and chlorophyll pigments, whereby he separated the pigments from one another by filtering the solution of mixed pigments through a column of dry $CaCO_3$, moistened first with the solvent being used (CS_2 or petroleum ether). The result was that as the pigments passed through the column with the aid of suction and a stream of the pure solvent, they differentiated themselves into various zones, depending upon the adsorption affinity of the $CaCO_3$ for each pigment. Carotin, being unadsorbed by the $CaCO_3$ passed through first as a rose or orange colored ring or zone, and could be collected by itself at the mouth of the glass tube, which is drawn out somewhat at the end. The arrangement for performing such a chromatographic analysis is shown in figure 1. The details of such an analysis may be found by referring to Tswett's work.

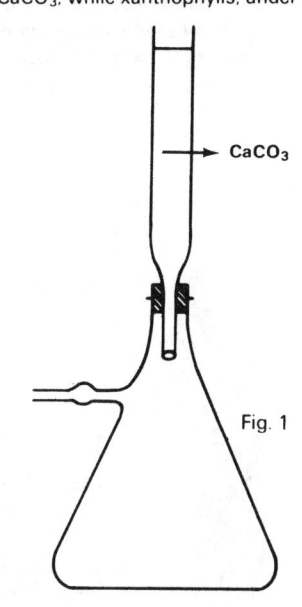

First description of Tswett's technique by American authors L. S. Palmer and C. H. Eckles (39)

validity of Kuhn's hypothesis that the "lutein" of egg yolk represents a mixture of "leaf xanthophyll" and zeaxanthin (just isolated by Karrer a year earlier from yellow corn). But how did one test this hypothesis? As told by Lederer, about that time he read Palmer's book (47) which contained the method of Tswett. He discussed this with Kuhn who gave him the copy of the translation of Tswett's book (26) originally obtained from Willstätter.

After reading the translation, in December 1930 Lederer prepared a small column packed with calcium carbonate and tried to determine whether lutein and zeaxanthin could be separated. (Based partly on this work, the nomenclature for xanthophylls was later modified: the term "xanthophyll" previously used for "leaf xanthophyll" was then considered the generic name for the C_{40} oxygen-containing carotenoids, whereas the term "lutein" previously used to express xanthophylls of egg yolk was then used exclusively for the main xanthophyll present, i.e., for a pure substance.) When this preliminary experiment proved successful, he made a larger column of 7-cm diameter and chromatographed the "lutein" solution prepared from 100 eggs (the white of which was previously "transformed to a delicious cake"). After development, the intermediate layer of the column packed was removed, suspended in the solvent, and the solution rechromatographed. This was repeated once. Investigation of the fractions proved the correctness of Kuhn's hypothesis. (In the paper the authors dutifully report that they used eggs from both Bulgaria and The Netherlands for the experiments.)

The experiments were repeated a couple of times, and on March 10, 1931, the manuscript of the paper "Zur Kenntnis der Xanthophylle" ("On the Knowledge of Xanthophylls") coauthored by Kuhn, Winterstein, and Lederer was submitted to *Hoppe-Seyler's Zeitschrift für Physiologische Chemie* (19) followed by a detailed paper one week later to the *Berichte der Deutschen Chemischen Gesellschaft* (49).

This work was soon followed by other publications from Kuhn's laboratory, and the technique was picked up within a very short time by many other laboratories. Particularly Winterstein became the unofficial envoy of the new technique, giving lectures and demonstrations at many places. [Martin remembers (10) that while a student at Cambridge

A happy day in Kuhn's laboratory in Heidelberg, sometime in 1931. Identified persons are: 1, R. Kuhn; 2, Mrs. Kuhn; 3, A. Wassermann (later immigrated to England); 4, H. Brockmann (became professor at Göttingen); 5, Miss G. Stein (became Mrs. Brockmann); 6, M. Hoffer (immigrated to USA and joined Hoffmann-La Roche); 7, Th. Wager-Jauregg (immigrated to USA but returned to Switzerland); 8, H. Roth (head of Kuhn's analytical laboratory; became professor at Greifswald and Braunschweig); and 9, E. Lederer

Richard Kuhn (right) and E. Lederer at a reunion in Heidelberg in 1963

Two xanthophylls first separated by Kuhn, Winterstein, and Lederer in 1930–1931, hereby starting rebirth of liquid column adsorption chromatography

University, he heard the lecture of Winterstein who visited them and also demonstrated the separation of a crude carotene solution on a chalk column.] Another important contribution to the universal acceptance of chromatography was the book of Zechmeister and Cholnoky first published in 1937. [Zechmeister's book on carotenoids published in 1934 (49) had already dealt in detail with the principles and application of Tswett's technique.]

László Zechmeister (1889–1972) and his work played an important role in making chromatography widely known and accepted as a scientific tool. He was born in Hungary and finished his studies in Zurich, at the ETH. After graduating from Willstätter's laboratory, he followed his professor to the Kaiser Wilhelm Institute for Chemistry in Berlin-Dahlem. At the outbreak of the First World War, he was drafted in the Hungarian Army, captured on the Russian front, and kept in a POW camp in Siberia until 1918 when he rejoined Willstätter in Munich. In 1923 Zechmeister became professor of medicinal chemistry at the University of Pécs in Hungary where he continued his work on cellulose derivatives and carotenoids started while working with Willstätter. Cholnoky (1899–1967) was Zechmeister's assistant and later successor at Pécs University. Their book became a scientific "bestseller," and within a year a second edition had to be prepared. This book was soon followed by Willstaedt's book and by four important chapters on chromatography in the most widely used methodology books of that time (50–53).

(Harry Willstaedt was born in 1904 in Switzerland to a German father and a French mother. The family moved in 1914 to Germany where he received his education, graduating in 1928 from the University of Berlin. In 1929 he spent a brief period in Stockholm working on carotenes and vitamin A but soon returned to Germany. In 1933 he also had to leave Germany; he went to the University of Uppsala where in continuing his studies on natural pigments he became involved in chromatographic work. In 1940 he moved to Stockholm where he worked first at the Weener-Gren Institute and then at the university. In 1947 during a trip to the United States, he became ill and died on New Year's Eve.)

The group at Heidelberg was able to remain together only for a few years. Less than two months after Hitler's coming to power in Germany, Lederer had to leave; being married to a French girl (a student at Heidelberg University), he went to Paris. Between 1935 and 1937, he was director of a laboratory at the Research Institute of Vitamines in Leningrad; since 1938 he has been active in France where he now serves as the director of the Institut de Chimie des Substances Naturelles in Gif-sur-Yvette, a suburb of Paris.

Winterstein stayed in Heidelberg until 1934 when he had to return to his native Switzerland where he joined Hoffmann-La Roche as their chief chemist. He had a very dis-

L. Zechmeister and L. Cholnoky (Chemical Institute, University of Pécs, Hungary): *Die chromatographische Adsorptionsmethode-Grundlagen, Methodik, Anwendungen.* Springer, Vienna; first ed.: 1937; second ed.: 1938.

H. Willstaedt (Medical Chemistry Institute, University of Uppsala, Sweden): *L'Analyse chromatographique et ses applications.* Hermann & Cie., Paris, 1938.

L. Zechmeister (California Institute of Technology, Pasadena, CA) and L. Cholnoky (Chemical Institute, University of Pécs, Hungary): *Principles and Practice of Chromatography.* J. Wiley & Sons, New York, 1941. (English translation of the second German edition).

H. H. Strain (Carnegie Institution of Washington, DC, and Stanford University, Stanford, CA): *Chromatographic Adsorption Analysis.* Interscience, New York; first printing: 1942; second printing: 1945.

G. Hesse (University of Marburg/Lahn, Germany): *Adsorptionsmethoden im chemischen Laboratorium mit besonderer Berücksichtigung der chromatographischen Adsorptionsanalyse (Tswett-Analyse).* W. de Gruyter, Berlin, 1943.

T. I. Williams (Oxford University, Oxford, England): *An Introduction to Chromatography.* Blackie & Sons, London; Chem. Publishing Co., Brooklyn, NY, 1947.

L. Zechmeister (California Institute of Technology, Pasadena, CA): *Progress in Chromatography 1938–1947.* Chapman & Hall, London, 1950.

H. G. Cassidy (Yale University, New Haven, CT): *Adsorption and Chromatography.* Interscience, New York, 1951.

E. Lederer (Institute of Biology and Physico-Chemistry, Paris, France): *Progrès récents de la chromatographie. Première partie: Chimie organique et biologique.* Hermann & Cie., Paris, 1949

M. Lederer (Radium Institute, Paris, France): *Progrès récents de la chromatographie. Deuxième partie: Chimie minérale.* Hermann & Cie., Paris, 1952.

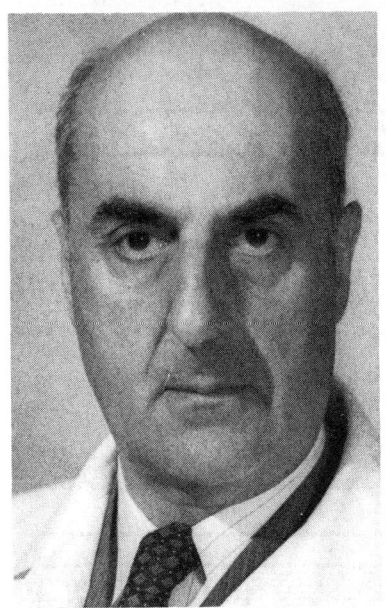

László Zechmeister in 1948

Monographs published in first 20 years of rebirth of chromatography

tinguished career at this company, one of the largest pharmaceutical houses in the world; he died in 1960 while attending a congress in Tokyo.

Kuhn remained in Heidelberg as the director of the institute until his death in 1967. He became the winner of the 1938 Nobel prize in chemistry "for his results in the field of carotenoids and vitamines" but was not permitted to accept the prize.

Another victim of the political events in Europe was Zechmeister who in 1940 immigrated to the USA and continued his work on carotenoids and other natural products as professor at the California Institute of Technology, until his retirement in 1959.

The second edition of Zechmeister's book was translated into English and published in 1941; it too became an instant success. It was followed within a year by Strain's book and then by monographs of others. The monographs published in the first 20 years of the rebirth of chromatography are listed separately. As seen, besides Zechmeister, Cholnoky, Willstaedt, and Strain (whose books have already been mentioned), Hesse in Germany, Williams in England, Cassidy in the United States, and Edgar and Michael Lederer belong to the pioneers from whose writings future chromatographers gained their knowledge. [The two French books by E. Lederer and M. Lederer served as the basis of a new book published a few years later in English (18).]

In 1941 chromatography received a new impetus through Martin and Synge of the Wool Industries Research Association's Laboratory in England, by their description of partition chromatography (20), soon followed by the development of paper chromatography by Consden, Gordon, and Martin (17) which revolutionized biochemical analysis. A further contribution of Martin to liquid chromatography was reversed-phase chromatography introduced in 1950 (53). It is understandable that in 1952, the Nobel prize in chemistry was awarded to Martin and Synge for their pioneering activities.

Another Nobel prize related to chromatographic work had been awarded four years earlier to Tiselius (1902–1967); the citation refers to his work on adsorption analysis and electrophoresis. Tiselius, Claesson, and coworkers at the University of Uppsala in Sweden contributed significantly to the development of liquid chromatography at its early stage (55–57). In 1940–1946 they systematized the method by pointing out the three variants such as frontal analysis (without a carrier), elution analysis (with a pure solvent as the carrier), and displacement development (using a more strongly adsorbed displacer). They also pioneered in adding an optical detector—based, e.g., on interferometry or refractive index measurement—at the column end to recognize the concentration changes when the individual sample components emerge, eliminating the need for the collection of the individual fractions (58). In 1952 Tiselius and coworkers further improved the efficiency and usefulness of liquid column chromatography by introducing gradient elution (59). Later, molecular exclusion chromatography ("gel filtration") was developed in Tiselius' laboratory, but this belongs to the most recent developments of the technique and is outside the scope of this article.

Two other chromatographic techniques, both belonging to liquid chromatography, can also be traced back to this period. *Thin-layer chromatography* has its origin in the work of Izmailov and Shraiber who in 1938 at the Institute of Experimental Pharmacy in Kharkov, USSR, first used layers of adsorbents on glass plates for the separation of tinctures of belladonna, digitalis, and rhubarb; they called the technique "spot chromatography" (60). In 1951 Kirchner and coworkers at the laboratory of the U.S. Department of Agriculture in Pasadena, CA, utilized glass strips coated with adsorbents ("chromatostrips") for the analysis of terpenes in essential oils (7, 61). However, the technique became widely accepted only after the work of Stahl in 1956–1958, who at that time was teaching at the University of Mainz in Germany (62, 63).

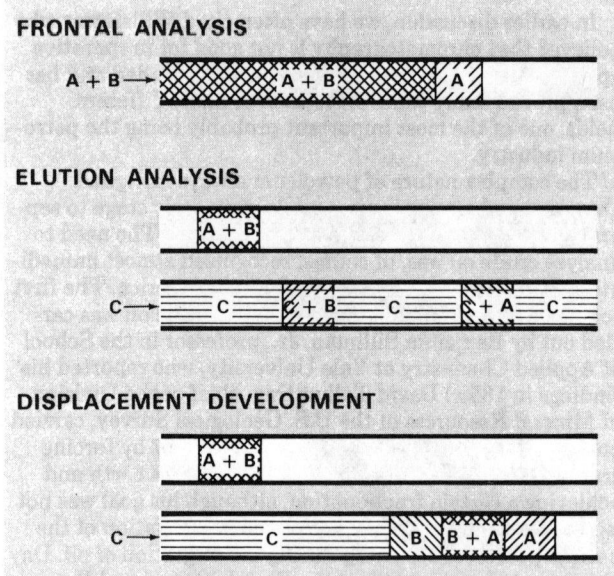

A. Tiselius in 1951

Principles of the three chromatographic techniques employed in the past. A and B are the two components to be separated, and C is the carrier or displacer. Today, elution chromatography is used most generally although recently introduced affinity chromatography employs displacement development

Another important chromatographic technique evolved in this period was based on the use of *ion exchange* for separation. In 1938 Taylor and Urey at Columbia University showed the spectacular separation of lithium and potassium isotopes using long columns packed with zeolite. As described in their paper (64), the longest column was made of a 100-ft long hard rubber pipe with a 1¼-in. diameter which was constructed in one of the stairwells of the Chandler Chemical Laboratory of the university. About the same time, Schwab and Dettler, in their third paper on the use of chromatography for the separation of inorganic ions (see below) indicated that the separation on aluminum oxide is also partly based on ion exchange (80).

The real development of ion-exchange chromatography started after ion-exchange resins became commercially available in the early 1940's and after Samuelson demonstrated their application in analytical chemistry (65). This technique was particularly important in the separation of rare earth and various transuranium elements in connection with the Manhattan project. Because of the nature of the work, it was published only after the Second World War (see, e.g., 66–68). This technique was also used in 1953 in the identification of elements 99 and 100 by examining fallout debris from a thermonuclear explosion in the Pacific in 1952 (69).

The work carried out on the use of ion-exchange chromatography in the field of atomic energy was then extended to problems in biochemistry. Cohn's pioneering work at Oak Ridge National Laboratory (70, 71) on the separation of nucleic acid constituents was practically simultaneous with the work of Moore and Stein at the Rockefeller Institute for Medical Research in New York City on the separation of amino acids on starch (72, 73); this was followed by separation on sulfonated polystyrene resins (74) which, a few years later, led to the development of the first high-performance liquid chromatograph, the amino acid analyzer (75). In 1972 Moore and Stein received the Nobel prize in chemistry as a recognition of their activities in the field of enzymes representing the logical continuation of their original work.

Ion-exchange chromatography was later also extended to paper chromatography by M. Lederer in Paris, who first succeeded in preparing ion-exchange papers by dipping the paper in the suspensions of colloidal resin aggregates (76). (M. Lederer is a cousin of E. Lederer; he left Vienna with his parents when Austria was annexed by Germany. He studied in Australia after which he moved to France where he was active at the Institut du Radium in Paris. Since 1960 Dr. Lederer has been associated with the University of Rome in Italy; he has also served as the editor of the *Journal of Chromatography* since its inception.)

Inorganic Chromatography. With the exception of the work of Taylor and Urey and those involved in the Manhattan project, we have so far only discussed the application of the chromatographic techniques for the separation of organic substances. It is, however, important to realize that a few years after the rebirth of chromatography, it was also extended to the separation of inorganic ions.

The pioneer in this field was Schwab, a professor at the University of Munich, associated with the institute of Wieland, the successor of Willstätter. On July 8, 1937, at the meeting of the Deutsche Bunsen Gesellschaft, he presented a short paper "Chromatography as a New Ancillary Method in Analytical Chemistry" (77) which was followed by five papers (78–82). We have quoted Schwab's remark in his first paper (78) about the "brilliant resurrection" of chromatography; he continued the evaluation of the development of chromatography by stating that it

> ... advanced again peculiarly. In five years of intensive chromatographic work, the methods remained completely restricted to organic substances and even the advance to adsorption from aqueous solution instead of using organic solvents was celebrated as an important step forward. However, the separation of inorganic ions—the problem known by every chemist as the basic question of his school years—remained completely untouched by the development in the adjacent field of organic chemistry.

In this series of articles, Schwab and his students demonstrated the use of liquid column chromatography for most inorganic anions and cations. Schwab emphasized that their success was only possible through the close and successful cooperation of those working in various disciplines in the Munich laboratory (which, of course, is in line with Willstätter's tradition). Also, contact with Winterstein is acknowledged, thus relating this pioneering activity to that in Heidelberg (even though at that time, Winterstein was already working in Switzerland).

Schwab opened the way for chromatography in inorganic analysis. This was continued and further advanced to paper chromatography by M. Lederer at the Radium Institute in Paris, F. H. Pollard (1906–1965) at the University of Bristol in England, and many others. [The second part of the French monograph by E. Lederer and M. Lederer (written by the latter) represents the first detailed review of the use of chromatography in inorganic analysis.]

Use of Chromatography in the Petroleum Industry

In earlier discussion, we have often cited Willstätter who believed that chromatography is not good for preparative applications. Of course, this belief was unfounded and has been proved many times since then in many different fields, one of the most important probably being the petroleum industry.

The complex nature of petroleum always intrigued chemists, and attempts were made at an early stage to separate some of the constituents in pure form. (The need to analyze crude oil was, of course, recognized almost immediately after the discovery of petroleum in America. The first scientific investigation of Pennsylvania crude oil was carried out by Benjamin Silliman, Jr., professor in the School of Applied Chemistry at Yale University, who reported his findings in 1855.) David Talbot Day, chief of the Division of Mineral Resources of the U.S. Geological Survey, carried out "chromatography" to some extent in 1897 by forcing crude oil fractions through pulverized Fuller's earth and achieving a certain fractionation, although his goal was not isolation or separation but rather the investigation of the natural processes occurring during the migration of oil. Day also reported on his work at the First International Petroleum Congress in Paris on August 20, 1900. His work was continued in Germany by Engler and Albrecht whose work can be considered as definitely superior to Day's. A few chemists in other countries were also working along this line at this time, but all of them were restricted in their efforts by unsatisfactory adsorbents and some lack of understanding of the real process involved. The outbreak of World War I marked an end to these activities. (For details on the work of Day, Engler, Albrecht, and others, see refs. 3, 4, and 83.)

In connection with this early period, the work of Russian scientists who achieved important results in the use of selective adsorption for the classification of hydrocarbon groups in petroleum fractions should also be mentioned. Between 1912 and 1926, Gurwitsch established some of the theoretical backgrounds used later in petroleum chromatography and also described different types of adsorption fractionations somewhat similar to various chromatographic techniques described later by others. Another Russian scientist, Tarasov, reported in 1926 on the use of silica gel for the quantitative removal of aromatics from petroleum samples. Some of Gurwitsch's papers were published in German journals, whereas Tarasov's results were published only in Russian. Thus, it is questionable how his work influenced the further use of adsorption chromatography in the petroleum industry. Camin and Raymond (83) summarize briefly the results of Gurwitsch and Tarasov.

API Research Project No. 6. In 1927, recognizing the need to identify and determine the chemical constituents of commercial petroleum fractions, the American Petroleum Institute established Research Project No. 6 at the National Bureau of Standards. In 1950 the project was moved to the Carnegie Institute of Technology in Pittsburgh, PA. The first director of the project was E. W. Washburn; after his death in 1934, F. D. Rossini took over this responsibility until July 1960 when B. J. Mair became the director of the project which was finally terminated in 1967.

In a summary paper (84), Rossini estimated that up to 1960, approximately 500 man-years of work had been put on the project. For the investigations, crude from Brett No. 6 well in the Ponca City field of Oklahoma was selected as "the representative petroleum".

At the beginning only systematic distillation was used, but around 1935 fractionation by selective adsorption on silica gel was also introduced. Around 1943 this work switched from simple adsorptive filtration to true chromatographic separation. Very complex and elaborate systems were developed for the project, e.g., five 53-ft long columns extending through five floors were built, and their length was later increased to 63 ft (85). After 1956 gas chromatography also came into use for separation.

The literature on the results of API Research Project No. 6 is extensive; summaries with detailed bibliographies have been published by the leaders of the project (84–87), and the paper of Camin and Raymond (83) places the whole project in its proper historical context.

The API Research Project No. 6 and associated investigations in the laboratories of oil companies, although carried out separately from the mainstream of organic chemistry and biochemistry where the application of liquid column chromatography was concentrated, greatly enhanced our knowledge in methodology. In this respect, the very detailed chapter written in 1961 by Mair, the last director of the project, on the liquid solid adsorption process in column chromatography (88) is worth mentioning. And let us not forget that the scientists who made a far-reaching contribution to the theory of chromatography (89) were associ-

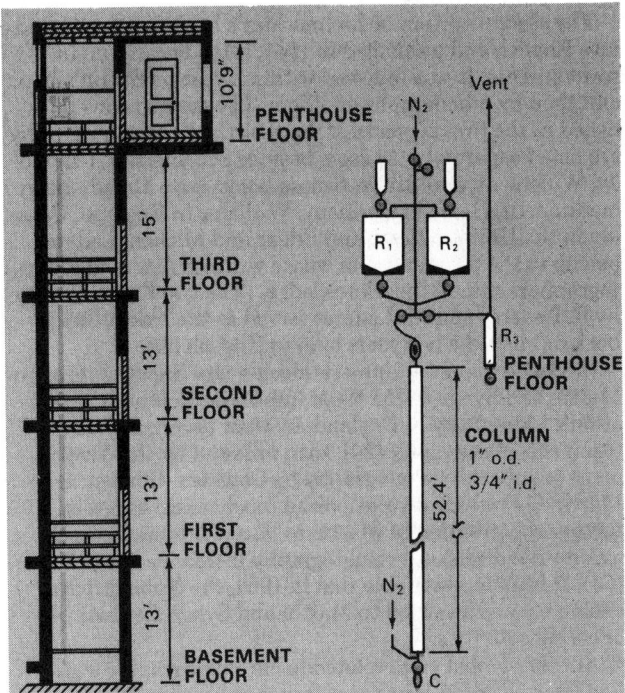

API Research Project No. 6: Figure on left shows arrangement of 52.4-ft long adsorption column in 62-ft shaft; figure on right is functional schematic of column assembly. Reservoirs R_1 and R_2 are for holding sample and desorbing liquid (alcohol), respectively; typical sample volume is 0.5 liter. Entire system can be placed under inert gas (nitrogen) pressure either from top or from bottom. Column proper is made of 1-in. o.d. ¾-in. i.d. stainless-steel tube and is filled with total of 3.7 kg silica gel. Insulation around column contains electric resistance heating wire. Receiver C at bottom of column is for withdrawal of fractions. After separation is finished, the desorbing liquid (alcohol) remaining on column is regenerated by introducing nitrogen at bottom and heating column to 200°C; evaporated alcohol is collected in receiver R_3 (from ref. 85)

ated with probably the largest petroleum research laboratory in the whole world.

One last remark is necessary. As mentioned by Rossini (84), by 1960, the isolation of 175 hydrocarbon compounds in the Ponca Crude was successful. This was the result of 33 years of work. Of course, "isolation" here means the preparation of limited quantities which increases the problems involved by at least an order of magnitude. Still, it is worthwhile to quote Desty who in 1961 reported (90) on the gas chromatographic analysis of the C_4–C_9 fraction of the Ponca Crude, representing about 30 wt % of the total crude oil. By using a 900-ft long × 0.006-in. i.d. glass open tubular column having over one million effective plates and operated at room temperature, 122 peaks were resolved in 24 hr; the length of the chart paper was about 20 meters. In contrast, in the API Project only 89 hydrocarbons were separated in this region.

Evolution of Theory of Chromatography

Until the early forties, chromatography remained an art based only on a body of empirical observations and an intuitive use of the underlying physicochemical phenomena. Chromatographers were only too happy to solve their particular separation problem and get on to more interesting things without asking themselves what was going on in the columns. Today, of course, chromatography rests on a sound theoretical basis. [The early theories of chromatography have been critically reviewed by Helfferich (91) and Giddings (92).] This is probably best indicated by the following quotation from the preface of a recent book by Aris and Amundson (93) on first-order partial differential equations:

> *When dealing with systems of equations we have focussed attention on the equations of chromatography, which give an admirable realization of reducible systems and theatre for the interplay of simple wave and discontinuous solutions.*

Therefore, one cannot but feel amused by reading the four paragraphs in the book of Willstaedt published in 1938 (94), constituting the entire theoretical section.

Shortly after the publication of this book, the situation rapidly changed, however. Recognizing that no quantitative treatment by which even an approximate prediction of the width or rate of development of the bands to be expected in a given chromatogram could be made, in 1940 Wilson (95) of the California Institute of Technology published the first theoretical paper. Although his quantitative model assumed complete equilibration and linear sorption isotherm (and therefore did not predict band spreading), the paper shows a remarkable qualitative understanding of the major phenomena involved in the chromatographic process: diffusion, rate of adsorption, and the nonlinearity of the isotherm.

The first breakthrough in the quantitative treatment of chromatographic peaks came with the theoretical plate model of Martin and Synge (20) in 1941. At that time, the concept of theoretical plates introduced by Peters (96) in 1922 was widely used in chemical engineering because it provided a convenient measure of system efficiency in distillation. Although the plate model later received strong criticism for representing a continuous process by a discontinuous one (which does not take into account mass transport and kinetic effects) and, thus, yielding no information on the factors affecting the actual chromatographic process, the introduction of the number of theoretical plates and HETP (height equivalent to one theoretical plate) as the measure of column efficiency had a stupendous impact on the further development of elution chromatography. Since the plate height could be readily evaluated from the

Theory of Chromatographic Adsorption. In order to formulate a theory of chromatographic adsorption, we assume the column divided in a large number of very thin parallel disks (1, 2, 3, 4, etc.) (see Fig. 8). The solution that enters the column at the top contains a given number of substances (A, B, C, etc.) which have different chemical affinities with respect to the adsorbent. These substances will compete for the free adsorbing sites. As a result the substance with the greatest affinity for the adsorbent—assuming it is A—will occupy all sites in disc 1. The solution, deprived of A but with an unchanged concentration of B, C, etc., enters the following disk 2, where the same competition occurs for the substances B, C, etc. Consequently, the substances will adsorb from the top to the bottom of the column in an order that corresponds to the degree of intensity of their affinity for the adsorbent.

When elution starts, pure solvent enters 1 and dissolves a small part of A. This solution then enters disk 2 which, so far, did not contain any substance A. The substance A is adsorbed in this disk and a part of substance B is consequently detached and transported to the following disk.

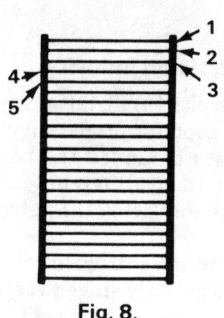

Fig. 8.

A series of such micro-adsorptions and micro-elutions will result in what we observe as the chromatogram.

Complete section dealing with theory of chromatography in Willstaedt's book ("l'Analyse chromatographique et ses applications," Hermann & Cie., Paris, 1938)

moments of the Gaussian peaks on the chromatogram, it provided for the first time a practical expression for column efficiency, even if the model itself did not take into account the phenomena actually causing band spreading which determined column efficiency.

In retrospect, the greatest merit of the plate model was, besides being the first viable theoretical approach to chromatography, that it provided a conveniently obtainable quantity, the HETP, to measure the overall effect of the various parameters in a highly complex system. Even if the meaning of the plate height has completely changed with the evolvement of theories closer reflecting the physical reality, its usefulness manifests itself in the fact that use of this expression is still more popular.

The complexity of the chromatographic process, which deterred earlier attempts to treat it theoretically, is rooted in the interplay of thermodynamic, mass transfer, and kinetic phenomena as shown in the triangle. It appears that analytical solution of the conservation equation which accounts for all phenomena involved is impossible.

The earliest theories were based on Wilson's assumption that there is complete equilibration of the solute between the mobile and stationary phases; consequently, dynamic effects were neglected. In 1943 De Vault (97) of Stanford University was the first to find analytical solutions to express the concentration profiles for linear and nonlinear isotherms and constant flow rate. His rigorous treatment clearly leads to the fundamental relationship:

$$V_R = V_M + V_S K$$

which established the relationship between the retention volume V_R, the equilibrium constant K (called the partition coefficient in partition chromatography), and the volumes of the mobile and stationary phases, V_M and V_S, respectively.

The relationship between the retention in chromatography and the thermodynamic equilibrium constant which could be related to the molecular structure of the sample components was of great importance in facilitating their identification and received great attention in the early fifties. The major contributions in this area were made by Martin (98) in 1949, James and Martin (99) in 1954, Le

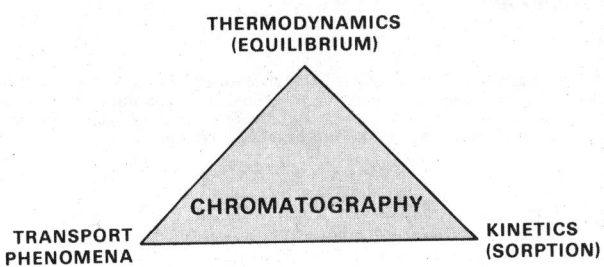

Triangle representing the three major areas of science which interact in chromatography. Consequently, complete theoretical description of chromatographic process has to take into account equilibrium, transport, and kinetic phenomena and therefore is extremely complex

Rosen and his coworkers at Louisiana State University (100) in 1951, and Pierotti and his coworkers at Shell Development Co., in Emeryville, CA (101), in 1956. On the other hand, the possibility of measuring the equilibrium constant in a convenient way with precision instrumentation later turned chromatography into a valuable tool for physiochemical measurements.

Of course, the equilibrium theory was far from being representative for the events in the chromatographic columns and was insufficient to account for band spreading. Although the existence of nonequilibrium was recognized, its first theoretical treatment was possible owing only to the brilliant contributions by Thomas at Yale University (102, 103). His second paper (103) entitled "Chromatography: A Problem in Kinetics" entails the fundamental assumption of the ensuing rate theories and introduces a remarkable transformation to linearize the equation of chromatography so that nonequilibrium conditions and also a nonlinear isotherm could be treated. (This paper was presented at the Conference on Chromatography held by the New York Academy of Sciences November 29–30, 1946. This meeting can be considered the precursor of the international symposia which, starting in 1956, played such a vital role in the advancement of chromatography and the exchange of ideas.) As a result, his formulae made it possible for the first time to obtain overall adsorption and desorption rates from the experimental concentration profiles. Nevertheless, the effect of transport phenomena per se was ignored by Thomas, and the first rigorous theory which accounted for mass transfer and longitudinal diffusion in linear chromatography was put forward by Lapidus and Amundson (104) at the University of Minnesota in 1952. Their paper, however, was too exacting to be useful directly in chromatography.

Both the plate model and Thomas' results implied an approximation that the flow is relatively slow; thus, near equilibrium conditions prevail and, in the case of linear isotherms, Gaussian peaks are obtained. This approximation played a crucial role in the further development of the theory of chromatography. In 1954, using the same assumption, Glueckauf (105) at the Atomic Energy Establishment in Harwell, England, derived the first comprehensive equation which established the relationship between the HETP and the major factors involved in ion-exchange chromatography such as particle size, particle diffusion, and diffusion through the film surrounding the particles. (E. Glueckauf is also a native of Germany who left the country in 1933. His interest in chromatography started when he was working in the British equivalent of the Manhattan project.)

It was, however, the introduction of gas chromatography by James and Martin (106) at the National Institute for Medical Research in London in 1952, which provided a theater for the development of chromatographic theories. [It is interesting to note that the preface of the 1955 book of E. Lederer and M. Lederer (18) contains the following statement: *In view of the fact that a book by A. J. P. Martin and A. T. James on the theory of chromatography will be published by the Elsevier Publishing Company in the near future, we thought it is unnecessary to encroach on this field; no detailed discussion of this subject is therefore included in this book.* Evidently, the book planned by Martin and James was never realized.] The rapid growth of this technique and the ease of obtaining quantitative results with the gas chromatographs provided strong stimuli for theoretical work. In 1956 van Deemter, Zuiderweg, and Klinkenberg (89) at the Koninklijke/Shell Laboratoriums in The Netherlands put forth their rate theory by simplifying the formulae of Lapidus and Amundson to a Gaussian distribution function, thus, expressing HETP as a function of the flow velocity and the most significant characteristics of the chromatographic system such as particle diameter and solute diffusivity. Since then the popular "van Deemter plot" greatly facilitated the optimization of chromatographic conditions.

Epilogue

In the 1950's it became clear that, by virtue of improvements in column efficiency and the sensitive detection methods, chromatography was not only a separation method but a quantitative microanalytical technique with unsurpassed versatility. At that time, a variety of novel column materials had also been successfully employed.

In 1957 the first panoramic view of chromatography was exposed in the book of Cassidy (107), professor at Yale University, who did a great deal of pioneering work in liquid chromatography. This book is still in print and makes fascinating reading because it shows that almost all ideas, on which present-day liquid chromatography is based, were put forth a long time ago. The publication of this book coincided with Keulemans' book on gas chromatography (108) which greatly helped to popularize that chromatographic technique.

The last chapters of the history of liquid chromatography are yet to be written. Undoubtedly, the development of

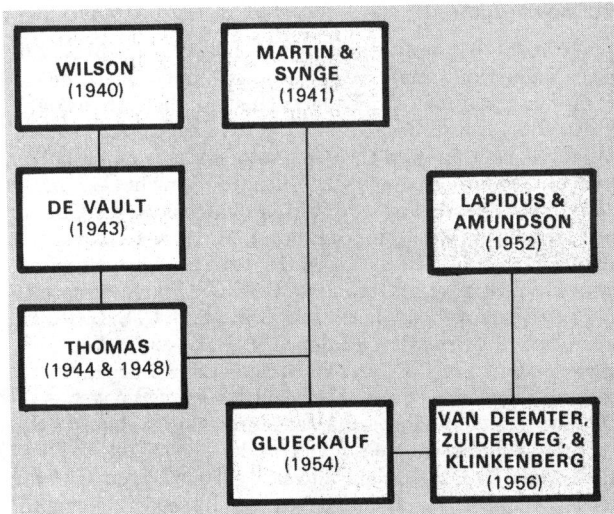

Milestones in early development of theory of band spreading in chromatography. Chart by no means complete since prior to 1956 there was a great deal of work which significantly contributed to refinement of the theory but, in the author's opinion, had lesser impact

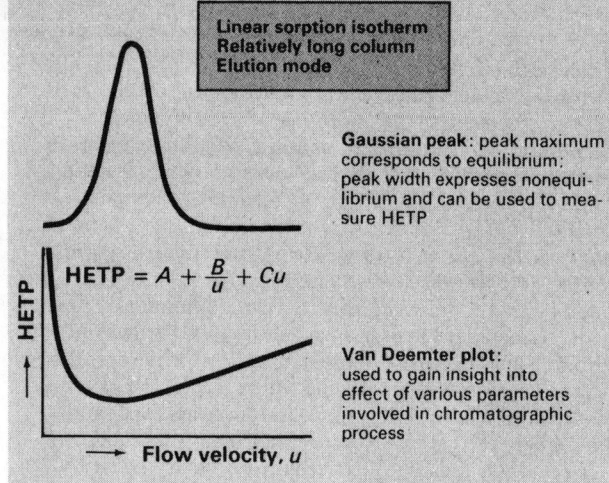

Elution mode of chromatography found widest application, and introduction of sensitive detectors made possible use of small samples so that sorption isotherm was sufficiently linear. Experimental data obtained with precision instruments, mainly in gas chromatography, thus could be linked to theory

gas chromatography greatly facilitated the metamorphosis of the art into science. The conflux of notions and methodology which had continuously evolved in liquid chromatography and the sophistication of gas chromatography in terms of theory and instrumentation had finally precipitated a groundwork for which a rapid development of modern liquid chromatography began in the sixties.

In many respects, this development followed the suggestion by Martin and Synge in 1941 (20), that

> ... the smallest HETP should be obtainable by using very small particles and high pressure difference across the length of the column

which was not reduced to practice at that time, because they felt that

> ... in striving for conditions for uniformity of flow, the high pressure and small particle size desirable for smallest HETP have to be abandoned.

A generation and many trials and tribulations later the pioneers' fancy is just about to become a triumphant reality.

Acknowledgment

We are very grateful to many who helped us by providing material and information for this article. First of all, we express our gratitude to E. Lederer (Institut de Chimie des Substances Naturelles, Gif-sur-Yvette, France) for his personal recollections and the two photographs from his collection. Thanks are due to D. L. Camin (Sun Oil Co., Marcus Hook, PA), W. Fritsche (Gesellschaft deutscher Chemiker, Frankfurt am Main, Germany), M. Lederer (University of Rome, Italy), R. E. Kaiser (Institut für Chromatographie, Bad Dürkheim, Germany), B. Kolb (Bodenseewerk Perkin-Elmer & Co., Überlingen, Germany), C. S. G. Phillips (Oxford University, England), F. D. Rossini (Rice University, Houston, TX), and W. Simon (ETH, Zurich, Switzerland). F. Wold, head of the Department of Biochemistry, University of Minnesota, sent information about L. S. Palmer, including his portrait. Thanks are also due to Jean-Marc Engasser (Yale University, New Haven, CT) for translating the theoretical section in Willstaedt's book and to the California Institute of Technology for the photograph of László Zechmeister.

We spent endless hours in the Kline Science and Sterling Chemistry Libraries of Yale University, studying their magnificent collection. Without the help of their polite and courteous staff, our article could never have been written.

References

References for which the original articles were checked contain the first and last pages. References taken from another source contain only the first page. Every important book (except notably Tswett's 1910 book published in Russian in Warsaw) was checked in the original edition.

(1) K. Sakodynskii, *J. Chromatogr.,* **49,** 2–17 (1970).
(2) K. Sakodynsky, *Chromatographia,* **3,** 92–94 (1970).
(3) V. Heines, *Chem. Technol.,* **1,** 280–85 (1971).
(4) L. S. Ettre, *Anal. Chem.,* **43** (14), 20A–31A (Dec. 1971).
(5) K. Sakodynskii, *J. Chromatogr.,* **73,** 303–60 (1972).
(6) I. M. Hais, *ibid.,* **86,** 283–88 (1973).
(7) J. G. Kirchner, *J. Chromatogr. Sci.,* **11,** 180–83 (1973).
(8) G. Zweig and J. Sherma, *ibid.,* pp 279–83.
(9) E. Heftmann, *ibid.,* pp 295–98.
(10) A. J. P. Martin, "Historical Background," in "Gas Chromatography in Biology and Medicine (1969 CIBA Foundation Symposium)," R. Porter, Ed., pp 2–10, Churchill, London, England, 1969.
(11) L. S. Ettre, *Am. Lab.,* **4** (10), 10–16 (October 1972).
(12) C. Dhéré, *Condollea (Geneva),* **10,** 23–73 (1943).
(13) G. Hesse and H. Weil, "Michael Tswett's First Paper on Chromatography," M. Woelm, Eschwege, 1954.
(14) L. Zechmeister, "Historical Introduction," in "Chromatography," E. Heftmann, Ed., 2nd ed., pp 3–10, Reinhold, New York, NY, 1967.
(15) A. J. P. Martin, "Past, Present and Future of Gas Chromatography," in "Gas Chromatography (1957 Lansing Symposium)," V. J. Coates, H. J. Noebels, and I. S. Fagerson, Eds., pp 237–47, Academic Press, New York, NY, 1958.
(16) A. T. James, "The Development of an Idea," in "Gas Chromatography (1959 Lansing Symposium)," H. J. Noebels, N. Brenner, and R. F. Wall, Eds., pp 247–54, Academic Press, New York, NY, 1961.
(17) R. Consden, A. H. Gordon, and A. J. P. Martin, *Biochem. J.,* **38,** 224–32 (1944).
(18) E. Lederer and M. Lederer, "Chromatography—A Review of Principles and Applications," quote from the preface, Elsevier, Amsterdam, The Netherlands, 1955.
(19) R. Kuhn, A Winterstein, and E. Lederer, *Hoppe-Seyler's Z. Physiol. Chem.,* **197,** 141–60 (1931).
(20) A. J. P. Martin and R. L. M. Synge, *Biochem. J.,* **35,** 1358–68 (1941).
(21) H. H. Strain and J. Sherma, *J. Chromatogr.,* **73,** 371–97 (1972).
(22) R. Willstätter, "Adolf v. Baeyer," in "Great Chemists," E. Farber, Ed., pp 733–47, Interscience, New York, NY, 1961.
(23) R. Willstätter and A. Stoll, "Untersuchungen über Chlorophyll; Methoden und Ergebnisse," Springer, Berlin, Germany, 1913.
(24) R. Willstätter, "Aus Meinem Leben—Von Arbeit, Musse und Freunden," A. Stoll, Ed., 2nd ed., (1st ed., 1949), quote from p 357, Verlag Chemie, Weinheim, 1958.
(25) R. Willstätter and A. Stoll, "Investigations on Chlorophyll; Methods and Results," quote from American ed., transl. by F. M. Schertz and A. R. Merz, a, p 14; b, p 142; c, p 212, Science Press, Lancaster, PA, 1928.
(26) M. Tswett, "Khromofilly v Rastitel'nom Zhivotnom Mire" (Chromophylls in the Plant and Animal Life), p 233, Warsaw Univ., 1910.
(27) R. Willstätter and W. Mieg, *Ann. Chem.,* **355,** 1–28 (1907).
(28) R. Willstätter, "Die Blattfarbstoffe," in "Abderhalden's Handbuch der biologischen Arbeitsmethoden," Abt. I, Teil 11.1; pp 1–70, Urban & Schwarzenberg, Berlin, Germany, 1924.
(29) E. Lederer, *J. Chromatogr.,* **73,** 261–66 (1972).
(30) E. Turba, "Chromatographische Methoden in der Protein-Chemie," Springer, Berlin, Germany, 1954.
(31) R. Willstätter, "Untersuchungen über Enzyme," Vol I, a, p 66; b, p 295, Springer, Berlin, Germany, 1928.
(32) L. Rosenthaler, review of P. Karrer's paper "Zur Frage der Reinheit und Wirkungsstärke von Vitamin-A-Präparaten," *Z. Vitaminforsch.,* **10,** 94 (1940).
(33) C. Dhéré and W. Rogowski, *C. R. Acad. Sci. Paris,* **155,** 653 (1912).
(34) C. Dhéré and G. Vegezzi, *ibid.,* **163,** 399 (1916).
(35) F. Czapek, "Biochemie der Pflanzen," Vol I, 2nd ed., p 802, Fisher, Jena, Germany, 1913.
(36) K. H. Coward, *Biochem. J.,* **18,** 1114 (1924).
(37) Th. Lipmaa, *C. R. Acad. Sci. Paris,* **182,** 867 (1926).
(38) R. Jennes and R. W. Luecke, *J. Nutr.,* **63** (1), 3 (1957).
(39) L. S. Palmer and C. H. Eckles, *J. Biol. Chem.,* **17,** 191–210 (1914).
(40) L. S. Palmer and C. H. Eckles, *ibid.,* pp 211–21.
(41) L. S. Palmer and C. H. Eckles, *ibid.,* pp 223–36.
(42) L. S. Palmer and C. H. Eckles, *ibid.,* pp 237–43.
(43) L. S. Palmer and C. H. Eckles, *ibid.,* pp 245–49.
(44) L. S. Palmer, *ibid.,* **23,** 261–79 (1915); **27,** 27–32 (1916).
(45) L. S. Palmer and H. L. Kempster, *ibid.,* **39,** 313–37 (1919).
(46) L. S. Palmer and C. Kennedy, *ibid.,* **46,** 559–77 (1921).
(47) L. S. Palmer, "Carotenoids and Related Pigments; The Chromolipids," Am. Chem. Soc. Monograph Series, Chemical Catalog Co., New York, NY, 1922.
(48) R. Kuhn and E. Lederer, *Chem. Ber.,* **64,** 1349–57 (1931).
(49) L. Zechmeister, "Carotinoide, Ein biochemischer Bericht über pflanzliche und tierische Polyenefarbstoffe," Springer, Berlin, Germany, 1934.
(50) A. Winterstein, "Fraktionierung und Reindarstellung organischer Substanzen nach dem Prinzip der Tswettschen Adsorptionsmethode," in "Abderhalden's Handbuch der biologischen Arbeitsmethoden," Abt. V, Teil 10.2; pp 247–63, Urban & Schwarzenberg, Berlin, Germany, 1938.

(51) G. Hesse, "Die chromatographische Adsorptionsanalyse," in "Berl-Lunge Chemisch-technische Untersuchungsmethoden," J. d'Ans, Ed., Suppl. vol. to 8th ed., Springer, Berlin, Germany, 1938.
(52) H. Vetter, "Die chromatographische Adsorptionsverfahren und seine Anwendung in der organischen Chemie," in "Physikalische Methoden der analytischen Chemie," W. Böttger, Ed., Part III, p 1, Akademische Verlagsgesellschaft, Leipzig, Germany, 1939.
(53) Ch. Grundmann, "Chromatographie und verwandte Methoden in der Enzymchemie," in "Die Methoden der Enzymforschung," E. Bamann and K. Myrbäck, Eds., pp 1452–66, G. Thieme, Leipzig, Germany, 1940.
(54) G. A. Howard and A. J. P. Martin, *Biochem. J.*, **46**, 532–38 (1950).
(55) A. Tiselius, *Ark. Kemi, Mineral. Geol.*, **14B** (22) (1940).
(56) A. Tiselius, *ibid.*, **15B** (6) (1941).
(57) S. Claesson, *ibid.*, **23A** (1) (1946).
(58) S. Claesson, *Ann. NY Acad. Sci.*, **49**, 183–203 (1948) (detailed summary of work carried out at Uppsala).
(59) R. S. Alm, R. J. P. Williams, and A. Tiselius, *Acta Chem. Scand.*, **6**, 826–36 (1952).
(60) N. A. Izmailov and M. S. Shraiber, *Farmatsiya*, **3**, 1–7 (1938); for English transl. of this article and of ref. 63, see N. Pelick, H. R. Bottinger, and H. K. Mangold, in "Advances in Chromatography, Vol 3," J. C. Giddings and R. A. Keller, Eds., pp 85–118, Dekker, New York, NY, 1966.
(61) J. G. Kirchner, J. M. Miller, and G. J. Keller, *Anal. Chem.*, **23**, 420–25 (1951).
(62) E. Stahl, G. Schröter, G. Kraft, and R. Renz, *Pharmazie*, **11**, 633–37 (1956).
(63) E. Stahl, *Chem. Ztg.*, **82**, 323–29 (1958).
(64) T. I. Taylor and H. C. Urey, *J. Chem. Phys.*, **6**, 429–38 (1938).
(65) O. Samuelson, *Z. Anal. Chem.*, **116**, 328 (1939).
(66) E. R. Tompkins, J. X. Khym, and W. E. Cohn, *J. Am. Chem. Soc.*, **69**, 2769–77 (1947).
(67) D. H. Harris and E. R. Tompkins, *ibid.*, pp 2792–800.
(68) B. H. Ketelle and G. E. Boyd, *ibid.*, pp 2800–12.
(69) S. G. Thompson, B. G. Harvey, G. R. Choppin, and G. T. Seaborg, *ibid.*, **76**, 6229–36 (1954).
(70) W. E. Cohn, *Science*, **109**, 377–78 (1949).
(71) W. E. Cohn, *J. Am. Chem. Soc.*, **71**, 2275–76 (1949).
(72) W. H. Stein and S. Moore, *J. Biol. Chem.*, **176**, 337–65 (1948).
(73) S. Moore and W. H. Stein, *Ann. NY Acad. Sci.*, **49**, 265–78 (1948).
(74) S. Moore and W. H. Stein, *J. Biol. Chem.*, **192**, 663–81 (1951).
(75) D. H. Spackman, W. H. Stein, and S. Moore, *Anal. Chem.*, **30**, 1190–1205 (1958).
(76) M. Lederer, *Anal. Chim. Acta*, **12**, 142–45 (1955).
(77) *Angew. Chem.*, **50**, 613–14 (1937).
(78) G. M. Schwab and K. Jockers, *ibid.*, pp 546–53.
(79) G. M. Schwab and G. Dattler, *ibid.*, pp 691–92.
(80) G. M. Schwab and G. Dattler, *ibid.*, **51**, 709–11 (1938).
(81) G. M. Schwab and A. N. Ghosh, *ibid.*, **52**, 666–68 (1939).
(82) G. M. Schwab and A. N. Ghosh, *ibid.*, **53**, 39 (1940).
(83) D. L. Camin and A. J. Raymond, *J. Chromatogr. Sci.*, **11**, 625–38 (1973).
(84) F. D. Rossini, *J. Chem. Educ.*, **37**, 554–61 (1960).
(85) B. J. Mair, A. L. Gaboriault, and F. D. Rossini, *Ind. Eng. Chem., Ind. Ed.*, **39**, 1072–81 (1947).
(86) F. D. Rossini, B. J. Mair, and A. Streiff, "Hydrocarbons from Petroleum," Reinhold, New York, NY, 1953.
(87) B. J. Mair, *Proc. Seventh World Petrol. Congr.*, **9**, 39 (1967).
(88) B. J. Mair, "Chromatography: Columnar Liquid-Solid Adsorption Processes," in "Treatise on Analytical Chemistry, Part I: Theory and Practice, Vol 3: Separation Principles and Techniques," Chapt. 34, pp 1469–1520, I. M. Kolthoff, P. J. Elving, and E. B. Sandell, Eds., Interscience, New York, NY, 1961.
(89) J. J. van Deemter, F. J. Zuiderweg, and A. Klinkenberg, *Chem. Eng. Sci.*, **5**, 271–89 (1956).
(90) D. H. Desty, A. Goldup, and W. T. Swanton, "Performance of Coated Capillary Columns," in "Gas Chromatography (1961 Lansing Symposium)," pp 105–35, N. Brenner, J. E. Callen, and M. D. Weiss, Eds. Academic Press, New York, NY, 1962.
(91) F. Helfferich, *Angew. Chem., Int. Ed.*, **1**, 440–53 (1962).
(92) J. C. Giddings, "Dynamics of Chromatography, Part 1, Principles and Theory," pp 13–26, Dekker, New York, NY, 1965.
(93) R. Aris and N. R. Amundson, "First-Order Partial Differential Equations With Applications. Mathematical Methods in Chemical Engineering," Vol 2, p xii, Prentice Hall, Englewood Cliffs, NJ, 1973.
(94) H. Willstaedt, "L'analyse chromatographique et ses applications," pp 15–16, Hermann & Cie., Paris, France, 1938.
(95) J. N. Wilson, *J. Am. Chem. Soc.*, **62**, 1583–91 (1940).
(96) W. A. Peters, Jr., *Ind. Eng. Chem.*, **14**, 376 (1922).
(97) D. De Vault, *J. Am. Chem. Soc.*, **65**, 532–40 (1943).
(98) A. J. P. Martin, *Biochem. Soc. Symp.*, **3**, 4–15 (1949).
(99) A. T. James and A. J. P. Martin, *Brit. Med. Bull.*, **10**, 170–76 (1954).
(100) A. L. Le Rosen, P. H. Monagan, C. A. Rivet, and E. D. Smith, *Anal. Chem.*, **23**, 730–32 (1951).
(101) G. J. Pierotti, C. H. Deal, E. L. Derr, and P. E. Porter, *J. Am. Chem. Soc.*, **78**, 2989–98 (1956).
(102) H. C. Thomas, *ibid.*, **66**, 1664–66 (1944).
(103) H. C. Thomas, *Ann. NY Acad. Sci.*, **49**, 161–82 (1948).
(104) L. Lapidus and N. R. Amundson, *J. Phys. Chem.*, **56**, 984–88 (1952).
(105) E. Glueckauf, in "Ion Exchange and Its Applications," pp 34–36, Soc. Chem. Ind., London, England, 1955.
(106) A. T. James and A. J. P. Martin, *Biochem. J.*, **50**, 679–90 (1952).
(107) H. G. Cassidy, "Fundamentals of Chromatography," Interscience, New York, NY, 1957.
(108) A. I. M. Keulemans, "Gas Chromatography," C. G. Verver, Ed., Reinhold, New York, NY, 1957.

Leslie S. Ettre

Csaba Horvath

Leslie Stephen Ettre is a senior staff scientist associated with the Chromatography Department at Perkin-Elmer Corp., in Norwalk, CT. He has a diploma in chemical engineering as well as a DSc in analytical chemistry from the Technical University, Budapest, Hungary. Prior to 1958, he was active in Hungary and Germany in industrial research and teaching. After coming to the United States in the fall of 1958, he was employed at the Perkin-Elmer Corp. and served as application chemist, product specialist, and chief applications chemist in gas chromatography. Between 1968 and 1972, he took over the executive editorship of the "Encyclopedia of Industrial Chemical Analysis," a 20-volume series. In 1972 he returned to Perkin-Elmer in his present position. Dr. Ettre is the author (and coauthor) of numerous books and publications. He has lectured widely in the United States, Canada, and Europe and has been active in the organization of several international symposia in chromatography. He was the chairman of the Anniversary Symposium on Chromatography held during the Fall National American Chemical Society Meeting in 1972 and cochairman of the Summer Symposium of the ACS Analytical Division in 1973. He is a member of ACS, ASTM, the New York Academy of Sciences, the British Society for Analytical Chemistry, the British Chromatography Discussion Group, and a fellow of the American Institute of Chemists.

Csaba Horvath is associate professor in the Department of Engineering and Applied Science at Yale University, New Haven, CT. He graduated in chemical engineering at the Technical University of Budapest, Hungary, in 1952 and obtained a PhD in physical chemistry at the University of Frankfurt am Main, Germany, in 1963. Prior to coming to Yale, he was at the Institute of Organic Chemical Technology at the Technical University of Budapest, the Farbwerke Hoechst AG in Frankfurt am Main-Hoechst, and at Harvard University. In the past 15 years Dr. Horvath has been interested in the development of both gas and liquid chromatography, and his contributions range from the introduction of support-coated open tubular columns for gas chromatography to pioneering in high-pressure liquid chromatography and the development of pellicular column materials. Although still active in chromatographic research, he is also exploring new areas in biochemical engineering, such as the study of biochemical reactors with immobilized enzymes and the behavior of heterogenous enzyme systems. Dr. Horvath is a member of ACS, AICE, American Ceramic Society, Sigma Xi, and the British Chromatography Discussion Group.

The Development of Chromatography

LESLIE S. ETTRE
Encyclopedia of Industrial Chemical Analysis
John Wiley & Sons, Inc.,
605 Third Ave., New York, N.Y. 10016

The years 1971–72 represent three important anniversaries related to the various chromatographic techniques: the 100-year anniversary of the birth of Tswett, the inventor of adsorption chromatography; the 30-year anniversary of the development of partition chromatography; and the 20-year anniversary of the first paper on gas-liquid partition chromatography.

The purpose of this report is, on the occasion of the Seventh International Symposium on the Advances in Chromatography, to note the importance of these anniversaries and summarize the results of the pioneers that led to the techniques used today.

The development of chromatographic techniques alone would assure a permanent place for the pioneers—M. S. Tswett, A. J. P. Martin, R. L. M. Synge, and A. T. James—in the history of chemistry. However, their merit is more than that; they not only described new techniques but also properly interpreted them, summarized the theoretical basis of the techniques, and showed the possibilities of practical applications. Thus, their work is an example of how scientific work should be carried out.

Tswett, Martin, Synge, and James represent the starting points in a line of distinguished scientists, all of whom contributed greatly to the advancement of chromatography. It is impossible to deal in this report with their activities; also, we cannot deal here with the development of ion exchange, thin-layer, and gas adsorption chromatographic techniques, each of which probably would merit a report in itself. However, it should be em-

On the occasion of the Seventh International Symposium on Advances in Chromatography, this Report deals with the importance of the work of Tswett and the circumstances which led to the development of liquid-liquid and gas-liquid partition chromatography

phasized that the pioneers only represent the start, and without the achievements of many known and unknown chemists and scientists, we would not be where we are today.

Adsorption Chromatography: The Life and Work of Tswett

For almost every invention, one can find persons who worked earlier and more or less carried out similar investigations but who are still not considered as the inventors of given techniques, processes, or machinery. Usually, the person recognized as the real inventor not only utilized or described a phenomenon but also could interpret it and apply it knowingly for some purpose.

The situation is not different in chromatography. Tswett certainly had his forerunners who described separation obtained by selective retardation on a column containing a solid substance. The most important person in this field was David Talbot Day (1859–1925), who was connected with the U.S. Geological Survey for 28 years and served as the director of its Division of Mineral Resources for 21 years. He demonstrated in 1897 that when crude oil fractions are pressed through pulverized Fuller's earth, a certain fractionation takes place and, in the forthcoming years, investigated this phenomenon in detail. He reported on his results in 1900 at the First International Petroleum Congress in Paris and then three years later at the 43rd meeting of the U.S. Geological Society in Washington, D.C. (For details on the activities of Day and his collaborators and their publications, see ref. *1* and *2*.) It should be noted, however, that, although Day recognized the analytical potential

of the process investigated, he and his co-workers interpreted incorrectly the physicochemical basis of the separation, calling it a capillary diffusion process. This is mainly the reason why today, Tswett and not Day is recognized as the inventor of adsorption chromatography.

Michael S. Tswett's life and activities are fairly well documented (3–6) although some of the statements in these summaries are contradictory. His life was a fascinating one, and a typical example of the troubled life encountered later by so many other European scientists who, due to the events beyond their control, had to settle at places far from their homes or find refuge after losing everything. His scientific work shows the activity of one of the clearest and most conclusive minds.

Michael S. Tswett was born on May 14, 1872, in Asti, Piemonte, Italy, as the son of Simeon Tswett, a Russian subject, and his wife Maria Dorozza, an Italian. He grew up in Switzerland, studied there, and graduated with a PhD; in 1896, he moved to Russia and, in January 1901, after spending a few years in minor jobs, he found a permanent position at the University of Warsaw (Poland was then part of Russia). He successively served as assistant, associate, and full professor at the University, the School of Veterinary Medicine, and the Institute of Technology of Warsaw until World War I interrupted his life. In 1915, he left Warsaw with the Institute of Technology which was evacuated to Nishi Novgorod (today: Gorkiy) before the advancing German troops. In 1917 he became ill and spent some time in the Caucasus recuperating, then accepted the chair of botany at the University of Jurjeff. [This was then the name of the city; its German name is Dorpat. After World War I, this area became Estonia, and the city's name was changed from the Russian Jurjeff to the Estonian name Tartu. When, in 1940, Estonia again became part of Russia (the Soviet Union), the Estonian name was kept.]

In the fall of 1918, the University was evacuated to Voronezh before the advancing German troops; he first decided to stay but, at the last minute, followed the University group. His illness at that time was already advanced, and he died in Voronezh on June 26, 1919, probably of heart disease. He was buried in the cemetery around a monastery which (and this is the last intervention of history in Tswett's life) was destroyed in World War II during the fighting with the German troops, and thus, his grave cannot be located anymore.

In practically his whole working life, Tswett dealt with investigations related to chlorophyll. Toward the end of the 19th century, many scientists showed interest in the pigments occurring in the leaves of plants, but there was no real way to separate them from each other and to check their identity in different plants. Since these substances are very labile, one could not be sure whether material obtained through chemical manipulation really corresponded to the form existing in the living plant.

Tswett's approach was different; he was looking for a physical method which would permit the separation of these pigments from each other and from closely related compounds that others felt to represent the same substances. For he was convinced that chlorophyll, as isolated by other researchers, was not a single substance.

In his work, he systematically checked a large number of solvents capable of extracting the pigments from vegetable matter, and more than 100 solid substances capable of selective retardation of the individual pigments through adsorption, and he also deducted a number of important rules for the adsorption phenomenon.

His first paper in which his preliminary work was summarized was presented on March 21, 1903, before the Biological Section of the Warsaw Society of Natural Sciences and published in Russian in the proceedings of the Society (7); it is also available in English translation (4). The title of the paper is "On a New Category of Adsorption Phenomena and Their Application to Biochemical Analysis," and in it, he, in essence, describes chromatography without yet naming it as such. Three years later, in his two remarkable papers entitled "Physico-Chemical Studies of Chlorophyll. The Adsorptions" (14), and "Adsorption Analysis and the Chromatographic Method. Application in the Chemistry of Chlorophyll" (9) published in the journal of the German Botanical Society, he very clearly defined the technique. It is worthwhile to quote from his first paper (8a):

"When a petroleum ether solution is filtered through a column of an adsorbent (I use mainly calcium carbonate which is tamped firmly into a narrow glass tube), then the pigments are resolved, according to the adsorption sequence, from top to bottom, into various colored zones, since the more strongly adsorbed pigments displace the more weakly adsorbed ones, and force them farther downward. This separation becomes practically complete when, after the pigment solution has flowed through, a stream of pure solvent is passed through the adsorbing column. Like light rays in the spectrum, the different components of a pigment mixture, obeying a law, are resolved on the calcium carbonate column and then can be qualitatively and quantitatively determined. I call such a preparation a chromatogram and the corresponding method the chromatographic method."

It is generally assumed that the word "chromatography" has its origin in the Greek words *chroma* (color), and *graphe* (writing). However, already Tswett (8b) emphasized that:

"It is self-evident that the adsorption phenomena described are not restricted to the chlorophyll pigments, and one must assume that all kinds of colored and colorless chemical compounds are subject to the same laws."

Actually, one may question whether Tswett really meant "color writing" when coining the name chromatography (10). The interesting fact is that the surname of

Tswett, in Russian, is identical with the Russian word for color (ЦВЕТ) and, as expressed by Purnell (11), "It would be nice to think that Tswett, whose name, in Russian, means color, took advantage of the opportunity to indulge his sense of humour."

In his later papers, Tswett developed even further his technique—the separation of substances by retardation through selective adsorption.

Tswett did not invent adsorption; this had been described well before him and also explained from the theoretical point of view (e.g., in the books of W. Ostwald from 1891–95). Also, scientists before him had separated plant pigments by selective solution. Tswett's merits are in the generalization of the technique as an analytical method and in the detailed investigations of the role of various adsorbents and solvents. As pointed out by Zechmeister (1),

"Tswett's achievement is superior to Day's in two respects. First, he recognized and correctly interpreted chromatographic processes; and second, he devised a useful laboratory method that includes as an important feature the development of chromatograms by pure solvents. The distances between the individual zones are thus increased and complete resolutions advised within minutes."

Although Tswett published his most important papers in German journals (German was then the general language of chemistry), his work and the importance of the chromatographic technique were probably not understood immediately. It is interesting to note that in 1922 in the U.S., Palmer (12) reported on a number of chromatographic experiments, giving credit to Tswett, but his work went largely unnoticed. Only 25 years after the publication of Tswett's two basic papers was his work taken out of oblivion by Kuhn et al. (13, 14) who applied it successfully to the separation of carotene and egg yolk xanthophyll into their isomeric compounds. From there on, the development of adsorption chromatography was straightforward.

Partition Chromatography

It is rare in the history of science that somebody can claim a number of major inventions. One of those men is A. J. P. Martin. Thirty years ago, together with Synge, he invented partition chromatography and demonstrated its use with a liquid carrier (15). Three years later, together with Consden and Gordon, he described paper chromatography as a simple variant of liquid partition chromatography (16). Then, 20 years ago, in cooperation with James, he showed the validity of their prediction in the first paper on partition chromatography that a gas could be used as well as the carrier instead of a liquid (17). With this publication, the unparalleled growth of gas-liquid partition chromatography began. Today, one could not even imagine chemistry and biochemistry without partition chromatography, and its importance is best demonstrated by the fact that Martin and Synge received the 1952 Nobel prize in chemistry for their work, thus joining the few scientists who have received the Nobel prize for a development in analytical chemistry.

The history of the development of the three techniques is fascinating and is described in detail by Martin (18) and James (19). As Martin explained, he early became interested in distillation columns. In 1931, when Winterstein (who, two years earlier with Kuhn and Lederer, brought back adsorption chromatography from oblivion) demonstrated at Cambridge University the separation of carotene on a chalk column, Martin realized that the processes involved in the separation are similar in both techniques. This early thinking resulted later in the expression of the efficiency of chromatographic columns by using terms established in distillation theory (HETP, number of theoretical plates).

At Cambridge University, when working on the separation of carotenes, Martin developed a very complicated countercurrent extraction apparatus and continued to utilize this technique for the separation of amino acids in wool when, in 1938, he moved to the Wool Industries' Research Association where Synge became his collaborator. However, the whole system was extremely complicated and difficult to operate, and therefore, he tried to develop some other technique which would do the job. As described by Martin (18):

"In 1940, it occurred to me that the crux of the problem was that we were trying to work two liquids in opposite directions simultaneously ... Then I suddenly realized that it was not necessary to move both the liquids; if I just moved one of them, the required conditions were fulfilled. I was able to devise a suitable apparatus the very next day, and a modification of this eventually became the partition chromatograph with which we are now familiar."

In their early work, chloroform containing a small amount of alcohol was used as the mobile phase, water as the stationary phase, and silica gel as the support; they could separate the monoamino monocarboxylic acids, and, according to Martin, "One foot of tubing in this apparatus could do substantially better separations than all the machinery we had constructed until then."

In their paper, Martin and Synge (15) emphasized that, by the selection of suitable mobile phase–stationary phase combinations, the technique can be used for many other separations, and they predicted that:

"The mobile phase need not be a liquid but may be a vapour. We show below that the efficiency of contact between the phases (theoretical plates per unit length of column) is far greater in the chromatogram than in ordinary distillation or extraction columns. Very refined separation of volatile substances should therefore be possible in a column in which permanent gas is made to flow over gel impregnated with a nonvolatile solvent in which the substances to be separated approximately obey Raoult's law."

It is interesting to note that, although this prediction clearly and unequivocally predicted the possibility of gas-liquid partition chromatography, nobody picked it up, and it took 10 years until Martin, then with A. T. James, proved its great potential.

Paper Chromatography

Liquid partition chromatography, as described originally by Martin and Synge, had superior separation power. However, in their original work, using water as the stationary phase and silica gel as the support, they were unable to separate the dicarboxylic and basic amino acids. They realized that the problem lay in the adsorptive power of silica gel and thus, were looking for some other support. As Martin explained, he had seen a "paper chromatogram" of dyes before and thus their first choice was paper. First, they used circles of papers in a Petri dish containing water; a drop of the amino acid solution was placed in the center of the paper, and water-saturated butanol was fed to the center of the paper; when it reached the edge, the paper was dried and sprayed with ninhydrin, a substance found by Gordon to give proper color reaction with the amino acids, enabling the detection of their spots. Later, they used paper strips in boxes, in an atmosphere saturated with water, and the edge of the paper was dipped into the solvent (the moving phase). They also learned to run the paper in two dimensions. This is how paper chromatography, which revolutionized biochemical analysis, was born (16).

Gas-Liquid Partition Chromatography

As mentioned earlier, Martin and Synge predicted, in their original paper, the possibility of using a gas as the moving phase in partition chromatography. However, nobody thought to test experimentally this prediction at that time although gas adsorption chromatography underwent an important development in the 1940's. (See, e.g., the works of G. Hesse, E. Cremer, S. Claesson, E. Glueckauf, and C. S. G. Phillips; for a listing of their papers, see Bibliography, ref. 20).

In 1948, Martin moved to the National Institute for Medical Research, where A. T. James, who had previously been working with Synge, joined him. They were engaged in research work which did not result in any success, and Dr. James became so discouraged that Martin suggested that they switch to a project that surely must work—to test the prediction of the possibility of gas-liquid partition chromatography. A request to try to develop a more advanced method for fatty acid analysis came as a good model for the investigations. Their work succeeded fairly rapidly, and soon the new technique was born; their paper was submitted for publication on June 5, 1951, and published in the first part of 1952 in *Biochemical Journal* (17).

In their original work, James and Martin added the sample by a pipet at the front of the column and determined the eluted fractions by titration, and it was fairly difficult to record the "chromatogram."

In the same year, D. H. Desty of British Petroleum, N. H. Ray of I.C.I., and R. P. W. Scott of Benzole Producers Assn. contacted Martin about the separation of hydrocarbons, and then the possibility of using a thermal conductivity detector (applied at that time by Claesson) in gas adsorption chromatography was raised. In the next three years, the applicability of the new technique to a wide variety of problems was shown, and in 1955, the first commercial instrument appeared on the market. The rest is history which those of us who "joined the club" in its early period will never forget.

Sometime ago I read about life in the Spanish ports in the years after Columbus' return from the New World, where every ship brought something new, interesting, and exciting in treasures, people, and tales about new discoveries. We who have participated in the development of gas-liquid partition chromatography since its beginning, probably felt similarly; every new issue of the journals, every meeting we have attended has brought something new and interesting that everybody wanted to try out in his own laboratory the next day. It is proper to quote Dr. Lipsky who finished one of his lectures (21) by paying tribute to Dr. Martin:

"He has twice made outstanding contributions to this field—in his discovery of partition chromatography and in his pioneering work on gas chromatography. He has thus altered, for the better, the lives of many of us. We, his scientific colleagues, thank him for allowing us to share with him this wonderful adventure."

Interaction of Chromatographic Techniques

It is interesting to note that, until fairly recently, some artificial classification barriers have been dividing the various types of chromatography, and scientists rarely deserted their own fields, Martin being a rare exception. A few years ago, however, these artificial barriers started to corrode. Scientists who, during the genesis of gas chromatography became identified with it, started to be active in the other branches of chromatography. As a result of this healthy development, chromatography is treated today more and more by a unifying approach.

This happening is a natural one, and the proper way development of a complex method is carried out. It starts in different channels which, for some time, look like separate techniques with nothing in common. But, sooner or later, it becomes clear that these channels are not parallel but approach each other and, at a given time, converge. From this point on, they are developed together, benefiting from each other's achievements.

Today, chromatography is the most widely used analytical technique, and there is practically no laboratory in the world where it is not practiced. It permits results which otherwise would be impossible. Present-day chemists who, in less than 30 min, can separate a complex multicomponent mixture and establish its concentration often do not realize that, until fairly recently, the solution of this problem would probaby have required weeks of hard work. All of us who are active in this field should therefore appreciate the genius of the pioneers of this technique.

References

(1) L. Zechmeister, "Historical Introduction," in "Chromatography," E. Heftmann, Ed., 2nd ed., pp 3–10. Reinhold, New York, N.Y., 1967.

(2) V. Heines, *Chem. Tech.*, **1**, 280–5 (1971).
(3) C. Dhéré, *Condollea* (Geneva), **10**, 23–73 (1943).
(4) G. Hesse and H. Weil, *Michael Tswett's First Paper on Chromatography*, M. Woelm, Eschwege, Germany, 1954.
(5) L. Zechmeister and L. Cholnoky, "Die chromatographische Adsorptionsmethode," Springer, Vienna, 1936; English ed., "Principles and Practice of Chromatography," Wiley & Sons, New York, N.Y., 1941.
(6) K. Sakodynsky, *Chromatographia*, **3**, 92–4 (1970).
(7) M. S. Tswett, *Proc. Warsaw Soc. Nat. Sci. Biol. Sect.*, **14**, minute No. 6 (1903).
(8) M. Tswett, *Ber. Deut. Bot. Ges.*, **24**, 313–26 (1906); the two quotations are from p 322 and 323.
(9) M. Tswett, *Ber. Deut. Bot. Ges.*, **24**, 384–93 (1906).
(10) L. S. Ettre, *Chromatographia*, **3**, 95–6 (1970).
(11) H. Purnell, *Gas Chromatography*, p 1, Wiley & Sons, New York, N.Y., 1962.
(12) L. S. Palmer, "Carotinoids and Related Pigments," Chemical Catalog Co., New York, N.Y., 1922.
(13) R. Kuhn and E. Lederer, *Ber. Deut. Chem. Ges.*, **64**, 1349–1357 (1931).
(14) R. Kuhn, A. Winterstein, E. Lederer, *Hoppe Seyler's Z. Physiol. Chem.*, **197**, 141–60 (1931).
(15) A. J. P. Martin and R. L. M. Synge, *Biochem. J.*, **35**, 1358–68 (1941).
(16) R. Consden, A. H. Gordon, and A. J. P. Martin, *ibid.*, **38**, 224–32 (1944).
(17) A. T. James and A. J. P. Martin, *Biochem. J.*, **50**, 679–90 (1952).
(18) A. J. P. Martin, "Historical Background," in "Gas Chromatography in Biology and Medicine," R. Porter, Ed., pp 2–10, J. & A. Churchill Ltd., London, England, 1969.
(19) A. T. James, "The Development of an Idea," in "Gas Chromatography." H. J. Noebels, N. Brenner, and R. F. Wall, Eds., pp 247–54, Academic Press, 1961.
(20) R. L. Pecsok, Ed., "Principles and Practice of Gas Chromatography," Wiley & Sons, New York, N.Y., 1959, pp 154–6.
(21) S. R. Lipsky, "Gas Chromatography; The Anatomy of a Scientific Revolution," in "Gas Chromatography in Biology and Medicine," R. Porter, Ed., pp 11–16, J. & A. Churchill Ltd., London, England, 1969.

Leslie S. Ettre *is executive editor of the 18-volume "Encyclopedia of Industrial Chemical Analysis," published by John Wiley & Sons, Inc. Mr. Ettre graduated in 1945 from the Faculty of Chemical Engineering of the Institute of Technology, Budapest, Hungary. He became interested in chromatography in 1957 while employed at LURGI-Companies in Frankfurt am Main, Germany. From 1958 to 1968, he was working in gas chromatography at Perkin-Elmer Corp., serving in the last years as chief applications chemist. He has over 70 publications, mostly in the field of gas chromatography, is author of the book, "Open Tubular Columns in Gas Chromatography," and editor of the books, "The Practice of Gas Chromatography," (with A. Zlatkis) and "Ancillary Techniques of Gas Chromatography" (with W. H. McFadden). He is a member of the editorial advisory board of the Journal of Chromatographic Science, one of the regional editors of Chromatographia, and serves as chairman of the Subcommittee on Nomenclature of ASTM Committee E-19 on Chromatography. Mr. Ettre is a member of ACS, ISA, the New York Academy of Sciences, the British Society for Analytical Chemistry and Gas Chromatography Discussion Group, and a fellow of the American Institute of Chemists.*

Leslie Ettre has cooperated with Albert Zlatkis in the organization of the International Symposia on Advances in Chromatography since their beginning.

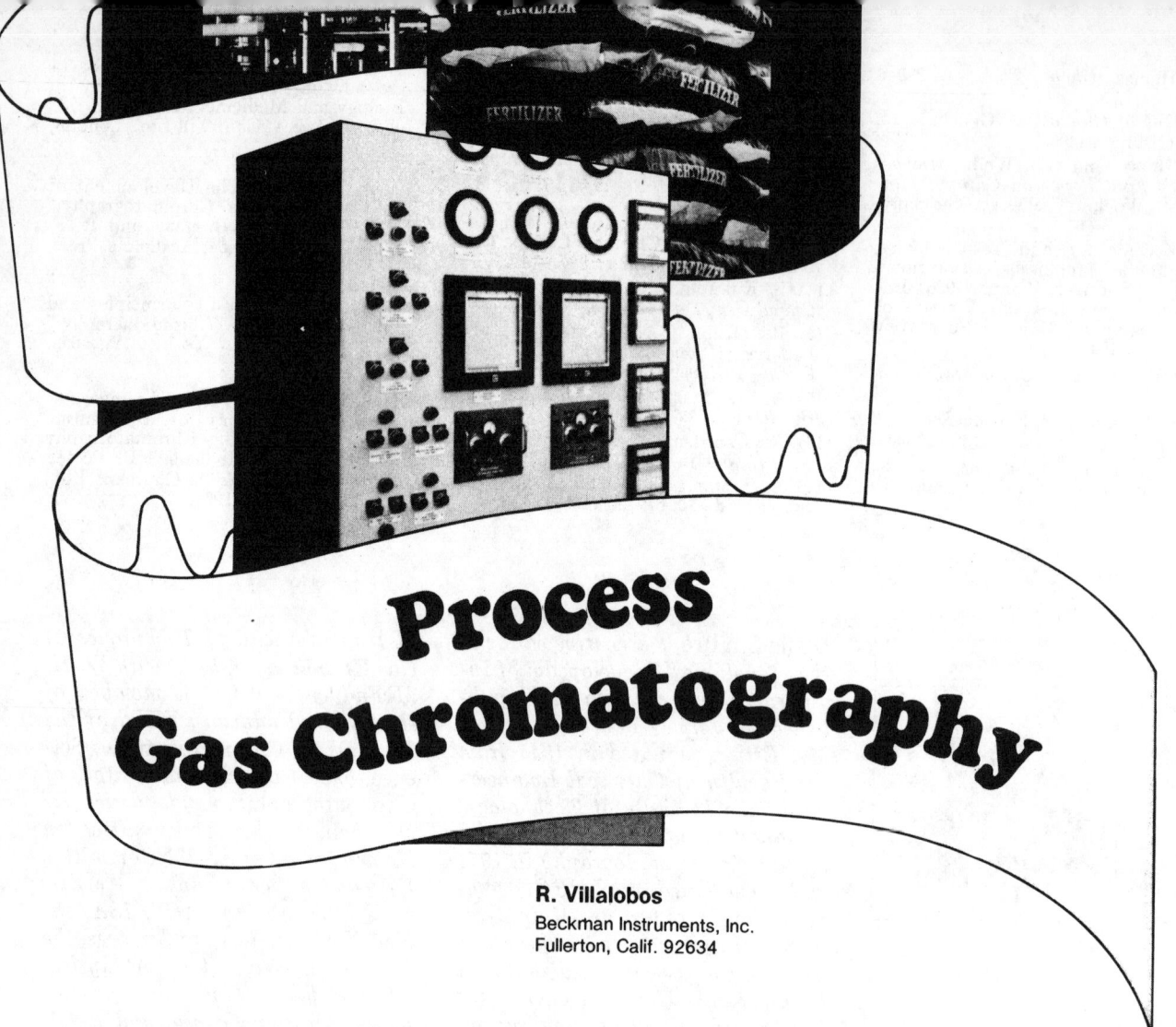

Process Gas Chromatography

R. Villalobos
Beckman Instruments, Inc.
Fullerton, Calif. 92634

Process gas chromatographs have been commercially available for almost 20 years. In that time they have become the most widely used process analyzer. A 1968 survey of U.S. refineries by the American Petroleum Institute (1) showed that chromatographs comprised fully 24% of all stream analyzers in use. This was considerably more than oxygen analyzers—the second most numerous—and 10 times as many as the number of pH systems. As in the laboratory, gas chromatography has proved to be the most versatile method to come along in the history of process analytical instrumentation.

While most analysts are familiar with laboratory gas chromatographs, few are familiar with the process versions and the considerable technology that has been developed with relation to their application to automated onstream monitoring. While the similarities between laboratory and process chromatographs would be readily apparent to most analysts, the differences are perhaps more important, though less well understood. These differences are not only with regard to the appearance of the instrument itself, but the manner in which the data are obtained, and perhaps most important, the way in which it is used.

What Is a Process Gas Chromatograph?

The aim in installing analyzers on-line is to obtain the analytical results with a speed of response that is comparable to process changes. The objective is to use the information to take corrective action.

Hence, a process gas chromatograph (GC) is an instrument which has been designed to meet this objective and operates continuously on-line, automatically analyzing a flowing process stream, in a cyclic and repetitive manner. In general, such an instrument is dedicated to performing a particular analysis on a single stream, or at most, a few liquid or gas streams (multistream). Moreover, it will usually be designed to measure only one or, at most, a few components in the sample.

A distinguishing characteristic of process chromatographs is that sample is transferred from the process sample point to the chromatographic column untouched by human hands. A supply of fresh sample is withdrawn continuously from the process and circulated to the sample valve, which injects a small volume into the column.

The sample valve, and the lines which connect it with the sample point, can be maintained hot. Therefore, the chromatograph can accept hot gaseous samples that contain large amounts of water vapor or other condensibles; samples which cannot be transported to the laboratory without drastically altering their composition. Indeed, many sampling situations which are difficult or impossible for laboratory analysis are routine for process chromatographs. A consequence of this is that the septum inlet, probably the most common component in a laboratory instrument, is unusable in a process analyzer. Instead, considerable emphasis has been placed on the development of sampling valves and column switching valves which are highly reliable.

How Are Process Chromatographs Used?

The uses of chromatographs in industry are varied. Table I lists some of the principal uses of on-line chromatographs. The most frequently encountered applications are for open- or closed-loop process control. In open-loop control the operator makes

Table I. Principal Uses of Process Gas Chromatographs

Process Control—Use information to adjust process through open- or closed-loop control

Process Study—Obtain information about process to improve yield or throughput. Correlate process variables with product quality

Process Development—Obtain information about process characteristics, as in pilot plants. Correlate process variables with reaction products and yields

Material Balance—Use information to calculate material balance for process units

Product Quality Specification Monitor—Monitor impurities in outgoing or incoming product for conformance to specifications

Waste Disposal Monitoring—Monitor liquid or gas effluent wastes for loss of valuable product or for presence of toxic compounds

Personnel Safety-Area Monitoring—Monitor ambient air for presence of toxic compounds

---

adjustments to the process conditions based on the results of the chromatograph. In closed-loop control the chromatograph data are converted to a continuous analog signal which is input to conventional control instrumentation to control the process automatically.

Components of Process Gas Chromatographs

Basic elements of a process GC are shown in Figure 1.

Analyzer (A). This contains all components of the analytical system—columns, sample valves, column switching valves, and detector—in a precisely thermostated oven compartment. For economy and simplicity, a single temperature zone is most frequently used. Multitemperature zone units with several columns at different temperatures have also been used, but are less common. Carrier gas flow controls, temperature controls, valve controls, and detector electronics are also located in the analyzer. The entire unit is located "on-line", as close to the sample point as possible. Supplies or carrier gas and other gases are also located in close proximity.

The analyzer is usually housed in a walk-in analyzer shelter or house, as shown in Figure 2. The house provides weather protection for analyzer and sample conditioning components as well as for maintenance personnel. Several types of analyzers are often mounted in the same analyzer house.

Sample Conditioning System (SH). This contains most or all of the components necessary to condition and maintain constant flow of sample to the analyzer (e.g., pressure reduction, filters, vaporizers, flow controls, etc.) plus sample switching or selector valves for multiple stream applications and for introducing calibration standard (B). It is usually mounted below or next to the analyzer in its own heated or unheated compartment or on a flat open-plate.

Some elements of the sample conditioning system, such as vaporizer, filters (F), and pressure reducers (R), may be located at the sample point (P) itself.

It should be emphasized that the sample system is probably the most critical part of the entire system. If the sample is not representative and properly conditioned, the entire system will fail in its objective. Hence, the sampling system must be designed as an integral part of the chromatograph and not as an afterthought. The interested reader is referred to the monograph by Houser (2).

Programmer-Controller (PC). This unit contains the program timer, power supply, signal conditioning electronics, and computer interface (where applicable). It controls all operations in the analyzer—sample injection, column switching—as well as housekeeping functions (auto zero), component gating and attenuating and data transfer to the appropriate readout channel. This unit is usually located in the control room (as much as 1,000 ft or more from the analyzer) but sometimes in a separate room near the control room.

Readout Devices (Recorders). A strip chart recorder for recording bar graph (BG) and for trend records (TR) is located in the control room. Additionally, the system may communicate with a computer by means of priority interrupt or long-term memories (not shown).

Considerations in Process Chromatograph Design

The techniques and methodology used in process GC are in a general sense the same as in laboratory GC; however, a far different emphasis is placed upon the use of various methods and accessory devices. Some methods are more widely used in process—notably the use of multiple columns and column switching valves—while others are less frequently or seldom used—for example, capillary columns and temperature programming. Beyond that, process hardware bears little resemblance to its laboratory counterpart.

Overall design and appearance are influenced by the following factors.

Purpose. The system's purpose is to obtain information to control the process. Hence, for the majority of applications, it is usually necessary to measure only one or a few components. (Exceptions are pilot plant applications, wherein it is desirable to measure *all* of the components to characterize the process under study.)

Location. The system is usually located in a hazardous area and requires explosion proof construction to satisfy National Electrical Code requirements (3).

Analyzers and sample conditioning systems are designed to meet Class I, Groups C and D, Division 1. These are locations in which hazardous concentrations of flammable gases are present under normal conditions.

Programmers and recorders are

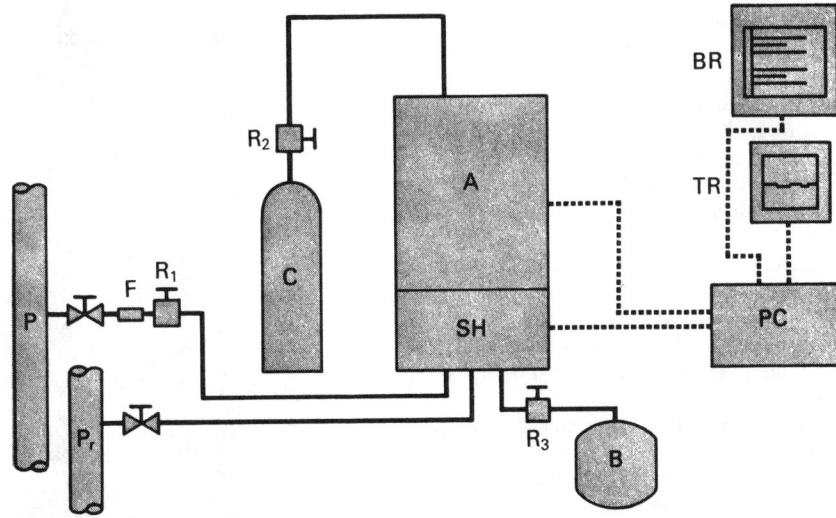

Figure 1. Basic elements of process gas chromatograph system

Sample withdrawn continuously from process line, P, filtered by filter, F1, pressure reduced by regulator, R1, circulated through sample conditioner, SH1, and returned to low-pressure point, Pr. Slip stream withdrawn and circulated to sample valve in analyzer, A, which also contains columns and detector in thermostated oven. Carrier gas supply, C, controlled by Regulator R2. Calibration blend, B, introduced by valves in conditioner, SH. Programmer, PC, controls functions in A and SH and converts signal for recording as bar graph, BR, or trend record, TR

Figure 2. Analyzer shelter showing field mounting of chromatograph and other types of analyzers. Chromatograph is at left
Courtesy of COMSIP-Customline Corp.

is the oven compartment which houses valves, columns, and detector. Heating is usually by forced air bath which uses air passed over an electrical heating element. Process GC's are usually limited to a maximum operating temperature of 225°C—a limitation imposed primarily by the material of construction of the sample valve. This has not been a serious limitation, since 99% of process analyses can be performed at below 160°C.

Electronics in the analyzer include the temperature controller, detector amplifier, and valve controls. The detector signal may be transmitted to the progammer at a high level (0–1 V) to minimize line loss and noise pickup.

Sampling Valves. Sampling and column switching valves have been the object of considerable development and refinement in recent years. Numerous types have been used, but the most widely used are the sliding plate (Figure 4, A), the diaphragm valve, and the spool and O-ring valve. Gases are usually metered by an external sample loop; liquids by the volume of an internal hole or channel in the

usually designed to meet General Purpose Classification or by modification with air purging, Class I, Groups C and D, Division 2. The latter are locations in which hazardous concentrations of flammable gases are not normally present.

Reliability. The entire system should operate continuously and without maintenance for periods of at least 4–6 months and preferably a year. Hence, a premium on simplicity, ruggedness, and dependability for all components.

Automatic Operation. Provision must be included for completely automatic and repetitive operation of all analytical functions.

Maintainability. When failure does occur, the design of the system must permit its rapid repair and return to service with a minimum of down-time. Troubleshooting aids must be included to aid in isolating and identifying the faulty components.

Use of Data. Provision must be included for presenting the data in a variety of ways depending upon its end use, from bar graph readout to input into a closed-loop control system to direct readout into a computer.

Analyzer Construction

A view of a typical analyzer is shown in Figure 3. The interior section

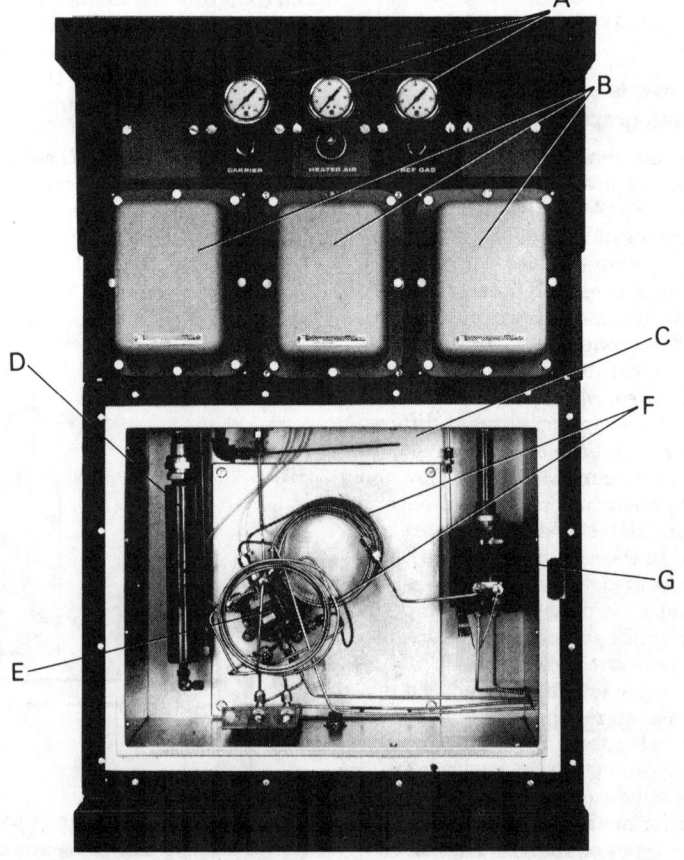

Figure 3. Process GC analyzer interior

A, carrier gas and heater air flow controls; B, electronics in explosion proof condulets—temperature control, valve drivers, detector amplifier; C, analyzer oven; D, heater; E, sample valve; F, columns; G, thermal conductivity detector

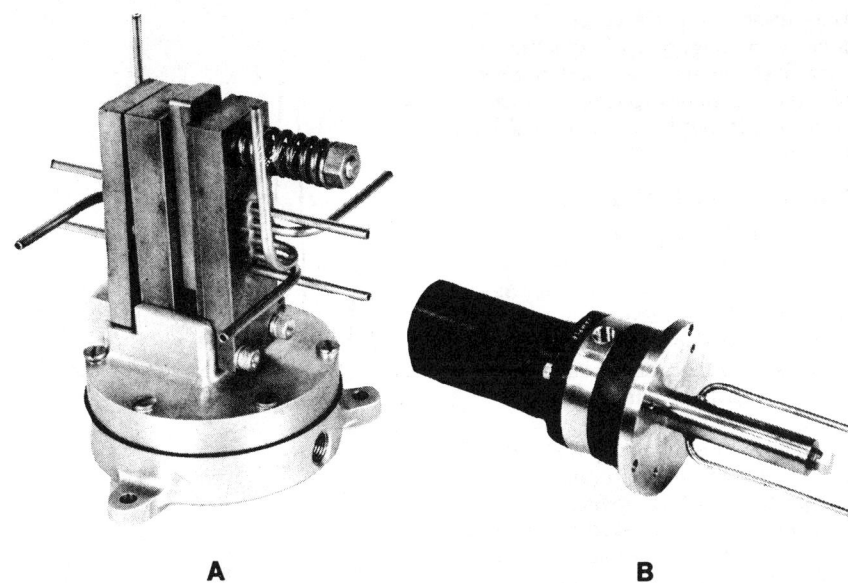

Figure 4. Sample valves used in process chromatographs
A, sliding plate valve, 10-port; B, vaporizing valve for injecting liquids through wall of analyzer

valve. Liquid sampling valves of this type have been used for pressures of up to 500 psig, though pressures under 200 psig are more common. Sample volumes as small as 0.5 µl are attainable, with a reproducibility of ±0.25%.

The choice of sample valve is a critical element in every application, since it must be compatible with the sample and operate reliably at elevated temperatures. Valves are available in corrosion resistant materials such as Hastelloy, Monel, or in all Kel-F and Teflon construction for corrosive samples. Other special designs permit injecting the liquid sample through the wall of the analyzer from the cold outside zone to the heated interior where the sample is vaporized by the heat (Figure 4, B). This is particularly useful for liquids which have high vapor pressures (such as propane or butane) and cannot be subjected to the analyzer temperature.

Process liquid sample injection techniques are in contrast to those used in the laboratory. In the latter, liquid samples are injected through a septum into a heated block where the sample is instantly vaporized and rapidly transferred to the column in plug flow fashion. The temperature of the injection block is frequently higher than the column—and usually higher than the highest boiling point of the substances being examined. In contrast, the process GC sampling valve is usually maintained at the temperature of the column. When the captured liquid volume is injected into the stream of carrier gas, the pressure is released and the liquid vaporizes under its own vapor pressure. The mass of the valve acts as a heat sink and provides sufficient heat to cause rapid vaporization. This requires that the flowing liquid sample be maintained at high enough pressure to prevent incipient boiling of the more volatile components which would form bubbles and affect reproducibility.

Detectors. Although a great number of different detectors have been used in gas chromatography, only a very few meet the stringent requirements of simplicity, ruggedness, and reliability necessary for process GC use. Detectors which have been used to any significant extent and the probable frequency of their useage are as follows: 85–90% for the thermal conductivity detector (TCD); 10–15% for the hydrogen flame ionization detector (FID); and less than 1% for all others [gas density balance, helium ionization detector (microcross section), and flame photometric detector].

A number of other detectors, such as the catalytic combustion (filament) for trace hydrocarbons and the electrolytic P_2O_5 hygrometer cell for trace moisture, have also been reported. However, their use has not been significant.

Programmer-Controller. Analytical and communication functions which are performed by the programmer include sample injection, column switching, component gating (select attenuator for correct scaling factor), integrator (start, stop, readout, and clear), rezero baseline (auto zero), peak picker (convert peak signal to steady-state signal), present signal to recorder, switch samples in sample conditioner, and communicate with computer (transfer signal, signal beginning of analysis, signal end of analysis, and identify stream being analyzed).

A timer mechanism is required to repeatably perform these functions. Early programmers used mechanical cam timers with 1 cam per function. Many forms of mechanical timers have been and are available: rotating plastic wheels with photocell actuation, turntables with adjustable pins on the periphery, etc. Present trends, however, are to solid-state electronic digital devices. The overall cycle time is selected by setting rotary switches which read directly in seconds or by patching in on a matrix board. Two sets of similar switches are associated with each function which is to be programmed. One sets the time "on," and the other the time "off". A decoding circuit decodes the pulses from the timer and activates the circuit at the time selected.

Programmers with microprocessors are also available. Timing for all functions is entered into microprocessor memory, which is then used to actuate each function at the proper time.

Scaling, or calibration, for each measured component is accomplished with a dedicated component attenuator. When a measured component elutes, the corresponding attenuator is activated or "gated" in by the timer, and the proper scaling factor is applied to that component.

Numerous other functions are located in the programmer. The "auto zero" and "peak picker" are two which have no equivalent in the laboratory chromatography, but are absolutely essential in a process GC.

Auto Zero. This circuit rezeros the baseline at selected times during the analysis, to compensate for short- and long-term detector drift or changes in baseline due to column switching transients. Like the other functions, it is programmable.

Peak Picker and Short-Term Memory. These circuits provide the capability of converting the transient peak signal to a continuous signal which can be stored in a long-term memory or transferred to a computer. The sequence of events is shown in Figure 5. A simple locking circuit (analogous to a diode) holds the peak maxima developed for a fixed period (e.g., 5 sec) after the component gate closes. As the gate closes, a transfer command is sent out, and the signal is transferred and stored in a long-term memory. The short-term memory is then cleared and is ready for a subsequent component peak. Hence, only one short-term memory circuit is required regardless of the number of peaks transferred. This method is widely used to communicate data to on-line process control computers via "priority interrupt".

Data Presentation

Peak height is most commonly used as a measure of component concentrations. When sample size, temperature, carrier gas flow rate, and other conditions are held constant, this relationship is accurate. Moreover, if sample size is such as to avoid column overloading and obtain symmetrical peaks, the relationship is linear. The simplest method of recording the peak height is the "bar graph" in which the component is recorded with the recorder chart stopped. The attenuator provides the correct scaling factor to give the desired full-scale range (e.g., 0–10%, 0–5 ppm, etc.). The relation of the bar graph to the chromatogram is shown in Figure 6.

Not all components in the chromatogram are recorded: only those which are "gated" by the programmer. At all other times the input to the recorder is simply shorted, and the recorder reads "zero," thus eliminating unmeasured peaks and baseline transients from the chart record.

If the recorder chart is advanced continuously during bar graph presentation, a time record of the gating for the measured components is obtained. This record is sometimes called an "elution time check" (Figure 6, b) and is useful as a maintenance check to detect shifting of the peaks within the fixed time gates.

The transient peak signal can be converted to a continuous analog signal suitable for closed-loop control. A peak picker is used in conjection with a long-term memory device which holds the signal from the peak picker until it is updated with a new value the subsequent cycle. A record of the long-term memory is called a "trend" record, shown in Figure 6, d. Such a signal can also be input to a conventional controller for use in closed-loop control.

Computer Interfacing

The number of process GC's which are tied into on-line computers has been increasing rapidly during the last decade. This requirement is now so common that for a number of years, process GC's have been available with standard options to make the interface.

The two most common methods of interfacing chromatographs to computers are long-term memory and priority interrupt (short-term memory). A third method is the computer-controlled chromatograph system, which eliminates the programmer entirely.

Long-Term Memory. The long-term memory and its operation were described above. During the analysis the information is stored in the long-term memories. At the end of the analysis cycle, a contact closure signals the computer that the analysis is complete and that the information is available. The computer then scans the signal in each long-term memory, digitizes it, and stores it in its own memory bank.

A disadvantage of this approach is that a separate long-term memory is required for each component. From a cost standpoint, it is less attractive than the priority interrupt approach and is now infrequently used.

Priority Interrupt. As computers have increased in speed and capacity, the priority interrupt method has become the preferred approach. In this method the long-term memories are eliminated altogether and replaced by a single short-term memory.

Sequence of events associated with the transfer of the information to the computer is identical to the peak picker function shown in Figure 5.

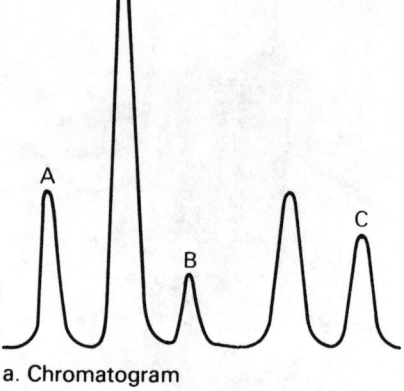

a. Chromatogram

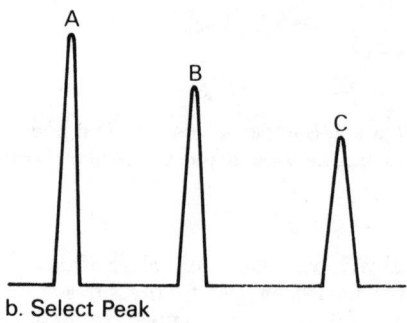

b. Select Peak Presentation

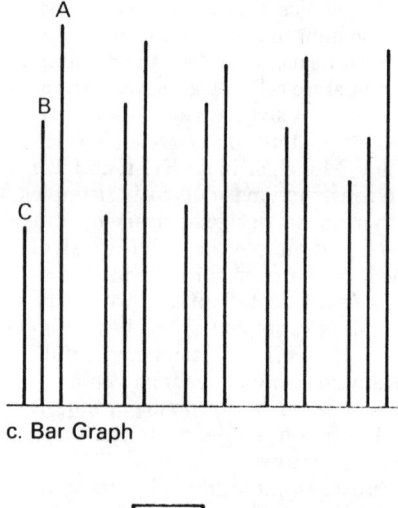

c. Bar Graph

d. Trend

Figure 6. Modes of data presentation
Chromatogram (a) and select peak presentation (b) are used in manual operation only. Presentation during automatic operation is bar graph (c) and/or trend record (d)

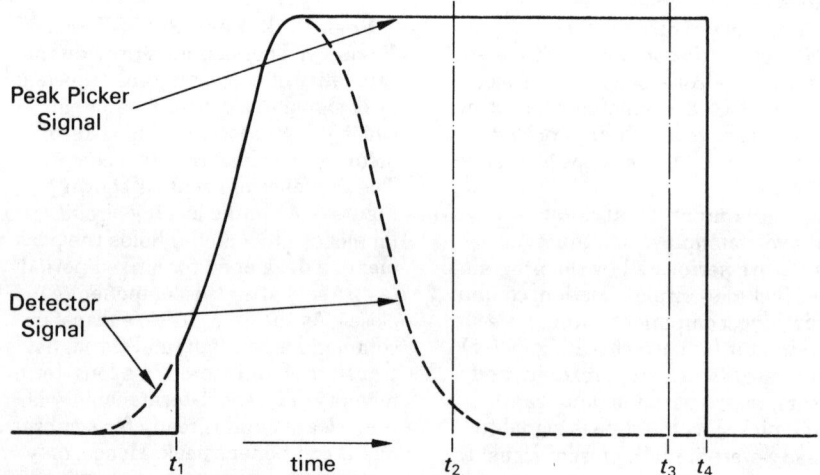

Figure 5. Peak picker operation
Sequence of events for transferring peak value to long-term memory or to computer by priority interrupt. $t1$, component gate comes ON and actuates component attenuator and peak picker which follows detector signal and holds peak value. $t2$, component gate turns OFF and initiates transfer of signal to long-term memory or to computer. $t3$, transfer of signal completed ($t3 - t2 = 5$ sec). $t4$, after additional 1-sec delay, peak picker output resets to zero

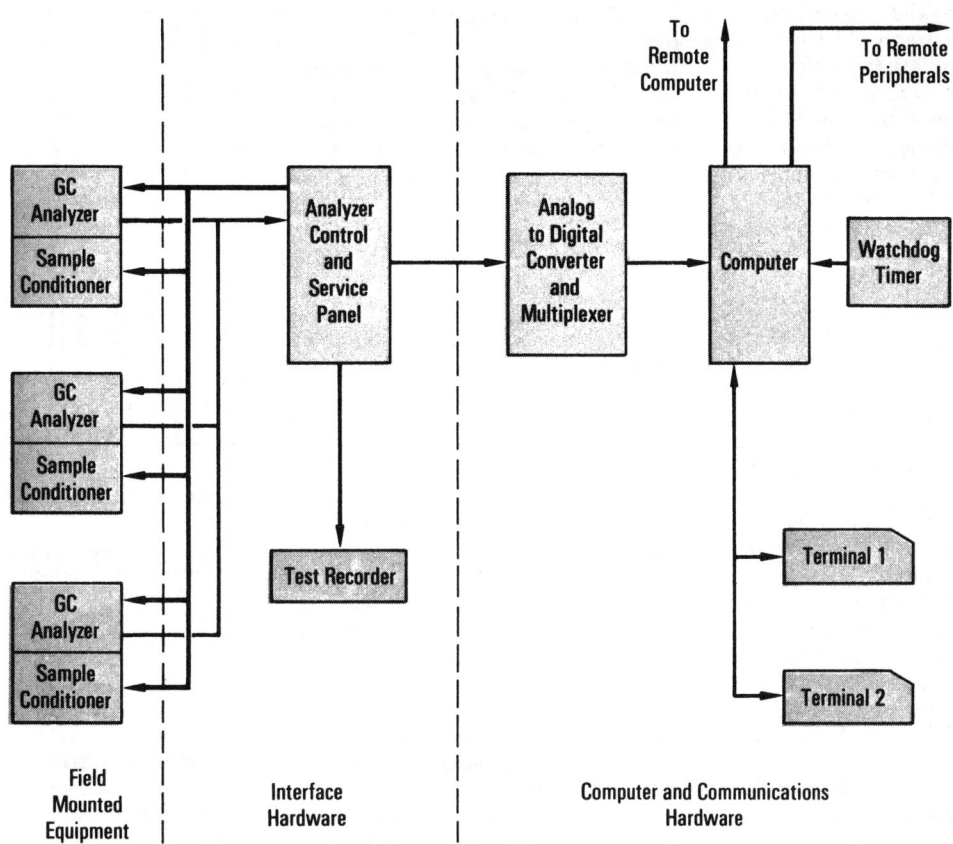

Figure 7. Computer-controlled process chromatograph system—block diagram
Dedicated minicomputer controls up to 32 chromatographs and performs all data acquisition and reduction functions. Conventional programmers are eliminated

As previously described, the short-term memory stores the scaled analog value for each component as it elutes from the column. As the component gate closes, a contact closure, or "come read" signal, signals the computer that a value is stored in memory. Duration of the closure is usually about 50 msec, but varies with the individual computer. Upon receipt of this signal, the computer interrupts its routine and "services" the interrupt. It scans the short-term memory output, digitizes it, and stores it in its own memory. All this occurs during the few seconds that the value is held by the short-term memory. The short-term memory is then available for the next component to be measured.

The obvious advantage of this approach is that only one memory is required regardless of the number of components which are input to the computer.

Computer-Controlled Chromatograph System

In this system a dedicated minicomputer replaces *all* the programmer controllers for an array of process chromatographs. The computer performs all functions associated with control of the analyzer units and all analytical data handling, namely:

- Control of analyzer sample inject and column switching valves
- Control of multiple stream and calibration sample valves in the sample conditioning modules
- Monitoring of detector output, integration of peak areas, component identification, and data reduction
- Automatic baseline correction for zero drift
- Separation of incompletely resolved peaks or of trace components riding on the tail of major components
- Automatic system monitoring and alarm for off-limits data and malfunctions
- Data presentation and transfer for control room operators and maintenance personnel
- Transfer of reduced analytical data in digital form to central supervisory computer.

A system of this type consists of three major hardware segments (Figure 7):

- *Field mounted equipment*—Consists of analyzer and associated sample conditioning modules
- *Interface hardware*—Provides a central terminal for all detector signals, as well as all analyzers and sample system electrical connections. Also includes provision for isolating the analyzer and sample conditioner from the computer during start-up and maintenance and for permitting manual control of all analytical functions during start-up and maintenance
- *Computer and communications hardware*—Consists of dedicated computer and associated teletypes for data logging, operator communications, program loading, and all other input functions. It may also include provision for analog output (pneumatic, voltage, current) to conventional control instruments.

A distinctive feature of this system is that the computer samples the output of each detector at up to 10 times each second, integrates the area under each peak, identifies the component, applies a response factor and computes the composition of the sample. This may be done by internal normalization or by comparison with calibration standards.

Systems of this type, like their laboratory counterparts, are able to correct for incomplete resolutions between adjacent peaks and for minor peaks which ride on the "tail" of a preceding major peak. This is a decided advantage over the conventional process chromatograph which cannot correct for incomplete resolution.

Data output is by means of a teletypewriter located in the control room for maximum data accessibility to the operators. The printouts may include a complete analysis or for selected components only, as required by the operator. A second teletypewriter is used for communicating with the minicomputer. Changes in operation or status of analyzer—such as removing from service for maintenance or reprogramming of events when columns are changed—are communicated by this means. The most significant advantage of this type of system is the greater accuracy which is potentially available. A second advantage is the more complete data which can be obtained by the computer.

Availability. Complete computer-controlled process GC systems are currently available from most manufacturers of process analyzers. The software package is tailored to the specific needs of the user. Large systems of this type with as many as 32 chromatographs are most frequently purchased for new plants or expansions where formerly no analyzers existed. More and more, however, such systems are being added to existing plants with numbers of analyzers already in service. Existing programmers may be replaced with the interface hardware described above. Alternatively, they may be left in service, with the computer performing only a data acquisitioning and reduction function and the programmed analytical functions remaining under the control of the chromatograph programmer.

Column Design

As in the laboratory chromatograph, the objective in designing the column is to separate all the components of interest in the minimum of time possible. Hence, good column design practices are dependent upon a thorough understanding of chromatographic column theory as developed during the last 20 years. However, the requirements of process chromatographs impose somewhat different constraints on the design of the columns. Some principles which must be observed are:

• *All components in the sample must be quantitatively accounted for and removed from the column system each cycle.* Components which are left on the column will either accumulate and change the characteristics of the column, or will elute during a subsequent cycle and interfere with a measurement.

• *The column separation should be designed not only for the normal composition, but also for the upset condition.* The chromatograph must continue to provide reliable data even when abnormal conditions prevail.

• *Columns must be protected from components which are irreversibly or too strongly absorbed* for the same reasons given in the first principle above.

• *The system should be as simple as possible.* Adjustment and maintenance, when necessary, should be easily and conveniently performed.

Emphasis on Valves and Column Switching. Because of these design constraints, process column technology has developed along lines which are in contrast to laboratory practice. A greater emphasis has been placed on isothermal, multicolumn techniques. On the other hand, programmed temperature has not been widely used, although commercial units are available. Numerous other techniques such as pyrolysis, derivatization, reaction chromatography, etc., have not found even limited application in on-line process analyzers.

With the emphasis on multicolumn methods, a great variety of schemes using valves to switch columns have been developed. While there are literally hundreds of possible column configurations, most are built up from only a few basic configurations. These configurations and how they are used are described in the following sections.

Stripper (Precut or Backflush to Vent). This is the most widely used. It consists of two columns, a stripper (or precut) and an analysis column in series, with provision for backflushing the stripper to a separate vent and providing carrier gas to the analysis column while the stripper is being backflushed.

The stripper column makes a partial separation and is then backflushed, rejecting the unwanted components. The remaining components are further separated on the analysis column while the stripper is being backflushed. A typical analysis is shown in Figure 8 of a full-range gasoline containing components as high as C_{12}.

Moisture, a contaminant often found in hydrocarbon samples, can be conveniently rejected with this configuration. In general, use of the stripper is good housekeeping practice; it ensures that unknown heavy components are completely removed from the system each cycle and will not appear unexpectedly during later cycles to interfere with a key measurement.

Back Flush to Measure. This is similar to the foregoing configurations, with the exception that the analytical column is backflushed to the

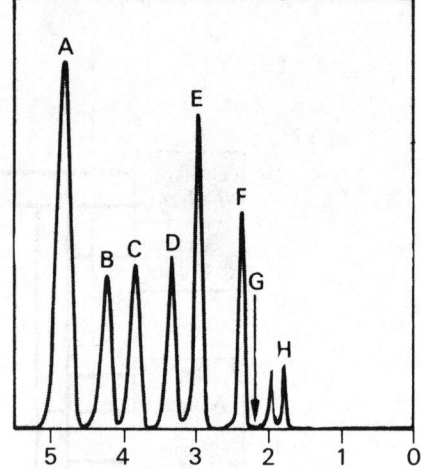

Figure 8. Example of use of stripper column

Analysis of light hydrocarbons in gasoline. *n*-Pentane and heavier backflushed to vent during 5-min cycle. A, *i*-pentane; B, *cis*-butene-2; C, *trans*-butene-2; D, *i*-butylene; E, *n*-butane; F, *i*-butane; G, backpurge; H, propane

detector permitting measurement of all components remaining in the column. A frequent use is to measure the total of material heavier than a particular carbon number, e.g., pentanes and heavier.

Dual Column. A valve is used to switch components so that components eluted from a first column are passed either into a second column or directly into the detector. A typical separation is shown in Figure 9.

Heart Cut (Cutter Column). This arrangement is used most frequently for trace analysis with the hydrogen flame ionization detector (FID). It is particularly useful in measuring a trace component which elutes on the "tail" of a major component. Two columns are so arranged that narrow cuts of effluent from the first (cutter) column can be taken into the second (analysis) column, the bulk of the sample being discarded through a separate vent. A trace component riding on the interfering tail is thus separated from the tail on the second column. Complete separation can be achieved with this arrangement in much less time than could be attained with a single column. This arrangement is essential when a sensitivity of 500 ppm full scale or less is required. A typical separation with a heart cut system is shown in Figure 10.

High-Speed Chromatography. Closed-loop control of processes with short response times may require analysis cycle times as short as 15–90 sec. Such rapid analyses may often be achieved with conventional packed columns (2 mm i.d.) as shown in Figure 11, particularly if only one or two components in the sample are to be

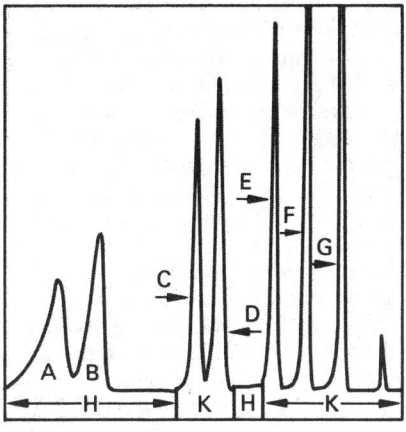

Figure 9. Example of dual-column analysis

Unresolved pair i-butene–1-butene is diverted to second column for further separation. A, i-butene (5%); B, 1-butene (5%); C, cis-2-butene (5%); D, trans-2-butene (5%); E, n-butane (5%); F, i-butane (5%); G, propane (10%); H, dual column; K, single column

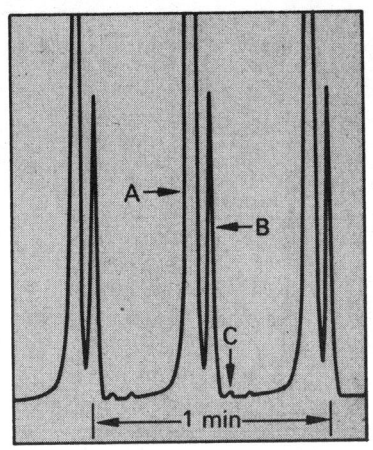

Figure 11. High-speed analysis with conventional packed columns

6 ft × 2 mm i.d. bis(butoxyethyl)phthalate on Chromosorb P. Sample size, 50 μl. A, n-butane; B, 2-butane; C, propane

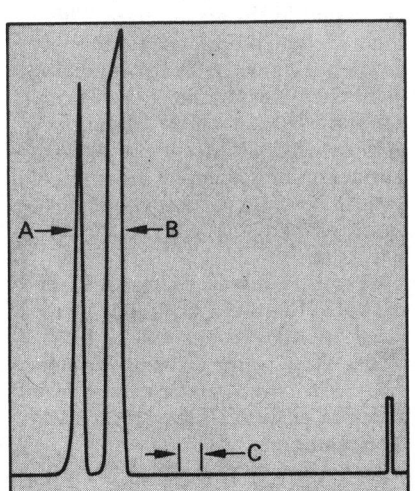

Figure 10. Example of heart cut analysis

Trace toluene impurity in benzene. Toluene trace from first column is cut into second column for separation from interfering benzene tail. A, toluene (20 ppm); B, benzene tail; C, cut for toluene

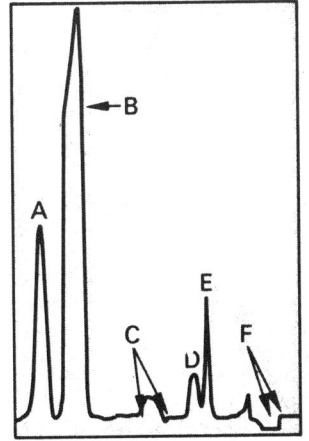

Figure 12. Determination of trace CO_2 in high-purity ethylene by methanation over nickel catalyst

Heart cut column system is used to separate CO_2 and acetylene from interfering tail from ethylene peak. A, acetylene (10 ppm); B, ethylene tail; C, cut for acetylene; D, ethylene tail; E, carbon dioxide (5 ppm); F, cut for carbon dioxide

Most widely employed is, of course, calibration with a known standard, in which a cylinder of calibration gas or liquid is periodically analyzed by the chromatograph. Component attenuators are then adjusted as required to give the correct reading. For this purpose, "certified" standards with guaranteed accuracies of 1% are available. Some users prefer to have the blend analyzed by their own laboratory and to use that value instead.

A further requirement of the calibration blend is that it be stable over periods of time, as long as a year or more. Reactive components or compounds which are strongly absorbed on cylinder walls add a dimension of uncertainty. At trace levels, even compounds which are not reactive in the percent levels can be troublesome. Fortunately, the technology of synthetic blends is currently undergoing major development and should within a short time result in improved quality of calibration standards, particularly in the parts-per-million range.

Analyzers may also be calibrated by having the laboratory analyze a sample taken manually at the same time that the analyzer injected the sample. This method is generally less satisfactory simply because it introduces all the uncertainties associated with manual sampling—possibility of fractionation during transfer, adsorption of components on container surfaces, condensation of heavy ends, etc. The API survey (1) indicated that the highest frequency of dissatisfaction with calibration methods occurred among users of this technique.

Internal normalization is of course potentially the most accurate method, but this requires integration of *all* components in the sample and access to an on-line computer. Hence, this method is used only when a dedicated computer is available for continuous on-line data acquisition and reduction, as in the computer-controlled chromatographic system previously described.

determined. Multicomponent analyses in short cycle times may be attained with micropacked columns (0.030–0.040 in. o.d.) packed with ultra fine diameter substrate and special low-volume detectors (4). Special low-volume valves are also required to obtain optimum performance.

Chemical Conversions. The measurement of carbon monoxide and carbon dioxide at trace levels cannot be done with the thermal conductivity detector because of its limited sensitivity. However, by methanating in the presence of hydrogen over a nickel catalyst, the oxides can be made measurable with the FID (5). This method is used most commonly for measuring low ppm levels of carbon monoxide and carbon dioxide in polymerization grade ethylene. By combining methanation with the heart cut technique, these compounds can be determined on the same system used to measure acetylene at the same low levels. An example is shown in Figure 12.

Quantitation and Calibration

Quantitation methods most commonly used are comparison with known standards (calibration blends), comparison with laboratory analysis (grab sample), and internal normalization.

Accuracy vs. Repeatability

A process chromatograph is capable of short-term repeatability of ±¼ to ½% of the full-scale range when in the conventional bar graph mode and using peak height measurements. This approximates a standard deviation of as small as 0.1%. In theory, this represents the limit of attainable accuracy. However, long-term effects, which include instrument drift, temperature effects on the calibration blend, and barometric effects on sample size and detector sensitivity, all conspire to degrade the repeatability which can be obtained in practice. Osborne (6) has

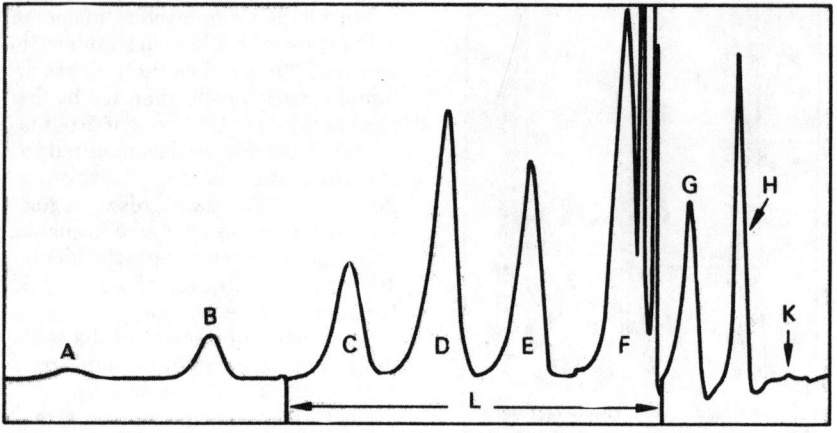

Figure 13. Determination of dissolved gases in transformer oil

Dual column with stripper configuration. Column 1: 3-ft Porapak N. Column 2: 3-ft Porapak Q + 4-ft Porapak N. Column 3: 6-ft Molecular Sieve 5A. Helium carrier 70°C. 1-ml oil sample is injected into column system where dissolved gases are removed and separated. Oil backflushed from system each cycle. A, CO (×10); B, methane (×10); C, acetylene (×10); D, ethane (×10); E, ethylene (×10); F, CO_2 (×10); G, N_2 (×1); H, O_2 (×1); K, H_2 (×1); L, single column

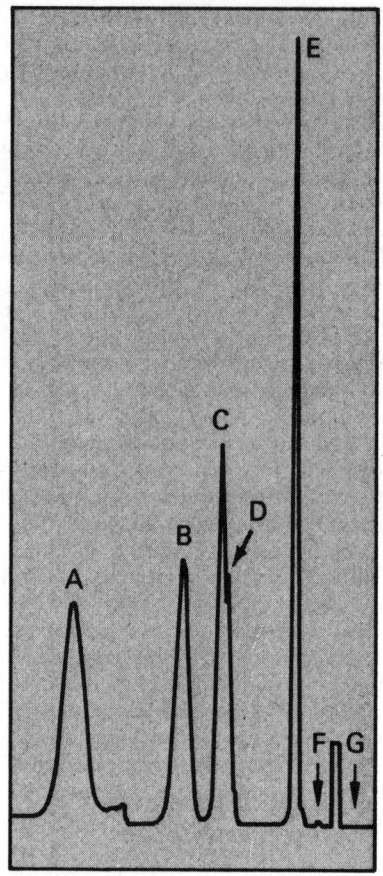

Figure 14. Detection of vinyl chloride in ambient air

Three-minute cycle time permits monitoring several sample points with a single analyzer. A, vinyl chloride (1 ppm, ×2); B, propylene (100 ppm, ×400); C, methyl acetylene (100 ppm); D, methyl chloride (100 ppm); E, ethane (100 ppm); F, air balance; G, start

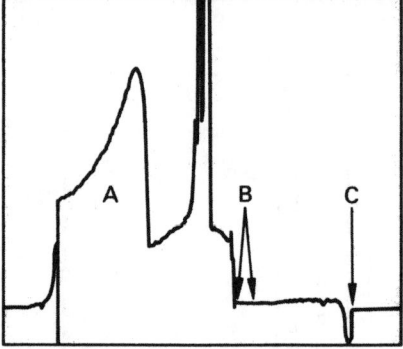

Figure 15. Determination of trace water in xylenes

Column, 3-ft × 3/16-in. Porapak-Q. Sample size, 0.5 ml liquid. Temperature, 105°C. A, water (2 ppm); B, sample inject; C, start analysis

nents in a single run, as shown in FIGURE [3/ A thermal conductivity detector permits detecting as little as 20–40 ppm wt/wt of each gas (30–300 volumes gas per million volumes oil). With argon carrier the sensitivity for hydrogen can be extended by a factor of 10. With an FID, as little as 0.1–0.2 ppb wt/wt (0.2 volumes gas per million volumes oil) can be detected.

Vinyl Chloride in Ambient Air Monitoring. Recent OSHA regulations require the monitoring of vinyl chloride in the atmosphere in the vicinity of plants which either produce it or use it. The analysis shown in Figure 14 illustrates this separation using a FID to detect vinyl chloride with a full-scale sensitivity of as little as 0–1 ppm. A single column stripper configuration is used to reject all components heavier than C_3. The design of the column is such that an interference-free measurement is produced regardless of the presence of any other chemical compound which is likely to be present in the atmosphere. This could include literally any other hydrocarbon found in a petrochemical plant or refinery. A short cycle time of 3 min is used to monitor 10 sample points within a total time of 30 min. The sampling system includes pumps to draw air from various points in the plant through 1/4-in. o.d. lines up to 400 ft long.

Trace Dissolved Water in Hydrocarbons. Dissolved water can be measured at the parts-per-million level with a TC detector by using porous polymer substrates (Porapak, Chromosorb Century Series) which permit rapid elution of water. By using liquid injection and large sample volumes (0.5–1 ml), water can be determined at the 1–5 ppm level, as shown in Figure 15.

Trace Hydrocarbons in Steam Condensate. Trace hydrocarbons in steam condensate return are undesirable in steam generation plants. The presence of hydrocarbons is indicative of leaks in heat exchangers or reboilers. A process chromatograph with FID can be used to monitor individual condensate return lines for hydrocarbons with sensitivities of as much as 0–5 ppm carbon. A porous polymer column separates the water from the hydrocarbons which are then backflushed from the column and measured by integration, as shown in Figure 16. This method is limited to hydrocarbons which have some volatility. Nonvolatile hydrocarbons such as polymers or organic salts require other analytical methods (e.g., catalytic combustion). The process GC method is best suited for chemical plants or refinery applications where the hydrocarbons are of known composition and volatility.

reported that records maintained over a period of several months show a long-term repeatability of ±8.2% relative at the 95% confidence level. Since the effect of these long-term variations can to a great extent be minimized through the use of on-line computer data acquisition and reduction, one can expect to see increasing use of on-line computers dedicated to process chromatographs.

Applications

The vast majority of applications are in the petroleum refining and petrochemical industries and for the most part are relatively straightforward and will not be discussed. A few which are of special current interest will be described below.

Dissolved Gases in Transformer Oil. The power industry has long been interested in the gases present in large oil-filled transformers. Gases of interest include hydrogen, carbon monoxide, carbon dioxide, methane, ethane, ethylene, and acetylene. A process chromatograph modified with a special sparger in the sample valve permits injecting a large volume of oil directly into the column system. A dual column with stripper configuration permits determining all these compo-

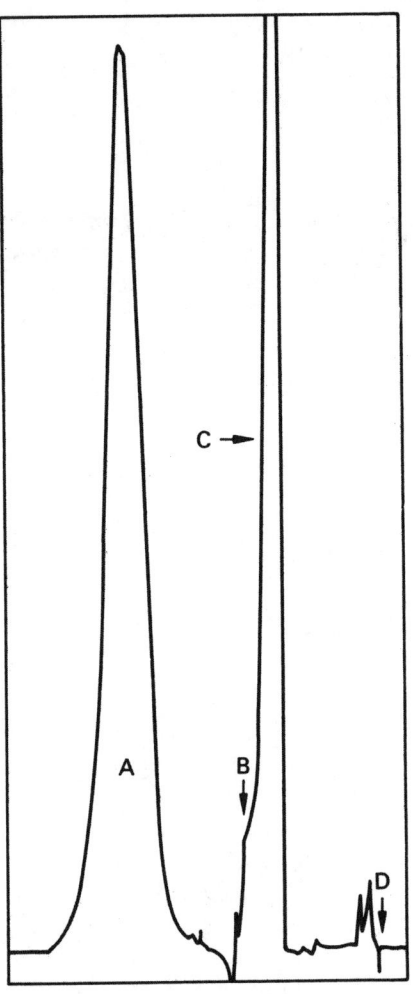

Figure 16. Determination of trace hydrocarbons in condensate

Backflush column configuration. Column, 5-ft Porapak Q, 120°C. A, Total hydrocarbons (45 ppm); B, backflush; C, water; D, sample inject

Summary

In the foregoing sections we have attempted to summarize the present state of the art and science of process gas chromatography—its hardware and technology—and to provide a few examples of their use. There is little doubt that process chromatographs are now indispensable instruments in the efficient operation of a refinery or petrochemical or chemical plant. Moreover, additional new applications are continuously being found in all areas of industry.

Chromatographs will also continue to increase in their capability and sophistication. Future efforts will be directed at increasing both reliability and accuracy to the maximum of which chromatographs are potentially capable. The recent explosive development in microprocessors undoubtedly will have an impact in that regard.

For the analytical chemist who is instrumentation oriented, this discipline will continue to provide an opportunity for creative efforts within the context of one of the more interesting fields of analytical instrumentation.

References

(1) P. R. Carl and J. G. Kerley, *Hydrocarbon Process.*, **47** (5), 127 (1968).
(2) E. A. Houser, "Principles of Sample Handling and Sampling Systems Design for Process Analysis," Instrument Society of America, Pittsburgh, Pa., 1972.
(3) "NFPA Handbook of the National Electrical Code," pp 425 ff, J. H. Watt, Ed., 3rd ed., McGraw-Hill, New York, N.Y., 1972.
(4) D. E. Durbin and J. Fruge, *Anal. Chem.*, **44** (8), 1502 (1972).
(5) R. Villalobos and R. L. Chapman, in "Air Quality Instrumentation," pp 114–28, J. W. Scales, Ed., Instrument Society of America, Pittsburgh, Pa., 1972.
(6) J. E. Osborne, *Chem. Eng. Progr.*, **70** (11), 76 (1974).

GC/MS/Computers

Francis W. Karasek, Department of Chemistry, University of Waterloo, Waterloo, Ont, Canada

Synergism effects are amply demonstrated by the combination of two powerful analytical instrumental techniques with computers. Application of these systems to chemical ionization ms and plasma chromatography can be expected

MANY YEARS AGO when one had to identify and analyze an unknown substance, there were well-established, workable laboratory procedures to use. Reaching the final answer, though, involved tedious, complicated, and time-consuming work. The first step was to determine if one were dealing with a mixture and, if so, to separate the components. Then would come considerable laboratory manipulation with the unknown and its reactions to determine the nature of the reactive entities in the molecule. Finally, the search would be narrowed by deduction to a certain type of compound; the validity of the deductions made needed proving by further reactions or by comparing properties with those of known pure compounds.

Today, for the single task of identification and analysis of organic compounds, the picture has been changed dramatically by combinations of new instrumental techniques. Analyses of unknowns that previously were impossible or extremely lengthy now yield to an afternoon's effort with microgram quantities. Such achievement has become so commonplace that the miracle occurs without fanfare. This situation has come about almost without conscious coordination of advances in the fields involved. Progress has been based largely on the development of instruments of great capabilities. Our purpose here will be to explore the capabilities of the techniques and instrumentation of gas chromatography, mass spectrometry, and the digital computer, when combined in various degrees and ways to reinforce one another. Although many cases involve the combination of two of these techniques, a remarkable synergism occurs when all three are integrated into a single unit. An unusual analytical performance emerges.

Over the recent past, work in these areas has been voluminous. Automation of the heavy routine analytical load of multichromatograph laboratories has been successfully accomplished by many different types of computer systems (1). However, the integrated gc/ms/computer is primarily a qualitative tool, and the key factor in its creation has been development of interfaced gc/ms systems. A detailed account of this development has been given by Watson (2). A most comprehensive review by Junk (3) of the massive information published on the gc/ms combination encompasses 850 references and organizes the reported results into topical summaries. Since the extent to which gc/ms has been used to solve analytical problems is adequately documented by Junk, only an outline need be presented here. It is the complex organic mixture that yields to the technique. In fact, the complexity of natural and biological mixtures provided much of the early incentive for development of gc/ms. It is not unusual for chromatograms of such mixtures to contain 100 or more peaks in which all components of greater than 0.5% concentration have been identified by gc/ms. Much early gc/ms work attacked the identification of odor and flavor components in food. Using high-resolution, high-capacity open tubular columns, compounds numbering in the hundreds have been identified (4). More recent work has involved identification of pesticides and other contaminants in the environment and the injurious organic components in complex engine exhaust mixtures. Applications in the biological and medical area are the most numerous of all: mixtures studied have included steroids, fatty acids, amino acids, carbohydrates, drugs, and drug metabolites.

Interfacing the Gc/Ms

In 1910 Thomson first separated masses of an element, and in 1919 Aston built his mass spectrograph. Mass spectroscopy as such has a long history, but it was not until the early 1940's that analytical mass spectrometers for the chemist appeared. Initially, these instruments were used for quantitative analysis until about 1952 when gas chromatography appeared and, with its great ability to separate and quantify components of a mixture, became the preferred method. Thus freed from routine analytical work, mass spectroscopists turned their attention to developing the qualitative, molecular structure determination aspects of mass spectroscopy. Such development is due to the pioneering work done on interpretation of mass spectra by many researchers, such as Bieman (5), McLafferty (6), and Budzikiewicz et al. (7). This interpretation is more empirical than theoretical so that structural assignments for completely unknown materials rest heavily on a strong background of empirical and semiempirical information or a comparison to established reference spectra.

Interfacing a gas chromatograph to a mass spectrometer is now a very widespread practice. There is a unique compatibility between these two instruments: the gas chromatograph separates the components of a mixture and delivers them one-by-one to the mass spectrometer for identification. This permits the identification of compounds present in quantities as low as 10^{-6} to 10^{-10} gram. Table I, comparing relative sensitivities of spectroscopic methods, shows the high sensitivity possible with the interfaced gc/ms.

The performance of the interface device determines to a large extent the type of results achieved by the entire gc/ms system (8). The device sits at a critical point and serves a single-minded purpose. It must remove as much of the carrier gas from a gc peak as possible and transport the maximum amount of the remaining organic material into the mass spectrometer ion source. How well a given device performs these functions is measured by an enrichment factor N, indicating the amount of helium removed from a gc peak, and a yield Y, indicating the percentage of sample that actually reaches the ms ion source. The interface must also perform the function of reducing the 760 torr pressure of the gc peak to the very low pressure (10^{-3} torr or less) required by the ms ion source.

Understanding the functions and performance of interface devices is important to the overall instrumental system. Although the many interfaces reported in the literature will vary greatly in detail and design, they can generally be classified into the four groups illustrated in Figure 1. The *direct coupled* uses a section of narrow-bore tubing to carry the gc column effluent into the ms ion source. Conditions must be adjusted to have the entire pressure drop (760 torr to 10^{-3} torr) occur across this connecting tubing. Various means are used to do this: splitter valves at the gc exit, adjustment of the gc to operate under specific flow conditions, restrictors, or use of a properly sized, auxiliary vacuum pump at the ms ion source. These direct systems in particular have a number of advantages of efficiency and simplicity that will be discussed in more detail later.

In the *effusive type* interface the gc effluent is forced to flow through a region where molecular flow conditions exist. This leads to selective removal of the carrier gas and creates the required pressure drop. Use is made of the fact that under molecular flow conditions the lighter molecules of the carrier gas move through orifices at faster rates. The Watson-Bieman example shown uses a porous glass fritted tube housed in an evacuation chamber. The pores in the frit, being less than 10^{-3} cm in diameter, provide the passages through which the carrier gas is selectively moved. Constrictions at the entrance and exit to the frit section are designed to give a viscous flow to the gc effluent.

Table I. Detection and Identification Limits for Analytical Methods

Method	Limit in grams	
	Detection	Identification
Gas chromatography	10^{-6}–10^{-12}	....
Infrared spectroscopy	10^{-7}	10^{-6}
Ultraviolet spectroscopy	10^{-7}	10^{-6}
Nmr (time averaged)	10^{-7}	10^{-5}
Ms (batch inlet)	10^{-6}	10^{-5}
Ms (direct probe)	10^{-12}	10^{-11}
Gc/ms	10^{-11}	10^{-10}

Other interfaces with this principle use porous stainless steel, silver, or Teflon in place of glass.

The *jet orifice* interface makes use of the properties of a jet of gas expanding as it moves toward an orifice. Mathematically and conceptually, it is very similar to the effusive type. The gc effluent flows through a nozzle opening from which it emerges into a lower pressure region as an expanding jet. A short distance from the nozzle creating the jet, and directly in line, is the exit orifice to the ms through which a gas stream enriched in the heavier component passes. Because of their lower forward momentum and greater diffusivity (according to Graham's law), the lighter carrier gas molecules are removed through the action of the vacuum pumps. Both single- and two-stage jet orifice interfaces have been used.

The *permeable membrane* interface employs a principle of separation quite different from that of the other types. Gc sample enrichment occurs because the organic component of the gc peak selectively passes through a thin elastomer membrane that exhibits a conductance to organic compounds 1000 times greater than it does to helium. The carrier gas flows over the membrane and on to the atmospheric exit, while the organic material passes through the elastomer film by a process of solution and diffusion to reach the ms ion source.

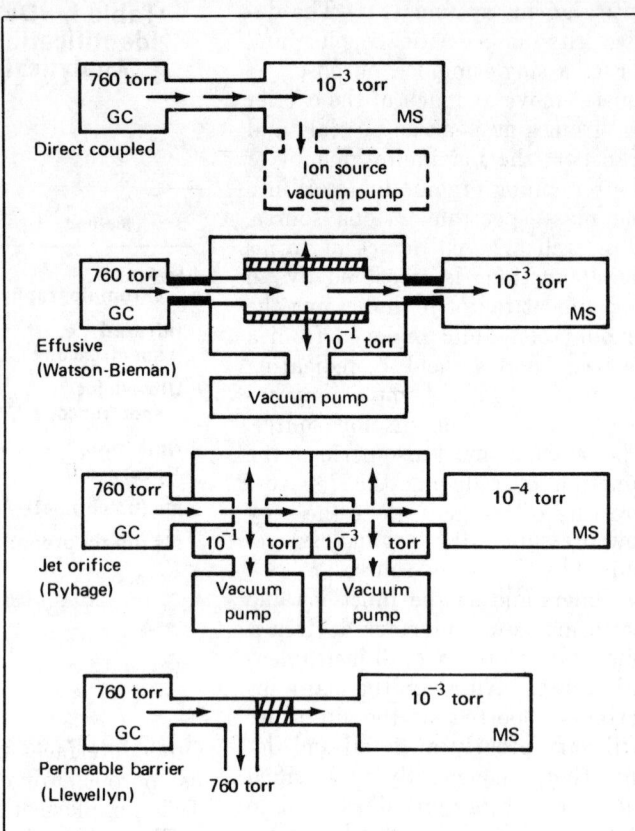

Figure 1. Gc/ms interfaces classified into four groups (8)

Courtesy Research/Development

Table II. Condensed Performance of Gc/Ms Interfaces

Classification	Efficiency N^a	Yield, % Y^b	Peak delay Sec	Peak distortion H^c	Effect on peak Inert	Carrier Gases
Perfect	∞	100	0	1	Yes	All
Jet	10^2	40	1	1–2	Yes	He, H_2
Direct	1	1–100^d	1	1	Yes	All
Effusive	10^2	50	1	1–2	Maybe	He, H_2
Permeable membrane	10^3	80–95	Variable	3	Maybe	Inorganic

$^a N = \dfrac{\text{conc in ms source}}{\text{conc in gc peak}}$

$^b Y = \dfrac{\text{quantity in ms source}}{\text{quantity in gc peak}} \times 100\%$

$^c H = \dfrac{\text{peak width in ms}}{\text{peak width from gc}}$

^{d} Equals split ratio.

parameters of different mass spectrometers and their coupled gas chromatographs. Successful use of any gc/ms interface system depends strongly upon a thorough understanding of vacuum technology. The variables of pressure and dynamic flow of gases are many and peculiar to each individual system. Chromatographs operate at a wide range of carrier flows, temperatures, and quantities of organic sample per peak. Mass spectrometers have different vacuum systems, flow conductances, ion sources, and mass selector designs. The connecting lines throughout the gc/ms system will contain valves, tubings, and orifices with individual effects on gaseous flow behavior.

A few selected principles and details that have often gone unappreciated in the gc/ms interface problem will reveal the limitations that a particular vacuum configuration imposes on various gc analytical situations. Exact computation of vacuum system behavior is difficult, but approximate calculations can be made that are adequate for the analysis of an individual gc/ms system (9, 10).

Mean Free Path. The mean free path of a gaseous particle is a measure of the average distance it will travel before it interacts with another particle. This interaction could be a direct collision with energy transfer, or particularly in ion-molecule systems, it may be a small deflection owing to two particles approaching within interacting distance. The value of mean free path is controlled by the parameters of velocity, collision diameter, and concentration and can be derived from kinetic molecular theory. Since molecules are moving at thermal velocities and ions are moving at velocities determined by accelerating potentials, and since collision diameters are different and largely unavailable, an exact calculation would not be practical. Instead, an approximate equation for mean free path of air,

$$\lambda = 5/p \quad (1)$$

where λ is in centimeters and p is the pressure in microns, gives results quite adequate for these calculations. The importance of estimating mean free path lies in the

High yields and efficiencies are possible with this simple interface. Single and double stages have been used.

Vacuum and System Considerations

Along with a description of principles by which these interface devices function, one needs an understanding of factors that affect the results produced. In addition to yield and efficiency, such factors as time delay, inertness, gc peak distortion and the carrier gases, and gc operating conditions, are important. Table II tabulates such factors.

A simple understanding of interface operating principles and performance is deceptive. Each interface has its own particular limitations which must be matched and balanced against the operating necessity of having it be significantly greater than any dimension in the ion source or the mass spectrometer ion flight path for proper performance. It also must be known to determine whether molecular or viscous flow conditions exist.

Viscous Flow. The exact nature of the flow of gas through a tube or orifice depends upon the pressure or molecular density of the gas. At high pressures the molecules undergo many interactive collisions with each other and hence move in a bulk fashion. Viscous flow requires a mean free path of the molecules much smaller than the diameter of the orifice or tube through which they are flowing and is related to the viscosity of the gas. This flow exists when $pD > 500$, where p is the pressure in microns, and D is the conduit diameter in centimeters.

Molecular Flow. When the molecular density is so sparse that few collisions occur between molecules, they will move independently of each other in a manner controlled by their thermal velocities and molecular weights. Molecular flow through an orifice or tube occurs when the mean free path of the molecules is much greater than the diameter involved. Then molecules only collide with the tube or orifice walls, not each other, and move independently at rates related to their molecular weights. Molecular flow will exist when $pD < 15$, where the terms are as defined for the corresponding rule for viscous flow.

Conductance. In vacuum terminology it is useful to express flow resistance as a conductance, which indicates the volume flow of gas that passes through an orifice or conduit. It appears in the equation

$$Q = C(p_1 - p_2) \quad (2)$$

where Q is the quantity of gas in l. torr/sec, C is the conductance in l./sec, and $p_1 - p_2$ is the pressure drop in torr.

In the high vacuum of a mass spectrometer, molecular flow conditions usually exist. A simple working equation for conductance C in l./sec for orifices in such systems is

$$C = 9.1(28.7/M \cdot T/273)^{1/2}D^2 \quad (3)$$

where M is the molecular weight, T is the absolute temperature, and D is the diameter of the aperture in centimeters.

The equation for conduits is

$$C = 12.1(28.7/M \cdot T/273)^{1/2}D^3/L \quad (4)$$

where L is the length in centimeters, and other terms are as defined in Equation 3. For an L/D greater than 20, the equation holds to better than 10%.

A calculation of conductance for viscous flow in a conduit can be made by using Poiseuille's law

$$C = 3.3 \times 10^{-5}D^4p/\eta L \quad (5)$$

where η is the viscosity in poises, p is the pressure in microns, and the other terms are as previously defined.

While it is not readily apparent from the equations, it is important to realize that for either viscous or molecular flow conductance is independent of pressure drop. This is best illustrated by Figure 2 which shows the conductance of tubes 1 m long with various diameters as a function of pressure. The behavior of the gas falls into the two regions of viscous and molecular flow separated by an intermediate region. At higher pressures the conductance is dependent upon pressure since molecules are pushed down the tube by collisions with each other. Since they are independent of each other at lower pressures, conductance is independent of pressure there.

Under molecular flow conditions conductance can never exceed the volume throughput permitted by thermal velocity of the molecules and the area of the conducting orifice or tube. A simple example will illustrate this important point. Suppose gas is flowing through a 6-cm^2 opening where complete removal is assumed once it moves into the infinite capacity pump. At room temperature, root mean square velocity of molecules (molecular weight equals approximately 100) will be close to 10^4 cm/sec, but since only a fraction of the molecules will have a component of velocity leading them out of the chamber, this factor must be modified by that fact. The volume of gas that will pass through the orifice is thus given by the product of this velocity factor and the area of escape. The conductance of the orifice can never

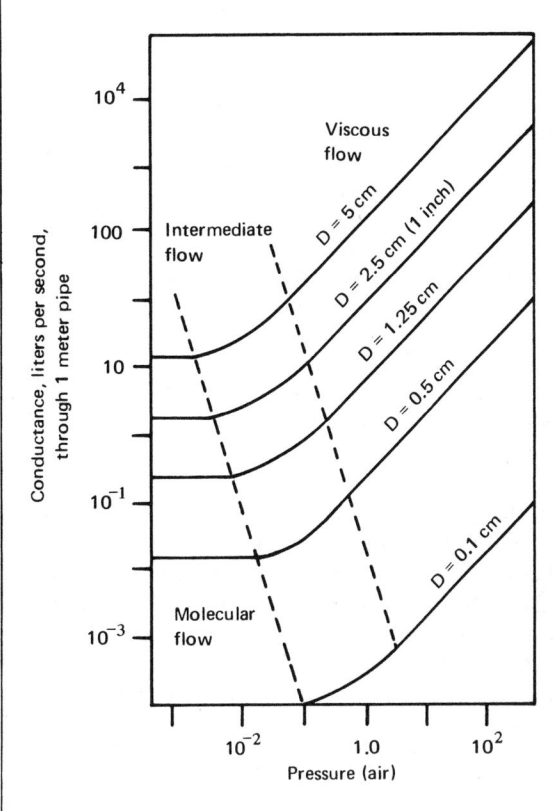

Figure 2. Conductance through tubes under viscous and molecular flow conditions (8)

Courtesy Research/Development

exceed 60 l./sec, irrespective of the pumping speed of the pump removing the gas. If the pump capacity were not sufficient to maintain a pressure significantly lower than that in the chamber containing the orifice, the actual conductance would be less by the amount of gas that diffused back into the chamber.

This molecular concept of conductance through an orifice can be used to estimate the pressure of sample in the ms ion chamber. Since the conductance is directly proportional to the area of the openings, it is only necessary to know the size of all openings in the chamber (electron beam slits, ion exit slit, solid probe entrance, etc.). Knowing the quantity of gas, Q, entering from the gc, the pressure in the ion chamber can be calculated from these conductances by the relationship $Q = Cp$.

The conductance or pumping speed of a system can be calculated from the individual conductances of each component. If components of conductances C_1, C_2, C_3, etc., are connected in series, the net conductance C will be given by

$$1/C = 1/C_1 + 1/C_2 + 1/C_3 + \ldots \quad (6)$$

Conductances in parallel are merely additive. It is also common practice to consider that a right-angle bend in a tubing is equivalent to increasing length by two diameters.

Another concept needed for analysis of vacuum systems is the speed of a vacuum pump. The mechanical and diffusion pumps now in use have a pumping speed relatively constant over a wide range of pressure. In a mechanical pump an eccentric cylinder sweeps out a volume of gas with each rotation. Its speed is the product of that volume and the rotational frequency. In a diffusion pump a constant speed exists which is determined by the velocity of the oil crossing the jets. The defining equation for pumping speed is $S = Q/p$, which resembles that for conductance. Pumping speed and conductances may be defined in the same units and be numerically equal, but they are never equivalent in meaning. Conductance implies a pressure gradient and is a geometrical property of impedance. Pumping speed is applied to any plane in the system that can be considered a pump with an ability to remove gas. Pumping speed can be seriously reduced if connecting elements have inadequate conductance. By referring to Figure 2 and the foregoing equations, one can see how seriously even a short section of narrow bore tubing will reduce the speed of a pump.

Application of Principles

The problem in the gc/ms technique is to place as much of the organic material into the ms ion source as possible without exceeding the vacuum requirements of the mass spectrometer. Vacuum requirements are twofold: those in the ion source chamber and those in the ion flight section of the spectrometer. Pressure must be kept below 10^{-2} torr as an upper limit in the ionization chamber to avoid the occurrence of ion-molecule reactions that lead to unrecognizable fragmentation patterns, to prevent formation of excessive numbers of charged particles that affect the ions being drawn into the spectrometer, and to avoid interference with passage and normal regulation of the electron beam creating ions. A most critical vacuum requirement lies in the analyzer section of the spectrometer because there the ions move down a fairly long path in a highly focused beam. Here, the mean free path must be significantly greater than 200 cm, corresponding to a pressure of 10^{-5} torr, to avoid scattering of the beam and degradation of ms performance. This factor is most demanding in determining the amount of gc effluent that can enter a mass spectrometer.

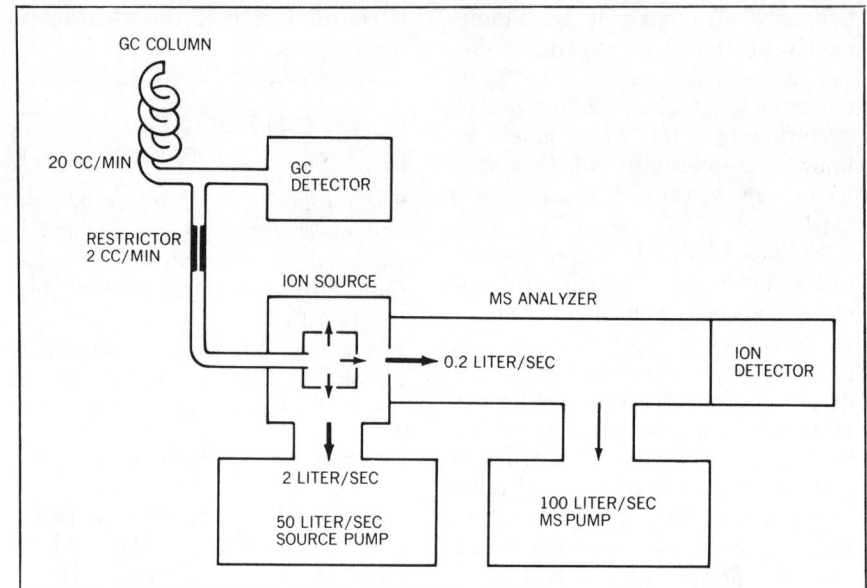

Figure 3. Pressure and flow analysis of gc/ms system shows existence of satisfactory conditions

Figure 3 illustrates a gc/ms system that can be appraised to see how well it meets the requirements outlined; the figure also serves to illustrate the application of these principles to other specific systems (11). A 20 cm^3/min flow from the gc is split by a restrictor to allow a 2.0 cm^3/min flow to enter the ms ion source. This gives a value for the quantity Q of 2.5×10^{-2} l. torr/sec. The analyzer section is being supplied gas only by the small exit slot from the ion chamber to the mass spectrometer with a conductance of 0.2 l./sec. The rest of the gas is removed via slits for the electron beam and other openings with a typical conductance of 2 l./sec. Calculations can be made according to the previously given equations to give the data presented in Table III to show that the operation of all components appears reasonably satisfactory. The ion source pressure is somewhat high. The source pump is operating near its limit and does not have capacity for much more gc effluent. Without the source pump, the gc effluent would have to be removed by the 100-l./sec pump in the mass analyzer tube, and only $1/10$ the amount of effluent could be admitted and still maintain the low pressure and 200-cm mean free path needed there.

Direct Coupled Gc/Ms. Recent systems have appeared that have been designed with differentially pumped ion sources to accept the total gc effluent (*12, 13*). They appear to provide many advantages: 100% yield, high efficiency, undistorted gc peaks, and tolerance to changing gc conditions. The case for differential pumping of the ion source is illustrated by the example discussed in Figure 3. If both the source pump speed and ion source conductance were increased by a factor of 10, the respective pressures would remain the same, and the gc effluent entering the ion chamber could be increased tenfold. This step requires altering the closed ion source design of most magnetic mass spectrometers and becomes easily feasible with the open ion source designs of such instruments as the quadrupole and time-of-flight.

When the ion source design has straight-through flows and proper conductances, an additional factor comes into play to increase the sample to helium ratio. Since the gc effluent enters a narrow bore tubing at atmospheric pressure and moves down this tube to the ms ion source, it must necessarily enter the ion source in a viscous flow fashion. Transition to molecular flow occurs quickly in its passage through the ion chamber and out the source pump. Since the helium carrier molecules move faster under these conditions, the ratio of an organic material of molecular weight 250 to helium in a gc peak will be increased on the average by a factor of 8. If conductances and the source pump speed are not designed adequately to keep the ion source pressure down to the molecular flow region (below 10^{-3} torr) at the maximum gc effluent admittance, then viscous flow conditions will exist and this increase will be lost.

Recent work with the Bendix TOF mass spectrometer indicates that it is also important to achieve a molecular beam flow at this point (*13*). In addition to a high capacity pump (500 l./sec) at the ion source, the gc effluent flow must be directed through the ionization region. High sensitivity is achieved for steroid hormones (*14*) and more recently for the common drugs indicated in Table IV.

Computerization of Gc/Ms

An operating gc/ms system generates an incredible amount of data. For a complex mixture yielding 100 gc peaks during an hour's run, 5 mass scans of 500 mass units for each gc peak result in 250,000 data points of mass position and ion intensity to be recorded and measured. Normalization of each spectrum for interpretation, correlation of the data with points on the chromatogram, and extraction of other interpretive data increase the effort required. This situation clearly points to the need for use of the data handling and computational capabilities of a digital computer. Without its use, only a small amount of this valuable information is acquired, and even then at considerable expense of time and manpower. A variety of approaches to computerization have been developed involving computers

Table III. Calculated Parameters for Gc/Ms System

	Q entering, l. torr/sec	Conductance out, l./sec	Pressure, torr	Mean free path, cm
Ion chamber	2.5×10^{-2}	2.2	1.1×10^{-2}	0.5
Source pump	2.2×10^{-2}	50	4.4×10^{-4}	Marginal
Ms analyzer tube	2.2×10^{-3}	100	2.2×10^{-5}	200

Table IV. Mass Spectral Data of Common Drugs by Means of Bendix MA-2/GC Gas Chromatograph/Mass Spectrometer System

Indexed by molecular weight

Compound name	Class[a]	Molecular wt	1st Peak	2nd Peak	3rd Peak	4th Peak	5th Peak	6th Peak	Molecular ion seen	Common name
Methyprylon		183	55	83	98	140	155	69	Yes	
Barbital	S	184	156	55	141	69	98	83	No	
Caffeine	St	194	67	55	109	194	82	42	Yes	
Glutethimide	S	217	117	132	91	189	115	51	Yes	Doriden
Butabarbital	S	224	156	41	141	57	69	157	No	
Amobarbital	S	226	156	41	141	42	55	157	Yes	
Pentobarbital	S	226	156	141	157	43	41	55	Yes	
Phenobarbital	S	232	204	117	118	51	232	77	Yes	
Secobarbital	S	238	168	167	43	41	97	124	Yes	Seconal
Mepiridine	N	247	71	70	57	247	91	103	Yes	Demerol
Promazine		283	58	86	283	199	198	237	Yes	
Diazepam	T	284	283	256	281	257	284	285	Yes	Valium
Pentazocine	An	285	217	285	70	69	110	202	Yes	
Imipramine		280	58	85	235	234	280	193	Yes	Tofranil
Chlordiazepoxide	T	299	282	281	283	284	77	220	No	Librium
Chlorpromazine	T	318	58	318	86	231	319	272	Yes	
Flurazepam		386	86	87	58	99	386	183	Yes	Dalmane

[a] N, narcotic; An, analgesic; S, sedative; St, stimulant; T, tranquilizer.

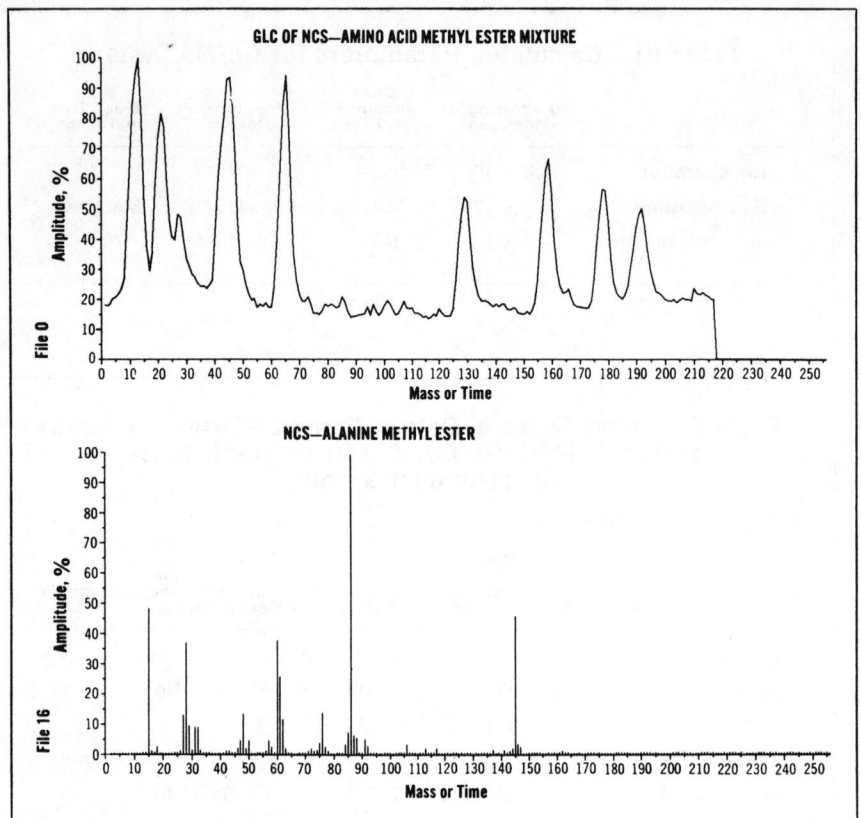

Figure 4. Data printouts from gc/ms computer-controlled system. For top curve, each unit on abscissa represents 5 sec. One spectrum was taken during each such unit. Ordinate is sum of all mass peaks or total ion current. Bottom curve is spectrum of 16th time slot, first glc peak

of all sizes, both off-line and dedicated completely to the system. Many of the sophisticated systems have originated in academic laboratories where the importance of the research and its exploratory nature permitted the expense involved. The more versatile and capable the system, the greater the cost, which limits its widespread use and forces compromises involving semiautomatic readout and data reduction. Computer hardware and software developed by computer companies economically tailored to different levels of gc/ms applications are now becoming available (3, 15).

The type of data produced by an integrated system is shown in Figure 4. Mass scans made as the chromatogram evolves are stored by the computer, which can then reconstruct the chromatogram by plotting the total ion current stored at each of the many points in the run. Identification of a gc peak is accomplished by printing and interpreting the normalized mass spectrum corresponding to the gc peak point on the chromatogram. Mass calibration procedures are eased, and the mass spectrum can be accurately plotted to ±0.2 amu. This single function removes much drudgery and inaccuracy involved in hand-counted mass spectra. Not only spectral plots, but plots with the background removed can be obtained. To further aid component identification, a mass chromatogram is produced which plots the points along the chromatogram at which an ion of a specific mass occurs, normalized relative to the largest intensity. Communication between operator and the system is through the teletype and plotter.

With all the aid a computer provides, interpretation and identification of unknown components in complex mixtures are still very difficult problems. Should the spectrum of an unknown correspond to that of a compound compiled in any of the several available catalogs of references, one is then faced with the practical problem of searching for a match and deciding on the criteria to use for positive identification. Computer search programs for this purpose have been developed, but reference spectra available are limited. The largest tabulation contains only 17,000 spectra of 8 peaks each (16). Mass spectral patterns are subject to many variables that must be considered when using computer or other search-match methods. Temperature of ionization, the type of mass spectrometer, and ionizing voltage all influence the pattern of ions observed for a compound (17). Consequently, efforts to program intelligent fragmentation rules into computers to deduce structure (18) or to conceive a matching system insensitive to ionic intensities are desirable. Grotch (19) reports such an approach to pattern recognition that involves the presence or absence of a peak above a specified threshold at each nominal mass over the range from 12–200. However, at present, interpretation and identification rest heavily upon the skill of the mass spectroscopist and his ability to extract the maximum amount of information from the combined instrumental components. Hence, computer systems must provide an easily executed dialogue between the operator and data.

Gc/Ms/Computer Units Commercially Available

All of the major manufacturers of mass spectrometers offer a gc/ms interfaced unit. Junk (3) tabulates a total of 19 different units available from nine companies at costs ranging from $20,000 to $125,000. Most of these companies also have an integrated gc/ms/computer system. In general, computerization adds about 50–100% to the cost.

The computerization of a quadrupole gc/ms developed by System Industries illustrates a number of aspects of these units. Since completion of their first system in 1968 (about 20 have now been installed), over 10 man years of computer and systems development have gone into making the unit (Figure 5) truly workable by the analytical chemist (20). This is not an uncommon effort and is in addition to the extensive work done by Reynolds et

Figure 5. Gc/ms/computer unit developed by System Industries with Finnigan 1015 quadrupole mass spectrometer and PDP-8 computer

Courtesy System Industries, Sunnyvale, CA

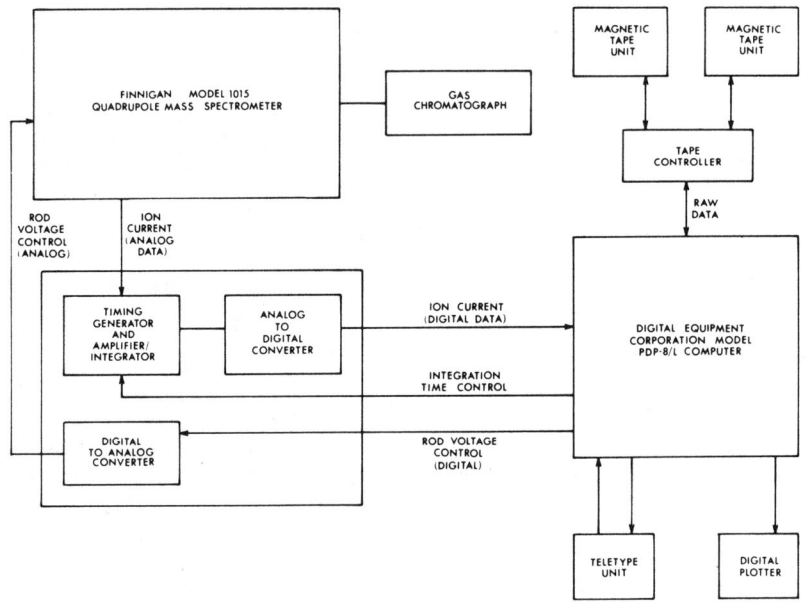

Figure 6. Functional diagram of computer-controlled system shown in Figure 5

Courtesy System Industries, Sunnyvale, CA

> **Table V. Programs and Functions Available to Operator for Gc/Ms/Computer System**
>
> **CALIBRATE:** Creates an accurate N Table. The N numbers which correspond to the peaks in the reference gas are used as the end points of a piecewise linear interpolation procedure for calculating a complete N Table.
>
> **COLLECT:** Is the primary data collection step. Here, the 750 N Table values are sent to the ms, and the 750 m/e intensities recorded. This operation can be repeated at 5-sec intervals as the data are filed on disk under an experiment name.
>
> **TYPE:** Allows the user to print out spectral data by indicating which spectra in a given file are to be reviewed. The user can request that the amplitudes of particular m/e positions be typed, that a given number of the highest amplitudes be typed, or that a consecutive number of them over a given range be typed.
>
> **PLOT:** Enables the user to have bar graphs produced by the computer-controlled digital plotter. The amplitudes to be plotted can be selected with the same flexibility as described in TYPE.
>
> **SUM:** Produces a plot of the total ion current over a series of gathered spectra. All responses of a spectrum are summed to produce one datum point on the plot. This plot corresponds closely with the glc output when running with the glc.
>
> **TRACE:** Produces a record of a spectrum similar to the normal chart recording output. The analyzer is sampled at all N values (about 10/amu) over a given range, and the result is plotted as a "broken line" (used for system check out).
>
> **MONITOR:** Provides for inspecting the peak profiles by sampling the spectrum around a given m/e position. The gathered data are then typed out (normally used for system service or service log).
>
> **DISPLAY:** Enables the user to display a given mass position (or N number) in the center of the console oscilloscope (used in the adjustment of the mass spectrometer).

al. (21) at Stanford University in originating and developing the concept through their use of three mass spectrometers, three computers, and two basic computer programs. The functions listed in Table V by Reynolds are characteristic of those found in many such systems. A significant aspect of this concept shown in Figure 6 is computer control of the mass spectrometer. A digital signal from the computer selects each mass position and allows the ion signal to be integrated there for a chosen period of time. By use of a variable integration period over a mass range, one is assured the best signal-to-noise ratio at each mass and enhanced sensitivity. A spectrum can be repeated every 5–10 sec and consists of all integer m/e values from 1–750 or any selected partial sequence.

The Du Pont (CEC) gc/ms uses its small double-focusing magnetic ms in a system based on a Hewlett-Packard computer with a digital tape storage unit and software worked out by spectroscopists (12). Across a chromatogram, 3000 mass scans can be run and stored. Ions reaching the mass spectrometer detector are counted and delivered as a digital signal directly to the computer. The result is a simpler system with a greater

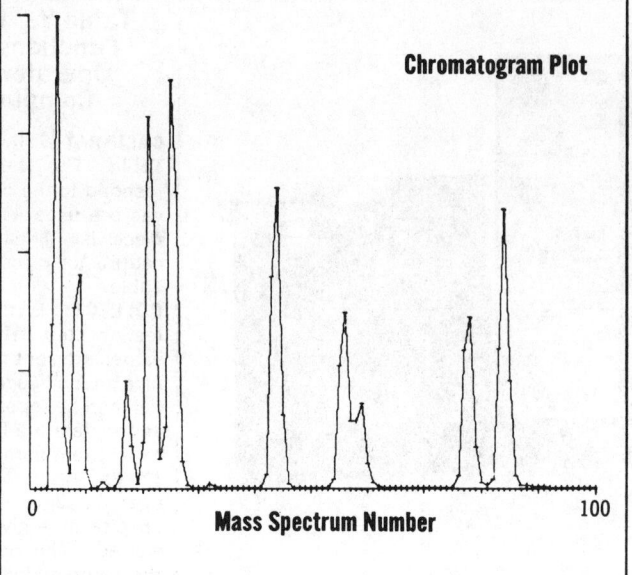
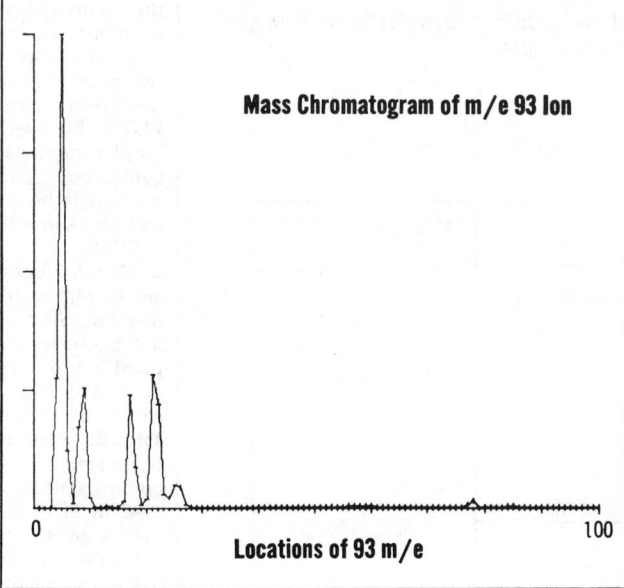

Figure 7. Data plotted by Du Pont gc/ms/computer system for mixture of natural products. Mass chromatogram (bottom) shows location of all m/e 93 ions, a diagnostic aid for monoterpene hydrocarbons

Courtesy Research/Development

plexity of the systems and increasing their reliability. Much yet needs to be done in compilation of readily accessible reference spectra and in developing schemes of automated structural retrieval from fragmentation patterns. We can expect the techniques of gc/ms/computer to be extended to include related methods, such as chemical ionization ms and plasma chromatography. These two comparable methods, based on ion-molecule reactions and high-pressure operation, provide complementary data to aid interpretation of mass spectra. We can look forward to these future developments with optimism.

dynamic range than the usual system of conversion from analog to digital from an electrometer output. Data from the system are illustrated by the reconstructed chromatogram and mass chromatogram shown in Figure 7.

Many integrated systems use one of the DEC (Digital Equipment Corp.) PDP-computers. Based on use of one of their PDP-12 computers, DEC engineers have devised a system of hardware and software generally applicable to any type of mass spectrometer. Entitled MASH (Mass Spectrometer Handler), it not only produces all of the gc/ms data processing needed but has several very significantly useful capabilities. Unlike many such dedicated computer systems, it is set up to work with other laboratory instruments such as gc, nmr, and epr. A most attractive function is the active CRT display of the data. Rather than using a teletype or plotter, the operator can call up any data onto the CRT display and then perform any desired programmed function. Once a needed result is achieved, it can be output on an X-Y plotter or teletype. The computer unit with the CRT display is shown in Figure 8.

Conclusions

The results of considerable activity in developing the analytical aspects of combining these instrumental techniques are now available. The analytical capabilities of a truly integrated system are impressive with the many solutions to complex problems reported. Much current effort needs to be expended toward reducing the cost and com-

References

(1) "Chromatography and Computers," Collected Papers Symp., 162nd National ACS Meeting, Washington, DC, September 1971; *J. Chromatogr. Sci.*, **9** (December 1971); *ibid.*, **10** (January 1972).
(2) J. T. Watson, "Ancillary Techniques of Gas Chromatography," L. S. Ettre, W. H. McFadden, Eds., Wiley-Interscience, New York, NY, 1969, p 145.
(3) G. A. Junk, *Int. J. Mass Spectrom. Ion Phys.*, **8**, 1 (1972).
(4) F. W. Karasek, *Res./Develop.*, **20** (10), 74 (1969).
(5) K. Bieman, "Mass Spectrometry: Organic Chemical Applications," McGraw-Hill, New York, NY, 1962.
(6) F. W. McLafferty, "Interpretation of Mass Spectra," W. A. Benjamin, Inc., New York, NY, 1966.
(7) H. Budzikiewicz, C. Djerassi, and D. H. Williams, "Mass Spectrometry of Organic Compounds," Holden-Day, Inc., San Francisco, CA, 1967.
(8) F. W. Karasek, *Res./Develop.*, **22** (9), 52 (1971).
(9) A. Guthrie and R. K. Wakerling, "Vacuum Equipment and Techniques," McGraw-Hill, New York, NY, 1949, p 1.
(10) R. B. Scott, "Cryogenic Engineering," Van Nostrand, Princeton, NJ, 1959, p 155.
(11) F. W. Karasek, W. H. McFadden, and W. E. Reynolds, "GC/MS/Computer Techniques," ACS Short Course Publication, December 1970.
(12) F. W. Karasek, *Res./Develop.*, **22** (5) 51 (1971).
(13) D. C. Damoth, Bendix Corp., Rochester, New York, NY, private communication, 1972.
(14) H. M. McNair, C. R. Dobbs, L. H. Aung, D. C. Damoth, and A. J. Luchte, "Analysis of Steroid Hormones by a New Direct GC/MS Interface," 19th Annual Conf. on Mass Spectrom., Atlanta, GA, May 1971.
(15) "Mass Spectrometer Handler (MASH), User's Manual, DEC-12-SQ2A-D," Digital Equipment Corp., Maynard, MA, 1971.

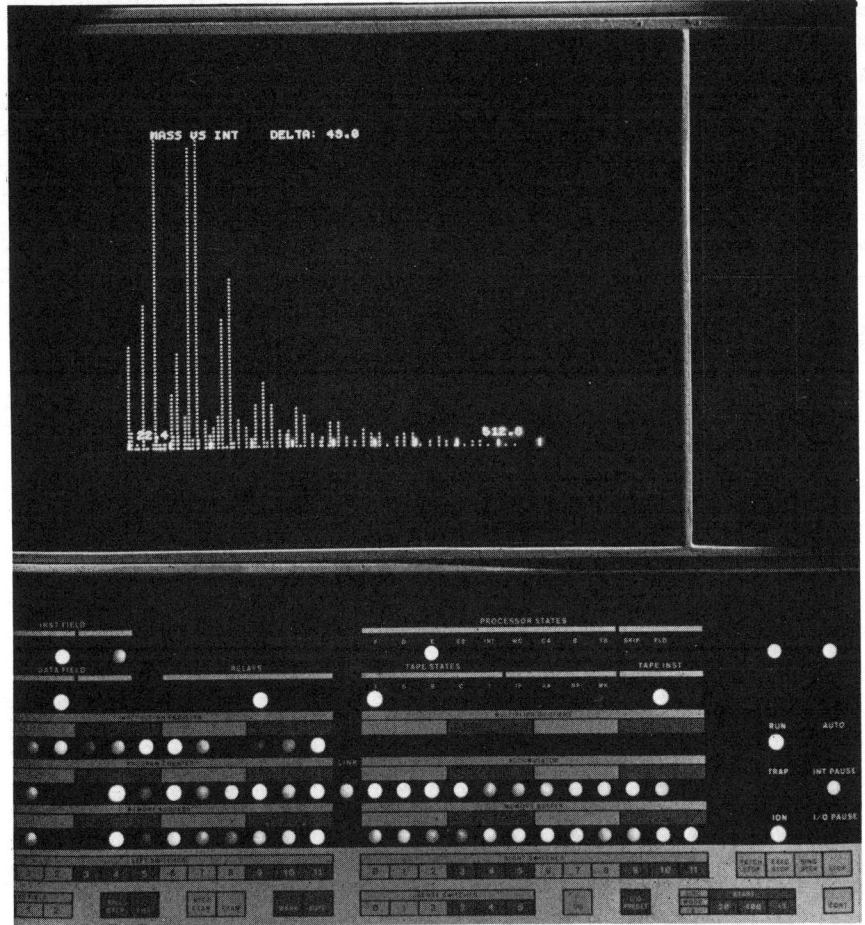

Figure 8. Control console of PDP-12 gc/ms/computer system showing CRT display

Courtesy Digital Equipment Corp.

Francis W. Karasek *is Associate Professor of Chemistry at the University of Waterloo in Waterloo, Ont, Canada. Dr. Karasek received his BS degree in 1942 from Elmhurst College and his PhD degree in 1952 from Oregon State University. For the next 16 years he led an automatic analytical instrumentation development group at the Phillips Petroleum Research Center. That work has resulted in many analytical instruments in direct control of chemical processes. In 1968 he joined the faculty at Waterloo, where his research interests are analytical instrumentation, gc/ms, and plasma chromatography. As an editorial associate for Research/Development, he writes a monthly article on new analytical instrumentation. He originated the Graduate Fellowship program of the ACS Analytical Division, is Councilor for the Division, and is Professor-in-Charge of the ACS Short Course on gc/ms/computers.*

(16) "Eight Peak Index of Mass Spectra," 1st ed., Vol I, II, British Information Services, 845 Third Ave., New York, NY (ASTM, Philadelphia, PA), 1970.
(17) F. W. Karasek, *Res./Develop.*, **21** (11), 55 (1970).
(18) F. W. Karasek, *ibid.* (4), 75 (1970).
(19) S. L. Grotch, *Anal. Chem.*, **43**, 1362 (1971).
(20) D. Kent Winton, System Industries, Sunnyvale, CA, private communication, 1971.
(21) W. E. Reynolds, V. A. Bacon, J. C. Bridges, T. C. Coburn, B. Halpern, J. Lederberg, E. C. Levinthal, E. Steed, and R. B. Tucker, *Anal. Chem.*, **42**, 1122 (1970).

Computerized

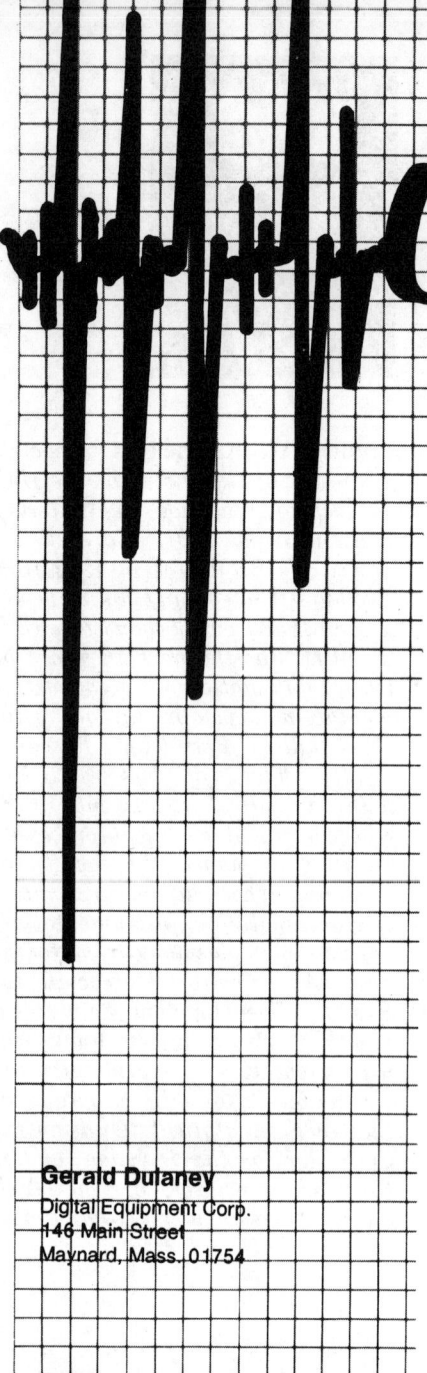

Gerald Dulaney
Digital Equipment Corp.
146 Main Street
Maynard, Mass. 01754

The design of computer systems for processing experimental data from chemical experiments is an area of growing significance. The wide variety of instrumentation and computer equipment available can cause uncertainties in the choice of optimum equipment for laboratory automation. This paper defines a common framework for signal processing and considers the requirements of computer interfacing to and processing of data from chemical instrumentation. Strong emphasis is placed upon cost effectiveness

The term signal processing is widely used to cover aspects of data handling, especially in connection with minicomputers. Here, signal processing shall be taken to be the reduction of measurable physical phenomena to a specific coordinate set. By extension it is assumed that the experimental data, whether analog or digital, produced by the experimental apparatus are capable of being accepted by the automation system under consideration and converted into a set of digital information by it.

Generally, any experiment can be considered as a superset of two-parameter measurements. Although a given experiment may embody multiple-parameter measurements, the resulting data can be converted to sets of x-y data; that is, two-parameter sets. These resulting x-y data sets can be presented and manipulated in a number of ways to obtain the desired multiparametric presentations and analyses. Therefore, one must consider the methodologies of converting the data output from the conventional laboratory instrumentation into a format acceptable to the minicomputers or microcomputers—the process of interfacing—and the subsequent processing of these data into specific coordinate sets.

For example, consider the possible automation of a mass spectrometer. Typical data would be ion-current intensity, indirect mass measurements (time), and the acceleration potential. This three-parameter coordinate set would ultimately be reduced to the two-parameter set of mass vs. intensity.

Other examples could be drawn from optical spectrophotometry, where analog data (y-axis) may be transmittance, absorbance, or optical density; and the x-axis measurements might be time, wavelength, frequency, etc. As before, the final data set will usually be defined in terms of some intensity parameter plotted against some x-axis parameter.

Thus, signal processing applications will best be defined in terms of readily automated two-parameter measurements. This will be taken as a design criterion throughout.

Common Framework for Computerized Signal Processing

Usually, automation systems are thought of as being composed of a data acquisition portion and a data reduction portion. However, this is too limited a division, and the framework presented here will permit definition of five serial phases of operation, incorporating all common aspects from analog signal acquisition and conversion through final manipulation and reduction. The five aspects usually required for the widest variety of instrument automation are: data acquisition and analog-to-digital conversion; digital preprocessing; procedural reduction; manipulation and transformation; and postoperative amalgamation and interpretation.

Data Acquisition and Analog-to-Digital Conversion. Data acquisition is the extraction of information from measurable physical phenomena and the conversion of that information into computer-compatible data. This aspect is what is classically considered to be the experiment, with the addition of the conversion of the output data into a form acceptable to the computer. The data source (instrument) may present the data in analog form (such as voltage), although it is becoming more popular in modern laboratory equipment to output the data in digital form. This is one instance of the impact of computers on instrumentation.

Aspects of Data Acquisition. The rate at which data can be acquired is an important consideration. The acceptable data rate for an automated system depends both upon the type of instrumentation and upon the characteristics of the computer system to be used. As a first approximation, data rates may be divided into four catego-

Signal Processing

ries:
- Slow data rates—fewer than 100 samples/sec. Such rates are typical of a wide variety of laboratory instrumentation.
- Medium data rates—from 100 to 1000 samples/sec. Such rates are associated with instrumentation such as that used for nuclear magnetic resonance (NMR), electron paramagnetic resonance (EPR), et al.
- High data rates—from 1000 to 50,000 samples/sec. Such data rates are encountered in mass spectrometry and Fourier transform NMR.
- Very high data rates—beyond 50,000 samples/sec. These are not relevant to the discussion.

The number of instruments that a computer system can handle depends upon the sample rate requirements. In low and medium data rate situations, a computer can handle several instruments at the same time. However, for high data rate applications, a computer system is usually dedicated to the servicing of the instrument producing the high data rates during the period of its operation. That is, the entire operation of the computer is restricted solely to that one instrument producing the high data rate samples. Without quite sophisticated (and usually costly) hardware, it is awkward and often impossible to require a computer system to operate simultaneously with both low and high data rate instrumentation.

Data rates can be specified from the frequency content of the output signal of the instrumentation. In many cases, this can be inferred from the experiment or the phenomenon being measured. For example, if the output of the instrumentation takes the form of peaks, the sample rate can be estimated as a function of the peak width. In general, the sampling frequency should be taken to be at least twice the frequency defined by the inherent characteristics of the analog output signal to prevent loss of information. Before the optimum sample rate can be adequately determined, it is desirable to understand exactly how the data being obtained will ultimately be processed. In the case of determining peak area or signal area, at least 20 samples over each significant peak or phenomenon would be required to make accurate calculations by means of numerical integration techniques.

Noise is another parameter that must be considered in the determination of the best data acquisition approach. There are a variety of noise types, several of which are always present. These may be categorized as follows:
- Line frequency noise (or mains frequency noise)—this is primarily the fundamental and harmonic frequencies of the alternating current used to power the instrument. It is almost always present in any analog output.
- Transmission noise (or pickup noise)—this is usually created by radio interference (RFI), electromagnetic interference (EMI), etc. When present, it is usually picked up in the cabling or circuitry between the output of the instrument and the computer or conversion device.
- System noise—it is reasonable to assume that the instrument itself generates an output noise, as a function of the phenomenon being measured and/or as a function of the instrumentation electronics. An important subset of system noise is interfacing or conversion noise; this can arise from the interfacing and/or conversion techniques used. Equipment component manufacturers have tried to insure that interface device quality will be sufficiently high to prevent the inclusion of noise from the interface. However, prospective users should always consider the possibility of this noise when planning an application setup.

There are several reasonable techniques for eliminating the effects of noise within a laboratory automation system. One technique that almost invariably produces improvements is the analog signal filter. The choice and design of such filters are much discussed in the literature. Such a filter will pass only signals with a frequency less than a specified amount (cutoff frequency). The cutoff frequency is typically chosen to be 2–10 times the sample rate being used. In practice, it is often convenient to have the actual filter accessible and modifiable by the user depending upon the application.

Additional advantage may be gained by using software filters—digital or numeric filtering techniques employed within the computer's program itself—in addition to analog filters. These software filters, which can assume many forms, have been discussed in the literature. Such filtering techniques will be considered in more detail subsequently.

Sometimes, a third technique can be employed to improve the signal-to-noise ratio (S/N). This technique can be used in situations where a series of equivalent (repeated) signals is obtained over a period of time. Here, the data set is acquired and stored temporarily. Another equivalent set is acquired, added to the first, and the sum stored. This add-and-store process is continued for some specified period of time. In the final set of data the random or noise element will have been lessened by the averaging of the ensemble of data sets. This technique is known as signal averaging or computed average transients (CAT) and has been widely employed in many applications. This technique can be performed by hard-wired nonprogrammable devices or can be implemented with computers.

Optimally, noise is minimized by a combination of the above discussed considerations. Choice of components with the least inherent noise is important. Good technique in analog signal handling and cabling is essential. Line frequency noise and RFI/EMI radiation can be decreased by the use of twisted-pair and shielded cable. This is further discussed in the section Data Transmission.

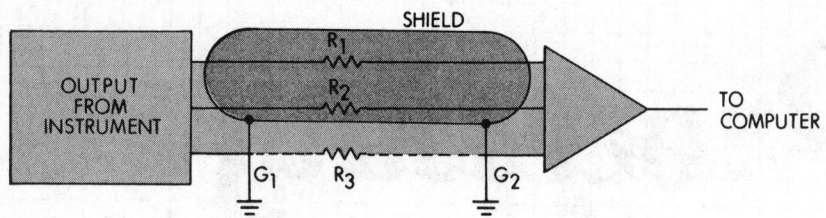

Figure 1. Ground loop phenomenon
Leads running from output from experimental instrumentation to interfacing input to computer can be subjected to ground loops. Let us consider a shielded two-line cable: actual lines have characteristic impedances of R_1 and R_2; effective impedance of shield is R_3. Instrument and computer system are grounded through G_1 at instrument and G_2 at computer input. R_3 is much, much lower than R_1 or R_2. If there is a potential difference between G_1 and G_2, there will be current flow through R_3, which can cause dc offset on results input to computer. It is also possible for loop between G_1 and G_2 to pick up alternating current signals, impressing noise onto input to computer

Dynamic range may be defined as the ratio of the largest desired observable signal (in voltage units) to the minimum desired observable signal (in voltage units). The dynamic range is extremely important when choosing a computer system for laboratory automation. The relevant parameters include signal resolution, amplifier (or instrument) signal linearity, and signal-to-noise performance.

Consider a simple curve such as might be generated by a thermocouple. The highest temperature at which the thermocouple will be used will generate an output (say A mV); the lowest temperature will similarly result in lower output (B mV). The ratio of these two readings, A/B, defines the dynamic range of the signal source.

The accuracy-of-reading and the resolution may be computed from the knowledge of the voltage range. For a desired accuracy in the readings, the resolution may be defined as the voltage range divided by the desired accuracy-of-reading. In the example above, suppose the voltage range (A–B, or effectively A since B is comparatively small) is 20 mV and the desired accuracy-of-reading is 1% (1 part in 100); then the resolution would be 20 mV/100 or 0.2 mV/unit.

Most chemical instrumentation today is capable of a dynamic range in excess of 1000. In certain cases, particularly chromatography, the dynamic range can exceed 10^6; but most applications do not require a dynamic range that would exceed several thousand. In these cases, all that may be required is amplification to increase the instrument's output signal to a level compatible with the computer system to be used. Moderately inexpensive amplifiers can be used for this purpose.

In situations where extremely large dynamic ranges are encountered, amplifiers with selectable gain are often used. The mode of gain selection can either be automatic in the amplifying device (autoranging amplifiers) or programmable (programmable gain-ranging devices). The choice between the two types is primarily determined by costs and by the programming (software) for the computer system.

Programmable gain-ranging amplifiers are used if it is known that the system's software can predict the desired gain range necessary for the dynamically changing system. Autoranging amplifiers can determine, within their own hardware elements, the appropriate gain range for the voltage level of the input system. Such autoranging devices typically tend to be fairly slow in operation and thus cannot be used with high data rate instrumentation.

Conversion of Analog Data to Digital Form. When automating a laboratory, it is always necessary to convert the analog output data from the instrument to a digital form compatible with the computer system. More precisely, if the data source produces analog information, this must be converted into parallel digital information. This is done by an analog-to-digital converter (ADC). These devices are commonly available from essentially all computer vendors as well as a large number of analog equipment manufacturers. There are many specifications for the ADC that must be considered, including dynamic range of the converter, its linearity, and its inherent noise. In addition, the speed of conversion (conversion rate) must be compatible with the desired data sampling rate.

ADC's themselves are rarely capable of processing low-level signals (less than 1 V) over their entire range. Thus, amplifiers must be used in conjunction with the ADC. Since a laboratory usually has several analog signals to be converted, the ADC can be preceded in the circuit by a device which allows the selection of the specific analog signal which is to be converted at any particular point in the analysis. The selections done by such a device—an analog multiplexer—are usually under the control of the program running in the computer.

The conversion rate is governed by the hardware rate (or throughput) of the ADC plus the multiplexer. There are two modes of operation of any computer-program-supported ADC. The simpler is program-controlled conversion; the computer program requests data from the ADC, manipulates them, stores them, etc. The other is the employment of hard-wired front ends and/or direct memory access/analog-to-digital converter (DMA/ADC) interfaces. Program-controlled conversion rates are typically 50 kHz or less, whereas hard-wired front ends can exceed 1 MHz acquisition rates. DMA/ADC's typically operate at rates below 200 kHz.

The entire interface between the instrumentation and the computer is considered in terms of analog filters, multiple-input channels (requiring a multiplexer), and the ADC. These must be specified with respect to accuracy, resolution, and signal-to-noise ratio.

Consider the specific example where the analog data consist of a series of peaks of definable width, amplitude, area, etc. The requisite dynamic range of the front end is determined from the maximum and minimum peaks of interest as previously described. The required accuracy for the minimum peak will then be determined primarily by the signal-to-noise ratio and the front-end resolution. If the required accuracy for the smallest peaks were to be 1%, then the front-end resolution must be at least 100 ADC units, and the signal-to-noise ratio 50:1 or more. (Here, the minimum signal is divided by the *peak-to-peak* noise). Thus, the dynamic range of the total front end for a specific resolution can be defined as the dynamic range of the input signals times the number of ADC units needed for the required accuracy of the smallest peak (Equation 1):

$$\text{DR (Front end)} = \text{DR (input)} \times \text{ADC resolution units} \quad (1)$$

If, in the above case of 1% resolution, the maximum and minimum peaks are

1 V and 1 mV, respectively, then the dynamic range of the front end is seen to be 100,000 ADC units. This corresponds to an ADC with a capability of a 17- or 18-bit result. Although such is available, this would be a very expensive item which suggests that other approaches may be more cost-effective.

Autoranging techniques can be used more economically than the direct approach just discussed. A gain-ranging amplifier, either autoranging or under computer control, is used in front of the ADC. The dynamic range required of the ADC can be reduced since low-level signals can be amplified sufficiently to be acceptable to this ADC. The ADC can then be chosen to have the resolution and accuracy required. For example, a measurement accuracy of 0.1% (resolution of 1 part in 1000) in the ADC range can be achieved with an ADC of 10 bits.

Data Transmission. An important decision is whether it is more cost-effective to convert the analog data to digital form at the instrument site, sending this digital information to the computer, or to transmit the analog signal directly to the computer location for conversion there. Converting analog data at the instrument site is becoming more prevalent. In this case, the facilities for data transmission become the limiting factor as will be seen below. Since a large number of older facilities are still extant, transmission of analog data directly to the computer is still the most widely used technique. Such analog-transmission techniques are usually restricted to distances under ¼ mile (400 m). Suitable transmission techniques involve use of moderate-cost analog front-ends incorporating preamplifiers, analog multiplexers, and the ADC.

Analog transmission is usually more expensive than digital, owing to the use of expensive shielded cable for increased noise immunity. For distances up to ¼ mile (400 m), it is possible to have the signal path be a direct connection from the instrumentation to the computer system's analog front-end. For greater distances a booster amplifier at the instrumentation site is strongly recommended to increase the signal strength sufficiently above the noise so that a good signal-to-noise ratio can be realized at the analog front-end.

Digital data transmission techniques require conversion of the analog signals to a digital signal or signals at or near the instrumentation site. Often, this conversion is in the form of binary-coded decimal (BCD). Once in the digital form, the information can be transmitted in either serial or parallel fashion. Serial transmission offers the highest noise immunity of any technique but requires the use of data communications hardware at both the instrument (data source) and at the computer site. Parallel transmission offers the highest speed in digital transmission, but many parallel cables may be required to satisfy the instrument resolution. The multiplicity of cables generally limits parallel transmission to distances under 100 ft (30 m), but this technique is very useful and inexpensive in smaller laboratory environments.

Ground loops should always be avoided. This phenomenon arises whenever the signal ground is different from the computer ground. This is shown for a common instrumentation setup in Figure 1. Here, there are multiple pathways through which current owing to signal can reach an earth ground that is different from signal ground. Avoidance of ground loops may mean modification of the signal source (the instrument) or slightly more sophisticated techniques of floating either the signal source or the signal sink (the computer).

Digital Preprocessing. Digital preprocessing is the initial, usually real-time, processing of data by the computer prior to the procedural processing routines. Such preprocessing often involves signal averaging, digital filtering, and real-time decision making.

Signal Averaging. Ensemble averaging is probably the oldest signal averaging technique used by minicomputer systems. Several sets of data (called an array) from a given experiment are collected. Based upon the assumption that the arrays are repetitive, they are averaged together point by point. The amplitude and phase of the noise relative to the data in the arrays should be random so that the noise should average to zero, allowing the extraction of the signal. The exact improvement made in the signal-to-noise ratio by this technique is given in Equation 2:

$$S/N(n) = 1/\sqrt{n} \times S/N(1) \quad (2)$$

Figure 2. Digital data before and after preprocessing by boxcar averaging

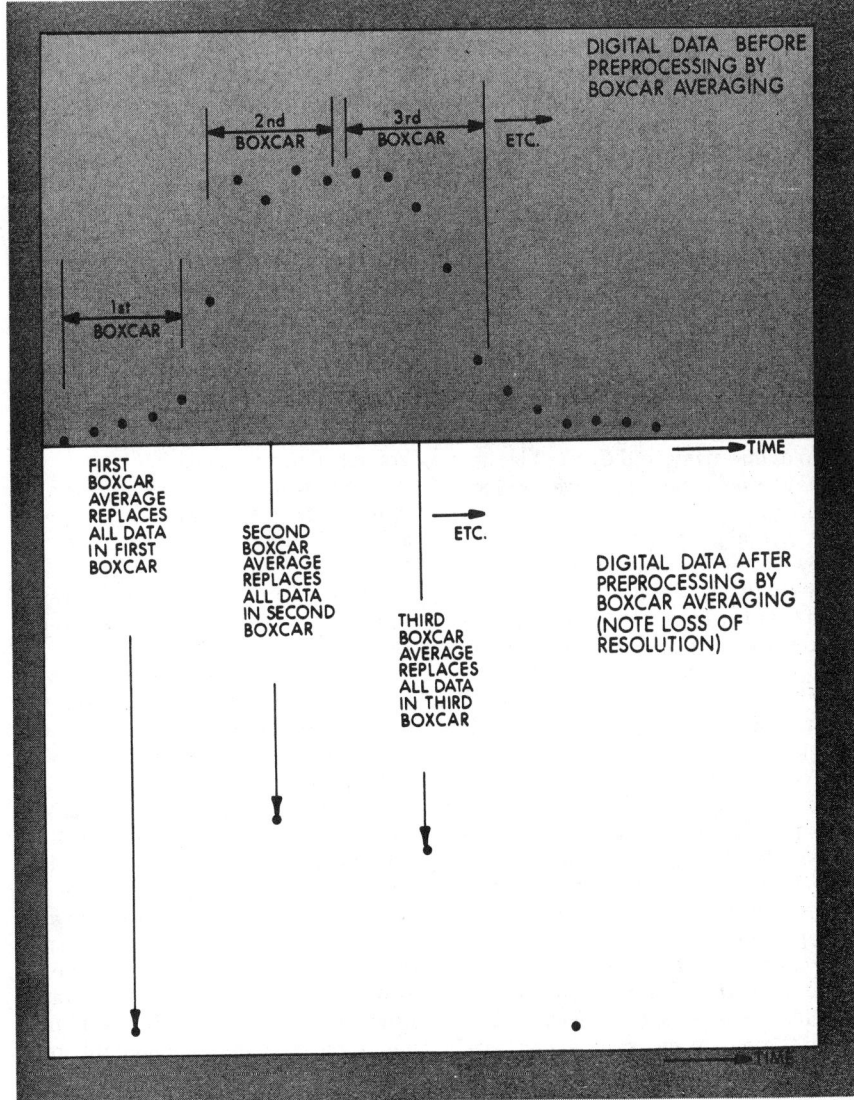

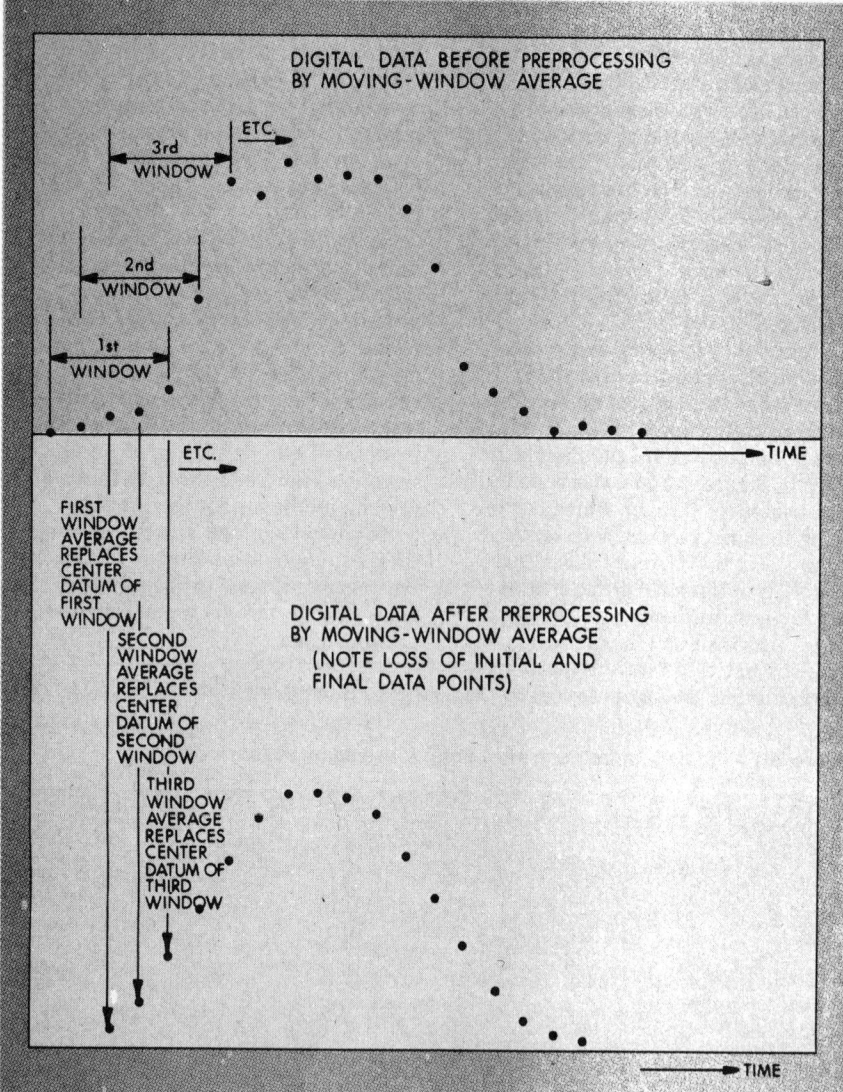

Figure 3. Digital data before and after preprocessing by moving-window average

where n is the number of arrays averaged, $S/N(1)$ is the signal-to-noise ratio of one array, and $S/N(n)$ is the signal-to-noise ratio of n averaged arrays.

A second form of averaging is known as boxcar averaging. The rationale for this technique is the assumption that the analog signal being sampled varies slowly with respect to the sampling rate and that an average of a small number of samples will be a better measure of the signal than a single sample since the S/N will be improved as discussed above (Figure 2). In practice, between 2–50 or so samples may be averaged together to generate a final datum. This lowers the effective sampling rate but removes noise, especially the high-frequency components imposed upon the signal. Boxcar averaging can be used in conjunction with ensemble averaging. Each array is boxcar averaged, and then these reduced arrays are ensemble averaged. This dual process reduces the number of arrays that would be required to get the same improvement in signal-to-noise ratio if ensemble averaging alone were used.

The moving-window average is a dynamic extension of the boxcar averaging technique. As in the boxcar technique, a subset of the array is averaged to form a new datum. However, this new datum does not replace this entire subset but only the subset's central datum. Subsequent subsets are then formed by dropping the first datum of the previous subset and adding the sample datum following the last datum of the previous subset, i.e., the first datum which was not included in the previous subset (Figure 3). In this manner, the window is said to be moved up one datum point. The average of this window is used to form a new datum to replace the new central point. The process is repeated. The averages generated form a new array which is then processed. It should be noted that only the original data are used throughout in calculating these averages. The new array is almost as large as the original, lacking only a number of data points equal to the width of the window, as it may be seen that a half-window width at the start and end of the array is lost. This technique has the noise reduction advantages of boxcar averaging without the concomitant significant reduction in effective sampling rate (and, hence, resolution) of that technique.

Digital Filtering. Digital filtering is another technique of data preprocessing performed by the computer; it has been widely discussed in detail. Most of the commonly used forms of digital filtering may be expressed by the following equation:

$$\overline{y}_{i+j} = (1/N) \sum_{j=1}^{N} [A_j y_{i+j} + B_j(y_{i+j})^2 + C_j(y_{i+j})^3 + \ldots] \quad (3)$$

where the y's are the data samples; N, the number of data points (called the span of the filter); the $\bar{y}$'s, the filtered or preprocessed points; and the $A, B, C, \ldots$ are the filter coefficients, the values of which determine the operation of the filter.

The moving-window average may also be regarded as a linear digital filter (or a linear coefficient digital filter). The moving-window average filter is obtained when all the A coefficients are unity, and the $B, C, \ldots$ coefficients are zero.

The least-squares derivation for a polynomial digital filter will generate nonequal coefficients with values weighted toward the center point of the span over which the filter is applied. The more strongly the coefficients give weighting toward the center value, the less effect the terminal points will have. High-frequency filtering with different cutoff frequencies can be obtained by varying the span. Other choices of coefficients can be used to create a high-pass digital filter to remove low-frequency information while retaining the high-frequency information.

Digital filtering is now widely used in most analytical instrumentation applications. In comparison to the digital techniques, analog filtering tends to be difficult, expensive, or both. By proper choice of coefficients and span, it is possible to create filter conditions which could not easily be obtained, if at all, in an analog manner. The ease with which computers allow the modification of filter parameters leads to great improvements in data enhancement techniques. Digital filtering is almost always required for low-level signal (less than 1 mV) applications, chromatography, and optical spectroscopy.

Real-Time Decision Making. Real-time decision making involves a variety of tests upon the preprocessed data with action to be taken in real-time based upon the results of those tests. The exact form of the decision-making processes depends upon time constraints. The data must be accepted, a decision made, and control functions implemented within the time frame of the external device(s). Such decisions might involve enabling or disabling certain instrumentation, modification of measurement ranges, or specifically timing the interval between special events. The speed of the computer in performing real-time decision making gives the experimenter considerably increased control over the experimental setup.

A choice can be made among microcomputers (based upon microprocessors), minicomputers, and larger general-purpose computers based upon memory capacity and the relative importance of different functions. When instrumentation automation emphasizes decision making and control functions rather than computation, it may prove more cost-effective to use microcomputers. However, if computational algorithms and data storage require memory capacities in excess of 3K, then minicomputers or larger computers are needed.

Procedural Reduction. Procedural reduction is a step toward the final two-parameter data set prior to user interaction, manipulation, interpretation, and report generation. It is the use of a fixed algorithm to process the data obtained from the acquisition and digital preprocessing stages; classically, this would be referred to as data reduction. The algorithm is self-sustaining, self-correcting, and capable of handling the entire data input stream and formulating it for later manipulation and transformation. These algorithms or techniques are usually computational in orientation rather than having real-time constraints. In certain cases advantages may be gained by performing the procedural reduction time-dependent as when the data rate of one instrument may be slow in comparison to others serviced by the computer or when several instruments with low data rates are used. A specific instance of this is the chromatograph in which such a procedural reduction in the form of peak analysis is a reasonable approach.

The nature of the reduction depends upon the data source, final data required, and the form and format of the final data presentation. This stage includes much of the system's analytical software and utilizes many of the advantages of the minicomputer, although in simpler computational cases, microcomputers may be used to advantage.

Procedural reduction may involve retention of the entire data input set or processing of each datum on-line in real-time. The former has the obvious advantage of retaining all information acquired during the experiment. The latter has the advantage of reducing the amount of storage needed if large amounts of data are required and of being more economical in experiments where retention of each datum is unnecessary.

The retention of the entire data set is desirable in some instances; for example, for spectral or frequency analyses, interactive cathode ray tube (CRT) or visual display handling, information plotting or comparative analyses among several data sets.

In terms of computer programming, storage of the entire data set from an experiment is the simplest type of operation. As the data are preprocessed or received, they are buffered temporarily and then stored, either within the core memory of the computer or on a mass storage device. The storage format should be planned so that information is readily retrievable and accessible to all programs of the computer system. Often, data will be gathered in real-time and then stored; data reduction may take place off-line later.

In more powerful computer systems, foreground/background operation offers the advantage of more effective and multiple use of the laboratory computer. In the foreground, data are accepted and stored, whereas in the background the user manipulates and transforms data acquired earlier.

On-line processing of each datum as acquired is usually performed only for moderately slow data rate (less than 2 kHz) experiments. Examples of such processing are unit conversions and limit testing. Unit conversions usually have a calibration table for the physical parameter which is compared against the ADC output. In the case of peak analysis, peaks present might be extracted and then reduced to a table containing such information as peak amplitude and its time-dependent position. More complex forms of peak analysis might involve other factors: area, width, or centroid. If input rates are moderate (less than 500 Hz), such analysis can be performed in real-time. However, very complex peak analysis such as the deconvolution of fused peak envelopes may require storage of the entire data set and off-line data reduction.

One example of procedural reduction is *deconvolution*. Deconvolution extracts accurate information concerning all parameters for each of the component peaks in a fused peak envelope. The deconvolution algorithms depend greatly upon the exact peak form and structure. Most such techniques for use in real-time minicomputer environments make simplifying assumptions concerning the peak shapes and approach the solution by approximation. Nonlinear iterative least-squares techniques are a specific useful example of these.

Many experiments generate a two-parameter data set (x-y) where the raw x-parameter is not meaningful physically; it is necessary to transform these data into a new coordinate set. *Frequency domain analysis* is a large area of procedural reduction where this is the case. The y- vs. time data are converted into an intensity vs. frequency set. Several newer forms of analytical instrumentation such as Fourier transform NMR and Fourier transform infrared (IR) require the use of frequency domain analysis to interpret the experimental data. Although frequency domain analysis typically retains the entire data set for off-line reduction, it is performed as a real-time function for certain applications. In vibration testing it is important to know, in real-time, the exact resonance frequencies as a function of some stimulus. By selecting limited bandwidths, real-time analysis of data from numerous (possibly a dozen or more) vibration sensors can be performed.

Input data can be sorted by *histogram formation*. In nuclear instrumentation pulse-height analysis (PHA), the input datum is a measure of energy. It is desired to know how many occurrences of that energy will appear in a given time; the data are represented as event-count vs. energy.

Correlation techniques are used when input analog data are extremely noisy and do not lend themselves to time-based signal averaging. The basic purpose of correlation is to determine an underlying repetitive signal, if any, in the analog data stream and to calculate its frequency(ies).

In autocorrelation the data set (or array) is time-shifted by an amount τ, multiplied by its unshifted values and a correlation coefficient calculated from the result. If this coefficient is plotted as a function of τ (autocorrelogram), local maxima indicate the frequencies ($1/\tau$) composing the signal.

Knowing the frequencies, the components of the signal can be found.

In cross correlation the data from one input channel are multiplied by the time-shifted data from another analog input channel, and correlation coefficients calculated. As above, a correlogram can be constructed. If it is assumed that the same frequencies are present in both arrays, then the phase shift between the two channels as well as the frequencies can be determined.

Manipulation and Transforma-

tion. This is the most general phase of laboratory system computing, and it requires the utmost flexibility to obtain the final presentation.

In many cases, the manipulation depends primarily upon the x-axis data being either time based or in the frequency domain. Time-based analyses consisting of intensity transformations and time-base conversions comprise the largest number of applications. Intensity transformations include base line corrections, peak intensity and area operations, deletion of unnecessary information, and comparative studies of several data sets. Time-base conversion is commonly done to such x-axis parameters as energy, wavelength, and frequency. Often, these conversions can be done directly, but they can also necessitate calibration tables and interpolation techniques. The use of an interactive computer system is extremely advantageous here since it allows quick, easy transformations and immediate presentation of the results.

Although frequency domain transformations were discussed under procedural reduction, it is a form of analysis amenable to flexible interactive manipulative techniques. The major frequency domain transformation is the Fourier transform. Commonly used fast Fourier transform (FFT) algorithms generate two arrays of data consisting of real (or absorption) mode and imaginary (or dispersion) mode coefficients. These coefficients are themselves often the desired end result of an experiment and its analysis. In addition, they are often manipulated further and transformed to obtain magnitude, power density, and correlation coefficients. From the Fourier coefficients it is possible to generate correlation coefficients equivalent to those described under procedural reduction. The use of the FFT allows derivation of information directly comparable to that obtained by real-time correlation.

While there have been a number of software (program) techniques for computing FFT data, hardware FFT computation techniques are now also available. Software techniques are governed by the transform program and the available computer; they require a smaller system expenditure than hardware techniques. They also are not as limited in the number of data points they can handle and so have better frequency resolution than the hardware FFT techniques. However, the hardware FFT techniques can typically process 1,024 data points in under 20 msec, whereas software transformations would average a second or more. In either technique it is important that the sampling frequency be at least twice that of the highest expected frequency in the analog signal to prevent the phenomenon of aliasing which is the contribution to low-frequency components of the signal from frequencies higher than the sampling frequency.

The final manipulative phase is the presentation of results, often involving coordinate axis calibration utilizing previously stored tables and interpolation techniques based upon finite difference theory. Information presentation is enhanced greatly by the use of interactive or visual computer techniques. The ability to display the information rapidly and modify it by use of a light pen or keyboard entry allows interactive procedures to be performed quickly and easily. The ability of the computer system to handle large amounts of data and to present them visually in a flexible manner aids considerably in the final interpretation of the data. Where printed or graphical output is required, the use of such output devices with general and flexible formats eliminates previous time-consuming manual writing, plotting, and drafting.

Postoperative Amalgamation and Interpretation. In this final phase of laboratory automation, the computer is no longer used primarily as an analytical tool but rather as a more general aid in the merging, sorting, storage, retrieval, and interpretation of data.

Data Merging. Data sets obtained from the previous stages of the experiment may be combined to form a superset of data. The elements of this superset need not necessarily be limited to those obtained from a particular instrument but can originate from a combination of data sets from a number of instruments serviced by the computer, from previous experiments performed with these instruments or even from other experiments. In the last instance, these data would be entered into the computer memory by the experimenter directly rather than through instrumentation.

Library Techniques. The computer can be used to store, for various analytical and interpretational techniques, standards and calibrations in the form of tables. Such tables may include standard spectra for matching, identification, and quantification of results. Such spectra might be in the two-parameter form of intensity vs. energy, intensity vs. frequency, or intensity vs. time.

File Storage/Retrieval. As a storage and retrieval device, the computer is unsurpassed. Data can be stored in a compact form and can be tagged in a number of ways so that only the desired data will be accessed and all the desired data will be accessed. It is therefore important that the proper format for the data and its labeling

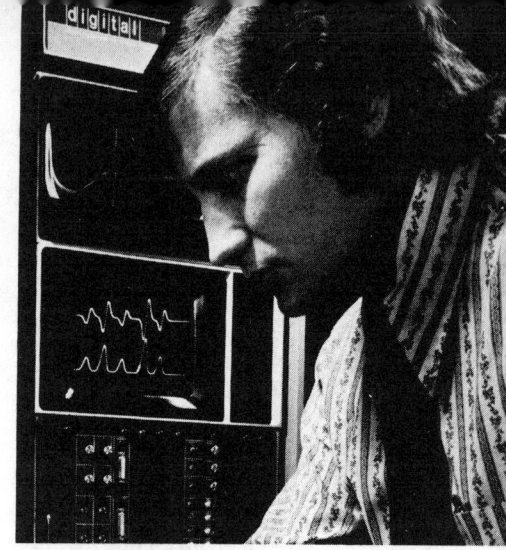

Gerald Dulaney is marketing supervisor for physical sciences within the Lab Data Products Group of Digital Equipment Corp. He earned a BA at Carthage college in chemistry and physics and undertook two years of graduate study at Purdue and three years at Virginia Polytechnic Institute and State University. His interests at that time were in rare earth Mössbauer spectrometry. Mr. Dulaney joined Digital Equipment Corp. in 1969 as senior applications programmer and for the last three years has worked in marketing as a specialist in instrument automation and lab product development.

should be chosen so that this primary function of a filing system can be obtained with a minimal amount of storage space and access time to optimize the cost-effectiveness of this section of the automation system.

Conclusions

The medium- and large-scale general-purpose computer, the microcomputer, and more particularly the interactive minicomputer, have brought about a new dimension in laboratory automation. By connecting the instrumentation to the computer through appropriate interfacing, analyses that frequently consumed hours to days of a researcher's time can now be completed in minutes. Other analyses, heretofore impractical, can be done routinely. However, to achieve these results, it is necessary to understand the processes involved in all areas of an automated laboratory system from the instrumentation output to the finished report. In this discussion each major area—techniques, interfacing, and hardware and software considerations—has been discussed. From this, it is hoped that a clearer perspective of the automated signal processing system for laboratories has been reached.

Figure 1. Typical microprocessor card set

Microprocessors
Part I: Bridging the Gap

Raymond E. Dessy, Peter Janse-Van Vuuren, and Jonathan A. Titus
Virginia Polytechnic Institute and State University
Blacksburg, Va. 24061

During the last year a new word has appeared in the ever-changing vocabulary of news and business magazines—microprocessors. These are the large-scale integrated (LSI) circuits that are at the heart of a revolution in the entertainment industry (PONG!), the large retail stores (point-of-sale or POS equipment), and more recently in automated analytical instrumentation (the digital self-computing gas chromatograph).

The significance and importance of this new technological development to the analytical chemist is that a microprocessor in conjunction with memory, control circuitry, and data input and output lines (the I/O bus) constitutes a microcomputer. The function of the LSI microprocessor chip(s) in this configuration is the same as that of the central processing unit (CPU) in larger computers. In other words, the complex circuitry necessary for the decoding of instructions and control of the resultant logical operations of the computer (the function of a CPU) has been micronized into a single (!) integrated circuit chip, hence, a microprocessor.

The extent to which LSI chips are currently incorporated into computers and in future designs makes microcomputers rather ill-defined devices since no criterion exists, or is likely to exist, which delineates the area where microcomputers leave off and minicomputers begin. Indeed, as many manufacturers begin to emulate their existing minicomputers, previously assembled from medium-scale integrated circuit components, with LSI chips the grey area will become even broader.

What is the significance of the microcomputer from the viewpoint of the analytical chemist?

Most minicomputer manufacturers are committed to the philosophy of developing machines with increasingly powerful computational capabilities, often to the exclusion of their less powerful and less expensive progenitors. In a decentralized configuration, minicomputers with up to 64,000 words of memory and 5 million words of disk storage and high-speed peripherals, such as line printers and CRT terminals, often provide an attractive logistical and economical alternative to a single massive data processing installation such as the proverbial IBM 360 or 370 systems. Fully equipped minicomputers offer all the advantages of higher-level languages such as BASIC or FORTRAN which are human oriented and which ease the cost of program generation.

Manufacturers provide programs called EXECUTIVES to supervise these high-level languages as well as data acquisition programs which are usually written in a language more closely related to the machine operations called ASSEMBLY or MACHINE Language. However, as the executive programs become more sophisticated, the vendor and the end user begin to spend more for the programs (software) to run the equipment than the equipment (hardware) itself, especially when a single large computer is to service the needs of a large number of instruments attached to it. At this point, the system becomes so complicated, interactive, and interlocked that the average chemist is more than happy to let the computing center personnel assume responsibility. But with this responsibility goes control of the design or modification of the system—often resulting in dissatisfaction at the bench level in system function, flexibility, and expandability.

A solution to this problem would be to devote a single minicomputer (with a minimum of memory and peripheral devices such as 2K of memory and a single teletype) to the control and data acquisition of a single instrument (a dedicated system) or a cluster of these dedicated systems as satellites to a larger CPU that is responsible for data manipulation and storage (a hierarchical system). However, current costs may preclude this approach. Alternatively, each instrument could have a hard-wired front-end or data reduction unit associated with it, similar to the peak detectors often found on gas chromatography equipment. For the average chemist this represents a formidable electronics design problem, and the end product is inflexible. Changing it to meet future needs is very difficult.

Microcomputers offer an interesting opportunity to implement such dedicated and hierarchical applications at relatively modest cost, but with complete flexibility. In the case of dedicated systems, microcomputers are installed in lieu of the more powerful but more expensive minicomputer systems where a careful analysis of needs indicates that the latter are an overkill to the specific problem. They will increasingly replace hard-wired front-ends to provide for low initial cost and future flexibility and expandability.

It is now possible to conceive of all moderately sophisticated analytical equipment being delivered with an inboard microcomputer for closed-loop control and optimization of the instrument conditions, as well as data collection and reduction. Completely operational microcomputers that are entirely adequate to service single instruments at data acquisition speeds of up to 1000–3000 digital data words/sec are available for about $1000. Multi-instrument installations in which relatively slow data rates place not too stringent speed requirements on control and data acquisition are also accommodated. Microcomputers thus bridge the gap that has existed between hardware implementation of automation and the use of minicomputers.

It is the purpose of this article to provide some perspectives and an overview of this exciting new development in analytical instrumentation. In Part I we will focus on:
• the most attractive feature of a microcomputer, i.e., the modular nature of the hardware components required to configure microcomputers with variable operational capabilities
• an outline of the basic principles of interfacing external devices, such as analytical instruments, to the input/output (I/O) bus of the microcomputer
• the powerful software instruction set utilized by the microprocessor and which is principally responsible for its operation as a full-fledged CPU.

In Part II we will:
• describe in detail a potential application in which the flexibility of the microcomputer as a control and data acquisition device will be illustrated
• spotlight some of the advantages and disadvantages of microcomputers
• summarize the application of microcomputers as relatively inexpensive instructional equipment for providing "hands-on" and "hands-in" experience in the teaching of digital techniques and applications in the analytical laboratory.

For those unfamiliar with digital electronics or interfacing principles in general, background material may be found in two review articles that have appeared in ANALYTICAL CHEMISTRY: "Computer Interfacing" (1) and "Instant Interfacing" (2).

Hardware

The description of hardware which follows is based on what is necessary to configure the Intel 8008 microprocessor into a microcomputer. This LSI chip is second-sourced, and four vendors can provide total systems for the end user. A typical system is shown in Figure 1. The different types of microprocessors are operationally very similar, and it is relatively easy to extrapolate the basic principles outlined below to other microcomputers and microprocessors. Many other manufacturers provide, or are claiming to provide, microprocessors.

The key components typically required to configure a microcomputer involve:
• the microprocessor or micro CPU—the "head-quarters" of all computer operations
and the following *external* components:
• memory—read/write and read-only for program and data storage
• external control circuitry—for synchronization and generation of control signals
• device decoder—for generation of special control signals for external devices
• interrupt input port—to force external control of the microprocessor.

These are illustrated in Figure 2a.

To complete the basic components required by the microcomputer to communicate with external devices, it is also necessary to include the following which will be added to our circuitry later in the article:
• output latches
• input ports—both to control and synchronize data transfer to and from the microprocessor.

The elements incorporated into the Intel 8008 microprocessor are:
• an instruction register associated with powerful and sophisticated internal decoding and control logic
• an accumulator and seven 8-bit general-purpose registers used in conjunction with flags, i.e., CARRY, SIGN, and ZERO
• a 14-bit program counter associated with a 14-bit, seven-register stack.

All these elements are present on the single microprocessor chip. Figure 2b represents a simplified diagram of these components.

CPU/Memory. The Central Processing Unit (CPU) is an arithmetic and logic unit that is associated with a memory. The memory consists of solid-state devices in which each "word" is made up of 8 binary digits (bits) of information. Part of the memory is *random access memory* (RAM) that the CPU can write data into, or read data from. Another portion of the memory is preprogrammed and can be read from only (*read only memory* or ROM). A series of commands (really binary numbers that are decoded by the CPU and cause certain operations to take place) is stored in memory and then executed one-at-a-time, usually sequentially. This is a program or software. The program may be stored in RAM or ROM; it is common to store the operating program in ROM and use RAM for data manipulation and storage.

Program Counter. When a program is to be executed, a 14-bit register in the CPU is loaded with the address of the first instruction to be executed, and it is "fetched" by the CPU from memory, decoded, and the CPU sets itself up to perform the desired operation. The program counter is automatically incremented at this point so that it is set to indicate the location of the *next* instruction to be executed.

General Registers. Typically, the initial instructions in a program would involve loading one of the seven general-purpose registers within the CPU (A, B, C, D, E, H, and L). The first of these registers is called the A register or accumulator. It is in this register that we perform the arithmetic (add, substract) and Boolean functions (and, or, compare).

Carry, Sign, and Zero CPU Flags. The A register has associated with it an extension bit, called the carry. Upon addition with carry, or subtraction with underflo, this bit gets set to a 1. The A register and its carry extension are really a circular register, and rotation of the contents is possible under software command. Such operations will be discussed in a later section. There are two other "flag" bits associated with the CPU. These are the sign and zero flag bits. Their contents will reveal information concerning the contents of the A register, or the results of Boolean operations. For example:

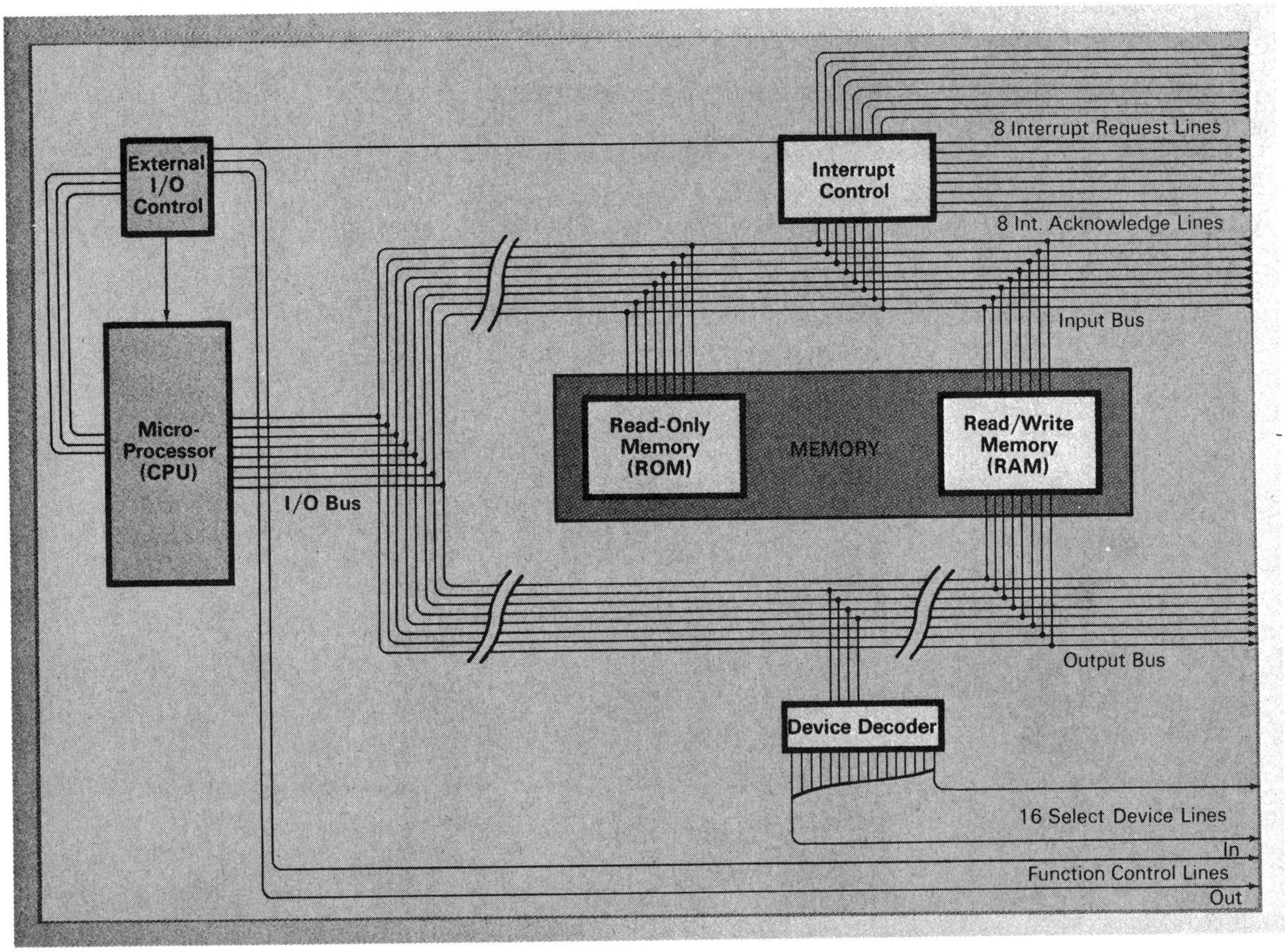

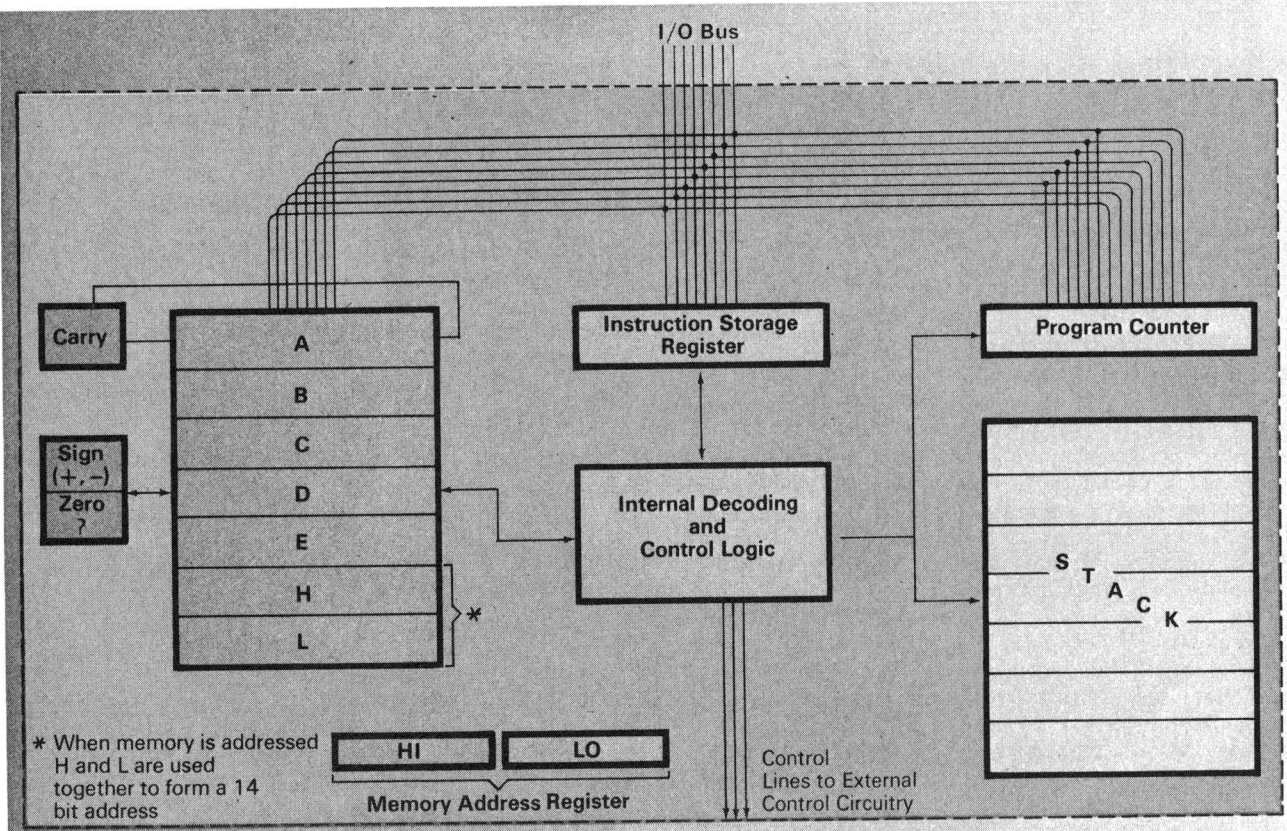

Figure 2a. Microcomputer-architecture external to microprocessor

Figure 2b. Internal architecture of microprocessor

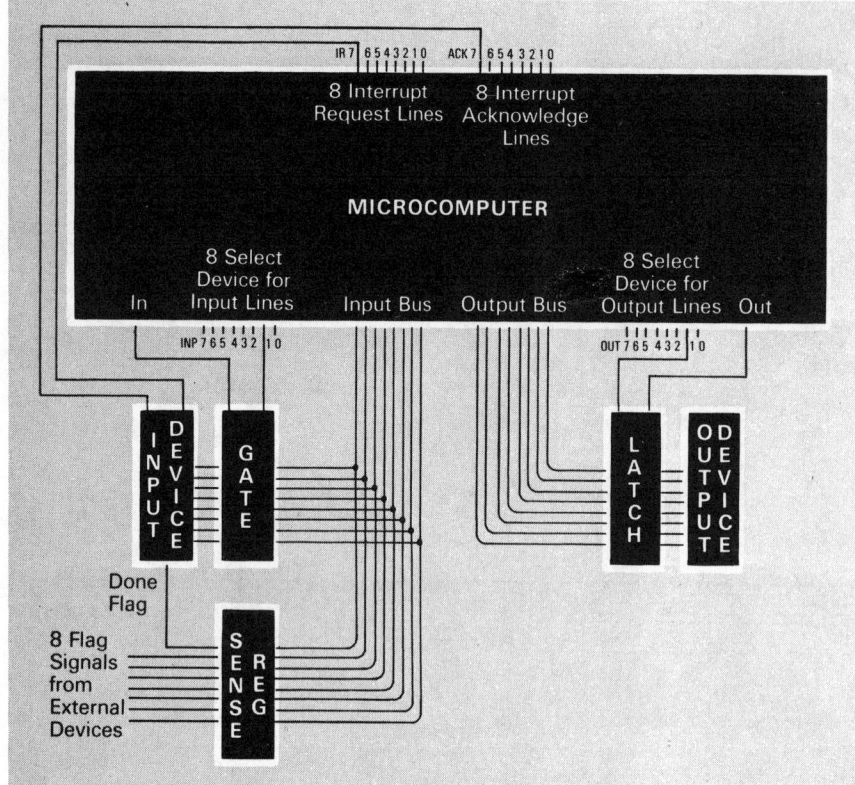

Figure 3. Input/output architecture

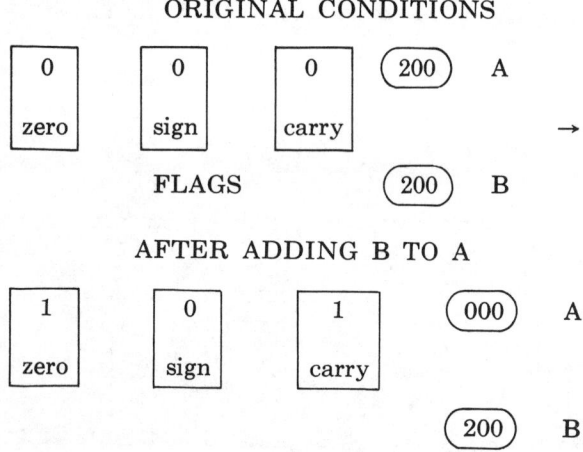

$200_8 + 200_8 = 400_8$. The largest 8-bit number is 377; thus, the result of the addition is zero, with a carry; just like 9 + 1 = 10, where the result is a zero with a carry. The appropriate flags are set as indicated.

The other six registers allow storage of constants and counters. These registers may be incremented or decremented by software commands. Data can be transferred from one register to another by software also.

High and Low Memory Address Registers. An operation that is important early in most programs is the ability to read data from memory or store data in memory. When this is desired, the location of that memory cell is loaded into the two registers called H and L. Two 8-bit registers have been used for this task; using one register would permit addressing of only 256 memory cells—(2^8), too small a memory for most applications. Only 14 bits of these registers are used, thus allowing direct addressing of 16K words of memory!

The Stack. Another common function in programming is a jump to a subroutine, implying that we want to branch off to perform a commonly used operation and return to the main program where we left it. In order to accomplish this, the contents of the program counter at the time of the jump-to-subroutine request is "pushed-down" into the stack, a 7-word × 14-bit set of registers associated with the CPU. The program counter itself now contains the address information pointing to the subroutine instructions. When we wish to return from the subroutine, the old address information is "popped-up" in the stack and back into the program counter by a software command called a RETURN.

The microprocessor has computational capabilities, and these will be discussed in subsequent paragraphs; however, it is the function of this article to emphasize the hardware necessary to input and output data into and out from the CPU (Figure 3).

Select Device/In and Out Pulses. The pathway in and out of the CPU is a set of bidirectional data lines, eight in number. The operational circuitry in the CPU accepts or places information on these lines only at certain times. (The complex multiplexing accomplishing this is not shown in the drawings for clarity; from this point on, the drawings will show separated INPUT and OUTPUT BUSSES). Because of this criticality in timing, and the fact that we usually want many input and output devices to be *attached* to the microcomputer, but only have a single input or output device *activated* at a time, the CPU has I/O control circuitry associated with it. When data are to be input to the CPU, a software command causes the external control circuitry to issue a pulse called IN, and the device decoder to issue a pulse unique to each device called SELECT DEVICE. These activate or enable a device called a GATE which momentarily places the desired information from the input device onto the bidirectional I/O lines. The microcomputer accomplishes this by means of tri-state logic. It is configured as shown in Figure 4.

Tri-State Gate (Input Port). Tri-state logic is constructed so that output is enabled only when the CON-

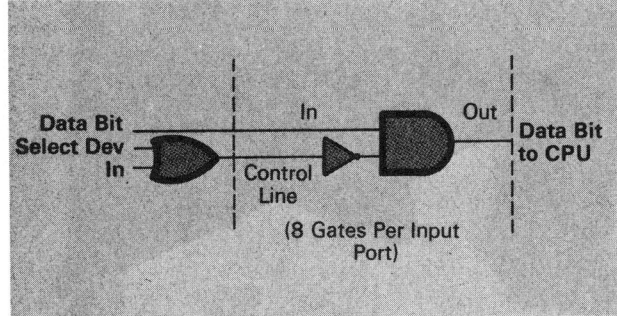

Figure 4. Tri-state logic configuration

TROL LINE is held at a logic 0. In this condition the DATA BIT at the input is "gated" to the output. When the control line is a logic 1, the output looks like a high impedence or open circuit.

Data bit	Control	Output
1	0	1
0	0	0
1 or 0	1	Hi Z

A set of 8 bits is called an INPUT PORT, and we can "strobe" data onto the Data Bus from them by momentarily grounding the CONTROL line associated with the PORT. This is done by ORing (OR gates and other integrated circuit components are discussed in ref. *1* and *2*) the SELECT DEVICE and IN signals together and connecting them to the CONTROL input of that port. Many devices may be attached to the CPU by using a *unique* SELECT DEVICE pulse for each port (Figure 5). Commonly, device decoders provide eight such pulses for input.

Data Latch (Output Port). When data are to be output (Figure 6), a software command causes the control logic to issue two signals, called SELECT DEVICE and OUT, which activate or enable a device called a LATCH, which accepts the data on the bidirectional I/O lines at that instant and holds them (examine Figure 6 for latch operation) for use by the output device. Since we wish to latch, or hold, data available on the Data Bus only during the period that OUT is 0, and since many different devices are connected to the Data Bus in parallel, logical NORing of a SELECT DEVICE pulse and OUT is used to cause the clock input on a DATA LATCH to go to a logic 1 momentarily, when and only when, the desired data are made available by the software and CPU action. Each device in the external world that requires latched data must have a data latch associated with it. Eight SELECT DEVICE pulses are available for output control in the configuration shown.

External Flags. It is necessary for the CPU to know when external devices require service, either for input or output. Depending on the timing criteria, one of two methods is ordinarily used to synchronize the operation of the CPU and external devices.

The first of these involves the external device creating a signal called FLAG which can be sensed by the CPU. This is done by having the CPU query a SENSE REGISTER (an input port to which the flags are attached) periodically, to ascertain whether any device requires service. If it senses such a condition, appropriate software will cause the issuance of the SELECT DEVICE and FUNCTION (IN or OUT) pulses appropriate to that device. Eight separate devices can be sensed using one gate input. Flags are usually built from flip-flops (Figure 7).

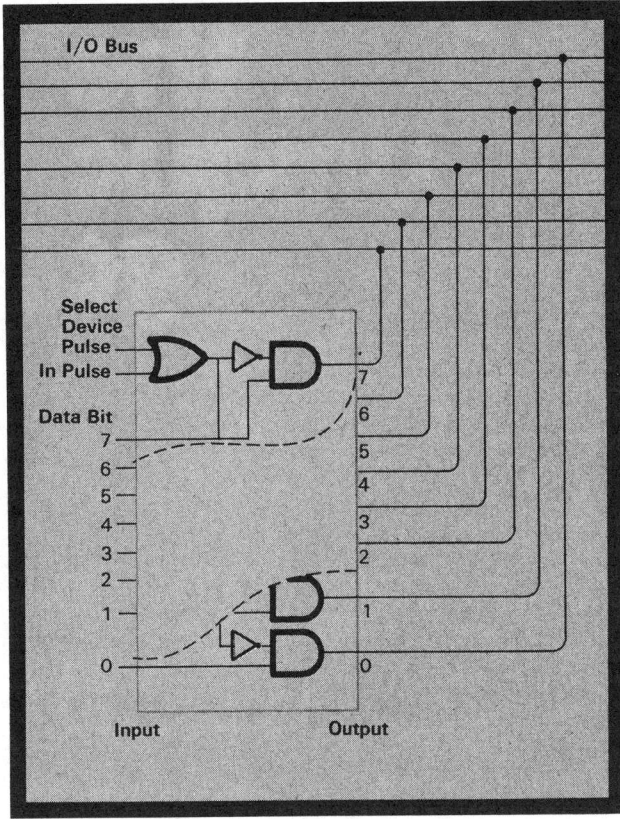

Figure 5. Connection of input ports to I/O bus

This method involves sufficient software to properly identify the device and direct the program to the service routine for that device. Under certain conditions, usually involving very rapid input or output, or long periods between I/O, such a method is inadequate or inefficient.

Interrupt Request and Acknowledge. To meet such needs, external devices can generate another signal called INTERRUPT REQUEST when I/O is needed (Figure 7). This interrupts the existing program, and *automatically* causes the following operations:

• stores the program counter in the stack (just like a call to subroutine)

• generates an INTERRUPT ACKNOWLEDGE signal that clears the flag, letting the device know its interrupt request has been accepted

• vectors (jumps) to the service routine for the device by "jamming" a jump command called a VECTOR instruction (created by external hardware) into the CPU for execution. These will be discussed in the Software section.

The microcomputer has the hardware to control eight devices in this manner. Those familiar with PDP-11 architecture will recognize that the microprocessor architecture is rather sophisticated.

The emphasis up to now has been on hardware. But the flexibility of the microcomputer lies in its ability to be programmed for a given task by software.

Software

A set of instructions called a program (software) is necessary to allow the microcomputer to accomplish anything. Programming for analytical instruments will be discussed in Part II of this paper. For the moment, we need to focus on the topic of input/output commands and the basic instruction set.

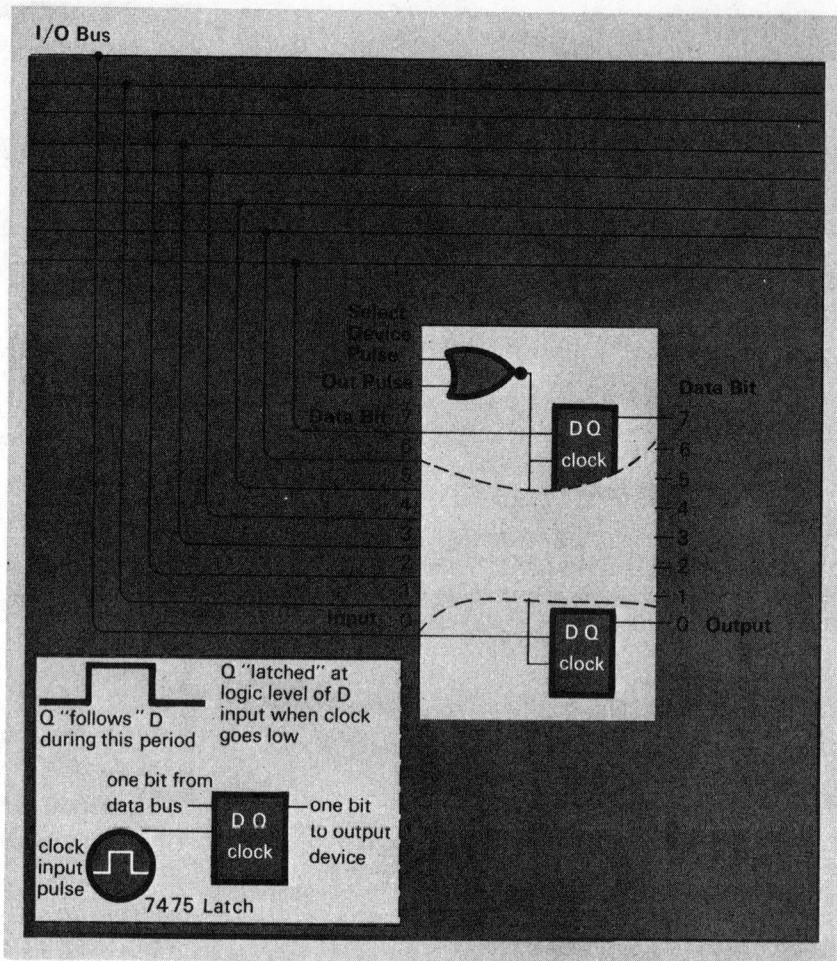

Figure 6. Connection of output ports to I/O bus

Input/Output Programming. All data flow to and from the microcomputer with respect to the outside world takes place via the A register. All data inputs must be activated during and only during the duration of a CPU generated pulse called IN. To distinguish between external devices, the CPU simultaneously produces a unique SELECT DEVICE pulse in response to the specific *software command* initiating data transfer; e.g.,

Octal code	Mnemonic equivalent
111	INP4 generates SELECT DEVICE INP4 and IN
113	INP5 generates SELECT DEVICE INP5 and IN

Similarly, data can be output from A during and only during the duration of a pulse called OUT. This and unique SELECT DEVICE pulses are generated by specific software commands: e.g.,

Octal code	Mnemonic equivalent
125	OUT2 generates SELECT DEVICE OUT2 and OUT
131	OUT4 generates SELECT DEVICE OUT4 and OUT
133	OUT5 generates SELECT DEVICE OUT5 and OUT

The SELECT DEVICE and FUNCTION pulse (IN or OUT) are, as shown previously, gated together to form a single pulse that activates the device, and causes data transfer.

Vector Instructions. Memory in a microcomputer might be configured as follows:

MEMORY ADDRESS

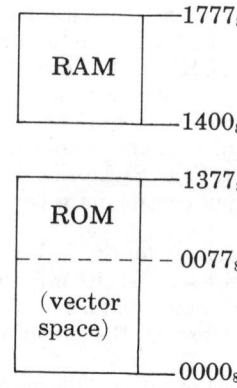

a total of 1024_{10} (1K) words of memory.

The microcomputer is configured to have available a number of instructions which will effectively cause a jump to certain specified ROM locations called VECTOR SPACE. These are the VECTOR INSTRUCTIONS:

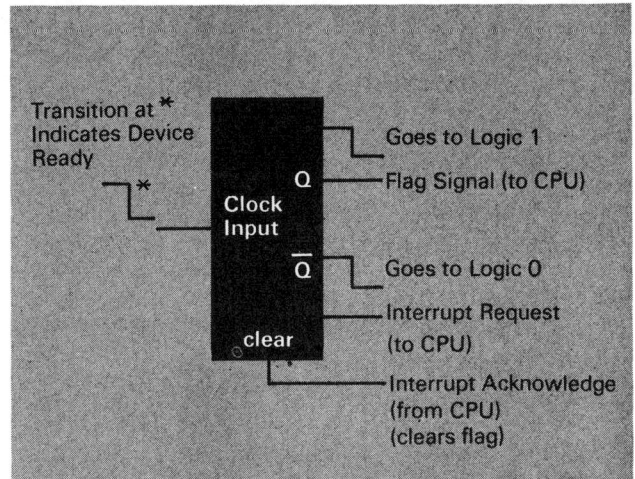

Figure 7. Flip-flop used as flag

Octal code	Operation performed	Hardware equivalent (Interrupt request)
075	Vectors to 0070	IR7
065	Vectors to 0060	IR6
055	Vectors to 0050	IR5
045	Vectors to 0040	IR4
035	Vectors to 0030	IR3
025	Vectors to 0020	IR2
015	Vectors to 0010	IR1
005	Vectors to 0000	IR0

VECTOR = Jump automatically

These actions may also be initiated by hardware action. As mentioned, available to the microcomputer user is a series of bus connections labeled INTERRUPT REQUEST. When *momentarily* grounded they effectively cause execution of the indicated vector command, as well as saving the current program counter in the stack and generating an acknowledge signal.

This allows one to write programs beginning in 00X0 and continuing through 00X7, which can initiate access to more lengthy service routines in ROM or RAM memory, wherever a device needs service. These would normally contain an appropriate input or output software command. A software RETURN command permits the CPU to return to the point in the main program where it was interrupted, after service is accomplished, by "popping" the stack.

In our microcomputer these interrupt requests can be divided into two classes determined by when they are recognized by the processor:
- IMMEDIATE requests: recognized at the end of execution of the instruction during which the request was made
- DEFERRED requests: these are recognized only at the time the software instruction OUT2 is executed.

This allows the user to divide his external devices into two groups: those that require almost immediate attention and those that may be checked periodically to see if action is necessary.

At each interrupt the register contents (A, B, C, D, E, H, and L) and the flag status bits must usually be saved. This permits the machine to be returned to the condition it was in at the time of the interrupt after exit from the service routine for the device requesting service. Since this is a tedious memory-consuming process, it is usually desirable to arrange for IMMEDIATE interrupts to be as uncomplicated as possible (in terms of register and flag alterations) and to use DEFERRED interrupt operation at times in the program when complex SAVE routines are not needed. This can be at the end of repetitively called subroutines when no meaningful data are in registers or the flag bits. Complex interrupt service should be avoided. It is time consuming in both execution and in program preparation for it.

Instruction Set. Once the data from an analytical instrument are read into the A register, it must be operated upon in some way. The instruction set in the 8008 microprocessor involves over 150 commands. Each of these commands is a number which can be decoded by the CPU and the appropriate operation performed. For human utilization each number has a short mnemonic equivalent associated with it. A few examples are given below using the mnemonic equivalents utilized in our laboratories.

Octal code	Mnemonic	Instruction performed
LOAD INSTRUCTIONS		
301	LDAB	Load A from B*
307	LDAM	Load A from Memory
310	LDBA	Load B from A
370	LDMA	Load Memory from A
MATH OPERATIONS		
201	ADDB	Add B to A
221	SUBB	Subtract B from A
010	INCB	Increment B
011	DECB	Decrement B
BOOLEAN OPERATIONS		
241	ANDB	Logical AND B with A
271	COMB	Compare B with A
251	XORB	Logical Exclusive OR B with A
261	IORB	Inclusive OR B with A
ROTATE A REGISTER		
002	RALT	Rotate A Left
012	RART	Rotate A Right
022	RALC	Rotate A Left Thru Carry
032	RARC	Rotate A Right Thru Carry

* Refer to registers. A = accumulator; B, C, D, E, H, and L = general-purpose registers.

The above commands do not affect the program counter. For branching two types of jump commands are available. Normal *JUMP* instructions JPxx) examine the flags, and if the condition specified in the command is met, alters the program counter to the desired address, BUT does not cause the old program counter to be stored in the stack. *JUMP TO SUBROUTINE* (JSxx) instructions DO store the old program counter in the stack. For this reason, they are associated with RETURN commands that "pop" the stack at the end of the subroutine. Let's examine jumps requiring a *True Carry* (=1): these are JPTC (*j*ump on a *t*rue *c*arry) and JSTC (*j*ump to *s*ubroutine on *t*rue *c*arry).

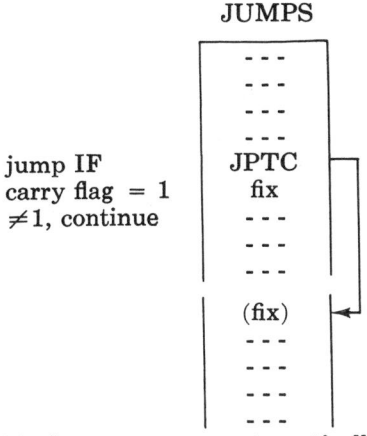

this does not return automatically

JUMPS TO SUBROUTINE

```
       ---
       ---
       ---
       ---
    JSTC      ┐ jump to subroutine IF
    fix       │ carry flag = 1
       ---    ┘ ≠1, continue
       ---
       ---
    (fix)
       ---
       ---
       ---
       ---
    RTUN      now return regardless of flag
              conditions (unconditional)
```
this does!

JUMP COMMANDS		FLAG CONDITION NEEDED
104	JPUN	*UNCONDITIONAL*
100	JPFC	*False Carry* (= 0)
110	JPFZ	*False Zero* (reg ≠ 0)
120	JPFS	*False Sign* (reg is +)
140	JPTC	*True Carry* (= 1)
150	JPTZ	*True Zero* (reg = 0)
160	JPTS	*True Sign* (reg is −)

JUMP TO SUBROUTINE		RETURN	COMMANDS
106	JSUN	007	RTUN
102	JSFC	003	RTFC
112	JSFZ	013	RTFZ
122	JSFS	023	RTFS
142	JSTC	043	RTTC
152	JSTZ	053	RTTZ
162	JSTS	063	RTTS

MISCELLANEOUS COMMANDS

250	CLRA	Clear A
300	NOOP	No Operation
000	HALT	Stop CPU

It is obvious with these commands that rather sophisticated data manipulation is possible. A number can be input, then examined to see if it falls between certain limits by *COM*pare commands, and then appropriate action taken by *JumP* commands depending upon the status of the flags called CARRY, ZERO, and SIGN. In fact, the instruction set is more powerful than that of the ubiquitous PDP-8 which has played such an important role in laboratory automation.

With this background, Part II of this article in next month's INSTRUMENTATION will present a typical application of microprocessors—the construction of a differential stripping electrochemical apparatus. The strong points and shortcomings of microprocessors will be examined, and some suggestions made on how to get started.

References

(1) R. E. Dessy and J. A. Titus, *Anal. Chem.*, **45** (2), 124A (1973).
(2) R. E. Dessy and J. A. Titus, *ibid.*, **46** (3), 294A (1974).

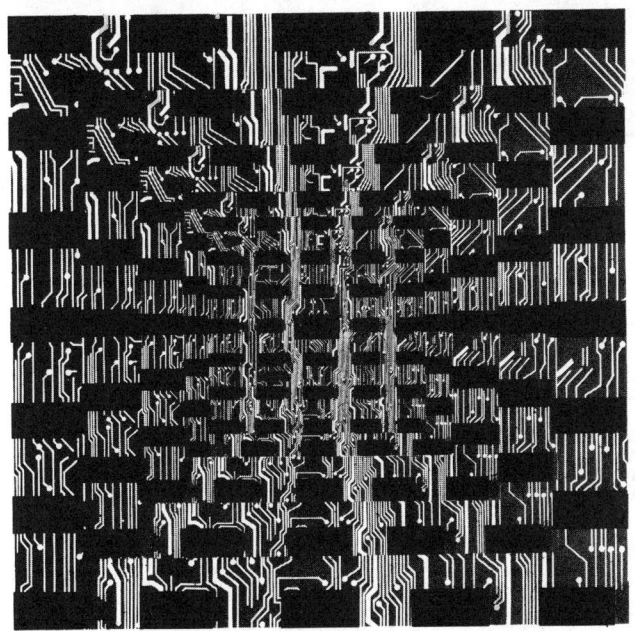

Microprocessors
Part II: Applications

Raymond E. Dessy, Jonathan A. Titus, and Peter Janse-Van Vuuren

Virginia Polytechnic Institute and State University
Blacksburg, Va. 24061

In Part I of this discussion of microprocessors, presented last month (*1*), the rather simple logical concepts of the use of microcomputers in the chemical laboratory were presented. Microprocessors are going to accelerate laboratory automation even more than the introduction of minicomputers did. They are bridging the gap between hardware implementation of control and the use of minicomputers for data acquisition. This month, applications of microprocessors to a "real world" problem are presented. The advantages and disadvantages of such systems are presented, as well as some suggestions on how to get started in the area.

Typical Application

Let us imagine that you want to construct a unit for repetitive analysis of water supplies for trace heavy metal contamination by use of stripping analysis. Hard-wired devices (consisting of logic elements alone) could be built, but most chemists are unfamiliar with their applications. In addition, once built, the units are rather inflexible, and their functions cannot be easily altered. Krause and Ramaley (*2*) have described the background of the method and appropriate circuitry. For the reasons given, the method is not often used. On the other hand, a programmable minicomputer system is a more sophisticated device than is really needed to do the job. A microprocessor is ideally suited to the task.

Since the example was chosen to cover the philosophy of microprocessor usage rather than focus on the method itself, let us first reduce the technique to the barest essentials for those unfamiliar with electrochemistry. An analog voltage signal varying in a complex manner with time is impressed on an electrode, and as the voltage is changed, the corresponding current that flows in the cell is measured.

All we need to add to our basic microcomputer to achieve the intended goal is an analog-to-digital convertor (ADC), a digital-to-analog convertor (DAC), and a clock, as indicated in Figure 1. These elements have been discussed in an earlier article entitled "Computer Interfacing" (*3*).

The analog voltage required is generated first by forming in the A register of the microprocessor its digital equivalent. This is latched out to the DAC by use of the software command OUT1, as described in Part I. (See Figure 1.) Whatever potential is output by the DAC controls a potentiostat made from operational amplifiers. Whatever voltage appears at the DAC output appears at the test electrode. The operational amplifier circuitry is necessary to boost the low current output of the DAC to useful levels. The current passing thru the cell, from test to counter electrode, is amplified to a suitable voltage by another operational amplifier. This voltage is input to an analog-to-digital convertor. The conversion process to digital form is started with the software command OUT6. When the conversion is finished, the ADC signals this fact to the CPU via interrupt request line IR2 by use of the ADC done flag. This is cleared with the corresponding acknowledge signal ACK2. A service routine is entered that acquires the digital equivalent of the current by an INP1 command which activates the gate, transferring the data into the A register. A real-time clock is serviced by IR1 and ACK1 in a manner similar to that of the ADC. It will be used to coordinate the actions of the microcomputer every 0.25 sec in real time.

The actual experiment requires electrolyzing the solution in the cell for a certain TIME at a given POTENTIAL. At the end of the plating time, the computer must generate a voltage ramp to more anodic potentials, upon which is superimposed a square-wave pulse of given AMPLITUDE. Current is sampled at the *bottom* and *top* of each square-wave pulse, and the *difference* measured and recorded [Figure 2 (a)]. If the results of such an experiment were plotted out at the end of a run, they would look like Figure 2 (b). Each peak represents the current necessary to "strip off" a plated metal. The voltage (E) identifies it. The integrated current (i) would give its relative concentration.

The software necessary to perform the operations desired is described in the accompanying flow chart (Figure 3). Five hundred and twelve words of ROM and 250 words of RAM memory would suffice.

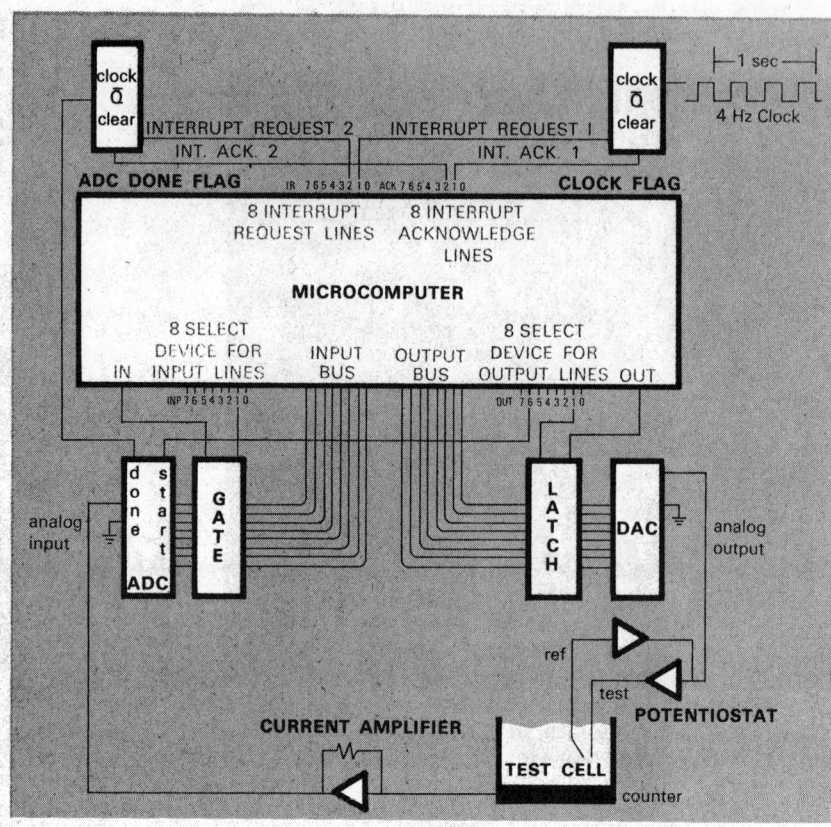

Figure 1. Repetitive stripping analysis unit using a microcomputer

The concept is to have the microcomputer ask for the following parameters, by use of a teletype:
- plating TIME—time plating potential is to be applied to test electrode
- plating POTENTIAL—potential at which solution is to be electrolyzed
- pulse AMPLITUDE—amplitude of pulse to be applied to stripping ramp.

The solution is electrolyzed at the indicated POTENTIAL for the desired TIME, and then a 8-mV/sec ramp to more anodic potentials is applied [Figure 2 (a)]. The scan range is 1.024 V anodic to the plating potential. The linear ramp is modulated by a 1-Hz pulse with the indicated AMPLITUDE.

Plate times of 256 min are accommodated. Stripping time is 2 min. These could be altered by simple changes in the software and the frequency of the clock used. (These parameter limits were chosen for clarity of presentation.)

Implementation of this experiment demonstrates both the computational and control capabilities of microcomputers.

As the plating TIME is typed on a teletype attached to the microcomputer, the data are accepted by the CPU into the A register. The ASCII code transmitted by the teletype is first converted to Binary Coded Decimal (BCD) format as shown. Some microprocessors are designed to handle numbers in BCD format. Others, such as the Intel 8008, require the number to be converted further to a straight binary representation by software (cf ref. 4). This number is stored in the B register.

No. typed	ASCII codes transmitted		BCD no.	Octal equivalent
23	262	263	23_{10}	27_8

As the plating POTENTIAL and the pulse AMPLITUDE are specified, e.g., 1.20 and 0.10 V, similar conversions are made. However, these numbers will be used in conjunction with a digital-to-analog convertor which has a maximum output of ±2.54 V—the exact output depending on the binary number input to the D/A convertor. It is a common practice in computers to let the most significant bit of a bi-

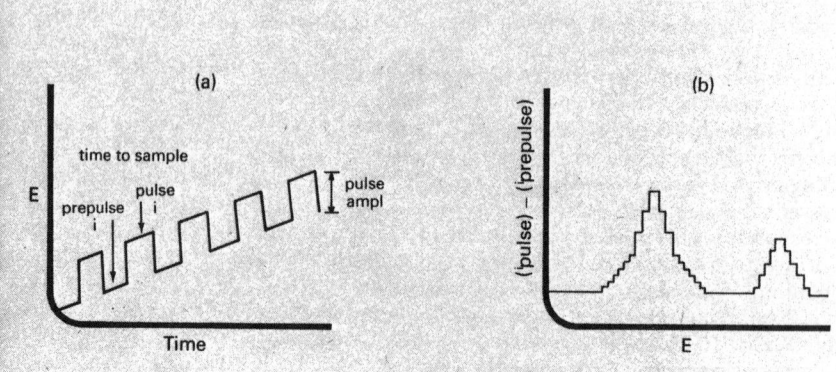

Figure 2. (a) Typical stripping analysis waveform. (b) Typical output current associated with (a)

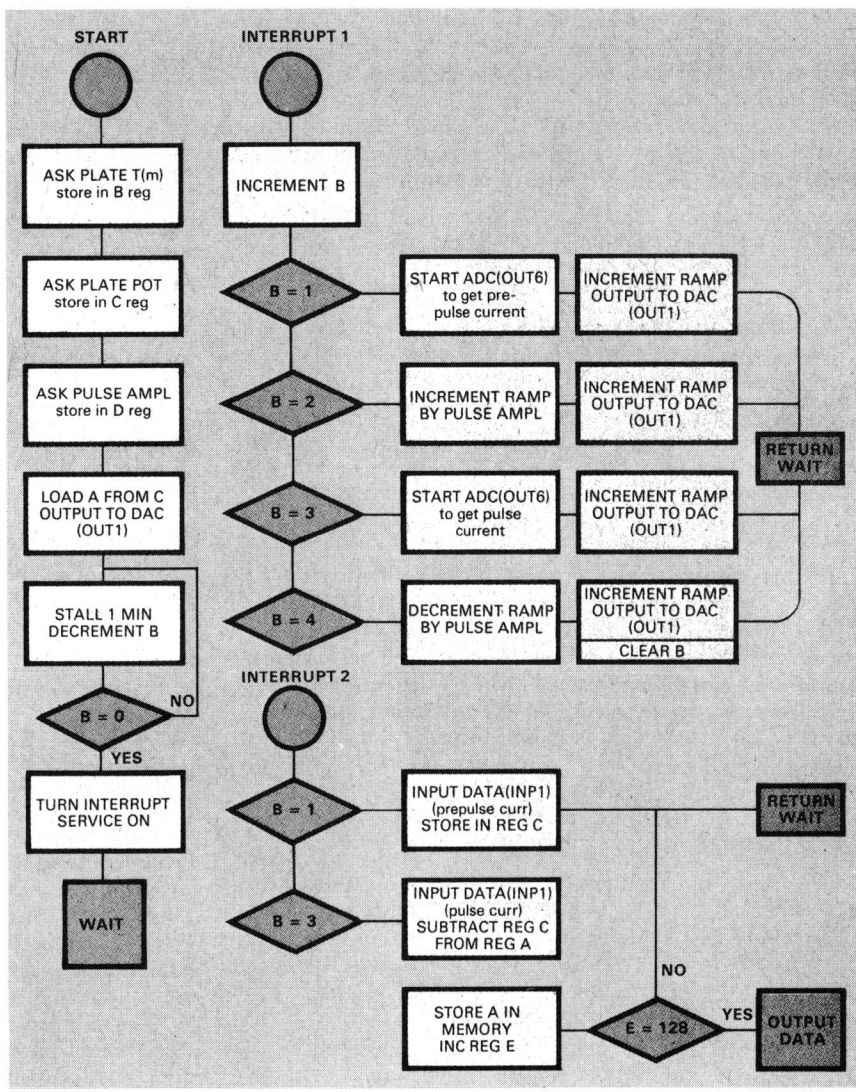

Figure 3. Stripping analysis flow chart

nary number indicate its sign; if it is set to a 1, the number is negative; if it is 0, the number is positive. This means that the following relationship exists between the numbers typed and the numbers to be stored in registers C and D, representing POTENTIAL and AMPLITUDE, respectively.

No. typed	Sign stored temporarily	Octal no. stored temporarily	No. finally stored
+2.54	+	376	177
+1.28	+	200	100
+0.02	+	002	001
0.00	+	000	000
−0.02	−	002	377
−1.28	−	200	300
−2.54	−	376	201

The microprocessor must sense the typed sign, accept the typed number ignoring the decimal point, and convert it to a straight binary representation. If the number is positive, this representation is divided by two since each increment of one in the digital number is equivalent to a 2-mV increase in the analog output of the D/A convertor. Division by two is easily accomplished by rotating the number one place to the right (this is the reason registers are designed to be rotated).

OCTAL 002_8 001_8
—ROTATE ONCE RIGHT→
BINARY $00\ 000\ 010_2$ $00\ 000\ 001_2$

If the number is negative, all of the bits are then complemented (all 1's become 0's; all 0's become 1's), and the result is incremented by one.

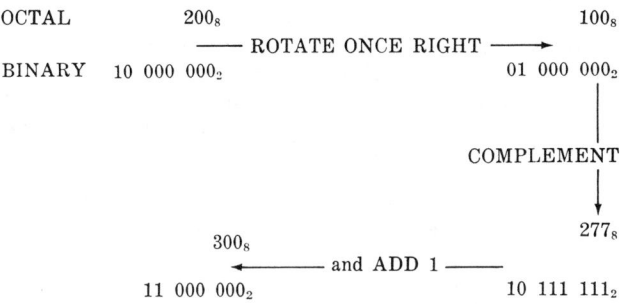

Try this on another value yourself!

With the initial dialog ended, the plate POTENTIAL is loaded from register C into the A register, output to the D/A convertor by use of an OUT1 command, and the electrolysis is started.

But how is time measured? One way is to use the fact that computer operations take known, fixed amounts of time. For example, the following simple program:

Octal command	Mnemonic equivalent	Operation
040	CNTR, INCE	/increment register E
110	JPFZ CNTR	/if E ≠ 0 jump back to CNTR
	(continue)	/if E = 0 perform next instruction instead
	======	

involving the incrementing of the E register—001 002 003 .... 376 377 000—requires 20 msec in the 8008 CPU. If we did this three times, 1 min in real time would have gone by, AND if we decremented the B register (containing TIME), every time this happens, we could use a similar software routine to determine when to terminate the electrolytic plating! (B would go to zero when the TIME was up.)

At this point, we could turn "on" a clock running at 4 ticks/sec by activating the interrupt service (allowing the CPU to accept interrupts). At each clock tick an interrupt occurs, and the service routine vectored to will perform different functions depending upon which "tick" in each 1-sec cycle is involved. At each clock tick the number output to the D/A convertor would be incremented by 1, increasing the output voltage by 2 mV. In addition, a special function would be performed at each clock tick:

CLOCK TICK	$1 + N*4$	$2 + N*4$
SPECIAL FUNCTION IMPLEMENTED	Start ADC to get i before pulse (OUT6 command)	Add pulse AMPLITUDE to number output to D/A conv.
CLOCK TICK	$3 + N*4$	$4 + N*4$
SPECIAL FUNCTION IMPLEMENTED	Start ADC to get i during pulse (OUT6 command)	Subtract pulse AMPLITUDE from number output to D/A conv.

The completion of every A/D conversion causes another interrupt of the microprocessor which reads the data in via an INP1 command. In every "4-clock tick cycle," the first value is the prepulse current. This is stored temporarily in C register, which is now free. When the next A/D interrupt occurs, and the data are read into A, representing the pulse current, it is an easy matter to subtract the two:

Octal command	Mnemonic equivalent	Operation
222	SUBC	/subtract C from A

This result is now stored in memory. Space is available for 128 data points, allowing a strip voltage span of 1.024 V. When 128 data points have been taken, the program can "plot" the data out on the teletype, or type the results out and produce a punched tape for later use by using the ASCII format. It could also plot the data out onto a Y/t plotter by using the same D/A convertor used in the voltage sweep. It would be a simple matter to integrate the areas under each peak—simply by adding together data values occurring above a chosen base line. If "voltage windows" were defined in the ROM operating program (e.g., Zn strips from −0.7 to −0.5 V), the program could also locate the maximum in each peak, compare it to the "look-up-table" of known metals, and simply provide a normalized report as in the first two columns:

Metal	Relative concn	Abs concn
Zn	1.00	0.93 E-08
Cu	0.28	0.26 E-08

A calibration run (with known Zn/Cu concentrations) could be run first, letting the microcomputer calculate and remember its own calibration factors, producing the absolute concentrations shown.

Of course, it would also be a simple matter to make a run on supporting electrolyte alone, generating a background file that would be subtracted point by point from the data before it was reported. This would be very difficult to do if a hardware approach had been used.

This system could have been designed to utilize control buttons (read by a SENSE REGISTER) to determine what function (software program) was to be executed. Input-parameters could be entered by a thumb-wheel switch, the BCD data being read into the A register via an input port.

Enter Time	Enter Potential	Enter Amplitude
Start Analysis	Start Calibration	Restart

A simple strip printer could be used for output.

Note that this application utilizes the power of a microcomputer properly. The program is contained in ROM; data entry from the teletype or switches uses RAM as a scratch pad and deposits all necessary operating parameters in the CPU's registers A–E. During stripping almost all of the operations involve these registers, accessing memory mainly for storage of data. The interrupt requirements are not complicated, and the mathematical operations are simple. The data base is small. Table I shows other areas in which microcomputers will find applications. If your area of interest is touched by any of them, the succeeding paragraphs discuss some of the problems in getting started.

Advantages/Disadvantages

Microcomputers should be considered when automation involves primarily control, with minor computation. Memory access in many of these devices is currently relatively slow and awkward. Software floating point packages (to perform four function arithmetic functions with high precision) are available, but they require approximately 1000 words of memory. Hardware math will come (such as the IC's used in calculators), but not for several years. Microprocessors should not be considered *currently* when massive data banks or programs above 3K are needed, because of the cost of the solid-state memories. The system is then approaching available minicomputer prices. Just what are the real advantages of microprocessor systems?

Primarily, the unique advantages in microprocessor systems, in comparison to their currently available competitors, are modularity, programmability, or cost. Most minicomputers, even in their stripped-down forms, are sold as complete systems with operating power and memory often in great excess of what is needed for the proposed task. Microprocessor systems can be purchased with memory increments (ROM or RAM) of 256 words. The interrupt structure can be added and augmented as needed. A control panel is usually not needed. Minicomputers usually are provided with a minimum of 4K of memory, a built-in program interrupt structure, teletype, and operator console. Thus, careful consideration of your actual needs and implementation via microprocessors can lead to considerable cost savings in the initial investment. Hardware implemen-

Table I. Potential Applications of Microprocessors

Electrochemical experiments
  Pulse polarography
  Stripping analysis
  Programmed triangular voltammetry
  Controlled potential electrolysis

Chromatography
  Liquid chromatography
  Gas chromatography
  Gel permeation chromatography

Optimization
  Mass spectrometers
  Magnetic resonance equipment

Process monitoring
  pH
  Pressure
  Refractive index

Sequencing
  Peptide synthesizers
  GC/LC programmer

Spectroscopy
  Control and data collection in X-ray fluorescence spectroscopy
  Scan control and report generation in densitometers
  Data reduction/correction in absorption/fluorescence spectroscopy

Data handling
  Format changing
    EBCDIC/ASCII
    Adapting data tapes to new protocols
  Data compression
    Peaks to area/position
    Report changes only
  Data reduction
    Generate derivative of data
    Apply calibration factors

Control and data acquisition
  pH-stats
  Automatic titrators
  DTA, TGA
  Autoanalyzers

---

tation of these needs might be cost-competitive, but any hard-wire implementation requiring 30 IC's or more should be reexamined for potential microprocessor utilization. The reprogrammability of such devices often becomes a deciding factor as future alterations in the system are envisaged.

Who/Where/What?

For those having some capability in-house in software development, hardware design, and troubleshooting (or those willing to develop the necessary skills), the microprocessor opens up the area of instrument development for the analytical chemist in the clinical, quality control, and research laboratories. Most chemists would rightly be hesitant about undertaking the hard-wired development of a wave-form generator suitable for the stripping analysis. But the logical nature of the software used by a microcomputer to accomplish such a task, augmented by the electronic power offered by the microprocessor, without the need to completely understand all of its intimate workings, should liberate the imagination of the analyst. Since the transistor/transistor logic (TTL) levels output by these devices can be current and voltage boosted with new solid-state relays, it is possible to control many items in the laboratory.

In chromatographic experiments, complex sequencing, temperature control, and report generation are obvious applications. Instruments should be capable of calibrating and optimizing themselves to yield the most accurate data and highest S/N ratio. Certainly the lack of imagination in instrument designers is most highly demonstrated in magnetic resonance and mass spectrometric equipment which have $40,000 data processing units attached but which do not automatically optimize themselves. Microcomputers will perform such tasks even in minimum configuration machines.

Correlative process monitoring, in which many transducer types continuously examine a number of system variables and produce a concentrated report, is within the scope of microcomputers. Availability of simple solenoid controlled valves and ion specific electrodes makes construction of closed-loop control devices utilizing microprocessors very simple. pH-stats for biological research controlling deviations in BOTH directions can be built at prices less than current commercial models utilizing hardware alone which control in only one direction. In many of these cases, the system function can be altered on a day-to-day basis by merely plugging in a new ROM program board.

Many of the tensile strength, deformation, and uniformity tests so common to the plastics industry's quality control laboratory will succumb to microprocessors since decentralization will reduce the signal transmission and program development problems associated with the single system. Clinical laboratories will utilize microcomputers in conjunction with autoanalyzers and autotitrators to not only control the data acquisition process but provide data compression before transmission to a central data base. The segregation here will also provide independence in case of system failures elsewhere.

Hospitals in British Columbia are already using them to provide inexpensive, portable data compression units for field monitoring of electrocardiograms. In a similar fashion, remote monitoring devices for effluent control by use of specific ion electrodes can be constructed and connected to a central computer for correlation of data via the telephone lines. A system based on this concept, by use of the respiration rate/cough rate of fish, has been designed at VPI.

Programming

At the present time, programming microcomputers appears to be a formidable problem to the beginner. He is faced simultaneously with the electronic aspect of interfacing AND learning assembly language programming and implementing it. There are cross-assemblers running on commercial time share equipment which can cross-assemble from a sub-set of PL-1 to machine language code that is acceptable to the microprocessor. In many environments such a facility is not easily assessible or cost justifiable if only a few programs are to be developed. Also, despite some vendor's claims, it creates code that uses costly memory inefficiently. On the other hand, production of machine code by an assembler resident in the microcomputer itself is also usually not cost justifiable if the user is involved with only a few microprocessors that are not continuously being altered. These programs are available, but too much ROM and RAM are required for the resident editor/assembler. One avenue that has been successfully employed at a number of installations is to prepare a cross-assembler that will run on available minicomputers. In some cases, these cross-assemblers have been written from scratch; these are available for the PDP-10 and PDP-8. Many laboratories have found it sufficient to take the assembler that is resident in their minicomputer and expunge its permanent symbol table, replacing it with the instruction set for the microprocessor. A few coding tricks and a translation routine to convert from the 12- or 16-bit output of the mini's assembler to the bit format of the microprocessor are all that are needed to generate code rather easily. Documentation and copies of a PDP-8 resident combined editor/assembler called TEACH which will generate code for the Intel 8008 are available from the authors. However, in program development it is also important to have available a set of ROM chip memories containing a basic set of subroutines, callable from the teletype, to aid in program prepara-

Figure 4. An interface teaching station at VPI&SU

tion and debugging. Some of these are available from vendors, but they vary in quality. A typical set that has been useful at VPI called OPERATE contains:
- paper tape reader routine
- paper tape punch routine
- memory examination/alteration routine
- break point (trap) routine
- single-step (trace) routine

and occupies 768 words of ROM. Documentation and copies are available from the authors.

Software resident in read/write memory can be used to collect data and altered to implement changes in the way data are manipulated or the way data are collected. But the volatility of these programs through operator error or power failure is often a concern, particularly where field work or technicians are involved. Microprocessors have been the focus for the utilization of *r*ead-*o*nly-*m*emories which have their programs "burnt" into them at the factory or in the field by ROM "blasters" which permanently fuse links in the IC device. These are truly nonvolatile memories, but they also cannot be changed. *E*rasable *p*rogrammable *r*ead-*o*nly-*m*emories (EPROMS) are floating gate-MOS circuits which can be biased by avalanche charging near the gate by high currents. Once biased, these devices remain a logic 1 until some conduction pathway is made available to the biasing electrical charge. This requires intensive UV irradiation at the biased junction. EPROMS are provided with quartz lids covering the memory area, and exposure to the UV sources used in TLC will erase their memories in 3–4 min. They may be reprogrammed in the same amount of time with a ROM blaster. Over 200 reprogrammings are possible with current devices. They thus provide FIRMware, less volatile than SOFTware, but more easily changed than HARDware. Vendors will program/reprogram your EPROM's for ~$10 if you provide a paper tape containing the program you want loaded.

Hierarchy

Connecting a microcomputer to a larger minicomputer system (a hierarchical or satellited configuration) is often desired; it might be for the purpose of acquiring a data base from the microcomputer and performing data reduction tasks on it, and then returning the concentrated data to the microprocessor for presentation or action. Alternatively, program preparation in the minicomputer could occur, with subsequent depositing of the program in the microcomputer and initiation of execution. This might be the result of the microprocessor sensing unusual conditions in the experiments it is controlling and requiring an altered operating program to meet these demands. Increasingly, the microprocessor is used to collect and compress a data base from several instruments for passing at a convenient time to the minicomputer at rates of 2–400 words (8 bits) per second. They can be accepted by the minicomputer by a serial interface and, to all intents and purposes, look like a TTY to the minicomputer. The concepts involved are discussed in more detail in the article entitled "Instant Interfacing" (4). For higher speeds and more sophisticated interrupt handling, parallel transmission can be used. Details of such a system operating in our VPI laboratories are available.

In the Teaching Lab at Virginia Tech

With their modern architecture, at small cost, microcomputers offer an alluring route to the low-cost introduction of computing philosophy and techniques to undergraduate, graduate, and postgraduate students. This can be accomplished only by a "hands-in" process, in contrast to "hands-on" experience which exposes the student to a teletype and turn-key laboratory data acquisition system. Hands-in experience can only be attained by exposure of the entire I/O bus of the CPU—a process often precluded by cost and extensive use of the computer available in the laboratory.

For $2,000 a complete interface teaching station has been provided, for Virginia Tech students and ACS Short Course participants, centered around microprocessors (Figure 4). The material learned can be extrapolated directly to minicomputers also.

Typically, the student examines the nature and time duration of SELECT DEVICE and IN/OUT pulses and uses them to input and output data. A program in ROM (OPERATE) allows the student to enter small programs into RAM bank and execute them, one step at a time. OPERATE reports the contents of registers A–E at each step.

For example, a typical TTY dialog might be as follows: human response is *underlined;* LF is a line-feed; CR is a carriage return.

TTY dialog	Operation
4100/713 006 LF	/Change memory location 4100 to 006
4101/700 014 LF	/To load A with 014
4102/000 010 LF	/Increment B
4103/777 201 LF	/ADD B to A
4104/121 320 CR	/Deposit sum in C
4100B CR	/Set break point at 4100
4100G CR	/Start program at 4100
000 000 000 000 000 S4100	/All registers (A–E) = 000 /Now do next step
014 000 000 000 000 S4102	/A = 014
014 001 000 000 000 S4103	/INC B
015 001 000 000 000 S4104	/ADD B to A
015 001 015 000 000	/And Store in C

No computer console is involved; therefore, the trauma of operating switches and thinking in binary are not encountered. In addition, the basic program OPERATE is in ROM memory where it cannot be destroyed by the student accidentally.

From this base the student examines sense switch and interrupt operation, printing out data from several sources every 10 sec. In this, as well as all succeeding experiments, the student must generate his own software and hardware.

Other typical experiments are:

Table II. Representative Microprocessor Suppliers

Company	Description
American Microsystems Inc. Santa Clara, Calif.	CK114 microprogrammable serial processor suitable for calculator operations. *Seven*-chip set. 75 instructions. True microprocessor expected mid-1974
Computer Automation Inc.	16-bit "naked mini" computer built around microprocessor chips is aimed strictly at OEM field. Mounted on single 15 × 17 board, it sells for around $1,000 in lots of 200
Control Logic Inc. Natick, Mass.	Complete microprocessor system available on 3 × 5 pc cards which mount in pre-wired card frame. Built around Intel 8008. Basic 3-card processor set is $335. Basic software available includes loader, ODT, assembler, and editor
Digital Equipment Corp. Maynard, Mass.	Complete microprocessor system on 11 × 8 pc cards which mount in DEC block. Built around Intel 8008. Basic 3-card processor set, plus some memory and interrupt capability, is $745. Cross assembler for PDP-8 family and standard software available
Fairchild Semiconductor Mt. View, Calif.	PPS-25 is a *seven*-chip parallel BCD set with 93 instructions. True microprocessor anticipated in 1974
General Automation Anaheim, Calif. (Microelectronics Division Rockwell International)	LSI 12/16 series with CPU/memory on 10 × 8 pc board. Software available. $495 in quantities of 1000
General Instrument Co. Hicksville, N.Y.	16-bit parallel processor and control chips available in 1974 as chips or cards
Intel Corp. Santa Clara, Calif.	4-bit (4-chip) parallel processors and 8-bit (1-chip) processors available as chips or systems in a variety of forms. Second generation version of 8008, the 8080, is being shipped. Software available
Motorola Semiconductor Products Phoenix, Ariz.	5-chip 8-bit processor expected 1974
National Semiconductor Corp. Santa Clara, Calif.	4-bit slice (4, 8, 12, or 16 bits) processors on boards or complete systems. IMP-8 card is $800; IMP-16 is $950. Software available
Pro-Log Corp. Monterey, Calif.	4- and 8-bit systems or cards built around Intel microprocessor chips
Rockwell/Microelectronics Anaheim, Calif.	PPS-4 is a 4-bit BCD-oriented system
Teledyne Systems Northridge, Calif.	TDY-52A is an 8-bit microprocessor chip; TDY-52B is a 16-bit version

Other vendors who have microprocessors in development are: RCA, Somerville, N.J.; Signetics Corp., Sunnyvale, Calif.; Western Digital, Newport Beach, Calif., and Raytheon Semiconductor, Mt. View, Calif.

- Digital Control—control 5-V motors (on/off; forward/reverse) for valve control
- Data Acquisition and Display—accepts analog data from GC, converts it to digital data; displays data as it is acquired on CRT; and punches final data file onto paper tape in ASCII format for utilization by a larger computer
- Interrupt Structure—monitor a number of external events (TGA, GPC) at the same time as CPU reports data from a previous run.

At the end of 30 hr of laboratory, the student is aware of the philosophies of interfacing and how he can utilize computers in his laboratory work. Details and courseware materials are available from the authors.

Reading Specifications

Keeping abreast of the microprocessors already available and those that have been announced (but often not even yet designed) is a serious problem. A list of vendors is given in Table II. The bit size (4, 8, 12, 16, etc.) and the number of general-purpose registers should be compared. The word length should easily accommodate your experiment's resolution [8 bits represents 1 part in 256—using two 8-bit words for data storage (double precision) will give a resolution of 1 in 64,000]. The more general-purpose registers there are, the easier programming will be. Current microprocessors can address 16 to 64,000 words of memory, but for most applications, memory limitation is not important, since cost-wise it would be hard to justify such a large microprocessor system in comparison to a minicomputer.

Some microprocessors, particularly the 4-bit versions, are BCD oriented and lend themselves readily to the human entry of data by touch-pads and keyboards. Their instruction set is limited in comparison to 8- and 16-bit machines which are binary.

Some microprocessors have both binary and BCD instructions. A vectored interrupt system is very desirable, since it reduces programming difficulties, but extremely sophisticated designs should be avoided because they lead rapidly to extremely sophisticated software, and another microprocessor might be cheaper in the long run.

Physical configurations are as varied as the internal architecture. Three distinct classes appear to be evident:

- Chips only—the integrated circuit CPU's and control IC's are available for the user to incorporate in his own printed circuit cards. This is definitely only for well-equipped and staffed electronic shops.
- CPU cards—the processor, control circuitry, and some memory are available on a printed circuit card. The user must add device decoding, advanced interrupt service, gates, and latches. A hardware background is essential to use them.
- Systems—the vendor provides a card frame and cards to fit it containing CPU memory in various forms, device decoding, interrupt service, gates, latches, and interfaces for teletypes, other computers (via serial data communication), relays, and A/D and D/A convertors. If you can specify your needs clearly, most chemists will find this route the best.

References

(1) R. E. Dessy, P. J. Van Vuuren, and J. A. Titus, *Anal. Chem.*, **46** (11), 917A (1974).
(2) M. S. Krause and L. Ramaley, *ibid.*, **41** (11), 1365 (1969).
(3) R. E. Dessy and J. A. Titus, *ibid.*, **45** (2), 124A (1973).
(4) R. E. Dessy and J. A. Titus, *ibid.*, **46** (3), 294A (1974).

AUTOMATED REACTION-RATE METHODS OF ANALYSIS

HOWARD V. MALMSTADT
EMIL A. CORDOS
COLLENE J. DELANEY

School of Chemical Sciences
Roger Adams Laboratory
University of Illinois
Urbana, Ill. 61801

Reaction-rate methods of analysis often offer advantages in selectivity and accuracy. New automated instrumentation encourages the development of specific procedures and the likely adoption of these methods

The first "automated" reaction-rate (kinetic) methods and instrumentation for the rapid selective determination of glucose and other constituents were reported over a decade ago (1–6). Through the 1960's, analytical methods based on the measurement of initial rates of chemical reactions grew in importance as many new sensitive and selective procedures were reported, and the instruments were improved. The frequent journal review articles and recent books attest to the interest in kinetic methods (7–12). It seems now that it will be the middle of the 1970's when reaction-rate methods become generally accepted and widely used on a routine basis.

It is certainly not coincidence that the conversion from research interest to general routine use of reaction-rate methods corresponds to the development of a new generation of elegant, completely automated, and computer-controlled chemical instrumentation. The rather recent concepts and advances in analytical instrumentation make it just as easy to obtain sensitive quantitative chemical results with reaction-rate methods as with conventional stoichiometric (equilibrium or endpoint) methods. By elimination of the barriers imposed by difficult laboratory rate techniques, it now becomes worthwhile to consider the inherent advantages of reaction-rate methods.

In this report the inherent advantages and possible limitations of reaction-rate methods as compared to equilibrium methods are reviewed, and the general concepts of encoding reaction-rate information are presented. However, the major discussion is focused on the automated systems that make it possible to perform hundreds of accurate, sensitive, and selective quantitative determinations per hour via rate data and to develop new methods more rapidly.

ADVANTAGES AND LIMITATIONS OF REACTION-RATE METHODS

With rate methods it is often possible to measure, immediately after mixing the reactants, the rate of change of some parameter P of the reactant whose concentration is to be determined, or other reactant or product of the reaction, and not wait for the reaction to go to completion (equilibrium). This is illustrated in Figure 1. The saving in time may or may not be significant depending on the specific reaction, but there are good examples (1–12) of obtaining quantitative rate results in seconds for selective reactions that would have required many minutes or hours to go to completion. This is especially true for many of the highly selective enzymatic reactions.

Because it is possible to obtain quantitative rate data shortly after the reagents are mixed, the measurement may be completed before interfering side reactions begin. This can be a distinct advantage in providing higher accuracy for some determinations.

One of the most important characteristics of the reaction-rate method is that it involves a relative measurement. The absolute value of the parameter (i.e., absorbance, cell potential, fluorescence) chosen to monitor the reaction does not have to be measured accurately, as shown in Figure 1. It is only necessary to measure the parameter's time rate of change with high precision and accuracy. Hence, even for extremely rapid reactions, the reaction-rate method can offer freedom from those interferences which contribute to the absolute value of the parameter (turbidity, dirty cells, junction potentials, and other fluorescing materials) but do not enter the chemical reaction and do not contribute to the rate of change of the parameter with time.

Of course, there are those applications where it is the rate per se, rather than the absolute concentration of a specific species, which is the important quantity to be determined. Most noteworthy in this category is the determination of *enzyme activity*.

Another possible advantage of kinetic methods is that they sometimes provide a means of determining the concentration of two or three constituents of closely related chemical properties without physical separation. As a rule of thumb, the successful development of differential rate methods requires that the first-order rate constants of the individual components differ by at least a factor of 10. For example, silicate and phosphate in mixtures have been determined by a differential rate procedure based on the formation reactions of the heteropolymolybdate and the reduced heteropolyblues (*13*).

There are some limitations in the general application of reaction-rate methods. The most important is that imposed by the reaction rate itself. The half-time of the reaction must be greater than the mixing time of the instrumental system available. Considering the other extreme, very slow reactions with half-times greater than a few hours are not too practical for routine analyses. Also, the accuracy and precision of the measurement depend upon good reproducibility (although not necessarily good accuracy) for all experimental conditions such as temperature, pH, ionic strength, size, and shape of reaction vessels.

GENERAL CONSIDERATIONS IN ENCODING CHEMICAL REACTION-RATE INFORMATION

Determination of Glucose. A few practical examples of reaction-rate methods are presented here to illustrate the general considerations involved in the encoding of rate information. The first example is the quantitative determination of glucose by use of the well-known selective oxidation of glucose in the presence of the enzyme, glucose oxidase, as illustrated by Equation 1.

$$\text{Glucose} + O_2 \xrightarrow{\text{Glucose oxidase}}$$
$$\text{Gluconic acid} + H_2O_2 \quad (1)$$

Encoding Indirectly Via Coupling Reaction. In our own laboratory we first became heavily committed to the development of rate methods more than a dozen years ago by devising an automated rate measurement system for the determination of glucose in blood serum (*1*). For expediency we first chose to determine the rate of change of glucose by indirectly following the rate of formation of H_2O_2. Since the H_2O_2 did not have a physical parameter that could be easily and directly measured, the relatively fast coupling reaction of H_2O_2 with an organic dye was used. This was conventional practice at the time for the end-point methods because the colored reaction product, as illustrated in Equation 2, provided high sensitivity by photometric measurements.

$$\text{Organic dye A} + H_2O_2 \xrightarrow[\text{Enzyme}]{\text{Peroxidase}}$$
$$\text{Colored product B} \quad (2)$$

Unfortunately, the peroxidase enzyme that catalyzed the coupling fast reaction was rather unstable and expensive, and the automated two-point, reciprocal time, rate-measuring system was at first relatively crude. But from this example, it became apparent that sensitive and precise quantitative measurements could be realized with the aforementioned advantages over end-point methods if suitable reactions could be selected and controlled; sensitive devices for converting (transducing or encoding) the concentration change for one of the reaction species to a measurable electrical signal could be developed; and reliable electronic systems for the rate measurement could be developed.

The glucose reaction then became a test system for demonstrating new encoding systems and rate measurement systems. By use of a different fast coupling reaction, as illustrated in Equation 3, it was possible to determine the rate of change of H_2O_2

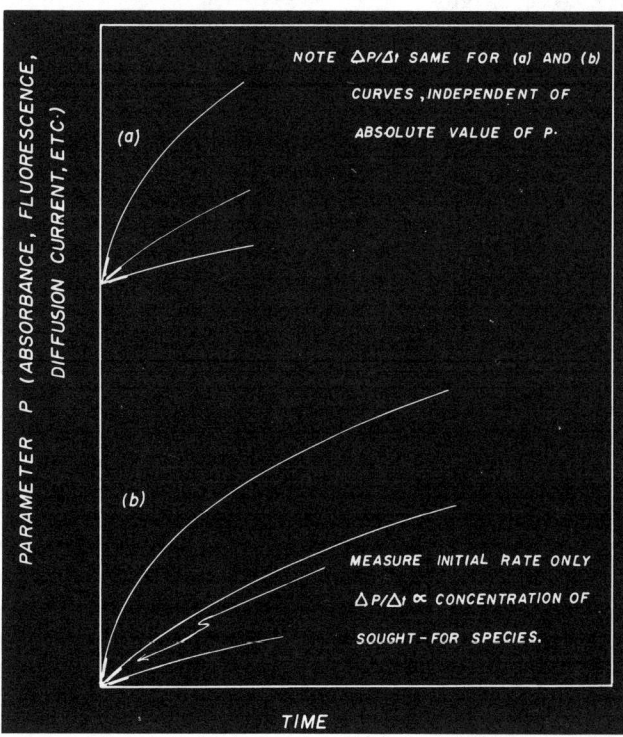

Figure 1. Initial rates of reaction proportional to concentration of sought-for species

$$H_2O_2 + 3I^- \xrightarrow[\text{Catalyst}]{\text{Molybdate}}$$
$$I_3^- + 2OH^- \quad (3)$$

by following the I_3^- with potentiometric (2), spectrophotometric (3), and amperometric (14) encoding (transducer) systems, and to eliminate the troublesome peroxidase enzyme.

In Equation 1 the rate of formation of gluconic acid could be used to obtain the initial rate information. Therefore, a sensitive digital pH-stat system (15) was devised wherein the pH was held constant by adding small increments of NaOH during the glucose reaction. The number of increments of base added during a short fixed period of time proved to be directly proportional to the glucose concentration. However, the system is not as sensitive as the methods based on the H_2O_2 coupling reactions.

Encoding Directly Via Primary Reaction. The obvious desire to eliminate any type of secondary or coupling reaction also led to the investigation of direct methods for following the rate of change of O_2. Voltammetric encoding systems that could measure O_2 directly (16–18) were developed, although again the method was not as sensitive as the indirect H_2O_2 color methods. The conversion of glucose or gluconic acid rate information to a directly measured physical parameter has not been reported.

Determination of Phosphate.
Encoding Indirectly Via Secondary Reaction. The determination of phosphate by a rate method illustrates similar considerations as those for the glucose determination. First, the classical molybdenum blue procedure could be developed into a sensitive rate method involving the primary reaction of phosphate with Mo(VI) to form 12-molybdophosphoric acid (12-MPA) and its subsequent reduction to form the heteropolyphosphomolybdenum blue, PMB (19), as shown in Equations 4 and 5.

$$H_3PO_4 + 6Mo(VI)_t \longrightarrow$$
$$12\text{-MPA} + 9H^+ \quad (4)$$

$$12\text{-MPA} + n\text{Red} \longrightarrow$$
$$PMB + nOx \quad (5)$$

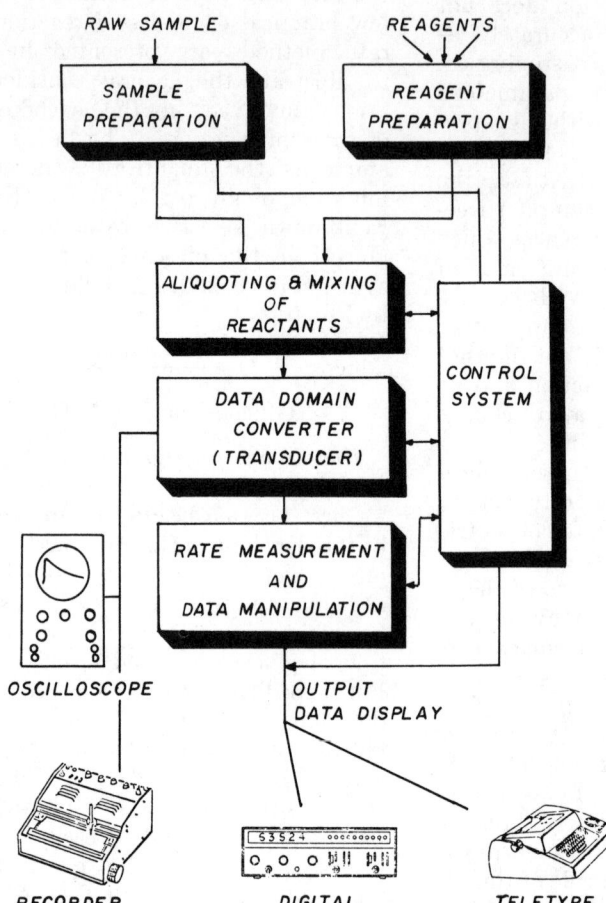

Figure 2. Block diagram of complete instrumentation system for reaction-rate methods

In this case, the measured reaction rate depends on the nature of the reductant, order of adding reagents, as well as acidity, etc. (19).

Encoding Directly Via Primary Reaction. It was again obvious that it would be advantageous to eliminate the reduction reaction (Equation 5) and convert one of the species in the primary reaction to a physical parameter that could be readily measured. Fortunately, the 12-MPA has a relatively high absorptivity at readily available wavelengths. However, for the first time in our development of quantitative reaction-rate procedures, there was a confrontation with a reaction whose half-life was so short that it was necessary to make rate measurements in milliseconds rather than seconds. A rather long development program was thus required to develop the new automated equipment that would provide rapid mixing and readout of rate information. After the automated system was developed, it was then possible to determine phosphate directly (20), utilizing the primary reaction shown in Equation 4. It is feasible to make as many as 3000 determinations of phosphate concentration per hour with this type of system.

General Conclusions. It is seen from these examples for glucose and phosphate that a sequence of chemical reactions is often used to obtain a chemical species that is related to the desired rate information and readily converted to a sensitive measurable signal, or to obtain a readily measured chemical species whose rate of formation is slow enough to be measured by available equipment. Although the methods utilizing multiple sequential reactions can be dependable under carefully controlled conditions, almost inevitably they will be less reliable than methods using measurements on one primary reaction.

Much research effort has been invested in providing more sensitive and selective encoding systems and high-speed automated rate-measuring instruments, and a rela-

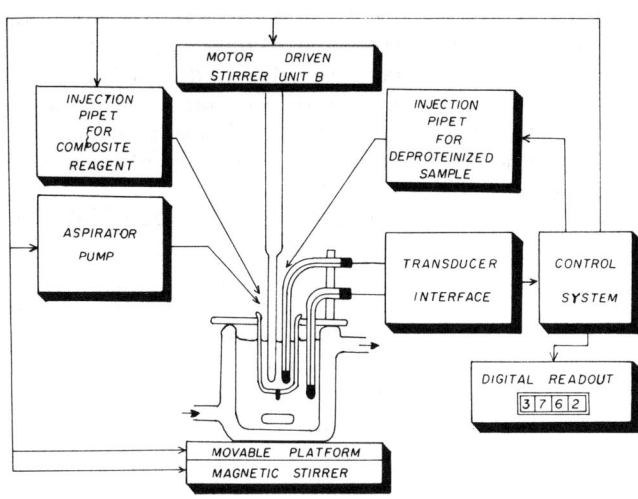

Figure 3a. Block diagram of automatic potentiometric reaction-rate analyzer (5)

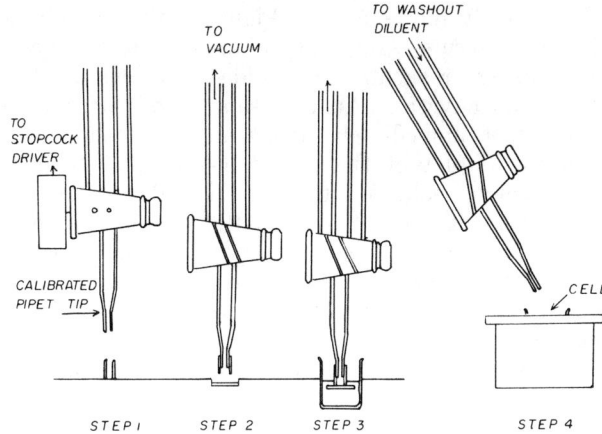

Figure 3b. Manipulations for filtering, measuring, and delivering deproteinized serum or plasma samples. Step 1: picking up polyethylene sleeve. Step 2: picking up filter paper. Step 3: filtering deproteinated sample. Step 4: delivering sample into reaction cell

tively large choice of methods is now available (1–21). The method of choice for encoding the rate information from either the primary or coupling reactions depends, of course, on several factors including sensitivity, freedom from interferences, simplicity, and dependability.

INSTRUMENTATION SYSTEMS FOR REACTION-RATE METHODS

The work on automating reaction-rate methods in the past decade primarily provided improved rate measurement devices (as described in this month's Instrumentation feature, page 79 A) and sensitive data domain transducers for converting the chemical rate information to measurable electrical signals. Today, the attention is being focused on complete systems that start with the raw samples and reagents and end with a formatted printed readout in the desired quantitative units. A general system for obtaining quantitative data is illustrated by the block diagram in Figure 2. By preliminary treatments (e.g., dissolution, dilution, filtration, ion exchange) the sample and reagent solutions are prepared as required for the specific procedures. Predetermined volumes of sample and reagents are then introduced, mixed, and transported to a vessel which serves as the reaction cell. The chemical reaction is monitored by a suitable transducer which converts the rate information about a specific species inherent in the reaction to a measurable signal in the electrical domain. Frequently, several interdomain conversions are required to obtain data in the preferred form. The control and rate measurement systems can be hardwired for specific applications, or they can be incorporated in a minicomputer-interfaced system that can provide through software much versatility in control of the measurement sequence and the processing and readout of data. Readout is visually displayed with digital lights or printed out on serial teletype or a high-speed parallel printer. When desired, a servo recorder or storage oscilloscope can display the parameter vs. time and rate curves.

The weakest links in reaction-rate instrumentation systems in the recent past have been the sample and reagent preparation and aliquoting and mixing systems. In fact, it becomes apparent that the greatest differences in completely automated reaction-rate instruments will probably be the methods of sample and reagent preparation, aliquoting, and mixing.

These operations are not only tedious, repetitious, and time demanding when done manually but are subject to human error and bias. It is not unusual that the sample-handling procedures require much more time than the rate measurement itself. This, of course, is not a problem exclusively associated with the development of reaction-rate methods; rather, it is an important consideration in the development of all automated analytical methods.

One of the first reaction-rate instruments that incorporated automated pipetting of sample and reagents and provided rapid deproteinization and filtration of serum samples was presented by Malmstadt and Pardue (5). The instrument was designed specifically for glucose in serum determinations. A potentiometric concentration cell, Figure 3a, was designed as the transducer to encode the initial rate information. One automated injection pipet (22) was used to deliver a 1-ml aliquot of composite reagent to the sample compartment. A second automated pipet delivered a measured aliquot of diluent to wash out the calibrated delivery tip that contained an accurate aliquot of deproteinized serum. The rate information for glucose was automatically measured within about 30 sec. The reaction mixture was then removed through an aspiration tube, and the cycle repeated.

The deproteinization, filtration, and pipetting of an accurate aliquot of sample into the reaction cell is schematically represented in Figure 3b. The serum sample and deproteinizing reagents are added to a small cuvette. Another sampling pipet immediately draws an accurate aliquot of the deproteinized sample solution through a glass–fiber filter which retains the precipitate. When the calibrated tip is filled with sample filtrate, the filter is knocked off, and the three-way valve is turned to connect the diluent injection pipet to wash out the sample and diluent into the reaction vessel. All of the operations illustrated by Figure 3b can be performed automatically.

Although it is, of course, preferable to develop a procedure that does not require filtration, the method does illustrate how classical deproteinization, filtration, and pipetting operations can be automated in discrete steps. Many of the newest commercial instruments operate in similar discrete steps. Considerable research, design, and engineering activities are currently being directed to automation of aliquoting and mixing of reactants, and many commercial models have only recently appeared on the market.

Automated Stopped-Flow Reaction-Rate System. The stopped-flow apparatus is widely used for kinetic studies (23, 24) of fast reactions, and it has been modified (20, 31) for rapid automated reaction-rate methods, as illustrated in Figure 4.

Many sample and reagent handling systems are only suitable for utilizing relatively slow reaction-rate systems. As pointed out earlier, there are advantages in selecting a fast reaction for analytical purposes. The limitation in making rate measurements on a fast reaction is that the time taken for mixing and observation must be shorter than the half-life of the reaction. Today, observation time can be reduced to the millisecond range by use of fast-responding detector-readout systems (25, 26) so that the real limitation becomes the time required for physical mixing of reactants and initiation of reaction. To minimize this time, mechanical systems are used which drive reactants rapidly enough to promote turbulent flow through the system and thereby insure rapid mixing and uniformity of solution composition in the observation cell.

Although a number of flow methods have been used for studying fast reactions, the stopped-flow method has been most widely applied for analytical purposes. This

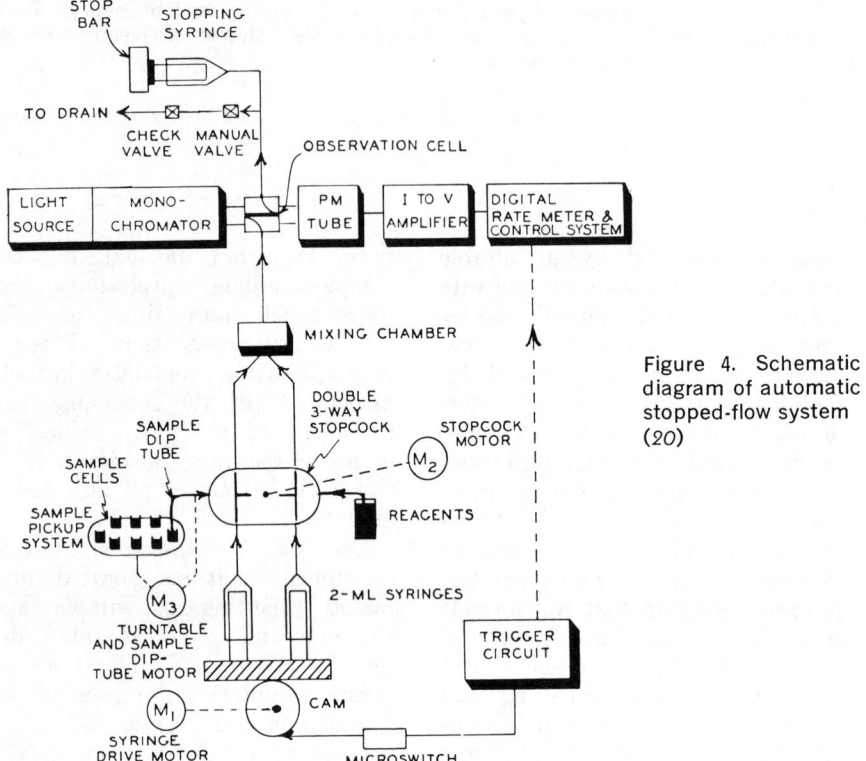

Figure 4. Schematic diagram of automatic stopped-flow system (20)

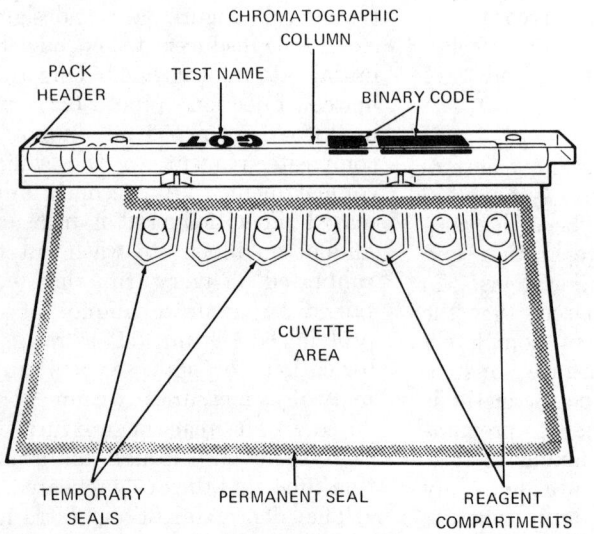

Figure 5. Analytical pack for Du Pont automatic clinical analyzer
Courtesy of E. I. du Pont de Nemours & Co., Wilmington, Del.

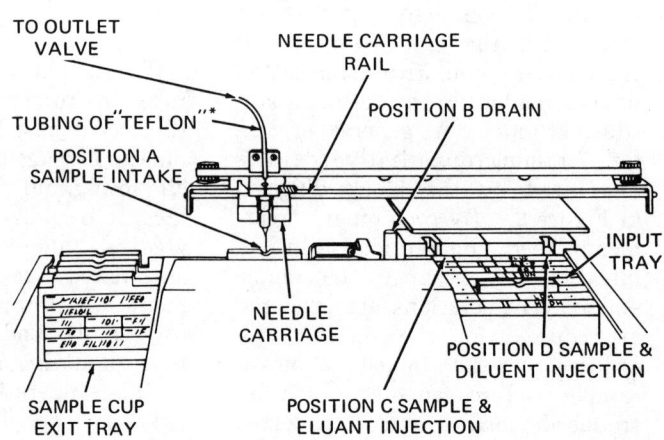

Figure 6. Filling station for Du Pont automatic clinical analyzer
Courtesy of E. I. du Pont de Nemours & Co., Wilmington, Del.

technique consists of rapidly mixing reactants by forcing the solutions through a mixing chamber and into an observation cell. The flow of solution is abruptly stopped, creating a back pressure which completes the mixing, and the rate measurement is rapidly made. Once experimental parameters have been selected, and samples and composite reagent have been loaded on the sample tray, the modified stopped-flow measurement system (*31*) operates automatically and continuously until reagents are exhausted. A logic circuit is used as a control system for a number of operations. At the end of a measurement cycle, the logic circuit provides a signal which activates the syringe drive circuit, thereby injecting sample and reagent automatically. The number of rate measurements per injection can be preset on the logic unit. Likewise, the number of injections per sample cup is programmed. A preselected number of injections are used to insure complete flushing of the flow system of previous reacted sample. During this flushing interval, the readout system is locked by the control circuit. After the preset number of injections and measurements have been made, the logic circuit activates the sample tray, and the next sample cup is brought into position. A teletype is used to log the data and to generate a paper-punched tape. The paper tape is subsequently used to input the data to a small computer (Digital Equipment Corp. PDP-8L). Results for standards and samples are computer averaged and corrected for blank if necessary. A least-squares routine is used to provide printout of concentration of the samples.

Automated System with Prepackaged Reagents. A novel method for automation of reaction-rate and equilibrium techniques minimizes the problem of laboratory reagent preparation and contamination by prepackaging all required reagents in stable form in a disposable plastic pack. The same pack also serves as the reaction chamber and observation cell. This approach has been initiated by Du Pont (E. I. du Pont de Nemours & Co., Wilmington, Del.) in the development of the automatic chemical analyzer (aca) system.

As shown in Figure 5, the reagents are contained in plastic bubbles near the top of the pack. The pack header contains the name of the test and a binary code which is interpreted by a small, ROM (read-only-memory) computer. The code specifies experimental conditions such as sample volume, reagent dilution, and processing cycle in addition to providing for automatic setting of specified instrumental parameters and for manipulation of the data so that the final printout is in desired units. Since the code is embossed on the top of the header, the packs may be loaded in any order to provide the desired sequence of analyses for a given sample. An additional feature is the inclusion of a gel filtration or ion-exchange column in those packs for determinations which require the elimination of interfering components.

Initially, the sample is placed either manually or by automatic sampler into a plastic cup to which an identification card is clipped. The cup assembly and test packs, after being manually loaded on the input tray of the analyzer, are automatically transported into the filling station shown in Figure 6. Here, the binary code on the individual test packs is decoded by the ROM computer, and the filling needle moves sequentially to perform the operations of flushing the delivery needle (position B in Figure 6), withdrawing the specified volumes of diluent (or elution diluent if a chromatographic column is used) and sample (position A in Figure 6), and injection of sample and diluent into the pack (at position D or C in Figure 6).

The fluid metering system consists of a series of reagent reservoirs which are connected to a piston pump through a series of precision microvalves. The pump is driven by a stepping motor. The volume of liquid drawn and delivered by the pump is proportional to the number of impulses applied to the stepping motor by the built-in computer. The operation of the microvalves is also computer-controlled.

At the conclusion of the filling cycle, each pack is automatically moved into the main processing section of the analyzer. As the pack proceeds on a transport chain, it is heated to and maintained at a constant temperature, and reagents are released and mixed at two breaker-mixer stations by rupture of the reagent bubbles and oscillation of the pack.

Upon completion of this cycle, the pack advances to the photometer station where a set of mechanical jaws close around the pack generating hydraulic pressure within the pack to form a precision cuvette between quartz windows. The initial rate is measured by use of the fixed-time technique, and the computer automatically supplies an appropriate conversion factor so that concentration or activity is printed out on a report sheet along with the sample identification number. The ROM computer is also used for surveillance of all operations required for processing the sample. Should a malfunction be detected, the system also prints out a code on the report sheet indicating the source of error.

Parallel Vs. Sequential Mixing of Samples and Reagents. In the sample and reagent handling approaches described in the above methods, the reactants are aliquoted and dispensed into the mixing chamber, and the process repeats sequentially for a series of samples. In most cases, the reagents are directly expelled from a pipet into the mixing chamber where the mixing operation is immediately carried out. An alternate approach can be taken by using a time separation of these two operations and then the simultaneous (parallel) mixing of reactants for a group of samples. All reactants are premeasured and introduced into reactant wells. The mixing operation takes place at a later time.

This principle of parallel mixing is applied in the unique GeMSAEC analyzer developed by Anderson (*27*) and Coleman et al. (*28*) at Oak Ridge National Laboratory (Oak Ridge, Tenn.). The latest system allows parallel mixing of reactants for up to 42 samples. Centrifugal force is used to transfer reactants into separate mixing chambers which also serve as the observation cells.

Samples and reagents are manually or automatically pipetted into individual holding compartments of the transfer disk as shown in Figure 7. The transfer disk is then manually locked into place in the

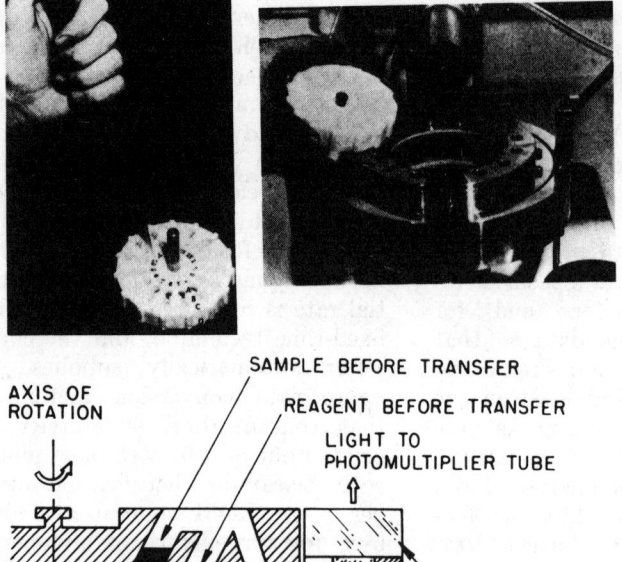

Figure 7. Sections of the GeMSAEC fast analyzer. Upper left: loading of transfer disk. Upper right: transfer disk and centrifugal analyzer section (27, 28)

Courtesy of International Scientific Communications, Inc., Green Farms, Conn.

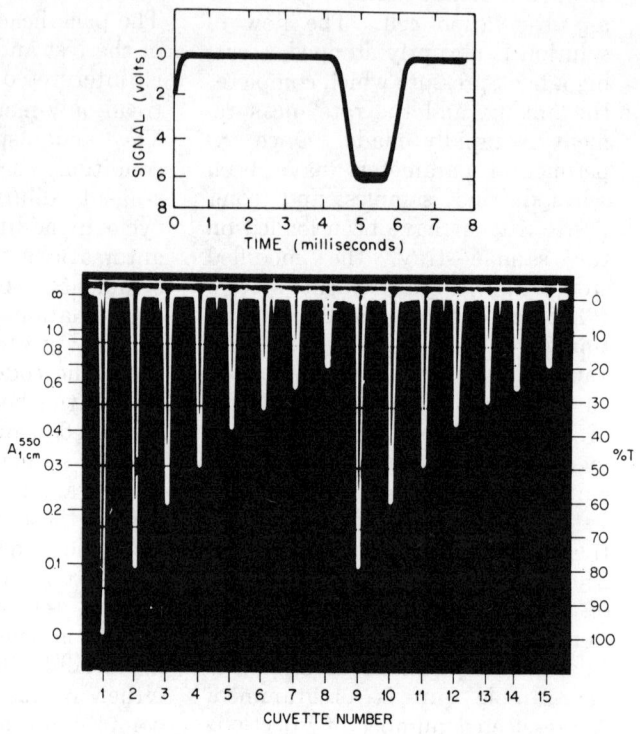

Figure 8. Oscilloscope traces from photometer output of 15-cuvette GeMSAEC analyzer. Upper: signal for a single cuvette. Lower: display for water blank and duplicate sets of protein standards (28)

Courtesy of International Scientific Communications, Inc., Green Farms, Conn.

rotor which also insures proper indexing of the samples. The holding compartments are arranged in two or more concentric circles in the transfer disk, with each circle containing the same number of compartments. The compartments are aligned radially and are angularly arranged so that samples and reagents are dumped through a transfer cavity into the corresponding reaction chamber (which also serves as a photometric cuvette) when the rotor reaches a sufficient speed (typically 400 rpm).

To prevent lateral splashing of the solutions in the transfer disk, the rotor is gradually accelerated, and then during a breaking interval, a pulsed vacuum is applied which draws a stream of small air bubbles through each cuvette. The resultant turbulence produces complete mixing of the reactants. The air bubbles are removed by rapid acceleration followed by deceleration to the normal operating speed of about 500 rpm. In present designs this entire process requires only a couple of seconds.

The cuvettes rotate past a stationary light beam completely interrupting it between cuvettes. The signal from the photomultiplier is continuously displayed on an oscilloscope which is synchronized with the rotor to provide an immediate visual monitor of the signals for all of the cuvettes. The photomultiplier signal is also passed through a buffer amplifier, directed to an analog-to-digital converter, and subsequently to a PDP 8/I computer. During each rotation of the rotor, the dark current value is obtained between each peak, and a blank peak value equivalent to 100% transmittance, in addition to a series of standard and experimental peak values, is determined, as shown in Figure 8 for a 15-place rotor. After the spectrophotometric rate or stoichiometric measurements are completed, the samples are dumped through drainage siphons located in each cell by application of an air stream to the center cavity of the rotor which extends radially to each cell. The rotor is then stopped, and wash water is added from a wash bottle to both the sample and reagent compartments of the transfer disk. The rotor operational sequence is repeated, allowing the wash water to be transferred to the cuvettes which are drained as described above.

To increase sample throughput rate, research is being conducted to provide automation of sample and reagent loading and of cleaning and drying of the rotor between runs. To extend the utility of the analyzer to more complex methodology, two additional research objectives reported (27) are incorporation of ion-exchange and gel filtration columns into the system and provision for posttransfer self-decanting of supernatant from a precipitate. Also, throwaway disks already loaded with reagents are being considered.

When an end-point determination is made, the readout is obtained by averaging the results for a number of rotor passes. For a reaction-rate procedure, a fixed-time technique is used. Two or more sets of digitized absorbance readings for each cuvette are stored simultaneously, and the difference is calculated and expressed as $\Delta A/\min$.

The commercial versions of GeMSAEC are the Centrifichem (Union Carbide, Tarrytown, N.Y.), the Rotochem (American Instrument Co., Silver Spring, Md.), and the GeMSAEC (ElectroNucleonics, Fairfield, N.J.).

ELLA. The use of a small computer with a spectrophotometric rate system can be extended to mechanistic investigations by programmed decision-making. Automation of such investigative experiments enhances the collection of data required for the development of new, reliable reaction-rate methods. This approach to automated instrumentation has been incorporated into a system called ELLA (29).

The hardware used for ELLA (Experimental LINC Laboratory Analytical System), excluding the computer I/O devices, is shown in Figure 9. The reactants are placed on the sample tray in the order of desired addition to the system. For example, for an enzymatic experiment the order is buffer, enzyme, buffer, substrate, buffer. The initial volume of enzyme to be aliquoted and incubation time required prior to rate measurement are specified via a teletype by the experimenter in response to the system's programmed request. The thermostated digital pipet sequentially draws reactants through a dip tube from the turntable and into a holding coil. The volumes of solutions to be picked up and delivered are determined by the number of clock pulses received from the accumulator buffer of the computer (DEC PDP-12). These impulses activate a stepping motor which allows delivery of 5 μl from the pipet per pulse. The reactants are expelled from the pipetting system into a thermostated mixing chamber, and the reaction mixture is transferred by vacuum into the flow cell where the spectrophotometric measurement is made.

The software used for ELLA permits performance of a kinetic experiment to a desired end point, making all necessary decisions and controlling all instrumentation via a special-purpose time-sharing system. This software consists of four subprograms called the initialization routine, the on-line operating system, the updating system, and the cleanup routine. The initialization routine requests the experimenter to enter the volume of enzyme and the incubation time to be used, sets up a blank reading tube and the first series of reaction tubes containing varying amounts of substrate, calibrates the reaction-rate reading system, and activates the sample-reagent introduction system automatically.

The on-line operating system takes readings of the rate of reaction, controls the reaction-rate interface, and performs a least-squares fit to the data. When a sufficient number of data points have been obtained which exhibit an error of estimation below the level programmed by the experimenter, a flag is set indicating the experiment is completed.

The updating program runs after each tube to process information obtained from the previous tube and to set up and activate the hardware for the next run. The program performs bookkeeping tasks and calculates the rate. The calculated initial rate is corrected for blank and printed along with other desired information. If the experiment is not completed, control is returned to the on-line operating system.

If the experiment is completed, the cleanup routine is called which stores the data in the file on magnetic tape for successive experiments. The system then prints out this material upon command.

Multimode Instrumental System. Recently, Deming and Pardue (30) have developed a computer-controlled instrumental system for characterization of chemical reactions. The system consists of four distinct elements: a programmable digital computer (controller), a spectrophotometer used to convert the chemical information into electrical information, an analog-to-digital converter, and a device for initiation and control of the experiment. The initiation and control device is composed of a number of electrical and mechanical components which enable the system to introduce into the spectrophotometer cell the proper volume of reagents and diluents. Four stepping motors are used to drive four calibrated syringes which pull the reagents from a stock reservoir and dispense them. The volume of delivered solution is controlled by applying a corresponding number of pulses to the stepping motors. The system is also provided with

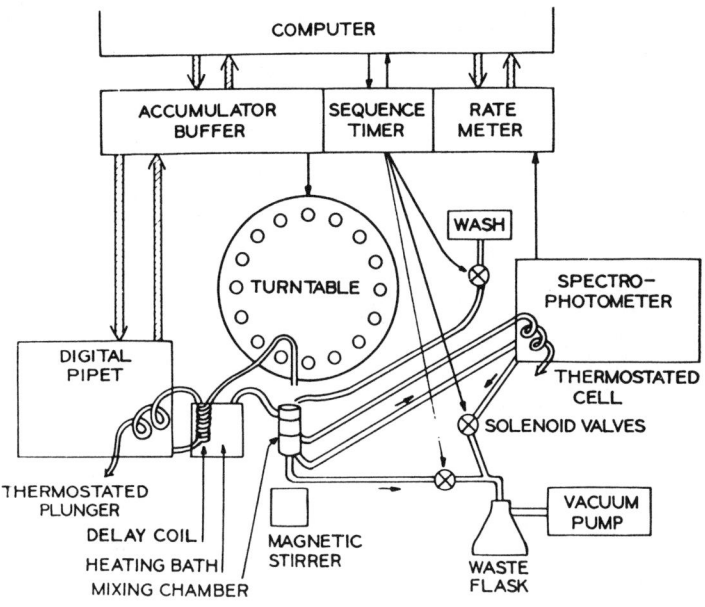

Figure 9. Block diagram of hardware used in ELLA (29)

the ability to remove the contents of the cell, rinse the cell, and mix the reagents.

The automated instrumental system has been evaluated for three types of operations: routine operations, experimental design, and data interpretation.

For routine operations, the experiments are preprogrammed into the information base, and data are plotted on the storage scope and printed out on the teletype. In the experimental design operation, the concentration of reagents and the limits over which they must be varied are introduced into the computer. Then the instrument plots the reaction rate vs. concentration of the species.

Data interpretation type of operation is accomplished by using the results of precision experiments. These results are interpreted, and new experiments are automatically designed to complete the characterization.

The automated instrumental system allows the time required for the operations involved in the experiment preparation to be decreased. The efficiency of the experimenter is greatly increased by the rapid availability of the results of a characterization and by more free time to accomplish other tasks that are more difficult to automate.

CONCLUSIONS

A new generation of automated and computer-controlled instrumentation makes it possible to perform reaction-rate methods of analysis at times per sample that are equivalent to the fastest equilibrium methods. The rate methods often have advantages in selectivity and accuracy that will encourage their adoption as the new automated instrumentation becomes available in laboratories, and more methods are consequently developed. The present interest in automating basic chemical reaction-rate studies should make it possible to develop specific new procedures many times faster than in the recent past.

Because of space, no attempt was made to review all of the new automated methods or the well-known Technicon continuous-flow analyzers. The methods presented illustrate that the major differences in chemical reaction-rate analyzers are in the sample and reagent-handling and mixing procedures. The functional units of the instruments are similar, and with suitable modular designs, it should be possible to easily modify analyzers for various general approaches to either reaction-rate or equilibrium analytical methods.

At present, only the stopped-flow analyzer has been applied for rapid quantitative procedures based on measurement of initial reaction rates for extremely fast reactions.

REFERENCES

(1) H. V. Malmstadt and G. P. Hicks, *Anal. Chem.*, **32**, 394 (1960).
(2) H. V. Malmstadt and H. L. Pardue, *ibid.*, **33**, 1040 (1961).
(3) H. V. Malmstadt and S. I. Hadjiioannou, *ibid.*, **34**, 452 (1962).
(4) H. V. Malmstadt and T. P. Hadjiioannou, *ibid.*, p 455.
(5) H. V. Malmstadt and H. L. Pardue, *Clin. Chem.*, **8**, 606 (1962).
(6) W. J. Blaedel and G. P. Hicks, *Anal. Chem.*, **34**, 388 (1962).
(7) G. A. Rechnitz, *ibid.*, **38**, 513R (1966); **40**, 455R (1968).
(8) G. G. Guilbault, *ibid.*, **42**, 344R (1970).
(9) K. B. Yatsimerskii, "Kinetic Methods of Analysis," Pergamon Press, New York, N.Y., 1966.
(10) H. Mark and G. Rechnitz, "Kinetics in Analytical Chemistry," Wiley, New York, N.Y., 1968.
(11) G. G. Guilbault, "Enzymatic Methods of Analysis," Pergamon Press, Oxford, England, 1970.
(12) H. L. Pardue, *Rec. Chem. Progr.*, **27**, 151 (1966).
(13) J. D. Ingle, Jr., and S. R. Crouch, *Anal. Chem.*, **43**, 7 (1971).
(14) H. L. Pardue, *ibid.*, **35**, 1240 (1963).
(15) H. V. Malmstadt and E. H. Piepmeier, *ibid.*, **37**, 34 (1965).
(16) H. V. Malmstadt and A. C. Javier, unpublished internal report, 1965.
(17) T. Kajihara and B. Hagihara, *Rinsho Byori*, **14**, 322 (1966).
(18) Kadish, E. Litle, and J. Sternberg, *Clin. Chem.*, **14**, 116 (1968).
(19) S. R. Crouch and H. V. Malmstadt, *Anal. Chem.*, **39**, 1084, 1090 (1967).
(20) A. C. Javier, S. R. Crouch, and H. V. Malmstadt, *ibid.*, **41**, 239 (1969).
(21) H. V. Malmstadt, C. Delaney, and E. Cordos, *CRC, Critical Rev. Anal. Chem.*, **2**, 559 (1972).
(22) H. V. Malmstadt and G. P. Hicks, *Anal. Chem.*, **32**, 115 (1960).
(23) B. Chance, *J. Franklin Inst.*, **229**, 455, 613, 737 (1940).
(24) Q. H. Gibson and L. Milnes, *Biochem. J.*, **91**, 161 (1964).
(25) E. Cordos, S. R. Crouch, and H. V. Malmstadt, *Anal. Chem.*, **40**, 1812 (1968).
(26) H. V. Malmstadt, C. J. Delaney, and E. Cordos, *ibid.*, **44** (12), 79A (1972).
(27) N. G. Anderson, *Amer. J. Clin. Pathol.*, **53**, 778 (1970).
(28) R. L. Coleman, W. D. Shults, M. T. Kelly, and J. A. Dean, *Amer. Lab.*, **3**, 7, 26 (1971).
(29) A. A. Eggert, G. P. Hicks, and J. E. Davis, *Anal. Chem.*, **43**, 736 (1971).
(30) S. Deming and H. Pardue, *ibid.*, p 192.
(31) C. J. Delaney and H. V. Malmstadt, unpublished report, 1971; or C. J. Delaney, PhD thesis, University of Illinois, Urbana, Ill., 1972.

Howard V. Malmstadt is Professor of Chemistry at the University of Illinois in Urbana. After receiving a BS degree in chemistry from the University of Wisconsin (1943) and serving as a radar officer in the Pacific, he returned to Wisconsin and earned his MS and PhD degrees in 1948 and 1950. He joined the faculty at the University of Illinois in 1951 following a postdoctoral year at Wisconsin. He is the author or coauthor of more than 90 technical publications including the well-known and widely used books: "Electronics for Scientists," "Digital Electronics for Scientists," and "Computer Logic." He was a Gugenheim Fellow in 1960, the 1963 ACS Chemical Instrumentation award winner, and recipient of the 1970 Eckman award in edutation (Instrument Society of America). Dr. Malmstadt was chairman of the ACS Analytical Division in 1964 and served on Analytical Chemistry's Advisory Board (1961–63). His major areas of research are in new spectroscopy methods and instrumentation; short-time phenomena in laser plumes, flames, and spark discharges; reaction-rate methods; and instrumentation automation.

Emil A. Cordos is an assistant professor in the Analytical Division of the University of Cluj, Romania, and is also a member of the Directing Council of the Center for Research in Analytical Chemistry, Cluj. Dr. Cordos received his BS and PhD degrees in chemistry from the University of Cluj in 1959 and 1969. In 1967–68, he studied at the University of Illinois under H. V. Malmstadt as an exchange visitor and returned to Illinois in 1970 as a two-year postdoctoral research associate. Dr. Cordos' research interests include the development of rapid methods of analysis, electrolysis in molten salts, inorganic ion exchangers, gas chromatography of inorganic gases, automated atomic absorption and fluorescence spectroscopy, and automated reaction-rate methods. He has published more than 20 papers on the above topics.

Collene J. Delaney is a postdoctoral trainee in clinical chemistry at University Hospital, University of Washington, Seattle, Wash. She received her BS in chemistry from Purdue University in 1965 and earned her MS and PhD degrees from the University of Illinois in 1967 and 1972. Her research interests include reaction-rate methods of analysis; flame emission, absorption, and fluorescence techniques; and automation in clinical chemistry. She is especially interested in developing in-service training and continuing education programs for clinical laboratory personnel. Dr. Delaney is coauthor of six papers on the above topics. She is a member of Iota Sigma Pi, the ACS, and AAAS.

Instruments for Rate Determinations

HOWARD V. MALMSTADT, COLLENE J. DELANEY, and
EMIL A. CORDOS

School of Chemical Sciences
University of Illinois at Urbana–Champaign
Urbana, Ill. 61801

Widespread commercial introduction of inexpensive general-purpose minicomputers and hardware interfaces is greatly changing the design concepts of chemical reaction-rate instruments. Circuits, integration systems, variable-time methods, hybrid analog-digital systems, and software systems for reaction-rate determinations are presented

THE DETERMINATION and utilization of the *rate of change* of one physical quantity with respect to another are frequently encountered in analytical methods. For example, one type of automatic titrator (1) electronically generates a signal proportional to the rate of change of transducer output (E) vs. volume (V) of titrant and utilizes the first derivative signal (dE/dV) to control the rate of titrant delivery. Another type of automatic titrator electronically computes a second derivative signal (d^2E/dV^2) during the course of the titration. This signal is ideally suited for the automatic termination of a titration at the endpoint, because it crosses the zero reference at the inflection point of the titration curve independently of the absolute magnitude of the transducer output (2, 3). A series of regular and derivative titration curves is shown in Figure 1a.

Recently, there has been increased interest in derivative spectrophotometry (4, 5), where the rate of change of absorbance (A) or emission (F) with respect to wavelength (λ) is determined and recorded vs. wavelength. The $dA/d\lambda$ derivative signal (Figure 1b) is especially useful in locating overlapping absorption bands (5, 6). The second derivative signal ($d^2A/d\lambda^2$), Figure 1c, is said to provide improved sensitivity in the determination of absorbing gases in the ultraviolet regions (4). The second derivative emission signal ($d^2F/d\lambda^2$), Figure 1d, has been used to provide increased sensitivity for trace constituents that have emission lines superimposed on strong background signals (7).

Of all the analytical examples where rate information is utilized, none has had so much attention focused on the various instrumental methods of encoding rate data as have the so-called kinetic or reaction-rate methods of analysis (8). In these methods the time rate of change of some property P that is proportional to the concentration of one of the reaction products is measured, as illustrated in Figures 1e and 1f. For most practical procedures the desired rate information is obtained shortly after the start of the reaction. The rate measurement system must be sensitive and capable of determining average initial rate amidst considerable background noise.

After the first automated reaction-rate methods and instrumentation for the rapid determination of glucose were developed in our laboratories about a dozen years ago (9–12), many further investigations were initiated—both of specific chemical reactions and of the instrumentation that would provide sensitive and selective quantitative rate procedures. It was apparent from the start of these studies that more instrumentation developments were essential if rate methods were to be widely accepted as routine quantitative chemical methods. Also, of course, reliable automated instrumentation was important for more rapid investigation of chemical reactions that might be used for quantitative rate procedures.

For over a decade there have been many published improvements from several laboratories for all major sections of reaction-rate instrumentation for quantitative analyses. Many of these developments, including computer-controlled instruments with automated sample and reagent handling systems, are described in this month's Report for Analytical Chemists, page 26 A (13). In this report the principles and characteristics of several electronic devices for rate determinations are presented which provide excellent noise immunity. Although the methods presented here were specifically developed for chemical reaction-rate procedures, several are generally applicable whenever the direct readout of the rate of change of some parameter with respect to another is required.

Classical Circuits for Rate Determinations

A differentiator circuit that provides an output proportional to the rate of change of an input signal can be easily assembled from a single operational amplifier (OA), an input capacitor, and a feedback resistor (14). However, relatively small noise components on the input signal can result in a high noise level at the output that often obscures the desired rate information. Even when the time constant of the differentiator is increased, the noise is usually prohibitive for chemical reaction-rate procedures (15).

An all-electronic rate comparison technique has been successfully used for some reaction-rate procedures. An

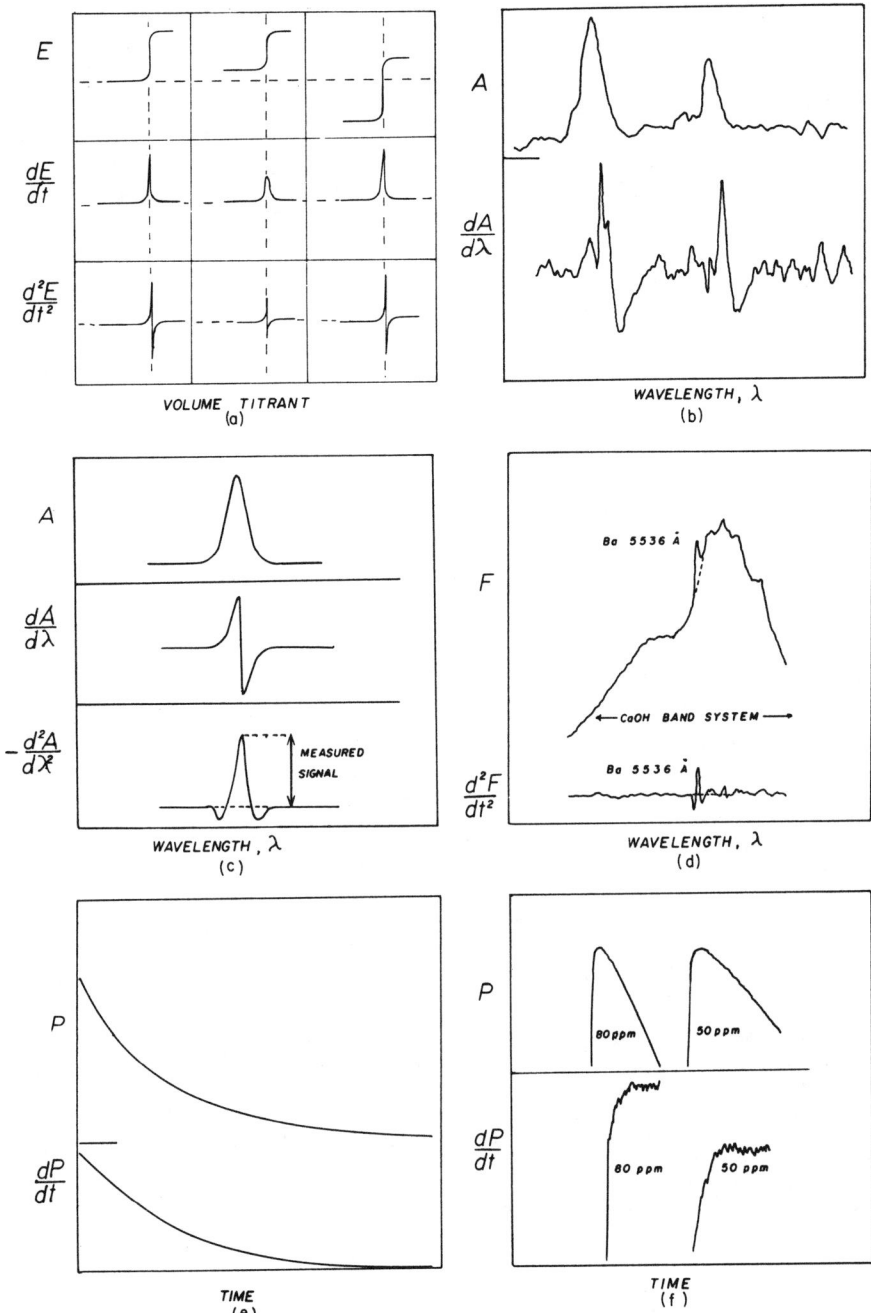

Figure 1. Analytical methods by use of rate determinations. (a) Titration curves illustrating first and second derivative curves. (b) Benzene absorption curves recorded by minicomputer-controlled spectrophotometer (6). (c) Atomic absorption line showing first and second derivative curves. (d) Flame emission spectra of barium in presence of calcium (7). (e) and (f) Reaction-rate curves

OA integrator is used to generate a reference rate curve that is continuously compared by an OA comparator and maintained equal to the unknown input rate curve (15, 16). Although the performance of the comparison system is superior to the simple OA differentiators, it does not give sufficient averaging of noise for many typical chemical reaction-rate measurements, especially where high sensitivity is required. Both the simple OA differentiator and the rate comparison circuits can have fast response to provide output data proportional to the "instantaneous" rate. Instruments capable of providing more or less instantaneous rate data for reaction-rate methods have been classified as *derivative* methods (8, 17, 18).

An early approach to combat noise on rate signals was to use one of the two-point methods as illustrated in Figure 2. The change in monitored parameter, ΔP (fixed-time), is measured over a preselected time interval, or the change in time, Δt, is measured over a preselected change in monitored parameter, ΔP (variable time). The so-called *fixed-time* and *variable-time* methods can also be considered as general categories for the classification of circuits for rate determinations (8, 17, 18). Pardue (17) and Ingle and Crouch (18) have indicated that the choice of method might be dependent on considerations other than noise, but in any event, it is important to modify the basic techniques to eliminate the noise problems illustrated in Figure 2. In making a fixed-time measurement (Figure 2a), the noise component causes uncertainty in determining the value of P_1 and P_2 and, therefore, in the measured value, ΔP. The values determined for P_1 and P_2 can fall anywhere within the dashed lines.

The same considerations apply to the variable-time measurement shown in Figure 2b. In addition to the uncertainty caused by the noise, an even greater error can be introduced by random spikes that can cause false triggering of the measurement system. For example, if the spike shown in Figure 2b is interpreted by the measurement system as an indication that the preselected value of P_2 has been reached, then the time interval $(t_2' - t_1)$ is erroneously taken as the measured value, Δt.

Greater noise immunity has been obtained for all three general methods by using hardware and software averaging and smoothing techniques (19–23). Because space restrictions prevent a detailed discussion of all three categories, it was considered of greater value and interest to show how only one general method, the fixed-time method, can be implemented to provide high noise immunity with analog and digital hardware or with software procedures. The variable-time and derivative methods are briefly outlined and referenced.

Integration (Fixed-Time) Technique for Reaction-Rate Determinations

Greater noise immunity has been obtained by applying an integration fixed-time technique rather than the two-point fixed-time method (16). The method involves the integration of two segments of the rate curve and subsequent subtraction of the resultant areas. As shown in Figure 3, the integration is made over two equal time increments, Δt, which are sequential, and the difference, ΔA, between the resultant areas A_1 and A_2 is the parallelogram $ABCD$. The area of the parallelogram is given by:

$$\Delta A = (\Delta t)a \qquad (1)$$

The slope is defined as:

$$S = \tan \alpha = \frac{a}{\Delta t} \qquad (2)$$

Substitution of Equation 1 into Equation 2 provides an expression relating

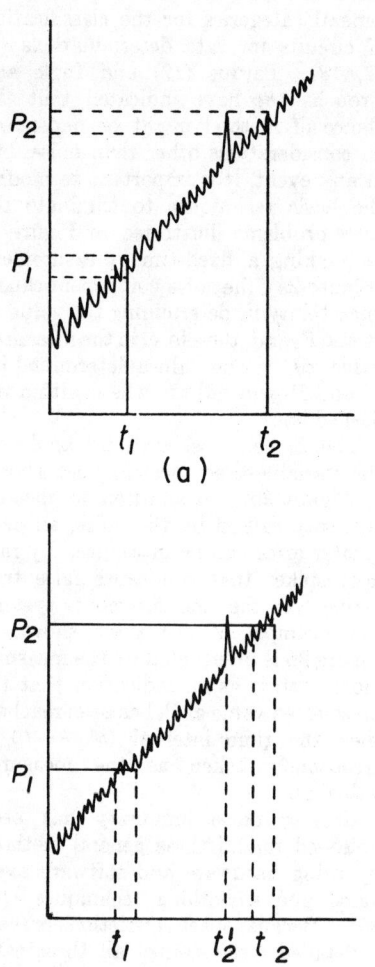

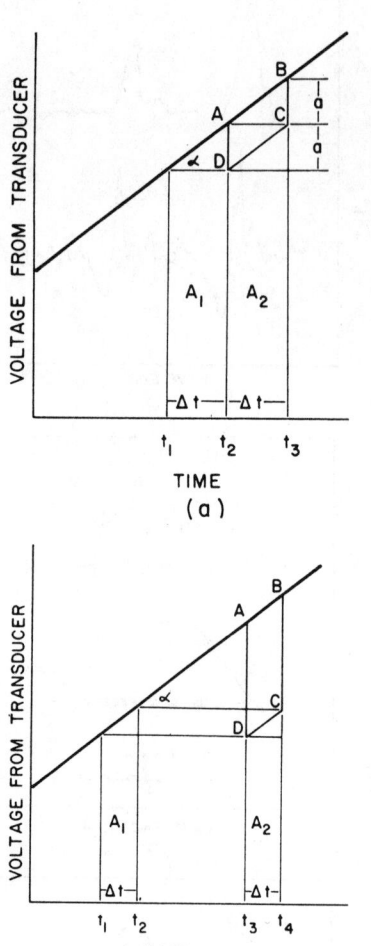

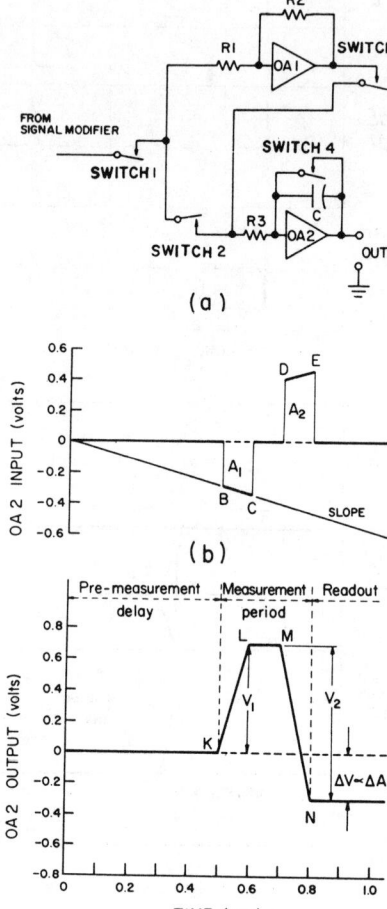

Figure 2. Illustration of noise susceptibility in (a) fixed-time measurement technique and (b) variable-time measurement technique

Figure 3. Principle of integration method for reaction-rate determinations. (a) Expanded section of typical rate curve illustrating integration and subtraction of two sequential areas A_1 and A_2. (b) Expanded section of typical rate curve illustrating integration and subtraction of two areas A_1 and A_2 separated by a measurement delay interval $t_2 - t_1$

Figure 4. (a) Integration and subtraction circuit of rate meter. (b) Oscilloscope trace of slope from ramp generator and input voltage of OA2. (c) Oscilloscope trace of output voltage of OA2

slope to the difference in area:

$$S = \frac{\Delta A}{(\Delta t)^2} \quad (3)$$

Since Δt is a constant the slope is directly proportional to the measured difference, ΔA. In Figure 3b the two time increments are not consecutive but are separated by a time interval $t_3 - t_2$. In this case Equation 3 can be rewritten as:

$$S = \frac{\Delta A}{(t_3 - t_2)\Delta t + (\Delta t)^2} \quad (4)$$

Since $(t_3 - t_2)$ is also kept constant, the slope remains directly proportional to the measured area. The total time for the measurement is $2\Delta t + (t_3 - t_2)$.

Rate Determinations with Analog Integration System

The integration and substraction operations illustrated in Figure 3 can be implemented by use of the relatively simple analog circuit shown in Figure 4a. This circuit uses two operational amplifiers (OA's) which are tied to one another and to a signal modifier by a switch network. During the first integration period, the signal is directed through switches 1 and 2 to the integrator (OA2). If the signal modifier output is as illustrated in Figure 4b, then line BC represents the signal applied to the integrator which charges capacitor C, causing the output of OA2 to rise to a voltage V_1 as shown in Figure 4c.

During the second integration period the signal from the modifier is first directed to the gain-of-one inverter (OA1) by switch 1 and then to the integrator by switch 3. The voltage represented by line DE is applied to the integrator input which discharges capacitor C, and the integrator output decreases in magnitude by the amount V_2. The voltage difference, ΔV, which is read out at the end of the measurement period, is proportional to the difference in area, ΔA, and is therefore proportional to the initial slope according to Equation 3.

The switches (S1–S4) are activated with pulses generated by a logic circuit. Thus, upon receipt of a suitable trigger pulse, the logic circuit controls the start of the sequence of delay time, the length of the integration periods, the subtraction, and the readout of the result. The integration period is selected to be as long as feasible to average the noise most effectively. The shortest practical integration period is governed by the switch times and OA response and at present is about 0.1 msec.

During the integration period, the output voltage, e_0, of the integrator changes as the square of the time, t, so that $e_0 = St^2/(R_3C)$. Thus, the rising and falling portions of the in-

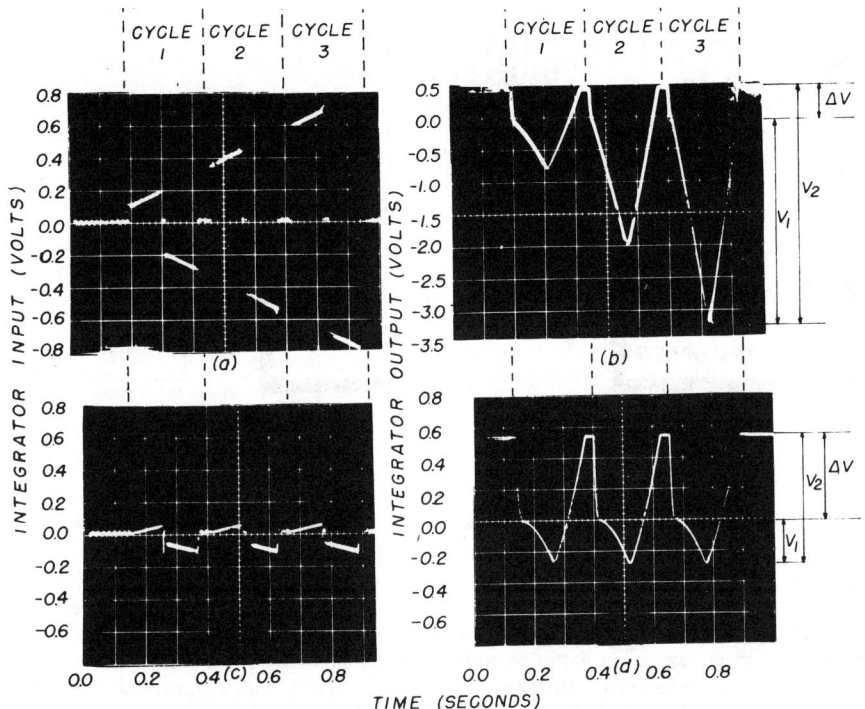

Figure 5. Oscilloscope traces of integrator input and output voltages for three successive measurement cycles. Integrator input (a) and integrator output (b) without curve-following suppression. Integrator input (c) and output (d) with curve-following suppression

tegrator output illustrated in Figure 4 should exhibit a curvature. This curvature is most pronounced when the integration operation is initiated at a time close to the moment when a slope is generated (t is small). In the case of the utilization of a long premeasurement delay interval (t is large), the curvature is quite small. This effect can be seen in the photograph in Figure 5.

The rate meter described above is capable of measuring slopes over the range 0.45 mV/sec to 450 V/sec. It has been evaluated with synthetic signals from a ramp generator simulating both slow and fast reactions. The results given in Tables I and II show from the proportionality constant that the output voltage is a linear function of the input slope, and that in both cases the relative standard deviations are typically about 0.2%.

Tables III and IV show results obtained with the rate meter for the enzymatic determination of glucose with the H_2O_2–I_3^- method (24) and for the determination of phosphate with the 12-MPA method (25). Relative errors and standard deviations were about 1% for glucose determinations at the µg/ml level and were less than 1% for phosphate determinations at the µg/ml level.

Modified Analog Integration System

Complex chemical systems often exhibit rate curves that require the rate measurement to be made on a segment of the curve where the absolute value of the input signal to the rate meter is high. The measurement of ΔV (Figure 4c) thus requires the determination of a small difference between two relatively large values, which can result in a decrease in the accuracy of the rate measurement. A similar situation occurs when successive measurements are made along the same rate curve, even with a simple chemical system, since the absolute value of the input signal to

Table I. Automatic Rate Measurements with Long Integration Time and Signals Typical of Slow Reactions

Input rate, mV/sec	Digital readout,[a] mV	Proportionality constant	Rel SD, %
2.50	111.8	0.0223	0.24
3.60	163.4	0.0220	0.24
5.30	239.7	0.0221	0.33
7.90	357.3	0.0221	0.15
10.80	485.6	0.0222	0.09

[a] Averages of 10 results; R_2 = 450 kilohms; integration time: 10 sec; premeasurement time: 30 sec.

Table II. Automatic Rate Measurements with Short Integration Times and Signals Typical of Fast Reactions

Input rate, mV/sec	Digital readout,[a] mV	Proportionality constant	Rel SD, %
180	81.8	2.20	0.24
385	175.3	2.19	0.20
590	268.8	2.19	0.14
795	362.8	2.19	0.11

[a] Averages of 10 results; integration time: 50 msec; premeasurement time: 100 msec; R_2 = 15 kilohms.

Table III. Automatic Results for Glucose

Direct concn readout[a]	Glucose concn in µg/ml		
	Taken	Rel error, %	Rel SD, %
5.0	5.0	0.0	1.6
10.0	10.0	...	1.0
15.1	15.0	+0.7	0.7
20.1	20.0	+0.5	0.8

[a] Averages of five results; 10.0 µg/ml standard used to set readout; integration time: 10 sec; premeasurement time: 30 sec.

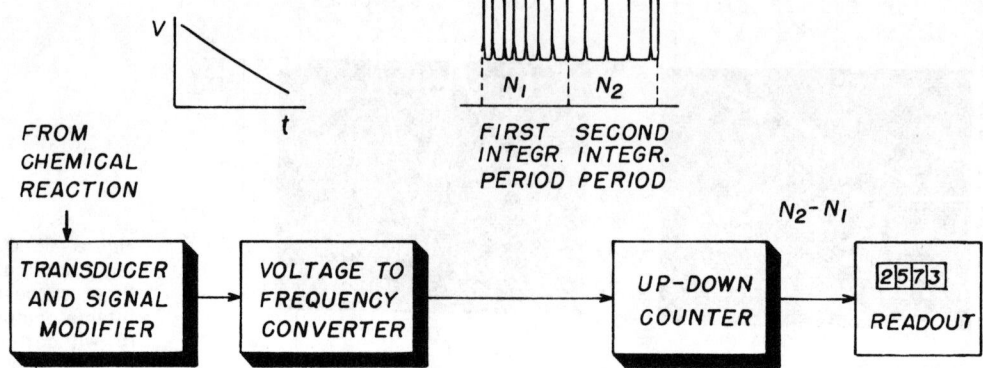

Figure 6. Block diagram of digital integration system for rate determinations

the rate meter increases after each measurement cycle. This case is demonstrated in Figures 5a and 5b, which are oscilloscope traces of the integrator input and output voltages obtained when three successive measurements were made on a 1.0 V/sec ramp signal. Whereas the slope remains constant for each measurement cycle, voltages V_1 and V_2 increase with each cycle. Not only is the accuracy of the measurement of ΔV decreased, but the integrator might limit during the measurement.

To prevent this problem, the basic analog integration system has been redesigned (26) to provide a curve-following suppression circuit which allows more accurate measurement of rate over a wider dynamic range. Also, provision has been made for automatic recycling of the measurement operations for multiple integrations along successive portions of the same rate curve and printout of the result after each measurement cycle.

At the beginning of each measurement cycle, the input signal is sampled and held by the curve-following suppression circuit. This voltage is directed to the integration and subtraction circuit during the measurement cycle in such a way that it is always in opposition to the signal which is integrated. Thus, for each measurement period the initial input signal to the integrator is essentially zero, as shown in Figure 5c. The effect of the curve-following suppression circuit on the readout for the multiple measurement operation can readily be seen from the oscilloscope trace for the integrator output voltage shown in Figure 5d. Now, regardless of the cycle, ΔV is a relatively large difference measured between two voltage levels which do not vary from cycle to cycle. The results tabulated in Table V demonstrate the improvement in accuracy and precision obtained during multiple measurement operation by use of the modified analog system.

Digital Integration System

Rather than performing the integration and subtraction operations with an analog circuit, these operations can be implemented with a digital circuit (27). A simplified block diagram of such a rate instrument is shown in Figure 6. The input signal to the rate meter is applied to a voltage-to-frequency converter which generates a pulse train. The frequency of the pulses is proportional to the input voltage. The pulse train is then applied to an up-down counter which is set to count up to N_1 during the first integration period and set to count down in magnitude by N_2 during the second integration period. The resultant output of the up-down counter, the number N_2-N_1, is decoded and read out.

The relative standard deviations reported for measurement of synthetic slopes with this rate meter were about 0.2%. These results are similar to those obtained with the basic analog rate meter described above. The primary reason is that although the integration and subtraction operations are

Table IV. Automatic Reaction-Rate Results for Phosphate

P, μg/ml	Av readout, mV	Rel SD,[a] %
0.50	54	1.01
1.00	103	0.79
1.25	125	0.85
1.50	149	1.05
1.75	172	0.97
2.00	200	0.54
2.25	226	0.40
2.50[b]	251	0.42
3.00	302	0.79
3.50	352	0.95
4.00	401	0.67
4.50	451	0.44
5.00	501	0.21

[a] The % relative standard deviation between averages of 10 results. [b] The 2.50 μg/ml standard was used to set the readout for direct digital concentration data.

Table V. Automatic Reaction-Rate Results for Successive Integrations with and Without Offset Suppression

Measurement condition[a]	Digital readout,[b] mV	Rel error,[c] %	Rel SD, %
First cycle with curve-following suppression	581.1	...	0.05
First cycle without curve-following suppression	582.5	0.24	0.21
Sixth cycle with curve-following suppression	582.1	0.17	0.15
Sixth cycle without curve-following suppression	566.9	2.45	0.39

[a] Input rate = 1 V/sec; integration time = 100 msec; integrator time constant = 20 msec. [b] Averages for 10 results. [c] Based on first-cycle measurement with curve-following suppression.

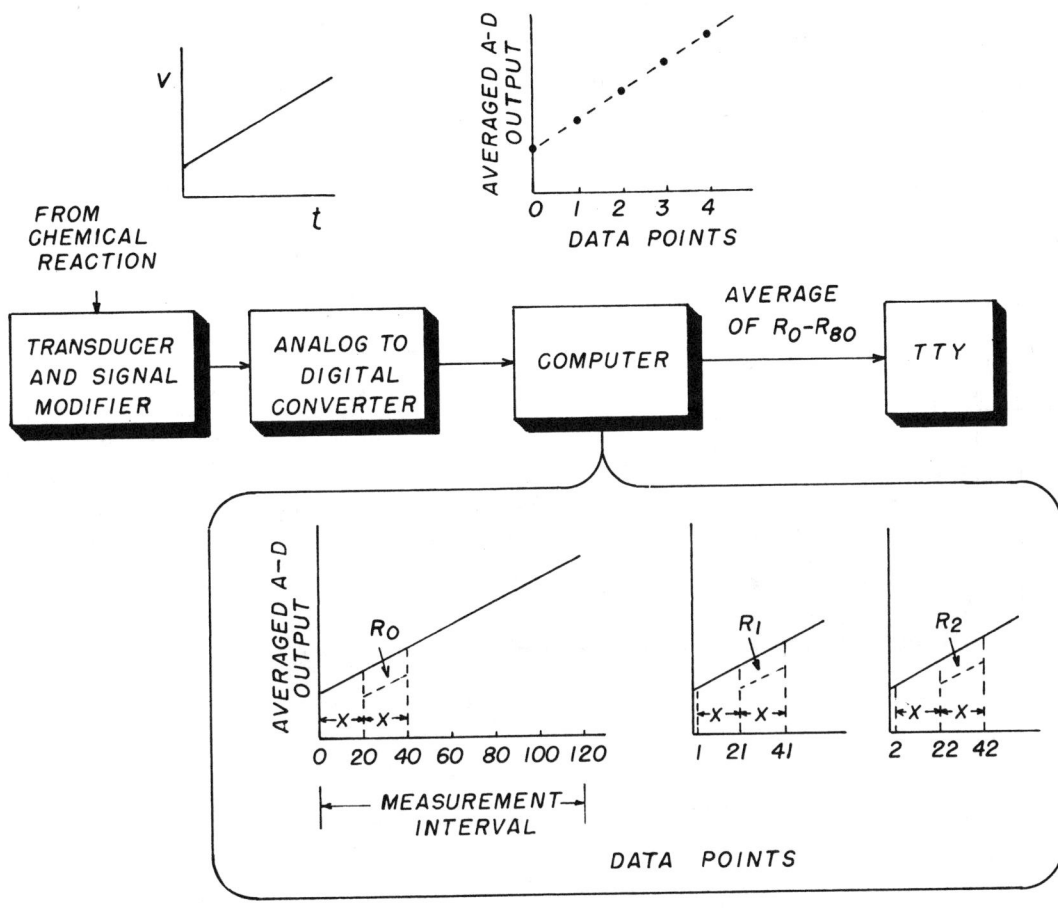

Figure 7. Block diagram illustrating minicomputer implementation of integration technique for rate determinations

performed with a digital circuit, the limiting operation is the analog-to-digital conversion. The same operation is the limiting factor when using the basic analog rate meter in conjunction with a digital readout device. The accuracy cannot be expected to change appreciably, since the primary difference is when the analog-to-digital operation is carried out. A truly digital rate measuring system would be one in which the monitored signal was obtained in a digital form. For example, in making spectrophotometric rate measurements, rather than using current-to-voltage and voltage-to-frequency converters, a pulse amplifier and signal level discriminator could be used in conjunction with the photomultiplier tube as a photon counting system.

Minicomputer Implementation of Integration Technique

The integration and subtraction operations can be performed by use of a minicomputer-interfaced software approach (23). The block diagram shown in Figure 7 illustrates that, as with the digital integration system, the analog signal obtained from the signal modifier system is fed to an analog-to-digital (A-to-D) converter, and the digitized data are operated on. The output of the A-to-D converter is first fed to a register where a preselected number of A-to-D conversions (e.g., 128) is added and averaged. Each resultant average is stored as a data point. The data points are obtained at a constant rate; the numbers on the data point axes in Figure 7 are proportional with time.

After a predetermined number of data points has accumulated (e.g., 121), the first X points (e.g., 0–20) are integrated, and this value is subtracted from the computed integral of the next X points (e.g., 20–40). The resulting number, R_0, which is proportional to the integral of the change in signal, is stored. The integration and subtraction processes are repeated by use of the same number of data points, with the X segments now taken between 1 to 21 and 21 to 41, and the resulting number, R_1, is stored. This process of sliding the integration intervals point-by-point is repeated until all data points have been included in the integration process. The stored R values are averaged, and the resultant value is printed out on the teletype. This approach exhibits excellent noise rejection, since the data have undergone smoothing three times (23).

Variable-Time Methods for Reaction-Rate Determinations

The variable-time method has been applied by use of one fundamental approach—the two-point technique. The method has been implemented in a variety of ways (16, 20–22, 28–30) by use of both hardware and software systems. The primary operations required in applying the variable-time method are measurement of the time interval, Δt, required for a preselected change in monitored parameter, ΔP, and computation of the reciprocal of the time interval, $1/\Delta t$.

Analog Circuits for Variable-Time Method

An analog variable-time rate meter has been developed by Stehl et al. (28) which measures the interval Δt, computes log Δt, and subsequently differentiates the expression to provide a readout voltage directly proportional to $1/\Delta t$.

A similar approach has been reported by James and Pardue (29). A computation circuit integrates the signal from a voltage interval detector for the measurement interval Δt. A log computation and exponential computation are then performed sequentially on the output of the integrator. The resultant of these operations is a voltage level which is directly proportional to $1/\Delta t$. The rate meter, when evaluated by a colorimetric procedure for the determination of alkaline phosphatase activity,

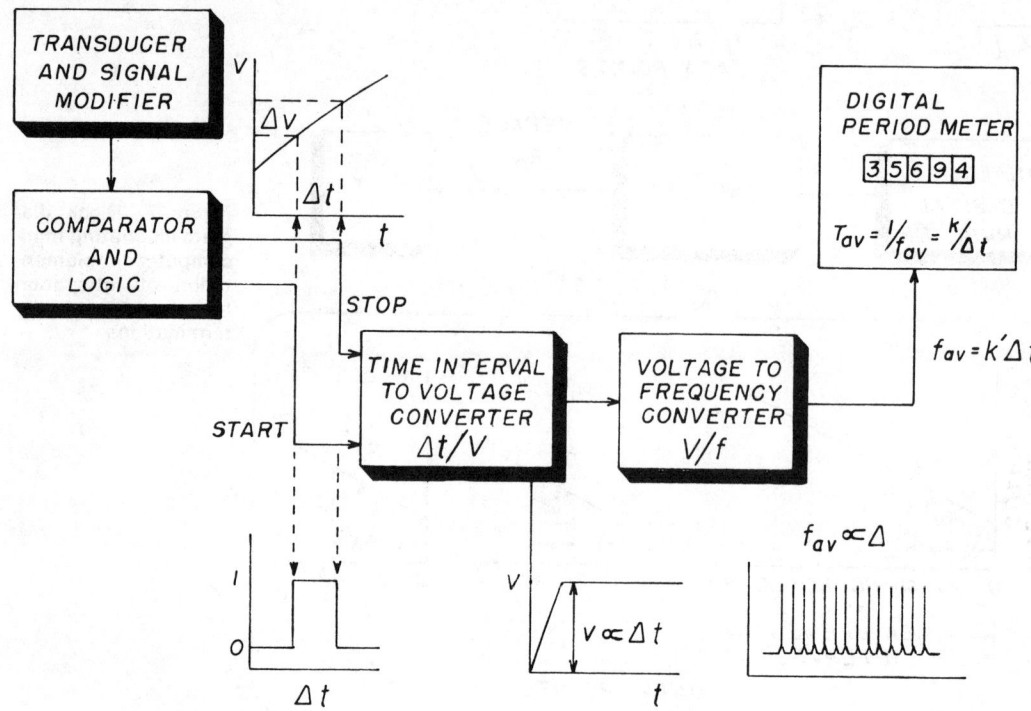

Figure 8. Block diagram of hybrid analog-digital variable-time system for rate determinations

gave results with relative standard deviations of about 1%.

Hybrid Analog-Digital Systems

One hybrid analog-digital approach which implements these operations (30) is outlined in Figure 8. The signal from the reaction monitor, after it is transduced and appropriately modified, is directed to a comparator and logic network. This circuit starts and stops a timing circuit when the transduced signal reaches two preselected voltage levels. The timing circuit is a part of a time interval-to-voltage converter which produces an output voltage, V, proportional to the input time interval, Δt. This output voltage is directed to a voltage-to-frequency converter which generates a pulse train whose frequency is proportional to the input voltage and, therefore, to the time interval, Δt. By measuring the period of this pulse train, the reciprocal of the time interval, $1/\Delta t$, is obtained.

The digital meter used as the readout device is operated in its multiple period averaging mode to provide a more accurate result. The output of the digital average period meter can then be fed to a computer or printed out. When this measurement system was evaluated by using synthetic slopes which varied from 5 mV/sec to 2 V/sec, the rate measurements were obtained with relative standard deviations which ranged from 0.15 to 1.8%.

A similar, yet complementary, hardware approach has been taken (20) by substituting a time-to-period converter for the time interval-to-voltage and voltage-to-frequency converters. In this system, it is the frequency which is measured to obtain a readout proportional to the reciprocal of the time interval, $1/\Delta t$. The time-to-period conversion is performed by using an entirely digital approach. The system also provides the possibility of noise averaging at either end of the measurement interval. This digital system yielded results with about 0.2% relative standard deviation when synthetic slopes were used as the input signals.

Software System

A software approach based on the variable-time method for on-line processing of reaction-rate data has been used (21). The transduced signal is directed to an analog-to-digital converter (ADC), and the digitized signal is entered in computer memory where it is stored as a function of time.

The acquired data points are processed in two parallel branches by the variable-time method and by a pseudo-fixed time method which provides some versatility to the system. To ensure that the available range of the ADC is not exceeded, a preliminary rate measurement is made by the variable-time program. The time constant of a digital filter is also adjusted on the basis of such preliminary rate measurements, allowing an analysis of rates over a wide dynamic range without making any changes in hardware. After the preliminary computations and adjustments are completed, a series of independent measurements are made for each reaction. The results are averaged and printed out along with individual results on the teletype. Data conversions (e.g., transmittance to absorbance) are performed either prior to or after the data acquisition process.

The computer-assisted system has been evaluated by using two chemical systems. The results for the determination of alkaline phosphatase activity indicate a relative standard deviation of about 0.3% and linearity (rate vs. activity) of 1%. Results from the catalytic determination of osmium by use of the Ce(IV)–As(III) reaction system gave relative standard deviations of about 5% at $10^{-11}M$.

Computer Implementation of Derivative Method

A software approach based on the derivative method has been reported by Willis et al. (22). As with the software system used for variable-time measurements, the transduced signal is directed to an analog-to-digital converter (ADC), and the digitized signal is entered in computer memory where it is stored as a function of time.

The acquired data points are averaged, and then a smoothing routine and computation of slope based on the methods of Savitsky and Golay (31) are performed. Several kinetic parameters are either displayed on an oscilloscope, or a permanent record is obtained via teletype printout or punched paper tape. These parameters include the rate curve (A vs. t) and the apparent first-order rate constant ($d \ln (A_\infty - A/dt)$).

General Considerations of Methods for Rate Determinations

Although the rate measurement methods in principle differ, there is a definite relationship between them. For example, in considering the integration technique (Figure 9a), the measurement interval is composed of the two fixed Δt integration periods. If the integration periods are made small, and the measurement interval is held constant, the result is the two-point fixed-time technique (Figure 9b). If the measurement interval is made small (Figure 9c), then the result is the derivative method.

This conclusion can be supported by the fact that any practical instrumentation system used requires that the measurement be made over a finite time interval as determined by the time constant of the instrument.

However, in practice it becomes necessary to consider the derivative method apart from the two-point methods, because the resultant of the measurement is an indication of the instantaneous rate, and the instrumental implementations are different.

Considerations in Selecting Method of Rate Computation

The factors that influence the noise immunity of the techniques are integration time, measurement interval, noise amplitude, and the period of the noise. For the integration technique, the most important factor is the relationship between the integration time and the period of the noise. A good noise average, and consequently a high noise immunity, are obtained when the integration time is much larger than the period of the noise. Given the same rate curve, no noise averaging is accomplished in applying the two-point fixed-time technique in the absence of software averaging and smoothing methods. In this case, the factor that has the greatest influence on the measurement error is the amplitude of the noise.

In the case of the derivative method, the measurement interval is much smaller than the period of most noise. Consequently, the measured value is the summation of the slope of the monitored signal and the slope of the noise. The slope of the noise can be either positive or negative and can be much larger than the slope of the monitored signal during the measurement interval.

If the rate instruments are to be applied to the determination of chemical reaction rates, then additional factors must be considered ($17, 18$). This is a result of the fact that the rate information is obtained on a segment of the reaction-rate curve which does not coincide with time $t = 0$. However, the ratio $\Delta P/\Delta t$ must be related to the tangent at time $t = 0$ by use of the basic kinetic equations relating concentration and time. Hence, the relationship between initial concentration and the measured ratio $\Delta P/\Delta t$ depends upon the reaction order, reaction mechanism, the kinetic role of sought-for species (substrate, catalyst), and the transfer function of the transducer used to monitor the reaction.

In discussing these factors, the kinetic equation in which they are included must be considered, and an explicit relationship between initial concentration C_0 and the remaining parameters must be obtained. Generally, the relationship between the initial concentration of sought-for-species C_0, the measured parameter P, and time t can be expressed as $C_0 = f(P)/\phi(t)$.

In choosing between the variable-time and fixed-time methods, one has to consider the complexity of functions $f(P)$ and $\phi(t)$. If $f(P)$ can be expressed in the simple form, $f(P) = a \cdot \Delta P$, where a is a proportionality coefficient, then a fixed-time method is advantageous since $\phi(t)$ is kept constant, and the readout value, ΔP, is linearly related to C_0.

The variable-time method is preferred when $\phi(t)$ can be reduced to $\phi(t) = b \cdot \Delta t$, where b is a proportionality coefficient. In this case the numerator, $f(P)$, is held constant, and C_0 is proportional to $1/\Delta t$.

When the two functions are complex, the problem is solved by making several approximations until one of the functions can be expressed as a first-degree equation.

For first or pseudo first-order kinetics, the fixed-time method yields the most accurate results ($17, 18$). For catalytic and enzymatic methods, a variable-time procedure could be the most appropriate method, although the rate equations must be individually considered for each case.

Conclusions

The widespread commercial introduction of inexpensive general-purpose minicomputers and hardware interfaces is greatly changing the design concepts of chemical reaction-rate instruments. Certainly, the specific-purpose electronic rate instruments will be useful and more economical in certain specialized applications. However, when the minicomputer is a part of a versatile reaction-rate system and is interfaced and programmed for control of instrument settings; sample and reagent handling operations; and smoothing, manipulation, and conversion of rate data, then it can provide economy and elegance in design of the total package and in operation. A block diagram of a complete system is shown in this month's Report for Analytical Chemists (13). After the rate curve is acquired under optimum computer-controlled conditions and is stored in memory, then the data can be smoothed, converted, and read out in any desired

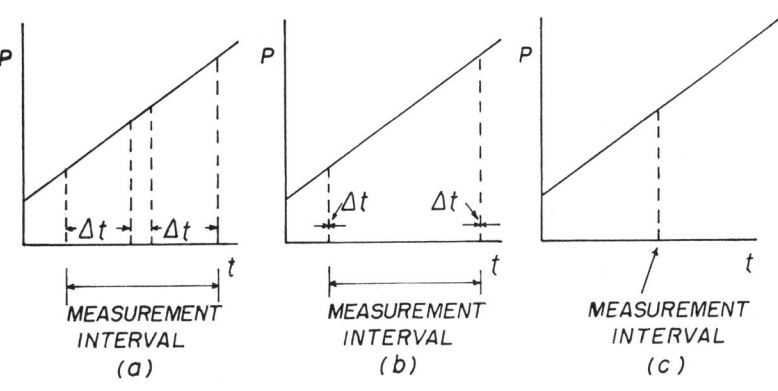

Figure 9. Illustration of relationship between (a) integration technique, (b) two-point fixed-time technique, and (c) derivative method

form. The absorbance and derivative curves of Figure 1b were obtained with such a minicomputer-interfaced system (6).

References

(1) H. V. Malmstadt, *Rec. Chem. Progr.*, **17**, 1 (1956).
(2) H. V. Malmstadt and E. R. Fett, *Anal. Chem.*, **26**, 1348 (1954).
(3) H. V. Malmstadt and C. B. Roberts, *ibid.*, **27**, 741 (1955).
(4) D. T. Williams and R. N. Hager, Jr., *Appl. Opt.*, **9**, 1597 (1970).
(5) F. Gorum, D. Paine, and L. Zoeller, *ibid.*, **11**, 93 (1972).
(6) R. Timmer and H. V. Malmstadt, submitted for publication, *Anal. Chem.* (1972).
(7) N. Snelleman, T. C. Rains, K. W. Yee, H. D. Cook, and O. Menis, *Anal. Chem.*, **42**, 394 (1970).
(8) H. V. Malmstadt, C. Delaney, and E. Cordos, *Critical Rev. Anal. Chem.*, **2**, 559 (1972).
(9) H. V. Malmstadt and G. D. Hicks, *Anal. Chem.*, **32**, 394 (1960).
(10) H. V. Malmstadt and H. L. Pardue, *ibid.*, **33**, 1040 (1961).
(11) H. V. Malmstadt and T. P. Hadjiioanou, *ibid.*, **34**, 455 (1962).
(12) H. V. Malmstadt and H. L. Pardue, *Clin. Chem.*, **8**, 606 (1962).
(13) H. V. Malmstadt, C. J. Delaney, and E. A. Cordos, *Anal. Chem.*, **44**, 26A (1972).
(14) H. V. Malmstadt and C. G. Enke, "Digital Electronics for Scientists," W. A. Benjamin, Menlo Park, Calif., 1969.
(15) H. V. Malmstadt and S. R. Crouch, *J. Chem. Educ.*, **43**, 340 (1966).
(16) H. L. Pardue, C. S. Frings, and C. J. Delaney, *Anal. Chem.*, **37**, 1426 (1965).
(17) H. L. Pardue in "Advances in Analytical Chemistry and Instrumentation," Vol 7, pp 141–207, C. N. Reilley and F. W. McLafferty, Eds., Interscience, New York, N.Y., 1969.
(18) J. D. Ingle, Jr., and S. R. Crouch, *Anal. Chem.*, **43**, 697 (1971).
(19) E. A. Cordos, S. R. Crouch, and H. V. Malmstadt, *ibid.*, **40**, 1812 (1968).
(20) R. A. Parker, H. L. Pardue, and B. G. Willis, *ibid.*, **42**, 56 (1970).
(21) G. E. James and H. L. Pardue, *ibid.*, **41**, 1618 (1969).
(22) B. G. Willis, J. A. Bittekofer, H. L. Pardue, and D. W. Margerum, *ibid.*, **42**, 1340 (1970).
(23) E. S. Iracki and H. V. Malmstadt, submitted for publication, *Anal. Chem.* (1972).
(24) H. V. Malmstadt and S. I. Hadjiioanou, *Anal. Chem.*, **34**, 452 (1962).
(25) A. C. Javier, S. R. Crouch, and H. V. Malmstadt, *ibid.*, **41**, 239 (1969).
(26) C. J. Delaney and H. V. Malmstadt, unpublished report, 1971.
(27) J. D. Ingle, Jr., and S. R. Crouch, *Anal. Chem.*, **42**, 1055 (1970).
(28) R. H. Stehl, D. W. Margerum, and J. J. Latterell, *ibid.*, **39**, 1346 (1967).
(29) G. E. James and H. L. Pardue, *ibid.*, **40**, 796 (1968).
(30) S. R. Crouch, *ibid.*, **41**, 880 (1969).
(31) A. Savitzky and M. J. E. Golay, *ibid.*, **36**, 1627 (1964).

Applications of Chemiluminescent Reactions to the Measurement of Air Pollutants

Development of automated instrumentation employing chemiluminescent reactions has led to low-cost, sensitive, specific methods for ozone, sulfur compounds, and oxides of nitrogen. Manufacturers of commercially available instruments are listed

**Robert K. Stevens
and J. A. Hodgeson**
Chemistry and Physics Laboratories
National Environmental Research Center
Environmental Protection Agency
Research Triangle Park, N.C. 27711

Air pollution monitoring instruments of the 1970's bear little resemblance to the classic techniques that have been used to measure the major contaminants in our atmosphere during the 1960's. The requirements for monitoring pollutants are largely based on the obligation of regional programs to protect our health and welfare and to a lesser degree to determine trends and sources of pollution. For this reason, the air pollution monitoring instruments of tomorrow must be extremely reliable and provide accurate measurements. In addition, the cost of acquiring, setting up, and maintaining such a network must be within the monitoring resources of regional directorates and also not be beyond the technical capabilities of manpower available to maintain and operate the instruments.

Over the past six years the Division of Chemistry and Physics of EPA has exploited the application of chemiluminescent methods of air monitoring to meet the restraints cited above. We chose this means of measuring gaseous pollutants because we believed successful application of chemiluminescent reactions would lead to air pollutant monitors which are low-cost, sensitive, and specific.

Most optical techniques used in the past for air pollutant monitoring were based on absorption spectroscopy. However, to measure sub part-per-million levels for such pollutants as sulfur dioxide or nitrogen dioxide by absorption spectroscopy, an absorption path length of 20-50 m is required. Multipass optical cells required for such measurements are complicated and contain a large volume and internal surface area, on which reactive pollutant molecules readily adsorb.

By contrast, chemiluminescent methods are based on emission spectroscopy. Trace concentrations of pollutants are readily detected in a small volume by modern low-level light measuring techniques. Chemiluminescent measurements are highly specific compared to conventional measurement techniques such as coulometry and colorimetry. For another molecule to be a positive interference, this molecule must react with the reagent added, and the reaction must be highly exothermic and yield chemiluminescence. Furthermore, the resultant emission must overlap the spectral region of the pollutant emission. A third body which quenches the emitting state is a potential negative interference. The predominant quenching agents in ambient air are molecular oxygen and nitrogen, which do not vary significantly in concentration, and the quenching effect, if present, is generally a constant.

The discussion that follows describes the development of automated air pollutant instrumentation employing specific chemiluminescent reactions for the measurement of ozone, sulfur compounds, and oxides of nitrogen. These instruments are now available from a number of manufacturers of air pollution monitoring instruments.

Ozone

Chemiluminescence is a characteristic feature of the reactions of ozone with many inorganic and organic materials *(1, 2)*. Bowman and Alexander *(3)* have employed ozone as a reagent in the chemiluminescent detection of organic compounds in solution. Organic dye molecules, such as rhodamines and safranin, produced quite in-

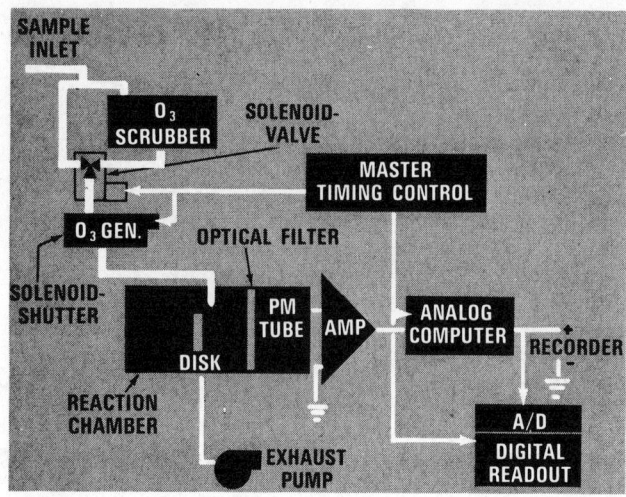

Figure 1. Automated gas-solid chemiluminescent ozone analyzer

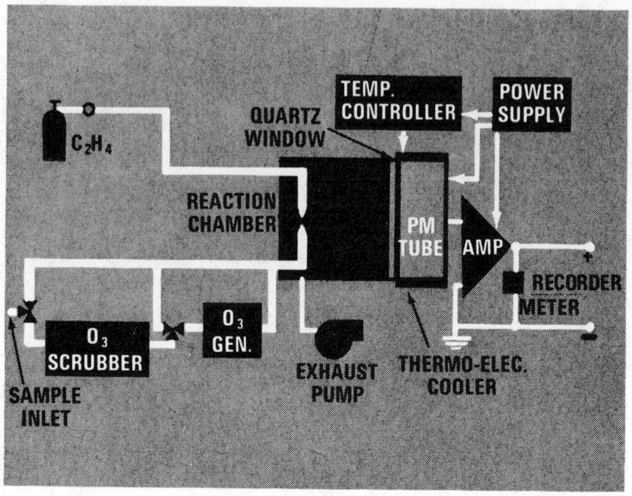

Figure 2. Automated gas-phase chemiluminescent ozone monitor

tense chemiluminescence when oxidized by ozone.

Regener *(4, 5)* utilized the intense emission from the reaction between ozone and Rhodamine-B, adsorbed on activated silica gel surface, in the first chemiluminescence detector for atmospheric ozone. Work was then initiated by the Environmental Protection Agency (EPA) on the development and application of Regener's technique for ozone monitoring in polluted atmospheres. The response characteristics of the detector were studied by Hodgeson et al. *(6),* and a chemiluminescence surface with improved lifetime and stability was developed. Figure 1 is a diagram of the prototype detector developed by EPA *(7).* An internal ozone source is necessary to provide frequent, periodic calibrations of the surface to compensate for sensitivity changes. An automated version of this analyzer is commercially available from Tritek Corp., Chapel Hill, N.C.

Ozone measurements by the Regener procedure have the advantage of requiring no support gases and only periodic replacement of the Rhodamine-B surface. Because the surface sensitivity changes, frequent recalibration is required, and the electronics and gas-flow arrangements are somewhat complicated.

A homogeneous gas-phase chemiluminescence technique for the detection of O_3 was reported by Nederbragt et al. *(8).* The detector employed an atmospheric pressure chemiluminescent reaction between ozone and ethylene. The reaction between ozone and ethylene yields chemiluminescent emission in the 300–600 nm region (λ max $\simeq$ 435 nm) *(9).* The intensity of the emission is directly proportional to ozone concentration. Response characteristics have been determined *(9)* by use of the detector cell and associated electronics shown in Figure 2. Several other geometries for mixing the ethylene and air streams have been tested, but no significant improvements in sensitivity have been observed. The detector responds linearly to ozone concentrations between 0.003 and 30 ppm, and no interferences have been observed *(9).* Modest improvements in sensitivity can be achieved by cooling the photocathode or by increasing the ethylene flow rate.

Good correlations have been obtained between prototype Regener and Nederbragt O_3 monitors in extensive field studies conducted in Los Angeles *(7)* and St. Louis *(10).* These studies were instrumental in establishing the Nederbragt approach as the reference method to monitor O_3, as cited in the *Federal Register (11).* The Nederbragt O_3 method is now commercially available from a number of instrument manufacturers both in this country and Japan. Table I lists the manufacturers of this device. In a period of only 18 months, the Nederbragt approach evolved through the stages of laboratory prototype construction and evaluation, field evaluations and comparisons, promulgation as a reference method, and commercialization.

The reaction between O_3 and NO in the gas phase yields chemiluminescence in the infrared region and may be used for detection of either O_3 or NO *(12).* This reaction, however, has been used almost exclusively in detectors for the measurement of NO and NO_x. Details of this analytical system will be discussed later.

Sulfur Compounds

The atmosphere contains a variety of sulfur compounds emanating from both natural and industrial activities. In most urban atmospheres, sulfur dioxide from the burning of fossil fuels is the predominant sulfur species. Malodorous sulfur compounds such as hydrogen sulfide, organic mercaptans, and sulfides are emitted by the kraft paper industry and oil refinery activities. Sulfuric acid and sulfates in the particle form result from photochemical oxidation of sulfur dioxide.

At present, flame chemiluminescence is the only approach being used routinely for monitoring atmospheric sulfur compounds. When volatile sulfur compounds are burned in a hydrogen-rich flame, an intense blue chemiluminescence occurs. This emission results from the recombination of sulfur atoms formed in the reducing flame environment by the following reaction sequence:

$$S + S = S_2^* \qquad (1)$$

$$S_2^* = S_2 + h\nu \ (350\text{--}450 \text{ nm}) \quad (2)$$

The chemiluminescence consists of a series of evenly spaced bands between 350–460 nm *(13).* Since the recombination involves two sulfur atoms, the intensity of the S_2^* luminescence is proportional to the square of the con-

Table I. Chemiluminescent Ozone Monitors

	Monitor Model No.
Bendix, Process Instruments Div. P.O. Drawer 477 Ronceverte, W.Va. 24970	8000
Beckman Instruments, Inc. 2500 Harbor Blvd. Fullerton, Calif. 92632	950
REM Inc. 2000 Colorado Ave. Santa Monica, Calif. 90404	612 B
McMillan Electronics Corp. 7327 Ashcroft Houston, Tex. 77036	1100-2 B
Meloy Labs, Inc. 6715 Electronic Dr. Springfield, Va. 22151	300
Kimoto Electric Co. Osaka, Japan	804

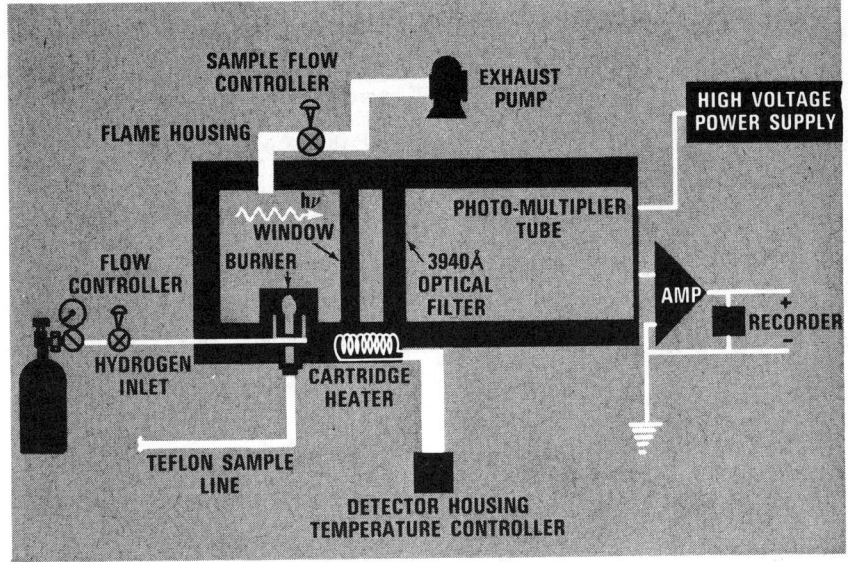

Figure 3. Typical commercial flame photometric detector

centration of sulfur compound in the flame.

The original flame photometric detector (FPD) for sulfur compounds was revealed in a patent by Draeger (14). Crider (15) described a version of this detector designed to measure sulfuric acid. Brody and Chaney (16) described the application of this detector to gas chromatography. The calibration and application of the FPD for detecting atmospheric concentrations of SO_2 were discussed by Stevens et al. (17).

The typical flame photometric detector uses a fuel mixture of 200 cm^3/min of air and 200 cm^3/min of hydrogen. The gases are usually mixed in a recessed barrel which functions as a light shield (Figure 3). The photomultiplier views the cool region of the flame afterglow through an interference filter centered at 394 nm. Meloy Labs and Bendix Corp. both offer the flame photometric detector as a sulfur dioxide monitor. In areas where hydrogen sulfide may coexist with SO_2, they offer as an accessory a scrubber, usually silver heated to 135°C, to remove hydrogen sulfide without removing SO_2.

Stevens et al. (18) have extended the specificity of the FPD by coupling a special gas-chromatographic column to this detector. The GC column consists of a 36-ft × 1/8-in. Teflon column packed with Teflon beads coated with polyphenylether and phosphoric acid. This column separates H_2S, SO_2, and CH_3SH quantitatively between 0.005 and 10 ppm. The system has been automated and evaluated in several urban areas (7, 10).

Table II lists the manufacturers of the FPD-GC combination that has been adapted to the measurement of ambient concentrations of sulfur dioxide, hydrogen sulfide, methylmercaptan, and dimethylsulfide.

Oxides of Nitrogen

The principal reaction which has been used in the detection of oxides of nitrogen is the chemiluminescent reaction between nitric oxide (NO) and ozone (O_3):

$$NO + O_3 = NO_2^* + O_2 \quad (3)$$
$$NO_2^* = NO_2 + h\nu \quad (4)$$

The mechanism and kinetics of this chemiluminescent reaction have been described by Clough and Thrush (19). The emission observed in this reaction is a modification of the air afterglow, which is observed in electrical discharges through air or nitrogen containing trace concentrations of oxygen (20). The emission is red shifted from the air afterglow and is a continuum extending from 0.6 to 3 μm (20). The direct chemiluminescent reaction is applicable only to the detection of NO. For detection of NO_x (NO + NO_2), a prior conversion of NO_2 to NO is required.

Another reaction which is applicable to the detection of NO is that with atomic oxygen:

$$NO + O + M = NO_2^* + M \quad (5)$$
$$NO_2^* = NO_2 + h\nu \quad (6)$$

This is the air afterglow (20) reaction which yields emission from 0.4 to 1.4 μm. Any NO_2 present is rapidly converted to NO via reaction with O atoms (20). This reaction may be applicable to the detection of NO_x at high concentrations, as may exist in source emissions.

The initial development work on an NO detector by use of O_3 as a reactant was initiated in 1968 by an EPA contract with Aerochem Research. The feasibility of detecting ambient concentrations of NO by chemiluminescence was demonstrated early in this work, and the results were published by Fontijn et al. (12).

The same chemiluminescence approach was also used in a NO detector developed at Ford Research Labs and reported by Stedman et al. (21). This detector was similar in many respects to the Aerochem monitor and was the prototype for commercial systems being applied largely to mobile source measurements.

A schematic of a typical NO-O_3 chemiluminescence detector is shown in Figure 4. Because of the known quenching effect of air on the NO + O_3 chemiluminescence, the earlier detectors used a small vacuum pump which pulled the air sample and O_3 stream through the reaction chamber at total pressures of 1-5 torr. Typical flow rates into the cell at STP were 50-100 cm^3/min sample air and 50-100 cm^3/min of oxygen containing approximately 0.5% O_3. In all NO detectors a Corning cut-off filter, which absorbs any emission below 600 nm, is placed between the reaction chamber and the photocathode. This filter is required to eliminate interfering emissions at shorter wavelengths from the reaction of O_3 with other molecules, e.g., olefins.

When the S-20 photomultiplier tubes employed are cooled to temperatures of -25°C or lower, the low-pressure chemiluminescent detectors are capable of detecting NO concentrations of 0.001 ppm. The detector response is linear from 0.001 to 10,000 ppm NO or a dynamic range of 10^7. The detector response time to changes in NO concentration can be 1 sec or less, depending on the electronic damping. No interferences have been observed in the chemiluminescent detection of NO when other pollutants are present at concentrations common to ambient air (12, 21). High concentrations of CO_2 and water vapor present in undiluted auto exhaust may cause some quenching of the chemiluminescence.

Table II. FPD-GC Sulfur Chromatographs

	Model No.
Analytical Instrument Development Co. 250 S. Franklin St. W. Chester, Pa. 19380	513
Bendix, Process Instruments Div. P.O. Drawer 477 Ronceverte, W.Va. 24970	8700
Tracor, Analytical Instruments Div. 6500 Tracor Lane Austin, Tex. 78721	270
Varian Aerograph 2700 Mitchell Drive Walnut Creek, Calif. 94598	1490-5

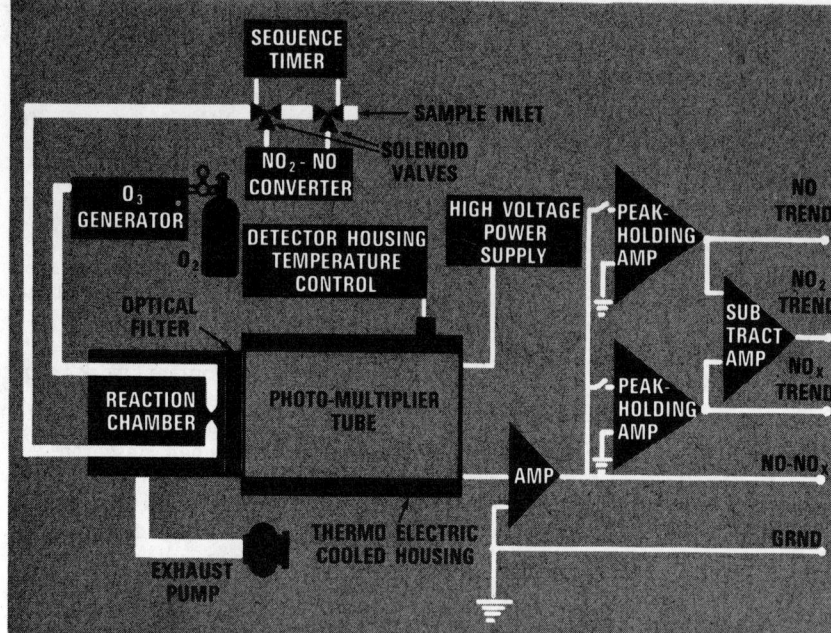

Figure 4. Automated NO, NO₂, NO$_x$, chemiluminescent analyzer

In a recent development the feasibility of detecting ambient concentrations of NO in a reactor cell operated at atmospheric pressure was demonstrated *(22)*. Quenching of the emission does occur as pressure is increased, but significant emission can still be obtained even at 1 atm total pressure. The decreased emission rate owing to quenching is at least partially compensated for by the use of higher sample flow rates (1–2 l./min, STP) and a small volume reaction zone which is closely coupled to the photocathode of the photomultiplier. The requirement for a mechanical vacuum pump is eliminated by higher pressure operation. Most of the commercial models being sold today for ambient air measurements operate at about 200 torr or greater and use small, diaphragm-type air pumps. When operating the detector at 200 torr, a sample flow of 100–200 cc/min is normally used in most commercial systems. At these flow rates and at this pressure, the minimum detectable level is between 0.002 and 0.005 ppm.

Present ambient air quality standards require only NO₂ measurements. NO$_x$ measurements are usually needed for automotive and stationary source emissions. Because of the enviable features of the NO chemiluminescence monitor, considerable effort has been expended during recent years on development of methods for the conversion of NO₂ to NO.

The earliest work on NO₂ conversion for application to chemiluminescence analysis was by Sigsby et al. *(23)*, who observed quantitative conversion of NO₂ to NO through a stainless-steel tube heated to temperatures of 600°C or greater. Under these conditions, however, ammonia (NH₃) is oxidized to NO and becomes a potential interference in both source and ambient measurements. Acidic scrubbers have been described which are capable of the quantitative removal of NH₃ without affecting NO₂ concentrations *(22, 23)*. The reduction of NO₂ at lower temperatures, where no oxidation of NH₃ occurs, is the method commonly being used today for eliminating the NH₃ interference.

In an earlier report *(22)*, a gold wool mesh surface heated to 240°C was used for the specific conversion of NO₂ at ambient concentrations. Over this type of surface, NO₂ is apparently reduced by surface impurities, and such convertors have unpredictable lifetimes *(22)*. Breitenbach and Shelef *(24)* have obtained conversion of NO₂ at source concentrations without NH₃ oxidation over various metals and carbon-impregnated metals.

Over the past 18 months, considerable work has been performed in our own EPA laboratories and in industry on the development, improvement, and evaluation of various carbon-based convertors. The use of carbon has been demonstrated to be a workable approach for the specific redox conversion of NO₂:

$$C + NO_2 = CO + NO \quad \Delta F°_{500°K} = -32.2 \text{ kcal} \quad (7)$$

Carbons of various forms, e.g., charred sucrose, graphite, and carbon black, are currently being used in commercial monitors.

Figure 1 is a schematic of an automated NO, NO$_x$, NO₂ monitor of the type being used in routine monitoring today. The detector operates in a cyclic mode in which the incoming air sample, at 30-sec intervals, alternately passes through the convertor or a convertor bypass line. Electronic subtraction of the NO signal from the preceding NO$_x$ signal is used to generate an NO₂ output. The detector shown has four analog outputs, a continuous signal corresponding to instantaneous NO or NO$_x$ concentrations, and three outputs for NO, NO₂, and NO$_x$ average concentrations. The latter three are updated once each minute. Current manufacturers of chemiluminescence NO$_x$ instruments are shown in Table III.

Conclusions

Chemiluminescent techniques have been applied by the Environmental Protection Agency to provide an effective method of monitoring atmospheric pollutants. Commercial manufacturers have capitalized on these developments and now offer a family of chemiluminescent monitors capable of monitoring sulfur dioxide, hydrogen sulfide, oxides of nitrogen, and ozone. On-going research at EPA and other laboratories on chemiluminescent reactions should, in the near future, add to the number of pollutants that can be measured by this powerful technique.

Acknowledgment

The authors acknowledge the assistance of Ralph Baumgardner and

Table III. Oxides of Nitrogen Monitors

	Model No.
AeroChem Research Labs, Inc. P.O. Box 12 Princeton, N.J. 08540	AA-5
Beckman Instruments, Inc. 2500 Harbor Blvd. Fullerton, Calif. 92632	952
Bendix, Process Instruments Div. P.O. Drawer 477 Ronceverte, W.Va. 24970	8100
McMillan Electronics Corp. 7327 Ashcroft Houston, Tex. 77036	1200
Meloy Labs, Inc. 6715 Electronic Dr. Springfield, Va. 22151	NA-520
REM Inc. 2000 Colorado Ave. Santa Monica, Calif. 90404	642
Thermo Electron Corp. 85 First Ave. Waltham, Mass. 02154	14B
Scott Research Labs P.O. Box D-11 Plumsteadville, Pa. 18949	225
Yanaco Yanagimoto Manufacturing Co., Ltd., Fushimi-ku-, Kyoto, Japan	ECL-7

Charles Jamerson in preparing the tables and figures and thank A. E. O'Keeffe and Thomas A. Clark for their consultation during preparation of the text.

References

(1) H. J. Bernanose and M. J. Rene, *Advan. Chem. Ser. 21*, pp 7-12, Amer. Chem. Soc., Washington, D.C., 1959.
(2) K. D. Gundermann, *Angew. Chem., Int. Ed.*, **4**, 566 (1965).
(3) R. L. Bowman and N. Alexander, *Science*, **154**, 1454 (1966).
(4) V. H. Regener, *J. Geophys. Res.*, **65**, 3975 (1960).
(5) V. H. Regener, *ibid.*, **69**, 3795 (1964).
(6) J. A. Hodgeson, K. J. Krost, A. E. O'Keeffe, and R. K. Stevens, *Anal. Chem.*, **42**, 1795 (1970).
(7) R. K. Stevens, J. A. Hodgeson, L. F. Ballard, and C. E. Decker, "Determination of Air Quality," G. Mamamtov and W. D. Shults, Eds., p 83, Plenum, New York, N.Y., 1970.
(8) G. W. Nederbragt, A. Van der Horst, and J. Van Duijn, *Nature*, **206** (4979), 87 (1965).
(9) J. A. Hodgeson, B. E. Martin, and R. E. Baumgardner, Environmental Protection Agency, Research Triangle Park, N.C., Eastern Analytical Symp., Paper No. 77, New York, N.Y., 1970.
(10) R. K. Stevens, T. A. Clark, L. F. Ballard, and C. E. Decker, 1972 Air Pollution Control Association Meeting, Paper No. 72-13, Miami, Fla., June 1972.
(11) Environmental Protection Agency, *Fed. Regist.*, **36** (228), 22384–97 (Nov. 25, 1971).
(12) A. Fontijn, A. J. Sabadell, and R. J. Ronco, *Anal. Chem.*, **42**, 575 (1970).
(13) R. W. B. Pearce and A. G. Gaydon, "The Identification of Molecular Spectra," 3rd ed., p 266, Chapman and Hall, London, England, 1965.
(14) B. Draeger, W. German Patent 1,133,918 (July 26, 1962).
(15) W. L. Crider, *Anal. Chem.*, **37**, 1770 (1965).
(16) S. S. Brody and J. E. Chaney, *J. Gas. Chromatgr.*, **4**, 42 (1966).
(17) R. K. Stevens, A. E. O'Keeffe, and G. C. Ortman, *Environ. Sci. Technol.*, **3**, 652 (1969).
(18) R. K. Stevens, J. D. Mulik, A. E. O'Keeffe, and K. J. Krost, *Anal. Chem.*, **43**, 827 (1971).
(19) P. N. Clough and B. A. Thrush, *Trans. Faraday Soc.*, **63**, 915 (1967).
(20) F. Kaufman, *Progr. React. Kinet.*, **1**, 1 (1961).
(21) D. H. Stedman, E. E. Daby, F. Stuhl, and H. Nikki, *J. Air Pollut. Contr. Ass.*, **22**, 260 (1972).
(22) J. A. Hodgeson, K. A. Rehme, B. E. Martin, and R. K. Stevens, 1972 Air Pollution Control Association Meeting, Paper No. 72-12, Miami, Fla., June 1972.
(23) J. E. Sigsby, F. M. Black, T. A. Bellar, and D. L. Klosterman, Publ. Prepr., Environmental Protection Agency, Research Triangle Park, N.C., 1972.
(24) L. P. Breitenbach and M. Shelef, Tech. Rep. No. SR 71-130, Scientific Research Staff, Ford Motor Co., Dearborn, Mich., 1971.

A portion of this material was presented at a joint session of the Water, Air and Waste Chemistry and Analytical Chemistry Divisions, 165th Meeting, ACS, Dallas, Tex., April 1973. It is possible that there are manufacturers of chemiluminescent equipment other than those cited in this text; any omission of these manufacturers is unintentional. Mention of a company or product name is not intended to constitute endorsement by the Environmental Protection Agency.

Chemiluminescence and Bioluminescence

W. Rudolf Seitz and Michael P. Neary

Department of Chemistry, University of Georgia,
Athens, Ga. 30602

Light accompanying a chemical reaction is known as chemiluminescence (CL). CL that occurs in a living system or is derived from one is known as bioluminescence (BL). Three conditions are required for CL to occur: The chemical reaction must release sufficient energy to populate an excited energy state; the reaction pathway must favor the formation of excited state product; and the excited state product must be capable of emitting a photon itself or transferring its energy to another molecule that can emit.

The efficiency of CL, ϕ_{CL}, can be defined:

$$\phi_{CL} = \frac{\text{number (or rate) of photons emitted}}{\text{number (or rate) of molecules reacting}}$$

It is equal to the efficiency of excited state production (number molecules going to excited state/number molecules reacting) times the efficiency of emission (number of photons emitted/number of molecules in excited state). The highest efficiencies are observed for BL. The firefly reaction has an efficiency close to unity. For nonbiological CL, ϕ_{CL} rarely exceeds 0.01 even for the brightest reactions.

CL reactions can be used for chemical analysis by adjusting concentrations so that the CL intensity, I_{CL}, is related to the concentration of the reactant to be determined. At any time, t, I_{CL} is given by the expression:

$$I_{CL}(t)\left(\frac{\text{photons}}{\text{sec}}\right) = \phi_{CL}\left(\frac{\text{photons}}{\text{molecule reacting}}\right)$$
$$\frac{dc(t)}{dt}\left(\frac{\text{molecules reacting}}{\text{sec}}\right)$$

where $dc(t)/dt$ is the reaction rate for the starting material forming an electronically excited state. I_{CL} can be measured as a function of time (kinetic analysis), or it can be integrated for a known time period. For chemical analysis, a convenient means of performing the integration is to carry out the reaction in a flow system observing steady-state CL intensity.

CL and BL offer three important advantages for chemical analysis:

CL and BL methods are extremely sensitive because it is easy to measure low levels of light emission. It is possible to calculate theoretical detection limits for CL and BL methods from ϕ_{CL} and the capability of modern instrumentation to measure low light levels. In practice, however, sensitivity is usually limited by reagent purity, rather than by light-measuring capability.

The only apparatus required for CL analyses is a light detector, a system to mix the reactants, and in some cases a filter to resolve the CL of interest from other sources of light. Because of the high intrinsic sensitivity of CL methods, the light detector often does not need to be particularly sensitive.

For many of the available CL and BL reactions, response is linearly proportional to reactant concentration over several orders of magnitude.

The first section of this article describes some CL reactions that have been applied analytically, and the second section deals with BL methods. The number of CL and BL methods is small because of a lack of available reactions. However, those methods that have been developed are quite successful because of the unique advantages of CL.

Chemiluminescence

Ozone and Nitric Oxide. Recently developed methods for atmospheric ozone and nitric oxide provide a good example of the advantages of CL-generating reactions for chemical analysis. Ozone can be determined either by its reaction with rhodamine-B adsorbed on an activated silica gel surface or by its gas-phase reaction with ethylene. The rhodamine-B method has all the advantages of CL: it is sensitive to less than 1-ppb ozone, response is linear up to 400 ppb, and the only required instrumentation is a gas flow system to pull the sample over the surface and a photomultiplier to measure CL intensity. The only problem with this method is that the sensitivity of the CL surface changes with time as rhodamine-B is consumed in the reaction, thus necessitating frequent recalibration.

Analysis based on the ozone–ethylene reaction avoids this problem. This reaction produces CL emissions in the 300–600-nm region with maximum intensity close to 435 nm. Like the rhodamine-B reaction, this reaction is specific for ozone so that no optical resolution is required. The method is sensitive down to 0.003-ppm ozone, and response is linear up to 30 ppm.

Ozone is also involved in the determination of NO by use of the CL reaction:

$$NO + O_3 \longrightarrow NO_2^* + O_2$$
$$NO_2^* \longrightarrow NO_2 + h\nu$$

CL emission is a continuum from 0.6 to 3.0 μ. In the presence of excess O_3, CL intensity is proportional to NO concentration. Greatest sensitivity is obtained at reduced pressure because of quenching effects at higher pressures. Nevertheless, ambient NO concentrations can be measured at atmospheric pressure. At reduced pressure with a

in Chemical Analysis

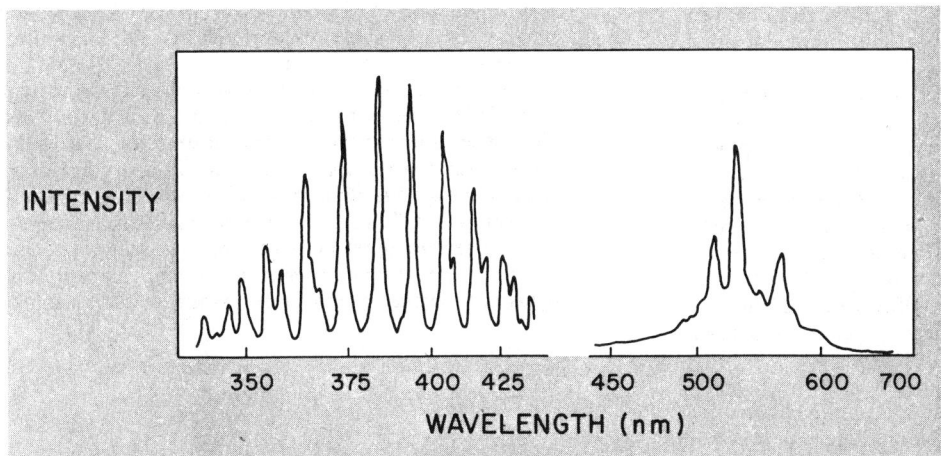

Figure 1. Chemiluminescence spectra for S_2 (left) and POH (right) recorded while aspirating aqueous SO_2 and phosphoric acid, respectively, in hydrogen–nitrogen diffusion flame

cooled photomultiplier tube, this method can detect 0.001-ppm NO. Response is linear up to 10,000 ppm, a linear dynamic range of 10^7. Because ozone reacts with other atmospheric contaminants to generate CL, a cutoff filter absorbing wavelengths shorter than 600 nm is included in NO monitors.

Total oxides of nitrogen ($NO + NO_2$) can be determined by reducing NO_2 to NO with carbon before reacting with ozone. The NO_2 concentration is equal to the difference between total oxides of nitrogen and NO concentration.

The instrumentation used for CL in air pollution monitors is described in detail in the April 1973 Instrumentation feature of *Analytical Chemistry*.

Sulfur and Phosphorus. Gas-phase CL can also result from the recombination of species generated in a flame. The most important analytical applications of flame CL have been to determine sulfur and phosphorus by observing the molecular emission that occurs when sulfur and phosphorus compounds are burned in a hydrogen-rich flame.

The sulfur emission comes from S_2 molecules, whereas phosphorus comes from POH. It has not been established exactly what chemical reactions are involved in producing CL. For sulfur, it may be the recombination of sulfur atoms, whereas for phosphorus, CL may come from the reaction between hydrogen atoms and PO.

Figure 1 shows the CL spectra for both S_2 and POH. For analytical applications the peak emission bands at 394 nm for sulfur and 526 nm for phosphorus can be resolved by use of interference filters. Sulfur and phosphorus can be detected simultaneously with two detection channels, each with the appropriate filter. Because two sulfur atoms are required to produce one excited molecule, CL intensity is proportional to the sulfur concentration squared.

The potential advantages of CL are not all realized when a flame is required to generate the reactants. Emission from the flame itself produces a background signal that limits sensitivity. This background can be reduced by shielding the flame. When the shield separates the burned gases above the flame from the outside air, CL from the recombination reactions is observed well above the flame itself. Thus, the CL can be viewed by the detector without looking directly at the flame.

Maximum efficiency for S_2 and POH CL is observed at temperatures below 400°C. This limits the temperature of the flame used to generate the reactive species. A hydrogen–oxygen flame is too hot and must be diluted with nitrogen to reduce the temperature. Hydrogen–air flames are satisfactory. Shielding of the flame helps to reduce the temperature of the burned gases and permits the use of hotter flames. Flame temperatures up to 1400°C have been achieved while maintaining efficient CL.

Flame CL analysis for phosphorus and sulfur works best with vapor-phase samples. Commercial detectors are available that use flame CL to selectively detect sulfur- and phosphorus-containing compounds as they elute from a gas chromatograph. They are sensitive down to minimum detectable levels of 0.04 ng of P and 0.2 ng of S. Response is linear up to 300 ng of P, and the square root of response is linear up to 100 ng of S. This application is discussed in more detail in the December 1973 Report in *Analytical Chemistry*.

With liquid-phase samples introduced as aerosols to the

Table I. Analytical Applications of Sulfur and Phosphorus Flame Chemiluminescence

Analysis	Performance
Sulfur- and phosphorus-containing gas chromatography effluents	Detection limits: 0.2 ng sulfur; 0.04 ng phosphorus
Sulfur-containing air pollutants	Detection limit: 5 ppb, H_2S, CH_3SH, and SO_2 can be resolved by gas chromatography
Sulfur in petroleum products	Detection limit: 0.5 ppm; requires combustion of sample
Phosphorus in detergents	Detection limit: 1.2 ppm; greater accuracy with prior ashing
Phosphorus in water	Detection limit: 0.003 ppm; measures dissolved phosphorus only

hydrogen-rich flame, there are several problems. Organic solvents cannot be used because they interfere directly with the CL reactions. For this reason, sulfur and phosphorus in oil cannot be determined without prior combustion of the sample.

It is possible to analyze for S and P in aqueous samples. However, the emission signal per unit sulfur or phosphorus is over 10 times smaller than for vapor-phase samples, even when ultrasonic nebulization is used to produce a finely divided aerosol. Emission per unit P or S increases with increasing volatility of the compound being analyzed. For example, P as triethyl phosphate produces emission 1.8 times greater than P as H_3PO_4 for an equivalent P concentration. In the case of alkali sulfides and sulfites, approximately a hundredfold increase in signal is observed upon acidification to produce a volatile species.

Because efficient CL requires a cool flame even when shielding is used, phosphorus and sulfur emissions are subject to chemical interferences. All metal ions depress emission intensity to a greater or lesser extent because the flame does not possess sufficient energy to break up the salt particles that form in the flame as the aerosol dries. Prior to an analysis, metal ions need to be removed by treatment with an ion-exchange resin.

Table I lists some applications of flame CL analysis for P and S. Gas-phase analysis of P- and S-containing GC effluents and sulfur-containing air pollutants has been much more widely applied than analysis of aqueous samples. However, improvements in sensitivity by using ultrasonic nebulization of aqueous samples may increase use of flame CL for solution analysis.

Luminol. The oxidation of luminol (5-amino-2,3-dihydrophthalazine-1,4-dione) in basic solution is one of the best known and most efficient CL reactions. The CL spectrum matches the fluorescence of the amino-phthalate oxidation product:

The most frequently used oxidant is hydrogen peroxide in the presence of a catalyst such as $Fe(CN)_6^{-3}$, Cu(II), and Co(II). Other CL-generating oxidants include hypochlorite, iodine, permanganate, and oxygen in the presence of a suitable catalyst. The optimum pH for CL varies somewhat with catalyst and oxidant. Most oxidizing systems have an optimum pH close to 11.

The luminol reaction differs from the CL reactions discussed above in that it occurs under a wide variety of conditions. Specific analysis using luminol requires that the chemistry be controlled so that CL is proportional only to the species of interest. This extends the unique advantages of CL to a wide variety of possible analyses rather than being restricted to only a couple of species.

Apparatus. It is possible to do CL analysis with luminol by simply injecting sample and reagents into a sealed con-

Figure 2. Diagram of flow system for making steady-state chemiluminescence measurements

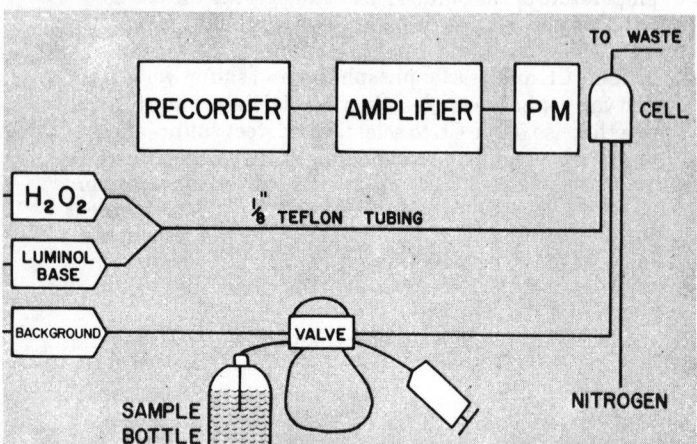

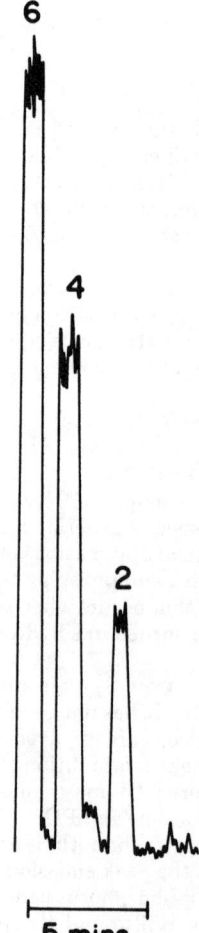

Figure 3. Typical data using flow system for chemiluminescence measurements. Peaks are for slugs of Cr(III) passing through cell

Conditions: $10^{-2}M$ H_2O_2, $10^{-3}M$ luminol, $10^{-1}M$ KOH–H_3BO_3 buffer, pH 10.5; peak 2 = $2.0 \times 10^{-8}M$ Cr(III); peak 4 = $4.0 \times 10^{-8}M$ Cr(III); peak 6 = $6.0 \times 10^{-8}M$ Cr(III)

Table II. Analytical Characteristics of Some Metal Ions That Catalyze Luminol Chemiluminescence[a]

Catalysts	Approx detection limit, M	Linear range, M	Remarks
Co(II)	10^{-11}	10^{-11}–10^{-7}	....
Cu(II)	10^{-9}	....	Nonlinear
Ni(II)	10^{-8}	10^{-8}–10^{-5}	....
Cr(III)	10^{-9}	10^{-9}–10^{-6}	....
Fe(II)	10^{-10}	10^{-10}–5×10^{-7}	Catalyst with oxygen
Mn(II)	10^{-8}	....	Requires amines to be a catalyst

[a] The data for Fe(II) were obtained by use of oxygen to stir the CL cell. All the other catalysts are effective only with H_2O_2. The conditions were $10^{-2} M$ H_2O_2, $10^{-3} M$ luminol, $10^{-1} M$ KOH–H_3BO_3 buffer and cell pH between 10 and 11.

Table III. Analytical Characteristics of Some Oxidants That React with Luminol to Produce Chemiluminescence

Oxidant	Approx detection limit, M	Linear range, M	Remarks
OCl^-	10^{-9}	....	Requires O_2
I_2	10^{-9}	10^{-9}–3×10^{-7}	Second- and third-order response also observed
MnO_4^-	10^{-10}	10^{-10}–10^{-7}	No O_2 needed
H_2O_2	10^{-9}	....	Excess Cu(II) catalyst

tainer surrounded by photographic film and measuring film exposure as a function of concentration. Alternatively, a stopped-flow spectrophotometer can be used to measure CL vs. time-after-mixing. However, for chemical analysis there are several advantages to performing the reaction in a flow system like that diagrammed in Figure 2. "Background" solution, usually the solvent of the analyte, is mixed with reagents in a cell positioned in front of a photomultiplier which measures CL intensity. This provides a reference level of light emission characteristic of the background solution. Slugs of samples are inserted into the background flow line with a sampling valve. As a slug of sample (catalyst or oxidant) passes through the cell, steady-state CL is observed with intensity proportional to sample concentration. Typical data are shown in Figure 3. The advantages of this system are: Background light emission from reagents provides a continuous reference level of CL rather than having to be subtracted out as a blank; the data come out in the form of peaks; a flow system can be readily adapted to continuous analysis, or if necessary it can be used as a detector for a chromatographic column; and the sample can be maintained at any pH in any eletrolyte until it enters the cell as long as it does not conflict with the requirement of a basic pH for luminol CL.

Either overhead stirring or gas bubbling can be used to mix the reactants in the cell. Gas bubbling makes it possible to use a gas as a reactant and has been successfully used for luminol oxidizing systems involving oxygen. However, overhead stirring reduces the noise level observed on steady state CL.

Analytical Performance. Table II lists some catalysts of luminol oxidation along with their analytical characteristics. The advantages of CL are apparent. Response is sensitive and linear over several orders of magnitude for most metals tested. The detection limits in Table II are imposed by background CL that is over two orders of magnitude greater than the PM dark current. Reagent purification should lead to even lower levels of detection.

Table III lists oxidants that can be determined by use of the luminol reaction along with their analytical characteristics. In the absence of peroxide, background CL is about 100 times less. Therefore, CL intensities are about 100 times less than for the catalyst systems in Table II for equivalent detection limits.

Indirect analysis based on luminol CL is also possible. For example, CL can be used to determine complexing agents by measuring the extent to which they solubilize metal ions from insoluble metal salts. Since several of the oxidants in Table III are commonly used as titrants, luminol CL can be used to follow oxidant concentration as a function of the quantity of oxidant added to a sample. This extends CL analysis to species that do not directly interact with luminol. Table IV lists analytical applications of luminol CL. Some of these applications have already been demonstrated, whereas others are feasible on the basis of present data. The chemical basis for obtaining selectivity for a particular species is included in the table.

Other CL Reactions. The CL reactions discussed above all have demonstrated analytical application. Several other CL reactions appear to have potential analytical applications but require further developmental work.

Ozone Reactions. Ozone reacts with a variety of compounds in the gas, liquid, and solid phases to generate CL. For example, the rhodamine-B method for ozone could be turned around to measure rhodamine-B concentration, and the ethylene–ozone reaction could be used to measure ethylene and other olefins. Since CL spectra are a function of olefin structure, it might be possible to use CL to characterize types of olefins.

Lucigenin. Lucigenin is similar to luminol in that it chemiluminesces upon oxidation by peroxide in basic solution in the presence of metal ion catalysts.

However, it has been reported in the Russian literature that lucigenin CL is catalyzed by Pb(II), Bi(III), Tl(III), and Hg(II), none of which catalyze luminol CL. If this is correct, then the lucigenin reaction could provide the basis for analytical applications not possible with luminol.

O Atom Reactions. Oxygen atoms undergo several gas-phase CL reactions that could be applied to trace air pollutant analysis. For example:

$$O + O + SO_2 \longrightarrow O_2 + SO_2 + h\nu \quad (\lambda_{max} = 200 \text{ nm})$$

is sensitive down to 0.001 ppm SO_2.

$$O + NO + M \longrightarrow NO_2 + M + h\nu$$

Table IV. Some Analytical Applications of Luminol Chemiluminescence

Type of analysis	Application	Basis for selectivity
Single catalyst	Fe(II) in water	Different oxidation potential from other catalysts of luminol oxidation by oxygen
	Cr(III) in water	EDTA used to quench other catalysts; Cr(III)–EDTA is kinetically slow to form
Multiple catalyst	Cu, Ni, Co, etc.	CL detection combined with ion-exchange separation
Organic analysis by measuring catalyst solubilization	Complexing agents in natural waters	Ion-exchange separation
Oxidant	Chlorine in waste water	Only CL-generating species present
	Protein-bound iodine	Different oxidation potential from other oxidants
	H_2O_2 generated by oxidase enzymes	Only CL-generating species present
Titrations	SO_2 in air titrated with I_2	Selective collection of SO_2
	Arsenic titrated with I_2	Distillation of $AsCl_3$

Table V. Luciferin–Luciferase Sources

Firefly	*Photinus*
	Photuris
	Luciola
Ostracod crustacea	*Cypridine*
	Pyrocypris
Bacteria	*Achromobacter fischerii*
	Photobacterium fischerii
Protozoa (dino flagellate)	*Gonyaulax polyedra*
Sea pansy	*Renilla reniformis*
Jellyfish	*Aequorea*

is an alternate method for NO which would have the advantage of producing emission partly in the visible, therefore requiring a less expensive photomultiplier tube. The problem in developing these and other methods is the lack of a suitably stable source of O atoms.

Flame CL. When NO or NO_2 is introduced into a hydrogen-rich oxy-hydrogen flame, CL from HNO is observed with a maximum at 690 nm. This reaction could be used to monitor NO and NO_2 emissions from gasoline engines. Conceivably, it could be adapted to develop a nitrogen-specific gas chromatography detector.

Riboflavin–H_2O_2. Riboflavin reacts with H_2O_2 to generate CL. CL is initiated either by adding a reducing agent or by irradiating with visible light ("photoinduced CL"). Because CL intensity is significantly enhanced by the presence of copper and is not greatly affected by other metal ions, this system can be used for copper analysis.

Bioluminescence

Background. Around 1885 Raphael Dubois, using the luminescing photogenic organ of the West Indian elaterid beetle, *Pyroporus,* discovered that when it was immersed in hot water until its light emisson ceased, a heat-stable compound was extracted. He also observed that when the cells of the photogenic organ were triturated in water at room temperature until the light emission ceased, a heat-labile compound was extracted. When the two oxygen-saturated extracts were mixed, light emission was immediately observed. The former extract was named Luciferin (LH_2) (*Lucifer* means light bearing in French) and is referred to as the substrate; the latter extract was named Luciferase (E) and is known to be an enzyme. It was at first thought that LH_2 was a protein; this is, in general, not true with the exception of the *Aequorea* bioluminescing system in which LH_2 is tightly bound to a protein matrix. There are many different sources for as many different luciferins and luciferases. Table V gives a few representative sources. The choice of a source for LH_2 and E is primarily governed by the chemistry involved, as will be discussed later; however, availability, stability, and cost are other important factors. Owing in part to these factors, certain BL systems have been favored for study. The firefly and bacterial systems have been extensively studied during the past 75 years; *Cypridina, Aequorea,* and *Renillia reniformis* are BL systems studied more recently.

Firefly Bioluminescence System. Of all the bioluminescing systems, that of the firefly is the most studied. The following scheme summarizes the firefly mechanism as presently understood.

$$LH_2 + E + ATP + Mg^{+2} \longrightarrow E \cdot LH_2 \cdot AMP + MgPPi$$

$$E \cdot LH_2 \cdot AMP + O_2(g) \longrightarrow [Oxyluciferin]^* + AMP + CO_2 + H_2O$$

$$[Oxyluciferin]^* \longrightarrow Oxyluciferin + h\nu \qquad \lambda_{max} = 562 \text{ nm}$$

where ATP = adenosine triphosphate, AMP = adenosine monophosphate, and PPi = inorganic pyrophosphate. A BL efficiency of one is routinely observed for this reaction.

When this reaction is carried out so that the analyte is ATP, 0.1 to 1.0 picomole is claimed for a minimum detection limit with linearity of response extending five orders of magnitude. Such a detection limit and linear range is obtainable from a so-called "crude" extract of ATP. Some investigators, after taking extreme care with their ATP extraction as well as with pH control and reactant purity, have been able to detect the ATP in a single bacterium, i.e., 2.0×10^{-5} to 1.0×10^{-2} picomoles. An accompanying increase in the linear range is also observed. The difference in the two minimum detection limits arises from contamination of the reactants with ATP or the presence of the interferent AMP in the ATP extract.

Specificity of the firefly LH_2 is high for ATP; however, studies have shown that both cytedine-5'-triphosphate (CTP) and inosine-5'-triphosphate (ITP) stimulate light production to the same extent as ATP. However, the contribution by these contaminants to the total light emission is small in a normal ATP-containing sample, since in natural systems the concentration of ATP is much greater than that of CTP and ITP. Other contaminants have been shown to act as inhibitors to the emission of light in the following order of activity Ca > K > Na > Rb > Li. Hg(II) at 2 ppm inhibits the emission of light, owing to its influence on E.

A typical ATP analysis utilizes firefly LH_2 (1.0 mg/ml), E (1.0 mg/ml) in 0.05M THAM buffer (pH 7.4), and 0.01M Mg(II). The mixture may be lyophilized and stored at $-65°C$. This is a convenient form for the reactants since they can be stored in this way indefinitely without losing their activity, and a simple addition of deionized water reconstitutes them for immediate use. For an analysis, no less than 0.5% dissolved O_2 must be present in the reaction solution. ATP is determined by adding it to excess

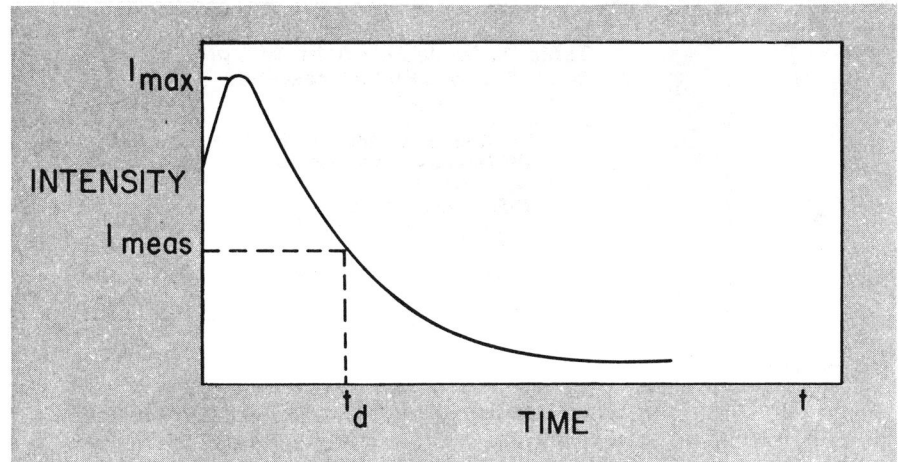

Figure 4. Typical bioluminescence vs. time-after-mixing curve observed for ATP analysis
I_{max} = maximum observed intensity
I_{meas} = intensity after fixed time interval, t_d

reactants and measuring the light intensity vs. time. A typical response curve is shown in Figure 4. Calibration curves are generated from measurements of either I_{max} or I_{meas} or the total integrated intensity as a function of ATP concentration.

ATP is present in *all* living cells regardless of whether the cell is photosynthetic or heterotrophic. ATP's phosphate bonds serve as an energy reservoir for the cell; thus, ATP becomes involved in many important metabolic reactions and is analytically important.

It was previously mentioned that it was possible to measure the ATP content in a single bacterium; with such knowledge, ATP analysis of a bacteria-containing sample would give information regarding their number. In practice, however, the ATP per bacterium is not determined on the basis of a single bacterium but rather on some large number of them. From various studies it is concluded that in any analysis of a particular bacterium for the concentration of ATP per bacterium, the investigator should compare the firefly–ATP analysis with plating of the same bacteria. The advantages of doing bacterial counting by ATP assay instead of plating include speed, accuracy, and expense. Moreover, certain filamentous microorganisms cannot be counted by conventional techniques and may best be counted by the ATP method. To illustrate these advantages, consider the study in which bacterial counts were made of samples of food, water, and urine. As few as 1000 bacterial cells could be measured in less than 5 min per sample. For such a measurement, log (ATP) vs. log (number of cells) exhibited a 0.93 positive linear correlation coefficient.

A typical extraction of ATP from the cell is accomplished by treating the triturated cells with five volumes of boiling ethanol for 1 min. After air drying, the extract is stored at $-20°C$ to be reconstituted later at the time of analysis. Table VI lists some applications of the BL assay for ATP. This reaction is so widely used that it has given rise to commercially available instruments specifically for this assay, such as Aminco's Chem-Glow photometer and Du Pont's biometer. Table VII lists some of the other species determinable by the firefly reaction along with their importance.

Bacterial Bioluminescence System. The bacterial system follows the firefly system in popularity for study. The following reactions schematically represent the bacterial system.

Table VI. Applications of Bioluminescence Assay for ATP

Application	Area of application
Monitoring of fermentation rates	Process control in food, beverage, and drug industries
Detection of bacterial contamination	Quality control in food, beverage, drug, and cosmetic industries
Measuring biomass in water	Control of activated sludge waste water treatment processes
	Fundamental studies in limnology and oceanography
Detecting presence of life	Extraterrestrial investigations
Determining viability of red blood cells	Medical
Measuring infection-causing bacteria	Medical

Table VII. Applications of Firefly Reaction to Species Other Than ATP

Species analyzed	Importance
Creatine phosphate	Energy reservoir for muscle activity
Cyclic AMP	Mediator of hormone activity
Dissolved oxygen	Measure of water quality Medical
Inorganic pyrophosphate	Starting material for several biological compounds

Table VIII. Sensitivity of Analytical Methods for FMN

Method	Min detectable concn, $\mu g/100~\mu l$
Paper chromatography	10^2
Cytochromic reductase	10^2
Lactic oxidase	1
Fluorometry	10^{-4}
Bacterial BL	10^{-5}

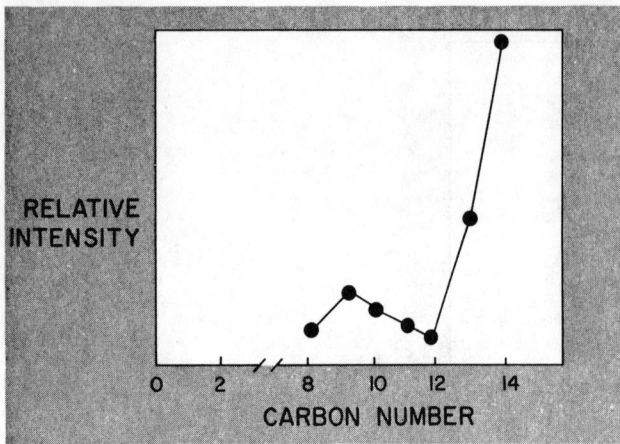

Figure 5. Relative intensity of bacterial bioluminescence as function of chainlength of added aldehyde

Table IX. Nucleotide Activators of *Renilla Reniformis* Bioluminescence

Activator	Rel activity, %
3′5′-Dephosphoadenosine	100
2′5′-Dephosphoadenosine	1
Coenzyme A	7
3′-Phosphoadenosine-5′-phosphosulfate (PAPS)	15–98[a]
ATP	0

[a] Depends on extent of acid hydrolysis.

The overall reaction is proposed to be:

$$2FMNH_2 + 2O_2 + RCHO + E \longrightarrow 2FMN + H_2O + H_2O_2 + RCOOH + h\nu$$

which is thought to be the sum of reaction series A and B.

Series A:

$$FMNH_2 + O_2 + E \longrightarrow E \cdot FMNH_2 \cdot O_2 \text{(I)}$$

$$\text{(I)} + RCHO \rightleftharpoons E \cdot FMNH_2 \cdot O_2 \cdot RCHO \text{(II)}$$

$$\text{(II)} \longrightarrow E \cdot FMN^* + RCOOH + H_2O$$

$$E \cdot FMN^* \longrightarrow E \cdot FMN + h\nu \qquad \lambda_{max} = 492 \text{ nm}$$

FMN = Flavin mononucleotide

Series B:

$$FMNH_2 + O_2 \longrightarrow FMN + H_2O_2$$

The BL efficiency for this reaction is about 0.05.

The oxidation of $FMNH_2$ proceeds with or without the aldehyde in the presence of O_2 to give 27 kcal/mol free energy, but emission from an excited FMN molecule of 492 nm would require 54.9 kcal/Einstein. The additional energy is provided by the oxidation of the long chain aldehyde to a carboxylic acid. The total free energy thus provided is approximately 95 kcal/mol. The intensity of the emission from this reaction depends on the chainlength of the aldehyde as shown in Figure 5.

Data in Table VIII show that the bacterial BL system exhibits a minimum level of detection for the analysis of FMN which is superior to other popular methods. It also shows that fluorometry might compete for the analysis of FMN; however, its lack of specificity vitiates its use. Compounds commonly found in samples of biosystems interfere with the fluorometric analysis of FMN.

Bacterial BL has high specificity for FMN. Some substituted FMN's and flavin adenine dinucleotide (FAD) react with the bacterial (E) to produce light; however, the level of emission is low enough so that it is of little analytical concern. The relationship between light output and $FMNH_2$ concentration is linear from 1.0×10^{-4} to 1.0 μg/ml.

The two most popular sources of the bacterial luciferase (E) are the bacteria Photobacterium fischerii and Achromobacter fischerii. The luciferin, FMN, may be obtained from virtually any living system. The luciferase is generally used at a concentration of 1.0 μg/ml in 0.05M THAM buffer (pH 7.4). Dodecylaldehyde complexed with bisulfite is widely used as the required aldehyde. The FMN may be extracted from the sample by treating it with a boiling solution of 6% butanol in 0.01M THAM buffer containing $10^{-3}M$ EDTA. The extraction is complete in less than a minute and is followed by filtration. The supernatant containing FMN and $FMNH_2$ is generally treated with either $NaBH_4$ with $PdCl_2$ as a catalyst to reduce FMN to $FMNH_2$, or FMN may be reduced by NADH (reduced form of nicotinamide adenine dinucleotide) in the presence of H^+. Following the above treatment, the extracted, reduced FMN is mixed with the enzyme (E) and the long chain aldehyde–bisulfite complex before a light sensitive detector, and the intensity of the emission is compared with a calibration curve. The noise or background in this analytical scheme is primarily endogenous light from the reactants and is eliminated by the calibration procedure since the background is generally constant for a given experiment.

The BL assay for FMN is applicable to many of the same systems as the ATP method. It has been used to monitor infectious bacteria and has been extensively investigated as a possible detector of extra terrestrial life.

Another type of application involves exposing the bacteria to organic vapors and observing the decrease in luminescence. This can be used to study the effect of anesthetic vapors or to detect the presence of various compounds such as alcohols, aldehydes, and ketones.

The dependence of the bacterial BL on $FMNH_2$, and FMN's participation in the following reaction leads to the possibility of analyses which do not depend directly on FMN but on some substrate being oxidized by NADH or reduced by NAD^+.

$$2NADH + FMN \xrightarrow{E'} 2NAD^+ + FMNH_2 \quad \text{where } E' = \text{NADH dehydrogenase}$$

For example, a method for NO_3^- could be based on the following reactions:

$$H^+ + NAD^+ + NO_3^- \xrightarrow{E''} NADH + NO_2^-$$

$$H^+ + NADH + FMN \xrightarrow{E''} FMNH_2 + NAD^+$$

$$E'' = \text{nitrate reductase}$$

The concentration of the resulting $FMNH_2$ is proportional to the initial concentration of nitrate.

Other BL Systems. *Aequorea.* The BL system of the hydromedusid, *Aequorea,* is unusual among BL systems in that it seems not to rely on either the typical LH_2 and E or O_2. The luminescence involves only the photoprotein aequorin in the presence of Ca(II) ions. It is thought that the protein provides a matrix for the LH_2, E and O_2 thus forming a complex which, when triggered by Ca(II), bioluminesces.

Early research proposed that the *Aequorea* system was specific for Ca(II); however, later research has shown that the photoprotein aequorin can be stimulated to bioluminescence by over a dozen other cations, such as Co(II), Pb(II), and Yb(III). However, since these cations are not normally present in significant amounts in biologically derived samples, the aequorin reaction may still serve as a means of Ca(II) trace analysis in samples of biological origin.

Renilla Reniformis. Renilla reniformis, commonly referred to as the sea pansy, produces blue-green bioluminescence in concentric waves across its surface. The emission wavelength maximum is 485 nm and follows the reaction scheme:

$$LH_2 + \text{nucleotide activator} + Ca(II) + E \longrightarrow \text{activated } LH_2$$

$$E = \text{sulfokinase (probably)}$$

$$\text{activated } LH_2 + O_2 \xrightarrow{E'} \text{products} + h\nu$$

$$E' = \text{Luciferase} \quad \lambda_{max} = 485 \text{ nm}$$

The feature of this BL system which is of analytical significance is the required nucleotide activator. Table IX shows some activators and their relative activity in the reaction. PAPS is of some interest in the study of brain metabolism.

Fungal. The fungal BL systems are pyridine-nucleotide linked and are thought to follow the path shown in producing light.

$$DPNH + H^+ + L \xrightarrow{\text{DPNH oxidase}} LH_2 + DPN^+$$

$$LH_2 + O_2 \xrightarrow{E} \text{products} + h\nu \quad \lambda = 528 \text{ nm}$$

$$E = \text{an enzyme (luciferase) bound to a surface}$$

The pyridine-nucleotide linkage of this system makes it potentially important analytically, owing to the broad range of reactions involving the pyridine-nucleotide.

References

General

M. J. Cormier, D. M. Hercules, and J. Lee, Eds., "Chemiluminescence and Bioluminescence," Plenum Press, New York, N.Y., 1973.

K. D. Gundermann, "Chemiluminescenz Organischer Verbindungen," Springer-Verlag, New York, N.Y., 1969.

W. D. McElroy and B. Glass, Eds., "Light and Life," Johns Hopkins Press, Baltimore, Md., 1961.

F. H. Johnston and Y. Haneda, Eds., "Biolouminescence in Progress," Princeton University Press, Princeton, N.J., 1966.

E. N. Harvey, "Bioluminescence," Academic Press, New York, N.Y., 1952.

A. C. Gese, "Photophysiology," Vols II, IV, V, Academic Press, New York, N.Y., 1970.

E. White and H. H. Seliger, "Bioluminescence," *Sci. Amer.*, **207** (12), 76-89 (1962).

B. L. Strehler, "Bioluminescence Assay: Principles and Practice" in "Methods of Biochemical Analysis," D. Glick, Ed., **16**, 99-179 (1968).

Air Pollutants

A. Fontijn, P. Golomb, and J. A. Hodgeson, "A Review of Experimental Measurement Methods Based on Gas-Phase Chemiluminescence," in "Chemiluminescence and Bioluminescence," M. J. Cormier, D. M. Hercules, and J. Lee, Eds., pp 393-426, Plenum Press, New York, N.Y., 1973.

R. K. Stevens and J. A. Hodgeson, "Applications of Chemiluminescent Reactions to the Measurement of Air Pollutants," *Anal. Chem.*, **45**, 443A (1973).

J. A. Hodgeson, K. J. Krost, A. E. O'Keefe, and R. K. Stevens, "Chemiluminescent Measurement of Atmospheric Ozone," *Anal. Chem.*, **42**, 1795 (1970).

G. J. Warren and G. Babcock, "Portable Ethylene Chemiluminescence Ozone Monitor," *Rev. Sci. Instrum.*, **41**, 280 (1970).

V. H. Regener, "Measurement of Atmospheric Ozone with the Chemiluminescent Method," *J. Geophys. Res.*, **69**, 3795 (1964).

G. W. Nederbragt, A. van der Horst, and J. van Duijn, "Rapid Ozone Determination Near an Accelerator," *Nature*, **206**, 87 (1965).

A. Fontijn, A. J. Sabadell, and R. J. Ronco, "Homogeneous Chemiluminescent Measurement of Nitric Oxide with Ozone," *Anal. Chem.*, **42**, 575 (1970).

P. N. Clough and B. A. Thrush, "Mechanism of Chemiluminescent Reaction Between Nitric Oxide and Ozone," *Trans. Faraday Soc.*, **63**, 915 (1967).

Sulfur and Phosphorus

A. Fontijn, D. Golomb, and J. A. Hodgeson, "A Review of Experimental Measurement Methods Based on Gas-Phase Chemiluminescence," in "Chemiluminescence and Bioluminescence," M. J. Cormier, D. M. Hercules, and J. Lee, Eds., pp 393-426, Plenum Press, New York, N.Y., 1973.

R. K. Stevens and J. A. Hodgeson, "Applications of Chemiluminescent Reactions to the Measurement of Air Pollutants," *Anal. Chem.*, **42**, 1795 (1970).

H. W. Grice, M. L. Yates, and D. J. David, "Response Characteristics of the Melpar Flame Photometric Detector," *J. Chromatogr. Sci.*, **8**, 90 (1970).

R. M. Dagnall, K. C. Thompson, and T. S. West, "Molecular-emission Spectroscopy in Cool Flames, Part I. The Behavior of Sulphur Species in a Hydrogen-Nitrogen Diffusion Flame and in a Shielded Air-Hydrogen Flame," *Analyst*, **92**, 506 (1967).

R. M. Dagnall, K. C. Thompson, and T. S. West, "Molecular-emission Spectroscopy in Cool Flames. Part II. The Behavior of Phosphorus Containing Compounds," *Analyst*, **93**, 72 (1968).

K. M. Aldous, R. M. Dagnall, and T. S. West, "The Flame-spectroscopic Determination of Sulfur and Phosphorus in Organic and Aqueous Matrices by Using a Simple, Filter Photometer," *Analyst*, **95**, 417 (1970).

A. Syty, "Determination of Phosphorus in Detergents by Flame Emission Spectrometry," *Anal. Lett.*, **4**, 531 (1971).

M. J. Prager and W. R. Seitz, "A Flame Emission Photometer for Determining Phosphorus in Natural Waters," submitted to *Anal. Chem.*

Luminol

W. R. Seitz and D. M. Hercules, "Chemiluminescence Analysis for Trace Elements," in "Chemiluminescence and Bioluminescence," M. J. Cormier, D. M. Hercules, and J. Lee, Eds., pp 427-49, Plenum Press, New York, N.Y., 1973.

W. R. Seitz and D. M. Hercules, "Determination of Trace Amounts of Iron(II) Using Chemiluminescence Analysis," *Anal. Chem.*, **44**, 2143 (1972).

W. R. Seitz, W. W. Suydam, and D. M. Hercules, "Determination of Trace Amounts of Chromium(III) Using Chemiluminescence Analysis," *Anal. Chem.*, **44**, 957 (1972).

In addition, there are many papers in Russian journals authored or coauthored by A. K. Babko that describe various luminol systems.

Other Reactions

R. L. Bowman and N. Alexander, "Ozone-Induced Chemiluminescence of Organic Compounds," *Science*, **154**, 1454 (1966).

B. J. Finlayson, J. N. Pitts, Jr., and H. Akimoto, "Production of Vibrationally Excited OH in Chemiluminescent Ozone-Olefin Reactions," *Chem. Phys. Lett.*, **12**, 495 (1972).

L. I. Dubovenko and E. Ya. Khotinets, "Catalytic Effect of Bi(III) on the Chemiluminescent Reaction of Lucigenin with Hydrogen Peroxide," *Ukr. Khim. Zh.*, **37**, 1154 (1971).

K. J. Krost, J. A. Hodgeson, and R. K. Stevens, "Flame Chemiluminescence Detection of Nitrogen Compounds," *Anal. Chem.*, **45**, 1800 (1973).

E. L. Wehry and A. W. Varnes, "Selective Determination of Copper(II) in Aqueous Media by Enhancement of Flash-Photolytically Initiated Riboflavin Chemiluminescence," *Anal. Chem.*, **45**, 348 (1973).

M. O. Stone and R. H. Steele, "Studies on the Chemiluminescence from an O_2 and/or H_2O_2 Adduct Derived from the Riboflavin-Copper(I) Chelate," *Biochem.*, **9**, 4343 (1970).

Firefly System

L. Dufresne and H. J. Gitelman, "A Semiautomated Procedure for Determination of ATP," *Anal. Biochem.*, **37**, 402-08 (1970).

E. Beutler and M. C. Baluda, "Simplified Determination of Blood ATP Using the Firefly System," *Blood*, **23** (5), 688-96 (1964).

N. Suzuki and T. Goto, "Studies of Firefly Bioluminescence II," *Tetrahedron*, **28** (15), 4075-82 (1972).

R. A. Johnson, J. G. Hardman, A. E. Broadus, and E. W. Sutherland, "Analysis of Adenosine 3',5'-monophosphate with Luciferase Luminescence," *Anal. Biochem.*, **35**, 91-97 (1970).

J. B. St. John, "Determination of ATP in Chlorella with the Luciferin-Luciferase Enzyme System," *Anal. Biochem.*, **37**, 409-16 (1970).

E. W. Chappelle and G. V. Levin, "Use of the Firefly Bioluminescent Reaction for Rapid Detection and Counting of Bacteria," *Biochem. Med.*, **2**, 41-52 (1968).

P. E. Stanley and S. G. Williams, "Use of the Liquid Scintillation Spectrometer for Determining ATP by the Luciferase Enzyme," *Anal. Biochem.*, **29**, 381-92 (1969).

J. W. Patterson, P. L. Brezonik, and H. D. Putnam, "Measurement and Significance of ATP in Activated Sludge," *Environ. Sci. Technol.*, **4** (7), 569-75 (1970).

O. Holm-Hansen and C. R. Booth, "The Measurement of ATP in the Ocean and Its Ecological Significance," *Limnol. Oceanogr.*, **11**, 510-19 (1966).

O. Holm-Hansen, "Determination of Microbal Biomass in Ocean Profiles," *Limnol. Oceanogr.*, **14**, 740-47 (1969).

Bacterial System

E. W. Chappelle, G. L. Picciolo, and R. H. Altland, "A Sensitive Assay for FMN Using Bacterial Bioluminescence Reaction," *Biochem. Med.*, **1**, 252-60 (1967).

P. E. Stanley, "Determination of Subpicomole Levels of NADH and FMN Using Bacterial Luciferase and Liquid Scintillation Spectrometer," *Anal. Biochem.*, **39**, 441-53 (1971).

D. E. White and C. R. Dundas, "Effect of Anaesthetics on Emission of Light by Luminous Bacteria," *Nature*, **226**, 456-58 (1970).

J. Lee, "Bacterial Bioluminescence," *Biochem.*, **11**, 3350 (1972).

Other Systems

O. Shimomura, F. H. Johnson, and Y. Saiga, "Microdetermination of Calcium by Aequorin Luminescence," *Science*, **140**, 1339 (1963).

E. B. Ridgway and C. C. Ashley, "Calcium Transients in Single Muscle Fibers," *Biochem. Biophys. Res. Commun.*, **29** (2), 229-34 (1967).

J. P. Hamman and H. H. Seliger, "The Mechanical Triggering of Bioluminescence in Marine Dinoflagellates: Chemical Basis," *J. Cell. Physiol.*, **80** (3), 397-408 (1972).

K. T. Izutsu, S. P. Felton, I. A. Siegel, W. T. Yoda, and A. C. N. Chen, "Aequorin: Its Ionic Specificity," *Biochem. Biophys. Res. Commun.*, **9** (4), 1034-39 (1972).

W. Rudolf Seitz received an AB in chemistry from Princeton University in 1965 and a PhD from MIT in 1970 where he studied with D. N. Hume. Dr. Seitz recently joined the faculty at the University of Georgia after working for three years at the Southeast Environmental Research Laboratory in Athens, Ga. His principal research interest is applying chemiluminescence to chemical analysis. Dr. Seitz was a coordinator of the symposium on Recent Advances in the Analytical Chemistry of Pollutants held in Athens in May 1973.

Michael P. Neary is a graduate student of Professor Hercules at the University of Georgia where he is carrying out research in analytical applications of chemiluminescence. Mr. Neary earned a BA degree in chemistry from the University of Colorado in Boulder and subsequently worked for five years with the atmospheric chemistry group at the National Center for Atmospheric Research in Boulder and five years with Beckman Instruments, Inc., in Fullerton, Calif.

Immobilized Enzymes: Analytical Applications

Howard H. Weetall
Corning Glass Works
Corning, N.Y. 14830

Enzymes can be immobilized or confined on or in a variety of water-soluble and water-insoluble matrices with little or no immediate loss of their catalytic activity. This is of interest and importance for theoretical reasons. Most enzymes are not found randomly dissolved in the cytosol or cytoplasm but are attached to membranes in or on various cell organelles. Further, the physicochemical characteristics of these "insolubilized" enzymes are affected by the microenvironments in which they are found. When enzymes are removed from their microenvironments, their behavior may differ. Immobilization of soluble enzymes on or in synthetic membranes or on organic or inorganic matrices permits researchers to study them in environments modeled after their natural states.

A second and more practical important reason for immobilizing enzymes is that immobilized enzymes allow researchers to add, remove, and reuse enzymes at will. This also means that industrial processors presently using soluble enzymes may, in some cases, economically substitute immobilized enzyme systems. Similarly, the ability to remove and reuse these materials makes them likely candidates for use in the analytical and clinical laboratory and even in the physician's office.

Not a new technology, enzyme immobilization can be traced back at least 50 years to the work of Nelson and Griffin (1). These workers adsorbed invertase to animal charcoal and observed that the "immobilized enzyme" retained biological activity and could be reused over a long period of time. For the most part, these studies went unnoticed until the 1950's when Grubhofer and Schlecth immobilized several enzymes on polyaminopolystyrene and on a chlorinated resin Amberlite XE-64 by covalent attachment (2, 3). It was around this time that other methods of enzyme immobilization also came into prominence.

Methods of Enzyme Immobilization

There are several methods of immobilizing enzymes. [For a general review of immobilization, see Zaborsky (4).]

Absorption. Absorption of an enzyme to a water-insoluble surface is the simplest method of immobilization. The subject has been reviewed in depth by McLaren and Packer (5).

Adsorption. The adsorption of enzymes onto surfaces is dependent on many variables including pH, type of solvent, ionic strength, temperature, and enzyme concentration (6). The stability of the adsorbed enzyme and its activity are dependent on assay and storage conditions, substrate concentration, ionic strength of storage solution and many of the same parameters which effect the adsorption phenomenon itself.

Commonly used adsorbents include: alumina, carbon, celluloses, clays, hydroxyapatite, and glasses including controlled-pore glass. Messing has shown that enzymes adsorbed to glass surfaces have extremely long half-lives and can be successfully used repeatedly without activity losses (7, 8).

Adsorption and Cross-Linking. Enzymes have been adsorbed to colloidal silica followed by cross-linking with glutaraldehyde (9). These enzymes retained their activity and could be reused several times. In a similar fashion Gaffield et al. and Olson and Stanley adsorbed lactase to a phenolformaldehyde resin and crosslinked the adsorbed enzyme with glutaraldehyde (10, 11). The technique is generally rapid and simple and produces a product of relatively good stability.

Cross-Linking. Enzymes can be immobilized by cross-linking with low-molecular-weight multifunctional reagents producing covalent bonds with intermolecular cross-links between the reagent and the enzyme. The activity of the cross-linked enzyme is dependent on many factors including: concentration of enzyme, reagent, pH, ionic strength, and number of cross-links produced. The overall apparent activity of the derivative is also dependent on the size of the substrate. Generally, high-molecular-weight substrates cannot come in contact with enzymes in the interstices of such immobilized enzyme supports. Multifunctional agents most commonly used include diazobenzidine and its derivatives and glutaraldehyde. Other less-used agents include carbodiimides, diisothiocyanates, diisocyanates, and disulfonic acids.

Ion-Exchange Resins. Enzymes can be immobilized through ion-exchange techniques. Many enzymes will bond to ion-exchange resins without significant loss of activity. Unless the pH or ionic conditions are changed to cause elution of the enzyme, it will remain attached and active. This approach has successfully been used in immobilizing L-aminoacylase on DEAE-Sephadex. This immobilized enzyme is used commercially for production of L-methionine at a rate greater than 20 metric tons per month.

Entrapment. Enzymes can be immobilized by entrapment within cross-linked water-insoluble polymers. The method of preparation usually involves the cross-linking of the polymer in the presence of the enzyme, physically entrapping the enzyme. The polymer lattice structure is such that the large enzyme molecule cannot diffuse out, but small substrate molecules can diffuse into the polymer. Gel entrapment was first successfully used by Bernfeld and Wan (12). Materials used for entrapment include: polyacrylamides, silicone rubber, silica gel, and starch.

Microencapsulation. Enzymes have been immobilized by encapsulation within semipermeable membranes. The encapsulation is carried out by an interfacial polymerization or coacervation of a preformed polymer. The first reports of encapsulating enzymes in permanent membranes were those of Chang (13). Enzymes have been encapsulated in membranes of collodian, polystyrene, and cellulose derivatives (4). The most common polymer used for encapsulation, however, is nylon.

Copolymerization. Enzymes can become immobilized by being covalently incorporated into polymers. The most common materials for copolymerization are maleic anhydride and ethylene, which were first reported by Bar-Eli and Katchalski (14). Copolymerized enzymes can be prepared as either neutral or charged derivatives. As with entrapped and microencapsulated enzymes, these derivatives show little or no activity with large macromolecular substrates.

Covalent Attachment. Chemical attachment of enzymes to water-insoluble carriers is the most commonly used method of immobilization. Theoretically, covalent coupling offers the most stable, most versatile method of immobilizing enzymes. Methods of covalent attachment are too numerous to be discussed in a review of this nature. However, some of the more common methods of attachment and carriers are as follows:

Carboxymethylcellulose Azide. The carboxymethylcellulose is converted by the reactions schematically shown in Figure 1. The method is relatively simple. The final coupling step is carried out at a slightly alkaline pH. The final product is an amide formed primarily to the ξ-amino group of lysine. Other possible attachment sites include cysteine, serine, and tyrosine.

Azo Linkage. Arylamines can be diazotized and coupled via azo linkage to proteins. The technique has been successfully used for covalent attachment of proteins to polyaminopolystyrene, *p*-aminobenzoylcellulose, and arylamine-glass (15). The coupling to protein occurs at a slightly alkaline pH. The reaction is schematically represented in Figure 2. Coupling occurs through an available tyrosine, although lysine, arginine, cysteine, and serine have been implicated in the reaction.

Isocyanates and Isothiocyanates. Arylamines and alkylamines can be converted to isocyanates and isothiocyanates. These activated derivatives will then react with amines on the protein, primarily the ξ-amino group of lysine, to form substituted ureas. The reactions are schematically represented in Figure 3. The coupling is carried out at slightly alkaline pH.

Carbodiimides. Water-soluble carbodiimides can be used for covalently attaching proteins to derivatives through amide linkage. The enzyme may be coupled under slightly acid conditions. The reaction is schematically represented in Figure 4. Amide derivatives can also be used for coupling to the carboxyl group of the protein with carbodiimides, although under these conditions, cross-linking is quite possible.

Cyanogen Biomide. The activation of cross-linked dextrans, including agarose and even cellulose, is a simple and attractive method of covalently coupling proteins to water-insoluble carriers. The reaction sequence is schematically represented in Figure 5. The coupling reaction works best at pH 9.0, although many workers prefer a more neutral pH. The ϵ-amino lysine group is the group through which coupling generally occurs to the protein.

Glutaraldehyde. The covalent attachment of proteins to water-insoluble carriers via glutaraldehyde is one of the simplest and most gentle of coupling methods. The reaction is carried out at neutral pH. Attachment is from the amine carrier to the ϵ-amino group of lysine in the protein. The reaction schematically represented in Figure 6 is oversimplified for purposes of this review. There are many other methods of covalently coupling enzymes to water-insoluble carriers. [For further details, see one of several available reviews (4, 16, 17, 18).]

Characteristics of Immobilized Enzymes

When an enzyme is immobilized—either within a matrix or on the surface of a carrier—several changes may occur in the enzyme's apparent behavior. The factors effecting this behavior are many. However, several of the observed changes occur quite commonly and will be considered here.

pH Profile. All enzymes have an optimal pH at which they show a maximum reaction rate. When the enzyme is immobilized, the optimal pH may shift, depending on the nature of the carrier. Goldstein and coworkers (19) studied this phenomenon in detail. In a nutshell, they report: If a carrier is negatively charged, then a high concentration of positively charged ions (H^+) will accumulate at the boundary layer between the carrier and the surrounding solution. This accumulation of hydrogen ions will cause the pH at the carrier surface to drop below that of the bulk solution. The enzyme, therefore, sees a pH below that of the bulk solution. In this manner, the apparent pH of the immobilized enzyme may be

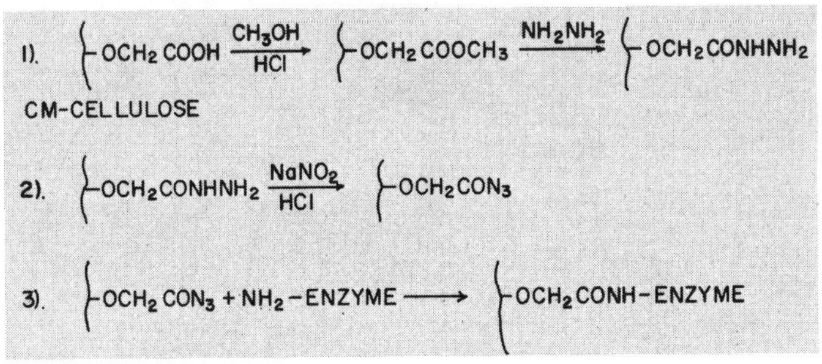

Figure 1. Preparation of carboxymethylcellulose azide and attachment of enzyme to active derivative

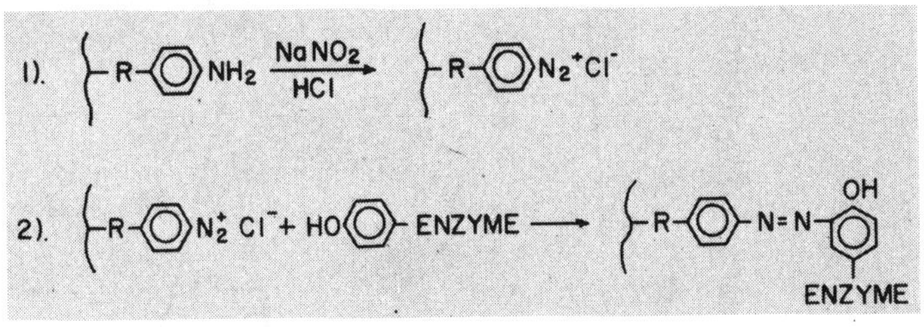

Figure 2. Preparation of diazonium chloride and attachment of enzyme through azo linkage

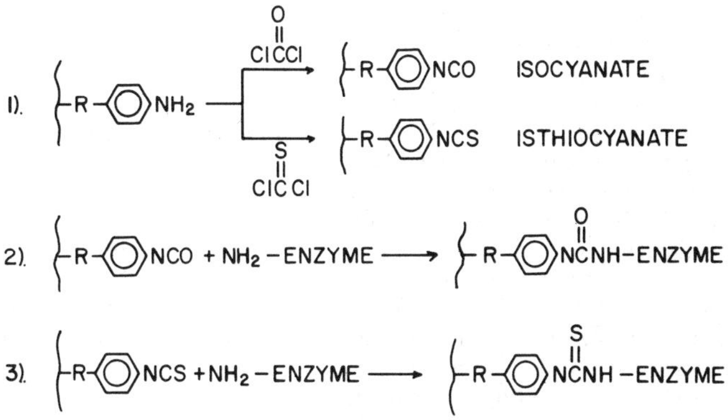

Figure 3. Preparation of isocyanate and isothiocyanate derivatives and their attachment to an enzyme

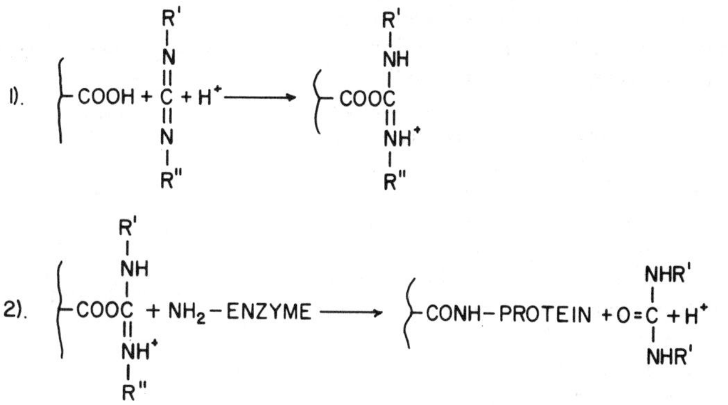

Figure 4. Preparation of an active pseudourea from a carboxyl and a carbodiimide and reaction with an enzyme forming amide linkage

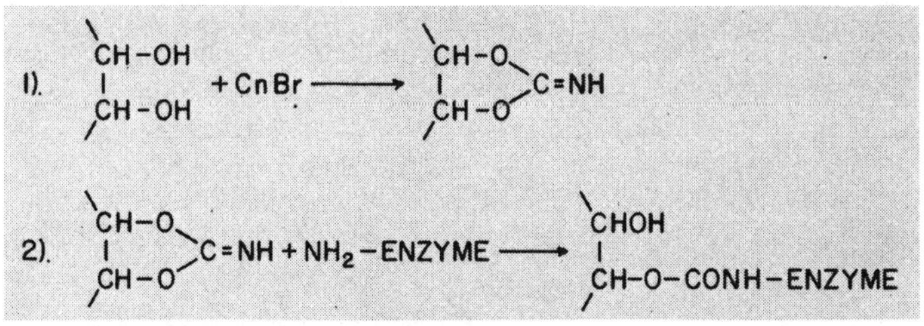

Figure 5. Preparation of cyanogen bromide-activated derivative and its attachment to an enzyme

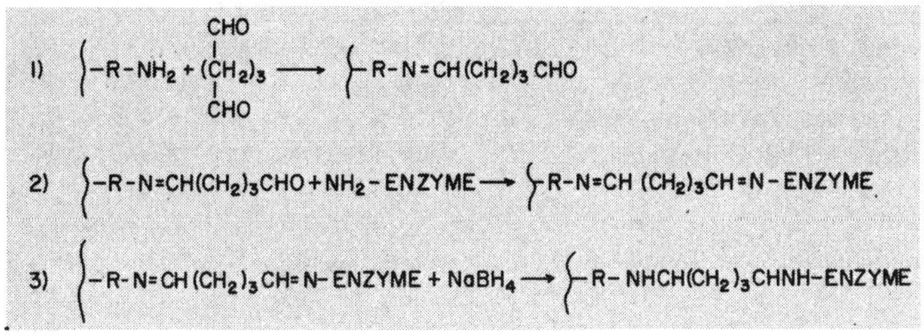

Figure 6. Preparation of an active aldehyde derivative and its attachment to an enzyme

increased. If the carrier is negatively charged, the opposite may occur (Figure 7).

Kinetics. For the most part, the kinetics of immobilized enzyme reactions are studied in terms of "apparent" values. Several excellent discussions of the kinetics of immobilized enzymes are available (16, 20, 21). When an enzyme is immobilized, one generally observes an increase in K_m. This increase is usually related to the charge on the substrate and/or carrier, diffusion effects, and in some cases, tertiary changes in enzyme configuration. However, in some cases, no change in K_m is observed (Table I).

Stability. Like all other proteins, enzymes are susceptible to thermal denaturation, whether they are immobilized or in the "free" state. In many cases, however, the rate of inactivation and denaturation of an immobilized enzyme is less than that of the free enzyme. The thermal stability of the enzyme papain is shown in Figure 8. Inactivation occurred at lower temperatures with the free enzyme. Similar results have been observed with many other enzymes. Table II gives the activation energies for several soluble and immobilized enzymes. Enzymes which show excellent thermal stability do not necessarily show excellent operational stability, because the operational stability of immobilized enzymes is not only a function of thermal stability, but of such factors as carrier durability and organic inhibitors and inhibitor concentrations, including that of heavy metals. The clogging of the carrier also effects half-life—that point where the amount of activity of the enzyme is 50% of what it was when initially used. Enzyme half-lives under controlled operating conditions should be temperature-dependent, but this does not mean that the activity decrease is denaturation dependent. Table III gives an example of the dependency of half-life on operating temperature for an immobilized enzyme.

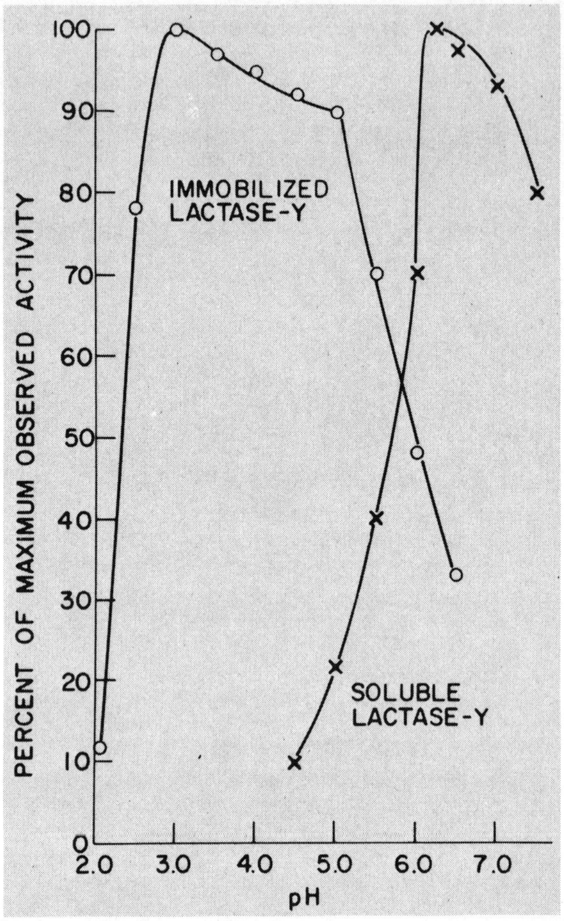

Figure 7. pH profile of yeast lactase in solution and covalently attached to ZrO_2-coated control pore glass, 550 Å pore diameter, 40/80 mesh. Substrate was 10% lactose solution. (From ref. 34)

Applications of Immobilized Enzymes

Analytical Applications—Enzyme Electrodes. An enzyme electrode is basically a combination of an enzyme and an electronic sensing device. Such an electrode possesses the specific properties of the enzyme and the adaptability and ease of readout of a clinical electrode device.

The first enzyme electrode as developed by Updike and Hicks (22) was prepared by entrapping the enzyme glucose oxidase in a gelatinous polymer coating over the surface of a

Table I. Comparison of K_m Values of Some Soluble and Immobilized Enzymes[a]

Enzyme	Substrate	K_m, mM	
		Soluble	Immobilized
Invertase	Sucrose	0.448	0.448
Arylsulfatase	p-Nitrophenyl-sulfate	1.85	1.57
Glucoamylase	Starch	1.22	0.30
Alkaline phosphatase	p-Nitrophenyl-phosphate	0.10	2.90
Urease	Urea	10.0	7.60
Glucose oxidase	Glucose	7.70	6.80
L-Amino acid oxidase	L-Leucine	1.00	4.00

[a] All enzymes were immobilized on ZrO_2-coated control porous 96% silica glass particles. K_m values were determined under identical conditions for both soluble and immobilized derivatives for comparison.

Table II. Comparison of Activation Energies of Some Soluble and Immobilized Enzymes[a]

Enzyme	Activation energy, kcal/gmol	
	Soluble	Immobilized
Papain (amide linkage)	13.8	11.0
Papain (azo linkage)	...	13.8
Glucose oxidase (azo linkage)	6.6	9.0
Glucoamylase (Shiffs base)	16.3	13.8
Yeast lactase (Shiffs base)	10.5	11.3
Microbial lactase (Shiffs base)	10.4	6.5

[a] Enzymes immobilized on inorganic supports. Comparative values were obtained between the same temperature ranges where possible.

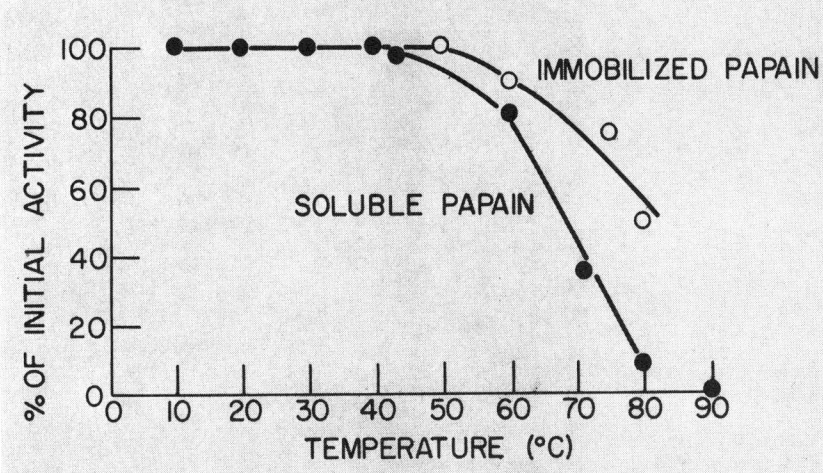

Figure 8. Thermal stability of an enzyme is in some cases increased by immobilization. This figure compares papain in solution and immobilized on control pore glass to storage at increasing temperatures for 30-min intervals. Soluble enzyme was dissolved while immobilized derivative was suspended in water. Residual activity was measured by assay with 1% casein at 25°C. (From ref. 32, Part I)

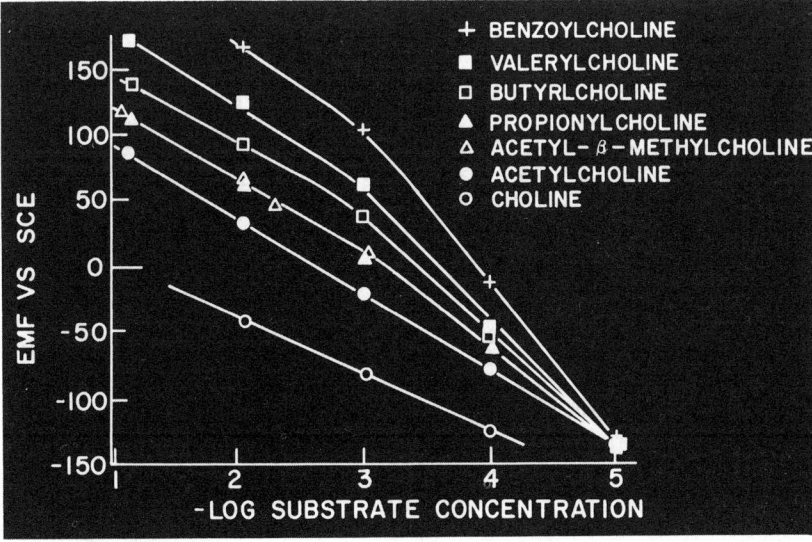

Figure 9. Selectivity of liquid organic ion-exchange electrodes toward choline and choline esters. (From ref. 25)

electrode using L-aminoacidoxidase, which deaminates L-amino acid as follows:

$$RCHCOO^- + H_2O + O_2 \xrightarrow{enzyme}$$
$$NH_3^+$$
$$RCOCOO^- + NH_4^+ + H_2O_2$$

The H_2O_2 can be removed by the addition of catalase.

Similarly, specific amino acids can be quantitated with specific amino acid deaminases. Rather than the oxygen and cation-specific electrode, Clark used polarographic oxidation of H_2O_2 to measure glucose and amino acids (24). He used a platinum electrode to which an applied voltage causes oxidation or reduction and results in producing a current. The platinum electrode is operated at a voltage so that the current produced is proportional to H_2O_2 concentration. Clark identified some 30 enzymes which use O_2 and produce H_2O_2. Using these enzymes with polarographic platinum electrodes, Clark estimated that the following compounds could be detected:

Acetaldehyde, D-mannose
D-alanine, methanol
Aliphatic nitro compounds, L-methionine
D-aspartate, 6-methyl-D-glucose
Benzaldehyde, N-methyl-L-amino acids
Diamines, NAD
2-Dioxy-D-glucose, NADH
Ethanol, oxalate
Formaldehyde, L-phenylalanine
L-galactonolactone, D-proline
D-galactose, purine
β-D-glucose, pyridoxamine phosphate
D-glutamate, pyruvate
Glycollate, saccosine
L-gulono-λ-lactone, spermine
Hypoxanthine, sulfite
D-lactate, tyramine
L-lactate, urate
Lactose, D-valine
(+) Mandalate, xanthine

polarographic oxygen electrode. When the electrode is placed in contact with a glucose solution, glucose and O_2 diffuse into the gelatinous layer around the electrode where the diffusing glucose is oxidized, producing gluconic acid. The resulting depletion of oxygen (O_2) is measured by the oxygen electrode. At glucose concentrations below K_m for the entrapped enzyme, Updike and Hicks found a linear relationship between glucose concentration and the measured O_2 depletion rate.

Guilbault (23) prepared a urea electrode by entrapping urease in a polyacrylamide matrix which is then placed over a cation-selective electrode. The urease hydrolyzes urea to ammonia and carbon dioxide. The rate at which ammonia is produced from the urea diffusing into the membrane is proportional to the quantity of urea present when operated at urea concentrations between 1.0 and 30.0 mg of urea per 100 ml of solution. Guilbault has shown that these electrodes are quite stable and can be operated continuously for at least three weeks.

Enzyme electrodes of many types and varieties are possible. For example, one can determine L-amino acid content with an ammonia-sensitive

Table III. Half-Life and Temperature Vs. Productivity for Immobilized Glucoamylase Covalently Coupled to ZrO$_2$-Coated Porous Glass[a]

Temp, °C	Half-life, days	Relative reaction rate, % of that at 60°C
60	13	100
50	100	70
45	645	30
40	900	25

[a] When plotted as half-life vs. 1/T (°K), a positive slope is observed. From the slope the deactivation can be calculated (33).

Though not all the above compounds are of interest to the clinical and analytical chemist, the list indicates the wide range of compounds capable of detection with particular types of enzyme electrodes.

Similarly, one can mass a list of compounds that produce or utilize oxygen or ammonia and are therefore quantitatable with an oxygen or cation electrode.

A third type of electrode system uses enzymes to quantitate either a substrate or a product. Developed by Baum and Ward (25), this is known as an organic liquid ion-exchange electrode. Baum's electrode specifically detected acetylcholine and thus could be used for the detection and quantitation of organophosphates which inhibit the activity of acetylcholinesterase. The reaction between the enzyme and product could be monitored by using the acetylcholine-specific electrode to determine the substrate depletion rate. Since the electrode could respond to a variety of choline compounds (Figure 9), it was useful with a variety of cholinesterase substrates. The electrode was used to perform studies on cholinesterase activity in blood. Results compared favorably with data obtained with established colorimetric and pH-stat techniques. Similarly, this electrode was successfully used to measure the presence of the organophosphate pesticides Paraoxon and Tetram by cholinesterase inhibition. Inhibitor concentrations of 10–100 ng/ml were quantitatable by this method.

Baum later combined his choline-sensitive electrode with immobilized cholinesterase. The immobilized enzyme was in the form of a disc which was simply added to the solution to be quantitated. Immobilized cholinesterase discs were used extensively over a 64-day period with excellent results.

Immobilized enzymes have also been used successfully for analytical purposes in the form of columns. Weibel et al. (27) developed a glucose analyzer that uses a column of immobilized glucose oxidase. As illustrated in Figure 10, besides the column of immobilized glucose oxidase, the system includes a Clark electrode to detect O_2 depletion, and a pump. Though the system operates in either a kinetic or end-point mode, the end-point mode is preferred. Weibel et al. found that complete conversion of glucose could be accomplished in less than 60 sec with immobilized glucose oxidase columns containing only 400–600 μl of porous glass.

Using glutaraldehyde, Inman and Hornby covalently attached urease and glucose oxidase to partially hydrolyzed nylon power and tubes (28).

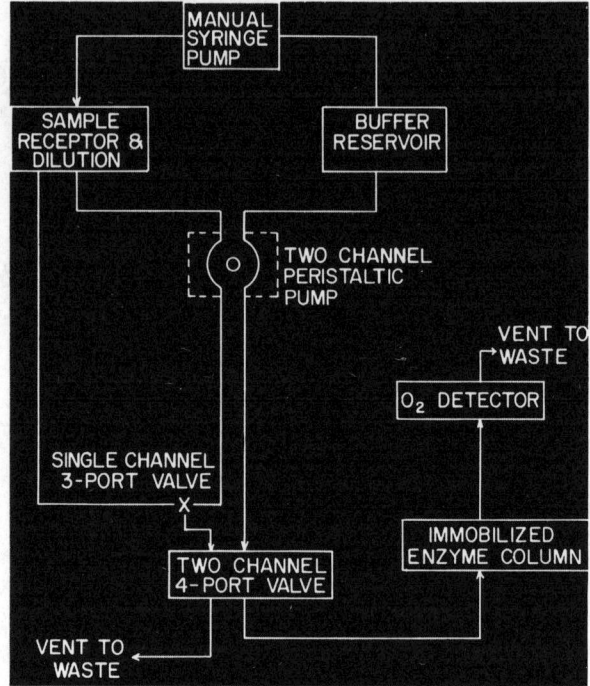

Figure 10. Schematic representation of glucose monitor showing principal components. This monitor uses immobolized glucose oxidase in a column as opposed to a membrane electrode. (From ref. 27)

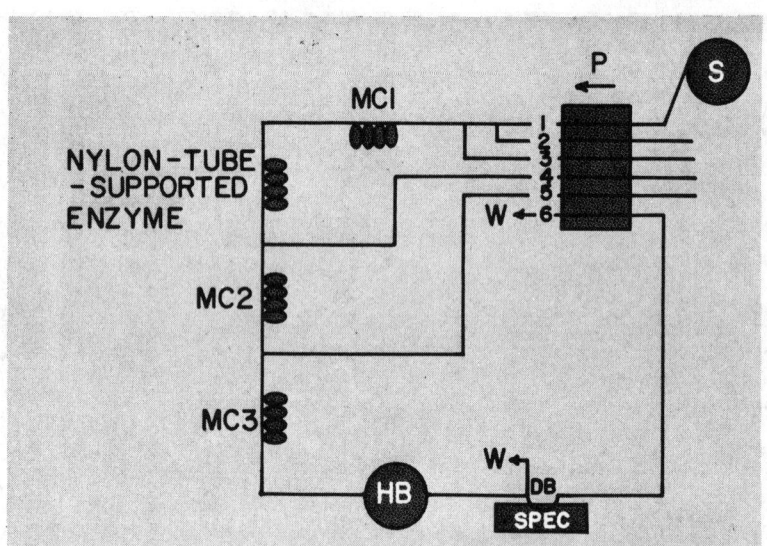

Figure 11. Flow diagram for use of nylon tube-supported enzyme automated system. For determination of glucose, lines 1–6 operate at predetermined flow rates. MC, mixing coil; HB, heating bath; S, samples; P, pump; SPEC, spectrophotometer; DB, debubbler; W, waste. (From ref. 28)

These derivatives have been successfully used in the automated analyses of glucose and urea. Glucose is determined spectrophotometrically as a function of the quantity of I_2 produced in reacting KI with H_2O_2. The urea is determined by ammonia formation according to the method of Chaney and Marbach (29). The flow system, using nylon tubes, is illustrated in Figure 11. The system can process 60 samples per hour without difficulty. The stability of the tubes was determined by performing 150 assays per day for 30 days. During this period, no activity losses were noted. Similarly, the urease derivatives through 5000 assays showed no detectable loss in activity. The use of immobilized enzymes in automated systems is under intensive investigation in Hornby's laboratory, where researchers are attempting to develop assays for a wide variety of clinically important substrates.

Immobilized enzymes can also be used for detection and quantitation of inorganic ions. This author has developed an immobilized enzyme system for detecting and quantitating inor-

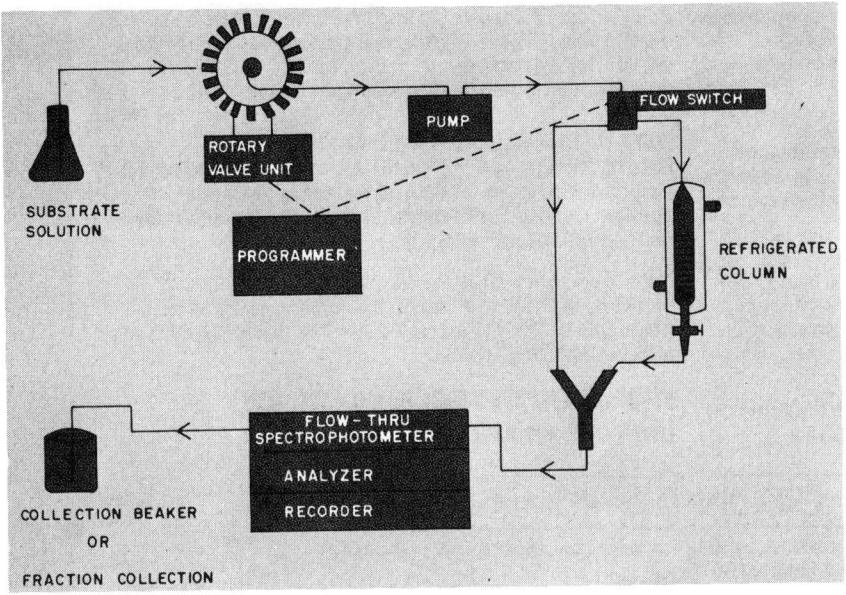

Figure 12. Schematic representation of apparatus for continuous monitoring of inorganic phosphate in solution. (From ref. 30)

ganic phosphate and sulfate with immobilized alkaline phosphatase (30) and immobilized arylsulfatase, respectively. The reactions involved are:

p-nitrophenolphosphate + $H_2O \longrightarrow$
(colorless)

p-nitrophenol + HPO_4^{2-}
(yellow)

p-nitrophenylsulfate + $H_2O \longrightarrow$
(colorless

p-nitrophenol + HSO_4^{2-}
(yellow)

The presence of an added inorganic anion causes a shift in the equilibrium between substrate and product. Moreover, the anion in both cases above is believed to act as a competitive inhibitor.

A schematic representation of the anion-monitoring system is shown in Figure 12. Because of the mode of operation, the system is relatively unaffected by column flow rate, temperature, or even a relatively wide range of molar concentrations of the substrate. Operating the system with a relatively high salt concentration eliminates the nonspecific anionic effects, and the system responds only to the anion of interest.

Immobilized enzymes in many different forms and systems can be used for a broad range of analytical purposes. For example, this limited report has covered: enzymes immobilized in membranes closely associated with ion-specific electrodes; organic ion-specific electrodes that monitor organic species, i.e., choline used with soluble and insoluble enzymes; immobilized enzymes in plug flow or tubular reactors interfaced with readout systems. In addition, there are many other types of immobilized enzyme analytical systems. For instance, Messing (31) reported an assay of glucose by using glucose oxidase and differential conductivity. Messing's method is based on the conversion of a relatively nonconducting species of glucose into a more conductive species—gluconic acid. Similarly, urease and differential conductivity could be used to measure the conversion of urea to ammonia and CO_2.

It is not within the scope of this review to discuss the range of industrial and therapeutic applications for immobilized enzymes. These ureas have been covered in depth elsewhere in the literature (4, 17, 20, 32).

Summary

Despite the rather extensive research effort being spent in enzyme technology, it is still regarded by many as "a solution in search of a problem." It is probably true that the real opportunities for immobilized enzyme technology lie 5 or 10 years ahead.

Certainly, the emphasis in this field today—both in government-funded research and in industry—is on its use in large-scale systems for the food, beverage, and pharmaceutical fields. With respect to analytical applications, several companies—notably, Corning Glass Works, Leeds, and Northrup, and Technicon—are developing instruments that utilize immobilized enzymes. Indeed, one, Yellowsprings Instrument Co., is already marketing a glucose analyzer which uses glucose oxidase trapped in a membrane on a platinum electrode—a system based on the approach recently described by Clark (24).

Owing principally to their high degree of reaction selectivity and their speed, enzymes, particularly immobilized enzymes, hold considerable promise in a range of process-related fields, not the least of which is analytical instrumentation.

References

(1) J. M. Nelson and E. G. Griffin, *J. Amer. Chem. Soc.*, **38**, 1109 (1916).
(2) N. Grubhofer and L. Schlecth, *Hoppe Seyler's Z. Physiol. Chem.*, **297**, 108 (1954).
(3) N. Grubhofer and L. Schlecth, *Naturwissenschaften*, **40**, 508 (1953).
(4) O. R. Zaborsky, "Immobilized Enzymes," CRC Press, Cleveland, Ohio, 1973.
(5) A. D. McLaren and L. Packer, *Advan. Enzymol. Relat. Areas Mol. Biol.*, **33**, 245 (1970).
(6) C. A. Zittle, *ibid.*, **14**, 319 (1953).
(7) R. A. Messing, *J. Amer. Chem. Soc.*, **91**, 2370 (1970).
(8) R. A. Messing, *Enzymologia*, **38**, 370 (1970).
(9) R. Haynes and K. A. Walsh, *Biochem. Biophys. Res. Commun.*, **36**, 235 (1969).
(10) W. Gaffield, Y. Tomimatsu, A. C. Olson, and E. F. Jansen, *Arch. Biochem. Biophys.*, **157**, 405 (1973).
(11) W. L. Stanley and A. C. Olson, U.S. Patent 3,736,231 (1973).
(12) P. Bernfeld and J. Wan, *Science*, **142**, 678 (1963).
(13) T. M. S. Chang, *ibid.*, **146**, 524 (1964).
(14) A. Bar-Eli and E. Katchalski, *Nature*, **188**, 856 (1960).
(15) H. H. Weetall, *Science*, **166**, 615 (1969).
(16) G. R. Stark, "Biochemical Aspects of Reactions on Solid Supports," Academic Press, New York, N.Y., 1971.
(17) H. H. Weetall and R. A. Messing, in "The Chemistry of Biosurfaces," M. L. Hair, Ed., Vol II, Marcel-Dekker, New York, N.Y., 1972.
(18) H. H. Weetall, *Separ. Purification Methods*, **2**, 199 (1973).
(19) L. Goldstein, Y. Levin, and E. Katchalski, *Biochemistry*, **3**, 1913 (1964).
(20) L. B. Wingard, Jr., Ed., "Enzyme Engineering Sym. 3," Wiley, New York, N.Y., 1972.
(21) I. H. Silman and E. Katchalski, *Ann. Rev. Biochem.*, **35**, 873 (1966).
(22) S. J. Updike and G. P. Hicks, *Nature*, **214**, 986 (1967).
(23) G. Guilbault, in "Enzyme Engineering," L. B. Wingard, Jr., Ed., pp 361–76, Wiley, New York, N.Y., 1972.
(24) L. C. Clark, Jr., Ed., *ibid.*, pp 377–94.

(25) G. Baum and F. Ward, *Anal. Biochem.*, **42,** 487 (1971).
(26) G. Baum, F. Ward, and S. Yaverbaum, *Clin. Chem. Acta*, **36,** 406 (1972).
(27) M. K. W. Weibel, W. Dritschilo, H. J. Bright, and A. E. Humphrey, *Anal. Biochem.*, **52,** 402 (1973).
(28) D. J. Inman and W. E. Hornby, *Biochem. J.*, **129,** 255 (1972).
(29) A. L. Chaney and E. P. Marbach, *Clin. Chem.*, **8,** 130 (1962).
(30) H. H. Weetall and M. A. Jacobson, Proc. IV IFS Ferment. Tech. Today, 361, 1972.
(31) R. A. Messing, *Biotech. Bioeng.*, in press.
(32) H. H. Weetall, *Food Prod. Develop.*, Part I, April 1973; Part II, May 1973.
(33) H. H. Weetall and N. B. Havewala, "Enzyme Engineering," Wiley, New York, N.Y., 1972.
(34) H. H. Weetall, N. B. Havewala, W. Pitcher, Jr., C. C. Detar, W. P. Van, and S. Yaverbaum, *Biotech. Bioeng.*, **16,** 295 (1974).

Howard H. Weetall is a senior research associate in biochemistry in Corning's Research and Development Division. He joined the company in 1967 as a research biochemist, was appointed a research associate in 1971, and was given his present position in 1973. Author of many technical articles, Weetall holds BA and MA degrees from the University of California at Los Angeles.

Howard H. Weetall (right) and L. S. Hersh are shown in the process of bonding heparin to polyester tubing at Corning's research laboratory

Ultrapurity in Trace Analysis

James W. Mitchell
Bell Laboratories, Murray Hill, N.J. 07974

Trace analytical techniques require special consideration to eliminate or minimize the effects of contaminants and impurities. Obstacles to extending the limits for trace element determinations and future needs are discussed

Although present needs for trace measurements and techniques for microanalysis are substantial, recent literature indicates that demands for these analytical capabilities will increase (1–6). The applications listed in Table I, for example, attest to the increasing need for analytical information on practical problems of national interest. For some of these challenges the realm of quantitative and reliable qualitative analysis must be extended beyond the limits presently existing for routine determinations of trace constituents.

As the need for analytical information pertaining to these problems becomes more critical, practical analytical scientists will encounter frontiers in which the development of new methods and techniques for highly sensitive and accurate analyses will be extremely important. These advances in measurement will demand highly skilled practitioners using state-of-the-art techniques for minimizing or eliminating the effects of contaminants and impurities.

Efforts by many investigators to perfect techniques for characterizing pure materials or determining trace constituents have generated the art of "Ultrapurity in Trace Analysis," an endeavor in which the expertise has been advanced significantly in recent years. General discussions of techniques and methods for preventing contamination have been given in several manuscripts and texts (7–11). Although considerable practical information and descriptions of useful techniques have recently appeared, these have been scattered widely throughout the literature.

In this report an attempt is made to provide a current reference for analytical chemists, materials scientists, and others responsible for the preparation or practical characterization of ultrapure materials. A summary of procedures used in this laboratory for trace analysis and other recent advances in methods and techniques are presented, major obstacles to extending the limits for determining trace elements are discussed, and developments required in the future are treated.

The Blank, a Persistent Problem in Trace Analysis

Masking of components in a sample, severe interferences with detection, uncertain qualitative analysis and unreliable quantitative measurements are all problems in trace analysis potentially connected through a common factor, the blank. Few analytical techniques are free from the influence of these difficulties. Even though the threshold for measuring trace constituents has been lowered significantly with the discovery and development of techniques with high sensitivity, the full potential for measurement by spectrophotofluorometry, polarography, anodic stripping, kinetic methods, and other techniques with sensitivities sufficient for measuring elements at the nanogram level is precluded in many practical applications.

In the case of the inability to reproducibly control the blank at levels insignificant in comparison with the constituent being determined, or where difficulties associated with quantitatively manipulating and recovering submicrogram quantities of trace elements exist, the limits at which trace elements can be measured by most techniques will be established by these restrictions. In the presence of constantly fluctuating blanks, the accuracy and precision of quantitative trace measurements are also influenced significantly. Thus, first-order improvements in the reliability of measurements at or below the ppm region depend greatly upon controlling and reducing the size of the blank and where possible, eliminating its effects.

Contamination from particulates in air, impurities in reagents, and trace elements from containers are considered to be primarily responsible for the blank. Hazards from less conspicuous sources, including instrumental noise interpreted as a signal from a component of the sample, must be considered as well (9, 12, 13). Since considerable attention by analytical chemists to these problems is necessary in trace analysis, state-of-the-art techniques for providing pure atmospheres and working conditions, for purifying and storing ultrapure reagents, and for performing routine analytical procedures under ultra-clean conditions are subsequently discussed.

Table I. Practical Applications for Trace Analysis Techniques

Characterization of ultrapure materials for technological advances (1)
Establishment of meaningful tolerance limits for pollutants (2)
Analysis of geological and celestial samples (3)
Determining the distribution and abundance of trace constituents to effectively use the ocean's resources (4)
Scientific evidence for forensic investigations (5)
Studying the physics and chemistry of semiconductors (5)
Illucidating the role of trace metals in biological functions (6)
Determining mechanisms by which heavy metals induce toxicity (6)

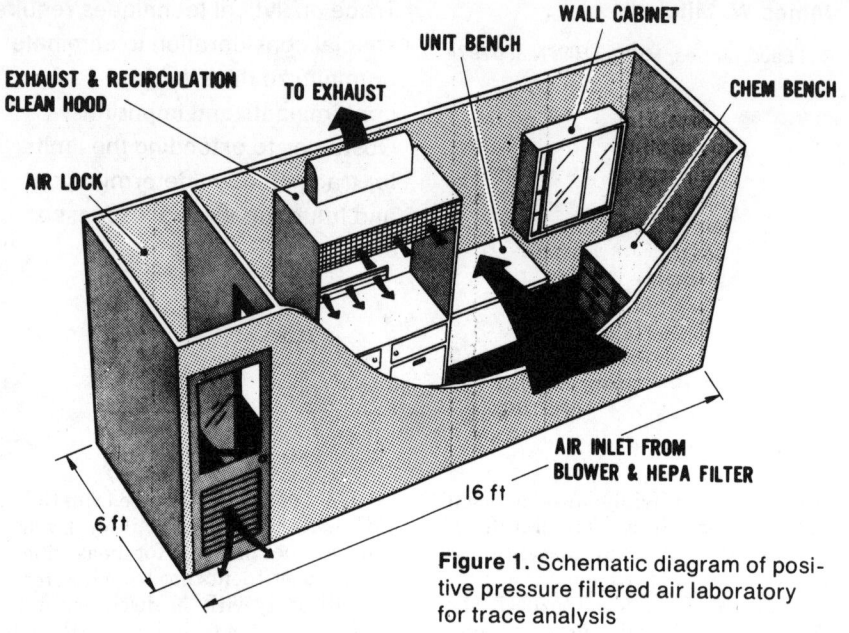

Figure 1. Schematic diagram of positive pressure filtered air laboratory for trace analysis

Essential Facilities for the Trace Analysis Laboratory

Controlled Atmosphere. Airborne contaminants in the form of volatile compounds, dusts, soots, and aerosols circulate in the atmosphere and enter laboratories through any channel permeable to air. Thus, the composition of the air in the laboratory will often approximate that of the surrounding atmosphere and will fluctuate with prevailing atmospheric conditions. Up to 200 μg of dust per liter has been collected by filtering the air in a noncontrolled laboratory (7). Analysis of this sample showed 10% Ca, 3% Fe, 1.5% Al, 0.5% Cu, 5% Si, 1.5% Ni, 1% K, 1% Mg, 0.5% Mn, and traces of other elements. Since this composition changes with atmospheric conditions, it is not surprising that the size of the blank has been correlated with humidity (9).

Influences of the purity of the air in the laboratory on analytical results have also been readily measured (9). Such analysis of HF, HCl, and HNO_3 for Al, Fe, Ca, Mg, Pb, Ti, and B showed an order of magnitude higher concentrations in samples open to the atmosphere than in samples analyzed in a closed system containing inert gas. Similar documentations of contamination that could only originate from the air have substantiated the need to control the air in the laboratory for characterizing ultrapure materials.

The major development to date for providing particulate-free air, the HEPA (high-efficiency particulate air) filter, was first used by the USAEC to prevent the discharge of radioactive dusts (14). Units containing these filters or similar modifications have been routinely employed for supplying air containing less than 100 particles per cubic foot of a size 0.5 μ and larger. Although standard clean air hoods or custom-designed work stations are commercially available from several suppliers, no critical evaluation or measurement of the efficiency of the variety of designs has appeared.

Descriptions of elaborate and expensive vertical or horizontal flow clean rooms by use of HEPA filtering systems have been reported in a federal bulletin (15). Similar closed-cycled, recirculated, filtered air systems providing atmospheres essentially free from particulates have been installed in industrial corporations and governmental agencies but have not been widely used in laboratories dedicated for trace analysis. Relatively inexpensive and efficient nonlaminar flow clean rooms can be used to provide the analytical chemist with the necessary environmental conditions for performing trace measurements.

A schematic diagram of one of the clean facilities that is used for trace analysis at this laboratory is shown in Figure 1. The floors, walls, ceiling, and windows in the clean area of the laboratory are sealed, and the room is continuously supplied with a positive pressure of filtered air. By using the air lock to separate the clean facility from the remainder of the laboratory and by following standard procedures for controlling people contamination, the introduction of particulates from the noncontrolled area is minimized. The vertical laminar flow clean work station, located inside the positive pressure room, further purifies the air and recirculates it over the work area inside the hood. Placing these units in positive pressure clean rooms provided excellent conditions for preventing contamination from airborne particulate matter.

Analyses of particulates in laboratory air and in air from laminar flow clean hoods indicate a thousandfold decrease in the amounts of Fe and Pb and a tenfold reduction in Cu and Cd per cubic meter of air (16). During routine use in this laboratory of clean work stations in the recirculating mode, no significant contamination of water samples has been detected by spark source mass spectrometry or neutron activation analysis. The additional precaution of enclosing samples in nitrogen-purged chambers has been taken to ensure noncontaminating conditions when stations were operated in the exhaust mode. For special analyses requiring limited work space, inexpensive, extremely clean working conditions can be attained by using Plexiglas, polyethylene, or plastic glove boxes which are continuously purged with pure air or nitrogen. Inflatable transparent polyethylene glove bags can also be used advantageously.

Materials for Construction. Materials routinely used in the construction of buildings are known to contribute atmospheric pollutants or produce particle fallout. Calcium in trace quantities can be constantly emitted from materials commonly used in walls and ceilings. Paints containing metallic pigments are chipped, flaked, or abraded from walls and furniture. Stainless-steel hoods, sinks, and other furniture, copper faucets and pipes, gas regulators, metal heating and air conditioning units, and other metallic objects usually corrode after prolonged use under normal laboratory conditions. Metallic dusts from these objects contribute significantly to particulates in the atmosphere of the laboratory.

Experience in this laboratory has shown that metallic objects should only be tolerated when it is impossible to substitute materials constructed from synthetic plastics. As guidelines for selecting materials for the trace analysis laboratory, most of the specifications for materials suitable for constructing radiochemical laboratories can often be adopted (7). Descriptions and evaluations of a number of special construction materials have been reported (10).

Ultrapure Water. An extremely pure and readily available supply of highly pure water is one of the most important facilities in the laboratory. The quality of this reagent, needed in large volumes for dissolving samples and preparing solutions, must be periodically monitored by quantitative analysis. Comprehensive reviews have been published on techniques for ultrapurification and analysis of water (17–19). Common conclusions reached by various investigators include: systems which employ metallic or pyrex vessels generally produce water containing higher levels of cationic impurities than are found in water purified in quartz or synthetic plastic apparatus (18, 19); storage of purified water, even in plastic or Teflon containers for periods exceeding 30 days, will result in an increase in cationic impurities (19, 20); and resistivity measurements may be used as survey techniques but cannot be relied upon for an unequivocal indication of water quality (17).

The system used in this laboratory for purifying water is shown in Figure 2. Distilled water is fed at a flow rate not exceeding 100 ml/min through tube A into a series of mixed-bed ion-exchange columns B_1 and B_2. The ion-exchange system is composed of two plastic columns with all-Teflon fittings and Teflon connecting tubing. After deionization the water passes through the micron filter (C) to remove particulates and then flows into the lower reservoir of the all-quartz still (E). Upon distilling from the lower chamber and condensing in the upper chamber (F), the water is redistilled, condensed, and passed from the collecting port (H) into Teflon tubing (I), which leads to a polypropylene holding tank.

Results from a semiquantitative spark source mass spectrometric analysis of water produced with this system are reported in Table II (21). By directly comparing samples of purified water with equal volumes of water doped with known amounts of trace elements, the maximum concentration of each element was determined to be less than 0.1 ppb.

Ultrapure Reagents from Commercial Suppliers. Developments in trace analysis have been impeded by limited supplies of high-quality reagents. Special techniques and specific efforts are presently needed to produce a variety of highly pure and well-characterized acids, bases, solvents, buffers, fluxes, oxidants, reductants, chelating agents, and other chemicals used routinely in analytical work. Commercial suppliers have focused some attention to this problem and have introduced special lines of ultrapure, superpure, or electronic-grade reagents and metals (22–24). A list of suppliers of such products is given in Table III. Careful characterizations to determine purity prior to commercial distribution apparently were made for some of these chemicals (25). Purity, however, is a dynamic quality fluctuating with the degree of contamination occurring during preparation, analysis, or containment and also depending on conditions of storage and distribution.

At this laboratory neutron activation and X-ray fluorescence spectroscopy were recently used to determine trace transition elements in commercially available CH_3COOH, HF, HNO_3, H_2O_2, Na_2CO_3, NaCl, NH_4Cl, and $NaOOCCH_3$ (26). Although these chemicals, described as ultrapure or electronic-grade products, were found to be much better than reagent grade, discrepancies were found between quotations of purity from the supplier and the concentration of impurities measured by these methods.

Other experiences indicate that it is incumbent upon the analyst to determine the suitability of commercially available products prior to using for application in which purity is critical. To simultaneously determine many trace elements in highly pure reagents, emission spectrometric survey analyses have been widely used (27). Where more accurate determinations of a limited number of impurities were important, neutron activation, polarography, and mass spectrometric isotope dilution methods are preferred (26, 28, 29).

Methods for determining particulate matter in ultrapure chemicals are immensely important. Light scattering techniques have been used to count particle contamination in inorganic salt solutions and to determine the efficiency of their removal by filtration (30). A sensitive X-ray fluorescence method developed in this laboratory was used to determine trace transition elements in particulate matter separated from dissolved

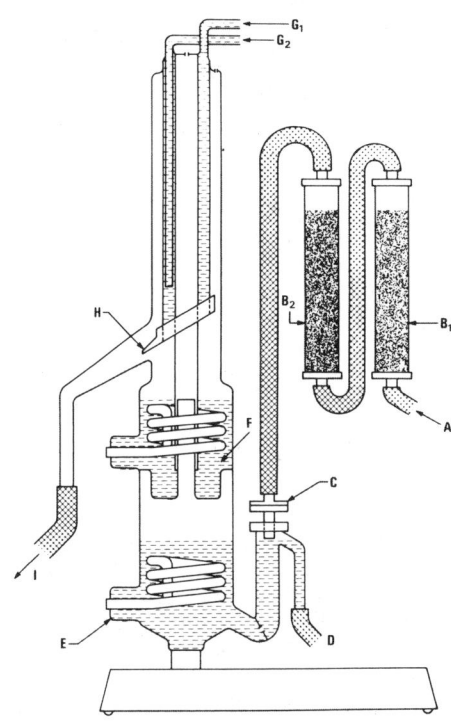

Figure 2. Apparatus for ultrapurification of water by ion exchange and double distillation in quartz

ultrapure salts (26). The role of particulate contamination in limiting the ultimate purity of reagents has not been fully investigated. Measurements of the relative amounts of elemental impurities in soluble and particulate form could produce valuable insight leading to the design of better procedures for ultrapurification.

Laboratory Preparation of Pure Chemicals. Several advances in techniques for the ultrapurification of reagents are changing this operation from a dreaded, arduous task to a promising laboratory procedure. Workers at NBS purified H_2O, HCl, HNO_3, $HClO_4$, and H_2SO_4 by non-boiling distillation, a technique in which infrared heaters vaporize the surface of a liquid without boiling. A modified Teflon still of NBS design was also used for the production of pure HF. Their analyses of the products by spark source mass spectrometric isotope dilution showed extremely pure reagents.

The apparatus used, a modified commercially available unit, makes the ultrapurification of liquids a practical batch or continuous laboratory operation (29). This commercial apparatus and a unit modified appro-

Table II. Mass Spectrometric Detection of Trace Impurities in Ultrapure H_2O

Cation	Blank + pure H_2O	Blank + distilled H_2O	Blank + standard[a]
Mn	N.D.[b]	M^{2+}	M^{2+}[c]
Ni	N.D.	M^{3+}	M^+[d]
Cu	N.D.	M^{2+}	M^{2+}
V	N.D.	(M^+)[e]	(M^+)
Co	N.D.	M^{2+}	(M^+)
Cr	N.D.	M^+	M^+
Zn	(M^+)	M^+	M^+
S	(M^+)	M^+	M^+
Fe	(M^+)	M^+	M^+

[a] 0.1 ml of H_2O with 1.0 ng of each cation/ml. [b] N.D., not detected. [c] M^{2+}, divalent cation detected. [d] M^+, monovalent cation detected. [e] (M), faint trace.

Table III. Suppliers of High-Purity Chemicals and Metals

Alfa Inorganics	Beverly, Mass.
Apache Chemicals	Rockford, Ill.
Asarco	S. Plainfield, N.J.
Atomergic Chemetals Co.	Long Island, N.Y.
Brinkman Instruments Inc.	Westbury, N.Y.
Cominco American Inc.	Spokane, Wash.
Engelhart Industries	Newark, N.J.
Fischer Scientific Co.	Pittsburgh, Pa.
Imanco	Monsey, N.Y.
J. T. Baker Chem. Co.	Phillipsburg, N.J.
Johnson Matthey	England
Kerr McGee Corp.	Oklahoma City, Okla.
Mallinckrodt Chem. Works	St. Louis, Mo.
Materials Research Corp.	Orangeburg, N.Y.
NBS	Gaithersburg, Md.
Princeton Organics	Princeton, N.J.
Texas Instruments, Inc.	Dallas, Tex.

priately for the purification of acids were described recently (31, 32). The apparatus shown in Figure 3 was used in this laboratory to purify 100-ml quantities of $HClO_4$ (33). A novel, inexpensive system based on a principle similar to nonboiling distillation has been described for the preparation of acids containing ultralow lead (34). These accomplishments indicate that nonboiling distillation is one of the best methods yet reported for the production of high-purity acids.

Reagent-grade NH_4OH, HCl, HBr, CH_3COOH, and HF have been isopiestically distilled into pure water to prepare high-quality reagents (35). Although extremely pure products have been prepared in this manner, the method is limited to the purification of volatile compounds and does not produce highly concentrated reagents. The latter problem can be circumvented by saturating pure water or isopiestically prepared reagents with pure gaseous compounds, NH_3, HCl, and HBr, for example (30, 36–40). A standard apparatus for preparation of reagents in this manner is shown in Figure 4.

Reports of quantitative determinations of trace elements by electrodeposition and anodic stripping imply general utility of electrochemical methods for ultrapurification of reagent chemicals (41, 42). Although a commercial apparatus for the removal of trace elements from NaCl, $NaOOCCH_3$, $NaNO_3$, and buffering and nonbuffering solutions by electrodeposition is available, no thorough investigation of this technique has been reported (43). Based on analytical data for quantitative analysis and on other more speculative information, it is expected that the apparatus will be capable of reducing Sb, Bi, Cu, Co, Ga, Au, In, Ir, Fe, Pb, Mn, Ni, Os, Pd, Pt, Rh, Ru, Ag, Tl, Sn, and Zn in sodium solutions to 0.05 ppb (44).

The techniques discussed here are generally applicable for ultrapurification of analytical reagents. Details of more specialized methods have been reported elsewhere (9, 10, 11, 45).

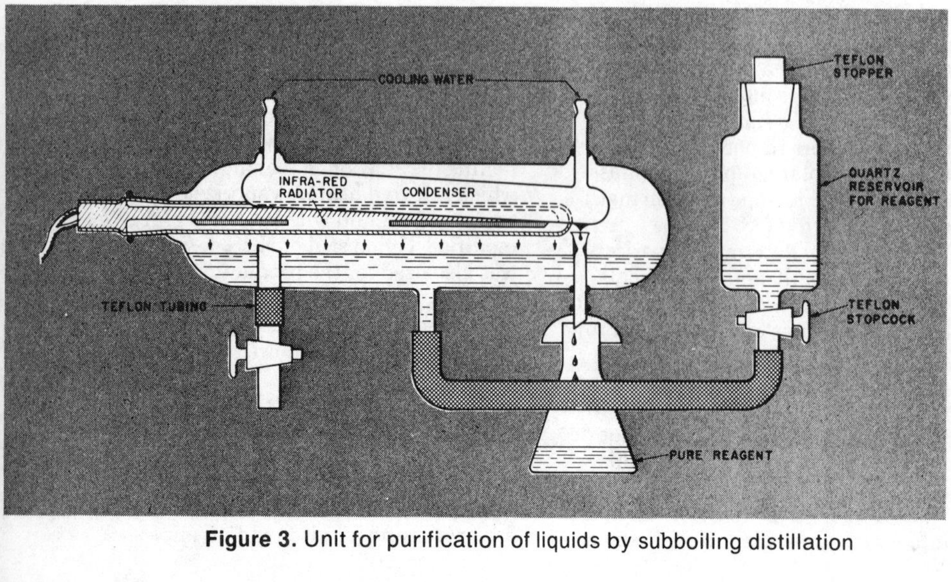

Containment and Storage of Pure Reagents. The purity of reagents is critically dependent upon conditions of storage, the duration, temperature, and container being particularly important. Adequately documented cases of contamination by leaching trace elements from the walls of vessels demonstrate the necessity to properly select storage vessels and to appropriately clean them to maintain the quality of pure reagents (8, 9, 13, 40).

Because of chemical inertness, polyethylene, Teflon, polystyrene, and other synthetic plastics have proved to be the most desirable materials for containing corrosive liquids. Fused silica, platinum, vitreous carbon, and Plexiglas are also quite popular. Presumably for economic reasons, borosilicate bottles have been investigated and recommended by commercial suppliers of pure chemicals (46). When criteria of nonpermeability to vapors and resistance to abrasion are important, the spectrum of suitable materials for a container is increased significantly to include other special materials (10, 11).

Typical concentrations of trace elements in various container materials have been reported (11, 12, 30). The

Figure 3. Unit for purification of liquids by subboiling distillation

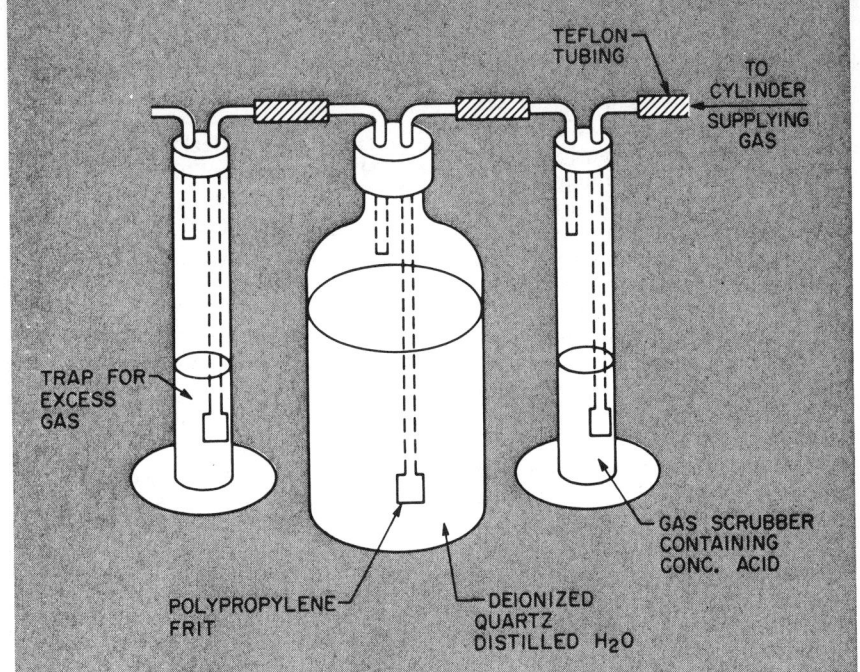

Figure 4. System for preparing ultrapure reagents by saturation of pure water with gases

important question of the extent or rate of leaching, diffusion, or adsorption of trace elements on these materials has not been investigated extensively. Neutron activation analyses were used to determine trace elements in fused silica and polyethylene tubes and to detect trace impurities leached from these materials by 1:1 HNO_3 at 80°C (47). The results reported in Table IV show several nonleachable contaminants in the polyethylene sample and large amounts of readily leached Sb in Vitreosil.

Preferred techniques for cleaning containers include leaching with HNO_3 at elevated temperatures for several days, rinsing with high-quality water or steam, and drying under heat lamps in laminar floor clean hoods (30, 48). Decontaminating the surface of containers by extracting with chelating agents is recommended as well (7). Quartz, Teflon, and plastic containers were routinely cleaned in this laboratory by rinsing repetitively in methanol and deionized water, followed by leaching for 2–4 days in 1:1 HNO_3 at 80°C. After leaching, containers were thoroughly rinsed with deionized quartz-distilled water and dried in a laminar flow hood under heat lamps. Droplets of the leach solution were not permitted to dry on the surface prior to the final rinsing. To evaluate this procedure, quartz and polyethylene tubes were cleaned, sealed, and irradiated with thermal neutrons. The interior of each tube was then rinsed with 1:1 HNO_3 and leached with this reagent at 80°C for 45 min.

Although the data in Table IV show easily detected traces of Cu, Mn, Na, and Cl in the polyethylene sample, none of these impurities was detected in solutions obtained by rinsing and leaching this material with HNO_3. The precaution, thoroughly rinsing cleaned vessels with the ultrapure reagent to be stored, is still warranted. Leaching impurities from the walls of containers and losing trace elements from solution by absorption and other processes occurring at the container interface are problems particularly important in trace analysis. Additional systematic investigations of the former under acidic and basic conditions are needed to accurately characterize materials for storing ultrapure reagents.

Performing Routine Analytical Operations Under Ultraclean Conditions

Techniques have continued to be developed for preventing contamination during the particularly susceptible procedures of sampling, fusing, ashing, dissolving, concentrating, and evaporating samples. Contamination has been reduced significantly by performing these operations in Teflon, quartz, or glass chambers that are supplied continuously with filtered air, nitrogen, or inert gas (9, 49, 50). Completely closed systems, Teflon-lined acid digestion bombs, for example, have been widely used for rapidly dissolving refractory materials (51–53). The apparatus in Figure 5 was designed and constructed at this laboratory and used to simultaneously dissolve six samples of ultrapure silicate materials (54).

The alternative to liquid dissolution, destruction of solid samples by solvent vapors, is highly effective for minimizing the introduction of impurities (55, 56). Trace elements in ultrapure fused silica, powdered silicon dioxide, and sodium lime silicate glasses have been concentrated into graphite at this laboratory by decomposing the samples in a Teflon cham-

Sample	Wt or vol	Concn. of trace elements, $\times 10^{-9}$ g							Other elements detected
		Cu	Mn	Na	Cl	Fe	Co	Cr	
Polyethylene	0.5576 g	52	8.9	208	66.7	...[a]	...	...	Ar
Rinse soln.	3.0 ml	N.D.	N.D.[b]	N.D.	N.D.	...	...	...	Ar
Leach soln.	3.0 ml	N.D.	N.D.	N.D.	N.D.				
Vitreosil		...	...	...	...	N.D.	N.D.	N.D.	Na, Sb
Leach soln.		...	...	...	...	N.D.	N.D.	N.D.	Sb
Spectrosil (1)		...	...	...	...	N.D.	N.D.	N.D.	Sb
Leach soln. (1)		...	...	...	...	N.D.	N.D.	N.D.	Sb
Spectrosil (2)	0.785 g	N.D.	N.D.	N.D.	4600	...	...	...	
Rinse soln. (2)	10 ml	N.D.	0.16	4.4	15	...	...	...	
Leach soln. (2)	5 ml	0.8	0.05	2.8	N.D.	...	...	...	

Table IV. Gamma-Ray Spectrometric Analysis of Fused Silica and Polyethylene

[a] Not determined. [b] Not detected, [Fe] < 100, [Co] < 3, [Cr] < 5 ppb.

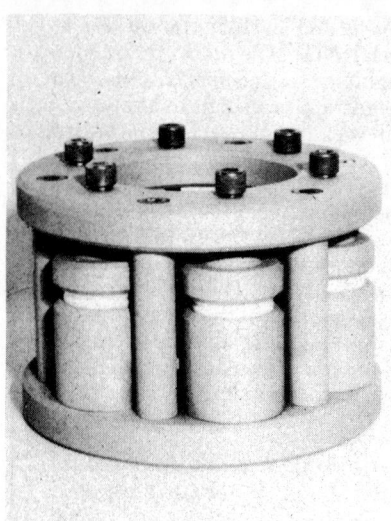

Figure 5. Teflon pressure vessel for dissolution of refractory materials

ber supplied with extremely pure HF vapors. Descriptions and discussions of the advantages of this apparatus (Figure 6) are available (57). Investigations of other chemical systems for vapor-phase dissolution of ultrapure materials should demonstrate conditions for preventing contamination that are superior to corresponding careful dissolutions in liquids.

Major advances in the development of ovens and furnaces for fusing or ashing materials without significant contamination have been lacking. Units lined completely with fused silica have potential applicability for low-temperature fusion or ashing of pure materials at temperatures below 1400°C. Radiofrequency induction heating in crucibles that introduce only tolerated impurities appears to be useful for high-temperature fusions; however, apparatus appropriately designed and priced for general analytical purposes are not yet readily available. Low-temperature ashing of biological samples in oxygen excited by a radio frequency discharge was first described in 1962 (58) and recently lauded as a noncontaminating technique (59).

Enhancing Analytical Capabilities for Accurate Trace Determinations

Although advances in methods and techniques that reduce the size and minimize fluctuations of the blank will be required to attain more precise measurements of trace constituents, improvements in the accuracy of these determinations will depend greatly on developments in other areas. The production and certification of practical standard reference materials, for example, must be accelerated even though noteworthy accomplishments including the NBS SRMS 610 to 619, 1571, and others have been reported recently (60–62).

Instruments and techniques for measuring microliter quantities of solutions must also be improved. In certain cases, it may be necessary to first measure the density of standard solutions and then subsequently weigh the amount of solution dispensed to measure a volume of solution small enough to contain an accurately known nanogram quantity of a trace element.

Reporting detailed descriptions of methods for calibrating instruments and procedures for trace determinations can be the key that determines whether other chemists are successful in achieving precise and accurate measurements by methods reported in the literature.

The following experience in this laboratory is probably illustrative of similar encounters of many analytical chemists. After several weeks of painstaking work, attempts to determine trace quantities of cations by a technique reported in the literature proved to be unsuccessful. Subsequent investigations with radiotracers showed that a significant portion of the trace element being determined had adhered to the walls of the plastic beakers used. (These containers were constructed of the same material as recommended.) By adjusting the pH between 2 and 3 instead of 4.0

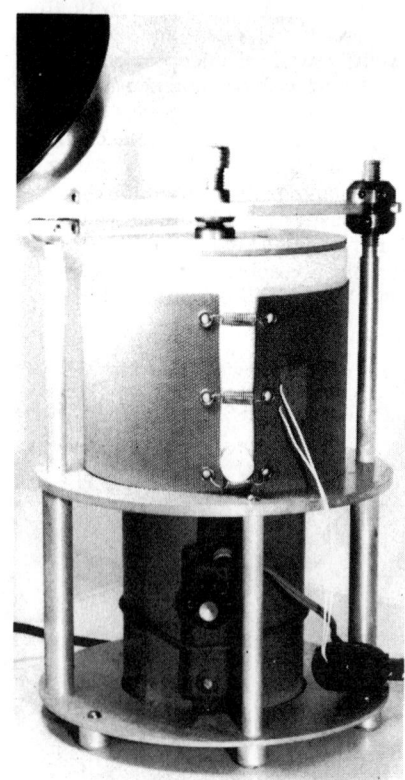

Figure 6. Teflon chamber for vapor-phase destruction of ultrapure silicate materials

---

Table V. Analytical Methods for Nanogram Quantities of Trace Elements

Neutron activation
Substoichiometric radioisotope dilution
Stable isotope dilution
Nuclear track counting
Anodic stripping voltammetry
Electron capture gas chromatography
Polarography
Spectrophotofluorometry
Carbon rod atomic absorption
Coprex X-ray fluorescence
X-ray excited optical fluorescence

(which was specified in the reported procedure), satisfactory results were obtained.

Although failures of methods practiced by investigators other than the originator may occur because of lack of skill, omitted details or erroneously reported procedures are particularly damaging. If analytical information on trace measurements is to be exchanged meaningfully by publication in the literature, full, accurate disclosures of procedural details are imperative.

Creations of new media for interchanging information between the original developer of an analytical technique and other chemists interested in acquiring the expertise to use the reported method are also necessary. Scheduling a single demonstration of new trace analysis techniques after disclosure by publication may prove to be valuable.

Promising Analytical Methods

Techniques potentially applicable for accurate determinations of submicrogram amounts of trace elements are listed in Table V. Although accuracy in measurement is an important criterion for many analyses, highly accurate methods are not essential for identifying trace elements responsible for degrading desired properties of materials, for pinpointing the location of impurities, and for following relative changes in the amounts of trace constituents in a series of similarly processed samples.

For these applications, analytical information of paramount importance has been obtained by solids mass spectrometry, emission spectroscopy, and numerous other methods which are available. Although the extra effort to obtain accuracy in these cases is not usually warranted, precautions to eliminate contamination are still necessary to obtain reliable qualitative information.

The list of techniques in Table V, though not inclusive, contains most of the methods potentially useful for accurate trace analysis. Since only one of these methods, nondestructive neutron activation analysis, is essentially free of contamination problems, the practice of and continued need for advances in techniques of ultrapurity are therefore obvious.

References

(1) A. D. Pearson and W. G. French, "Bell Laboratories Record," p 106, Bell Telephone Laboratories, Inc., Murray Hill, N.J., April 1972.
(2) Chem. Eng. News, 49, 29–33 (July 19, 1971).
(3) G. H. Morrison, Anal. Chem., 43 (7), 23A (1971).
(4) G. Thompson, "The Speaker, XVI (2)," Apex Industries Inc., Metuchen, N.J., 1971.
(5) Chem. Eng. News, 49, 40 (July 5, 1971).
(6) H. A. Laitinen, Anal. Chem., 43, 809 (1971).
(7) J. Ruzicka and J. Stary, "Substoichiometry in Radiochemical Analysis," pp 54–58, Pergamon Press, New York, N.Y., 1968.
(8) D. N. Hume, "Analysis of Water for Trace Metals," Advan. Chem. Ser., 67, 30–44, American Chemical Society, Washington, D.C., 1967.
(9) I. P. Alimarin, Ed., "Analysis of High-Purity Materials," pp 1–31, Israel Program for Scientific Translations, Jerusalem, Israel, 1968.
(10) B. D. Stepin, I. G. Gorshteyn, G. Z. Blyum, G. M. Kurdyumov, and I. P. Ogloblina, "Methods of Producing Superpure Inorganic Substances," Joint Publications Research Service JPRS 53256, U.S. Government Printing Office, Washington, D.C., 1971.
(11) M. Zief and R. Speights, Eds., "Ultrapurity Methods and Techniques," Marcel Dekker, New York, N.Y., 1972.
(12) D. E. Robertson, Anal. Chem., 40, 1067 (1968).
(13) D. E. Robertson, in "Ultrapurity," M. Zief and R. Speights, Eds., pp 208–50, Marcel Dekker, New York, N.Y., 1972.
(14) H. Gilbert and J. H. Palmer, "High Efficiency Particulate Air Filter Units," TID-7023, USAEC, Washington, D.C., August 1961.
(15) "Federal Standards," Circular 209a, GSA Business Service Center, Boston, Mass., 1966.
(16) J. K. Taylor, Ed., NBS Tech. Note 545, p 53, U.S. Government Printing Office, Washington, D.C., December 1970.
(17) R. C. Hughes, P. C. Müran, and G. Gundersen, Anal. Chem., 43, 691 (1971).
(18) R. C. Hughes, paper presented at 165th National ACS Meeting, New York, N.Y., August 1972.
(19) V. C. Smith, in "Ultrapurity," M. Zief and R. Speights, Eds., pp 173–91, Marcel Dekker, New York, N.Y., 1972.
(20) G. Healy, J. Morgan, and R. Parker, J. Biol. Chem., 198, 305 (1952).
(21) J. W. Mitchell and D. L. Malm, unpublished procedures.
(22) M. Zief, Ind. Res., 13, 36 (April 1971).
(23) M. Zief, Amer. Lab., 2, 55–57 (October 1969).
(24) H. G. Griffin and T. D. George, paper 72 presented at the Pittsburgh Conference, Cleveland, Ohio, March 1972.
(25) N. A. Kershner, E. F. Joy, and A. J. Burnard, Jr., Appl. Spectrosc., 25, 542 (1971).
(26) J. W. Mitchell, C. L. Luke, and W. R. Northover, Anal. Chem., 45, in press (1973).
(27) A. J. Barnard, E. F. Joy, K. Little, and J. D. Brooks, Talanta, 17, 785 (1970).
(28) J. K. Taylor, Ed., NBS Tech. Note 545, pp 49–52, U.S. Government Printing Office, Washington, D.C., 1970.
(29) E. C. Kuehner, R. Alvarez, P. J. Paulsen, and T. J. Murphy, Anal. Chem., 44, 2050 (1972).
(30) D. H. Freeman and W. L. Ziclinski, Jr., Eds., NBS Tech. Note 549, pp 62–69, U.S. Government Printing Office, Washington, D.C., 1971.
(31) "Highly Pure Water Generator," Leaflet 51-5A, Quartz Products Corp., Plainfield, N.J.
(32) K. D. Burrhus and S. R. Hart, Anal. Chem., 44, 432 (1972).
(33) J. E. Kessler and J. W. Mitchell, unpublished work.
(34) J. M. Mattinson, Anal. Chem., 44, 1715 (1972).
(35) H. Irwing, J. J. Cox, Analyst, 83, 526 (1958).
(36) E. Haberli, Z. Anal. Chem., 160, 15 (1958).
(37) H. Stegeman, ibid., 154, 267 (1957).
(38) R. E. Thiers, in "Trace Analysis," J. H. Yoe and H. J. Koch, Eds., pp 637–66, Wiley, New York, N.Y., 1957.
(39) R. E. Thiers, in "Methods of Biochemical Analysis," D. Glick, Ed., Vol 5, pp 274–309, Interscience, New York, N.Y., 1957.
(40) M. Knizek and J. Provaznik, Chem. Listy, 55, 389 (1961).
(41) H. E. Allen, W. R. Matson, and K. H. Mancy, presented at the 41st Conference of the Water Pollution Control Federation, Chicago, Ill., September 1970.
(42) W. R. Matson, R. M. Griffin, and G. B. Schreiber, in "Trace Substances in Environmental Health-IV," D. Hemphill, Ed., pp 396–406, University of Missouri, Kansas City, Mo., 1971.
(43) ESA brochure, Environmental Sciences Associates, Inc., Burlington, Mass., 1972.
(44) W. R. Matson, University of Michigan, Ann Arbor, Mich., private communication, July 1972.
(45) M. Tatsumoto, Anal. Chem., 41, 2088 (1969).
(46) N. A. Kershner, E. F. Joy, and A. J. Barnard, Appl. Spectrosc., 25, 542–49 (1971).
(47) J. W. Mitchell, J. E. Riley, and W. R. Northover, Tech. Memorandum, Bell Telephone Laboratories, Inc., Murray Hill, N.J., 1972.
(48) E. C. Kuehner and D. H. Freeman, in "Purification of Inorganic and Organic Materials," M. Zief, Ed., Marcel Dekker, New York, N.Y., 1969.
(49) R. E. Thiers, Symposium on Trace Analysis, New York, N.Y., 1955.
(50) C. I. Zilberstein, M. M. Piryutko, O. Nikitina, O. N. Fedorev, and F. Ju, Zavod. Lab., 28, 680 (1962).
(51) F. J. Lanmyhrand and P. E. Paus, Anal. Chim. Acta, 49, 358 (1970).
(52) B. Bernas, At. Absorption Newslett., 9 (2), 52 (1970).
(53) "Uni-Seal Decomposition Vessels," Uni-Seal Decomposition Vessels LTD, Haifa, Israel, 1972.
(54) J. W. Mitchell and S. S. DeBala, unpublished work.
(55) F. A. Pohl, K. Koker, and W. Bonsels, Z. Anal. Chem., 174, 6 (1960).
(56) F. A. Pohl, Chem. Eng. Tech., 30, 347 (1958).
(57) J. W. Mitchell, D. L. Nash, and S. S. DeBala, to be published.
(58) C. E. Gleit and W. D. Holland, Anal. Chem., 34, 1455 (1962).
(59) J. W. Mair, Jr., and H. G. Day, ibid., 44, 2015 (1972).
(60) D. H. Freeman, L. A. Currie, E. C. Kuehner, H. D. Dixon, and R. A. Paulson, ibid., 42, 203 (1970).
(61) T. E. Gills, W. F. Marlow, and B. A. Thompson, ibid., 1831 (1970).
(62) H. L. Rook, T. E. Gills, and P. D. LaFleur, ibid., 44, 1114 (1972).

James W. Mitchell is a supervisor in the Analytical Chemistry Department of Bell Laboratories, Murray Hill, N.J. After receiving a BS degree in chemistry from the Agricultural and Technical State University of North Carolina at Greensboro (1965), he entered Iowa State University at Ames and earned the PhD degree in analytical chemistry in 1970. He joined Bell Laboratories in March of that year. Dr. Mitchell's major areas of research include the analysis of highly pure materials by neutron activation and radiotracer techniques, the development of procedures for ultrapurification of analytical reagents and research chemicals, and investigations of the chemistry of solvent extraction processes. He has seven manuscripts pertaining to these topics, four published and three submitted recently. He is a member of Phi Lambda Upsilon, Sigma Xi, the American Chemical Society, and the Analytical Chemistry Division of the ACS.

The Other Face

Richard W. Roberts
National Bureau of Standards
Washington, DC 20234

It is a pleasure for me to be here today, and somewhat of an historic occasion. The National Bureau of Standards has long contributed to the fields of spectroscopy and analytical chemistry. We have been closely allied with the Pittsburgh Conference since the first meeting back in 1950. Over the years many outstanding Bureau scientists have presented papers before this forum. But never before has an NBS Director addressed the Conference, and I consider it a privilege to be here today. The timing is most appropriate, as March 3 is the anniversary of the Bureau's founding in 1901.

Analytical chemistry has a vital role to play in the years ahead. As we search for new materials, new energy sources, a cleaner environment, and better health, the central questions of composition and quantity will be raised almost daily.

As you know, trace element analysis plays an important role in many fields and has at least two major health impacts. Better measurements are providing new information on the necessary role of trace elements in vital metabolic processes. And trace element analyses, both of biological specimens and of the environment, are revealing the level of and potential danger of many pollutants.

Analysis has a major role to play in the materials area, since performance is so intimately tied to composition. As shortages of traditional materials accelerate the search for substitutes, as we seek to recycle more materials, and as requirements for stronger, tougher, longer lasting materials arise, the role of the analyst will become even more important.

Analysis affects the economy in several ways. Quality control based on valid analytical results can improve productivity, reduce warranty claims, and help American goods compete in world markets.

Analysis has an especially important role to play in the energy field. As we edge our way closer to controlled fusion devices, spectral characterization of the plasma becomes an increasingly important diagnostic tool. As new sources of coal and oil are developed, analysis of the fuel itself or of the combustion products is necessary in meeting environmental standards. And the development of new converters of solar energy will require accurate data on composition.

I could go on with other examples, but I think the point is clear: rapid, accurate, inexpensive analyses will play a major role in determining the type of world, and the quality of life, that we achieve in the years ahead. In some cases, using tried and proven methods of analysis, the answers will come easy. In others, totally new approaches will have to be developed. But no matter what quantity is being measured, no matter how simple or sophisticated the technique, accuracy will remain the very foundation of chemical analysis. An accessible system of standards is necessary if accuracy is to be attained through the measurement chain.

You chemists and spectroscopists provide the accurate measurement base that is of fundamental importance to our scientific and technological society. We at NBS provide what I call the other face of that measurement base: the standard reference materials, reference data, reference methods, calibration services, even the base units themselves, that help all of you achieve the accuracy you seek.

NBS is no stranger to measurement standards, analytical chemistry, or spectroscopy. Congress created NBS in 1901 to meet a national need for a unified measurement system, a need that had many roots. The rapidly growing electrical industry, the growth of industrial research, the campaign for fairness in trade, all created demands for accurate measurement.

The Act of Congress which established the Bureau assigned us responsibility for "custody, maintenance, and development of the national standards of measurement." Recognizing that having standards was only half

Metre bar, composed of 90% platinum–10% iridium, served as the Nation's fundamental standard of length from 1893 to 1960

of the Measurement Base

the job, we were also charged with providing the "means and methods for making measurements consistent with those standards."

As might be expected, NBS started small, with a staff of 11 and with borrowed quarters in the Coast and Geodetic building. Work began immediately on fundamental measurements in electricity, photometry, and thermometry. While a chemistry division as such wasn't formed until 1905, the first annual report of the Bureau recognized the need of chemists for calibrated burets and graduates, and a program was started in this area. We began early in spectroscopy, and the fourth paper ever published by NBS was by Perley Nutting on the spectra of mixed gases.

In 1903 William Noyes, one of the most distinguished chemists of his time, joined NBS. He started a line of contributions to the field of chemical analysis that has continued unbroken to the present time.

As new staff and new programs were added, we soon outgrew our temporary quarters. By March of 1903 work had begun on a permanent NBS facility on what were then the outskirts of Washington; and within a few years the facilities, staff, and work of NBS were recognized on a par with the great national laboratories of Great Britain and Germany.

In the 1960's NBS moved again, this time to a magnificent new site in Gaithersburg, MD. Here, on 576 acres, are located 23 major buildings, special facilities such as a nuclear reactor, an echoic chamber, a testing machine that can apply 12 million pounds in compression, and accommodations for conferences that bring over 20,000 scientists to our site each year. At Boulder, CO, is located a somewhat smaller lab that concentrates on cryogenics, time and frequency, and electromagnetics. Our total staff is about 3,600, and our budget this year is a little over $100,000,000.

Measurement Language

Today, environmental concerns, the oil crisis, the unique perspective provided by recent space flight, all have created a growing awareness of the need for international cooperation. While we are many nations, we are but one people. For the mutual benefit of mankind, we must recognize and foster the sharing of knowledge, materials, and products. No nation can close its eyes—or its borders—to the needs and contributions of its neighbors.

But there are barriers to the cooperation that is so vital to progress. One major hurdle is that of different languages, a distinct impediment to the interchange of knowledge and the creation of new understanding. Fortunately, we in science have a measurement "language" that provides a uniform, universal basis for the exchange of information.

This "language," the International System of Units (SI), was formalized by the 11th General Conference on Weights and Measures in 1960, and provides a set of definitions that is accepted and used world wide. Basically, the SI language is the essence of simplicity: seven base quantities provide the foundation from which are derived all other quantities. The seven are mass, length, time, temperature, luminous intensity, electric current, and amount of substance. In principle, these seven units are conceptually quite simple, but their realization requires continuing research at the frontiers of science and technology.

Before I touch on some of this research, let me provide just a brief framework of history and procedure. While the metric system had its origins in the reforms precipitated by the French revolution, it made little headway except in scientific circles until 1875. At that time a Metric Convention was signed by the United States and 16 other nations. Under the agreement, prototype standards for length and mass were to be constructed and delivered to each member nation, an International Bureau of Weights and Measures was established near Paris

In 1960 the metre was redefined by international agreement in terms of wavelength of light emitted by ^{86}Kr. To reduce thermal effects, lamp is operated inside Dewar flask at triple point of nitrogen

to coordinate the measurement system, and a General Conference on Weights and Measures was to be held every six years to settle important questions.

In 1890 after a drawing by lots, the United States received two metre bars (numbers 21 and 27), and two kilograms (numbers 4 and 20), which in 1893 were declared to be the Nation's fundamental standards of length and mass. Since that time the yard and the pound have been defined in terms of metric units, a situation that makes us in many ways a metric nation whether or not we realize it.

Length

The metre bars that came to this country provided the fundamental standard of length for over 70 years. These bars are composed of an alloy of 90% platinum, 10% iridium, and have two parallel, microscopic scratch marks that define the metre. There are, of course, fundamental difficulties associated with such a standard. Unknown changes in length, owing to crystal structure changes, could take place; a bar could be dropped or damaged in some other way; and most important, only NBS could have the nation's primary length standard.

As early as 1893 Michelson had proposed a wavelength definition of the metre, based on an emission line of cadmium. As time went by, other scientists championed other wavelength standards, including Meggers of NBS, who developed a mercury-198 lamp that was both convenient and accurate. Finally, in 1960 the metre bar was officially supplanted by a wavelength definition. The General Conference agreed that a metre was 1,650,763.73 wavelengths, in vacuum, of the orange–red spectral line of ^{86}Kr.

While this was a giant step forward, in that any well-equipped laboratory could operate its own primary length standard, the definition had some inherent problems. First, to reduce thermal effects, the krypton discharge lamp is operated at the triple point of nitrogen, an inconvenient arrangement. Next, ^{86}Kr can only produce interference fringes over a length of 50 centimetres or so, meaning that longer lengths must be "stepped off," with a resulting loss of accuracy. Finally, the emission line itself was discovered to be asymmetric, limiting the accuracy with which measurements could be made.

In 1960, the year in which ^{86}Kr was adopted as the definition of the metre, operation of the first laser was announced. An immediate interest was shown at NBS in using the laser as a potential length standard. NBS measurements, published in 1964, demonstrated the production of bright interference fringes over a path length of 100 metres, a distance that was dictated by the length of an available measuring room and not by the coherence of the He–Ne laser light.

But progress rarely comes easy. Studies at NBS and elsewhere revealed instabilities in the laser wavelength large enough to limit its usefulness as a length standard. Various stabilization schemes were tried, but the one found most effective was developed by Barger and Hall of our Boulder laboratories. Their approach was to lock the laser wavelength to a molecular absorption line in a cell of methane gas. This approach provided a basic line width of a part in a billion, and a reproducibility of a part in 100 billion for two independent lasers.

Again, there was a minor problem, for the laser line stabilized by methane is at 3.39 μm, in the infrared portion of the spectrum. A search was made for a suitable molecule with which to stabilize the 633-nm visible line of the same laser, and iodine was found to be the answer by Schweitzer, Kessler, Deslattes, Layer, and Whetstone at the Gaithersburg labs.

But stabilizing the laser is only half of the job: its wavelengths also had to be carefully determined in terms of the primary ^{86}Kr standard. This was done both at NBS and at other national laboratories which followed our lead, and in 1973 two laser wave-

Speed of light was determined by measuring both frequency and wavelength of same He–Ne laser line. (Shown is Kenneth Evenson of NBS-Boulder, who was responsible for frequency determination)

lengths were adopted by the General Conference as secondary wavelength standards. This may have been the first step toward adoption of a laser wavelength as a replacement for ^{86}Kr.

Recently, another elegant experiment tied the X-ray scale to SI units for the first time. In this work, the lattice repeat distance of a nearly perfect silicon crystal was measured by means of simultaneous X-ray and optical interferometry. X-rays were passed through three parallel silicon crystals, one of which was moved relative to the others, creating X-ray fringes as the moving crystal lattices passed or blocked the X-rays. The number of X-ray fringes was counted automatically. Attaching one mirror of a Fabry-Perot interferometer to the moving crystal, a simultaneous measurement was made of the distance moved, permitting calculation of the lattice spacing. Once the lattice spacing was known, a silicon sample was used to diffract Cu Kα_1 and Mo Kα_1 X-rays, and wavelengths were obtained by double-crystal spectrometry. Since the lattice spacing was known, the X-ray wavelengths could be expressed directly in terms of SI units.

Time

A similar story of research and progress applies to the definition and measurement of the second. The second was first officially defined in the

NBS-5, primary standard of frequency, is maintained by NBS, Boulder, CO, labs. Since this standard also measures base unit of time, the second, it is popularly known as the "atomic clock." Incredible accuracy of this standard approaches one part in 10 trillion (1×10^{-13})

metric system as $1/86{,}400$ ($24 \times 60 \times 60$) of the mean solar day. This was at best an unwieldy definition, depending as it did on precise astronomical observations. As evidence accumulated concerning unexpected changes in the rate of the earth's rotation about its axis, another, more regular natural phenomenon was substituted in 1956 as the definition. What was selected was a fraction of the time it takes the earth to make a complete orbit around the sun. To be specific, the second was redefined as $1/31{,}556{,}925$ (approx. $365 \times 24 \times 60 \times 60$) of the time of this orbit. While in principle this was a very exact definition, again only well-equipped and dedicated observatories could make such measurements. Obviously, a more accessible definition was needed.

Fortunately, work on a new clock was well under way. In 1949 a team led by Lyons at NBS devised a means for using the invariant vibrations of the ammonia molecule to control an oscillator with which to drive a clock. This was the first atomic clock ever built, and while its accuracy was limited, it opened the door to a whole new field of timekeeping. Subsequent work at NBS and other laboratories led to development of a clock based on the transition frequency of cesium atoms. Such clocks have achieved fantastic accuracies—on the order of 1 part in 10^{13}, or a deviation of 1 sec in about 300,000 years. In 1967 the second was redefined as the duration of 9,192,631,770 periods of the radiation corresponding to the transition between two hyperfine levels of the ground state of ^{133}Cs. Atomic clocks are now available commercially and are widely used in military, space, navigation, astronomy, and atomic/nuclear research applications.

Let me give just one example of the need for really precise time measurements. Better knowledge of movements of the earth's crust could lead to prediction of potential earthquakes, and a new and sophisticated approach to measuring earth movement is now under way. Antennas on either side of the San Andreas fault—one at Pasadena and the other at Barstow, CA— are positioned to receive radio signals from a quasar deep in the galaxy. The difference in arrival times of these signals allows calculation of the positions of the two antennas, and *changes* in the relative arrival times can reveal antenna movement. To resolve movements as small as 2.5 centimetres, arrival time differences of one ten-billionth of a second must be detected. Atomic clocks are used at the antenna sites to provide the necessary timing accuracy.

Volt

In addition to the SI base quantities, NBS also maintains dozens of *derived* quantities, such as the volt. Here again advanced research is applied to upgrade our capabilities. For years the volt was maintained by a large reference group of saturated cadmium sulfate standard cells, a cumbersome arrangement that produced a value known to drift with time.

On July 1, 1972, we adopted a new procedure for maintaining the legal volt. In this procedure the voltage produced by two Josephson junctions, operated in series, is compared to that of a standard cell. The comparison, made through a special fixed ratio (100:1) potentiometer, is accurate to two parts 10^8. The junction used in our work is a sandwich of lead–lead oxide–lead, cooled below its superconducting transition temperature. This junction converts an applied microwave frequency to an output voltage that is dependent on knowledge of the frequency alone. Since frequency can be measured with high accuracy, the junction's output voltage is well characterized and independent of time. Thus, the Josephson junction provides a stable reference against which the drift in voltage of the standard cells is monitored and corrected.

Atomic Weights

Over the years, mass spectrometry has been used at NBS to determine atomic weights with unprecedented accuracy. For example, the weights of silver, chlorine, bromine, copper, and rubidium, among many others, have been determined by NBS and accepted as "best values" by the International Union of Pure and Applied Chemistry. Uncertainties ranged from ±0.001 for silver to ±0.0026 for rubidium. While such work is of fundamental importance to chemists, it also can have practical implications. Knowledge of the atomic weight of gallium is a limiting factor in the fabrication and analysis of gallium–arsenide–phosphide light emitting diodes, and industry would like to know the value to a part in a million if possible, rather than the part in 3,000 now available. We have a program under way to provide a new value, aiming at, but of course not guaranteeing, a part in a million.

Other Constants

In addition to measuring atomic weights, which are in essence constants of nature, NBS has long been involved in determining other natural constants. Recent work of special interest involved direct measurements of both the *frequency* and *wavelength* of the same laser line. Since the speed of light is defined as frequency times wavelength, the unique results achieved at our Boulder labs have special significance. In fact, the value they derived for the speed of light, 299 792 458 m/sec, is 20 times better than any previous number, and has been provisionally accepted by the General Conference as the current best value. The acceptance is provisional because no other lab has measured both the frequency and length of the same wave.

Another fundamental constant whose value has been upgraded at NBS is the Avogadro number. This work is the result of three separate, very demanding experiments: determination of the atomic weight, density, and lattice spacing of the cubic cell of silicon. The new value, 6.0220943×10^{23}, is 20 times more accurate than earlier values.

Standard Reference Data

NBS is deeply involved in a program to provide the technical community with what we call standard reference data. Such data are numbers extracted from the literature, critically evaluated, and published by NBS as current "best values." Critical evaluation is the process by which recognized experts subject data to independent scrutiny as to their reliability and relevance to user needs.

The need for such data arises from many factors: 650,000 papers containing quantitative information are published each year, making it almost impossible to keep up in most fields. Next, some of the published data are definitely in error as indicated by the conflicting numbers that can be found for many quantities. Finally, up to 50% of the published data are unusable, not because of error, but because insufficient details are given with which to assess the work.

At the request of the President's Office of Science and Technology, a program to provide critically evaluated data was established at NBS in 1963. Briefly, NBS coordinates a group of about 40 data centers, located at NBS, in industry, universities, and other government agencies, which do the compilation and evaluation of data in the world's literature.

Our output to date has been 160 compilations containing 28,000 pages, with quantitative information on more than 30,000 materials. Over 200,000 copies of these compilations have been sold, and much of the information has been incorporated in various reference works and handbooks. In cooperation with the American Institute of Physics and the American Chemical Society, we publish the *Journal of Physical and Chemical Reference Data,* an outlet that greatly expands our impact. We feel that this effort makes a substantial contribution to the efficiency of research and measurement, in that data are readily accessible, and to accuracy, by providing the best possible information.

Standard Reference Materials

Another approach to measurement reliability is provided through the use of NBS standard reference materials (*1*). These materials, whose chemical or physical properties have been measured by NBS, are sold for use in standardizing measuring techniques and apparatus of all kinds. This program has a long history, since we sold our first SRM's, a group of four cast irons of certified composition, in 1906. Then, as today, we depend heavily on industry to provide much of the candidate material and to assist in its characterization. Without this close cooperation the SRM program would not have the broad impact it now has. Today, over 800 different materials are offered and widely used, and 32,111 individual items were sold in 1974.

The preparation of an SRM is not a trivial matter. First, a strong need must be established, as our resources are limited. We often conduct a survey to determine the potential impact of a proposed SRM.

Once the decision is "go," the preparation of the material itself can be a major undertaking. In most cases, we let a contract for the preparation of the basic material, and the requirements can be quite exacting. For example, one steel sample calls for controlled amounts of C, Mn, P, S, Si, Cu, Ni, and Cr, which really tests the foundry's skill.

Finally comes the process of careful measurement, first to assure homogeneity and then for certification. Sometimes this is done exclusively at NBS, sometimes it is done in collaboration with participating laboratories, and oftentimes it involves us in the development of new or improved measurement techniques.

I would like to mention a few new standards of major impact. One of the latest is the NO_2 permeation tube, a standard of considerable importance in pollution studies. In this rather simple device, NO_2 in a glass reservoir permeates through a Teflon plug at a rate that is temperature dependent. Our calibration consists of a simple but accurate weight loss determination over a period of about two months. This rather long period is dictated by the fact that tubes lose but about 1.4 mg per day, and we needed a larger weight loss to ensure accuracy at equilibrium. This device, developed under a cooperative program involving our Office of Air and Water Measures, the Analytical Chemistry Division, and the Environmental Protection Agency, has had an impact on upgrading the accuracy of NO_2 measurements.

Our SRM's are also having a major impact on clinical measurements. There are over 15,000 clinical labs in this country, making an estimated three billion measurements yearly. Some medical authorities feel better measurements will assist the physician in diagnosis or treatment. At the request of the College of American Pathologists and the American Association of Clinical Chemists, NBS turned its attention to SRM's for the clinical lab.

The first standard we produced, in response to a need for improved accuracy, was cholesterol of 99.4% purity. Other samples were soon to follow, including urea, uric acid, creatine, bilirubin, cortisol, and mannitol.

Reference Methods

But more than SRM's were needed to really upgrade clinical measurements. Many ASTM test methods call for use of an SRM in conjunction with a *specified measurement procedure;* the same approach was decided upon for some clinical measurements.

The first "reference method," as they are called, was developed for the accurate measurement of calcium. First, a standard of $CaCO_3$, 99.9% pure, was produced and issued as an SRM. Then, a method of extra high accuracy was selected as an absolute analytical technique. This method was isotope dilution mass spectrometry, and measurements in our own mass spec lab gave results accurate to within 0.2%, relative to serum level concentrations. With this as a base, work then proceeded on a reference method that could be used in clinical labs, not so much as a routine procedure, but rather as a check method of known accuracy. What we zeroed in on, in close

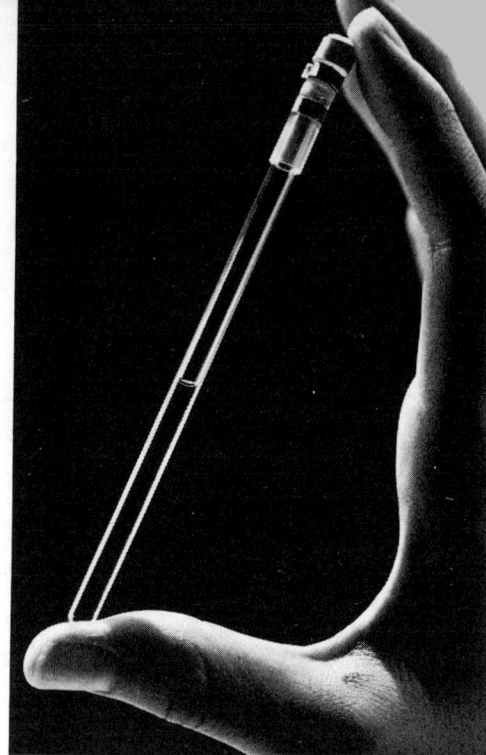

NO_2 permeates through Teflon plug in end of glass tube at constant, calibrated rate. Known rate of air flow past tube is used to produce desired NO_2 concentration for instrument calibration

cooperation with clinical pathologists and the Center for Disease Control, was the technique of atomic absorption spectrometry. By use of the $CaCO_3$ SRM and the carefully specified reference procedure, clinical labs can with confidence measure calcium in serum to within 2%. Already the reference method has been used to evaluate nine field methods for determining calcium. Of the nine, four were found adequate, four need improvement, and one should be discarded.

This concept of a reference method is being extended into other areas of medical interest, such as the determination of sodium, potassium, magnesium, and lead. We feel that the combination of SRM's with reference methods will greatly upgrade the reliability of clinical measurements and improve the system of health care in this nation.

Calibrations

Another approach to measurement accuracy is provided through our calibration services (*2*). NBS offers literally hundreds of tests by which we will, in our own laboratories, measure such devices as voltmeters, standard resistors, refractive index standards, spectral transmittance filters, and gamma ray sources. Although we have provided calibration services for over 70 years, and last year processed 7,000 items, the system has obvious limitations. Oftentimes the user cannot afford to have his standard out of service for the time it takes to have it calibrated; there is the obvious chance of outright damage or undetected change during transit; and most important of all, merely having an NBS-calibrated standard on hand in no way assures that a lab is *making* accurate measurements.

To help remedy that last point, NBS has evolved an approach that we call our Measurement Assurance Program. The essence of MAP is that participating laboratories can determine, on a continuing basis, the actual performance of their entire measurement process. Let me illustrate how this works in mass, the area to which MAP was first applied.

At NBS we had accumulated data, over a long period of time, on comparison weighings of two kilograms. A plot of the difference between the two kilograms as a function of time gave a clear indication of what the entire measurement process was doing. With such a record on hand, two things became obvious: future comparisons of these two kilograms could reasonably be expected to fall within the same limits, and points that fell outside the limits would immediately alert us to something wrong in the system. For example, a balance could be malfunctioning, or an operator could be having a bad day. Permanent changes in one of the weights would become obvious if a new cluster of data points accumulated. Through this process NBS develops assurance that its measurements are correct.

We felt that a similar approach would be valid for other labs as well and initiated our first Measurement Assurance Program. In broad terms, NBS consults with new participants, trains their staff if necessary, calibrates a set of starting standards for them, provides measurement protocols and data analysis services, helps evaluate the participant's process, and ties it to national mass standards.

The participant creates his own control chart for the range in which he intends to work by repeated measurements on standard weights or other appropriate test items. How much data are required to establish a chart and how often checks must be made depend on the degree of accuracy needed by the user. Once a control chart is established, we provide the means of evaluating their system for compatibility with the national system by having them measure a standard weight or other object that we had well characterized in our own lab. If their value plus their uncertainty (which is established by their control chart) overlaps the NBS value, they are in good shape. If it doesn't, we will consult with them and help resolve the difficulty.

The beauty of this system is that the user himself establishes a running record of the actual performance of his measurement *process,* a record that provides solid evidence of the reliability of measurements made on unknowns. No such evidence would exist if he merely had a set of NBS-calibrated weights which were used sporadically to check working masters or, as we often found, were considered "sacred" and not used at all.

Today we have 19 participants in our mass program. For 12 of the groups we actively provide data reduction and process evaluation services. Other MAP programs are available for those laboratories who calibrate standards of DC voltage, resistance, capacitance, and voltage ratio, and programs are being developed in length, temperature, laser power, RF power, impedance, and attenuation. We feel MAP is a significant step toward providing accurate measurements backed by a firm data base.

Statistical Considerations

Analytical chemists are aware of the importance of the precision of methods and the assessment of measurement uncertainty. At NBS we have a special group of statisticians whose main job is to help in the design of experiments and analysis of data, to search for systematic errors, and to realistically evaluate precision and accuracy of measurement results. Many of the accomplishments which I have mentioned, such as the Measurement Assurance Program, the atomic weights and physical constants work, and the development of standard reference materials, all have had substantial statistical analysis.

New trends in measurement technology are presenting new challenges to the statisticians, such as helping to plan efficient data acquisitions in automated experiments. For example, statistical techniques for exploratory data analysis uncovered unsuspected auto-correlation in instrumentation and led to modification of procedures in automated high-precision spectrophotometry. The development of analytic methods for use outside the laboratory has broadened our concern with measurement uncertainties to include sampling error owing to inhomogeneity of source material, or in the case of environmental measurements, owing to time and space variations. As the precision of measurement systems improves, the treatment of sampling errors and the planning of measurement schedules to control them will become increasingly important for all analysts.

Measurement Techniques

We have also made advances in measurement techniques. For example, Okabe, a specialist in molecular fluorescence, has developed a simple, rapid, accurate method for measuring SO_2. By use of light from a zinc vapor lamp, fluorescence is induced in any SO_2 contained in a gas sample, and the intensity of the fluorescence is measured with a phototube. The method is linear from less than a part per million to 1600 ppm and is insensitive to most other contaminant gases. Several firms are planning commercial instruments based on this effect.

Again in the pollution field, Marinenko of NBS has developed a chlorine monitor for continuous field use that is sensitive in the parts per billion range, a level so low that it is missed by most other methods. Chlorine, widely used to treat sewage, often combines with ammonia in waste water to form chloramines. The chloramines are highly toxic to both freshwater and marine organisms and have been linked to massive fish kills. In fact, EPA recommends that trout streams contain no more than 2 ppb residual chlorine, but until now no reliable method existed to measure such small amounts.

The NBS approach is based on the principle that chlorine, chloramines, and hypochlorites will oxidize iodide to iodine at the proper pH. An electrochemical system is used to monitor the iodine concentration, with the current resulting from the iodide–iodine reaction serving as the output signal of the device. This instrument has already been used to detect and correct excess chlorine levels in waste water and should find widespread application in both pollution control and biological research.

Since NBS has a high-flux nuclear reactor, it is natural that we are quite active in neutron activation analysis of trace elements. We have done pioneering work in careful characterization of the irradiation facility itself and have made intensive studies of the effects of such factors as configuration, sample positioning, density, and homogeneity on the precision and accuracy of counting.

Specific analytical techniques have been developed for the activation analysis of mercury in biological matrices at the nanogram (10^{-9} gram) level and then extended to the determination of selenium, arsenic, zinc, and cadmium. Recently, a technique was developed for the simultaneous determination of tellurium and uranium, both potent toxins, at the level of a few parts in 10^{-9}.

We have used activation analysis in characterizing a variety of SRM's. For example, trace amounts of copper, iron, lead, potassium, and other elements have been certified in bovine liver samples. Still in activation analy-

Richard W. Roberts, the seventh Director of the National Bureau of Standards, was named to this post in February 1973. Dr. Roberts earned his bachelor's degree at the University of Rochester in 1956 and his doctorate in physical chemistry from Brown University in 1959. After a postdoctoral year as a National Academy of Sciences Fellow at NBS, he joined the General Electric Co. and moved from laboratory research through successive management positions until becoming manager of Materials Science and Engineering. Dr. Roberts is the holder of three patents and has earned recognition for his studies in ultrahigh vacuum, surface chemistry, and lubrication of space-age metals. He is a coauthor of "Ultrahigh Vacuum and Its Application" and an associate editor of "Annual Review of Materials Science."

sis, a water sampler has been developed that is intended to introduce zero contamination into the sample it takes. It goes under water in the closed position, is opened at the sampling depth, then closed, and raised. As soon as a thorough evaluation of its performance is completed, and so far it seems excellent, it will be used to evaluate the contaminating effects, if any, of other samplers now in use.

We recently developed a new system for detecting vinyl chloride at levels below 1 part per million. In this technique a pulsed electrical field of about 2000 volts per centimeter is used to Stark shift an absorption line of vinyl chloride into and back out of coincidence with an emission line from either a CO or CO_2 laser. A photomultiplier is used to determine, by difference, the amount of absorption, which is proportional to the vinyl chloride concentrations. This development is especially important in view of proposed Federal regulations to limit vinyl chloride concentrations to 0.5 ppm.

We have, in the last few years, developed expertise in Raman spectroscopy. For example, we have demonstrated the feasibility of analyzing single particles in the micrometer range; we have analyzed the contents of bubbles in glass; and we have worked with geologists to identify minute inclusions in rock specimens. We are concerned with the provision of accurate reference spectra and are considering the production of an SRM that will help standardize measurements made in different labs.

In preparing this talk, the major problem I faced was what had to be omitted because of time limitations. Looking at the session titles of this symposium brought to mind, in almost every case, a matching NBS project. We are active in fluorimetry, gel permeation chromatography, ion selective electrodes, polarography, thermal analysis, ion probe work, and much more. What I hoped to leave with you was an impression of the breadth of our work and our broad impact on the practical measurement problems you face.

Spectroscopy

I would be remiss, however, if I didn't include some mention of spectroscopy. Coblentz came to NBS in 1905, and Meggers in 1914. Between them they laid the foundations for modern, quantitative spectroscopy, a story with which most of you are familiar. But the passing of these two giants and the recent retirement of Scribner, well known for his work on spectral analysis of uranium, do not mean we're out of the spectroscopy business. Far from it. Moore continues to evaluate data on atomic spectra and energy levels, part of our ongoing National Standard Reference Data System. Other elements of the System include the Atomic Energy Levels Data Center run by Martin and Hagan, who are picking up the work of Moore, and the Atomic Transition Probability Data Center under Wiese.

In 1961 we published Monograph 32, which reports the results of measurements of the spectral intensities of 39,000 spectral lines. This has become a standard reference for spectroscopists, and an updated, extended version is due to be published in the next month or two.

And we are continuing our laboratory work in spectroscopy. One of the latest results, now in press, provides data on the 5th ionization level of tantalum, with data on high ionization states of tungsten and molybdenum to follow. This information is especially important in the diagnosis of high-temperature plasmas. So our work in spectroscopy goes on at a very rapid pace.

Summary/Conclusion

I've covered a great deal of territory this afternoon, and I've left even more untouched. The impression I wish to leave with you is the importance we at NBS place on accurate, meaningful measurements, not just within our own laboratories but *throughout* the measurement system.

To achieve this goal takes many elements. First, it takes the universal language provided by the International System of Units. Next, it takes methods of tying the measurement system to those units, which we do through calibrations, Measurement Assurance Programs, SRM's, and standard reference data. It also takes a reservoir of knowledge and skill, both at NBS and elsewhere, to meet the new measurement challenges that arise. Perhaps most important, it takes communication of needs, ideas, problems, and solutions if progress is to be maximized. I hope that you think of and use NBS as what it is: a national resource devoted to measurement excellence.

References

(1) SRM's are described in *Standard Reference Materials—1973 Catalog, NBS Special Publication 260,* available from Superintendent of Documents, U.S. Government Printing Office, Washington, DC 20402. Use Catalog #C13.10:260 when ordering. Price: $1.25.
(2) NBS Calibrations are described in *Calibration and Test Services, NBS Special Publication 250,* available from GPO, Catalog #C13.10:250. Price: $2.00.

MEASUREMENT ANALYSIS BY PATTERN RECOGNITION

Applications should offer advantages over existing techniques to solve chemical problems or open new fields for investigation. Computer pattern recognition methods extend the ability of human pattern recognition, but, in the end, it is the chemist who must do the chemistry

Bruce R. Kowalski
Laboratory for Chemometrics
Department of Chemistry
University of Washington
Seattle, Wash. 98195

Analytical chemists are among many scientists who rely heavily on graphics for the interpretation of experimental results. Except for convenience the use of laboratory computers has not altered this reliance significantly. A primary output of computerized experimentation is still graphical representation of numerical data. There are at least two reasons why graphical presentation is not only prevalent but desirable. First, many chemists distrust computers and are not willing to allow a computer to reduce the results of experimental toil to a few numbers. Second, the human has a well-developed visual ability to recognize pictorial patterns and deviations from patterns. For the latter reason, the interactive role of the chemist in data analysis should be strengthened rather than reduced as originally proposed in the early days of computers.

If an analytical chemist is seeking the relationship between a physical measurement and the concentration of a species in a sample, a simple plot of the measurement vs. concentration usually leads to the correct functional relationship. The computer can be used to fit the function and report goodness-of-fit and any significant deviations from the relationship. The modern analytical chemist, however, is becoming more involved in experimental design and measurement systems. Therefore, data analysis often means finding relationships not only between a measurement and sample composition but multiple measurements and sample function or origin as well. In these cases the relationships are not only multivariant but may be nonlinear with respect to the measurements. The solution is simple; since the scientist is the best pattern recognizer, an examination of an N-dimensional plot where each measurement is represented by one of N axes should lead to the necessary mathematical relationship. While true in principle, the practical impossibility is obvious. Nevertheless, the need to deal with N-space plots cannot be avoided. Plotting each measurement vs. every other leads to $N(N-1)/2$ plots, but the pair-wise examination is neither effective nor practical. How then does one attack this dilemma? The historical approach is to solve the problem experimentally. This approach has evolved due to a lack of computers and a love for experimentation. But as computer resources become more available, experimentation becomes more expensive, and the complexity of problems which science must solve grows, advances in computer science and other fields can provide new approaches.

In 1969 a series of papers concerned with applications of learning machines (1) to the determination of molecular structural features directly from spectral data (2) appeared in ANALYTICAL CHEMISTRY. Shortly thereafter, a review described early results of such applications (3). Thus began a search for new mathematical methods to solve multivariant problems in chemistry. While it can be argued that recognition of the importance of multivariant data analysis methods in chemistry started much earlier with linear free energy relationships (4) and other studies, there is little doubt that a considerable amount of interest in new mathematical methods evolved after these early applications of pattern recognition. Although the scope of all chemical applications is large at this time and potentially even much greater, emphasis will be on the analysis of data generated by the analytical chemist. After a brief section on pattern recognition methods, the strengths and limitations of pattern recognition in chemistry will be discussed along with future uses and a perspective on how it fits into a much larger discipline the author identifies as chemometrics. Chemometrics includes the application of mathematical and statistical methods to the analysis of chemical measurements.

Methodology

For a detailed understanding of the many methods of pattern recognition, the interested reader should refer to the engineering and computer science literature (5, 6). A number of reviews on methods and applications which provide stepping stones to the more mathematical literature have also appeared in the chemical literature (7–9).

It is convenient to divide the operations or types of pattern recognition methods into four categories. These operations—display, preprocessing, supervised learning, and unsupervised learning—will be presented with the aid of a simple example. There are several methods from which to choose within each of the four categories. The choice depends on an understanding of the method's limitations, criteria, etc. Additionally, there are many useful chemometric methods that complement and enhance those of pattern recognition. A broad knowledge of chemometrics is necessary to ensure proper utilization of methods and optimal solutions to data analysis problems.

If N measurements are made on a collection of objects, and the goal is to learn something about the objects, it is useful to study the similarities and dissimilarities among the objects. One way of studying the objects is to represent them as labeled points in an N-dimensional plot. Each of the N axes would then correspond to a measurement, and the value of the jth measurement for the ith object would serve to position the ith point along the jth axis. This representation for two measurements reduces to the simple two-dimensional graphs, an example of which is shown in Figure 1. Here, the concentrations of two enzymes in blood are plotted for a collection of 55 patients, each suffering from one of two different liver disorders. If three clinical measurements are plotted, computer graphics can be used to analyze the three-space plot. Chemists are often concerned with the analysis of much higher dimensional spaces (often >300) which, of course, cannot be seen but can be manipulated with mathematical methods that follow direct analogies to the analysis of low-dimensional spaces. For example, the familiar two-space euclidean distance $d_{\alpha\beta}$ between points α and β is

$$^2d_{\alpha\beta} = \left[\sum_{k=1}^{2} (\chi_{\alpha k} - \chi_{\beta k})^2\right]^{1/2} \quad (1)$$

which is derived from the N-space euclidean distance,

$$^Nd_{\alpha\beta} = \left[\sum_{k=1}^{N} (\chi_{\alpha k} - \chi_{\beta k})^2\right]^{1/2} \quad (2)$$

which in turn can be derived from the N-space Mahalanobis metric

$$^NM_{\alpha\beta} = \left[\sum_{k=1}^{N} (\chi_{\alpha k} - \chi_{\beta k})^\rho\right]^{1/\rho} \quad (3)$$

It is important to recognize that by using extensions of mathematics commonly used in analyzing two-space plots, we can analyze N-space plots with the euclidean or some other metric.

If some of the objects are tagged with a known property, the corresponding points in N-space are referred to as members of a *training set*. In Figure 1 the category A patients are distinguished from the category B patients and collectively constitute the training set. Another set consisting of patients with known property, the *test set*, is frequently used to test the results of supervised and unsupervised learning. Also seen in the example are five points representing patients with unknown classification where it is perhaps the goal to use relevant clinical measurements to determine the classification to which an unknown belongs. This set can be called the *evaluation set*.

Preprocessing

If the measurements for each object are assembled in a matrix X with elements X_{ij} (ith object and jth measurement), the process of operating on X to change the N-space data structure is known as preprocessing. The data in Figure 1 were range scaled so that the extremes within the two mea-

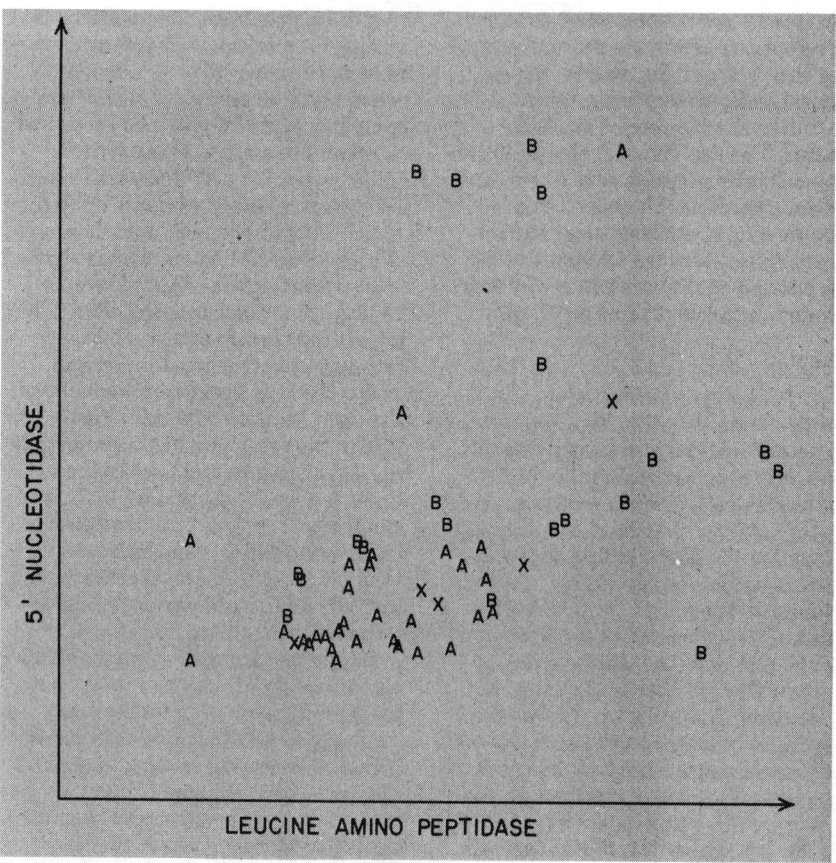

Figure 1. Concentrations of two enzymes in blood of patients suffering from two (A and B) liver disorders

surements form the limits of the plot. If this simple form of preprocessing was not used, the variation in one measurement would receive undue emphasis due to its dominant range. This simple form of preprocessing is familiar to anyone who has ever plotted two measurements with dissimilar ranges.

Preprocessing is usually applied to reduce the number of measurements used in subsequent analysis, enhance the information representation of the measurements, or both. When measurements are preprocessed, *features* are generated from the measurements for each object. The number of features can be greater than, equal to, or less than the number of measurements. The features can also be linear or nonlinear combinations of the measurements. Since the selection of a preprocessing method is dependent upon subsequent operations, further discussion on preprocessing will be included in the next three sections.

Display Methods

If the goal in the analysis of the clinical data shown in Figure 1 is the separation of the two known categories, it is clear that although the two measurements contain some discriminatory ability, they are not sufficient to satisfy the goal. Adding additional measurements to the study should obviously be done, but, unfortunately, the addition increases the dimension of the plot, and the unique pattern recognizing abilities of the scientist are no longer applicable. Are we then forced to blindly accept the results of the mathematical analysis methods in the form of tables of numbers so easily generated by computers? Experience in our laboratory has shown that the display methods (*10*) provide a useful but unfortunately only partial solution to this problem. These methods attempt to project or map the points in N-space down to two- or three-space with the criteria that the data structure in the N-space be preserved as much as possible. When six additional clinical measurements (Table I) are obtained on the patients in Figure 1, and a linear projection method is used to project the points from eight-space to two-space, Figure 2 results.

The eigenvector method, used here, attempts to retain total variance by forming two orthogonal projection axes which are linear combinations of the eight measurements. In Figure 2, 65% of the total variance is preserved. All eight measurements combine to provide a better separation of the two categories. It must be remembered that Figure 2 is an "approximate view" into eight-space, and the need to apply additional pattern recognition methods still exists.

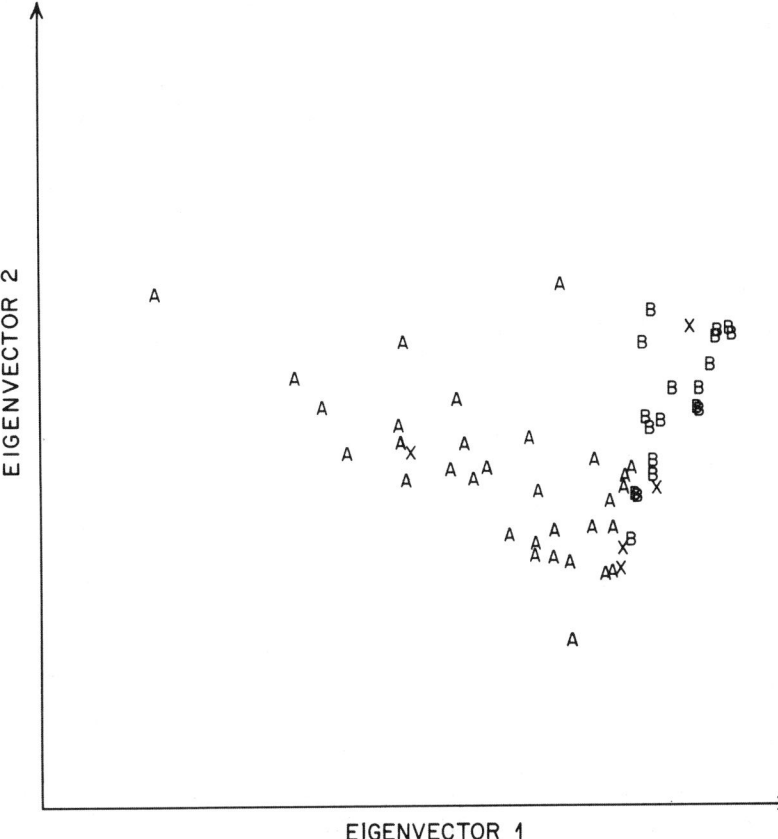

Figure 2. Eigenvector projection of eight-measurement data structure (65% information retention)

Before leaving display methods, two additional points can be made. First, by the definition of preprocessing methods, the display methods can be included as special cases. Dimensionality reduction is accomplished by the projection to two-space, and the information content of the two new features (coordinates of the points in Figure 2) is greater than in any two of the original clinical measurements. This is because the new features are combinations of all of the measurements. Eigenvector projections generate optimal linear combinations where nonlinear mapping methods preserve more information but form nonlinear combinations of the original measurements. Second, preprocessing can be applied to the measurement space prior to the application of the display methods. When the scaled measurements were weighted by variance weighting (*7*) (Table I) so as to increase the importance of measurements that are good individual discriminators of the two categories, the plot in Figure 3 results. The projection shows a greater degree of separation, but the analysis of the eight-space must continue because some information was lost in projection.

Supervised Learning

The result in the clinical data example thus far has been to investigate the separation of categories with display methods while applying some simple preprocessing methods. To attack this application more directly, supervised learning (*1, 5, 6*) methods can be applied that develop classification rules using the examples in the training set. If the results of classification attempts on the training and test sets are acceptable to the chemist, the rules can be applied to classify unknowns in the evaluation set via the measurements.

Table I. Blood Constituents

Measurement	Variance wt[a]
Leucine aminopeptidase	1.7
5'-Nucleotidase	1.5
Glutamate oxaloacetate transaminase	1.6
Glutamate pyruvate transaminase	1.8
Ornithine carbamoyl transferase	1.3
Guanine deaminase	1.8
Isocitrate dehydrogenase	1.2
Alkaline phosphatase	3.5

[a] Weight = 1.0 means no discrimination information.

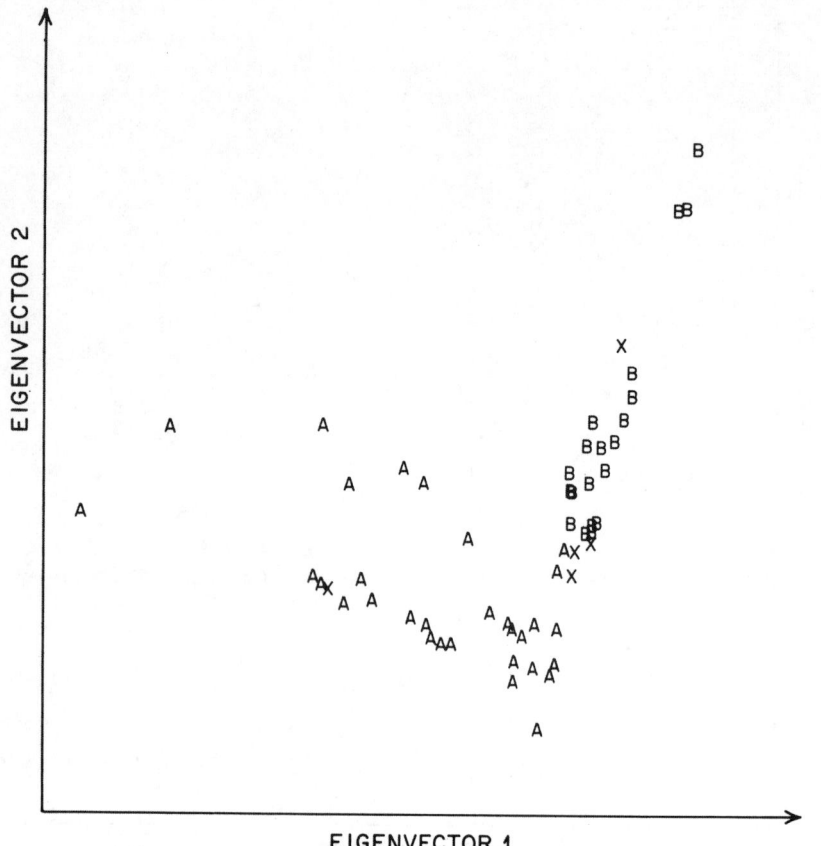

Figure 3. Eigenvector projection of variance weighted eight-measurement data structure (80% information retention)

Supervised learning methods attempt to do in N-space what we have done by examining the first three figures; namely, determine the separability of the categories in the measurement space (or feature space if preprocessing has been applied).

When the eight-space data in the clinical measurement example are analyzed by any of the several supervised learning methods used routinely in our laboratory, classification rules that separate points into the two categories are easily determined, and the features that are most useful for separation are reported. For example, the linear learning machine (1) iteratively moves a decision surface in the measurement space until either all of the points in one category lie on one side of the hyperplane and the points in the other category lie on the other, or a limit of time or iterations is exceeded. In the former case, the training set is said to be linearly separable. The separating surface was found for our example in a very few iterations. In the latter case, it can only be said that perfect separability was not attained.

Several other supervised learning methods are known, and a few have found application in chemistry. The k-Nearest Neighbor Classification Rule (7) which classifies an unknown point according to a majority vote of its k-nearest neighbors is conceptually simple and has the advantages of being a nonlinear, multicategory classifier. The learning machine and k-nearest neighbor rule are among several supervised learning methods that can be classed as nonprobabilistic because they do not determine or use the underlying probability distributions in the training set measurements. These distribution-free methods have a possible advantage when small training sets make the determination of the parameters of distributions uncertain. Their severe disadvantage is that a probability of correct classification, or the classification risk, cannot be reported.

If the measurements can be seen to fit a probability density function, then the parameters of the appropriate distribution can be determined from the training set and the preferred Bayes Classification Rule (5) can be used. This is illustrated in Figure 4 which represents two overlapping distributions in a single measurement space. In this simple example, points with measurement A and C would be classified as category one and two, respectively, with a probability of 1.0 associated with the classification. Point B would have an equal probability (0.5 for category one and 0.5 for category two) of coming from either category. Probabilistic classification rules have generally been avoided in chemical applications most probably due to a lack of sufficient data to establish the parameters and the simplicity and general availability of nonprobabilistic methods.

A new supervised learning method called statistical isolinear multicomponent analysis (SIMCA) developed by Wold (11) has recently found application to chemical data analysis. SIMCA combines the simplicity and general applicability of the nonprobabilistic methods with the robustness of probabilistic methods. The axes of greatest variance for each category are determined by principal component analysis, and least squares is used to fit the training set members to their respective principal component models. In most cases, the number of components is quite small compared to the number of original measurements. Experience in our laboratory has shown SIMCA to be the supervised learning method of choice for many chemical applications.

There are usually two reasons for poor classification results in a particular application. Either the necessary discriminating information is not contained in the measurements or the information representation is not adequate. In the former case, it is obvious that better measurements are needed. This was seen in the clinical data example when additional measurements led to 100% discrimination. In the latter case, preprocessing can be applied to change the representation. Scaling and measurement weighting have already been discussed, but the number of preprocessing methods from which to choose is greater than all of the rest of the pattern recognition methods combined. Several preprocessing methods have been used to enhance the results of supervised learning in chemical applications. Most prominent are those studies that use transform domains such as Fourier (12) and Hadamard (13) on spectral data prior to the determination of molecular structural features. The important point to note is that preprocessing is applied to reduce the number of measurements or features and/or to improve the information representation. Selection of a preprocessing method depends upon a knowledge of the ideal measurement representation for the supervised learning method of choice.

In our example, the property of the objects is a category or class membership. This is because the classic pattern recognition application involves classification. If the property of interest is continuous or semicontinuous, such as the ozone concentration over a

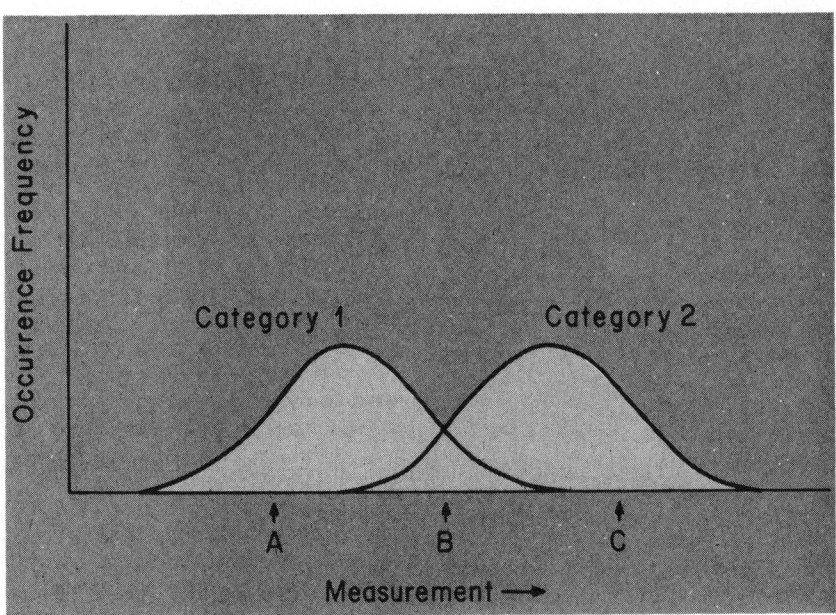

Figure 4. Overlapping distributions in single measurement

city, or a biological response to a chemical measured on a continuous scale between two limits, the objects can either be forced into classes based on ranges of the property or other chemometric methods can be used to relate the measurements to the property. There are actually several cases where classification according to ranges is desirable. For instance, atmospheric contaminant measurements are often conveniently divided into ranges associated with appropriate actions to be taken: low concentration requiring no action, moderate levels leading to a curtailment of industrial activity, and extreme levels demanding industrial shutdowns. The application of probabilistic supervised learning methods to predict future levels from current atmospheric measurements could complement current modeling efforts and provide more confidence in implementing suggested actions. Since sample collection and measurements represent the expensive parts of most studies, the application of several complementary data analysis methods is highly recommended for multimeasurement studies.

Unsupervised Learning

When a training set of labeled data points is not available and the goal of a study is to gain an understanding of the N-space data structure in the hope that a new property of the objects will be detected, the unsupervised learning methods of pattern recognition can be applied. The majority of unsupervised learning methods look for clusters of points (clusters are defined by a variety of criteria) in the N-space. When the eight-space data in the clinical example is subjected to an unsupervised learning method [Zahn minimal spanning tree method (5)], five distinct groups or clusters of points are detected. These groups are identified in Figure 5 with the aid of the eigenvector projection to two-space (Figure 2). Ignoring the classifications, Figure 5 is an excellent beginning to the analysis of the eight-dimensional data structure. Neither the display method nor the cluster analysis method was given the classifications of the points. Figure 5 is the combination of two complementary analyses of eight-space, each surfacing the inadequacies of the other. The two single-point clusters represent two patients with clinical measurements quite different from each other and the rest of the patients. From the positions of the three large clusters, it might be postulated that there are three different types of disorders represented within the collection of patients. Returning to the classifications, one cluster is purely "A", another purely "B", and the third is predominantly "A". The supervised learning methods had little difficulty in separating the two groups, but it is obvious from Figure 5 that no great void exists between the "A" group and the "B" group. In fact, the mixed cluster even indicates that a few "B"'s are more like the "A"'s.

Finding clusters is a simple task for the computer. Assigning a meaning to the separation of the groups of patients cannot be done by the computer and is by far the most difficult task. In our example, unsupervised learning suggested the discrimination of three liver disorders by the measurements. If, however, the data analysis began with the theory that only one disorder was involved, the theory would certainly be challenged by the results in Figure 5.

Unsupervised learning methods have not found extensive application in analytical chemistry. Nevertheless, when used with display methods, they can aid in understanding N-dimensional data structures and information representations. It is expected that as interactive pattern recognition programs become commonplace in the

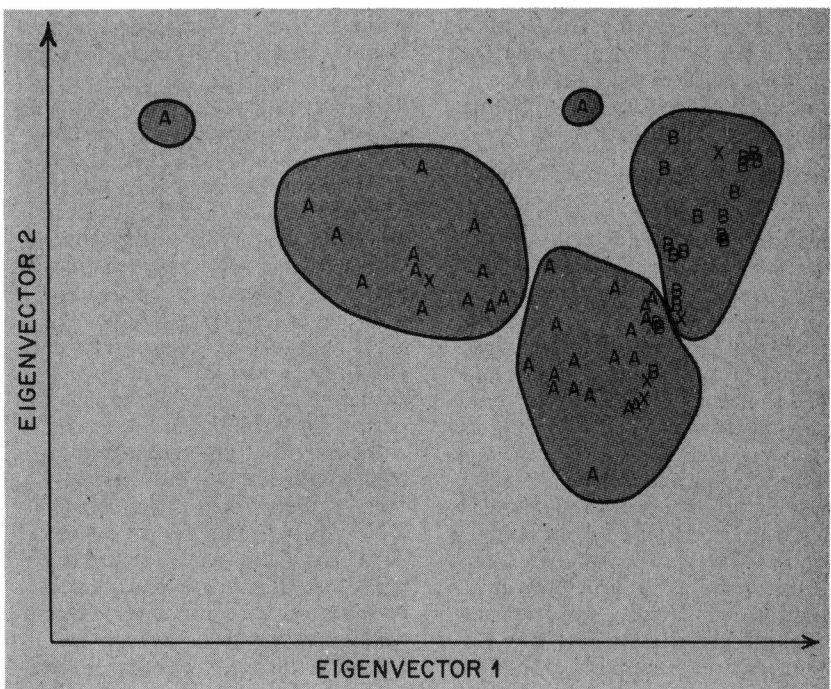

Figure 5. Unsupervised learning results presented with eigenvector projection (same as Figure 2)

analytical laboratory, their usefulness will increase.

Applications

Early applications of pattern recognition to chemical data involved the determination of molecular structural features from low-resolution mass spectra using the learning machine. Since then, several other pattern recognition methods have been applied to chemical data analysis, and the applications have seen an increase in both breadth and depth. The increase in breadth comes from the analysis of data obtained from other areas of molecular spectroscopy as well as atomic spectroscopy and even nonspectral sources. The increase in depth comes from the application of improved preprocessing and supervised learning methods to the analysis of spectral data. Justice and Isenhour (14), for example, compared six supervised learning methods for classification accuracy and cost of implementation.

The identification of structure features from spectral (molecular spectroscopy only) data can be represented by path 1 in Figure 6. Molecules or mixtures of molecules are therein represented by points in a high-dimensional spectral data space and concurrently in a high-dimensional structure or composition space. For example, the axes in the spectral space can be mass-to-charge ratios (m/e), and a point representing a molecule is positioned by the ion intensity at each m/e from its mass spectrum. In structure space, the axes can correspond to several molecular structural features (carbonyl groups, primary amine nitrogens, etc.), and the representative point is positioned by the frequency of occurrence of each structural feature in the molecule. Therefore, these early chemical pattern recognition applications really attempted the determination of rules that were trained on known structures to provide a mapping of molecules from spectral space to structure space. The objects under study need not be pure molecules but may be complex mixtures about which one might wish to map to a composition space or even to an average structure space. An excellent example of the latter is the work of Tunnicliff and Wadsworth (15) which mapped gasoline samples from mass spectral space to an average structural space for hydrocarbons (commonly called type analysis). The reverse mapping of 1 in Figure 6 is, of course, done by the spectrometer but is also represented by work done by Schechter and Jurs (16) which uses pattern recognition to generate mass spectra from a list of molecular structural features. In this study, candidate structures are used to generate spectra which are matched

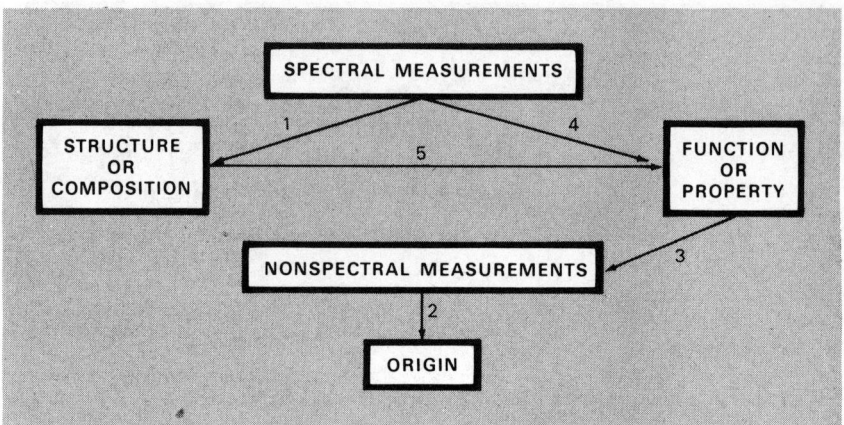

Figure 6. Chemical applications of multidimensional mapping using pattern recognition

to the measured spectrum for conformation.

Following this mapping perspective, samples (pure molecules or complex mixtures) can be positioned in N-dimensional spaces where each space represents either what is known about the samples (types of measurements that can be obtained) or what is to be learned. Pattern recognition methods can then be used to map these samples from one space to another. Traditionally, the mapping is from one N-space to only one dimension at a time in another N'-space. However, this should not be seen as a limitation because, for instance, the method of nonlinear mapping (10) and the method of Schiffman (17) can be used to map points from N-space to N'-space ($N \geq N'$). Further, information spaces can be combined, thereby allowing the mapping of points in two or more spaces to another space.

Pattern recognition has been applied to all of the mappings shown in Figure 6. The determination of the geographical origin of archaeological artifacts from the concentrations of trace elements (18) and the determination of the position in the heart from which a tissue sample was extracted (19), also by elemental composition, are just two examples of using pattern recognition for path 2 mapping. In these examples, the trace element data are considered as nonspectral data because the data need not be obtained by atomic spectroscopic measurements.

Solving material production problems (20) with pattern recognition is potentially a most rewarding area of application. These studies are represented by path 3 in Figure 6 and involve the discrimination of acceptable quality material from unacceptable quality material using chemical and physical properties of the material. These studies do not usually end with the mapping but involve dimensionality reduction of the measurement space to identify the measurements related to the quality of the material. The practical goal in most production processes is the manufacture of only quality material. The mapping, therefore, is only a means to an end and is seldom the end in itself.

Paths 4 and 5 represent exciting but difficult mappings recently attempted by a few chemical pattern recognizers. In these studies the function is the biological activity of molecules. The determination of differences in the biological function of a class of similar molecules from spectral data is an exciting prospect. Although the first attempt (21) suffered from an unfortunate selection of molecules, the idea is sound, and future research could lead to some very exciting results. The understanding of structure activity relationships (SAR) has long been of interest to chemists (22). Most SAR research uses multiple or stepwise regression analysis to relate physical and chemical properties of molecules to their biological activity (path 3). Others use regression analysis to fit structural information to biological activity (path 4). Recently, pattern recognition has been used to relate structural information to biological activity (22), and the future will no doubt see an increase in this activity.

In describing the philosophy and current chemical applications of pattern recognition, I was forced, for reasons of brevity, to ignore several very interesting and high-quality studies conducted by chemists. Additionally, there are even more excellent applications being conducted at the time of this writing.

Future

In some ways, chemistry has fallen behind other fields in the application of new methods to extract useful information from raw data. Analytical chemists and spectroscopists buried under the burden of interpreting the enormous quantities of data produced by the laboratory computer are well aware of this lag, and many are currently working on solutions. Chemists using pattern recognition methods, as well as optimization, factor analysis, and several other chemometric methods, are demonstrating that the computer is capable of aiding significantly in providing better and more useful chemical information with less effort expended by the chemist. To many chemists, it is painfully obvious that mathematicians, statisticians, and computer scientists cannot and will not solve our problems for us. It is the chemist who must accept the responsibility of infusing the tools of these information scientists into chemistry and even contributing to method development when and if necessary. This interfacing task is clearly a service to chemistry if the end result is a demonstrated enhancement in the acquisition and extraction of chemical information. With this in mind, what research in pattern recognition is needed in the future?

New and improved chemometric methods are needed to perform the mappings indicated in Figure 6. The greatest emphasis in method development should be on preprocessing. The chemical literature will no doubt see improved methods of display, supervised and unsupervised learning that will enhance the results of the various mappings shown in the last section. Preprocessing method improvements will provide better information representations and will therefore provide features more related to the information desired. The scientist is somewhat biased in thinking that if a spectrum or other datum is optimal for human analysis, then it should be ideal for computer analysis. This may not be the case. The scientist seeks simplified spectra to make analysis easier. The computer can take advantage of complex spectra. For instance, many of the successful applications of pattern recognition to spectral analysis use preprocessing methods that transform spectra into waveforms that look much more complicated and are of little use to the chemist.

Probably the most important area of future development is the incorporation of pattern recognition methods into on-line measurement systems. In this way, the power of the methods can be used for real problems, and the methods will become familiar tools to chemists and spectroscopists. Data from several instruments can be combined to attack more difficult problems involving high-dimensional data spaces. In much the same way as the chemist routinely uses various spectrometric tools, pattern recognition and other chemometric methods should be available to the chemist on the same basis. The use of interactive computer graphics, the cost of which has been falling sharply in recent years, will be an integral part of on-line measurement acquisition and analysis systems.

As pattern recognition programs become more available to chemists, the breadth and depth of applications will certainly increase. This process is already under way as more than 50 chemical laboratories have received pattern recognition systems from our laboratory (ARTHUR) or from C. F. Bender of the Lawrence Livermore Laboratory (RECOG).

Applications should either demonstrate a clear improvement over existing techniques to solve chemical problems or open new fields of investigation which were previously untouched because the tools were not available.

Pattern recognition methods operate with defined criteria and attempt to distill useful information from raw data. If the criteria used by the methods and their limitations and pitfalls are not clearly understood by chemists, the dangers are incorrect interpretation and a misuse of costly measurements. It is the author's opinion that they should be used to extend the ability of human pattern recognition and rely heavily on graphics for the presentation of results. The computer can assimilate many more numbers at one time than can the chemist, but it is the chemist who, in the end, must do the chemistry.

References

(1) N. J. Nilsson, "Learning Machines," McGraw-Hill, New York, N.Y., 1965.
(2) P. C. Jurs, B. R. Kowalski, T. L. Isenhour, and C. N. Reilley, *Anal. Chem.*, **41**, 1949 (1969) and references therein.
(3) T. L. Isenhour and P. C. Jurs, *ibid.*, **43**, (10), 20A (1971).
(4) N. B. Chapman and J. Shorter, Eds., "Advances in Linear Free Energy Relations," Plenum, London, England, 1972.
(5) H. C. Andrews, "Mathematical Techniques in Pattern Recognition," Wiley-Interscience, New York, N.Y., 1972.
(6) E. A. Patrick, "Fundamentals of Pattern Recognition," Prentice-Hall, Englewood Cliffs, N.J., 1972.
(7) B. R. Kowalski, "Pattern Recognition in Chemical Research," in "Computers in Chemical and Biochemical Research," Vol 2, C. E. Klopfenstein and C. L. Wilkins, Eds., Academic Press, New York, N.Y., 1974.
(8) T. L. Isenhour, B. R. Kowalski, and P. C. Jurs, *Crit. Rev. Anal. Chem.*, **4**, 1 (1974).
(9) P. C. Jurs and T. L. Isenhour, "Chemical Applications of Pattern Recognition," Wiley-Interscience, New York, N.Y., 1975.
(10) B. R. Kowalski and C. F. Bender, *J. Am. Chem. Soc.*, **95**, 686 (1973).
(11) S. Wold, Technical Report No. 357, Dept. of Statistics, University of Wisconsin, Madison, Wis.
(12) L. E. Wangen, N. M. Frew, T. L. Isenhour, and P. C. Jurs, *Appl. Spectrosc.*, **25**, 203 (1971).
(13) B. R. Kowalski and C. F. Bender, *Anal. Chem.*, **45**, 2234 (1973).
(14) J. B. Justice and T. L. Isenhour, *ibid.*, **46**, 223 (1974).
(15) D. D. Tunnicliff and P. A. Wadsworth, *ibid.*, **45**, 12 (1973).
(16) J. Schechter and P. C. Jurs, *Appl. Spectrosc.*, **27**, 30 (1973).
(17) S. S. Schiffman, *Science*, **185**, 112 (1974).
(18) B. R. Kowalski, T. F. Schatzki, and F. H. Stross, *Anal. Chem.*, **44**, 2176 (1972).
(19) J. Webb, K. A. Kirk, W. Niedermeier, J. H. Griggs, M. E. Turner, and T. N. James, *J. Molec. Colln. Cardiol.*, **6**, 383 (1974).
(20) B. R. Kowalski, *Chem. Technol.*, 300 (May 1974).
(21) K. H. Ting, R.C.T. Lee, G.W.A. Milne, M. Shapiro, and A. M. Guarino, *Science*, **180**, 417 (1973).
(22) G. Redl, R. D. Cramer IV, and C. E. Berkoff, *Chem. Soc. Rev.*, **3**, 273 (1974).

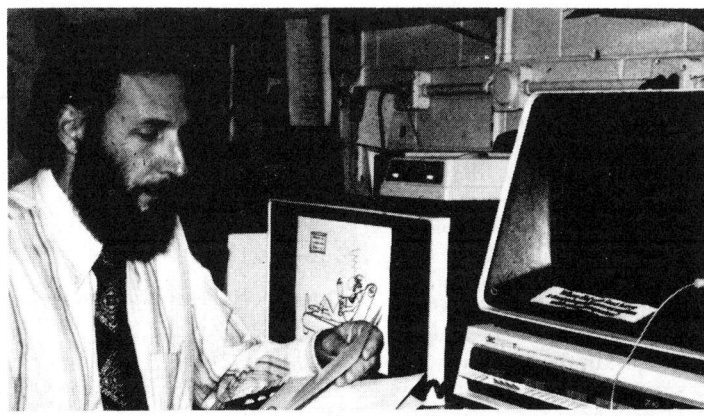

Bruce R. Kowalski is associate professor of chemistry at the University of Washington. After receiving his PhD from the University of Washington in 1969, he applied pattern recognition techniques to several areas of petroleum research for the Shell Development Co. in Emeryville, Calif., and Houston, Tex. In 1971 he joined the General Chemistry Division at Lawrence Livermore Laboratory and then, in 1972, became an assistant professor at Colorado State University. He then moved to Seattle where he has been since December 1973. His research interests include the development and application of chemometric methods primarily for the analysis of analytical data and to the relationship between molecular structure and biological activity. Additionally, he is investigating the use of improved chelating agents for trace metal analysis.

Comparison of Analytical Techniques for Inorganic Pollutants

R. F. Coleman
Laboratory of the Government Chemist
London, England

The interest in the concentration of pollutants has never been greater. An increasing number of laboratories are involved in the measurement of a wide range of substances in environmental samples, and much of the current research in analytical chemistry is directed toward improved methods of analysis for this purpose.

The main aim of this paper is to provide some background data which will assist in the selection of the appropriate technique for current and future problems. This is a very wide ranging subject, and I must draw some boundaries, albeit artificial ones, because a pollutant may sometimes be a nutrient, just as to a gardener a weed is merely a plant in the wrong place rather than something inherently undesirable. I shall limit the discussion to inorganic pollutants, but I shall ignore radioactive and also gaseous pollutants such as sulphur dioxide and oxides of nitrogen which require specialised techniques. In effect, this paper will discuss analytical methods which are applied to inorganic pollutants found in food, water, or collected on air filters from the atmosphere. The choice of the appropriate technique depends upon many factors; in addition to the inherent characteristics of that technique, it is very much dependent upon the nature of the requirement for analysis as well as the availability of particular kinds of samples. I propose, therefore, to start off with the kinds of operational requirements which one is likely to meet, follow this by a brief description of sampling techniques and problems which are common to all kinds of analysis, and then discuss the analytical techniques themselves in relation to these requirements and sampling problems.

Operational Requirement

It is of vital importance to try and understand the complete problem. Sometimes analysis is not required at all in the initial stages because the analysis of a few samples may be misleading if the problem has not been properly formulated. All too often, the analysis of samples is requested without any thought being given to the value of the data and the valid conclusions which can be drawn from it.

There are essentially three different kinds of operational requirement in which analysis can play an important part. The first might be the unexpected problem when a known or unknown pollutant is affecting the health of workers or the external population. The analytical chemist is then expected to provide some data rapidly so that necessary appropriate corrective action can be taken. In such circumstances, the appropriate technique should be capable of giving a rapid answer, be comprehensive and wide ranging, and initially, only qualitative or semiquantitative answers may be required. The second typical requirement is to study the effects of a single or, at the most, a few elements or compounds on the environment, for example, lead effluent from an industrial plant or on a larger scale, the lead content of the diet of the whole population. In such cases, one can select a technique which is appropriate for that particular element. If there are a large number of samples, automation may be necessary. Alternatively, several laboratories may collaborate in the study. In this case, it is essential that intercomparison of particular samples is undertaken at the beginning of the survey to ensure compatibility of analytical data. Careful planning of the sampling programme is also essential. The third requirement involves the analysis of a large number of elements simultaneously. These are often long-term background studies designed to monitor the environment and identify changes which have a health or economic significance. Naturally, multielement techniques are preferred in such surveys, and often some form of automation is necessary to handle the large number of samples generated.

Sampling

Sampling is a difficult and complex problem, and very often insufficient attention is paid to it. The results obtained in many surveys bear little resemblance to the true state of affairs because of changes occurring in the samples, between the source and the laboratory. In some cases, it is possible to avoid sampling altogether by making the measurements in situ. For example, selective ion electrodes may be used for the determination of certain pollutants in flowing streams of liquid. The sampling of water has been reviewed by Spencer and Brewer (1), but general solutions to all the problems are not known at the present time. Contamination and loss of trace elements during storage has been responsible for much erroneous data on the concentration of trace elements in the literature. Robertson (2) has shown major losses of some elements from unacidified seawater stored in polythene and glass. If it is important to distinguish between ions in true solution and suspended matter, filtration should be carried out as soon as possible. Hamilton and Minski (3) showed very significant differences between river waters analysed as sampled and after the addition of acid. The increase in concentration of many elements is attributed to the dissolution of particulate matter. The same authors found that soft water tends to extract more elements from containers than hard water.

Cawse and Pierson (4) have described a remote sampling station for airborne dust and rainwater which has been used satisfactorily for several years. Rainwater is collected monthly in a polythene funnel and bottle, while airborne dust is collected by drawing air through a filter paper in a polypropylene duct.

While it is relatively easy to sample in order to measure the concentration of a constituent in a particular foodstuff, sampling of the normal diet of a population is much more difficult. Harries et al. (5) have described the organisation of a total diet study which involved local purchase of foodstuffs and cooking in about 20 colleges in different areas of England and Wales in order to give a good geographical coverage of population centres and allow for regional variations in the diet.

Sample Preparation

The sample preparation will naturally depend upon the analytical technique to be used; however, there are some guiding principles which are common. Obviously, care must be taken to avoid contamination of the sample from the laboratory atmosphere, containers, and reagents. Tolg (6) reviews procedures involved in sample preparation and also methods of decomposing and dissolving samples for trace analysis. Hamilton et al. (7) considers the laboratory environment, sample preparation, and ashing procedures specifically for multielement analysis of biological materials.

Analytical Methods

The analytical techniques which have found the widest application in the study of inorganic pollutants are compared below on the basis of sensitivity, accuracy, precision, multielement capability, and range of application. In addition, where possible, the future trends in each technique will be discussed.

Microscopy. Although neglected by many analytical chemists, optical microscopy is particularly suited to the examination of particulate matter. It is a comprehensive technique capable of detecting as little as 10^{-12} gram quickly and with relatively simple apparatus. In addition, it provides information, not only on the elemental composition, but also the compounds present and in many cases the crystalline form. Identification of particles is greatly assisted by using the classification system of McCrone and Delly (8). Each particle is characterised by six parameters, e.g., colour, shape, and birefringence and can be compared with over 1000 substances in "The Particle Atlas." The groups of particles with the same classification can be differentiated on the basis of additional attributes such as density and refractive index and by direct comparison of the unknown with photomicrographs.

The next logical step is to use the scanning electron microscope. It shows the shape and surface of relatively large particles (1–100 μ) better than the polarising microscope because of the superior depth of field, and with magnifications of up to 30,000×, it is suitable for the identification of submicron particles also. A third advantage is the ability to determine the chemical composition of tiny single particles by energy dispersive X-ray analysis. "The Particle Atlas" contains electron micrographs of 100×, 1000×, and 10,000× and the X-ray spectrum of many particles. With transmission electron microscopy diffraction patterns of individual submicron particles may be obtained. With this technique my laboratory identified the presence of traces of talc in maize starch. The latter is used as a lubricant for surgeons' gloves, and talc is forbidden because of the possible deleterious effects of talc inside the body. This is a formidable analytical problem since direct X-ray diffraction on the bulk powder does not reveal talc, and magnesium oxide is also present to improve the free running characteristics of the powder and would greatly exceed the magnesium concentration due to talc. Optical and electron microscopy are particularly suited to troubleshooting and for a preliminary examination of many types of samples.

Atomic Spectroscopy. It is convenient to discuss atomic absorption, atomic fluorescence, and atomic emission together because of their similarities and the complementary nature.

Atomic absorption is particularly suited to the analysis of trace pollutants in studies involving one or a few elements. More than 60 elements can be determined by the technique, and it has become the most widely used technique for such applications. Winefordner et al. (9) have written an excellent review of theoretical and experimental aspects of atomic absorption, fluorescence, and emission flame spectroscopy. In addition to comparing the sensitivity of the three techniques, the review discusses the theoretical limitations of different types of excitation and the possible interference effects. Stone and Warren (10) review the modern instrumentation available for atomic absorption and give practical advice to assist in the selection of spectrometers from the wide range now available.

Although electrodeless discharge lamps and lasers have been used for atomic absorption, the hollow cathode lamp is still used in the vast majority of applications. The latter are available for over 60 elements and now have a fairly long life and adequate stability. The flame is the most widely accepted method of generating atoms for absorption. A variety of fuels and oxidants have been investigated, but for most purposes, air–acetylene (temperature, 2400°K) and nitrous oxide–acetylene (temperature, 3200°K) are satisfactory. The latter is particularly satisfactory for refractory compounds.

As with any technique, atomic absorption suffers from various types of interference. However, these are not generally serious, and in flames the mechanisms are well understood so that they can be reduced or eliminated. Spectral interferences rarely occur in atomic absorption spectroscopy because of the simple absorption spectra and narrow lines emitted by hollow cathode lamps. Variations in the relative proportions of ground state atoms and ionised states can occur if the matrix contains easily ionised elements such as potassium. This is fairly readily avoided by adding a sufficiently large quantity of the element to standards in order to swamp the matrix effect. Chemical interferences owing to compound formation between the element being determined and interfering substances do exist but are often broken down by the use of high-temperatures flames or by adding a releasing agent which preferentially binds the interferent. For example, lanthanum is widely used for this purpose as it will bind a variety of oxy-anions. Physical differences in samples such as viscosity can influence the formation of aerosols and cause errors if the samples and standards are not adequately matched. Samples containing a high salt content can cause scattering of radiation; thus, an absorption effect is detected other than that owing to the element to be measured. Many modern instruments contain a second light source which can eliminate background absorption and substantially increase precision and limits of detection in such samples.

An indication of the sensitivity of current flame atomic absorption procedures is given in Figure 1. Typically, the precision of analysis is 1–3% for concentrations more than 10 times the limit of detection. In principle, the accuracy should be of a similar order because of the limited interference in the technique; however, in practice the sampling and sample preparation procedures often introduce larger errors.

A variety of techniques have been developed to increase the applicability of atomic absorption. For example, the determination of mercury by flame atomic absorption is rather insensitive. However, by reducing mercury to the elementary form, it can be vapourised in a current of air into an absorption tube and then the absorption measured in the cold state. This technique is widely used to determine the mercury level of foods and other biological materials (11). The measurement of arsenic can be improved by production of arsine from the solution (12), which can then be led directly into the flame and the absorption integrated over time, or by collection in a balloon and subsequently discharged into the flame. Either of these two techniques can determine submicrogram quantities of arsenic. Similarly, the reduction of selenium to hydrogen selenide prior to the introduction into the flame has improved the sensi-

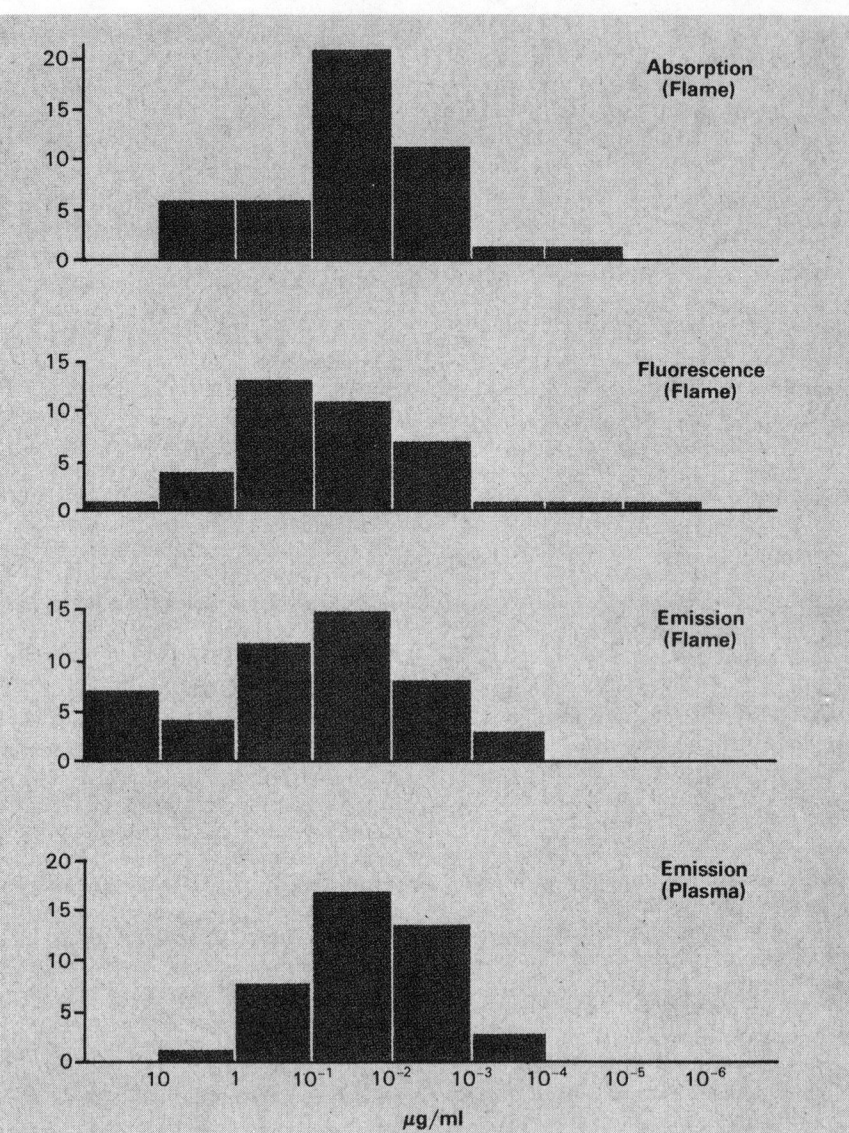

Figure 1. Limits of detection in atomic spectroscopy

tivity of the technique (13). A variety of flameless cells are now commercially available, such as graphite tube furnaces, graphite rod atomizers, and tantalum ribbon atomizers. These devices are electrically heated and produce temperatures up to 3000°K and are suitable for the less volatile refractory elements. The marked improvement in sensitivity of the graphite atomiser methods over flame sources is indicated in Table I. The precision of analysis is much worse than for flame methods, typically 5–25% depending on the type of atomiser and type of sample. The interferences have not been adequately studied, and insufficient information is available on the interaction between solution and gaseous state with carbon. Various physical interferences have been noted in some applications, apparently inhibiting the volatilisation of the sample.

Atomic absorption is widely used within the Laboratory of the Government Chemist for trace analysis of foods, water, and dust samples when the requirement is for one or a few specific elements such as lead, arsenic, cadmium, and mercury. Simultaneous multielement analysis is difficult by atomic absorption, but various

Table I. Comparison of Flame and Graphite Tube Sources

	Detection limits, ng/ml							
	Cd	Co	Cu	Mg	Mn	Ni	Pb	Zn
Flame	2	10	1	0.1	2	2	10	1
Furnace	0.0001	0.004	0.001	0.0001	0.001	0.01	0.006	0.0006

schemes have been reported (14) for the rapid sequential analysis of many elements.

Since 1964 when Winefordner and Vickers (15) demonstrated analytical applications of atomic fluorescence, many papers relating to this technique have appeared. A review by Browner (16) provides a good background description of the technique and allows the reader to make a reasoned choice between this and atomic absorption and flame emission spectrometry. At the present time, there are no commercial instruments available for atomic fluorescent spectrometry which must limit its general availability. It offers considerable improvement in detection capability for many elements, particularly those with principal resonance lines below 320 nm, but the instrumental drawbacks inherent in its nature have delayed its wider acceptance. It is essential to have a high-intensity stable source of the required wavelength. Many workers are now using thermostatic electrodeless discharge lamps, and these seem to offer the best solution at the present time. Atomic fluorescence has the same limited interference problems as absorption; in addition, quenching processes could cause problems. Most of the development work on fluorescence spectrometry uses nonflame cells since this reduces the possibility of quenching and still maintains good atomisation efficiency. Multielement analysis is more readily achieved with atomic fluorescence than absorption, and a spectrometer for the simultaneous determination of up to six elements has been described. As with most multielement techniques, the operating conditions will be a compromise.

Atomic emission has developed significantly in the last decade and is now applied to a large range of elements. The sensitivity, compared with absorption and fluorescence techniques, is indicated in Figure 1. Busch and Morrison (17) have reviewed the multielement capability of flame spectroscopy and conclude that emission is the technique most easily adapted for this purpose, although it is not the most sensitive of the three techniques for many elements.

The high-frequency plasma sources for spectroscopy first used by Greenfield et al. (18) are now becoming widely accepted, and commercial sources coupled with direct reading spectrometers are available. Scott et al. (19) have investigated plasma sources extensively and have developed a compact system ideally suited to practical multielement analysis. A pneumatic nebulizer feeds the aerosol produced from the solution directly into a plasma. The high temperature (7500°K) eliminates most chemical interferences observed with other emission sources; thus, simple standard solutions can be used to calibrate complex mixtures. The stability of the system is good, and precisions of about 1% (relative standard deviation) for sub-ppm solutions have been achieved. The detection limits with a plasma source for most elements are lower than flame atomic absorption or fluorescence. The technique seems well suited to the multielement analysis of biological samples, following removal of organic matter and dissolution of the sample.

Mass Spectrometry. The attraction of mass spectrometry for the analyst is the simplicity of the spectra. In theory, a spectrum of a mixture will contain only discrete lines for each isotopic species present, but in practice, some interference from compounds and hydrocarbons is always present. The spark source is usually used for ionisation for elemental analysis because of the relatively uniform ion formation for all elements. For analysis of environmental samples, it is usually necessary to remove the organic material and mix the residue with graphite in order to prepare conducting electrodes for the spark source. Hamilton et al. (20) have studied the procedure extensively for biological materials, and Brown and Vossen (21) have applied it to air pollution samples on filters.

Several instruments are available commercially with two-stage magnetic and electrostatic separators and ion detection by photographic plate or electrically by using photomultipliers. The method is most sensitive using photographic plates with exposure times of about 1 hr, and almost all elements can be detected down to ppb levels. Electrical detection is less sensitive but more rapid; in the scanning mode typical sensitivity limits are 0.1 ppm, taking about 10 min for examination of the full mass range, and 0.01 ppm by peak switching to examine specific elements. Direct semiquantitative analysis is possible without standards with an accuracy of about a factor of three. For quantitative analysis it is essential to measure the relative sensitivity of ionisation and detection of each element by using standards prepared in a matrix similar to the sample. In this way, the accuracy can be about ±10–25%—lower for electrical detection and higher for photographic methods.

Although the spark source has been used for many years for elemental analysis, it has serious limitations, and Gray (22) has shown that a plasma source has many desirable features for ion formation prior to mass separation. A solution of the sample to be analysed is sprayed directly into a capillary arc plasma, and the gases from the plasma plume directly enter the vacuum system through a small orifice in a thin wall. The ions are focused and directed into a second chamber with a quadrupole filter. The quadrupole mass filter permits only one mass number to pass through it, but by scanning, a complete mass range can be examined and recorded. Currently, the low mass end of the spectrum contains a complex collection of peaks owing to nitrogen, oxygen, and hydrogen compounds, but above mass 55 the background is very small.

Mass spectrometry gives the best coverage of elements at the ppb concentration range (Table II) and is well suited to studies requiring a large number of elements. The careful sample preparation necessary for spark sources limits the rate of analysis, but this may improve with the development of the plasma source.

Isotope dilution procedures are capable of giving some of the most accurate determinations for trace elements. Although not likely to be used on a routine basis for analysis of pollutants, they do offer the possibilities of providing an absolute method suitable

Table II. Typical Detection Limits in Mass Spectrometry (ng/g)[a]

Element	Spark source	Plasma source
Ag	2	2
As	0.6	10
Cd	7	5
Co	0.5	0.2
Cu	2	0.4
K	0.3	0.5
Mg	0.2	0.5
Mn	0.4	0.5
Pb	3	0.2
Zn	2	5

[a] For spark source a 10-g sample was ashed and mixed with carbon to form electrodes; for plasma source residue after ashing was dissolved in 2 ml and sprayed into plasma.

Table III. Comparison of Analytical Methods for Pollutants

Technique	Advantages	Limitations
Microscopy	Excellent for particulate matter Identification of compound and crystalline forms possible Very sensitive Rapid	Not suitable for pollutants in solution Not readily quantified Requires specialised skills and training
Atomic absorption and fluorescence	Applicable to more than 60 elements Simple spectra and instrumentation Sensitive Rapid	Simultaneous multielement analysis difficult
Atomic emission	Multielement analysis feasible Sensitive Rapid Plasma source eliminates most chemical interferences	Flame sources have significant matrix effects
Mass spectrometry	Multielement technique Excellent sensitivity Plasma source promises rapid analysis	Specialised sample preparation techniques for spark source Plasma source still in development phase
Neutron activation analysis	Freedom from contamination Excellent sensitivity for some elements Applicable to wide variety of matrices	Limited availability for routine analysis Total time for multielement analysis may be long
X-ray fluorescence	Direct examination often feasible May be nondestructive Multielement technique Rapid	Limited sensitivity Particle size effects must be carefully controlled
Anodic stripping voltammetry	Simple technique and equipment Excellent sensitivity for some elements	Not suitable for wide range of elements

for the standardisation of secondary techniques. The principle of the method is very simple; the concentration of an element is determined from the change produced in a natural isotopic composition by the addition of a known quantity of the same element, the isotope ratio of which has been artificially altered. A given amount of sample is equilibrated with a known quantity of an isotopically enriched spike; the element is extracted, and its new isotopic ratio measured by mass spectrometry. Since only isotopic ratios are involved, the extraction does not have to be quantitative. Isotope dilution techniques are capable of accuracies of the order of 0.1% for many elements at trace concentrations. The National Bureau of Standards has already used the technique to standardise the determination of calcium in blood and also lead in blood. It is hoped that through certified reference materials and methods such as isotope dilution, it will be possible to obtain compatible data for these elements throughout many laboratories.

X-ray Spectrometry. X-ray fluorescence spectrometry is well suited to the determination of the elemental composition of pollutants because in many cases, limited sample preparation is required, and the technique is nondestructive. All elements of atomic number greater than 11, i.e., sodium, can be analysed. Traditionally, the technique has used wavelength dispersion as a method of separating the X-rays emitted from the sample, but with the recent availability of silicon and germanium X-ray detectors of high resolution, energy dispersive spectroscopy is possible and often simpler. Gilfrich et al. (23) have reviewed X-ray spectrometry for particulate air pollution samples and conclude that there are two problems with solid-state detectors which are limiting in that application. First of all, the resolution is not adequate to separate the K-alpha line of one element from the K-beta line of the next lower atomic number in the region of sulphur to nickel; thus, all these elements will require mathematical unfolding to determine X-ray intensities. Also, the solid-state detectors are limited to about 10^4 counts/sec at their best resolution, and long count times would be necessary to get adequate counting statistics for multielement analysis. Because of these limitations, they conclude that large-scale quantitative analysis requires the resolution obtainable from crystal spectrometers, and for routine analysis the only practical approach is the use of multichannel wavelength spectrometers, because of the time required for the wavelength scanning mode.

At our own laboratory we have found X-ray spectrometry very useful for the analysis of food samples; the sample preparation is simple, merely requiring freeze drying followed by grinding at liquid nitrogen temperatures in order to produce a fine powder suitable for presentation to the X-ray beam. It has been particularly useful for light elements, e.g., sulphur and phosphorous, which are difficult to determine by most spectrographic techniques.

The limit of detection for most elements by wavelength dispersive spectroscopy is in the range 0.1–1 µg, with concentration limit depending on the sample size, shape, and physical form. X-ray spectrometry is mainly used for multielement analysis but often does not have adequate sensitivity for all the elements of interest in biological samples.

Neutron Activation Analysis. The radiation of a sample with neutrons transforms some of the atoms of each element present into a radioactive form. These radioactive atoms can be separated and detected more easily than stable atoms. For many years activation analysis was probably the most sensitive analytical technique, but this is no longer the case, and the merits of the technique must be assessed on other properties such as cost, availability, and range of elements analysed. It is not easy to sum up activation analysis because its performance depends upon many different attributes—first of all, the specific nuclear properties of the element concerned, the equipment available for measurement of induced activities, and the nature of the sample itself. One of the prime advantages of activation analysis is its freedom from contamination, provided the sample is not contaminated prior to irradiation; then, one does not have to worry about subsequent contamination from reagents.

After irradiation the measurement

of induced activity may: use direct instrumental methods for discriminating between the radiations emitted from each element present; rely on complete radiochemical separation with simple nonspecific detectors; or use group separations, e.g., Goode et al. (24), to reduce the complexity of each sample to more manageable proportions. The improvement in the sensitivity and resolution of germanium detectors over the last five years has encouraged more use of the direct method. However, for biological samples the intense radiation from sodium and chlorine usually makes some preliminary separation essential if a wide range of elements is to be determined.

An example of how the technique can be applied to analysis of pollutants is given by Pierson et al. (25) in their work on the composition of rain and dust in the atmosphere. The samples of filter paper or rain were irradiated for 20 sec; then the gamma spectrum of each sample was measured a few minutes after irradiation and at further times over the subsequent 24 hr, with sodium iodide detectors. The spectra were compared with those from standards by the method of linear regression analysis with a computer; about eight elements could be determined in this way. Samples were then irradiated for one week, and gamma spectra obtained over a period of several weeks with germanium detectors, and a further 20 elements determined. This example indicates direct access to a reactor for short and long irradiation; a variety of detection systems and computer data processing are necessary for successful routine multielement analysis by activation techniques. If these requirements cannot be met, then other techniques are likely to offer a better solution to the analytical problem.

Electrochemical Methods. Polarographic methods of analysis have been used for trace analysis of specific trace metals for many years, but the technique has many limitations and has not been so widely used as many spectrographic techniques. However, the related technique of anodic stripping voltammetry has been applied, increasing in recent years to the analysis of water samples and other solutions. The sensitivity of the technique is obtained by initially concentrating the trace elements into a small volume of mercury by electrolysis, at a more negative potential than the reduction potential of the trace element, over a period of 10–60 min. The element is then anodically stripped from the mercury in a few seconds with a linearly varying potential. The current peaks for the metal oxidation of the elements in the mercury are displayed on an X-Y recorder. The technique is particularly useful for elements such as copper, lead, zinc, and cadmium and can be extended to about 12–20 elements, although it is unrealistic to expect more than about six elements to be determined simultaneously in a solution. Multicell units are available commercially and, of the order of 100 samples per day, can be analysed with a precision of 5–10% at the 10–100 nanogram level.

Conclusions

In this review the most practical methods of analysis for pollutants have been discussed. Table III summarizes the advantages and limitations of the various technqiues. There may have been a bias toward multielement techniques, because I believe that most laboratories in the future will need to consider a wider range of elements. A variety of techniques of adequate sensitivity for the trace analysis of bulk samples are available, and it is important to relate the analytical problem to the overall requirement so that a rational and economical choice can be made.

Acknowledgment

The author acknowledges useful discussions with many members of the staff of the Laboratory of the Government Chemist but particularly R.G. Stone, D.E. Henn, S. Isherwood, and K.L.H. Murray.

References

(1) D.W. Spencer and P.G. Brewer, *CRC, Crit. Rev. Solid State Sci.,* **1,** 401 (1970).
(2) D.E. Robertson, *Anal. Chim. Acta,* **42,** 533 (1968).
(3) E.I. Hamilton and M.J. Minski, *Environ. Lett.,* **3** (1), 53 (1972).
(4) P.A. Cawse and D.H. Pierson, AERE Report R-7134, 1972.
(5) J.M. Harries, C.M. Jones, and J.O'G. Tatton, *J. Sci. Food Agr.,* **20,** 242 (1969).
(6) G. Tolg, *Talanta,* **19,** 1489 (1972).
(7) E.I. Hamilton, M.J. Minski, and J.J. Cleary, *Sci. Total Environ.,* **1,** 1 (1972).
(8) W.C. McCrone and J.G. Delly, "The Particle Atlas," Vols 1–4, 2nd ed., Ann Arbor Sci. Publ., Ann Arbor, Mich., 1973.
(9) J.D. Winefordner, V. Svoboda, and L.J. Cline, *CRC, Crit. Rev. Anal. Chem.,* **1,** 233 (1970).
(10) R.G. Stone and J. Warren, *Lab. Equip. Digest,* 49 (April 1974).
(11) W.R. Hatch and W.L. Ott, *Anal. Chem.,* **40,** 2085 (1968).
(12) E.F. Dalton and A.J. Malanoski, *At. Absorption Newslett.,* **10** (4), 92 (1971).
(13) E.N. Pollock and S.J. West, *ibid.,* **12** (1), 6 (1973).
(14) L.E. Ranweiler and J.L. Moyers, *Environ. Sci. Technol.,* **8,** 153 (1974).
(15) J.D. Winefordner and T.J. Vickers, *Anal. Chem.,* **36,** 161 (1964).
(16) R.F. Browner, *Analyst,* in press (1974).
(17) K.W. Busch and G.H. Morrison, *Anal. Chem.,* **45,** 713 (1973).
(18) S. Greenfield, I.L.W. Jones, and C.T. Berry, *Analyst,* **89,** 713 (1964).
(19) R.H. Scott, V.A. Fassel, R.N. Kniseley, and D.E. Nixon, *Anal. Chem.,* **46,** 75 (1974).
(20) E.I. Hamilton, M.J. Minski, and J.J. Cleary, *Sci. Total Environ.,* **1,** 341 (1972/73).
(21) R. Brown and P.G.T. Vossen, *Anal. Chem.,* **42,** 1820 (1970).
(22) A.L. Gray, *Proc. Soc. Anal. Chem.,* **11,** 182 (1974).
(23) J.V. Gilfrich, P.G. Burkhalter, and L.S. Birks, *Anal. Chem.,* **45,** 2002 (1973).
(24) G.C. Goode, C.W. Baker, and N.M. Brooke, *Analyst,* **94,** 728 (1969).
(25) D.H. Pierson, P.A. Cawse, L. Salmon, and R.S. Cambray, *Nature,* **241,** 252 (1973).

Ronald F. Coleman was appointed head of Research and Special Services at the Laboratory of the Government Chemist in 1973. The laboratory provides analytical service and advice on chemical matters to government departments in the United Kingdom, particularly in the fields of revenue and environmental protection. He obtained a BSc degree from the University of London in 1953 and then spent nearly 20 years at the Atomic Weapons Research Establishment Aldermaston. His research has been mainly connected with nuclear chemistry but he has wandered into many different areas in collaboration with other workers. In the late 1950's, he was studying fast neutron and heavy ion reactions and simultaneously used fast neutrons for activation analysis. In 1963 he started using thermal neutron activation techniques to solve many problems in forensic science, geochemistry, and medicine. He has published about 50 papers in nuclear and analytical chemistry.

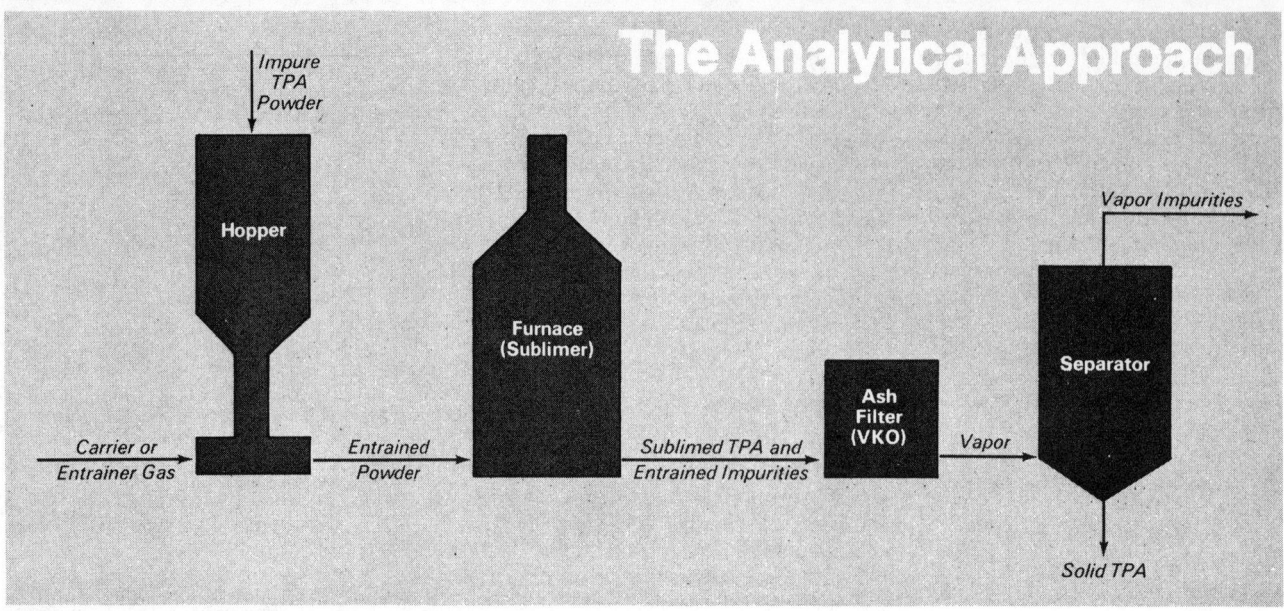

Figure 1. Terephthalic acid sublimation purification process

Urgent Production Problem Solved by Unique Capabilities of Analytical Chemists

Claude A. Lucchesi
Contributing Editor

Analytical chemists at the Mobil Chemical Co. were confronted with a challenging problem during the start-up stages of the company's terephthalic acid (TPA) pilot plant. The Mobil TPA process consisted of two parts, an oxidation section and the purification section which is illustrated in Figure 1. Dry, impure TPA powder from the oxidation section was fed into a hopper where it became entrained in a high-velocity carrier gas and passed into a furnace where substantially all of the solids were vaporized. The effluent from the furnace was passed through an ash filter where entrained solids, including catalyst residues from the oxidation section, were removed from the vaporized material. The vapor then entered a condenser for fractional condensation of the solid TPA product.

The TPA was being produced as a potential replacement for dimethyl terephthalate (DMT), then the only source of the dicarboxylic acid in many polyester films and fibers. TPA had an expected cost advantage over DMT, provided it could be produced at an equivalent level of purity which was judged by a set of specification tests. One specification test involved the color of a 5% solution of TPA in dimethylformamide (DMF). And this is where the problem started. Its solution, in time to be of practical value to the company, required the integrated efforts of a team of specialists in IR, X-ray diffraction, X-ray spectrography, arc-spark spectrography, thermal methods, gas chromatography, and solution chemistry. In addition, a literature searcher was needed to find a synthesis procedure for the tentatively identified material to "cinch" an identification.

An early pilot plant run in Beaumont yielded TPA product which, not only failed the color test, but also did not completely dissolve in the DMF. The question was, "What is the DMF-insoluble material?" This was the question I was asked to answer when I was with Mobil Chemical as manager of the Analytical and Physical Chemistry Department. The then vice president of the Research and Development Division phoned and told me to go to Beaumont and find out what that DMF "turbidity" was. Although the immediate question was the identity of the turbidity, the critical question was how to prevent it, whatever it was, from forming. The problem had top priority, and for several weeks most of the specialists in the department did little else. Five pounds of TPA product which failed the DMF test was requested for the Research Laboratory in Metuchen, N.J., and I went to Beaumont, Tex., to become familiar with the pilot plant operation and to obtain test samples for study.

Nondestructive Tests Used to Survey Test Samples

Because only a few milligrams of the DMF turbidity could be isolated from several pounds of TPA product, initial tests on the DMF-insoluble material were limited to the nondestructive techniques readily available in our lab: X-ray diffraction, X-ray spectrography, and infrared spectroscopy. The same measurements were made with the Beaumont test samples, and the material removed from the ash filter and the DMF turbidity from the TPA product were almost identical (Figure 2). The two materials had identical infrared spectra and nearly identical X-ray diffraction patterns. Also, the X-ray fluorescence spectrographic measurements showed that the two materials contained the same metallic elements in roughly the same concentration ratios. These findings enabled us to do subsequent work with the pound or so of ash filter material rather than with the limited amount of DMF-insolubles.

Identification of Inorganic Part

The nature and concentration of the inorganic materials in the ash filter sample were established to the extent justified by the nature of the sample as illustrated in Figure 3. From all the data shown, it was estimated that about half of the inorganic material was $CaSO_4$. Thus, at most, only 3 or 4% of a metal was available to form a salt or chelate with TPA. Consequently, the data on the inorganic part of the ash filter sample supported the IR conclusion that the main constituent of the ash filter sample (and the DMF insolubles) was not primarily a salt or chelate of TPA. (For example, Ca(TPA) contains 11% Ca.)

Cobalt, iron, and calcium were found in the DMF-insolubles by X-ray spectrography. The ratios of the three metals in the DMF-insolubles and in the ash filter sample and the concentrations in the ash filter sample suggested that less than 0.1% cobalt was in the DMF-insolubles. Consequently, the DMF-insolubles could not have been a cobalt salt or chelate. The same can be said for iron.

Identification of Organic Part

Having convinced ourselves that only a minor part of the DMF-insolubles could have been of an inorganic

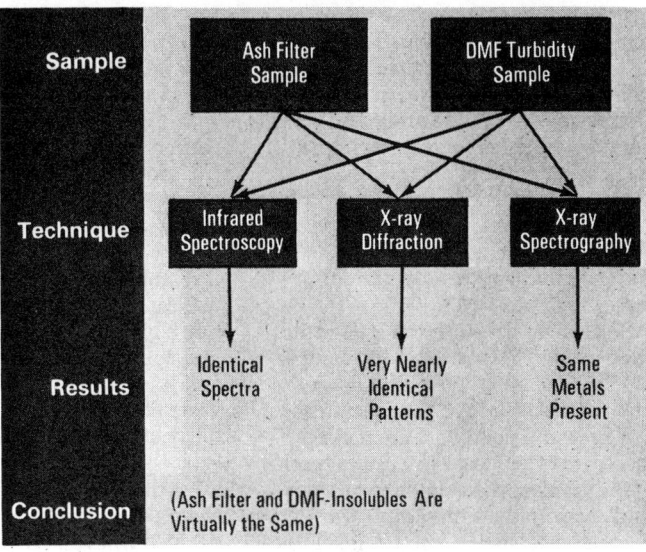

Figure 2. Nondestructive methods used to show DMF turbidity and ash filter material were virtually the same

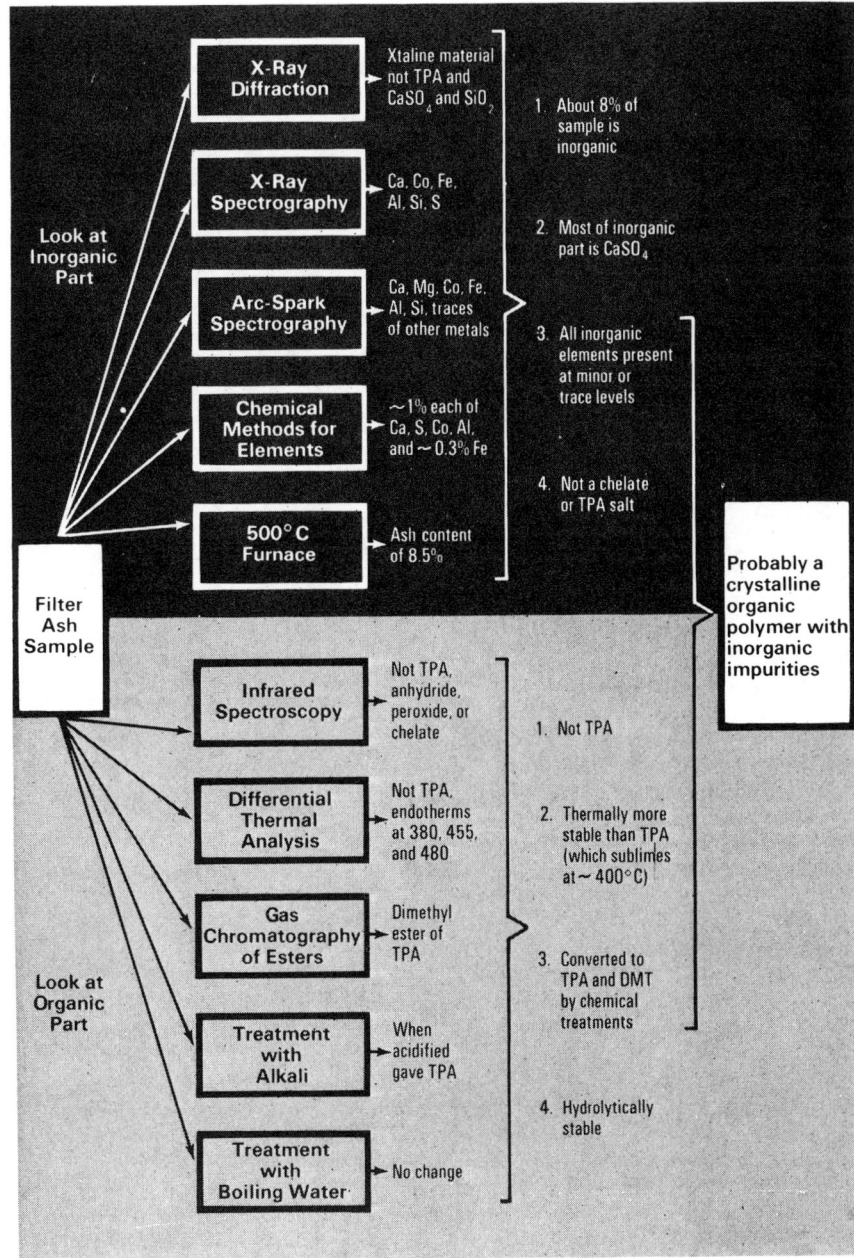

Figure 3. Techniques for inorganic and organic substances used in combination to show what material could and could not be

nature, we concentrated on the organic part of the problem. The ash filter sample and several TPA samples were treated with methanol to produce methyl esters and other volatile substances which were extractable with chloroform and could be passed through a gas chromatograph. Although only about 20% of the sample was esterified, almost all of the material that got through the chromatograph was DMT.

Because of the gas chromatographic observation that the ash filter sample yielded only about 20% TPA ester when carried through the esterification procedure and because the IR spectrum of the residue was the same as the starting material, a hydrolysis study was made. The sample was treated with NaOH, and the insoluble residue was separated by filtration, dried, and weighed. When the filtrate was acidified, it gave a precipitate which was identified as TPA. About 62% TPA was recovered with the first treatment. After four treatments, 83% was recovered. This gave a residue of 17% which was in the ball park with the 8.5% ash figure obtained earlier.

Unfortunately, we did one more experiment that caused unnecessary confusion. We boiled the ash filter sample in water for at least 2 hr, and it remained insoluble. Nevertheless, we came to the conclusion that we had a highly crystalline polymer that was hydrolyzed with base or acid to give TPA and could be partially converted to DMT by a conventional esterification procedure.

At this point, I had a meeting with all of the specialists who worked on the problem, and we systematically decided what the DMF-insolubles could *not* be. The only possibilities we had left was an anhydride or a peroxide, and we concluded that the TPA more likely had formed an anhydride. But it was difficult to convince our organic chemists that TPA most likely had formed an anhydride. Consequently, one of our literature searchers went to the Chemists' Club Library in New York City and found a 1959 German article describing the synthesis of TPA anhydride. We translated the article and followed the recipe. The IR spectrum of the synthesized material matched the IR spectra of both the ash filter sample and of the DMF-insolubles. The X-ray diffraction patterns also matched.

When we were convinced that the DMF-insolubles indeed were poly-(TPA anhydride), we suggested to the engineers that steam be used as a carrier gas instead of nitrogen to prevent the formation of the anhydride. This solved the problem: no more DMF-insolubles and no more color.

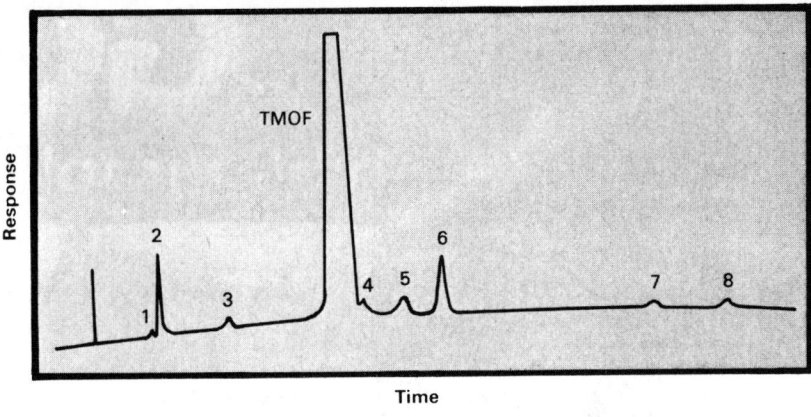

Figure 1. Gas chromatogram of trimethylorthoformate (TMOF)

Effects of Raw Material Change in Manufacturing Process Resolved

John Mitchell, Jr.
E. I. du Pont de Nemours and Co.
Plastics Department
Experimental Station
Wilmington, Del. 19898

Current shortages of many chemicals present a new challenge to the analytical chemist. Product quality in many industrial processes requires complete knowledge of compositions of chemical intermediates, including type and concentration of trace impurities. Active species may seriously affect product yield as well as product quality. This problem becomes acute when a plant finds it necessary to change the source of supply of a raw material, because maintenance of production requires rapid assessment of the nature of process stream impurities. A special emergency task force may be necessary to bring together the analytical expertise vital in resolving the effects of a raw material change on the manufacturing process. A plant faced such a problem in looking for another source of trimethylorthoformate used in a complex manufacturing process. Identification and determination of low levels of impurities were essential, requiring a multitechnique approach. Complete analysis required gas chromatography, infrared and ultraviolet spectrometry, nuclear magnetic resonance, mass spectrometry, distillation, and chemical methods.

Preliminary GC studies (Figure 1) indicated at least eight impurities. As indicated in Figure 2, GC/MS served to identify GC peaks 1, 2, and 3 as methyl formate, 2-methoxyethanol, and methanol, respectively. Compo-

Table I. Analytical Data on GC Peak No. 7

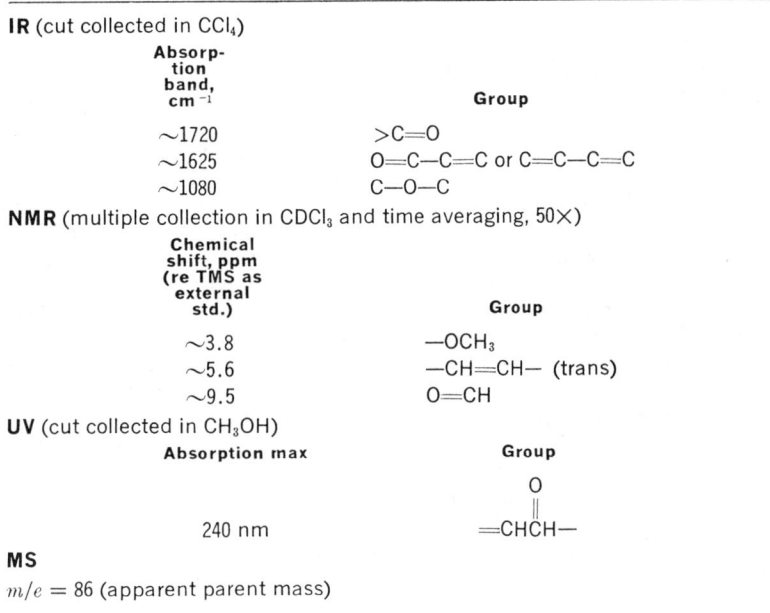

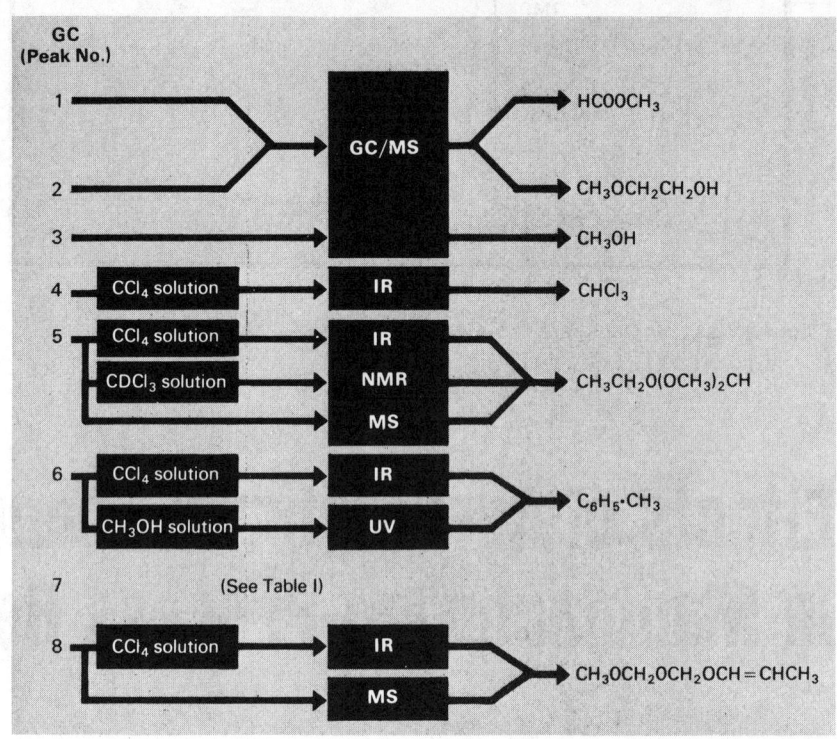

Figure 2. Identification of components separated by gas chromatography

nent 4 was identified as chloroform with the aid of IR. Fractions representing component 5 were collected in CCl_4 and in $CDCl_3$ for IR and NMR studies and MS was obtained directly. The spectra indicated ethyl dimethylorthoformate. Component 6 was identified as toluene. From GC, MS, and IR data, component 8 was identified as methoxymethoxymethyl-1-propenyl ether.

Component 7 proved to be the most challenging. Collection of analytical information is shown in Table I. Infrared spectra indicated C=O groups, conjugated unsaturation, and ether groups. NMR (with time averaging) showed methoxy, unsaturated methylene, and aldehyde group protons. Mass spectrometry indicated a molecular ion of 86; ultraviolet, an unsaturated aldehyde. The 2,4-dinitrophenylhydrazone (2,4-D) derivative, precipitated from ethyl alcohol solution, had a molecular weight of about 432 (Table II). It was orange in color and decomposed on heating, indicative of a dialdehyde. Elemental analysis of the 2,4-D supported the compound malondialdehyde. However, the data were inconsistent with direct results by IR, NMR, UV, and MS. Repeat of the direct analyses of the GC fractions verified the original results, including a methoxyl group. The suggested structure was $C_4H_6O_2$ (MW = 86), $CH_3OCH=CHCHO$ (3-methoxyacrolein). Formation of the di-2,4-D-derivative could be explained from cleavage of the methoxyl group in the strongly acid reagent to form the enol. The enol, in turn, would be expected to rearrange to the dialdehyde as the derivative was formed:

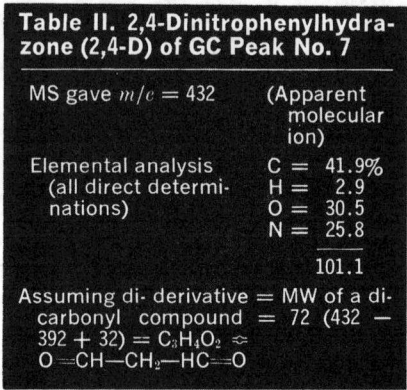

In summary, the eight compounds shown in Table III were identified. They are arranged in order of elution from the GC column.

In conclusion, the progressive analytical group associated with an industrial research organization combines expertise with versatility. Emphasis is on problem solving by the most efficient and effective means. Specificity is all important in detecting substances in concentrations varying from ppb to major amounts. The analytical chemist is presented with the challenge of contributing to special chemical structure needs in research, production, and marketing. This challenge is now all the more exciting as we must provide reliable results with respect to EPA, OSHA, and FDA requirements.

Table II. 2,4-Dinitrophenylhydrazone (2,4-D) of GC Peak No. 7

MS gave $m/e = 432$	(Apparent molecular ion)
Elemental analysis (all direct determinations)	C = 41.9% H = 2.9 O = 30.5 N = 25.8 ——— 101.1
Assuming di- derivative = MW of a dicarbonyl compound = 72 (432 − 392 + 32) = $C_3H_4O_2 \Leftrightarrow$ O=CH—CH$_2$—HC=O	

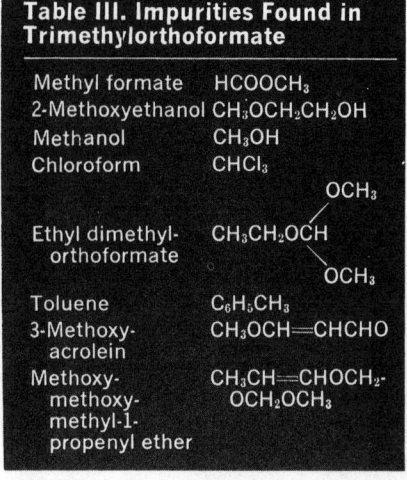

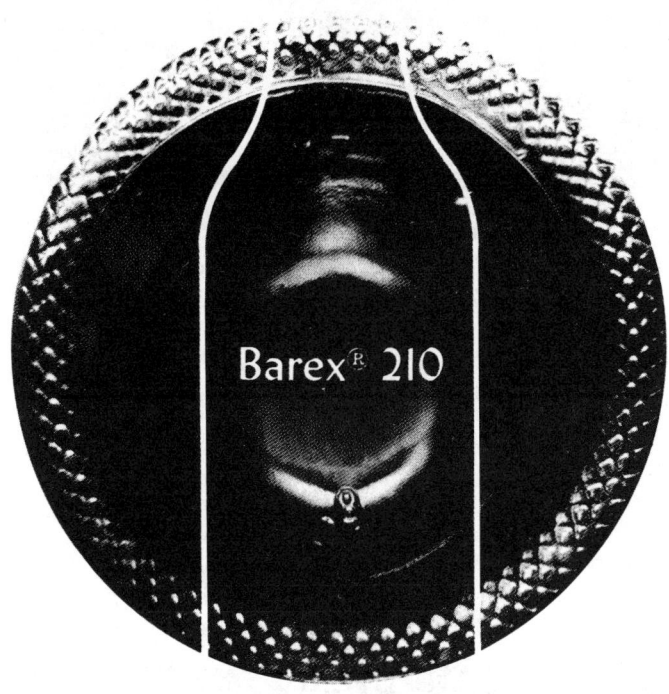

Analytical Chemists Vital in Commercialization of New Food Packaging Material

V. F. Gaylor

The Standard Oil Company (Ohio)
4440 Warrensville Road
Cleveland, Ohio 44128

Discovery and development of a new family of barrier resins several years ago plummeted Sohio analytical chemists into an unusual problem-solving area. The thermoplastic, impact-resistant resins developed by the polymer research chemists were very effective barriers against transmission of oxygen, carbon dioxide, and most other vapors. Food packaging was thus a logical marketing goal for Sohio's first commercialized resin. Concurrent with this marketing decision, management recognized the need for obtaining FDA approval for the new resin, trade named Barex® 210, and the importance of analytical chemistry in obtaining it. Analytical research on this problem was therefore initiated early in the development program and became a vital part of the whole commercialization process as shown in Figure 1.

The extent of the analytical work is indicated by the following figures. At least 1,632 Barex 210 bottles were ex-

Figure 1. Role of analytical research in commercialization of new food packaging material

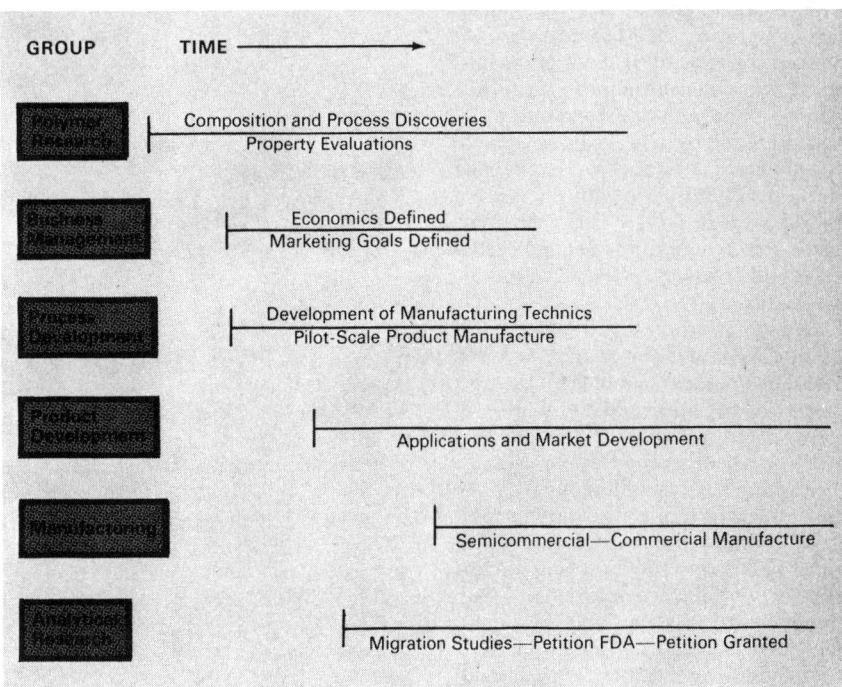

tracted, 302 liters (80 gal) of ultrapure water was used, 168 liters (44 gal) of extract was slowly evaporated from 100-ml evaporating dishes, and 1,053 analytical determinations were made.

Organization

Responsibility for obtaining an FDA regulation for Barex® 210 was delegated to a team representing three different disciplines. The team and its interaction with FDA and with the appropriate parts of the company are shown in Figure 2. The three team members represented a spectrum of expertise in administrative law; resin composition, properties, and processing characteristics; and instrumental and chemical analysis technics. Each member of this multidiscipline team had access to the total scientific resources of the R&D organization; thus, good two-way communication with all the various scientific and business groups involved in the resin development system was insured. Additionally, the team took advantage of advice and help available from FDA officials in the Petitions Control Branch of the Bureau of Foods. Invaluable advice on the required analysis program and on the supporting documentation requirements was received. The information developed in these joint meetings also helped the team guide process development pertaining to specific ingredients of the resin, i.e., potential migrants, and associated limitations.

Requirements for Food Packaging Regulation

Before regulating a new food packaging material, the FDA must be convinced that no harmful materials migrate from the container to the food. Migration levels are determined experimentally by contacting or extracting the packaging material with food or food simulating solvents. The exposed foods or solvents are then analyzed for any migrants, i.e., indirect food additives, extracted from the packaging material.

Migration studies on our food packaging candidate, Barex® 210, were carried out in bottles made from the new resin and with the food simulating solvents listed in Table I. The solvents were "cooked" in the resin bottles at 125° or 150°F to equilibrium, i.e., until migrant levels measured in the solvents showed no increase with time.

The complete program consisted of the sequential steps outlined in Figure 3. Exploratory extraction experiments defined temperatures and approximate equilibrium times for each food simulating solvent. Nonvolatile mi-

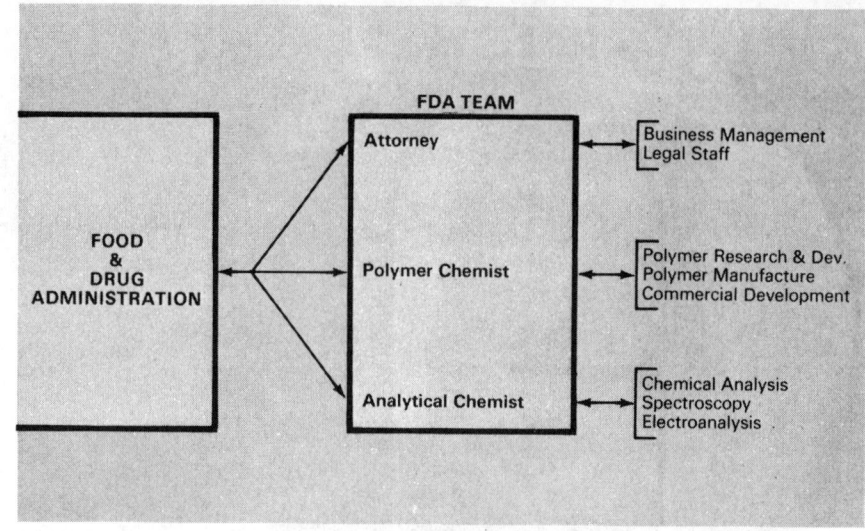

Figure 2. Organization approach for obtaining FDA approval

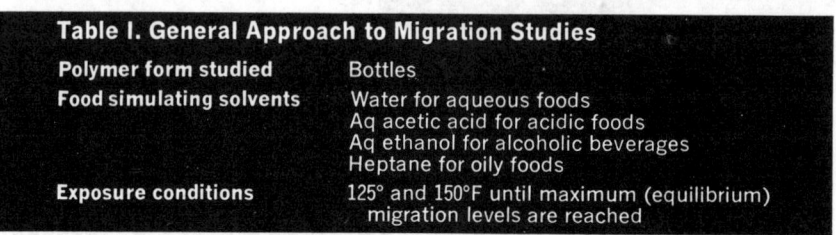

Table I. General Approach to Migration Studies

Polymer form studied	Bottles
Food simulating solvents	Water for aqueous foods Aq acetic acid for acidic foods Aq ethanol for alcoholic beverages Heptane for oily foods
Exposure conditions	125° and 150°F until maximum (equilibrium) migration levels are reached

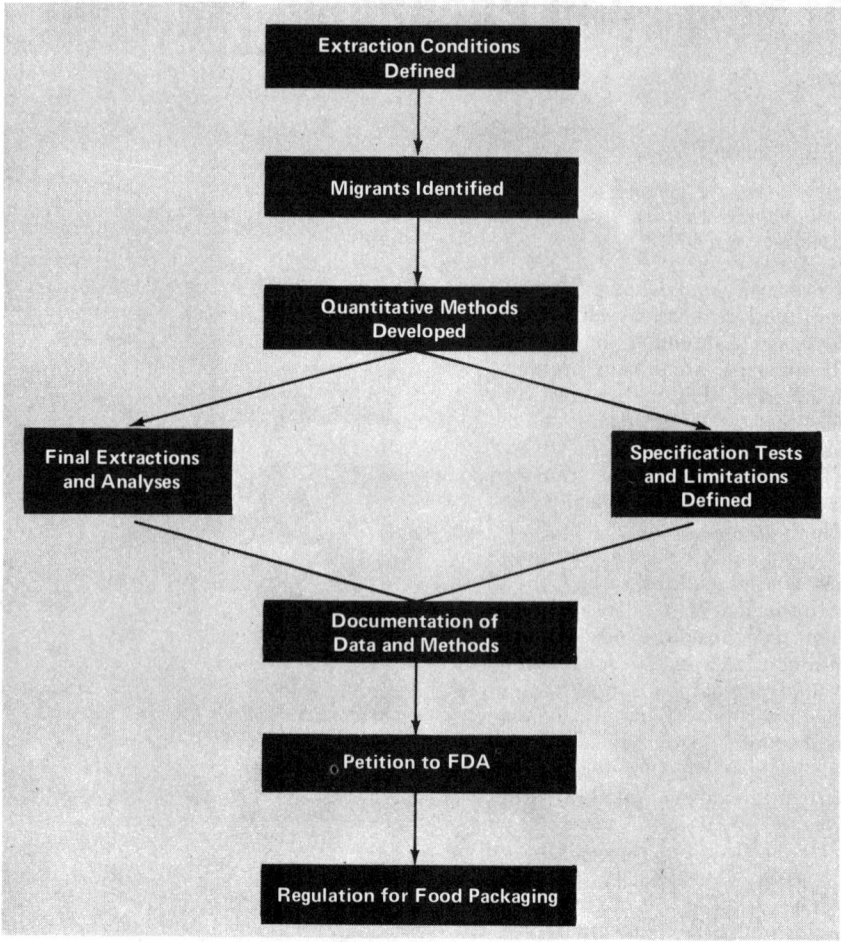

Figure 3. Sequence of FDA approval project

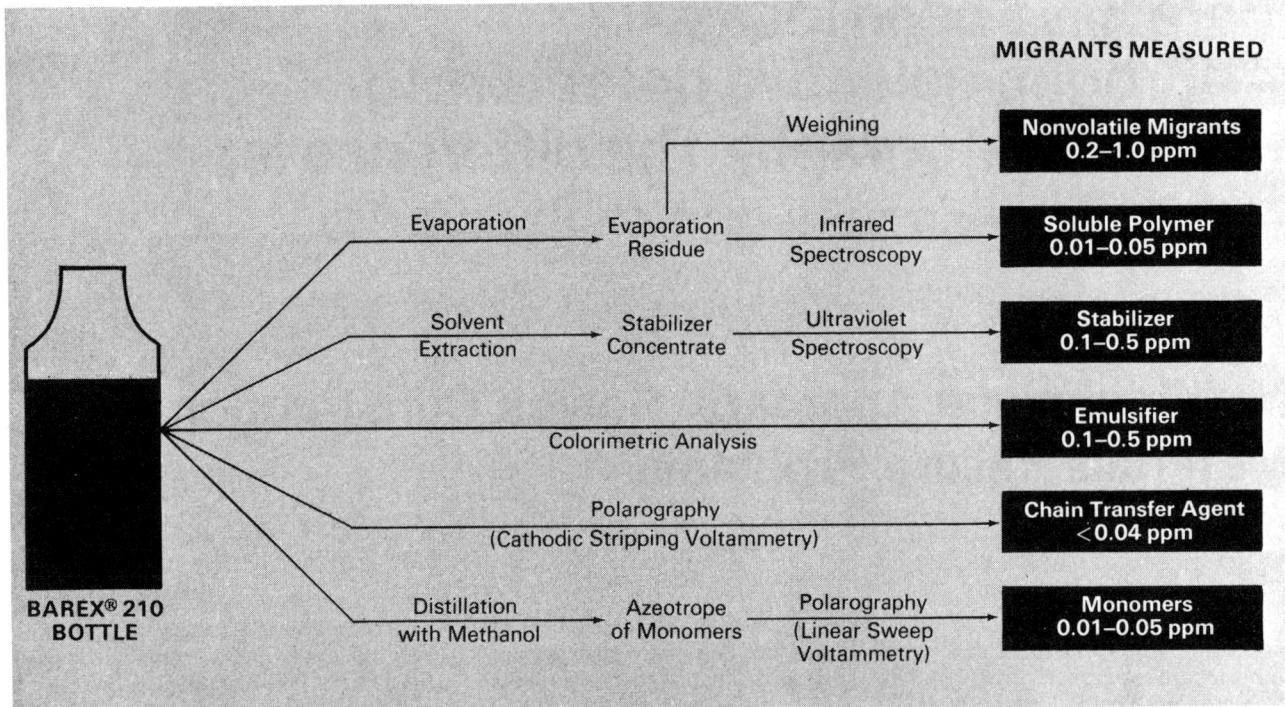

Figure 4. Analytical methods and results

grants were identified qualitatively by IR and UV spectrometry inspection of evaporation residues. Methods for quantitative measurements of both total and single migrants were then developed, in preparation for equilibrium extraction studies on several different resin batches. Quality control tests and specifications for a food-grade resin product were defined concurrently by relating bulk properties to results of the migration studies. The formal petition for the food packaging regulation contained the complete results of all the analytical studies, along with written procedures and copies of original records.

Analytical Methods and Results

Requirements for analyzing extracts of a new food packaging material partly depend on composition. As a minimum, the FDA requires measurement and identification of total nonvolatile extractables. Nonvolatile extractables of Barex® 210 were primarily emulsifier and stabilizer, both FDA-regulated food additives. Each of these was measured quantitatively by spectrophotometric procedures. We were also required to analyze the extracts for monomers, polymer, and chain transfer agent.

The complete analysis system is outlined in Figure 4. Total nonvolatile migrants were measured by weighing evaporation residues, as required by the FDA. Nonvolatile migrants from Barex® 210 totaled less than 1.0 ppm in most cases. Gravimetric measurement of these low levels required the highest standards of solvent purity, clean room handling technics, and a controlled humidity atmosphere for tare and final weight measurements.

The amount of polymer in the evaporation residue was measured by infrared spectrometry. A considerable amount of technic development was needed to develop a quantitative IR method. The evaporation residues were often invisible to the eye and, at best, looked like stains in the platinum evaporating dishes. Quantitative transfer for IR analysis was achieved by redissolving the "stains" and evaporating the solutions on KBr. Analysis of the resulting KBr pellet for the low levels (<0.05 ppm) of migrating polymer required 10× scale expansion on a high-resolution, grating spectrophotometer.

The higher levels (0.1–0.5 ppm) of stabilizer and emulsifier were comparatively easy to measure. Some method development was required to obtain adequate sensitivity for stabilizer measurement by UV spectrometry. This problem was solved by a solvent extraction step in which the stabilizer was concentrated by a factor of ten.

Polarography was employed to measure both chain transfer agent and monomers. The chain transfer agent is a mercaptan, and we used the polarographic anodic depolarization wave of the sulfhydryl group at a mercury electrode. No (<0.04 ppm) chain transfer agent was detected in any of the bottle extracts. Migration levels of the two monomers, acrylonitrile and methyl acrylate, were less than 0.05 ppm in every case. Measurement of these low levels by polarography required a concentration step. Both monomers formed azeotropes with methanol and were concentrated in the first few milliliters of distillate. In the distillation step the monomers also are isolated from possible traces of interfering alkali metals.

The migration studies showed that Barex® 210 is a safe food packaging material. Regulation 121.2614, entitled "Nitrile Rubber Modified Acrylonitrile–Methyl Acrylate Polymers," was published in the Federal Register on June 11, 1970, and is now part of Title 21 of the Code of Federal Regulations. Publication of the regulation cleared the way for commercial production and large-scale market testing of the new resin, and both events followed shortly thereafter.

Acknowledgment

Direct contributions as part of the "FDA team" were made by J. F. Jones, attorney, and B. F. Vincent, Jr., polymer chemist. Members of the larger analytical team who made equally important contributions include J. G. Grasselli, M. C. Helms, I. P. Horner, N. J. Meyer, C. Paxton, and M. K. Snavely.

William G. Hime
Erlin, Hime Associates
811 Skokie Boulevard
Northbrook, Ill. 60062

Multitechnique Approach Solves Construction Materials Failure Problems

Microscopists and analytical chemists at the Erlin, Hime Associates Laboratories are regularly confronted with problems involving failures of construction materials: cement, concrete, metals, paints, and coatings. The general approach to the solution of these problems involves initial study of the material by the techniques of petrographic microscopy to discover the mechanism of the failure, followed by the application of analytical techniques to ascertain the causative agents. The techniques typically used and the kinds of information obtained are listed in Table I.

It is estimated that over 90% of construction materials failure problems can be solved by the approach suggested in Table I, provided the microscopist–chemist team is expert enough in the chemistry and behavior of construction materials to know what to look for. For example, the presence of very large quantities of many substances has little effect on the properties of cement or concrete, but very small quantities of others cause enormously deleterious effects. To illustrate, silica in the form of quartz can be present as the major concrete component. But silica in the form of opal must be limited to a few percent. Even more powerful in their immediate effect are certain organic substances which at a thousandth of a percent level affect the setting, workability, or strength of concrete. The following three examples of "failure analyses" illustrate the approach.

"Unset" Concrete—Getting the Lead Out

When concrete forms were removed on a large construction project in New York, everyone held their breath. Occasionally, the concrete came pouring out. The uncertainty finally dictated that the multimillion dollar project be halted until the cause for the failure-to-set problem was determined and corrected. By the time a sample of the concrete was received in the laboratory, the "unset" concrete had already hardened. Microscopical analysis revealed unusual, thin rims on the cement particles, suggesting an excessive amount of a cement set-retarder.

Since sabotage had been suspected, the concrete was analyzed for sugar—a known set-retarder. (A cup of sugar can delay the set of yards of concrete for weeks.) Colorimetric methods did not detect sugar; therefore, other extracts of the concrete were then analyzed by infrared and ultraviolet spectroscopy for known cement hydration retarders, such as other polysaccharides and lignosulfates. These results also were negative.

X-ray fluorescence measurements were then made. Trace quantities of lead and zinc were detected. Since experience has indicated that quantitative analyses of quite varied materials are made more accurately by atomic absorption, AA determinations for lead and zinc were performed, and about 0.03% of each was found. Such quantities, when present as alkali-soluble compounds, are known to delay severely cement hydration.

Further work resolved the mystery. A dredged river gravel was being used

Table I. Techniques Used for Hardened Concrete

Technique or method	Information obtainable
Light microscopy	Air-void system Aggregate-composition, texture, classification, reaction rims Proportions of aggregate and paste Cracking patterns Identification of solid admixtures Extent of cement hydration Composition, fineness, and dispersement of relic cement particles Identification of hydrated cement compounds Identification and location of secondary compounds Detection of "unaccommodative" chemical reactions Physical properties of the paste such as hardness, granularity, porosity, density
Atomic absorption	Quantitative analyses of "oxides" present in cement and concrete
Infrared spectroscopy	Identification of organic admixtures (air-entraining, set-retarding, and workability agents)
Wet-chemical analysis	Cement content Chemical composition of aggregate Chemical composition of paste Chemical composition of secondary compounds Detection of some organic substances
X-ray diffractometry	Aggregate mineralogy Identification of secondary compounds Identification of hydrated and unhydrated cement compounds
X-ray fluorescence	Identification and relative proportion of elements present in aggregate

as the aggregate for the concrete. A thin band of the "New Jersey" lead and zinc deposit passed across the river. The dredging operation thus accounted for the sporadic occurrence of these elements in the concrete.

Holey Concrete

The quality of concrete is usually monitored by compression tests of samples taken during the "pour." Unfortunately, a lot of construction may take place before the initial (usually three-day) results become available. Thus, when tests on a large road paving project indicated strengths of 50% below requirements, all work was stopped while the laboratory team worked on the problem. The general approach taken in a problem of this type is illustrated in Figure 1.

The microscopists quickly determined the failure mechanism—15% air in the concrete. About 5% air by volume is frequently specified because it provides great protection against freeze–thaw deterioration. Such a quantity does not significantly affect strength, but each additional percent of air leads to a loss of about 5% strength. A photomicrograph of a polished section of the holey concrete is shown in Figure 2.

Samples of the concrete were extracted with a number of solutions, and the extracts were prepared for analysis by absorption spectroscopy as detailed in Figure 3. Infrared revealed the presence of two commercially available admixtures, materials added to concrete to produce special properties. One admixture, a triethanolamine salt and tall oil soaps, is added to cause entrainment of air. The other admixture identified (triethanolamine, polysaccharides, and lignosulfonate) is sold to increase the "workability" of plastic concrete. The concentrations of these admixtures were determined by infrared and ultraviolet–visible spectrophotometry by comparison of the sample extracts with extracts from "known" concretes containing the identified admixtures.

But one admixture singly was known to entrain only 5% air, and the other only 2%. A concrete mix containing both admixtures in the determined

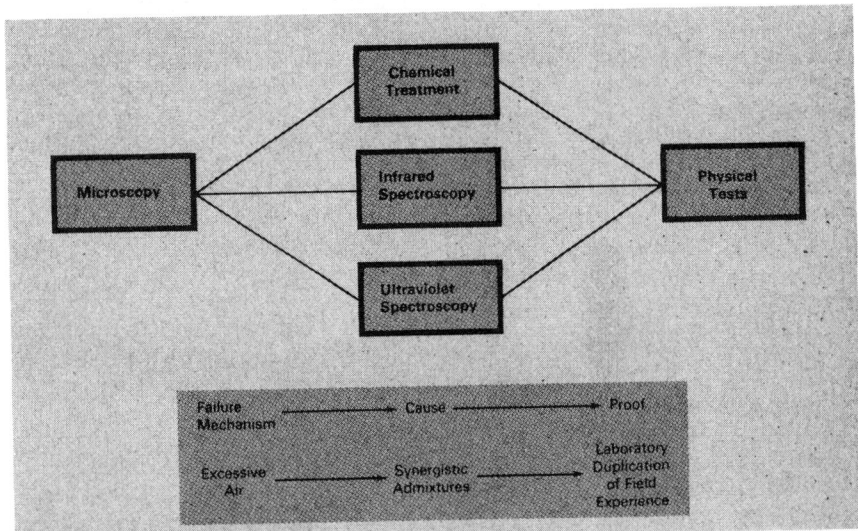

Figure 1. General approach to study of admixture problem

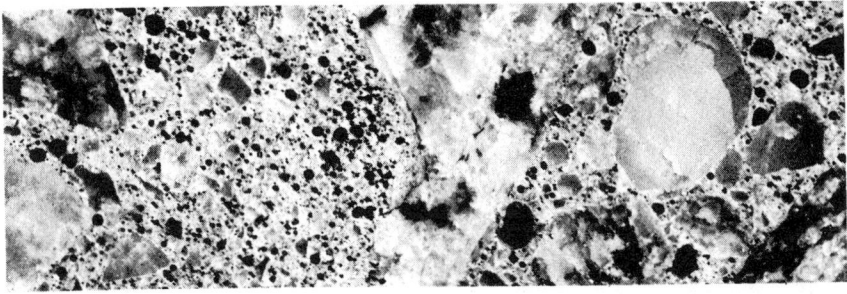

Figure 2. Photomicrograph of polished section of highly air-entrained concrete. Black "holes" are sectioned microscopic bubbles of air

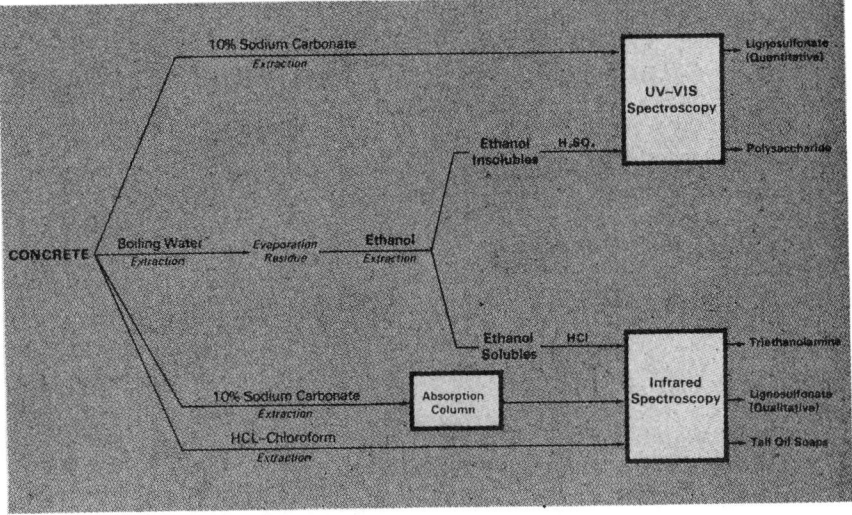

Figure 3. Analytical methods and results for admixture problem

dosages was prepared, and this particular combination proved synergistic—over 15% air had been entrained, and 50% loss in strength had resulted. Unfortunately, there was no way to save the mile of concrete pavement (about 10 million lb) that had been placed, but succeeding pavement set satisfactorily by controlling the amounts of the admixture in the concrete.

Concrete Scales

Many owners of wood frame houses recognize that the peeling of paint is a major economic headache. Concrete may experience the same distress. But when a concrete block building began losing not only its paint but also a ¼-in. layer of underlying concrete (Figure 4), consternation really abounded.

Microscopic analysis of the received "scales" revealed thin delamination layers within the spalled section received at our laboratory. This effect is characteristic of freeze–thaw damage to critically saturated concrete.

Analysis of the paint by infrared spectroscopy, solvent extraction, and pyrolysis techniques disclosed an alkyd type of paint. Such a paint is classified as "nonbreathing."

With this analytical data, the team speculated that moisture entering the building walls was prevented from escaping by the paint. The concrete near the paint surface became saturated and during winter froze because of its exposure at outside temperatures.

A site visit revealed a flashing detail error that allowed entrance of rain water into the walls. The corrective measures suggested were the elimination of the flashing detail error by properly redoing the flashing and the replacement of the nonbreathing paint with a breathing paint.

Figure 4. Underside of spalled concrete "flake" that originally extended across three masonry blocks. Section is about ¼-in. thick. Opposite side is painted

Daniel W. Vomhof
U.S. Customs Laboratory
610 South Canal Street
Chicago, Ill. 60607

Analytical Chemists in the U.S. Customs Service

The U.S. Customs Service is the oldest Federal law enforcement agency; it is celebrating its 186th anniversary this year. The 2nd Act of the First Congress of the United States provided for the collection of customs duties by the Secretary of the Treasury. By 1799 laws establishing the tariffs on several commodities, defining the customs districts, listing the functions, and regulating the employment of customs officers had been enacted. As more commodities were included in the Tariff Act, it became necessary to obtain expert opinion on such questions as purity and composition. The Congressional Act of June 26, 1848, authorized the appointment of special examiners of drugs, chemicals, and other commodities. The first special examiners were political appointments held by pharmacists and physicians. These special examiners were continued until the late 1870's when they were gradually replaced by chemists. In 1901 the Bureau of Standards was established in the Treasury Department. Its primary purpose was to develop a uniform set of weights and measures to be used by customs. The Bureau of Standards was transferred to the Department of Commerce in 1903 and later renamed the National Bureau of Standards. By 1900 customs had laboratories established in New York, Philadelphia, New Orleans, Chicago, and San Francisco.

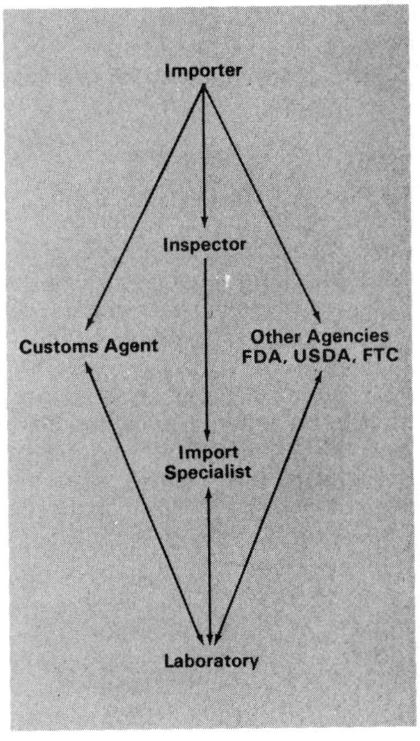

The purpose of the customs laboratories today is much the same as it was in 1848—the technical examination of imported merchandise to insure compliance with a variety of Federal laws, to detect fraud, and to provide general technical support for law enforcement activities. As a consequence, the customs laboratory activities are a combination of quality control, technical service, criminalistic, and analytical research and development. The commodities tested run the gamut from alabaster and alfalfa through human hair and marijuana to plastics and zinc. These commodities come from the importer, customs officials (inspectors, import specialists, special agents), and other Federal agencies as indicated in Figure 1. Because the staff of the customs laboratory is small relative to the quantity and breadth of work, it is essential that the analytical approach be used to obtain the maximum of information in the minimum of time. The general scheme employed is shown in Figure 2.

Three problems are presented to show the scope and variety of customs laboratory activities. One involves marijuana and a possible conspiratorial relationship. The second has to do with a question of the tariff classification of a fruit juice, and the third problem involves determining whether an item is in the antique classification and consequently not subject to tariff.

Marijuana Conspiracy

The first problem appeared when a customs agent brought two samples to the laboratory for examination. The first was plant material wrapped in a piece of red cloth taken from a suspect at one location, and the second sample was a sitar case found in the apartment of another suspect. The question was whether the marijuana in the possession of the first suspect came from the apartment of the second suspect. In other words, was there a conspiratorial relationship between the two suspects? Such a relationship is treated by the courts much more strictly than simple possession of marijuana.

The approach taken by the chemist assigned to the examination of the two samples is outlined in Figure 3. Examination of the plant material under the microscope revealed the presence of characteristic cystolithic hairs on the plant fragments, as well as a few seeds having the characteristic veination of marijuana seeds. A few plant fragments were subjected to the Dequenois–Levine spot test which resulted in a purple chloroform layer, indicating the plant material was marijuana. This preliminary result was checked by analyzing the plant material along with a known sample of marijuana and of tetrahydracannabinol (THC) by thin-layer chromatography (TLC). The developed TLC plate showed that the plant material was indeed marijuana, both by the comparison of the "fingerprint" of the spots of the known marijuana sample and by the presence of the THC spot. Fragments of plant material found in the sitar case were subjected to the same tests and gave the same results, indicating that marijuana also was present in the sitar case.

Figure 1. Origin of typical customs laboratory problem

The question of whether the two seizures were in fact related was pursued in two ways, a detailed study of the compositions of the marijuana found at each location and a careful comparison of the red cloth containing the plant material with the red fabric lining of the sitar case. The data obtained in this phase of the investigation are indicated in Figure 3.

Spot size measurements of the TLC plates indicated that the compounds cannabinol, cannabidiol, and THC were in about the same ratios in both samples. Next, portions of the two samples were both carefully ashed. The ash was then run on the emission spectrograph. Although this test is often more conclusive in showing dissimilarity, in this case, traces of two particular elements were found in both samples. The presence of these two elements had *not* been observed in a previous study of over 30 marijuana samples from several parts of the world. Thus, it was concluded that it was quite probable that all of the marijuana had originally been in the sitar case.

The sitar case was lined with a red fabric from which a large piece had been cut out. This lining was similar in appearance to the cloth containing the plant material. Chemical examination showed that both fabrics were cotton. Both fabrics had the same density and the same number of threads per inch. The size of the plant wrapper was identical with that of the missing piece of case lining. Thin-layer chromatography showed the same pigment patterns in the red dye of the two samples.

The probability of two samples having so many characteristics in common and yet not being related is so slight that the inevitable conclusion was reached that the marijuana sample had originally been in the sitar case. The technical evidence supported the theory of a conspiratorial relationship

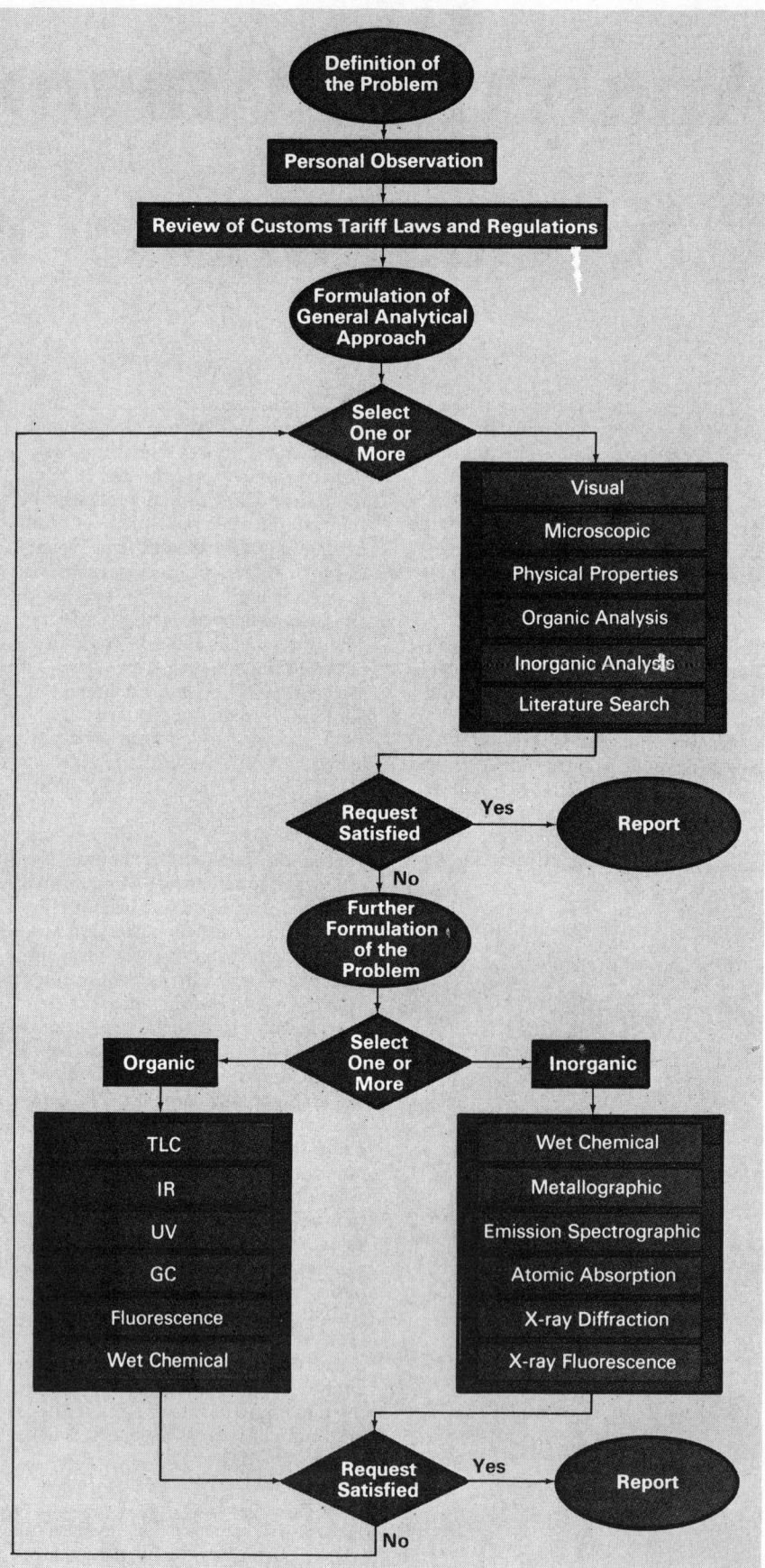

Figure 2. General analytical process for typical problem

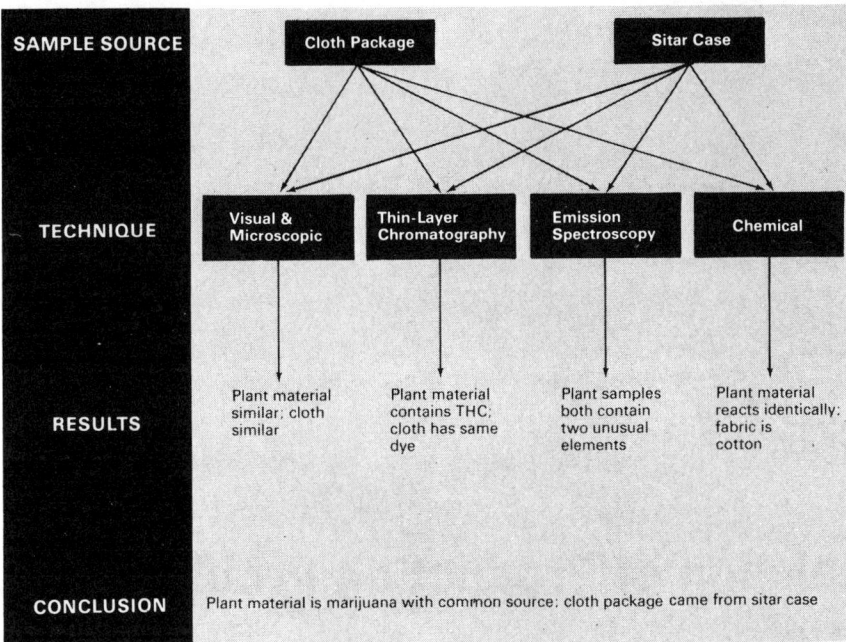

Figure 3. Approach to identification of physical evidence in criminal conspiracy

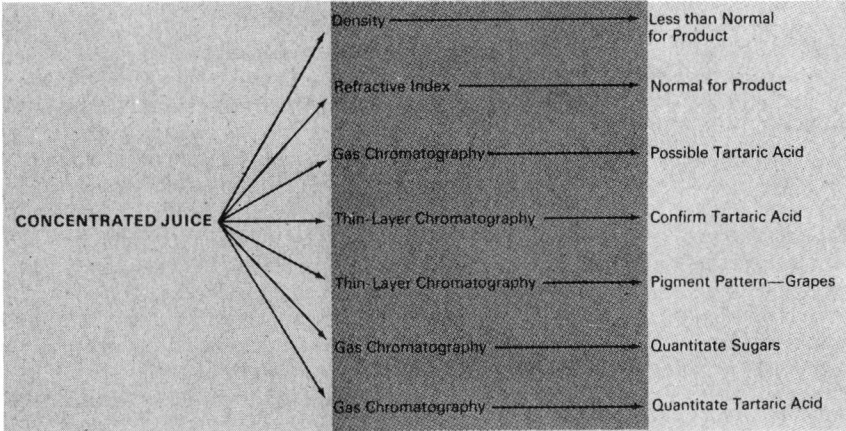

Figure 4. Testing concentrated fruit juice

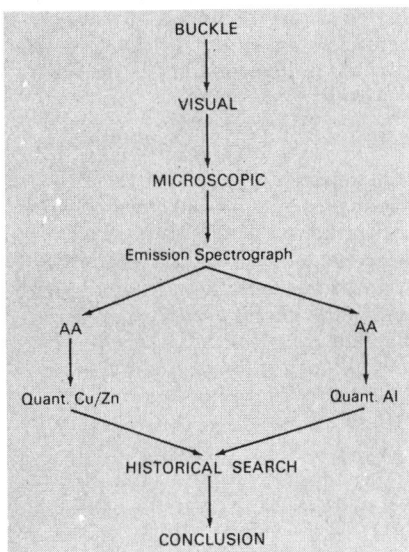

Figure 5. Analytical process in examination of metallic buckle invoiced as antique

taric acid, and this was confirmed by thin-layer chromatography. Gas chromatography was also employed to quantitate the amount of glucose, fructose, and sucrose. The ratios of these three sugars to each other and to tartaric acid were in agreement with that normally encountered in such concentrates. The chemist concluded that the sample was indeed a natural grape juice which had not been concentrated to the usual degree. With this information the import specialist was able to make the appropriate tariff classification so that the correct amount of duty could be collected. The techniques used in solving this problem are depicted in Figure 4.

"Antique" Belt Buckle

A tarnished belt buckle with the words "Wells Fargo" on it was presented to the laboratory by a port director who suspected fraud. He asked that the item be examined to determine whether it actually was an antique. How does one show if a belt buckle is antique? The approach taken is shown in Figure 5.

Customs regulations state that an item is an antique—and free of tariff duty—if it is over one hundred years old. The Wells Fargo belt buckle certainly looked like it could have been worn by a pony express rider. Under a low-power microscope, however, the patina and several of the scratches did not appear quite as they should have. They were a little too uniform, and it was decided that further testing was in order. Wiping the back of the buckle with an organic solvent quickly removed much of the tarnish. The conclusion was that the surface treatment, at least, was much less than the required age. At this point, shavings of metal were removed from the back side of the buckle. Care was taken to obtain more than just a surface sample while not irreparably damaging the article in case it was genuine. Portions of the sample were prepared for analysis by both emission spectrographic and atomic absorption (AA) techniques. The AA results showed a copper–zinc ratio which was not typical of that employed in brass works in the 1850's. The emission spectrograph plate was quite interesting in that the presence of aluminum was indicated. Quantitation on the AA determined that the alloy contained almost 7 wt % aluminum.

The presence of that much aluminum could hardly be due to an impurity. A review of the history of aluminum ascertained that it was not commercially available prior to 1890. Thus, the belt buckle was not legally an antique, and the importer was liable for duty on the shipment.

between the two suspects and was used by the government attorney to obtain convictions against the two suspects.

Tariff Classification of Fruit Concentrate

An import specialist submitted a sample of a viscous liquid invoiced as "grape juice concentrate" with a request to determine whether the commodity had been invoiced correctly. This information was required before the correct tariff classification could be made. The density of the liquid was about two-thirds that of most imported grape juice concentrates. The refractive index was also outside the range normally encountered. Thin-layer chromatography, however, indicated a pigment pattern typical of grape juice. Gas chromatography showed the probable presence of tar-

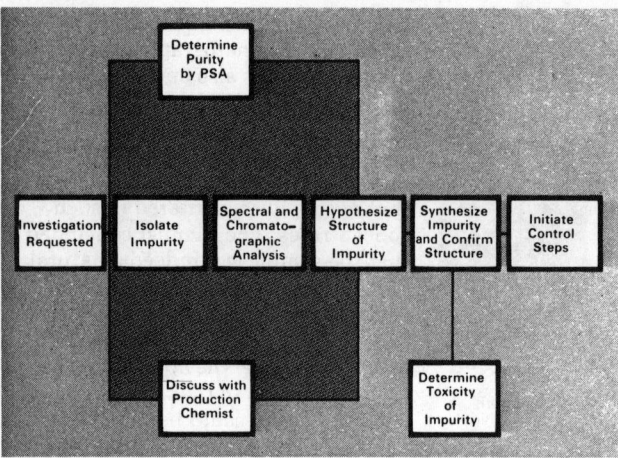

Figure 1. Analytical approach

Recognition and Solution of Production Problem in Vitamin Manufacture

E. MacMullan, R. Hagel, and R. Gomez
Quality Control Department
Hoffmann-La Roche Inc.
Nutley, NJ 07110

It is a truism that the first step in solving an analytical problem is recognizing that you have a problem. This was demonstrated recently in the Roche Control laboratories during the testing of a lot of thiamine mononitrate.

Thiamine (vitamin B_1) is an essential factor in human nutrition found naturally in rice hulls, cereal grains, yeast, liver, eggs, milk, and green leaves. It was originally isolated from rice bran; however, the thiamine used for dietary supplements is almost entirely derived from chemical synthesis. Thiamine mononitrate (I) synthesized at Roche must pass a rigid battery of chemical, physical tests to assure its purity and equivalence to the natural vitamin before being released for sale.

In the course of this testing, the analyst observed that one lot did not pass the solution clarity test. At a concentration of 2% in water, the solution was hazy whereas it should have been clear. In all other respects, the sample met specifications.

At this point, the analyst could have simply failed the lot and returned it to Production for reprocessing where an additional recrystallization would have undoubtedly brought the lot within specifications. But he recognized this as an unusual result worth a deeper investigation; therefore, he requested that the Analytical Research section look into the situation.

Defining the Problem

The general approach taken to define the problem was to:
- Determine the level of impurity in the sample
- Isolate some of the impurity
- Consult with the production chemist on likely impurities
- Subject the impurity to spectral and chromatographic analysis to determine its structure.

These steps are shown schematically in Figure 1.

The purity of the lot was measured by phase solubility analysis with procedures similar to those described by Webb (1) and MacMullan (2). Phase solubility analysis is a technique for determining the purity of materials based on a careful measurement of their solubility behavior. The method has its theoretical origin in the Gibbs Phase Rule in which absolute purity is defined as single component behavior. A solid material in equilibrium with a saturated solution which does not vary in composition as a function of the amount of excess solid in equilibrium with the solution is a pure solid. A solid whose saturated solution composition increases with increasing amounts of solid in equilibrium with the solution is impure to the extent of the increase in solution concentration.

The phase solubility analysis indicated an impurity level of 1.3% (Figure 2). This changed the picture significantly. What had been considered a trace amount of insoluble impurity was in fact a significant amount of a slightly soluble impurity. This was a serious problem and made the identification of the impurity even more urgent.

Identification of Impurity

To obtain more of the impurity, 11 grams of the thiamine mononitrate was shaken with 550 ml of water for 30 min. The resulting suspension was filtered, and 49 mg of dried impurity was recovered for study by UV, IR, NMR, and mass spectrometry. At the same time, discussions with production chemists led to the information that the impurity might be the immediate precursor in the synthesis (3), thiothiamine [3-(4-Amino-2-methyl-5-pyrimidinyl methyl)-5-(2-hydroxyethyl)-4-methyl-4-thiazoline-2-thione] (II), which is only slightly soluble in water.

This apparently was confirmed when thin-layer chromatography of the suspect lot showed an impurity spot at the

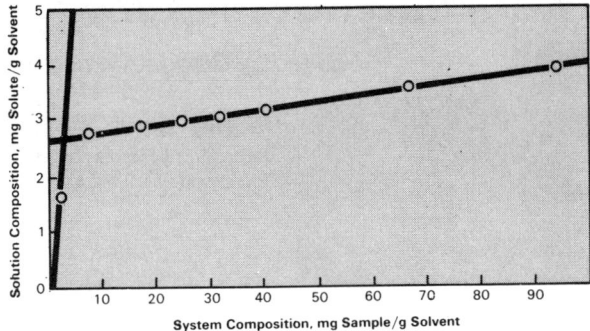

Figure 2. Phase solubility analysis of thiamine mononitrate
Sample: thiamine mononitrate lot A
System: methanol, 20 hr @ 25°C
Slope (computed by least squares with 95% confidence): 1.30 ± 0.08%
Extrapolated solubility: 2.63 ± 0.04 mg/g

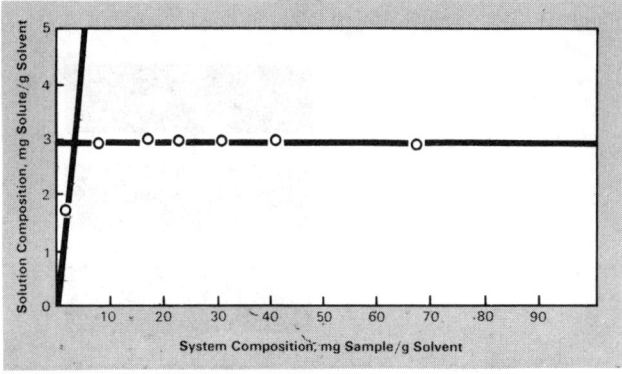

Figure 3. Phase solubility analysis of purified thiamine mononitrate
Sample: thiamine mononitrate lot A (phase purified)
System: methanol, 20 hr @ 25°C
Slope (computed by least squares with 95% confidence): −0.08 ± 0.3%
Extrapolated solubility: 2.92 ± 0.1 mg/g

same R_f as thiothiamine. However, when the plate was sprayed with thiochrome, the thiothiamine standard gave a fluorescent spot under longwave UV, but the impurity spot did not. Furthermore, the UV, IR, and NMR spectra were not consistent with the thiothiamine. The IR showed a carbonyl stretching band between two strong electronegative groups, and the NMR showed chemical shifts and splitting patterns consistent with a thiamine derivative containing a carbonyl in the 4-position of the thiazole ring. The identification data are summarized in Table I.

A hypothetical structure of the impurity was obtained by replacing the sulfur in thiothiamine with an oxygen. A search of the chemical literature showed that this compound had been made and was known by the trivial name thiamine thiazolone [3-(4-Amino-2-methyl-5-pyrimidinyl methyl)-5-(2-hydroxyethyl)-4-methyl-4-thiazoline-2-one] (III). Furthermore, it had been prepared from thiothiamine by oxidation in alkaline medium (4).

III

An authentic sample of thiamine thiazolone was prepared by the literature procedure, and the structure was confirmed by IR, NMR, MS, and elemental analysis. In all cases, the spectra of the insoluble impurity were identical to the authentic thiamine thiazolone.

Table I. Summary of Spectral and Chromatographic Results

Technique	Results	Conclusions
Thin-layer chromatography	Similar R_f to thiothiamine but different reaction to thiochrome	Impurity not thiothiamine
UV spectrum	Differs from thiothiamine	
IR spectrum	Carbonyl band between two strong electronegative groups	Hypothesized impurity as thiamine thiazolone
NMR spectrum	Consistent with oxygen analog of thiothiamine	
Mass spectrum	Molecular ion m/e 280 Fragmentation consistent with oxygen analog of thiothiamine	

Corrective Action

Once the impurity had been identified, the next step was to institute procedures to insure that it would not appear in subsequent lots. Further discussion with the production chemists indicated that the pH had probably gone up during the conversion of thiothiamine to thiamine mononitrate, allowing the formation of a small amount of thiamine thiazolone. The control steps instituted were:

- Better pH control of the reaction
- An in-process TLC done by the production chemist
- Purity determination of the finished lot by phase solubility analysis.

Another necessary step was to determine the toxicity of the thiamine thiazolone. The LD_{50} in mice was greater than 4000 mg/kg. This is even less toxic than thiamine which has a reported toxicity of 3000 mg/kg, and this represents a very low level of toxicity on an absolute basis (5). While we have instituted steps to eliminate this impurity, we have the added assurance of knowing it is a very nontoxic substance.

In retrospect, the critical steps in the solution of this problem were:

- Recognition of the problem by the control chemist
- Purity determination by phase analysis
- Close cooperation between analytical and production chemists
- Search of the chemical literature.

Given the success of these four steps, the successful laboratory solution of the problem was almost inevitable. These steps in some form or other are fundamental to the solution of any production problem.

Phase analysis had one more role to play in the process. This method can also be used as a separation technique since at solution equilibrium all the impurities end up in the solution phase, and the undissolved solids are essentially pure. This "phase purification" was run on a 50-gram sample of the impure lot and yielded 47 grams (94%) of material with a purity of 100.0% ± 0.1%. The analysis of the purified material is shown in Figure 3. This procedure could be scaled up for use in the production area.

References

(1) T. J. Webb, *Anal. Chem.*, **20**, 100 (1948).
(2) E. A. MacMullan, paper presented at Land-O-Lakes Conference on Pharmaceutical Analysis, August 1969.
(3) Maxion, U.S. Patent 2,844,579.
(4) T. Matsukawa and H. Hirano, *J. Pharm. Soc., Jap.*, **73**, 379 (1953).
(5) "The Merck Index," 8th ed., p 1037, 1968.

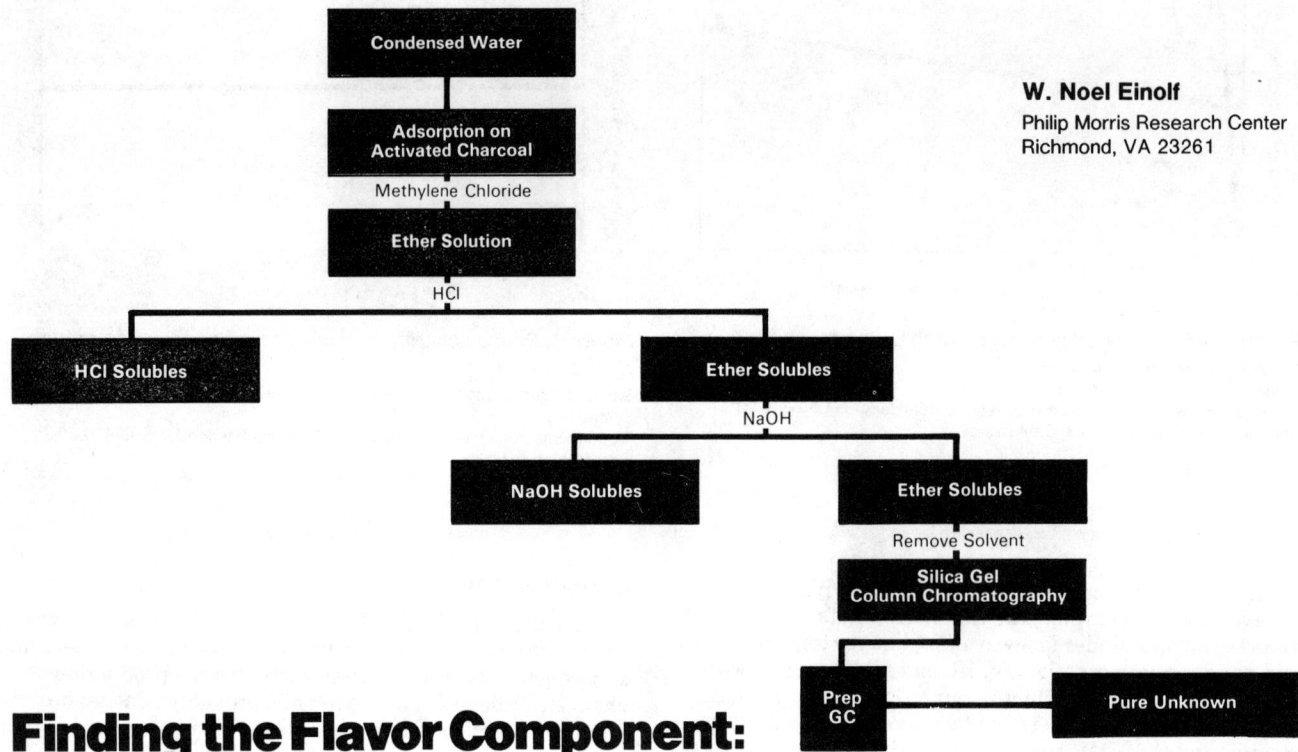

Figure 1. Steps in isolation of tobacco flavor component

W. Noel Einolf
Philip Morris Research Center
Richmond, VA 23261

Finding the Flavor Component:
A Job for the Analytical Chemist

In the tobacco industry "total utilization" requires that all stages of manufacturing be reviewed to determine which material classified as waste actually has some value or contains compounds of value. Since millions of pounds of tobacco are processed annually at Philip Morris, the problem of recovery of valuable waste is particularly important. It is this continuing review of the manufacturing processes that prompted a team of analytical chemists to investigate the source of a potentially valuable compound. Isolation and identification of compounds are accomplished only through the combined efforts of chemists with expertise in spectral interpretation, wet chemical methods, and synthetic organic chemistry.

The problem presented here originated when the condensed water from steam-treated tobacco was observed to have a tobacco-like aroma. This water was collected in a process used to make the tobacco leaf more pliable, and the presumption was that the steam process was removing some of the volatile flavor components from the tobacco. Thus, compounds isolated from the condensed water should be present in the tobacco itself. The condensed water was passed over a carbon bed, and the carbon extracted with methylene chloride. The methylene chloride residue was dissolved in

Table I. Analytical Data for Unknown Flavor Component and Two Synthetic Compounds

UNKNOWN AND COMPOUND I		COMPOUND II	
IR			
Absorption band, cm^{-1}	Description	Absorption band, cm^{-1}	Description
1761	>C=O α,β-unsat. lactone	1802	>C=O lactone
1637	C=C	1701	C=C
UV			
Absorption max	Description	Absorption max	
208 nm	α,β-unsaturated γ-lactone	None	
NMR: 16 protons observed			
Chemical shift, ppm	Description	Chemical shift, ppm	Description
1.22	—CH$_3$	0.09	—CH$_3$ singlet
1.27	—CH$_3$	1.87	—CH$_3$ singlet
1.50	—CH$_3$	1.20	—CH$_3$ singlet
5.52	H—C=C	2.37	Singlet (2H)
2.0–1.5	Complex pattern	4.97	Singlet (1H)
		1.59	Multiplet (2H)
		1.69	Multiplet (2H)
MS (high resolution)			
m/e	Composition	m/e	Composition
180	C$_{11}$H$_{16}$O$_2$	180	C$_{11}$H$_{16}$O$_2$
165	C$_{10}$H$_{13}$O$_2$	165	C$_{10}$H$_{13}$O$_2$
152	C$_{10}$H$_{16}$O	123	C$_8$H$_{11}$O
137	C$_9$H$_{13}$O		
111	C$_6$H$_7$O$_2$		
Elemental analysis (calculated for C$_{11}$H$_{16}$O$_2$)			
Calcd, %	Found, %		Found, %
C 73.30	72.90, 73.07		73.11
H 8.95	9.16, 9.31		8.88

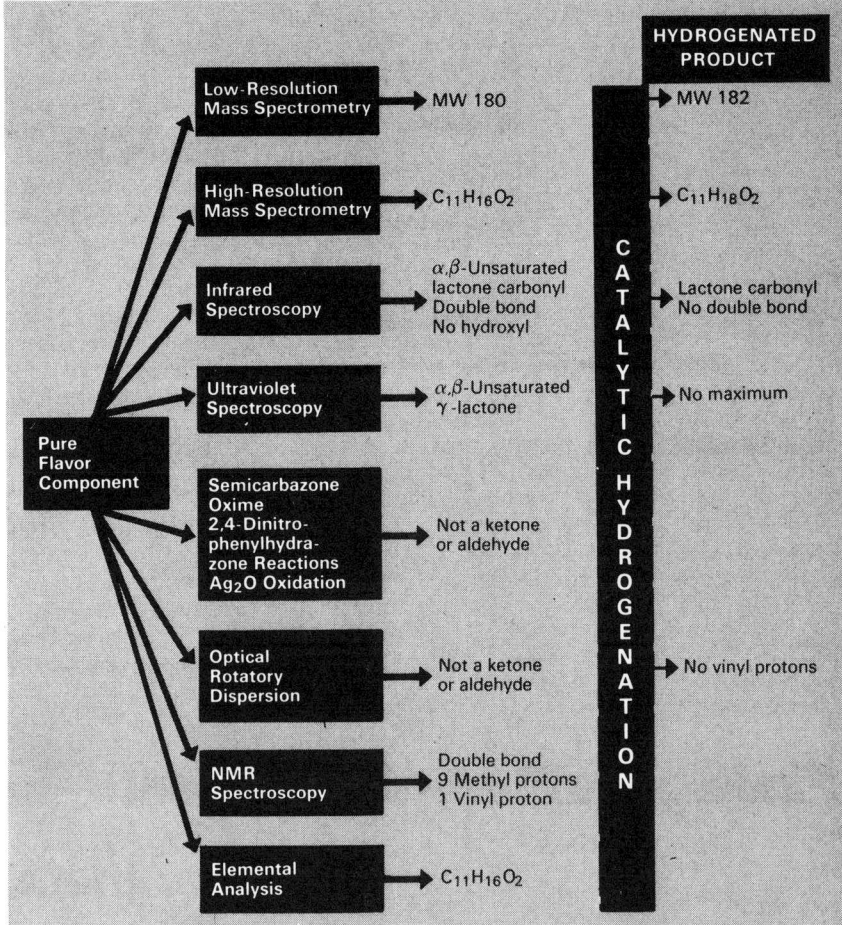

Figure 2. Techniques used in analysis of unknown flavor component

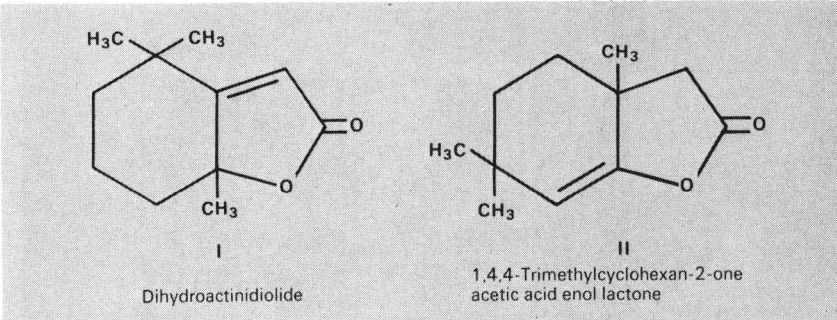

Figure 3. Possible structures of isolated flavor component

ether, and the ether solution was then treated with acid and base, and subjected to both column and gas chromatography to isolate a neutral, oxygen-containing component. The isolation procedure is summarized in Figure 1.

Nondestructive Techniques Used First

Nondestructive analytical techniques were first used; i.e., IR, UV, and NMR spectroscopy, followed by the destructive techniques of MS and elemental combustion analyses. The low-resolution mass spectrum showed a loss of 43 mass units from the molecular ion, which could have been due to the loss of either an isopropyl or acetyl group. To determine which of the two moieties was involved, precise mass measurements were necessary to obtain the elemental composition of the major ions present in the mass spectrum. These and other results are given in Table I. Based on these results, the loss of 43 mass units was found to correspond to the acetyl moiety, and the compound was found to have a molecular weight of 180 with a molecular formula of $C_{11}H_{16}O_2$.

A carbonyl function was suggested by the absorption band in the IR spectrum at 1761 cm^{-1}, but resistance of the compound to silver oxide oxidation and the inability of the compound to form 2,4-dinitrophenylhydrazone, oxime, or semicarbazone derivatives ruled out the possibilities of an aldehyde or a ketone. The lack of a Cotton effect in a qualitative optical rotatory dispersion experiment (700–300 nm) also supported the latter conclusion. A double bond was suggested by the IR and NMR spectra and confirmed by catalytic hydrogenation, followed by the observation of an increase of two mass units in the mass spectrum of the hydrogenated product (MW 182). The UV spectrum of the original compound gave an absorption maximum at 208 nm, characteristic of an α,β-unsaturated γ-lactone. After hydrogenation the absorption maximum disappeared, and the IR carbonyl band shifted from 1761 to 1770 cm^{-1}. The original compound was resistant to hydrolysis in methanolic potassium hydroxide, ethylene glycol–potassium hydroxide, or sulfuric acid, even at elevated temperatures. In addition, no hydroxyl function was present by IR analysis. The techniques used and the conclusions drawn from each are shown in Figure 2.

Synthesis Clinches Identification

At this point, it was decided to synthesize two likely compounds, based on the physical data that had been obtained. These are Compounds I (dihydroactinidiolide) and II (1,4,4-trimethylcyclohexan-2-one acetic acid enol lactone), shown in Figure 3. A literature search complicated the problem somewhat because there were three reports covering the isolation and identification of Compound I, with each giving a different set of physical properties. Two of the three reports involved isolation of a compound from natural sources. Since large discrepancies existed between the chemical data obtained for the isolated compound and that reported in the literature for the same proposed structure, the best approach was to synthesize Compound I by at least two different routes to obtain an unambiguous structure. This was done, with the product from each synthetic route giving spectral characteristics identical to those of the unknown compound; the mass spectrum is shown in Figure 4a. The product was a viscous oil, which soon crystallized to give a mp 42–43°C. Hydrogenation of Compound I also gave a product identical in physical properties to the hydrogenated unknown.

Compound II was synthesized to conclude the investigation. As suspected, the physical properties of this compound, given in Table I, were considerably different from those of Com-

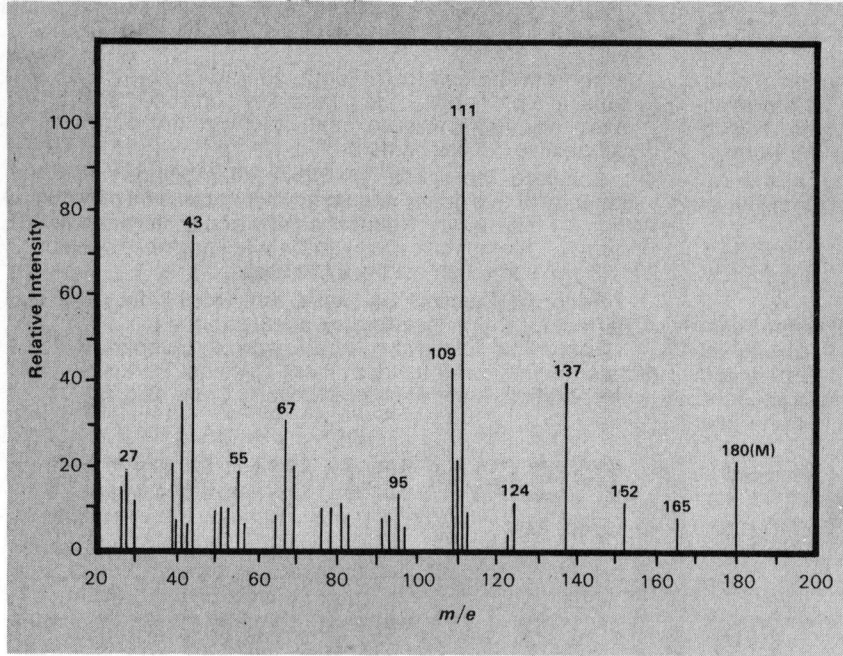

Figure 4a. Mass spectrum of Compound I, later identified as dihydroactinidiolide

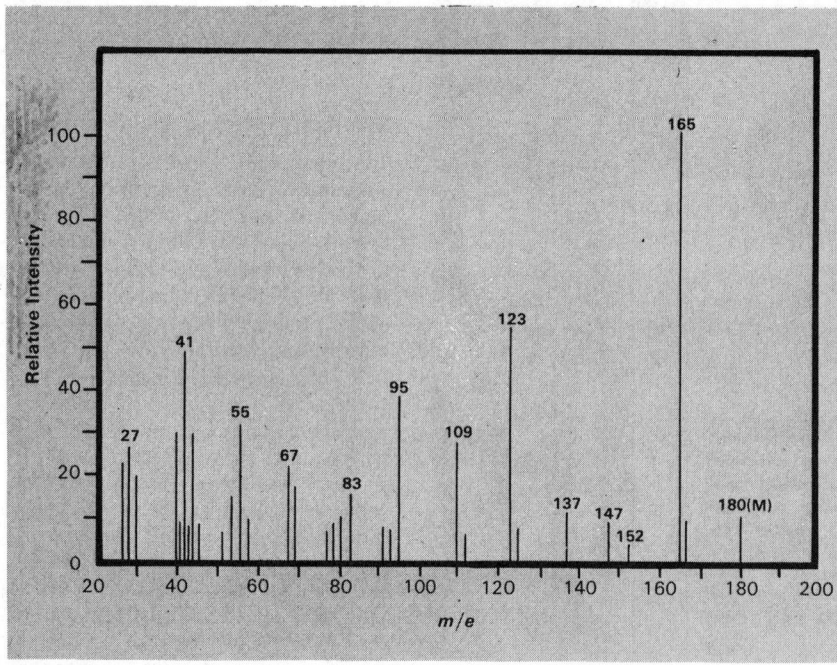

Figure 4b. Mass spectrum of Compound II, 1,4,4-trimethylcyclohexan-2-one acetic acid enol lactone

pound I. The mass spectrum of Compound II is shown in Figure 4b for comparison with that of Compound I.

The discovery of dihydroactinidiolide in the condensed water as well as in bright and Oriental tobaccos has added to the knowledge of tobacco chemistry in addition to providing a potential source for a recoverable flavor component. Problems similar to this one and frequently of a proprietary nature occur on a continuing basis, providing the challenge which only a group of skilled and knowledgeable analytical chemists is able to meet via the analytical approach.

Acknowledgment

The solution of this problem is the result of the efforts of the following people. Details appear in these two papers: William C. Bailey, Jr., Ajay K. Bose, Robert M. Ikeda, Richard H. Newman, Henry V. Secor, and Charles Varsel, *J. Org. Chem.*, **33**, 2819 (1968); and Paul H. Chen, William F. Kuhn, Fritz Will, III, and Robert M. Ikeda, *Org. Mass Spectrom.*, **3**, 199 (1970).

Figure 1. Structurally related amphetamines

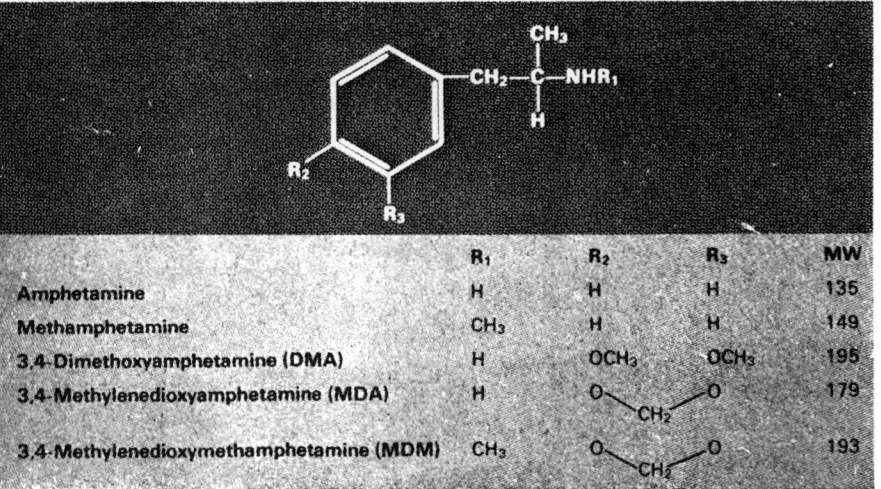

Larry S. Eichmeier and Michael E. Caplis
Northwest Indiana Criminal Toxicology Laboratory
540 Tyler Street
Gary, Ind. 46402

The Forensic Chemist
An "Analytical Detective"

The forensic chemist has the acute problem of constantly being required to make a rapid assessment of the nature of suspected illicit materials or to demonstrate the presence of poisons, drugs, and toxic chemicals in biological samples. Society demands that samples be completely analyzed prior to accusation and prosecution of any violation of its laws. Thus, the forensic chemist must be both swift and absolutely reliable in his assessment. To this end, it is his responsibility to identify and determine, without a reasonable doubt, illicit or toxic substances within a relevant time.

By virtue of this responsibility, the forensic chemist becomes an "analytical detective" who must deal with complex mediums from pharmaceutical and illicit preparations to samples of body fluids and tissues. Determination of toxic compounds in a biological matrix is complicated by the low concentration levels of the compounds and by their possible conversion to metabolites. Complete analysis of such materials requires a multitechnique approach. Techniques utilized include thin-layer chromatography, gas chromatography, high-pressure liquid chromatography, fluorospectrophotometry, gas chromatography–mass spectrometry, UV–VIS spectrophotometry, and infrared spectrophotometry, as well as chemical methods. Equipped with this armamentarium, there can be little justification for an incorrect analysis from the forensic laboratory.

Two examples in which a judicial combination of techniques was utilized are cited. One involved the identification of a new street drug closely resembling a known illicit preparation, and the other involved a drug overdose which led to a homicide investigation.

Identification of Suspected Illicit Drug

A major abused class of drugs is amphetamine and its derivatives. In an attempt to combat this problem, undercover narcotic agents are constantly striving to determine the source and trafficking route of these drugs. In one particular instance, agents purchased capsules suspected to contain amphetamine or methamphetamine from a major distributor (Figure 1). These capsules containing a pink powder were submitted to the forensic laboratory for analysis. Preliminary screening of a $1M$ sulfuric acid solution of the powder by UV spectrophotometry indicated a dioxyamphetamine derivative, such as 3,4-dimethoxyamphetamine (DMA) or 3,4-methylenedioxyamphetamine (MDA). Amphetamine and methamphetamine were ruled out since the indicative pattern for monosubstituted benzenes (252, 257, 263 nm) was absent. Thin-layer and gas chromatographic screening after chloroform extraction from a basic solution showed chromatographic properties similar to 3,4-methylenedioxyamphetamine (MDA).

Differential chemical visualization following thin-layer chromatographic separation was used to determine if the substance was a primary, secondary, or tertiary amine (Table I). The material did not react with ninhydrin–acetone, but reacted with ninhydrin–isopropanol–acetic acid and iodoplatinate, indicative of a secondary or tertiary amine.

On-column gas chromatographic derivatization with acetic anhydride and benzaldehyde was employed to confirm the amine structure. The on-column derivatization of the amine with acetic anhydride gave a change in peak retention time indicative of the acid amide formation expected with a primary or secondary amine.

However, on-column derivatization with benzaldehyde resulted in loss of any gas chromatographic activity, indicating formation of the chromatographically inactive substituted secondary amine.

Table I. Summary of Tests and Conclusions for Suspected Illicit Drug

Examination	Results	Conclusion
UV (1M H_2SO_4)	252 nm, 257 nm, 263 nm — Absent	No monosubstituted benzene present
	287 nm, 233 (max) — Present	Dioxyamphetamine-like derivative
TLC		
Ninhydrin–acetone	No reaction	
Ninhydrin–isopropanol–acetic acid	Reaction	Secondary or tertiary amine
Iodoplatinate	Reaction	
GC		
Acetic anhydride	Peak shift	Secondary amine
Benzaldehyde	No peak	
Mass spectrometer		
193 m/e	Weak—recognizable (molecular ion)	Secondary amine
135 m/e	Strong	Aryl methylenedioxy substituent
58 m/e	Intense (base peak)	Alkyl amine
151 m/e	Absent	No aryl dimethoxy substituent present
91 m/e	Absent	No monosubstituted benzene present

$$R-CH_2-\underset{R^1}{N}-H + \phi CHO \rightarrow R-CH_2-\underset{R^1}{N}-\underset{H}{\overset{OH}{C}}-\phi$$

Mass spectral analysis showed fragments at 58, 135, and 193 m/e. Even though the ion 193 m/e was recognizably weak, it was determined to be the molecular ion. Characteristically, a weak parent ion is common to aryl-substituted secondary amphetamine-like derivatives. The base peak at 58 m/e can be attributed to either of two fragments:

(A) $(CH_3)_2C=NH_2^+$

(B) $CH_3-CH=N^+(H)(CH_3)$

Greater significance can be assigned to fragment B because previous data indicated a secondary amine. Additionally, the absence of any ion at 151 m/e negated the possibility of a dimethoxy substituent, whereas the presence of the 135 m/e ion indicated a methylenedioxy substituent:

(151 m/e) — dimethoxybenzyl cation

(135 m/e) — methylenedioxybenzyl cation

Inspection of the analytical data allowed identification of the material as 3,4-methylenedioxymethamphetamine (MDM). The data and conclusions are summarized in Table I. This substance, similar in structure and general analytical properties to MDA, is exempt from federal control, whereas MDA is controlled. Thus, the rapid utilization of general and specific techniques provided valuable information on a suspected illicit preparation and averted embarrassment and unwarranted prosecution.

Drug Homicide

Blood from a woman found by a relative was submitted to the forensic laboratory for toxicological analyses. There was no apparent cause for death. The woman and two of her male friends had been drinking heavily outside her apartment complex earlier that evening. Police were called to quell a disturbance, and the trio had retired to her apartment. Later a witness testified that cocktails had been forcibly administered to the woman in an attempt to subdue her. Medical reports showed the woman was under a doctor's care with a prescription for phenobarbital.

Examination for blood volatiles involved gas chromatography and UV spectrophotometry (Figure 2). Screening was accomplished by injecting 10 μl of whole blood into a gas chromatograph. Results indicated the presence of ethyl alcohol. The ethanol concentration was determined by an enzymatic procedure with alcohol dehydrogenase (ADH).

$$CH_3CH_2-OH \xrightarrow[NAD \quad NADH\cdot H]{ADH} CH_3-\overset{O}{\underset{H}{C}} \xrightarrow{Semicarbazide} Semicarbazone$$

In this procedure a mole of ethanol is oxidized to a mole of acetaldehyde, which is removed as the semicarbazone, with the concurrent reduction of a mole of nicotine adenine dinucleotide (NAD). Monitoring of the reduced NAD at 340 nm with a UV spectrophotometer gave an ethanol concentration of 0.23% (0.1% = intoxication).

Preliminary screening with a chloroform extract of an acidified blood sample (Figure 2) gave conflicting results. Gas liquid chromatography suggested the presence of three barbiturates: amobarbital, secobarbital, and phenobarbital. However, thin-layer chromatography indicated only the presence of amobarbital and secobarbital. Gas chromatography–mass spectrometry resolved the inconsistency with identification of amobarbital, secobarbital, and n-dibutylphthalate. Dibutylphthalate has the identical gas chromatographic properties of pheno-

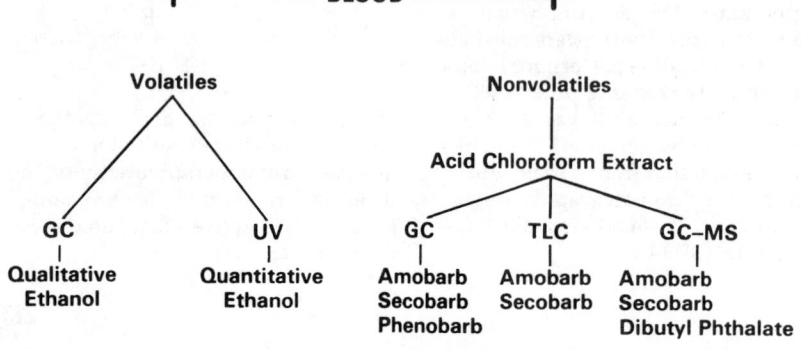

Figure 2. Isolation and identification scheme for volatiles and nonvolatiles in blood

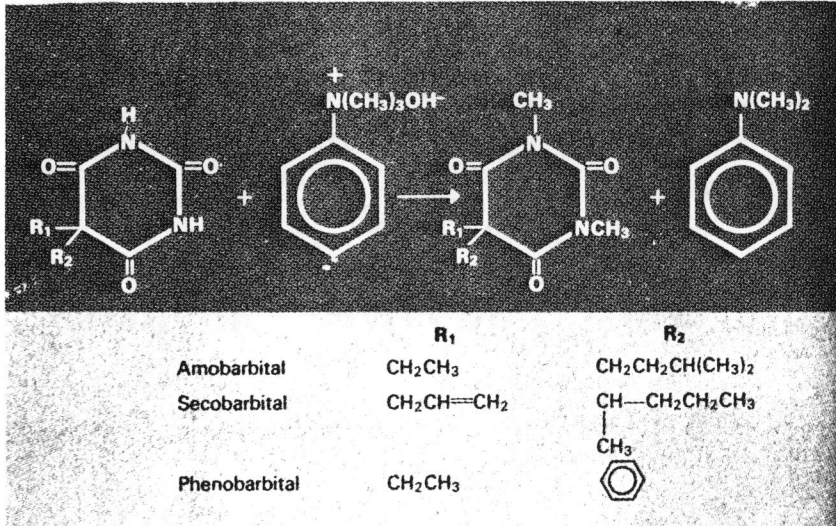

Figure 3. Methylation of barbiturates

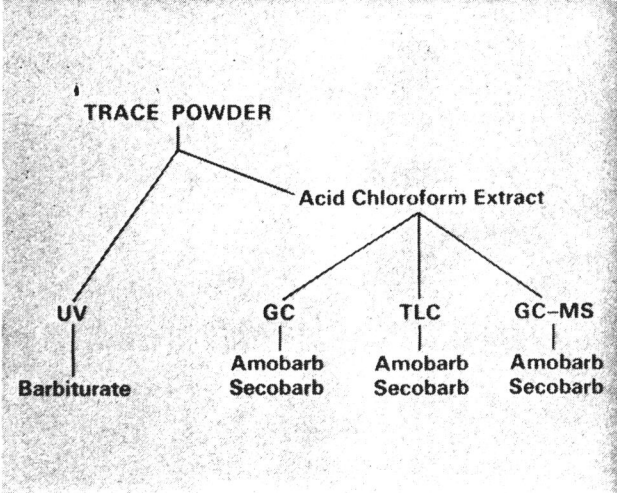

Figure 4. Identification scheme for unknown powder

barbital but without the chemical visualization utilized in thin-layer chromatography.

The secobarbital and amobarbital were quantitated by gas chromatography as their dimethyl derivatives via an on-column reaction with trimethylanilinium hydroxide (Figure 3). This provides sharper chromatographic peaks with a minimum amount of tailing. The secobarbital and amobarbital levels determined were greater than each drug's minimum toxic level, 1.0 and 1.5 mg/100 ml, respectively.

Each of the drugs [alcohol (0.23%), secobarbital (1.1 mg/100 ml), and amobarbital (1.6 mg/100 ml)] was present at toxic levels and individually could have caused intoxication. The synergistic effects of alcohol and barbiturates are well documented, and in this case, an acute intoxication and overdose death were quite probable. The forensic chemist's analysis indicating adverse blood barbiturate levels, to which the subject had no known access, prompted further investigation.

Apprehension of the male companions showed one to have traces of white powder in the pockets of the pants worn on the fatal evening. Preliminary screening of the trace powder with UV spectrophotometry indicated the presence of a barbiturate (Figure 4). Further examinations of the powder by gas chromatography, thin-layer chromatography, and gas chromatography–mass spectrometry showed the powder to contain both amobarbital and secobarbital.

Thus, a suspected overdose death was transformed into a drug homicide investigation and eventual criminal trial as a consequence of the forensic chemist's analyses.

Industrial Analytical Chemists and OSHA Regulations for Vinyl Chloride

S. P. Levine, K. G. Hebel, J. Bolton, Jr., and R. E. Kugel
Stauffer Chemical Co., Eastern Research Labs, Dobbs Ferry, N.Y. 10522

In 1971 the newly created Occupational Safety and Health Administration (OSHA) with the advice of its technical arm, the National Institute for Occupational Safety and Health (NIOSH), adopted a 500 parts-per-million (ppm) by volume permissible level for worker exposure to vinyl chloride monomer gas (VCM). In 1974 a review of this standard was prompted by reports of several deaths from a rare form of liver cancer called angiosarcoma among polyvinyl chloride (PVC) plant employees. This suggested a possible relationship to PVC production. Animal studies and epidemiological surveys indicated that exposure to VCM might be a causative agent involved in the development of angiosarcoma in humans (1). These facts led OSHA to issue in April 1974 a temporary emergency standard of a 50-ppm ceiling exposure. This standard also provided for regular monitoring of the work space by personnel monitoring systems able to assay 5-ppm VCM with a relative precision of ±20% (average for a 10-min air sample). This was to have a profound effect on the vinyl chloride and the polyvinyl chloride industries which employ approximately 360,000 workers in over 7,500 plants.

Following public hearings, OSHA published a final standard for VCM in October 1974 (Figure 1) (2, 3). However, the emergency temporary standard remained in effect until April 1, 1975. The final standard sets a maximum permissible level of 1 ppm for an 8-hr time-weighted average exposure. A ceiling limit of not more than 5-ppm VCM over a 15-min period has also been set. In addition, an action level of 0.5 ppm was set up; exposures above the action level require periodic monitoring, medical examinations, and training (4).

The emergency and the final standards called for vastly different approaches from an industrial hygiene point of view. Under the temporary standard, a survey was made to determine areas of emission, and only "grab" sampling was performed. It was only necessary to ensure that work areas did not exceed 50 ppm of VCM in the air. This sampling approach was changed with the advent of the permanent standard. Now areas must be regulated by both the ceiling (maximum) value and by the time-weighted average exposures of workers in those areas. These requirements call for classifying the areas and types of jobs in plants and for monitoring the actual exposures of workers over a typical workday. The combined sampling and analytical methods used have to be capable of determining VCM down to the 0.25-ppm level with a precision at the 95% confidence limit of ±50%.

Analytical Approach

To aid industry in monitoring programs designed to comply with this standard, NIOSH published a preliminary procedure for VCM sampling and analysis that was classified as "operational, but not thoroughly characterized" (5). This method calls for collection of VCM in glass adsorption tubes containing one of the specific NIOSH-approved lots of activated charcoal. Air from the breathing zone of the worker is drawn through the adsorption tube with the aid of a small low-flow battery-operated pump. After sample collection is completed, the tube is capped and sent to the laboratory for analysis. VCM is desorbed from the charcoal with CS_2, and the resulting solution is injected into a gas chromatograph (GC) for analysis. Separation of VCM from other components is performed with an SE-30 column. Since NIOSH realized that this procedure had not yet been thoroughly characterized, the final standard allowed this method or any equivalent method to be used.

Preliminary testing by our laboratory, as well as by others, indicated that

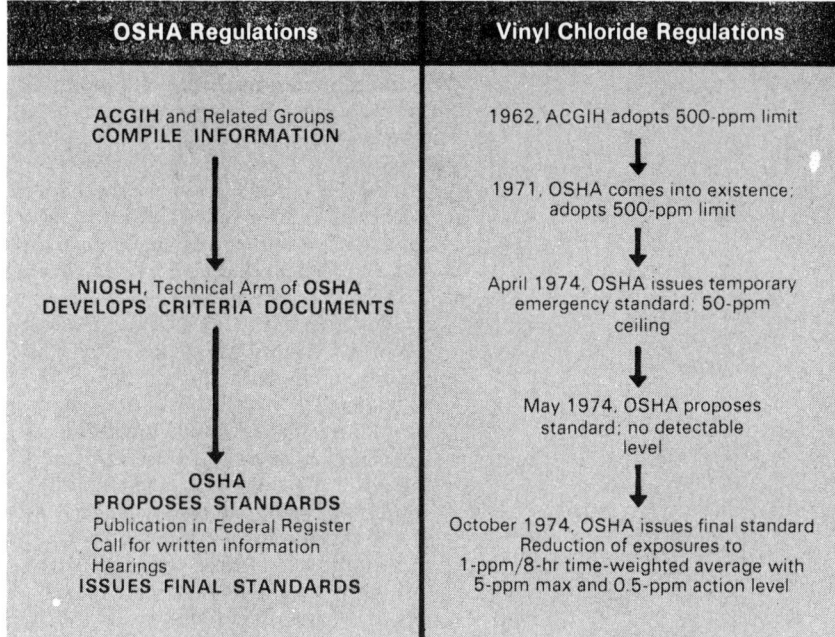

Figure 1. Genesis of OSHA regulations
ACGIH = American Conference of Government Industrial Hygienists. NIOSH = National Institute of Occupational Safety and Health. OSHA = Occupational Safety and Health Administration

there were some disadvantages with the recommended method. These were: poor storage stability of VCM on the charcoal tubes, lot-to-lot variations in charcoal, low and variable desorption efficiency of VCM with CS_2, the inadequacy of the SE-30 column to resolve VCM from other components (of the plant air) and/or CS_2 impurities, and the toxicity and flammability of CS_2. In addition, the volatility of CS_2 made it difficult to prepare stable standard solutions of VCM in CS_2.

Because of these problems, our laboratory, the Analytical Section of Stauffer Chemical Co.'s Eastern Research Center, sought to develop an improved method capable of VCM personnel monitoring for Stauffer Chemical Co.'s PVC resin and fabricating plants. In addition to the requirements set forth by OSHA, we had several other considerations to include when deciding which analytical approach to use:

• The sampling device had to be capable of storing VCM with no losses for periods of up to one week to permit shipping of samples from several plant locations to a central laboratory for analysis.

• The GC column must cleanly resolve VCM from interfering substances that might be found in plants employing VCM or VCM-containing materials used in a wide variety of synthetic and/or fabricating formulations.

• The analytical procedure should exceed in both accuracy and precision the stated OSHA requirements so that the number of personnel monitoring samples could be minimized. This requirement, plus a well-designed sampling program, was needed to ensure the validity of the resulting VCM exposure data, because the variations due to personnel, work shift, process, and even day of the week are not always controllable.

• Due to the effective date of the standard (January 1, 1975, delayed to April 1, 1975, by court order) and the time required to train personnel, strict time limits were imposed on the analytical method development stage of this project. This timing precluded the use of semiautomated VCM analysis systems that have since appeared on the market. Although many of these commercial systems are perfectly satisfactory, their precision, reliability, and delivery date were all unknown at the time that the VCM personnel monitoring surveys were started by Stauffer Chemical Co.

In an industrial environment, method development frequently involves more method adaptation than actual invention. The development and adaptation problems for this project involved two categories, the sampling system and the analysis system.

Sampling System

We have investigated the utility of two types of personnel sampling systems for organic gases. The first involves concentrating the sample in an adsorbent tube, such as that used in the NIOSH procedure. Although certain drawbacks have been noted in the NIOSH procedure, variations of adsorbent tube design, adsorbent, and/or VCM desorption techniques have been applied successfully by several groups. These variations involve the use of modified reusable charcoal tubes, heat desorption devices, head space analyzers, and desorption with CS_2 at Dry Ice temperatures. Advantages of the adsorbent tube approach are the small size of the sampling apparatus and the fact that large volumes of air can be drawn through the tube, thereby concentrating the VCM by several orders of magnitude. A drawback in the heat desorption and head space analysis procedures (which are applied to the adsorbent tubes) is that gas chromatographic analysis can be performed only once; repetitions are not possible since the sample is either totally consumed in a single determination or its concentration has been substantially changed. The use of Dry Ice baths to minimize losses of VCM and/or CS_2 during desorption from charcoal tubes was developed by Dow Chemical Co. (6). This procedure is a variation of the NIOSH-developed method and has been tested by our

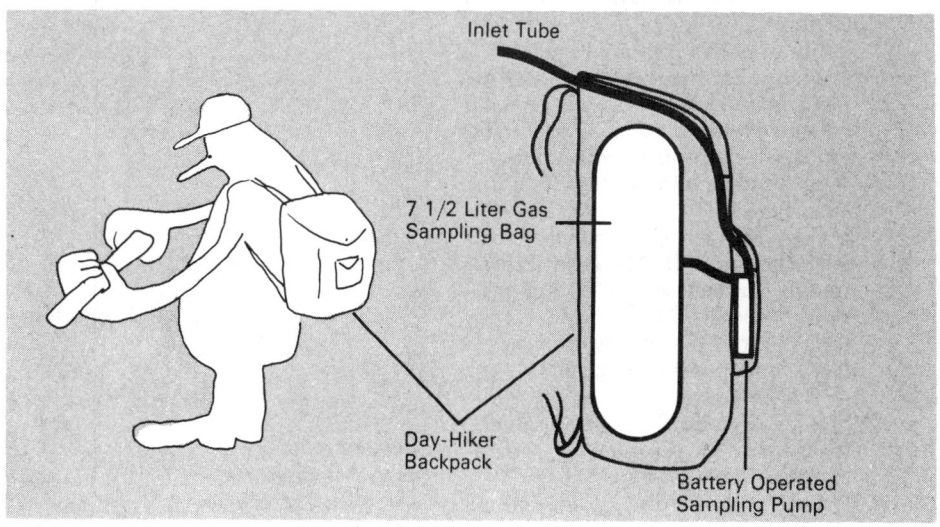

Figure 2. Vinyl chloride personnel sampling unit

277

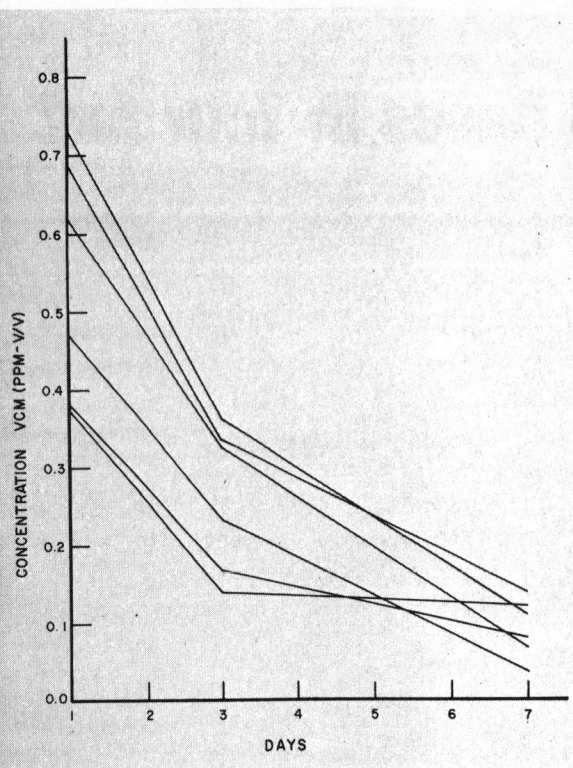

Figure 3. Loss of vinyl chloride gas from Teflon gas sampling bags

laboratory. Although somewhat time-consuming, it is more reliable than the original NIOSH procedure.

A second basic sampling procedure involves the collection and storage of the sample gases without concentration. The method of choice for personnel monitoring involves the use of gas sampling bags. A battery-operated pump is used to draw air from around the worker's breathing zone and exhaust it into the bag. The contents of the bag are then analyzed directly by gas chromatography or any other suitable analytical technique. The pump and the bag are placed in a small dayhike backpack which is then worn by the worker for a complete workshift (Figure 2).

Gas sampling bags are commercially available and are usually fitted with a metal twist-lock valve, although some are also equipped with a permanent or replaceable septum or with a filling snout. Although the storage stabilities of a wide variety of volatile materials in these bags have been summarized in the literature (7–11), none of these reports has dealt with 0.2–1.0 ppm concentrations of VCM in air. Therefore, the storage stability, memory effect (from previous samples), and losses of VCM in two commercially available gas sampling bags were studied. In addition, the precision and accuracy of the total sampling system (pump, bag, and tubing) were defined. Chosen for this study were a Teflon bag equipped with a replaceable septum and a twist-lock valve and an aluminized Scotchpak three-layer bag equipped with a valve.

Figure 3 shows the loss of VCM from Teflon bags to be in the range of 20% per day. It was not determined whether this loss resulted from the permeability of Teflon or from mechanical problems. There is really little need for a septum on a gas sampling bag since maximum GC precision can more easily be achieved by using gas sampling valve injection rather than gas syringe injection techniques.

Figure 4 illustrates the storage stability of VCM in aluminized Scotchpak bags. There is no detectable loss of VCM for a period of one week over the concentration range of 0.1–1.1 ppm VCM in air. Because of the possibility of leaks in gas sampling bags, it is recommended that they should be leak tested with clean compressed air for a period of several hours before use or reuse. In actual field use, we find about a 10% "mortality" rate for aluminized Scotchpak bags when they are used repeatedly.

All further studies were carried out on only the aluminized Scotchpak bags. Bags experimentally filled with between 1.0 and 10 ppm VCM had no detectable amount (<0.03 ppm) of VCM remaining after two repetitions of vacuum pumping of the bags and refilling with compressed air. It is, therefore, our practice to perform three pump-and-fill cleaning cycles before each reuse of a sampling bag.

The performance of the entire sampling and analytical system was checked by pumping 1.0 ppm of VCM in air from a full gas sampling bag through connecting tubing and a sampling pump into a second bag which had been evacuated prior to the experiment. The lengths of Teflon-lined neoprene tubing used and the pump were the same as would be used in the field. This was a simulation of the

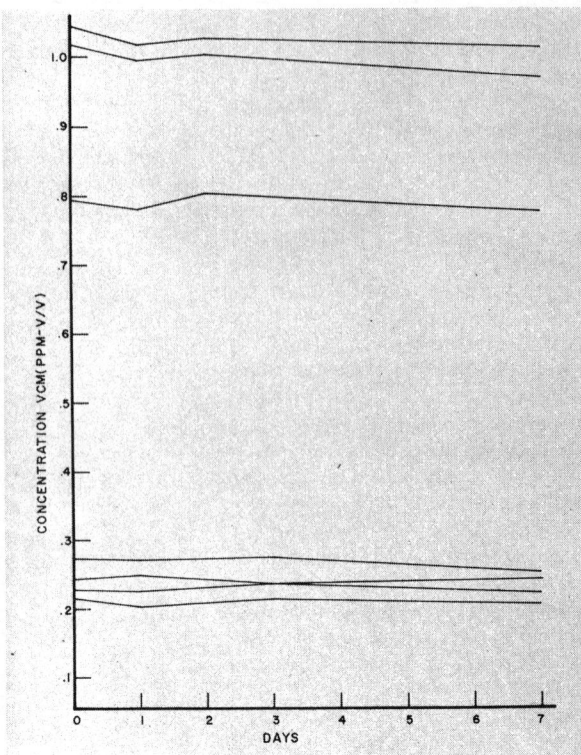

Figure 4. Stability of vinyl chloride gas in aluminized Scotchpak gas sampling bags

losses that could be expected under actual use conditions. The data showed a mean recovery of 96–97% of the VCM.

Analytical System

Because of the inadequacy of the NIOSH-recommended SE-30 gas chromatographic column, we have chosen to use a Durapak (Carbowax 400 on Porasil) packing. The use of Durapak requires a low inlet pressure which results in only minor baseline disruption during sample injection with the gas-injection valve. Sample loops of up to 5.0 ml can be used without significant loss of resolution. We have found nominal analysis times of between 3 and 5 min to yield more than adequate resolution and results with relative precisions (95% confidence level, single injection) of ±3% at the 1.0-ppm level and ±11% at the 0.25-ppm level. Since the precision increases as 1/square root of the number of injections, we routinely inject each sample two or more times. This system can detect less than 0.03-ppm concentrations of VCM in air.

Several valving configurations have been developed by other laboratories which incorporate the function of column backflush and injection, and cut-and-backflush and injection in a single automated valve. These valves minimize the time necessary to clean the GC column of constituents from the air sample having longer retention times than VCM. A cut-and-backflush system presently in use by many laboratories incorporates a precolumn of Durapak (n-octane on Porasil) with porous polymer and a short analytical column of porous polymer. An automated 10-port valve is used (12).

In conclusion, the industries which use VCM for resin production and PVC for fabrication of finished vinyl products were faced with developing, field proving, and installing a complete VCM personnel monitor system in a short time. Sampling and analysis guidelines supplied by NIOSH were not adequate for the goals of our program, thus necessitating extensive method adaptation and development. The method outlined above represents a viable analytical approach for an industrial setting.

The long-range goal for determining worker exposure to VCM must be, however, the utilization of continuous monitors. These VCM monitors must have the capability of correlating variable ambient air VCM concentrations with the results of corresponding personnel monitoring samples.

References

(1) H. Falk, J. L. Creech, Jr., C. W. Heath, Jr., M. N. Johnson, and M. M. Key, *J. Am. Med. Assoc.*, **230**, 59 (1974).
(2) "Threshold Limit Values for Chemical Substances in Workroom Air Adopted by ACGIH," American Council of Governmental Industrial Hygienists, Cincinnati, Ohio, 1973.
(3) G. Clack, *Job Saf. Health*, **3**, 4 (1975).
(4) *Fed. Regist.*, **39**, 194 (1974).
(5) "NIOSH Manual of Analytical Methods," U.S. HEW 75-121, pp 178, 1–178, 9, 1974.
(6) A. A. Allemang, L. W. Severs, and L. K. Skory, "Monitoring Personnel Exposed to Vinyl Chloride and Other Chlorinated Solvent Vapors in an Industrial Work Environment," American Industrial Hygiene Conference, Minneapolis, Minn., June 4, 1975.
(7) R. T. Maykoski and C. Jacks, "Review of Various Air Sampling Methods for Solvent Vapors," NTIS AD-752-525, 1970.
(8) F. J. Schuette, *Atmos. Environ.*, **1**, 515 (1967).
(9) "Methods of Air Sampling and Analysis," Interscience Committee, American Public Health Assoc., pp 7–8, 138, Washington, D.C., 1972.
(10) F. B. Higgins, Jr., "Sampling for Gas and Vapors," in "Source Sampling of Atmospheric Contaminants, Symp. Proc.," H. G. McAdie, Ed., Chem. Inst. Canada, Ottawa, Canada, 1971.
(11) R. E. Dilgren, Shell Chemical Co., Analytical Method HC-604-74, 1971.
(12) F. Zado, Western-Electric Co., Princeton, N.J., personal communication. This system is not marketed by Western-Electric Co.

PART I

Art Conservation: Culture Under Analysis

BEN B. JOHNSON and THOMAS CAIRNS

Conservation Center, Los Angeles County Museum of Art
Los Angeles, Calif. 90036

CONSERVATION is a relatively new discipline which has attained maturity only in the last few decades. It evolved in the late 19th century through a synthesis of analytical science and restoration, which by the first decades of the 20th century had developed into a new philosophy emphasizing respect of the material and aesthetic integrity of the original. Although there is no extensive history of restoration, there is evidence that man was concerned with preserving objects of both utilitarian and aesthetic interest several millennia before Christ. The Chou Chinese (1028–1256 B.C.) employed ingenious techniques in repairing their ceremonial bronzes. A Nishapur bowl (11th century A.D.) illustrates a crude type of ceramic repair utilizing bronze brackets to strengthen firing cracks or breaks (Figure 1). As early as the 16th century, accounts of restoration to paintings and sculpture indicate concern for restoring the original quality of the object.

By the 18th century specialized techniques for treatment of objects had developed which were beyond the capabilities of the general artist or repairman. Transfer of paintings from original support to an entirely new one was practiced as early as 1741 by Frederic Dumesnil (1710–91) (1). That Dumesnil had learned the technique from an Italian named Riario indicates it was already in use by the early 18th century in Italy. In the early 19th century, transfer of paintings from panel support to canvas became a common practice as did frequent cleaning. However, the picture cleaning campaign initiated at the National Gallery in London between 1846 and 1853 triggered the convergence of restoration and scientific analyses. Negative public reaction to the cleaned paintings prompted the House of Commons to form the Select Committee of Inquiry to investigate the National Gallery.

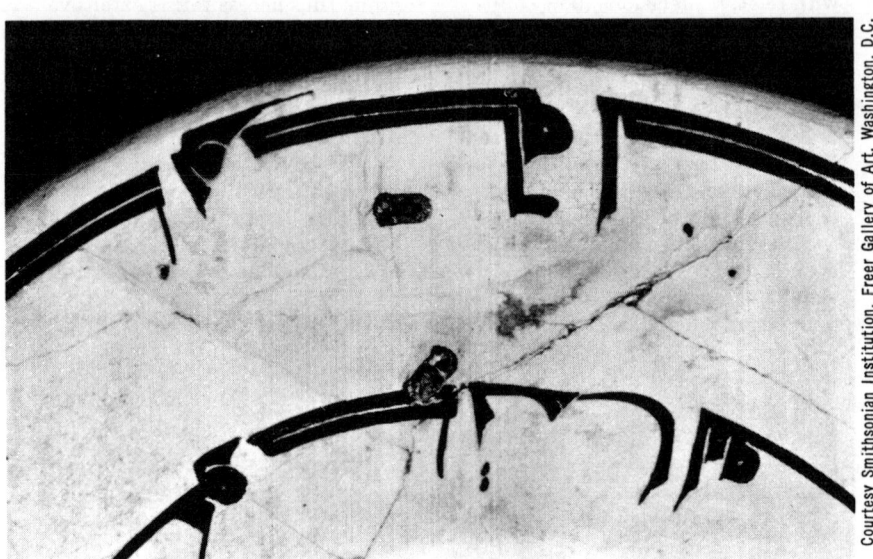

Figure 1. Ceramic plate, Nishapur, 11th century A.D.
Metal brackets applied prior to burial indicate early repair technique

The main purpose of this committee was to study the management of the National Gallery with special attention to its restoration practices. As a result, the committee published two reports which included interrogation of Gallery personnel and various experts covering such subjects as climatic conditions in the Gallery, cleaning practices, solvents, varnishes, relining and transfer procedures, and the formal training of restorers. The man mainly responsible for this cleaning campaign and Keeper of the National Gallery at that time, Sir Charles Locke Eastlake, had published an important work on the technical history of paintings (2). Thus, around 1850 the essential ingredients of Conservation had emerged: a study of the history of technology, awareness of environmental factors, examination of restoration practices and materials, and a concern for preventive and preservative methods.

Throughout the latter half of the 19th century, interest in the problem of preserving and restoring art objects steadily increased. Berger and Eibner in Germany, Russel and Abney (3) in England, plus many

280

Art conservation is a new discipline which utilizes modern analytical chemical techniques in the study and preservation of unique objects of artistic and cultural importance. Today's public consciousness of cultural heritage has elevated conservation to a new significance in the museum world

others concerned themselves with understanding not only the materials employed but also the effect of environmental agents on these materials. In this century knowledge has rapidly increased, and new materials and technology exist which are continually being evaluated by conservators for their potential application to the preservation of art objects. In the United States, first at the Boston Museum of Fine Arts in 1928 and then at the Fogg Art Museum in 1932, scientific laboratories devoted solely to the study of art objects were established. In the last 30 years many conservation laboratories have been established —i.e., Instituto del Restauro, Rome; Institut Royal du Patrimoine Artistique, Brussels; and the Los Angeles County Museum of Art in January 1967.

Conservation can be defined as the application of science to the examination and treatment of objects of art and to the study of the environments in which they are placed. Art restoration is that portion of conservation which deals primarily with the treatment of objects. It should be understood that restoration does not imply an attempt to return the object to its original state but rather to prevent deterioration of the original materials while respecting their integrity.

Environment and Art Object

Environment, as related to the art object, is defined as the aggregate of all the external influences on the object. Light, humidity, and atmospheric pollution are the three major elements which concern the museum conservator. The effects of light on art objects have been studied since the late 19th century—first by George Field who studied the fading and darkening of pigments and then McIntyre and Buckley who noted the effect of humidity on the rate of fading. More recent studies have revealed that short wave ultraviolet in daylight and fluorescent lighting is the most dangerous to museum objects (4). By far the most susceptible objects to fading are textiles, watercolors, pastels, inks, and colored prints which often have organic pigments and dyes subject to photodecomposition.

A similar phenomenon can also occur in oil or tempera paintings when a fugitive pigment is present, although the protection afforded by the oil medium will reduce the overall kinetic rate of such processes (Figure 2). Cellulosic materials, especially papers and textiles, undergo degradation such as discolor-

Figure 2. Detail, figure of standing saint by Antonio Crivelli, Italian, 15th century

Dotted line shows area protected by frame. To right of line, deep red color of garment has been preserved. To left, faded color reveals underdrawing

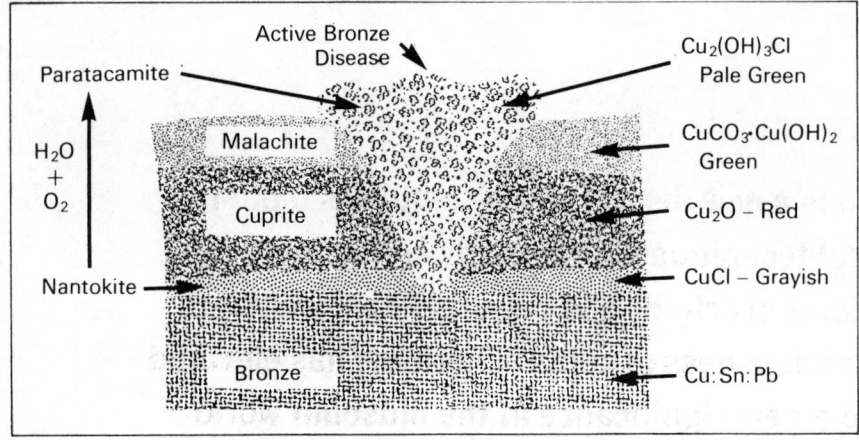

Figure 3. Schematic drawing of patination layers on ancient bronze shows active bronze disease

ing or tendering on exposure to ultraviolet light. Such harmful effects through exposure to strong ultraviolet light can be reduced considerably by the use of screening agents. Plexiglas is commonly employed as glazing in picture frames since it contains ultraviolet absorbers. Various plastic coatings containing absorbent materials are available to coat windows and skylights, and ultraviolet absorbing sleeves are manufactured to slip over fluorescent tubes (4).

The control of relative humidity in the museum is of tremendous importance because it directly influences the dimensional stability of certain materials. Objects may also develop fungi, effloresce salts, or otherwise deteriorate if strict humidity control is not maintained. The lower limits of relative humidity are dictated by objects such as panel paintings and furniture which have complicated wood structures and which respond dimensionally to humidity changes. A relative humidity below 50% may cause paint to flake, furniture and wooden sculpture to develop cracks, and Oriental scrolls to curl. On the other hand, relative humidities above 65% can foster mold growth in paper, canvas, and textiles. For these reasons, museums attempt to maintain a relative humidity at some point between 50–65% at a temperature between 68–72 °F.

Often individual objects require special attention with respect to relative humidity because of their chemical make-up or because of unusual structural problems. For example, bronzes buried for a long time acquire a protective coating or corrosion layer consisting usually of basic copper carbonate (malachite) and cuprous oxide (cuprite). Sometimes a thin layer of cuprous chloride is also present (Figure 3). Loss of the overlying malachite and cuprite accelerates the conversion of this cuprous chloride to cupric chloride—a light green powdery deposit (Figure 4). This so called "bronze disease" eats away at the bronze unless arrested by treating or by maintaining a low relative humidity.

Yet another unusual problem arose when Rembrandt's famous "Self-Portrait" of 1638 (see cover) was purchased in London about two years ago and brought to Los Angeles. Although the Los Angeles County Museum of Art has well controlled relative humidity at 52% ($\pm 4\%$), the painting had adjusted to the damper conditions in Great Britain. The panel on which the self-portrait is painted is a complex structure which when slightly warped has tremendous stress imposed upon it by the distortion of the horizontal elements at the top and bottom, much the same as a bow drawn back under tension. By experimentation in an environmental chamber, the panel, warped badly at 52% RH (museum conditions), was totally relaxed at 60 $\pm$ 2% relative humidity (Figure 5). A plexiglas case designed to fit into the ornate Louis XIV antique frame was filled with silica gel previously conditioned in a chamber to 60% RH and sealed with the painting inside (Figure 6). The special case with appropriate hygrometer and thermometers visible on the reverse maintains the internal humidity required to relieve tension within the panel support. From the normal front view, the painting appears simply as if glazed for protection against light and dust (Figure 7).

Atmospheric pollutants are somewhat more complex and difficult to overcome than light and humidity. In the urban atmosphere such as Los Angeles, sulfur dioxide, hydro-

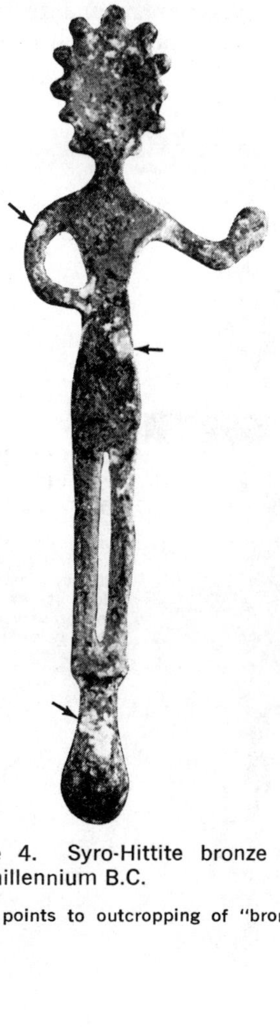

Figure 4. Syro-Hittite bronze figure, 2nd millennium B.C.

Arrow points to outcropping of "bronze disease"

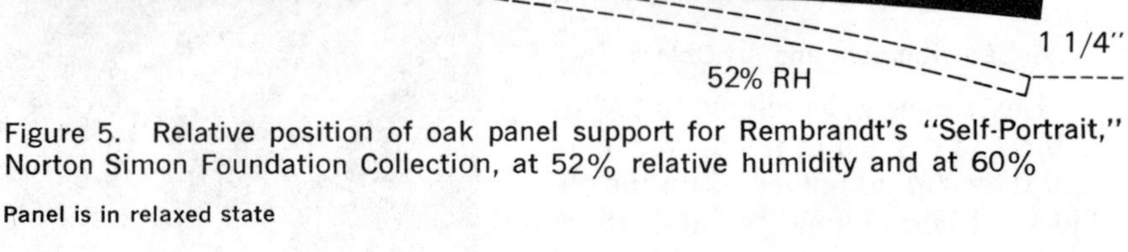

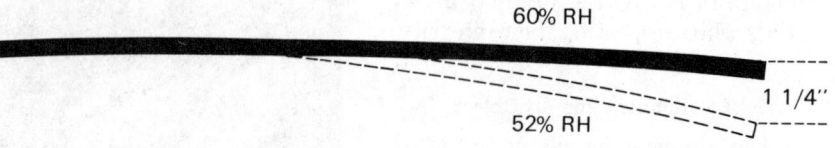

Figure 5. Relative position of oak panel support for Rembrandt's "Self-Portrait," Norton Simon Foundation Collection, at 52% relative humidity and at 60%

Panel is in relaxed state

Figure 6. Humidity control case during assembly

Painted panel (in small padded frame) is facedown in plexiglas case; silica gel framework inserts over panel back; neoprene tubing rests in groove visible on upper edge of framework, back plexiglas cover

Figure 7. Rembrandt's "Self-Portrait" as it appears on exhibit

Humidity control case aids in preservation of painting, yet does not impede viewer's enjoyment of the great masterpiece

gen sulfide, and ozone can cause rapid defacement of works of art.

Examination of Objects of Art

When an object enters the Conservation Laboratory, it undergoes an examination to determine its state of preservation. Information on the original materials and structure, former restorations, and deterioration is collected and used in determining the best technique for preserving the original. Of equal importance, the examination fulfills the purely academic function of providing knowledge of earlier cultures and technology. The conservator, curator, or collector often suffers the embarrassment of not knowing how to answer the laymen's simplest questions, "What is it made of? How was it done?" Through documentation and analysis, the conservator is only recently beginning to provide a few of the basic answers on technology and materials of earlier cultures.

The examination of an art object generally begins with a simple visual inspection by use of special lighting to enable the conservator's trained eye to distinguish differences in color, gloss, texture, or other material irregularities which may indicate problems in the structure. Based on the results of the initial observations, photographic techniques are used to document and further study the condition of the object. Raking light photographs (a strong light is placed at an oblique angle to the surface) show irregularities or deterioration in the structure because of the emphasis on surface relief. When a strong photo light is placed behind the painting, light transmitted through the canvas and paint structure can show old damages and losses and occasionally give a clue to the painting technique (Figure 8). Examination by ultraviolet light sometimes enables differentiation between the restored and original materials because of a difference in fluorescence.

The fluorescence of some pigments and varnishes provides information which may be important during treatment of the painting. Infrared photographs can reveal damages or a preliminary drawing (5) under layers of dark varnish or brownish paint which may be transparent to invisible infrared. Recently, color infrared photographs have been used by the authors to identify various pigments, both organic and inorganic, on Indian miniature paintings dating from the 10th through the 19th centuries (6). X-rays (first applied to paintings

283

Figure 8. Transmitted light photograph of early American painting is used to document hundreds of small losses of paint

in the paint structure. White lead is usually more abundant and thicker than most pigments, but others such as vermillion (HgS) can also be recorded on the X-ray film. In addition, old repairs and retouchings seldom have the same densities as the original paint and are detectable in radiographs. Rembrandt, Van Gogh, and other artists made frequent compositional changes which when studied in radiographs give insight into the understanding of the artist's technique (Figure 9). Radiographs are extremely useful for three-dimensional objects where inner structures, joints, cores, etc., can be seen (Figure 10). Neutron radiographs have proved useful for studying the organic matter trapped in metal objects, specifically the core material in ancient bronzes which contains carbon (Figure 11).

The stereobinocular microscope, which magnifies up to about 40×, is used to make more exacting observations on paintings. One can distinguish retouching from original paint, age cracks, individual pigment particles, brushwork, etc. During treatment the stereomicroscope is used for taking minute samples (less than a milligram) which can be used for analysis by wet chemical methods utilizing the compound microscope under numerous types of light conditions—i.e., polarized, fluorescent, transmitted. Cross sections of paint structures studied under 100–200× magnification show clearly the stratography of the painting—i.e., how it was built up from the support and of what pigments the layers are composed. For example, a rich luminous maroon from a 15th century Flemish painting may owe its beauty to an underlayer of fiery vermillion. Cross sections of ceramics, bronzes, and even papers give extremely valuable information in determination of what treatment should be applied.

Although every conservation laboratory relies heavily on traditional analytical methods, most have specialized in some area of instrumental analysis. More and more, conservation chemists are tackling specific research problems, either organic—i.e., media, varnishes, dyes and adhesives—or inorganic pigments, metals, ceramics, and stones.

Because of space limitations, it is possible to mention only a few of

by Röentgen and Curie) can reveal old damages and artist's changes and may help in the study of the structure of the painting.

Since the major white pigment used in European easel painting until the early 19th century was lead white ($PbCO_3$), exposure to X-rays results in a film record of densities

Figure 9. Left, "L'Hiver" by Van Gogh, Norton Simon Foundation Collection. Normal photograph and radiograph (30 kV, 5 mA, 1 min, 30-in. distance). Right, Van Gogh used one of his earlier canvases of a "Woman Spinning" to paint winter scene

Radiograph is so clear because first painting was done with lead white pigment, whereas in top of winter scene, zinc white, a less dense pigment, was used

 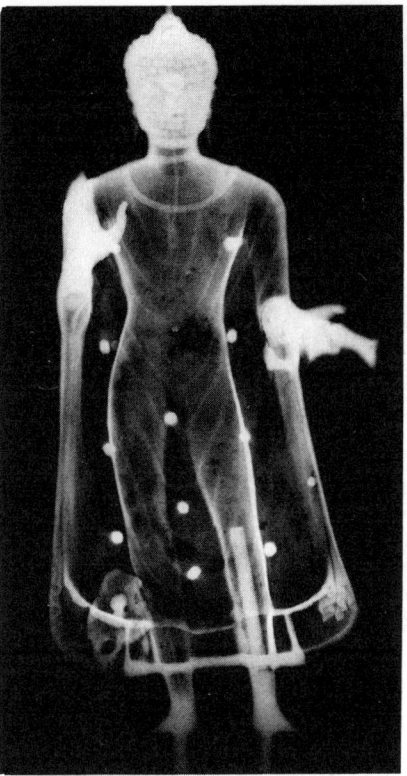

Figure 10. Left, normal photograph and right, radiograph of bronze "Buddha," Gupta period, ca. 6th century, Indian, Los Angeles County Museum of Art

Radiograph (250 kV, 6 mA, 10 min, 40-in. distance) shows modern repair in left leg proper. White dense areas are part of internal sprue system used in casting hollow bronze

phase of study on the Indian miniature research at the Conservation Center is the identification of organic pigments.

More progress has been made in the field of inorganic analysis, partly because any organic materials introduced during modern treatment do not confuse the issue and partly because of the statistical abundance of metal and ceramic objects from important ancient cultures. Gettens at the Freer Gallery of Art has pioneered the study of ancient Chinese bronzes (11). His definitive work combines wet chemistry and emission spectrometric analyses to determine major, minor, and trace elements of bronze ceremonial vessels. Metallographic studies on etched sections and microscopic surface studies along with radiographs and X-ray diffraction analysis have enabled Gettens to determine composition, fabrication methods, surface decoration techniques, and patination types. The monumental nature and classical quality of Getten's research have established the standard for future research on art technology.

the instrumental analytical techniques which have proved most useful. The work of Stolow, utilizing the technique of gas chromatography to separate linseed oil components, has been most rewarding (7). Essentially, the triglycerides of three of the basic fatty acids (oleic, linolenic, and linoleic) in linseed oil are converted by the process of transesterification into methyl ester molecules volatile enough to be separated on a gas chromatographic column. Not only has Stolow's work been most enlightening in understanding the essential processes involved in drying of linseed oil, but with further work it may prove useful in the study of modern paintings of doubtful authenticity. Stolow's earlier research on the effects of various solvents on linseed oil films has contributed greatly to the evaluation of picture cleaning processes.

Although used to identify media and adhesives, infrared spectroscopy has proved most useful in studies of varnishes used as surface coatings on paintings (8). In addition to fundamental research on media and adhesives, the identification of varnish type films often becomes essential to practical problems in cleaning. Robert Feller of the Mellon Institute has done invaluable work on varnish materials and has defined their physical as well as chemical properties. He has led a campaign to develop the perfect picture varnish (9). Modern plastics such as acrylics and vinyl acetates have proved to be tough and enduring replacements for the traditional mastic and damar resins which discolor and embrittle badly with age.

Recently, work has been carried out at the Los Angeles County Museum of Art with the MS902 mass spectrometer for analysis of media in Indian miniature paintings (Figure 12). The medium, usually an exudate from a tree such as gum arabic, is first hydrolyzed into its basic components—i.e., galactose, arabinose, mannose. A refinement of the chromatographic technique as described by Mme. Flieder (10) enables identification of the various gums. Individual components are further studied through mass spectrometric analysis. The next

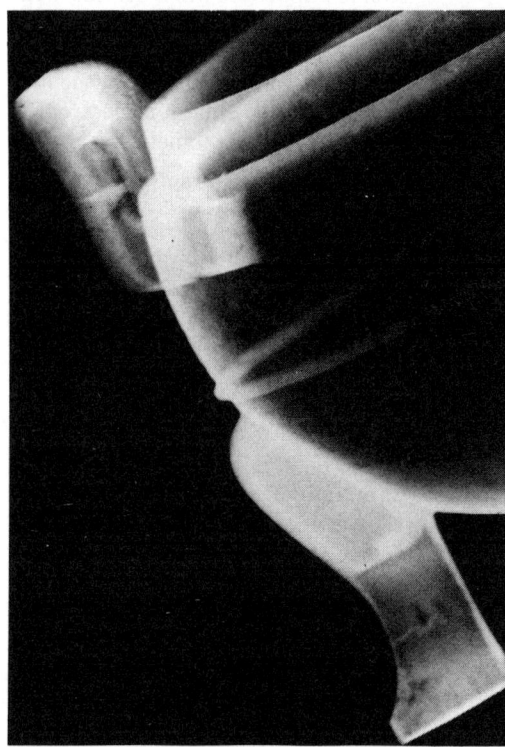

Neutron radiograph courtesy Atomics International, North American Rockwell

Figure 11. Neutron radiograph of Chinese bronze ceremonial vessel, "Ting," Chou dynasty

Note clay core in leg

Figure 12. MS902 mass spectrometer at Los Angeles County Museum of Art, Conservation Center

Instrument was gift of Mr. and Mrs. Stanton Avery

X-ray diffraction techniques have become important for pigment studies not only because of their sensitivity but also because of the differentiation of crystal structure (*12*). This is important when dealing with materials chemically similar but with different crystal designs—i.e., azurite vs. malachite or chrysocolla. Combined with emission spectrography or mass spectrometry, this provides a thorough and accurate analysis.

Inorganic mass spectrometry has been successfully applied to art objects, especially for metals and pigment analysis (*13*). Currently, the Los Angeles County Museum of Art is using mass spectrometry to study a large group of Luristan bronzes approximately 3000 years old. The small sample size (milligram range) and the sensitivity to trace elements provide potential for studies on geological sources for materials of art. This avenue of study has previously scarcely been considered because of the large quantity of material required to obtain accurate trace analysis with other techniques. Mass spectrometry has also been applied to pigment studies (European easel paintings, Indian miniatures, etc.); currently, a complete collection of pigments (Forbes Collection) provided by the Conservation Center of New York University is being fingerprinted in Los Angeles.

Recently, X-ray fluorescence has been used at the Winterthur Museum by Hanson for studies of silver objects. The technique is extremely useful as it is not necessary to remove a sample of the object but simply to focus on a surface zone. He has so far been able to establish differences in British sterling and American silver of contemporaneous dates.

William Young at the Conservation Laboratory of the Boston Museum of Fine Arts has applied the laser microprobe to the analysis of art objects with great success (*14*). Advantages of this technique are its removal of a sample only about 50–80 μ in diameter and the easy analysis of nonmetallic objects. It is also extremely useful on paintings since it combines the sampling and analysis in a single stage.

The increasing importance of analysis in evaluation, authentication, and treatment has prompted the conservator to rely heavily on new, sensitive techniques and to refine sampling and sample handling so that he can deal with extremely small specimens. The techniques briefly reviewed in the foregoing paragraphs are but a few of the many promising analytical techniques in current use (*15*).

Treatment of Art Objects

The most difficult and demanding task facing the Conservator is the treatment of an art object. It is impossible to discuss here all types of deterioration which art objects undergo, each of which may require a unique treatment process if preservation is to be ensured. Art objects are individuals by their very definition, and each must be studied to determine its makeup and properties. Only after a thorough study and much testing can the proper treatment be developed.

The treatment of paintings requires considerable experience and knowledge of painting techniques. The nature of the painting support, whether canvas, panel, metal, glass, paper, or other, will greatly influence the decision as to what should be done to preserve it (Figure 13). On the support is a rather complex structure of ground and design layers. These may be any one of numerous types of paint—i.e., oil, egg tempera, glair (white of egg), watercolor (gum arabic), distemper (animal glue), or a synthetic resin (acrylic or vinyl). Often the ground layer is one type of medium —i.e., glue with drying oil or egg tempera on top. To further com-

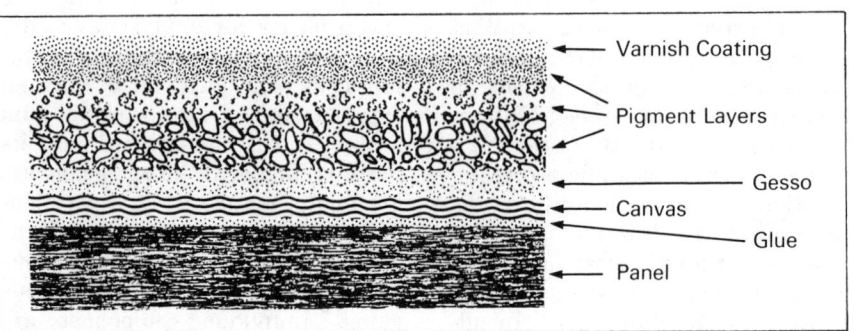

Figure 13. Schematic drawing of structure of early Italian painting shows complex structure from panel support to surface coating

Figure 14. "Portrait of Viscount Barrington" by Gilbert Stuart, American, late 18th–early 19th century, Los Angeles County Museum of Art

Painting is shown during cleaning process with varnish removed from left side and small rectangle in right center area. Effect of discolored varnish can be seen most easily in face but is also noticeable in cloak where details and highlights have appeared. Photograph was made without applying wetting agent or varnish in cleaned area

plicate the matter, a surface coating is usually present of natural soft resin (damar, mastic), wax, glue, natural hard resin (copal, amber), or synthetic resin (acrylic, vinyl). Not only are we faced with a complex structure of materials with vastly different chemical and physical properties, but the technical idiosyncrasies of each artist must be considered.

Perhaps the most complex process is safe removal of varnish and retouching from old master paintings. Varnish removal from an oil painting has both physical and aesthetic value. A vast majority of oil paintings or egg tempera paintings were covered with soft resin varnishes sometime within the last century and a half. Although virgin surfaces which predate the 19th century are rare, they are encountered occasionally. Soft resin varnishes not only discolor badly, but they also embrittle and crack with age and lose their protective value.

The "Portrait of Viscount Barrington" by Gilbert Stuart (Figure 14) illustrates the detrimental effects of varnish on the color qualities of the painting. In this case, the painting was covered with a soft resin varnish which contained a small percentage of drying oil. Acetone diluted with naphtha proved effective in removing the varnish film without disturbing the original oil medium. Since the linseed oil polymers in the paint medium cross-link with age, they become more and more resistant to the solvents necessary for removal of natural resin varnishes. Certain periods in the history of painting present problems in cleaning—i.e., early English paintings by artists such as Reynolds, Morland, and Constable often contain oleo-resinous media which have such a high resin content that normal cross-linking of the oil does not occur. The problem of removing varnish from oil-resin films can usually be solved by utilization of a technique called "reforming" (9). The surface film is sprayed with a slow evaporating solvent mixture and left for a period ranging from a few hours to a week. The reformation of the varnish in this manner makes it soluble in much milder solvents, such as toluene, which generally do not affect an oil-resin film. It is believed that the swelling caused by the initial spray solvent mixture breaks secondary linkages in the resin molecules which would require months to be reestablished.

Two other examples will illustrate more complex problems where large areas of the original surface have been repainted by an early restorer to the extent that the composition and quality of the original design have been altered. A painting by Jusepe de Ribera, a 17th century Spanish painter, representing an old woman gesturing with her right hand and entitled "A Sibyl" (Figure 15), came to the laboratory for examination. Radiographs of the lower left quadrant revealed a child's head under the visible surface. Ultraviolet examination indicated gross overpainting in the same area. Solvent tests were made on the retouched area, and dimethyl sulfoxide used in conjunction with acetone was applied to dissolve the repaint, which was oil, but had a different solubility from the original. The child's figure was uncovered in good condition, and the original composition regained. Of course the picture must now be restudied by art historians to correctly identify the subject matter.

The second example is an egg tempera painting on poplar panel by Lorenzo Monaco representing "The Martyrdom of Pope Caius" (Figure 16). The gold background in the upper left portion and the gold decoration on the garments of the figures, as well as the helmets and gloves of the figures in armor, have been repainted by a restorer. The gold retouching could not be recorded by using photographic techniques, but a cross section made from a sample taken in the gold area clearly revealed two layers: the thick gold leaf and the thinner modern gold paint. After removal of the gold repaint, the original surface was in unusually good condition. In the armor the areas which had been overpainted a dark blue color show traces of silver leaf which would have given a rich metallic luster in contrast with the surrounding gold.

Structural work on paintings often requires lining of the original canvas, or on panels, applications of moisture barriers, or sometimes transfer of the original to a new support. Lining is generally carried out on a vacuum hot table which utilizes atmospheric pressure to hold the original canvas in contact with the lining canvas while a thermoplastic material adheres the two together. Lining adhesives require

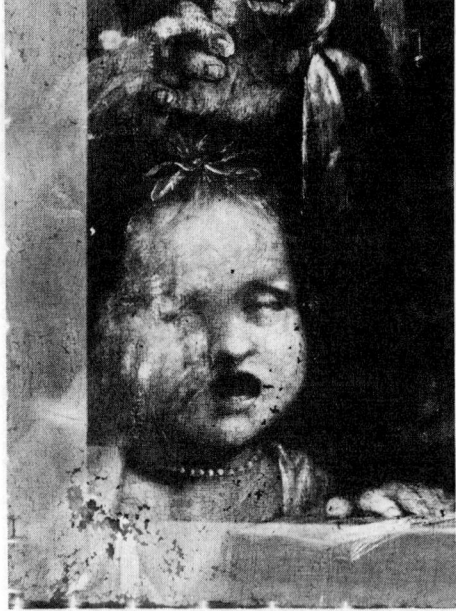

Figure 15. "A Sibyl" by Jusepe de Ribera, Spanish, 17th century, San Diego Museum of Fine Arts
Left, before treatment; center, radiograph (30 Kv, 5 mA, 1 min, 30-in. distance) of lower left portion shows child's head under visible surface; right, after treatment

Figure 16. "The Martyrdom of Pope Caius" by Lorenzo Monaco, Italian, 15th century, Santa Barbara Museum of Art
Left, before treatment; right, after treatment

more research. Traditional waxes or mixtures of waxes, resins, and balsams have been used. These adhesives have been favored because the process is reversible simply by reheating, and they also provide a certain amount of protection from humidity changes. Modern synthetic waxes or mixtures will probably yield a more neutral and durable product once enough research is done by conservators on their aging properties.

Cleaning of metals is another exacting process since their natural patinas, usually carbonates or oxides of copper on bronzes, often must be preserved. Surface dirt and resinous accretion from an Indian bronze was removed mechanically with brushes and scalpels without affecting the green malachite surface. The base of the bronze which had a heavy accretion was first bombarded with ultrasonic waves while in a detergent bath to soften the impacted dirt layer; then it could be easily removed.

If it is necessary to convert or remove the patina in a cleaning process, electrolysis can be employed. The Anatolian bull (Figure 17) shows the effectiveness of electrolysis in revealing a silver surface. The piece was first wound with copper wire and then hung as a cathode in a bath with an electrolyte of 2% sodium hydroxide. Iron anodes were hung on either side of the bull,

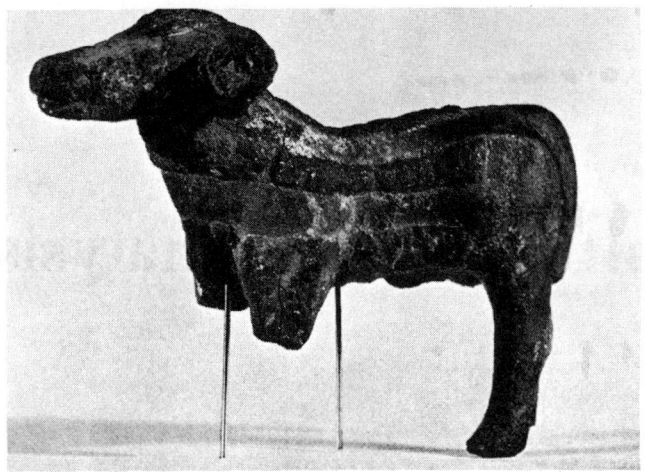 

Figure 17. "Anatolian Bull," Boston Museum of Fine Arts
Left, before treatment; right, after electrolytic treatment. Note especially well-preserved silver surface

and a 0.5 ampere current was applied. The electrolytic reduction took about five days with washing and brushing after three days.

Small areas of "bronze disease" (see Figure 2) can be treated by excavation of the area and packing with silver oxide. The silver oxide combines with the copper chlorides to form silver chloride which is insoluble and will stop further deterioration. Where entire objects require treatment for "bronze disease," they can be soaked in a solution of sodium sesqui carbonate (5%). The piece is washed frequently until chlorides can no longer be detected in the wash water.

Graphic arts such as prints and drawings have often been damaged owing to negligence and poor framing practices. Pulpwood cardboard, extremely acidic (pH ≈ 4.5), was for many years the most popular type of mounting and matting material. Of course, the acidic property of the surrounding materials migrates into the original paper and causes embrittlement and discoloration. Unlike easel paintings where damage to support does not always mar the design itself, the paper is usually an integral part of the graphic expression, and deterioration can cause a loss in quality. In addition to removal of the acidic materials and replacing them with chemically neutral matt boards, the original paper often requires bleaching (*16*) to remove stains and discoloration, and deacidification to prevent further deterioration. Deacidification is accomplished by application of a magnesium bicarbonate to the paper to neutralize the pH (*17*).

In next month's issue the authors will discuss some of the more significant conservation research projects and techniques of analysis.

References

(1) D. Coekelbergs, "Precisions sur la vie et L'oeuvre du Peintre-Restaurateur Bruxellois Frederic Dumesnil (vers 1710–91," *Bulletin, Institut Royal du Patrimoine Artistique*, Brussels, Belgium, **11**, 174 (1969).

(2) C. L. Eastlake, "Methods and Materials of Painting of the Great Schools and Masters," Vol I, II, Dover Publications, New York, N.Y., 1960.

(3) E. Berger, "Beitrage zur Entwickelungs-Geshichte der Maltechnik," 4 vol, Munich, Germany, 1901–2; A. Eibner, "Entwickelung und Werkstoffe der Wandmalerei," Munich, Heller, 1926; W. J. Russel and W. L. Abney, "Report to the Science and Art Department of the Committee of the Council of Education on the Action of Light on Watercolors," H.M. Stationary Office, London, England, 1888.

(4) R. L. Feller, "Control of the Deteriorating Effects of Light upon Museum Objects," *Museum*, **XVII** (2) (1964).

(5) J. R. J. Van Asperen de Bier, "Infrared Reflectography—A Contribution to the Examination of Earlier European Paintings," J. F. Duwaer, Amsterdam, Holland, 1970.

(6) B. B. Johnson, "The Technique of Indian Miniature Paintings," Los Angeles County Museum of Art Symposium on Indian Art, to be published, 1971.

(7) N. Stolow, "The Application of Gas Chromatography in the Investigation of Works of Art," in "Application of Science in Examination of Works of Art," Boston Museum of Fine Arts, Boston, Mass., 1967, pp 172–83.

(8) R. Kleber and F. Tricot-Marckx, "Identification d'une Vernis Moderne Recouvrant la Decente de Croix de Rubens," *Bulletin, Institut Royal du Patrimoine Artistique*, Brussels, Belgium, **6**, 63 (1963).

(9) R. Feller, N. Stolow, and E. Jones, "On Picture Varnishes and Their Solvents," rev. ed., Case Western Reserve University, Cleveland, Ohio, 1971.

(10) F. Fleider, *Stud. Conserv.*, **13**, 49 (1968).

(11) R. J. Gettens, "Freer Chinese Bronzes, Vol II, Tech. Studies," Smithsonian Inst., Washington, D.C., 1969.

(12) H. Barker, "Spectrographic and X-ray Diffraction Methods in the Museum Laboratory," in "Application of Science in the Examination of Works of Art," Boston Museum of Fine Arts, Boston, Mass., 1967, pp 218–21.

(13) T. Cairns, "Spark Source Mass Spectrometry," IIC-AG Technical Papers 1968–70, Conservation Center, NYU, New York, N.Y., 1970, pp 47–58.

(14) W. Young, "The Laser Microprobe and its Application to the Analysis of Works of Art," in "Application of Science in the Examination of Works of Art, Boston Museum of Fine Arts, Boston, Mass., 1967, p 230.

(15) For further reading, see "Studies in Conservation," Aberdeen University Press, Aberdeen, Scotland, published quarterly, 1955–present; also, "Recent Advances in Conservation," G. Thompson, Ed., Butterworths, London, England, 1963.

(16) For various bleaches, see H. J. Plenderleith, "The Conservation of Antiquities and Works of Art," Oxford University Press, London, England, 1956.

(17) W. J. Barrow, Spray Deacidification Permanence/Durability of the Book III, W. J. Barrow Research Laboratory, Richmond, Va., 1964.

PART II

Art Conservation: Culture Under Analysis

BEN B. JOHNSON and THOMAS CAIRNS
Conservation Center
Los Angeles County Museum of Art
Los Angeles, Calif. 90036

With the advent of modern methods of chemical analysis, such as polarography, atomic absorption, neutron activation analysis, energy-dispersive X-ray emission spectroscopy, laser microprobe analyzer, and spark source mass spectrometry, a great deal of attention has been focused recently on detailed analysis of art objects (1) with two objectives in mind. Firstly, there are the systematic programs of analyses of a large number of closely related objects in an attempt to fingerprint them by a good statistical profile, i.e., trace elements in cast bronzes. Such programs deal with research on the history of technology answering such questions as source of raw materials and techniques in fabrication. Secondly, there are specific analyses for practical purposes. This often involves the analysis of a single sample or even a small group of samples to answer specific questions usually relating to treatment.

In this respect the attention of the chemist is drawn to relate chemistry to authenticity. Until recently anachronisms in the use of materials were a good guideline to follow, but latter-day forgers try hard to mimic the original artist and/or technique. An outstanding example in the use of materials lies in the chronology of white pigment—lead white (PbO) was used since classical times; zinc white (ZnO) made its appearance around 1810; titanium white (TiO) was commercially available around 1920. Evidence of zinc white or titanium white in a 17th century style painting would certainly raise questions as to its authenticity. At this point it must be emphasized that the sample taken for analysis by the chemist in the museum laboratory is from an original area and not from a recently restored area where later materials might well have been employed. Without a prior physical examination of a painting, for instance, by uv, ir, or X-rays, mistakes might easily be made by a novice to the conservation–chemistry field. Frankel (2) in a recent article demonstrated the potential use of X-ray fluorescence (3) in the so-called detection of art forgeries via pigment identification. Such publications can, in this instance, give false impressions that chemistry alone tackles the job whereas interdisciplinary approaches are really necessary. In matters of authenticity, therefore, three points finally decide the outcome: scientific data, stylistic criteria, and conservator's experience and availability of related comparative material.

Instrumentation

Conservation–chemistry covers an extremely wide spectrum of materials for analysis. In the field of inorganic chemistry one encounters materials from metals, pigments, and stones to ceramics. On the other hand, organic materials encountered are mainly natural products, i.e., gums, glues, resins, oils. This conglomeration prompts the utilization of sophisticated instrumentation capable of handling such diverse materials.

Chemists have long been aware of the restrictions imposed on their studies by the limits of sensitivity of many of those currently available analytical methods which permit a number of elements to be determined simultaneously in one sample. This has led to the adoption of more sensitive methods for a restricted number of elements. e.g., neutron activation analysis of paper (4). Recently, spark source mass spectrometry has provided the chemist with the ability to cover the full range of elements in any sample in a single determination and the ability to detect those elements down to very low concentrations, i.e., parts in 10^6. The AEI MS702 spark source mass spectrometer in the Conservation Center of the Los Angeles County Museum of Art is used to tackle a variety of problems encountered in the conservation–chemistry field. Today, spark source mass spectrometry has found its way into even more new avenues of research. It has been used alongside atomic absorption and neutron activation analysis by Morrison and Kashuba (5) in the analysis of returned lunar samples (6). Harrison et al. (7) have reviewed the forensic application of spark source mass spectrometry in analyzing such products as hair and glass. Recently, the FBI laboratory in Washington, D.C., has acquired a spark source mass spectrometer. The technique owes its widespread success to its unique capability of multielement determinations at low concentrations.

With such diverse materials as gums and oils, separation into ma-

Conservation–chemistry represents a new and fascinating applied scientific discipline harnessing chemical knowledge to unlock the secrets of the history of technology. Particular attention is paid to ancient metallurgy and pigment and media studies

jor and minor components is of primary importance, followed by identification. To achieve this ultimate goal on small samples, a combined gas chromatograph–mass spectrometer system must be employed. The AEI MS902 high resolution mass spectrometer at the Conservation Center is used in conjunction with a Perkin-Elmer F-11 capillary gas chromatograph.

Reasons for Analysis

Concentration on chemical analysis of art objects is generally prompted by the curator, collector, and art historian. He appreciates the beauty and craftsmanship of an ancient object. However, he is unable to look into the historical study of technology since there are very few texts on the subject (8). Each object can be considered unique and has locked into its structure and fabrication some knowledge of the properties of the materials used. From the standpoint of his materials, the artist was rarely concerned about any scientific investigation into their nature—his concern was solely directed to physical properties. The aesthetic enjoyment of an object is greatly enhanced when its original creation is fully understood. In the case of metallic objects, technical analyses can often answer questions as to the source of raw materials (mines, trade), processing (smelting, cupellation), models, casting (molds, alloy composition), finishing (cold working, techniques such as chasing, incising, engraving, joining), and how these various stages are interrelated, if at all.

Were the processes involved indigenous achievements or results of cultural exchange?

Ancient Metallurgy

Sampling. Ancient metals and their alloys are sometimes either homogeneous or heterogeneous, and truly representative sampling is difficult to obtain (9). In particular, the existence of more than 4 wt % of lead (Pb) in a cast bronze (Cu/Sn) is clearly demonstrated (polished cross section) by the appearance of globules of Pb of widely varying size irregularly distributed throughout the Cu/Sn matrix (10), i.e., Pb is insoluble in the Cu/Sn alloy. It is strongly recommended, therefore, that both X-ray examination and, if possible, cross sections be prepared before sampling a structure. Knowledge that such a structure is under investigation aids interpretation of analyses.

One of the greatest problems today is comparing results from two different sources by two different techniques. At the moment a comparative study organized by the International Council of Museums (ICOM) between a number of museum laboratories is underway (11). The same samples are being analyzed by different techniques to document standard deviations and accuracy of results by the various techniques available in the conservation field throughout the world. For instance, Caley (12) has pointed out that the determination of tin by wet chemical methods is known to yield high values by about 10%. On the other hand, spark source mass spectrometry boasts of $\pm 10\%$ accuracy for "trace elements" (13).

With spark source mass spectrometry the question of homogeneity is a critical one since the actual sample size consumed during a typical analysis can be of the order of a few milligrams. The introduction of the ion beam chopper (selective sampling of the positive ion beam) has greatly improved the reproducibility (14) of analysis for both homogeneous and heterogeneous materials by use of the integrating properties of the Q2 photoplate. However, spark source mass spectrometry relies heavily on the use of standard reference materials, i.e., Cu doped with various trace elements at known concentration levels. The few existing bronze standards are somewhat inadequate and usually too heterogeneous for use in spark source mass spectrometry. In particular, standards prepared via a dilution technique are extremely heterogeneous (15).

Trace element analysis must be treated with some caution when interpretation is attempted. Two main sources of trace elements are the ores from which the metals were smelted and the smelting technique. To date, detailed trace element analysis by spark source mass spectrometry and neutron activation analysis is fragmentary, and any definitive conclusions are only tentative.

Historical. In an attempt to understand the development of metallurgy in ancient times, it is necessary to outline the chronological or-

Figure 1. Typical Luristan bronze bar bit depicting a wild mountain goat in profile with its head seen frontally with a bird on its rump (L.2567.67-241, 12.5 × 11.5 cm, bit 21 × 1 cm, wt 613 grams)

der of events that led man to casting and alloy formation. Admittedly, native gold and silver were among the first metals to be used, but native copper is reported to have been first discovered about 5000 B.C. in the Sinai Peninsula. The importance of copper (and of bronze) lies in the fact that it was extensively used, and its corrosion in soil generally forms a stable patina. Persia, rich in ores of all kinds, was logically a center of early ancient metallurgy (16).

By 4000 B.C. the discovery of the reduction of oxide ores by smelting with charcoal took place. Copper ores employed were probably the two oxides, cuprite (Cu_2O) and tenorite (CuO) and the two carbonates, malachite [$CuCO_3 \cdot Cu(OH)_2$] and azurite [$2CuCO_3 \cdot Cu(OH)_2$]. Carbon, coal, or coke was heated in the presence of oxygen to form carbon monoxide which in turn reacted with the ore to give molten metal and carbon dioxide. This discovery enabled man to develop the art of smelting and casting. To cast a pure copper artifact presented some difficulties. Pure copper melts at 1083°C. It is a very sluggish viscous liquid. Such high temperatures would certainly have taxed the furnace capabilities of the artisans at this early date. However, a number of pure copper castings do exist, mainly from Northern India and Tibet but usually of a much later date.

The restrictions imposed by attempting to use pure copper for direct casting resulted in the discovery of adding tin ore to the copper ores during smelting. Tin was utilized from around 1800–1600 B.C. in North West Persia in the form of cassiterite or stannic oxide (SnO_2). Bronze was therefore born. The addition of tin to copper in the ratio 25:75 reduces the melting point from 1083° to 795°C—a considerable drop. Such an alloy is more fluid and presents fewer problems in casting. Pure copper has the tendency to shrink greatly on cooling and thereby lose any fine detail of the original mold. Bronze, on the other hand, shrinks less and is more malleable and easier to cold work. Techniques of metal working in the Far East also developed as the Chinese during the Shang dynasty (ca. 1523–1028 B.C.) and the Chou dynasty (ca. 1027–222 B.C.) achieved the zenith of practical perfection in their bronze castings (10).

Another important factor in bronze technology is the use of lead in the Cu/Sn alloy especially where the casting was to be extensively cold worked. The principal lead ore is galena, lead sulfide (PbS), and might well have been known before cassiterite and bronze since smelting of this particular ore was primarily undertaken to separate out the silver content by a process commonly known as cupellation. The dross that forms on the surface of the crucibles containing the melt is continually removed until a shining surface is obtained. This dross contained all the base metal impurities. Separation of gold and silver is achieved by a modification of this process whereby salt is added to the melt to remove the silver as chloride. This so-called chlorination modification has been in practice since the second millennium B.C. In this way both gold and silver were refined to a fairly high degree of purity. Rarely, however, was cupellation carried to the ultimate extreme. The addition of lead, therefore, to bronze was a practical suggestion and found widespread usage in casting imparting to the alloy desirable extra qualities.

The entrance of zinc (Zn) into bronze technology might well have been extremely early especially if either the copper ore or the tin ore contained a natural amount of Zn as an impurity. During the smelting of such ores, however, the metallic Zn vaporizes extremely easily

Figure 2. Krishna Rajamannar Bronzes, South Indian, early 12th century. From left to right, Rukmini, Krishna, Satyabhama, and Garuda (M70.69.1;2;3;4.)

and would have been lost to the atmosphere save trace amounts. It was not until around 800 B.C. that brass (Cu/Zn) started to occasionally appear (17). Finely ground calamine $[Zn_4Si_2O_7(OH)_2 \cdot 2H_2O]$, charcoal, and granulated copper are placed in sealed crucibles and heated. The reduced metallic Zn vaporizes and then alloys with the copper since the reaction takes place in a sealed system. Hence the appearance of substantial quantities of Zn in ancient cast bronzes would suggest questionable authenticity.

Luristan Bronze Study. A carefully selected group of Luristan bronze horse bits from the Foroughi collection (18) are at the moment under investigation by spark source mass spectrometry. Such horse trappings are in a very distinctive decorative style (Figure 1) and belong to the period during the 7th and 8th centuries B.C. They have no counterparts outside their point of origin—the mountainous province of West Persia. Beginning early in this century, excavations have revealed more and more of these beautiful ancient bronzes belonging to the Luristan civilization. Such prerequisites made them very suitable candidates for a study of the potential of trace element profiles in characterizing artifacts of an ancient culture (19).

Drillings were taken from both the cheek plates and bit of one such trapping. In essence, to date, the results may be summarized briefly since the study has not yet been fully completed. The major constituents (Cu/Sn/Pb) vary from piece to piece, but the presence of the same trace elements (Bi, Sb, Ag, Se, As, Ni, Fe, Co) were noted in each analysis within certain concentration levels. A number of the bits, however, were manufactured from native copper and are excluded from the above generalization. Work is continuing to achieve the number of analyses needed to draw reliable conclusions.

The noticeable absence of zinc is worthy of mention. A number of forgeries, declared such on stylistic grounds, were also examined. The presence of high percentages of zinc (above 2 wt %) together with other trace elements certainly damned them, scientifically speaking. Such a combination of science and stylistic criteria is the most effective method in the detection of fakes and forgeries.

South Indian Bronzes. Quite recently an interesting bronze group of the early 12th century A.D. depicting Krishna and his two wives, Rukmini and Satyabhama, together with his messenger, Garuda (Figure 2), were acquired by the Los Angeles County Museum of Art through the generosity of Mr. and Mrs. Hal Wallis. The occurrence of such a group of high quality and importance is unique. On religious grounds they were probably cast and fabricated (cold worked after casting the rough shape) as a group with Sutras dictating proportions, attributions, etc. This prerequisite, therefore, permitted a detailed study (20) of the chemical composition from piece to piece in an attempt to ascertain if an exact science was operative or was each composition by chance. It turned out that all four pieces resembled each other fairly closely. Drillings were taken from various locations in each piece and analyzed several times, then averaged. Table I illustrates the typical results obtained for Satyabhama (Figure 3). Many other South Indian bronzes are now under study to determine whether any further correlations can be established.

Mughal Indian Miniature Painting

Although there are several excellent studies of the technique of Indian miniature painting which are based purely on literary sources and tradition passed from generation to generation, until present day very little has been published which deals with the technical examination of actual paintings. Chandra's treatise (21) is a wealth of information but leaves many basic chemical questions unanswered, e.g., whether lead white, which is poisonous and can turn black in the presence of sulfur, was used by Mughal painters in the 16th and 17th centuries or whether zinc white, which was preferred more recently, was adopted. The common appearance of ultramarine in Mughal paintings raises a natural doubt as to whether lapis lazuli $(3Na_2O \cdot 3Al_2O_3 \cdot 6SiO_2 \cdot 2NaS)$, a costly stone mineral, yielded the

Table I. Analysis Results for Satyabhama (M70.69.3)

Ht, 28 in.; base diam, 8.5 in.; wt, 58 lb

Wt %	Hip	Neck	Heel	Av	Base
Cu	93.05	92.65	92.45	92.71	93.31
Sn	2.25	2.21	2.66	2.37	1.62
Pb	3.50	3.96	3.53	3.66	4.33
Bi	0.01	0.01	0.04	0.02	0.03
Sb	0.09	0.09	0.13	0.10	0.09
Ag	0.08	0.08	0.11	0.09	0.09
As	0.022	0.026	0.04	0.02	0.032
Zn	0.08	0.11	0.09	0.09	0.05
Ni	0.50	0.45	0.55	0.50	0.40
Co	0.02	0.02	0.02	0.02	0.01
Fe	0.39	0.39	0.38	0.39	0.03

Figure 3. Satyabhama, one of Krishna's two wives (M70.69.3.)

Figure 4. Page from a Ragamala: Todi Ragini series. Mid-18th century (M71.1.42). Woman with vina stands in grove of trees (9.5" × 6.25")

blue or whether it was obtained from azurite [$2CuCO_3 \cdot Cu(OH)_2$], a much cheaper material.

Agrawal's paper (22) is the most comprehensive report to date on materials of Indian painting. It is mainly a survey of historical references but includes scientific examination of manuscripts as well.

To answer some of the questions regarding pigments, a study was undertaken (23) of the miniature paintings (Figure 4) from the collection of Nasli and Alice Heeramaneck (24).

Pigment Identification. The limited number of pigment analyses has precluded any final conclusions at this stage, but some interesting patterns have evolved. The analysis of the blues during the early Mughal period (mid-15th to mid-16th century) has revealed that genuine lapis lazuli was the most preferred and most frequently used blue, i.e., lack of evidence for the suggested use of azurite as a cheap substitute. In addition Mughal artists preferred to use their pigments in pure form in overlapping layers or in mixtures with a second pigment such as lead white.

For white pigments Mughal artists of the first half of the 17th century preferred lead white whereas in earlier paintings belonging to the 15th and 16th centuries kaolin ($Al_2O_3 \cdot 2SiO_2 \cdot 2H_2O$) was dominant.

Vermilion (HgS) was the most widely used red pigment. Second to vermilion, minium (Pb_3O_4) was employed, especially in paintings of the 16th and 17th century. Orpiment (As_2S_3) and realgar (As_2S_2), yellow and orange, respectively, were identified on 16th and 17th century paintings but not in significant patterns to be of diagnostic value. In a study of the green pigments, both malachite and copper resinate (25) were found.

A number of organic pigments were encountered and research is in progress via GCMS to elucidate their molecular structures.

Media Studies

Paint is the mixture of a suitable medium plus finely ground pigment. Medium is the descriptive terminology applied to the binding agent or vehicle for the pigment particles. Traditionally three major classes exist which can be chemically differentiated: plant gums (gum arabic) which are polysaccharides, glues of animal origin (size) and egg-tempera which are proteins, and drying oils (linseed) which are mixtures of various triglycerides. Historically both gums and glues (including egg products) were used in earliest times. It was not until the 15th century that drying oils such as linseed prevailed and quickly replaced the use of egg-tempera. Gums, however, are still used extensively today as the medium for commercially available water colors.

Gums. Starch and cellulose are the most ubiquitous of the plant polysaccharides known. Plant gums also belong to this general class as demonstrated as early as 1929 by Butler and Cretcher (26) who identified various hexoses ($C_6H_{12}O_6$) as products from acid hydrolysis of gum arabic. In essence, plant gums are high-molecular-weight polysaccharides built up by repeated condensation of various monosaccharides (both hexoses and pentoses). The hydrolysis products of such gums from various botanical sources (27) are related to their taxonomic origin (Table II).

Besides the pigment identification of the Mughal Indian miniatures described previously, media samples were also taken (about 0.5 mg) and hydrolyzed with 3% HCl under vacuum at 105°C for 24 hr (28). The hydrolysis products were then neutralized with 200 mg Amberlite IRA 68, filtered and evaporated (12 hr at 45°C) to dryness. The residue was then examined by thin-layer chromatography according to Stahl (29) and later by gas chromatography to obtain quantitative results. In almost every case studied, gum arabic was the preferred medium. One or two cases, however, were of protein origin. Flieder (30) has successfully used this technique in the identification of both sugars (from polysaccharides) and amino acids (from proteins) in a large number of illustrated manuscripts.

Glues. Before the advent of dry-

Table II. Composition of Various Plant Gums

Common name	Source	D-glucuronic acid	D-galactose	D-mannose	L-arabinose	Rhamnose	Xylose
Gum arabic	Acacia senegal	16	52		19	14	
Cherry gum	Prunus cerasus	12	21	10	55		
Peach gum	Prunus persia	7	36		43		14

Table III. Composition of Various Proteins

Amino acid	Casein, %	Glues, % Gelatin	Glues, % Elastin	Ovalbumin, 64.9%	Conalbumin, 13.8%	Ovomucoid, 9.2%	Lysozyme, 3.4%	Avidine, 0.1%
Glycine	2	27.25	26.7	3.05	5.7	3.8	5.7	4.6
Alanine	3.2	11.23	21.3	6.72	4.4	2.3	5.8	
Valine	7.2	2.78	17.7	7.05	8.2	6.0	4.8	4.2
Leucine	9.2	3.45	9.0	9.2	8.8	5.1	6.9	4.9
Isoleucine	6.1	1.53	3.8	7.0	5.0	1.43	5.2	5.5
Serine	6.3	3.73	0.85	8.15	6.3	4.2	6.7	4.5
Threonine	4.9	2.36	1.12	4.03	5.9	5.5	5.5	10.5
Phenylalanine	5.0	2.5	6.2	7.66	5.7	2.91	3.12	5.9
Tyrosine	6.3	0.24	1.5	3.68	4.6	3.18	3.58	0.88
Tryptophane	1.2			1.2	3.0	0.3	10.6	5.4
Proline	11.3	15.47	13.5	3.6	4.9	2.72	1.4	1.64
Hydroxyproline		13.24	1.6					
Cystine	0.34		0.35	0.51	3.8	6.7	6.8	0.47
Cysteine				1.35				
Methionine	2.8	0.63		5.2	2.03	0.95	2.06	1.41
Aspartic acid	7.1	6.7	1.1	9.3	13.3	13.0	18.2	9.7
Glutamic acid	22.4	11.56	2.4	16.5	11.9	6.5	4.32	6.6
Histidine	3.1	0.7		2.35	2.57	2.15	1.04	0.96
Arginine	4.1	9.04	1.3	5.72	7.6	3.7	12.7	6.5
Lysine	8.2	4.37	0.5	6.3	10.0	6.0	5.7	6.2
Hydroxylysine		0.76						

ing oils in the 15th century, glues and egg-tempera were widely used, although isolated examples exist of the use of oils as early as the 13th century. This takes the conservator–chemist into the realms of protein and amino acid chemistry. Here again the hydrolysis products (amino acids) have a direct bearing on the origin of the protein used (Table III). One can easily distinguish (31) casein (from milk), from animal glues (gelatin and elastin), and from egg-white. The two observed cases of protein as media in Mughal Indian miniature painting were both of animal glue origin.

Drying Oils. Vegetable drying oils consist largely of triglycerides (glycerol ester of fatty acids) of five fatty acids: palmitic, stearic, oleic, linoleic, and linolenic (Table IV). The analysis of triglycerides by gas chromatography is still somewhat in the development stage (32). To date, analysis of such molecules (mol wt approx. 950) has been achieved by transesterification of the triglyceride entity into the methyl esters of the three parent fatty acids which are then volatile enough to undergo facile gas chromatography on nonpolar columns. Table V lists the composition of a number of the more important vegetable oils in terms of fatty acid content. The iodine value is the percentage of iodine chloride, calculated as iodine, which is capable of being absorbed by the oil and is a direct measure of the total amount of unsaturation in the oil.

Gunstone and Padley (33) have recently proved that the distribution of the fatty acids as esters on the glycerol backbone conforms to a modified random distribution (Table VI).

A novel new way of rapid and sensitive analysis of triglycerides in oils has recently been developed by

Table IV. Principal Fatty Acids in Drying Oils

Name	Molecular formula	Double bonds, no.	Location
Palmitic	$C_{16}H_{32}O_2$	0	
Stearic	$C_{18}H_{36}O_2$	0	
Oleic	$C_{18}H_{34}O_2$	1	9
Linoleic	$C_{18}H_{32}O_2$	2	9, 12
Linolenic	$C_{18}H_{30}O_2$	3	9, 12, 15

Table V. Average Component Characteristics of Some Drying Oils

Oil	Source	Iodine value	Unsaturated entities palmitic + stearic, %	% Oleic	% Linoleic	% Linolenic
Perilla	Perilla ocimoides	198	7	20	5	68
Linseed	Linum usitatissmum	180	10	20	16	53
Candlenut	Aleurites moluccana	164	13	10	49	28
Soyabean	Glycine hispidu	132	14	23	55	8
Sunflower	Helianthus annus	136	11	16	74	0

Hites (34) on the basis that a molecular-weight distribution can simply be measured by the resultant mass spectrum of the oil, i.e., measurements on the various M^+ and $(M-18)^+$ ions. Positional isomers, however, cannot be distinguished. In spite of this, the rapidity of such measurements overcomes the other conventional and tedious methods of esterification followed by chromatography. A method to locate ultimate double bonds in triglycerides has been developed by Serck-Hanssen (35) using a rapid isothermal gas chromatographic determination of the main and end carbon chains split off as monocarboxylic acids by permanganate oxidation of the oils in acetone during a few minutes at room temperature.

Autoxidation of Drying Oils. The mechanism by which conjugated species polymerize in an autoxidation process is not yet fully understood, but it is certain that a system involving the reaction of free radicals is responsible for the construction of a three-dimensional macromolecular structure (36).

Table VI. Triglyceride Composition of Linseed Oil

Triglyceride[a]	Wt %
333	22
332	15
331	18
330	10
322	4
321	8
320	5
310	6
300	1
Others	11

[a] 3, 2, 1, and 0 refer to the acids linolenic, linoleic, oleic, and saturated species, respectively.

Farmer (37) was the first to demonstrate that oxidation may occur at reactive methylene groups in the fatty acid entities with the formation of a hydroperoxide. In the past, classical theory had demanded that oxidation involve the production of a cyclic peroxide via the double bonds themselves. Oxidation is now regarded as a free radical chain reaction as follows involving only the reactive allylic methylene groups in both linoleate and linoleneate chains:

Initiation: $\quad RH \rightarrow \dot{R} + H$

Propagation: $\quad \dot{R} + O_2 \rightarrow \dot{R}O_2$
$\dot{R}O_2 + RH \rightarrow ROOH + \dot{R}$

Termination: $\quad 2\dot{R}O_2 \rightarrow R-O-O-R$
$\dot{R} + \dot{R} \rightarrow R-R$
$\dot{R}O_2 + \dot{R} \rightarrow R-O-O-R$

Dimerization can therefore occur between unsaturated fatty acid chains on separate triglycerides (interpolymerization) or between fatty acid chains on the same glycerol backbone (intrapolymerization). The initial number of unsaturated sites available within these triglycerides is the key to the drying process. Polymerization proceeds at room temperature via an autoxidation process involving uptake of atmospheric oxygen. Via such a mechanism a network of bonded triglycerides is formed retaining the pigment particles within the polymeric framework. In addition to this polymeric structure, some triglycerides may remain unchanged owing to lack of sufficient unsaturation to participate in the autoxidation process, i.e., palmitic and stearic. Decomposition may also occur (possibly with the hydroperoxide formed) with the production of a number of smaller molecules.

Stolow (38) has already undertaken a detailed study of the mechanism of this polymerization process and a number of the controlling factors affecting the rate of polymerization in an attempt to relate the chemistry of such a polymerized system to real time. It has been demonstrated successfully that such tentative correlations do in fact exist, but much more experimental data must be collected to evaluate the program statistically. Stolow (38) encountered a number of compounds by gas chromatography that he was unable to identify by normal procedures. It is hoped that the use of the combined gas chromatograph–mass spectrometer system at the Conservation Center will help solve these problems. At the moment, however, research is underway at the Conservation Center to further the basic work on the gas chromatography of triglycerides.

New Materials

It is often necessary for the conservator–chemist to study the properties of new materials not only from the point of view of their application to conservation practices but also as a potential material for use by contemporary creative artists. Such a case is polyurethane elastomers which have already been used by prominent artists such as Claes Oldenberg in his composite molded relief of the Chrysler airflow car (39). After the first limited edition was published, the polymeric material discolored badly, and studies were initiated at the Conservation Center to determine causes.

Research into the basic components of the resin used revealed that an aromatic diisocyanate had been employed together with a suitable polyol. Polyurethane elastomers based on conventional aromatic diisocyanates are prone to yellow (i.e., oxidation) and lose gloss on exposure to sunlight. Replacement of the aromatic component by an aliphatic saturated hydrocarbon molecule such as hexamethylene diisocyanate removed the ability of the elastomer to undergo further reaction after casting (40). Tests on this new suggested formulation (i.e., prolonged exposure to ultraviolet) indicated that no such discoloration was likely to occur in the future. These discoveries necessitated a republication of the art work with new castings made of the more stable formulation.

Role of Analytical Chemistry

Today's public consciousness of cultural heritage has elevated Conservation to a new significance in the museum world. However, lack of suitably qualified analytical chemists has somewhat hindered rapid advancement in the field. The marriage of the two disciplines, chemistry and art, has not yet been formally conceived academically although it has existed to some extent in certain talented individuals throughout the world. Rapid development of more scientific laboratories in the U.S.A. devoted to this topic is under discussion by Congress and will necessitate qualified staff which do not formally exist. Fresh manpower oriented to this new discipline and committed to the principles of Conservation will be in demand in the very near future.

References

(1) Symposium on "Application of Spectrographic Techniques in the Museum Laboratory," 10th National Meeting of Society of Applied Spectroscopy, St. Louis, Mo., 1971.
(2) R. Frankel, *Isotop. Radiat. Technol.*, **8**, 1 (1970).
(3) R. S. Frankel and D. W. Aitken, *Appl. Spectrosc.*, **24**, 557 (1970).
(4) R. L. Brunelle, W. D. Washington, C. M. Hoffman, and M. J. Pro, *J. AOAC*, **54**, 920 (1971).
(5) G. H. Morrison and A. T. Kashuba, ANAL. CHEM., **41**, 1842 (1969).
(6) K. M. Reese, *ibid.*, **42**, 26A (1970).
(7) W. W. Harrison, G. G. Clemena, and C. W. Magee, *J. AOAC*, **54**, 929 (1970).
(8) "Art and Technology—A Symposium on Classical Bronzes," S. Doeringer, D. G. Mitten, and A. Steinberg, Eds., MIT Press, Cambridge, Mass., 1970.
(9) E. R. Caley, "Analysis of Ancient Metals," Pergamon Press, New York, N.Y., 1964.
(10) R. J. Gettens, "The Freer Chinese Bronzes—Technical Studies. Vol II," Smithsonian Institution, Washington, D.C., 1969, p 124.
(11) R. M. Organ, *Archaeometry*, **13**, 27 (1971).
(12) E. R. Caley, "Critical Evaluation of Published Analytical Data on the Comparison of Ancient Metals" in "Application of Science in the Examination of Works of Art," Boston Museum of Fine Arts, 1967.
(13) R. M. Elliott and P. Swift, *Appl. Spectrosc.*, **21**, 312 (1967).
(14) P. G. T. Vossen, ANAL. CHEM., **40**, 632 (1968).
(15) R. Brown, MS702 Users Meeting, St. Louis, Mo., 1971.
(16) H. E. Wulft, "The Traditional Crafts of Persia," MIT Press, Cambridge, Mass., 1966.
(17) R. J. Forbes, "Extracting, Smelting and Alloying" in "History of Technology," C. Singer, A. R. Hall, and E. J. Holmyard, Eds., Oxford, England, 1954, p 572.
(18) "7000 Years of Iranian Art," Smithsonian Institution, Washington, D.C., 1964.
(19) P. R. S. Moorey, *Archaeometry*, **7**, 72 (1964).
(20) B. B. Johnson, "Krishna Rajamannar Bronzes: An Examination and Treatment Report" in "Krishna: The Cowherd King," P. Pal, Los Angeles County Museum of Art, to be published, 1972.
(21) M. Chandra, "The Technique of Mughal Painting," The U. P. Historical Society, Lucknow, India, 1949.
(22) O. P. Agrawal, "A Study in the Technique and Materials of Indian Illustrated Manuscripts," paper presented at ICOM Symposium, Amsterdam, Holland, 1969.
(23) B. B. Johnson, "The Technique of Indian Miniature Painting," paper presented at Symposium in Indian Art, Los Angeles County Museum of Art, to be published, 1972.
(24) "The Arts of India and Nepal: The Nasli and Alice Heeramaneck Collection," Boston Museum of Fine Arts, 1966.
(25) R. D. Harley, "Artists' Pigments c. 1600–1835," Butterworths, London, England, 1970.
(26) C. L. Butler and L. H. Cretcher, *J. Amer. Chem. Soc.*, **51**, 1519 (1929).
(27) F. Smith and R. Montgomery, "Chemistry of Plant Gums and Mucilages," New York, N.Y., 1959, p 106.
(28) L. Masschelein-Kleiner and F. Tricot-Marckx, *Bulletin Institut Royal du Patrimoine Artistique*, Brussels, **8**, 180 (1965).
(29) E. Stahl, "Thin-Layer Chromatography," Academic Press, New York, N.Y., 1965.
(30) F. Flieder, *Stud. Conserv.*, **13**, 49 (1968).
(31) "Traite de Biochemie Generale," Masson, Paris, France, 1952.
(32) R. Watts and R. Dils, *J. Lipid Res.*, **9**, 40 (1968).
(33) F. D. Gunstone and F. B. Padley, *J. Amer. Oil Chem. Soc.*, **42**, 957 (1965).
(34) R. A. Hites, ANAL. CHEM., **42**, 1736 (1970).
(35) K. Serck-Hanssen, *Acta Chem. Scand.*, **21**, 305 (1967).
(36) G. H. Hutchinson, *J. Oil Color Chem. Ass.*, **41**, 474 (1958).
(37) E. H. Farmer, *Trans. Faraday. Soc.*, **38**, 340 (1942).
(38) N. Stolow in "Application of Science in Examination of Works of Art," Boston Museum of Fine Arts, 1967.
(39) "Profile Airflow," Gemini G. E. L., Los Angeles, Calif., 1969.
(40) Modern Plastics Encyclopedia, 1970–71, p 224.

Ben B. Johnson, *Head of the Conservation Center, Los Angeles County Museum of Art, received his BA in mathematics at the College of William and Mary and his MA in art history at the Institute of Fine Arts, NYU. He received the Certificate in Art Conservation from the Conservation Center of NYU. In Italy he studied at the Uffizzi with Lionetto Tintori and later received a Diploma in Art Conservation from Ghent University. In 1964 he became Conservator of European Paintings at the Freer Gallery of Art, Smithsonian Institution. In 1967 he established the Conservation Center of the Los Angeles County Museum of Art and was appointed Lecturer in Art Conservation in the Graduate Art History Department at UCLA. A Fellow of the Institute for the Conservation of Antiquities and Works of Art and a Consultant Fellow of the Conservation Center, Institute of Fine Arts, NYU, he has lectured widely in museums on the West Coast. He has published a small book entitled "Introduction to Art Conservation," several articles on conservation, and at the moment has two definitive articles in press, "Technique of Indian Miniature Paintings," and "South Indian Bronzes."*

Thomas Cairns, *Conservation Chemist for the Los Angeles County Museum of Art, received his PhD degree in chemical spectroscopy from the University of Glasgow, Scotland, in 1965. He joined Heyden and Son, Ltd., as publishing director in 1965 and served in this capacity until his appointment at the Museum in June 1968. Dr. Cairns was editor and author of the series, "Spectroscopy in Education," Heyden, 1967, and is the author of numerous papers dealing with hydrogen bonding in natural products. He was a summer session lecturer in chemistry at UCLA, 1968–71, and is a science advisor to the Food and Drug Administration in Los Angeles. Currently, Dr. Cairns is engaged in the application of mass spectrometry in an attempt to establish the origin of art objects via their trace analysis.*

"Report" Articles in ANALYTICAL CHEMISTRY, 1970-1975

Items in **bold face** *are reprinted in this volume.*

An Analysis of Teaching. W. E. Harris, **47** (*12*), 1046A (1975).
The Analytical Chemist and Multielement Chemical Testing in Preventive Medicine. Ronald H. Laessig, **43** (*8*), 18A (1971).
Analytical Chemistry — A Fading Discipline? Arthur F. Findeis, M. Kent Wilson, and W. Wayne Meinke, **42** (*7*), 26A (1970).
Analytical Chemistry and Consumerism in the Automobile Industry. Lynn L. Lewis, **46** (*11*), 866A (1974).
Analytical Potential of Photoelectron Spectroscopy. D. Betteridge and A. D. Baker, **42** (*1*), 74A (1970).
Analytical Toxicology. Irving Sunshine, **47** (*2*), 212A (1975).
Applications of Mass Spectrometry to Trace Determinations of Environmental Toxis Materials. Fred P. Abramson, **44** (*14*), 28A (1972).
Art Conservation: Culture Under Analysis. Part I. Ben B. Johnson and Thomas Cairns, **44** (*1*), 24A (1972).
Art Conservation: Culture Under Analysis. Part II. Ben B. Johnson and Thomas Cairns, **44** (*2*), 30A (1972).
Atomic Absorption Spectroscopy — Stagnant or Pregnant? Alan Walsh, **46** (*8*), 698A (1974).
Atomic Fluorescence Spectrometry. J. D. Winefordner and R. C. Elser, **43** (*4*), 24A (1971).
Automated Reaction-Rate Methods of Analysis. Howard V. Malmstadt, Emil A. Cordos, and Collene J. Delaney, **44** (*12*), 26A (1972).
Biopharmaceutics: The Role of the Analyst. Glenn A. Brewer, **45** (*8*), 702A (1973).
Carbon 13 Nuclear Magnetic Resonance Spectroscopy. George A. Gray, **47** (*6*), 546A (1975).
Chemical Analysis of Moon Samples. K. M. Reese, **42** (*6*), 26A (1970).
Chemical Ionization Mass Spectrometry. Burnaby Munson, **43** (*13*), 28A (1971).
Chemiluminescence and Bioluminescence in Chemical Analysis. W. Rudolf Seitz and Michael P. Neary, **46** (*2*), 188A (1974).
Chemistry for Consumers. Helen L. Reynolds, Hyman P. Eiduson, John R. Weatherwax, and Donald D. Dechert, **44** (*13*), 22A (1972).
The Coblentz Society Specifications for Evaluation of Research Quality Analytical Infrared Spectra (Class II). **47** (*11*), 945A (1975).
Columns for Modern Analytical Liquid Chromatography. J. J. Kirkland, **43** (*12*), 36A (1971).
Comparison of Analytical Techniques for Inorganic Pollutants. R. F. Coleman, **46** (*12*), 989A (1974).
Computer Interfacing. Raymond E. Dessy and Jonathan A. Titus, **45** (*2*), 124A (1973).
Computerized Signal Processing. Gerald Dulaney, **47** (*1*), 24A (1975).
Computers in Clinical Chemistry. Ronald H. Laessig and Thomas H. Schwartz, **46** (*4*), 398A (1974).
Crimialistics: Educational and Scientific Progress. Goeffrey Davies, **47** (*3*), 318A (1975).
Development of Analytical Chemistry as a Science. Izaak M. Kolthoff, **45** (*1*), 24A (1973).
The Development of Chromatography. Leslie S. Ettre, **43** (*14*), 20A (1971).
Electron Spectroscopy. David M. Hercules, **42** (*1*), 20A (1970).
Element Selective Detectors in Gas Chromatography. David F. S. Natusch and Thomas M. Thorpe, **45** (*14*), 1184A (1973).
Evaluation of Lunar Elemental Analysis. George H. Morrison, **43** (*7*), 22A (1971).
Forensic Science in Criminal Prosecution. Joseph M. English, **42** (*13*), 40A (1970).
Forensic Science — The Present and Future. Ray L. Williams, **45** (*13*), 1076A (1973).
Forensic Toxicology of Drug Abuse: A Status Report. Bryan S. Finkle, **44** (*9*), 18A (1972).
Foundations of Modern Liquid Chromatography. L. S. Ettre and C. Horvath, **47** (*4*), 422A (1975).
GC/MS/Computers. Francis W. Karasek, **44** (*4*), 32A (1972).
Immobilized Enzymes: Analytical Applications. Howard H. Weetall, **46** (*7*), 602A (1974).
Inductively Coupled Plasma-Optical Emission Spectroscopy. Velmer A. Fassel and Richard N. Kniseley, **46** (*13*), 1110A (1974).
Industrial Problem Solving by Infrared Spectroscopy. Peter R. Griffiths, **46** (*14*), 1206A (1974).
Inferences from Observations: Graphical Intuition to Bayesian Probability. Patrick C. Kelly, **44** (*11*), 28A (1972).
Instant Interfacing. Raymond E. Dessy and Jonathan Titus, **46** (*3*), 294A (1974).
International Standardization for Water-Quality Evaluation. Marvin W. Skougstad, **46** (*12*), 982A (1974).
Ion Microprobe Analysers: History and Outlook. Helmut Liebl, **46** (*1*), 22A (1974).
Ion Microscopy. George H. Morrison and Georges Slodzian, **47** (*11*), 932A (1975).
The Manpower Crisis in Clinical Chemistry. William C. Purdy and Robert S. Melville, **42** (*12*), 32A (1970).
Measurement Analysis by Pattern Recognition. Bruce R. Kowalski, **47** (*13*), 1152A (1975).
"Meets A.C.S. Specifications." Story of ACS Committee on Analytical Reagents. Samuel M. Tuthill, **42** (*3*), 30A (1970).
Met and Unmet Needs of the Automated Clinical Laboratory. S. Raymond Gambino, **43** (*1*), 20A (1971).
Multielement Flame Spectroscopy. Kenneth W. Busch and George H. Morrison, **45** (*8*), 712A (1973).
The Need for Analytical Chemistry in Clinical Chemistry Training Programs. Merle A. Evenson, **42** (*14*), 53A (1970).
Neutron Activation Analysis, a Tale of Three Meetings. W. S. Lyon, **45** (*4*), 386A (1973).
Nuclear Magnetic Resonance with Superconducting Magnets. LeRoy F. Johnson, **43** (*2*), 28A (1971).
Operations Research in Analytical Chemistry. Desire L. Massart and Leonard Kaufman, **47** (*14*), 1244A (1975).
Organic Pollutant Analysis. 1973 Analytical Chemistry Summer Symposium. Roy A. Keller and Leslie S. Ettre, **45** (*11*), 892A (1973).
The Other Face of the Measurement Base. Richard W. Roberts, **47** (*7*), 648A (1975).
The Pains and Pleasures of Industrial Analytical Chemistry. Vernon A. Stenger, **43** (*3*), 36A (1971).
Picasso the Chemist. Philip W. West, **46** (*9*), 784A (1974).
Picosecond Spectroscopy. P. M. Rantzepis and C. J. Mitschele, **42** (*14*), 20A (1970).
Plasma Chromatography. Francis W. Karasek, **46** (*8*), 710A (1974).
Primary Analytical Chemists. Sidney Siggia, **47** (*2*), 207A (1975).
Quantitation in Elemental Analysis. Part I. H. Kaiser, **42** (*2*), 24A (1970).
Quantitation in Elemental Analysis. Part II. H. Kaiser, **42** (*4*), 26A (1970).
Radioactive Inert Gases. Tool for Analysis of Gases, Liquids, and Solids. Vladimir Balek, **42** (*9*), 16A (1970). Correction. **42** (*11*), 32A (1970). Adendum. **42** (*14*), 32A (1970).
Radioimmunoassay and Related Methods. Charles D. Hawker, **45** (*11*), 878A (1973).
Reaching Students with Analytical Chemistry. Sidney Siggia, **42** (*14*), 49A (1970).
The Relevance of Graduate Research in Analytical Chemistry. Charles V. Banks and Robert F. Sieck, **42** (*14*), 56A (1970).
Remote Chemical Analysis During the Apollo 15 Mission. Isidore Adler, Jacob I. Trombka, and Paul Gorenstein, **44** (*3*), 28A (1972).
Selection of an Optimum Analytical Technique for Process Control. Frank A. Leemans, **43** (*11*), 36A (1971).
Simplex Optimization of Variables in Analytical Chemistry. Stanley N. Deming and Stephen L. Morgan, **45** (*3*), 278A (1973).
Solute Band Spreading in Liquid Chromatography: Causes and Importance. Eli Grushka, **46** (*6*), 510A (1974).
Some Chemical Applications of Machine Intelligence. Thomas L. Isenhour and Peter C. Jurs, **43** (*10*), 20A (1971).
Some Ideas on Analytical Chemistry Courses for Chemistry Majors. Roland F. Hirsch, **42** (*14*), 42A (1970).
Standard Reference Materials for Clinical Measurements. W. Wayne Meinke, **43** (*6*), 28A (1971).
Surface and Thin Film Compositional Analysis: Description and Comparison of Techniques. Charles A. Evans, Jr., **47** (*9*), 818A (1975).
Survey of Analytical Spectral Data Sources and Related Data Compilation Activities. Lewis H. Gevantman, **44** (*7*), 30A (1972).

Survey of Graduate Education in Analytical Chemistry. G. A. Rechnitz, 43 (*4*), 51A (1971).

Teaching Analytical Chemistry. A Need for Objective Data. W. E. Harris, 42 (*13*), 53A (1970).

Teaching of Analytical Chemistry — The Problem in Perspective. H. A. Laitinen, 42 (*14*), 37A (1970).

Trace Metals in Atmospheric Particulates and Atomic Absorption Spectroscopy. Jae Young Hwang, 44 (*14*), 20A (1972).

Tunable Lasers in Analytical Spectroscopy. James R. Allkins, 47 (*8*), 752A (1975).

Tunable Organic Dye Lasers. J. Pierce Webb, 44 (*6*), 30A (1972).

Ultrapurity in Trace Analysis. James W. Mitchell, 45 (*6*), 492A (1973).

Using Integrated Circuits in Chemical Instrumentation. John S. Springer, 42 (*8*), 23A (1970). Correction. 42 (*9*), 16A (1970).

Vacuum Technique in Analytical Chemistry. Peter F. Varadi, 42 (*11*), 28A (1970).

Water Quality Surveillance. Vernon T. Stack, Jr., 44 (*8*), 32A (1972).

Weighed in the Balance. John T. Stock, 45 (*12*), 974A (1973).

Workshop on Mass Spectrometric Analysis of Solids. 43 (*3*), 45A (1971).

Workshop on Mass Spectrometric Analysis of Solids. Committee VII — Study of Solids, American Society for Mass Spectrometry, 47 (*12*), 1059A (1975).

Workshop on Surface Analysis and Secondary Ion Mass Analysis, 45 (*4*), 398A (1973).

X-Ray Energy Spectrometry. David E. Porter and Rolf Woldseth, 45 (*7*), 605A (1973).

X-Ray Photoelectron Spectroscopy. William E. Swartz, Jr., 45 (*9*), 788A (1973).

"Instrumentation" Articles in ANALYTICAL CHEMISTRY, 1970-1975

Items in **bold face** *are reprinted in this volume.*

Anodic Stripping Voltammetry. T. R. Copeland and R. K. Skogerboe, 46 (*14*), 1257A (1974).

Application of a Silicon-Traget Vidicon Detector to Simultaneous Multielement Flame Spectrometry. D. G. Mitchell, K. W. Jackson, and K. M. Aldous, 45 (*14*), 1215A (1973).

Applications of Chemiluminescent Reactions to the Measurement of Air Pollutants. Robert K. Stevens and J. A. Hodgeson, 45 (*4*), 443A (1973).

Atomic Spectrochemical Measurements with a Fourier Transform Spectrometer. Gary Horlick and W. K. Yuen, 47 (*8*), 775A (1975).

Computer System for Structural Identification of Organic Compounds from Spectroscopic Data. P. R. Naegeli and J. T. Clerc, 46 (*8*), 739A (1974).

Continuous Flow Analysis. Morton K. Schwartz, 45 (*8*), 739A (1973).

Correlation Spectroscopy. J. H. Davies, 42 (*6*), 101A (1970).

Countercurrent Chromatography. Yoichiro Ito and Robert L. Bowman, 43 (*13*), 69A (1971).

Current Instrumentation for Continuous Monitoring for SO . Craig D. Hollowell, Glenn Y. Gee, and Ralph D. McLaughlin, 45 (*1*), 63A (1973).

Data Domains — An Analysis of Digital and Analog Instrumentation Systems and Components. C. G. Enke, 43 (*1*), 69A (1971).

Derivative Spectroscopy with Emphasis on Trace Gas Analysis. Robert N. Hager, Jr., 45 (*13*), 1131A (1973).

Design and Performance of a Mass-analyzed Ion Kinetic Energy (MIKE) Spectrometer. J. H. Beynon, R. G. Cooks, J. M. Amy, W. E. Baitinger, and T. Y. Ridley, 45 (*12*), 1023A (1973).

Direct Analysis of Stable Isotopes with a Quadrupole Mass Spectrometer. R. M. Caprioli, W. F. Fries, and M. S. Story, 46 (*4*), 453A (1974).

Double-Wavelength Spectroscopy. T. J. Porro, 44 (*4*), 93A (1972).

Electrofocusing in Gels. Daniel Wellner, 43 (*10*), 59A (1971).

Field Ionization Mass Spectrometry: A New Tool for the Analytical Chemist. Michael Anbar and William H. Aberth, 46 (*1*), 59A (197).

Fourier and Hadamard Transform Methods in Spectroscopy. Alan G. Marshall and Melvin B. Comisarow, 47 (*4*), 491A (1975).

Fourier Transform Approaches to Spectroscopy. Gary Horlick, 43 (*8*), 61A (1971).

Gas Chromatography Detectors. C. Harold Hartmann, 43 (*2*), 113A (1971).

Hadamard-Transform Spectrometry: A New Analytical Technique. John A. Decker, Jr., 44 (*2*), 127A (1974).

The Hall Effect. E. D. Sisson, 43 (*7*), 67A (1971).

High and Low Temperature Microscopy. Walter C. McCrone, 47 (*14*), 1279A (1975).

High-Speed Current Measurements. Pieter G. Cath and Alan M. Peabody, 43 (*11*), 91A (1971).

Image-Analyzing Microscopes. Philip G. Stein, 42 (*13*), 103A (1970).

Inductively Coupled Plasmas. Velmer A. Fassel and Richard N. Kniseley, 46 (*13*), 1155A (1974).

Insturmentation for the Study of Rapid Reactions in Solution. Richard M. Reich, 43 (*12*), 85A (1971).

Instrumentation for Water Quality Monitoring. Sidney L. Phillips, Dick A. Mack, and William D. MacLeod, 46 (*3*), 345A (1974).

Instruments for Rate Determinations. Howard V. Malmstadt, Collene J. Delaney and Emil A. Cordos, 44 (*12*), 79A (1972).

Interferometry in the Seventies. Peter R. Griffiths, 46 (*7*), 645A (1974).

Ion Cyclotron Resonance Spectrometry: Recent Advances of Analytical Interest. Michael L. Gross and Charles L. Wilkins, 43 (*14*), 65A (1971).

Ion-Sensitive Field Effect Transistors and Related Devices. J. N. Zemel, 47 (*2*), 255A (1975).

Ion Specific Liquid Ion Exchanger Microelectrodes. John L. Walker, Jr., 43 (*3*), 89A (1971).

Ionization Sources in Mass Spectrometry. E. M. Chait, 44 (*3*), 77A (1972).

Isotope Excited X-Ray Fluorescence. Theo. J. Kneip and Gerard R. Laurer, 44 (*14*), 57A (1972).

Laser Optoacoustic Spectroscopy — A New Technique of Gas Analysis. L. B. Kreuzer, 46 (*2*), 235A (1974).

Low-Angle Laser Light Scattering. Wilbur Kaye, 45 (*2*), 221A (1973).

Measuring Fast Optical Signals: Amplifiers, Displays, and Transmission Lines. F. E. Lytle,, 46 (*9*), 817A (1974).

Measuring Fast Optical Signals: Detectors. F. E. Lytle, 46 (*6*), 545A (1974).

Metastable Ions in Mass Spectra. J. H. Beynon, 42 (*1*), 97A (1970).

Microprocessors. Part I: Bridging the Gap. Raymond E. Dessy, Peter Janse-Van Vuuren, Jonathan A. Titus, 46 (*11*), 917A (1974).

Microprocessors. Part II: Applications. Raymond E. Dessy, Jonathan A. Titus, and Peter Janse-Van Vuuren, 46 (*12*), 1055A (1974).

A Miniature Fast Analyzer System. Charles D. Scott and Carl A. Burtis, 45 (*3*), 327A (1973).

A Miniature Mass Spectrometer. Peter H. Dawson, J. W. Hedman, and N. R. Whetten, 42 (*12*), 103A (1970).

Modern Aspects of Air Pollution Monitoring. R. K. Stevens and A. E. O'Keeffe, 42 (*2*), 143A (1970).

A Modular Approach to Chemical Instrumentation. Richard G. McKee, **42** (*11*), 91A (1970).

Henneberg and Gerhard Schomburg, **42** (*9*), 51A (1970).

Muhlheim Computer System for Analytical Instrumentation. Engelbert Ziegler, Dieter Henneberg and Gerhard Schomburg, **42** (*9*), 51A (1970).

Photoacoustic Spectroscopy: A New Tool for Investigation of Solids. A. Rosencwaig, **47** (*6*), 592A (1975).

Photon Counting for Spectrophotometry. Howard V. Malmstadt, Michael L. Franklin, and Gary Horlick, **44** (*8*), 63A (1972).

Preparative High-Performance Liquid Chromatography. Part I. J. J. DeStefano and J. J. Kirkland, **47** (*12*), 1103A (1975).

Preparative High-Performance Liquid Chromatography. Part II. J. J. DeStefano and J. J. Kirkland, **47** (*13*), 1193A (1975).

Process Gas Chromatography. R. Villalobos, **47** (*11*), 983A (1975).

Prospects for Molecular Microscopy. J. Wendell Wiggins and Michael Beer, **44** (*1*), 77A (1972).

Pulsed and Fourier Transform NMR Spectroscopy. T. C. Farrar, **42** (*4*), 109A (1970).

Pumps and Injectors for Modern Liquid Chromatography. Laverne Berry and Barry L. Karger, **45** (*9*), 819A (1973).

Quantitative Analysis of Solid Surfaces by Auger Electron Spectroscopy. Paul W. Palmberg, **45** (*6*), 549A (1973).

Radiation Sources for Optical Spectroscopy. August Hell, **43** (*6*), 79A (1971).

Rapid Scanning Spectroscopy: Prelude to a New Era in Analytical Spectroscopy. Robert E. Santini, Michael J. Milano, and Harry L. Pardue, **45** (*11*), 915A (1973).

Real Time Clocks for Laboratory-Oriented Computers. Brian K. Hahn and C. G. Enke, **45** (*7*), 651A (1973).

The Renaissance in Polarographic and Voltammetric Analysis. Jud B. Flato, **44** (*11*), 75A (1972).

Secondary Ion Mass Analysis: A Technique for Three-Dimensional Characterization. Charles A. Evans, Jr., **44** (*13*), 67A (1972).

Semiconductor Light-Emitting Diodes. Richard A. Chapman, **42** (*8*), 69A (1970).

Signal-to-Noise Enhancement Through Instrumental Techniques. Part I. Signals, Noise, and S/N Enhancement in the Frequency Domain. G. M. Hieftje, **44** (*6*), 81A (1972).

Signal-to-Noise Enhancement Through Instrumental Techniques. Part II. Signal Averaging, Boxcar Integration, and Correlation Techniques. G. M. Hieftje, **44** (*7*), 69A (1972).

Surface and Thin Film Analysis. Charles A. Evans, Jr., **47** (*9*), 855A (1975).

Temperature Compensation Using Thermistor Networks. Ray Harruff and Charles Kimball, **42** (*7*), 73A (1970).

Thin-Layer Densitometry. Morton S. Lefar and Arnold D. Lewis, **42** (*3*), 79A (1970).

Time-Sharing Minicomputer Data Acquisition-Processing System. Mack W. Overton, Larry L. Alber, and Donald E. Smith, **47** (*3*), 363A (1975).

TLC Programmed Multiple Development. John A. Perry, Thomas H. Jupille, and Louis J. Glunz, **47** (*1*), 65A (1975).

Total Internal Reflection Enhancement of Photodetector Performance. Tomas Hirschfeld, **42** (*14*), 87A (1970).

TV-Type Multichannel Detectors. Yair Talmi, **47** (*7*), 697A (1975).

Two-Directional Immunoelectrophoresis. Leslie M. Shaw and Dean A. Arvan, **44** (*9*), 57A (1972).

When the Computer Becomes a Part of the Instrument. Marvin Margoshes, **43** (*4*), 101A (1971).

INDEX

A

Absorption.. 219
   spectra of NO_2 86
   spectroscopy... 83
Ac polarography.. 114
Acetylene flame source, nitrous oxide...................... 31
Accelerators... 100
Accuracy vs. repeatability................................. 151
Acknowledge.. 175
Activation analysis, neutron............................... 252
Activation energies of enzymes............................. 222
Activators nucleotide...................................... 216
Address regesters.. 174
Admixture problem.. 263
Adsorption... 219
   chromatography...................................... 134, 139
Aequorea... 216
AES (auger electron spectrometry)...................... 91, 100
Air pollutants... 205
Air, vinyl chloride in ambient............................. 152
Aldehyde, chainlength of................................... 216
Aldehyde derivative, active................................ 221
Amalgamation, postoperative................................ 170
Amide linkage.. 221
Analog
   circuits for variable-time method....................... 201
   data to digital form.................................... 166
   digital conversion...................................... 164
   digital systems, hybrid................................. 202
   integration system................................. 198, 199
   interferograms....................................... 64, 66
Analysis
   automated reaction-rate methods of...................... 186
   in quality control, trace................................ 71
   speed of... 98
Analyte, atomization of..................................... 42
Analytical
   chemists.. 259
      capabilities of..................................... 254
      industrial.. 276
      in the U.S. Customs Service......................... 265
   chemistry as a science, development of................... 1
   methods... 249
      detection and identification limits for............. 155
      for nanogram quantities of trace elements........... 232
      operations under ultraclean conditions, routine..... 231
   performance... 213
   spectroscopy, tunable lasers in.......................... 82
   techniques for inorganic pollutants..................... 248
Analyzers.. 145, 146
Anodic stripping voltammetry.......................... 115, 118
Antique belt buckle.. 267
API Research Project....................................... 133
Apparatus.. 212
Aromatic carbon in hydrocarbon mixtures..................... 20
Art
   conservation.. 280, 290
   objects
      environment and.................................... 281
      examination of..................................... 283
      treatment of....................................... 286
ASV applications... 123
Atmosphere controlled...................................... 228
Atomic
   absorption... 31
      spectrometry.. 49
      spectrophotometer............................... 26, 27
      spectroscopy.. 23
   clock... 237
   emission, multielement................................... 30
   emission spectrometry................................... 49
   flame fluorescence spectrophotometer, nondispersive... 28
   flame spectrometry...................................... 48
   fluorescence... 30, 41
      flame spectrometry, for............................. 46
      instrumentation..................................... 44
      fluorescence, mechanism of.......................... 41
      fluorescence, radiance of........................ 41, 42
      spectrometry........................... 40, 44, 47, 85
      spectroscopy.. 45
   gas... 41
   spectrochemical... 61
   spectroscopy.. 249, 250
   weights.. 237
Atomization of analyte...................................... 42
Atomizers for atomic fluorescence spectrometry......... 45, 46
ATP, applications of bioluminescence assay for............ 215
Auger Electron Spectrometry (AES)...... 91, 100, 102, 103
Auto zero.. 147
Automated
   reaction-rate methods of analysis....................... 186
   stopped-flow reaction-rate system....................... 190
   system with prepackaged reagents........................ 191
Automatic
   operation... 146
   potentiometric reactionrate analyzer.................... 189
   rate measurements....................................... 199
   reaction-rate results for phosphate..................... 200
   results for glucose..................................... 199
Autoxidation of drying oils................................ 296
Availability... 150
Azo linkage... 220, 221

B

Backscattering spectrometry (BS)............................ 92
Back flush to measure...................................... 150
Bacterial bioluminescence system........................... 215
Balance, double pan... 52
Band spreading chromatography, theory of.................. 135
Bandwidth, spectrometer.................................... 55
Barbiturates, methylation of............................... 275
Big-beam ion probes.. 108
Binding energies................................... 76, 78, 79
Bioluminescence.. 214
   assay for ATP... 215
   in chemical analysis.................................... 210
   Renilla Renifermis.................................... 216
   system.. 214, 215
Blanks... 227
Blood constituents... 243
Blood, volatiles & nonvolatiles in......................... 274
Boxcar averaging... 167
Broadband operation dye laser system........................ 85
Bronze
   disease, active... 282
   Indian South.. 293
   Luristan.. 293
   patination of ancient................................... 282

Buckle, antique belt 267
Buckle, metallic 267
Bulk detection limits 95

C

Calibration 151, 161, 238
Carbaryl, nitrosated 114
Carbodimides 220, 221
Carbon 76
   1_s electron binding energies and functional groups.... 76
   1_s photoelectron spectrum..................... 74
   13 (^{13}c) 11
      Chemical Shifts........................ 13, 15
      NMR proton decoupling in 12
      NMR spectrum 12
   carbonyl 21
   degree of protonation of 14
   dioxide in high-purity ethylene trace 151
   in hydrocarbon mixtures, aromatic 20
Carbonyl carbon of methacrylate polymer 21
Carboryl, active pseudourea from 221
Carboxymethylcellulose azide 220, 221
Central Processing Unit (PU).................... 172
Cesium, interferograms 64, 65
Chainlength of aldehyde 216
Charged particle excited x-ray energy fluorescence
   spectrometry 72
Chemical(s)
   analysis, chemiluminescence and bioluminescence in . 210
   analysis ESCA, electron spectroscopy for 74
   applications of multidimensional mapping......... 246
   conversions 151
   information 97
   laboratory preparation of pure................. 229
   shifts 15, 74
   suppliers of high purity 230
Chemiluminescence........................... 210
   in chemical analysis 210
   Luminol 213, 214
Chemiluminescent ozone monitors 206
Chemiluminescent reactions to the measurement of
   air pollutants 205
Chemists(s)
   analytical 259
   capabilities of analytical 254
   forensic 273
   industrial analytical 276
   in the U.S. Customs Service, analytical 265
Chemistry relationships, physical 5
Chemistry as a science, development of analytical 1
Choline 223
Chromatogram of trimethylorthoformate, gas......... 257
Chromatography design, process 145
Chromatograph system, computer controlled 149
Chromatographic
   adsorption 134
   results 269
   techniques 142
Chromatography
   adsorption 139
   development of 138
   elution mode of 136
   evolution of theory 134
   gas 258
   gas–liquid partition 142
   high-speed 150
   inorganic 132
   modern liquid 125
   paper 142
   partition 141
   in the petroleum industry 133
   process gas 144
   rebirth of 129
   theory of bank spreading in 135
Circuits for rate determinations, classical 196
Circuits for variable-time method, analog 201
Clinical analyzer, Du Pont automatic 190
Coherent single-frequency off-resonance decoupling 15
Coherent spectrometers 55
Collect 161
Column
   cutter.................................... 150
   design 150
   dual..................................... 150
   stripper 150
   switching 150
Commercialization of new food packaging material.... 259
Complexes 9
Compositional analysis, surface and thin film 90
Computer(s)
   Controlled chromatograph system 149
   GC/MS 154
   implementation of derivative method 202
   interfacing 148
Computerization of Gc/Ms....................... 159
Computerized signal processing 164
Concrete
   hardened................................. 262
   highly air-entrained 263
   Holey 263
   scales.................................... 264
   spalled................................... 264
   "unset" 262
Condensed performance of Gc/Ms interfaces 156
Conductance 157
Conservation, art 280, 290
Construction materials 228, 262
Containment of pure reagents 230
Continuous monitoring of inorganic phosphate
   in solution 225
Continuum source of excitation 44
Conventional wavelength-dispersive spectrometry....... 67
Conversions, chemical 151
Copolymerization 220
Coupling reaction, encoding indirectly via 187
Covalent attachment 220
Coverage 94
CPU (Central Processing Unit) 172
Cross-correlation, ^{13}C ^{1}H chemical shift 17
Cross-linking 220
Cutter column 150
Customs service, analytical chemists in the U.S........ 265
Cyanogen biomide 220
   activated derivative........................ 221

D

Data
   acquisition 69, 101, 164
   handling 101, 105
   latch (output port) 175
   Presentation............................... 148
   standard reference.......................... 237
   structure, eight-measurement 243
   use of 146
DC pulses.................................... 57
Decision making, real-time 168
Decoupling
   in ^{13}C NMR, proton........................ 12
   coherent single frequency off-resonance 15
   gated 15
   selective proton 17
Detection limits............ 42, 48, 94, 95, 124, 155
Detection systems 33, 37, 100
Detective, analytical 273

Detector(s) 147
   multichannel 36
   systems, multiple slit-multiple 35
Deuteration 17
Deproteinized serum or plasma samples 189
Derivative method, computer implementation 202
Diazonium chloride 221
Differential pulse
   anodic stripping voltammetry 115
   polarography 111
   scan 120
Digital
   filtering 168
   form, analog data to 166
   integration system 200
   systems, hybrid analog 202
Digitized interferograms 65
Dihydroactinidiolide 272
Dinitrophenylhydraizone 258
Direct
   coupled Gc/Ms 159
   multicannel spectrometers 53
   stripping voltammetry 116
Displacement development 132
Display 161
   methods 243
Distillation, ion exchange and double 229
Distillation, subboiling 230
Distributions in single measurement 245
Double pan balance 52
Dormant period, Tswett's technique in 128
Dual column 150, 151
Duplex excitation 71
DuPont automatic clinical analyzer 190
Drug(s)
   homicide 274
   identification of suspected illicit 273
   mass spectral data of common 159
Drying oils 295, 296
Dye laser
   sodium D_2 line recorded with CW 84
   system for broadband operation 85
   tunable 82

E

Eigenvector projection 243, 245
Electrodes, enzyme 222
Electrodes, ion-exchange 223
Electrolytic treatment 289
Electron
   binding energies 76, 78, 79
   spectrometer 101
   spectrometry Auger (AES) 91, 100, 102, 103
   spectroscopy for chemical analysis
   (ESCA) 74, 92, 100, 104, 105
Elemental sensitivity 94
ELLA 193
Elution analysis 132
Elution mode of chromatography 136
Encoding chemical reaction-rate information ... 187, 188
Energetic ion beam techniques 91, 100
Energy
   level diagram for Raman process 87
   spectrometers 69
   spectrometry 67
Entrance optics used in atomic fluorescence
   spectrometry 47
Entrapment 220
Environment and art objects 281
Enzyme(s)
   activation energies of 222
   to active derivative attachment of 221
   automated system, nylon tube-supported 224
   through azo linkage, attachment of 221
   electrodes 222
   immobilized 219, 220
   immobilization, methods of 219
   thermal stability of 223
Equilibrium, phenomena 134
ESCA (electron spectroscopy for
   chemical analysis) 74, 92, 100, 104, 105
Ethyltrifluoroacetate 74
Ethylene, trace carbon monoxide 151
Evidence in crimal conspiracy, identification
   f physical 267
Excitation 100
   line and continuum source of 44
   sources requirements for atomic fluorescence
      spectrometry 44
   system 67
   waveform, potential 112
Failure problems, construction materials 262
Fast analyzer, GeMSAEC 192
Fast linear sweep voltammetry 114
Fatty acids in drying oils, principal 295
FDA approval project 260
Film
   analysis, surface and thin 99
   compositional analysis, surface and thin 90
   saturation of 122
Filter photometers 35
Filtering, digital 168
Firefly bioluminescence system 214
Firefly reaction 215
Fixed-time technique for reaction-rate determinations .. 197
Flag, flip-flop used as 177
Flags, external 175
Flame
   atomizer shapes 46
   cells 30
   characteristics for multielement flame spectrometry ... 32
   fuel-rich oxyacetylene 30
   methods 37
   photometric detector 207
   source, nitrous oxide–acetylene 31
   spectrometry, multielement 32, 33
   spectroscopy, multielement 30
   techniques, multielement capability of 37
   tube sources 250
Flavor component 270
Flip-flop used as flag 177
Fluorescence
   atomic (see Atomic fluorescence) radiance
      for a dilute atomic gas 41
   spectroscopy absorption 83
Food packaging material 259
Food packaging regulation 260
Forensic chemist 273
Fourier transform
   ion cyclotron resonance spectroscopy 58
   NMR, pulsed 12
   spectrometer 61
   spectroscopy 52, 56
Frequency domain represenations of DC pulses 57
Frontal analysis 132
Fruit concentrate, tariff classification of 267
Fruit juice, testing concentrated 267
Fuel-rich oxyacetylene flame source 30
Functional groups, correlation chart for C(1s) 76

G

Gamma-ray spectrometric analysis of fused silica
   and polyethylene 231

Gas(es)
　chromatography 144, 257, 258
　fluorescence radiance for a dilute atomic 41
　–liquid partition chromatography 142
　–phase chemiluminescent ozone monitor 206
　phlogiston theory of 2
　sampling bags, Scotchpak 278
　–solid chemiluminescent ozone analyzer, automated . 206
　in transformer oil, dissolved 152
Gated decoupling 15
GC/MS
　computers.............................. 154, 159
　direct coupled 159
　interfaces 155, 156
GeMSAEC fast analyzer 192
Gibbs contributions 6
Glow discharge mass spectrometry (GDMS) 93
Glow discharge optical spectrometry (GDOS) 93
Glucoamylase, immobilized 223
Glucose
　automatic results for 199
　determination of 187
　monitor..................................... 224
Glues.. 293
Glutaraldehyde 220
Graphite substrates on peak currents 122
Graphite tube sources 250
Ground loop phenomenon 166
Gums 293, 295

H

^{1}H chemical shift cross-correlation, ^{13}C 17
H_2O_2, riboflavin................................. 214
Hadamard transform encoding-decoding ("Multiplex")
　spectrometers................................ 54
Hadamard transform methods in spectroscopy 52
Hardware 172
Heart cut analysis 151
High purity chemicals and metals, suppliers of 230
High-speed chromatography 150
Holey concrete 263
Homicide, drug 274
Humidity control case 283
Hybrid analog-digital systems...................... 202
Hydrocarbon(s)
　mixtures, aromatic carbon in 20
　in steam condensate, trace..................... 152
　trace dissolved water in 152

I

ICR spectrometer, (ion cyclotron resonance) 56, 58
Identification limits for analytical methods 155
Identification scheme for unknown powder 275
Image-dissector photomultiplier 35
Immobility glucoamylase.......................... 223
Immobilized enzymes 219, 220
Impregnation on peak currents 122
Impurities
　identification of............................... 268
　spectrometric detection of trace 230
　in trimethylorthoformate 258
In and out pulses/select device 174
Incoherent spectrometers 55
Indian bronzes, South 293
Indian miniature painting, Mughal 293
Industrial analytical chemists 276
Industry, chromatography in the petroleum.......... 133
Infrared laser source spectrometer, hypothetical 56
Inorganic
　chromatography 132
　pollutants, analytical techniques for 248
　substances, techniques for 255
Input/output architecture 174

Input/output programming 176
Input port 174
Instruction set 177
Instrumental system, multimode 193
Instrumentation............................. 121, 290
　atomic fluorescence 44
　future 108
　systems for reaction-rate methods 189
Instruments for rate determinations 196
Integration
　(fixed-time) technique for reaction-rate
　　determinations........................ 197, 198
　system, analog........................... 198, 199
　system, digital 200
　technique, minicomputer implementation of 201
　times 199
Interface teaching station 184
Interfaces, condensed performance of Gc/Ms 156
Interfacing the Gc/Ms 155
Interferences 43, 48
Interferograms, analog......................... 64, 66
Interferograms, digitized 65
Interpretation, postoperative...................... 170
Interrput request and acknowledge 175
Intracavity absorption spectra of NO_2 86
Ion(s)
　beam techniques, energetic 91, 100
　that catalyze Luminol chemiluminescence, metal 213
　cyclotron resonance (ICR) spectrometer........... 56
　detection and data handling 105
　exchange
　　electrodes, liquid organic..................... 223
　　purification of water by 229
　　resins.................................... 220
　impact radiation (SCANIR), surface composition by
　　analysis of neutral and...................... 93
　-inuced x-rays 92
　mass spectrometry (SIMS) 93, 101
　microprobes........................ 101, 105, 106
　microscope 101, 106, 107
　probes, big-beam............................ 108
　scattering spectrometry (ISS) 93, 101
　sources 99, 101
　sputtering 90, 100
Isocyanates................................. 220, 221
Isothiocyanates 220, 221
ISS (ion scattering spectrometry)............... 93, 101
Italian painting, structure of early 286

J

Juice, testing concentrated fruit 267

K

Kinetic phenomena............................... 134
Kinetics... 222
Kuhn, Richard 130

L

Laboratory, positive pressure filtered air 228
Laboratory preparation of pure chemicals 229
Large-beam SIMS systems 101, 106, 107
Laser(s)
　in analytical spectroscopy 82
　-excited atomic fluorescence flame spectrometry...... 85
　sodium D_2 line recorded with CW dye............. 84
　sources, tunable.............................. 87
　system for broadband operation 85
Lateral (x,y) analysis 95
Lattice relaxation (T_1), spin- 17
Lead in aqueous solution, spectrum of 71
Learning results, unsupervised 245
Leaves, orchard 71, 72
Lederer, E....................................... 130

Length 236
Light source for multielement flame spectrometry 32
Light, speed of 236
Line source of excitation 44
Linseed oil, trialyceride composition 296
Liquid chromatography, modern 125
Liquids, purification of 230
Lithium, interferograms of 64
Logic configuration, tri-state 175
Luciferin-luciferase sources 214
Lucigenin 213
Luminol 212–214
Luristan bronze study 293

M

Maintainability 146
Manipulation 169
Manufacturing process, raw material change in 257
Mapping, chemical applications of multidimensional ... 246
Marijuana conspiracy 265
Masks 55
Mass
   Spectral data of common drugs 159
   spectrometer 104
   spectrometric detection of trace impurities 230
   spectrometry 251
   spectrum, ICR 58
Matrix effects 96
Mean free path 156
Measurement
   analysis of pattern recognition 241
   base 234
   language 235
   overlapping distributions in single 245
   techniques 239
Media studies 294
Memory
   address registers, high and low 174
   /CPU 172
   long-term 148
   peak picker and short-term 147
Mercury, solubilities of elements in 122
Metal ions that catalyze Luminol chemiluminescence .. 213
Metals, suppliers of high purity 230
Metallurgy, ancient 291
Methacrylate polymer, carbonyl carbon of 21
Methodology 242
Methyl group by polar substituent, replacement of 15
Methylation of barbituates 275
Metre 236
Microanalysis 72
   nuclear 92
Microcomputer 180
Microencapsulation 220
Microprobe mass spectrometers, ion 101
Microprobes and ion microscope, ion 105–107
Microprocess card set 171
Microprocessors 171, 179, 185
Microscope, ion 101, 106, 107
Microscopy 249
Migration studies 260
Minicomputer implementation of integration
   techniques 201
Monitor 161
Monochromators, wavelength tracking for scanning 34
Moving-window average 168
MS/computers, GC 154
Mughal Indian miniature painting 293
Multichannel
   detectors 36
   devices 33, 35
   resonance randioeters 36

Multicomponent radiation beam 34
Multidimensional mapping 246
Multielement
   analysis 36
   capability of flame techniques 37
   falme spectrometry 32, 33
   falme spectroscopy 30
   techniques 38
Multimode instrumental system 193
Multiple slit-multiple detector systems 35
Multiplex advantage 62
Multiplex spectrometers 54

N

N(1s) electron binding energies 78
Neutral and ion impact radiation (SCANIR), surface
   composition by analysis of 93
Neutron activiation anaylsis 252
Neutron radiograph 285
Nitric oxide 210
Nitrogen 77
   oxides 86, 207
Nitrous oxide–acetylene flame source, separated 31
Nitrosated carbaryl 114
Noise susceptibility 198
Nondestructive tests used to survey test samples 254
Nondispersive atomic flame fluorescence
   spectrophotometer 28
Nonflame atomizers for atomic fluorescence
   spectrometry 46
Nonvolatiles in blood 274
Nuclear magnetic resonance spectroscopy (NMR) 11
   nuclei 11
   spectrum characteristics of ^{13}C 12
   proton decoupling in ^{13}C 12
   pulsed Fourier transform 12
Nuclear microanalysis 92
Nucleotide activators of *Renilla Reniformis*
   bioluminescence 216
Nuclei, NMR 11
Nylon tube-supported enzyme automated system 224

O

O-atom reactions 213
Off-resonance decoupling, coherent single-frequency 15
Off-resonance spectra 16
Offset suppression 200
Oil(s)
   drying 295, 296
   linseed 296
   transformer 152
Operational requirement 248
Optical methods 6
Optical spectrometry (GDOS), glow discharge 93
Optics, entrance 47
Optics, primary source and 101
Organic
   analysis, elementary 4
   functional groups 78
   substances, techniques for 255
Organization 260
Oscilloscope traces 199
OSHA regulations 276, 277
Output port 175
Overhauser spectrum, suppressed 15
Overlapping distributions in single measurement 245
Oxidants that react with Luminol to produce
   chemiluminescence 213
Oxyacetylene flame source for multielement atominc
   emission 30
Ozone 205, 210
   analyzer, automated gas–solid chemiluminescent 206

monitors, chemiluminescent 206
reactions 213

P

Packaging material, commercialization of new food ... 259
Packaging regulation, food 260
Painting, Mughal Indian miniatures 293
Painting, structure of early italian 286
Palmer, Leroy Sheldon 129
Paper chromatography 142
Parallel mixing of samples and reagents 191
Partition chromatography 141, 142
Patination layers on ancient bronze 282
Pattern recognition 241, 246
Peak
adsorption, sharp line source to measure 25
currents 122
picker 147
Personnel sampling unit, vinyl chloride 277
Petroleum industry, chromatography in the 133
pH profile 220, 222
Phase solubility analysis 269
Phlogiston theory of gases 2
Phosphate
automatic reaction-rate results for 200
continuous monitoring of inorganic 225
determination of 188
Phosphorus 78, 211
Photoelectron spectroscopy, x-ray 74
Photoelectron spectrum for ethyltrifluoracetate,
carbon 1_s 74
Photograph of early American painting,
transmitted light 284
Photometers, filter **35**
Photomultiplier, image dissector 35
Photon source 69
Physical chemistry relationships 5
Pigment identification 294
Plant gums 295
Plasma samples 189
Plot 161
Polar substituent 15
Polargraphic analysis 109
Polargraphy, Ac 114
Polargraphy, differential pulse 111
Polyethylene 231
Pollutants, air 205
Pollutants, inorganic 248
Pollution 81
Polymer(s) 81
carbonyl carbon of methacrylate 21
solid 21
Positive pressure filtered air laboratory 228
Postoperative amalgamation and interpretation 170
Potassium interferograms 64, 65
Potential excitation waveform 112
Potential-time waveforms 119
Potentiometric reaction-rate analyzer, automatic 189
Potentiostatic control 110
Potentiostatic three-electrode system 110
Powder, identification scheme for unknown 275
Preprocessing 242
Pressure vessel for dissolution of refractory materials;
Teflon 232
Probes, big-beam ion 108
Procedural reduction 169
Process chromatograph design 145
Process gas chromatography 144
Production problems 254
in vitamin manufacture 268
Program counter 172
Programmer-controller 145, 147

Programming 183
Projection, eigenvector 243, 245
Proteins, composition of 295
Proton decoupling in 12, 17
Proton shifts 16
Protonation of carbon 14
Pseudourea from a carboxyl and a carbodiimide 221
Pulse scan, differential 120
Pulsed Fourier transform NMR 12
Pulsed stripping voltammetry 116
Purification of liquids 230

Q

Quality control, trace analysis in 71
Quantitation 96, 151
Quantitative analysis 18, 72, 80
Quartz, ion exchange and double distillation in 229

R

Radiance of atomic fluorescence 41, 42
Radiation
beam, multicomponent 34
SCANIR (surface composition by analysis of
neutral and ion impact) 93
source spectrometer bandwidth 55
Radiometers, multichannel resonance 36
Raman process, energy level diagram for 87
Raman spectroscopy 86
Rate
computation, selecting methods of 203
determination
with analog integration system 198
classical circuits for 196
instruments for 196
methods for 203
Radiograph, neutron 285
Raw material change in manufacturing process 257
Reaction
primary 188
rate
analyzer, automatic potentiometric 189
determinations for 197, 198, 201
information, encoding chemical 187
methods of analysis 186, 189
results for phosphate, automatic 200
system, automated stopped-flow 190
secondary 188
Reading specification 185
Readout devices (recorders) 145
Reagents
automated system with prepackaged 191
containment and storage of pure 230
parallel vs. sequential mixing of 191
preparing ultrapure 231
Real-time decision making 168
Recognition, measurement analysis by pattern 241
Recorders 145
Reference materials, standard 238
Reference methods 238
Refractory materials 232
Registers, general 172
Reliability 146
Renilla Reniformis 216, 217
Repeatability 151
Repetitive stripping analysis unit 180
Resonance radiometers, multichannel 36
Riboflavin-H_2O_2 214

S

S(2p) electron binding energies 79
Sales of atomic absorption spectrophotometers,
world 27

Sample(s)
  chamber and manipulation 100, 101
  consumption and alteration 96
  mounting and manipulation 100
  parallel vs. sequential mixing of 191
  preparation 249
  thick vs. thin 71
Sampling 249
  bags, gas 278
  unit, vinyl chloride personnel 277
  valves 146
SCANIR (surface composition by analysis of neutral
  and ion impact radiation) 93
Scanning
  detectors 34
  monochromators 34
  spectrometers 33
Scattering spectrometry (ISS), ion 101
Science, development of analytical chemistry as a 1
Scotchpak gas sampling bags 278
Secondary ion mass spectrometry (SIMS) 93, 101
Select device/in and out pulses 174
Selective proton decoupling 17
Sensitivity, elemental 94
Sensitivity variations 94
Sequential mixing of samples and reagents 191
Serum samples, deproteinized 189
Sharp line source to measure peak absorption 25
Signal-to-noise improvement 13
Signal-to-noise ratios 15
Signal processing, computerized 164
Silica and polyethylene, fused 231
Silicate materials, ultrapure 232
Silicon-germanium, spectrum of 71
Silicon vidicon tube 36
SIMS (secondary ion mass psectrometry) 93, 101
  instruments, large-beam 101
  systems, large-beam 106, 107
Single-frequency off-resonance decoupling, coherent 15
Single-slit scanning absorption spectrometer 53
Slit-multiple detector systems, multiple 35
Sodium D_2 line recorded with CW dye laser 84
Software 175, 202
Solid polymers 21
Solution analysis 7
South Indian bronzes 293
Spatial multichannel devices 35
Specific labeling 17
Specificity 94
Spectral data of common drugs, mass 159
Spectral results 269
Spectral scanning 68
Spectrochemical measurements with a Fourier
  transform spectrometer, atomic 61
Spectrometer(s) 19
  Auger electron 102, 103
  bandwidth 55
  direct multichannel 53
  electron 101
  Fourier Transform 61
  Hadamard transform encoding-decoding
    ("multiplex") 54
  hypothetical infrared laser source 56
  incoherent and coherent 55
  ion cyclotron resonance (ICR) 56
  ion microprobe mass 101
  mass 104
  scanning 33
  signals, time domain 57
  single-slit scanning absorption 53
  wavelength dispersive, x-ray 68
  wavelength and energy 69

Spectrometric analysis, gamma ray 231
Spectrometric detection of trace impurities, mass 230
Spectrometry 251
  atomic absorption and atomic emission 49
  atomic fluorescence 40, 44, 47
  Auger electron, (AES) 91, 100
  backscattering (BS) 92
  charged particle excited x-ray energy fluorescence 72
  conventional wavelength-dispersive 67
  detection limits in mass 251
  flame atomizer shapes for atomic fluorescence flame . . 46
  glow discharge (GDMS) 93
Spectrometry (GDOS), glow discharge optical 93
  ion scattering (ISS) 93, 101
  laser-excited atomic fluorescence flame 85
  limits of detection in atomic flame 48
  multielement flame 32, 33
  nonflame atomizers for atomic fluorescence 46
  secondary ion mass (SIMS) 93, 101
  x-ray 252
    energy 67
Spectrophotometer, atomic absorption 26, 27
Spectrophotometer, nondispersive atomic flame
  fluorescence 28
Spectroscopy 240
  absorption and fluorescence 83
  atomic 249
    absorption 23
  atomizers used in atomic fluorescence 45
  for chemical analysis ESCA, electron 74
  Fourier transform 52, 56
    ion cyclotron resonance 58
  limits of detection in atomic 250
  multielement flame 30
  nuclear magnetic resonance 11
  Raman 86
  tunable lasers in analytical 82
  x-ray energy 68
  x-ray photoelectron 74
Spectrum acquisition 100
Speed of analysis 98
Spin-lattice relaxation (T_1) 17
Sputtering chamber 28
Sputtering, ion 90, 100
Stability 222
Stainless steels 72
Standard reference data 237
Standard reference materials 238
Statistical consideration 239
Steam condensate, trade hydrocarbons in 152
Steels, stainless 72
Solubilities of elements in mercury film 122
Solubility analysis, phase 269
Stopped-flow reaction-rate system 190
Storage of pure reagents 230
Stripper column 150
Stripper precut or backflush to vent 150
Stripping analysis unit using a microcomputer,
  repetitive 180
Structure determination 76
Structure, sensitivity of ^{13}C shifts to chemical 14
Subboiling distillation 230
Sulfur 78, 206, 211
Sum 161
Supervised learning 243
Suppliers of high-purity chemicals and metals 229, 230
Suppressed Overhauser spectrum 15
Suppression, offset 200
Surface
  analysis 80, 97
  composition by analysis of neutral andion
    impact radiation (SCANIR) 93

and thin film analysis..........................90, 99
Susceptibility, noise 198

T

Tariff classification of fruit concentrace 267
Teaching lab at Virginia Tech 184
Teaching station, interface 184
Teflon
    chamber for vapor phse destruction of
      ultrapure silicate materials 232
    gas sampling bags 278
    pressure vessel for dissolution refractory materials... 232
Temporal multichannel devices 33
Test samples, nondestructive tests used to survey 254
Thermal stability of an enzyme 223
Thiamine monoitrate, purified 269
Thick vs. thin samples 71
Thin film compositional analysis, surface and 90, 99
Three-electrode system potentiostotic 110
Time 236
    averaging 12
    domain representations of DC pulces 57
    domain spectrometer signals 57
Tiselius, A.. 132
Titration methods.................................... 4
Trace... 161
    analysis in quality control 71
    analysis, ultrapurity in 227
    elements 232
    impurities, mass spectrometric detection of 230
Transform methods in spectroscopy, Fourier
    and Hadamard 52
Transformation 169
Transformer oil, dissolved gases in 152
Transient ICR signal 58
Transmitted light photography of early
    American painting 284
Transport, phenomena............................. 134
Triglyceride composition of linseed oil 296
1,4,4-Trimethylcyclohexan-2-one acetic acid
    enol lactone 272
Trimethylorthoformate 257, 258
Tri-state gate 174
Tri-state logic configuration 175
Tswett 126, 128
Tube-supported enzyme automated system, nylon 224
Tube sources, flame and graphite 250
Tunable lasers in analytical spectroscopy 82, 87
Type 161

U

Ultraclean Conditions·............................. 231
Ultrapure
    reagents from commercial suppliers.............. 229
    reagents, preparing 231
    silicate materials 232
    water 229
Ultrapurity in trace analysis 227
"Unset" concrete 262

V

Vacuum 99, 156
Valves and column switching 150

Valves, sampling 146
Vapor-phase destruction of ultrapure silicate
    materials..................................... 232
Variable-time method, analog circuits for 201
Variable-time methods for reaction-rate
    determinations................................ 201
Vector instructions 117
Vinyl chloride
    in ambient air 152
    gas from Teflon gas sampling bags 278
    OSHA regulations for 276
    personnel sampling unit 277
Virginia Tech, teaching lab at...................... 184
Viscous flow 157
Vitamin manufacture, production problem in 268
Volatiles in blood 274
Volt... 237
Voltammetric analysis 109, 110
Voltammetry
    anodic stripping 118
    differential pulse anodic stripping 115
    fast linear sweep 114
    pulsed and direct stripping 116

W

Water
    in hydrocarbons, trace dissolved 152
    ultrapurification of 229
    in xylenes, trace 152
Waveform, potential excitation 112
Waveforms, potential-time 119
Wavelength
    dispersive spectrometry, conventional 67
    dispersive x-ray spectrometer 68
    range 43
    spectrometers................................. 69
    tracking for scanning monochromators 34
Wax impregnation 122
Weights on a balance............................. 52

X

X-ray(s)
    analysis system 67
    energy fluorescence spectrometry, charged
      particle excited 72
    energy spectrometry 67
    energy spectroscopy 68
    ion-induced 92
    photoelectron spectroscopy 74
    source 100
    spectrometer, wavelength dispersive 68
    spectrometry 252
(x,y) analysis, lateral 95
Xylenes, trace water in........................... 152

Y

Yeast lactase 222

Z

Zechmeister, Laszlo 131

QD
75.2
M63
v.2